ALLEN COUNTY PUBLIC LIBRARY

P9-ECP-000

Developmental Biology

Nelson T. Spratt, Jr.
The University of Minnesota

Wadsworth Publishing Company, Inc.
Belmont, California

Original drawings by Darwen Hennings, assisted by Vally Brown Hennings.

ISBN-0-534-00010-X

L.C. Cat. Card No. 79-164996

Printed in the United States of America

1 2 3 4 5 6 7 8 9 10—76 75 74 73 72 71

Preface

This book presents a synthesis of animal and plant development, stressing the basic principles that unite the field. It also treats the various levels of development—molecular, cellular, and organismic—in a balanced way.

The broad problems of development and the relationship of development to genetics are presented first. Then the diverse patterns of development are surveyed, including not only reproductive and growth patterns but also patterns of continuous development and repair and regeneration in adults. I feel that students can better understand the problems of how developmental processes are controlled after they know something about the nature and diversity of developmental patterns. Thus, although control mechansims are discussed throughout the text to unify much of its diversity, the detailed explication of such mechanisms is reserved for Part Three. The last part of the book reviews and further unifies all the preceding material: it deals with such problems as the emergence of order, the role of cell products, and pattern aspects of development.

This book reveals to students the challenge developmental biology presents to human intellect and creativity. Though the field is fraught with unanswered questions about developmental processes, I have tried to present the material and problems as simply as possible, while trying not to mislead the student into thinking that these problems can be easily solved. I have tried to present in a balanced way what we do and do not know about development, hoping that students will be stimulated to fill in the gaps in our knowledge.

I wish to thank Professors Aron Moscona, William Jensen, and Cleon Ross for their very helpful reviews of the manuscript.

Contents

Part One. The Problem and Processes of Development

14 *Alternation of Asexual and Sexual Patterns 316*

15 *Mosaic and Regulative Patterns of Development 326*

Part Three. Universal Control Mechanisms of Development

16 *Control of the Basic Processes of Development 340*

17 *Control of Gene Activity and the Program of Nucleoprotein Synthesis 358*

18 *Extranuclear Genetic Control: Cytoplasmic Inheritance 374*

19 *Hormonal Control in Animal and Plant Development 390*

20 *Nucleo-Cytoplasmic Interactions 414*

21 *Induction: Cellular Interactions 444*

22 *Environmental Control of Development 462*

Part Four. Special Aspects of Development

23 *The Role of Extracellular Products 486*

24 *Developmental Basis of Behavior 498*

Part One:

The Problem and Processes of Development

1 *The Problem of Development*

What is development? Development may be defined as the action of genes in: (1) creating a new organism from some part of a parent organism, (2) maintaining or increasing the size of a fully formed mature organism, and (3) repairing accidental defects or losses in an organism. More broadly stated, development is any transformation from one state, condition, composition, or function to another that is progressive and relatively permanent under natural conditions. The growth of a tree or a man, the transformation of a caterpillar into a butterfly, the replacement of an amputated salamander arm, the transformation of an egg into an adult, the germination of a seed are all developmental transformations.

With such a broad definition it is not surprising that there are many kinds or patterns of development and that development is exhibited by all levels of biological organization from the molecular to the organismic and even the organism population level. These diverse processes and patterns of development might appear to have so little in common that nothing would be gained by considering them as one distinct biological phenomenon, but all developmental processes do have one basic common feature: *every developmental process is characteristically a progressive series of transformations—a sequence of orderly changes or stages.* The transformations appear to be programmed from the beginning. The program consists of an inherited set of instructions in the form of a chemical code, packaged for the most part in the nucleus of the cell. The problem is how the inherited instructions in the genome are used to direct the program of development. The ways in which the instructions are used may not be the same for all developmental phenomena, but probably many of them are similar. In all patterns of development in nature the program of changes seems clearly predetermined.

It is also necessary to distinguish developmental activity from homeostatic cyclical processes (which maintain a steady state by negative feedback actions that tend to reverse any transformation). As a progressive process, development is a perpetuation and stabilization of transformations (Davis, 1964). Homeostatic processes, which are also controlled by inherited instructions, are equally important, however, to the survival of the organism and species and are thus basic to developmental processes.

Control of Development

The fundamental and most interesting problem that confronts the student of development is how development is controlled. It appears that both information and guidelines for the use of the information are inherited. But are all the necessary guidelines preformed—present from the beginning—or are some of them epigenetic—progressively acquired as a consequence of the developmental process itself?

For reasons that will become apparent as we study developmental patterns, we shall hold the view that both preformed and progressively formed guidelines (directive influences) control the course of development. The relationship and interplay of the two types of guidelines that operate in the development of a new individual are diagramed in Figure 1-1 (Spratt, 1966).

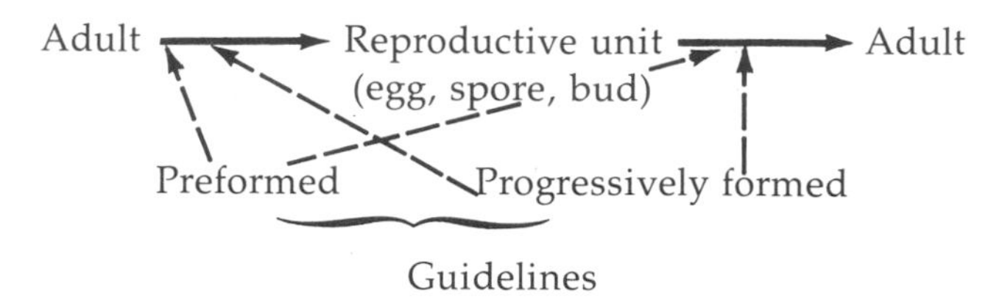

Figure 1-1. *Interplay of preformed and progressively formed guidelines in the formation and development of a reproductive unit such as an egg, spore, or bud.*

Two aspects of reproductive development—the formation of the reproductive unit and the development of this unit into the new adult—are indicated in the diagram of Figure 1-1. Because it is relatively more difficult to study, less is known about the formation of reproductive units than about their development. However, many reproductive units, especially those that become separated from the mother organism, contain all the information necessary to produce the new individual. Thus, understanding the nature of this information and how it is put into the unit is just as important as understanding how the information is used. The diagram also attempts to show that preformed guidelines (for example, coded information in the genes) and progressively formed guidelines (the changing internal environment of the developing reproductive unit) control both the formation and development of the reproductive unit.

The formation of a reproductive unit such as an egg is controlled by the genotype of the mother organism. In this process, probably as a consequence of the combined action of genes in the egg nucleus and in the cells of the mother surrounding the forming egg, an egg cytoplasm of specific composition is built up. When the egg is fertilized or artificially activated it begins to divide repeatedly to form a population of cells. As it is unlikely that the cytoplasm of any egg is truly homogeneous, some of the nuclei in the increasing number of cells will lie in cytoplasmically different environments. Consequently, it is reasonable to expect the nuclei to differentiate in the different cytoplasmic environments surrounding them. It is well known that the composition of the cytoplasm has an influence on the use of the genome in the nucleus. The diverse cytoplasmic regions built into the egg during its formation thus constitute preformed guidelines for further gene action. With the addition of different gene products in response to the diverse cytoplasmic environments of the egg, the cytoplasmic composition of the cells will progressively change. In this way, further guidelines are progressively built up. The changing cytoplasmic composition of the cells elicits further differential and selective gene action as development progresses.

With the increase in the number of cells produced by division of the egg, an intercellular environment made up of products released by the cells begins to develop. Because some of the cells at this stage probably would already be different in their patterns of gene action, a differential release of gene products would occur, resulting in the development of a heterogeneous intercellular environment. In turn, the composition of this environment would be expected to undergo further progressive changes.

The crux of this concept of development is the idea that the progressively formed guidelines are a consequence of the use of the instructions packaged in the preformed guidelines. The genome controls the development of its environment, which in turn controls further activity of the genome. Implicit in our concept of the orderly control of development is the existence of a mechanism by which inherited instructions or information can be sequentially utilized to transform the reproductive unit into an adult. As we proceed with our analysis of development we shall have many occasions for testing the validity of this suggested mechanism for the control of development.

Plan of the Text

As development is not a single process but a complex of processes, we begin to examine these processes in the next chapter. Appreciation of the significance of the patterns of development would be difficult without a knowledge of the underlying basic processes. A chapter summarizing the role of genes in the control of protein synthesis follows. Succeeding chapters review some of the many patterns of development. The purpose of the first part of the book is to introduce the material of developmental biology. As the patterns are surveyed, the underlying mechanisms (where known) and the simplest possible explanations and interpretations are presented. This does not mean that all aspects of development are simple to explain. The interpretations presented may be either correct or incorrect, but to suggest no explanation of a phenomenon would be inexcusable. Most developmental phenomena are not well understood, many not even sufficiently described or studied to permit a critical treatment of our understanding of them. However, if an example of development is interesting or remarkable, the fact that it is not well studied is no reason to exclude it from our consideration. Knowledge of the diversity of developmental patterns and processes is just as important as knowledge in depth of a few patterns. Both kinds of knowledge are necessary if progress in our comprehension of these phenomena is to continue.

After we have become familiar with examples of the material and special mechanisms of developmental biology, we will be in a better position to examine some of the generally applicable control mechanisms. By doing this, we obtain a more unified view of developmental phenomena than could be achieved by considering mechanisms only in relation to specific phenomena. Understanding the diverse patterns of development will eventually depend on our knowledge of these underlying mechanisms, which are used again and again to control and direct development. An examination of the general control mechanisms constitutes the second major portion of the text. Next we examine special aspects of development, such as the emergence of order in developing systems and the problem of pattern formation. We conclude with a formulation of the general principles of development.

Conclusion

A word of advice to the student may be helpful. The student will gain little understanding of the phenomena of development if he allows himself to be "carried away" by the beauty and apparent mystery of the endless patterns of development, as satisfying as this might be for some. Nor will he gain much insight by studying only one kind of development, be it the development of a frog gastrula, an onion root tip, or a protozoan. He will understand more by looking for a common design in molecules, cells, organisms, and societies. As he expands his knowledge and understanding, the exceptions he finds to the common design will contribute to his interest.

A list of selected general references, mostly textbooks, reviews, and publications of symposia dealing with aspects of development, is provided. Specific references are listed at the end of each chapter. Much selection has been necessary in the citation of literature, but it is hoped that these citations will introduce the student to the literature in areas of his special interest.

General References

Alston, B. E. 1967. Cellular Continuity and Development. Scott, Foresman, Chicago. This paperback deals largely with intracellular mechanisms regulating heredity and development. It contains a good discussion of mitosis and meiosis.

Balinsky, B. I. 1970. An Introduction to Embryology. W. B. Saunders, Philadelphia. An excellent textbook primarily concerned with the embryology of vertebrates.

Bell, E. 1965. Molecular and Cellular Aspects of Development. Harper & Row, New York. A collection of outstanding original papers mostly in the area of animal embryology.

Berrill, N. J. 1961. Growth, Development and Pattern. W. H. Freeman, San Francisco. A detailed treatment of invertebrate embryology and regeneration, especially of coelenterates and tunicates.

Bertalanffy, L. von, and J. H. Woodger. 1933. Modern Theories of Development. Oxford University Press, Oxford. This presentation of theories of development attempts a formalization of principles in terms of symbolic logic.

Bonner, J. T. 1952. Morphogenesis. Princeton University Press, Princeton, N. J. A delightful treatment of ideas and concepts of development.

Bresler, S. E. 1970. Introduction to Molecular Biology. Academic Press, New York. A comprehensive introductory text including molecular genetics and biochemical regulatory mechanisms.

Child, C. M. 1941. Patterns and Problems of Development. University of Chicago Press, Chicago. A detailed discussion of axial gradients and their role in development.

Clowes, F. A. L., and B. E. Juniper. 1968. Plant Cells. Blackwell, Oxford. An excellent review of the structure of plant cells and their function in development.

Davidson, E. H. 1968. Gene Activity in Early Development. Academic Press, New York. An excellent review of the program of nucleoprotein synthesis in early animal development.

DeHaan, R. L., and H. Ursprung. 1965. Organogenesis. Holt, Rinehart and Winston, New York. A collection of original reviews of current knowledge of the development of animal organs and organ systems.

Ebert, J. D. 1965. Interacting Systems in Development. Holt, Rinehart and Winston, New York. A paperback primarily concerned with mechanisms of development.

Flickinger, R. A. 1966. Developmental Biology. Wm. C. Brown, Dubuque, Iowa. A collection of original papers in the field of development similar in scope but briefer than the book edited by Bell.

Florkin, M., ed. 1967. Comprehensive Biochemistry, Morphogenesis, Differentiation and Development, Vol. 28. Elsevier, New York. An excellent collection of papers concerned with the biochemical basis of developmental processes.

Goss, R. J. 1969. Principles of Regeneration. Academic Press, New York. An excellent discussion of the problem of regeneration and the principles that have emerged from experimental analysis.

Haggis, G. H., D. Michie, A. R. Muir, K. B. Roberts, and P. M. B. Walker. 1964. Introduction to Molecular Biology. Longman, London. An excellent review of the molecular basis of biological phenomena.

Harris, H. 1968. Nucleus and Cytoplasm. Clarendon Press, Oxford. An interesting discussion of nucleo-cytoplasmic interactions.

Hay, E. D. 1966. Regeneration. Holt, Rinehart and Winston, New York. A small book that concisely reviews aspects of regeneration in animals, especially the problem of dedifferentiation.

Kerr, N. 1967. Principles of Development. Wm. C. Brown, Dubuque, Iowa. A paperback that summarizes and discusses principles.

Kühn, A. 1965. Entwicklungsphysiologie. Springer, Verlag, Berlin. A superior textbook in the field of developmental physiology of plants and animals in which pattern aspects are emphasized.

Loewy, A. G., and P. Siekevitz. 1963. Cell Structure and Function. Holt, Rinehart and Winston, New York. A brief but excellent summary of current knowledge of the molecular machinery of the cell.

Picken, L. 1960. The Organization of Cells. Clarendon Press, Oxford. An unusual book in which usually neglected aspects of cell biology are presented.

Raven, C. P. 1964. An Outline of Developmental Physiology. McGraw-Hill, New York. A brief outline of experimental embryology of animals.

Saunders, J. W., Jr. 1970. Patterns and Principles of Animal Development. Macmillan, New York. A brief but excellent presentation of much of the material of animal development with emphasis on principles.

Sinnott, E. W. 1960. Plant Morphogenesis. McGraw-Hill, New York. An excellent treatment of many aspects of plant development.

Sinnott, E. W. 1963. The Problem of Organic Form. Yale University Press, New Haven, Conn. Discusses problems of symmetry and pattern in the development of plants.

Spratt, N. T., Jr. 1964. Introduction to Cell Differentiation. Reinhold, New York. A small paperback presenting various aspects of cellular specialization and the mechanisms involved.

Steiner, R. F., and H. Edelhoch. 1965. Molecules and Life. D. Van Nostrand, Princeton. A book for the nonscientist explaining the genetic code.

Thompson, D. W. 1942. Growth and Form. Cambridge University Press, Cambridge. A classic book in which the role of growth in the development of organic form is extensively developed.

Torrey, J. G. 1967. Development in Flowering Plants. Macmillan, New York. An excellent review of the processes of development in higher plants.

Wardlaw, C. W. 1955. Embryogenesis in Plants. John Wiley & Sons, New York. The most comprehensive treatment of the problems and nature of plant embryology.

Wardlaw, C. W. 1965. Organization and Evolution in Plants. Longman, London. A discussion of the evolutionary relationships of plants and the developmental basis of these relationships.

Wardlaw, C. W. 1968. Morphogenesis in Plants. Methuen, London. A summary of the experimental analysis of the shoot apex in various groups of plants.

Weber, R., ed. The Biochemistry of Animal Development, Vols. 1 and 2. Academic Press, New York. An excellent collection of review papers concerning the biochemical aspects of developmental processes.

Wilkins, M. B., ed. 1969. The Physiology of Plant Growth and Development. McGraw-Hill, New York. An excellent review of the role of plant hormones in development.

Willier, B. H., P. Weiss, and V. Hamburger, eds. 1955. Analysis of Development. W. B. Saunders, Philadelphia. A collection of original reviews concerning aspects of experimental animal embryology.

Wilson, E. B. 1925. The Cell in Development and Heredity. Macmillan, New York. A large (1,232 pages) classic and still useful book, which emphasizes the cellular aspects of heredity and animal embryology.

Yčas, M. 1969. The Biological Code. Frontiers of Biology, Vol. 12. Eds. A. Neuberger and E. L. Tatum. North Holland Publishing Co., Amsterdam. A collection of excellent papers discussing the genetic code.

2 *Basic Processes of Development*

Although the basic processes in developmental systems, from the renewal of cells in an adult to the formation of a new adult, are closely integrated in normal development, it has been possible to dissociate them both conceptually and experimentally (Needham, 1950). We can recognize five basic processes of development at the cell level of biological structure, and all special processes can be classified within these basic categories.

a. Growth—increase in living mass (cell size and number)
b. Differentiation—cell specialization (specific biosynthesis)
c. Cellular interaction—influence of one cell or one cell group on another
d. Movement—changes in position of single cells or tissues (morphogenetic movements)
e. Metabolism—the source and use of energy; the basic cost of living and developing

These processes can be separated in experiments, but our major concern is how the processes and our concepts of them merge as we change our frame of reference from one level of organization (for example, the cell) to another (tissue, organ, organism). For example, cell movement may result in the growth of a cell population, as in the formation of an aggregate by migration of cells into the aggregate. Conversely, growth, particularly when one part grows more rapidly than another, can result in movement of one part relative to the other, as in the spreading of germ layers during gastrulation in some animals. The influence of one group of cells on another can result in differentiation of one or both groups. At the subcellular level, growth (synthesis) of cytoplasmic components—particularly when it occurs at a different rate in different cells or in different parts of a single cell—may be a manifestation of cellular differentiation. The nature and importance of the basic processes are briefly outlined below.

Growth

Although it is difficult to define "growth" precisely, we shall employ the word in its most general application to living organisms. Growth is the addition of material to a living organism—a continuous process in all living organisms (Needham, 1964). This does not mean that an organism continues to increase in size indefinitely, but net biosynthesis occurs at all levels of complexity from the molecular to the organismic. Thus, synthesis of cellular components, increase in cell size, increase in cell number, and increase in organ or organism size all constitute growth.

The phenomenon of growth has always fascinated man. Growth can be measured easily. If we can measure it we tend to think we understand it, but this never removes the aboriginal pleasure of simply measuring (paraphrased from Bonner, 1952). We measure growth in units of increase in size or amount or as a rate of growth. The S-shaped growth curve in Figure 2-1 clearly shows that growth is limited. The relationship for the S-shaped curve is expressed mathematically by the equation:

$$\frac{dy}{dt} = ry(b - y)$$

y = amount (size)
t = time
r = rate constant
b = limiting factor (for example, food supply)

Thus, in a given time t the food supply b will have decreased to $(b - y)$, giving the S-shaped curve. If the slope of the S-shaped curve is plotted against time, the result is a measure of the rate of increase or decrease of growth. It is doubtful that unlimited growth occurs in nature, but presumably continuous (but not uncontrolled and unlimited) growth occurs in hydra and other hydroid coelenterates (Lenhoff and Loomis, 1961) and in the giant redwoods.

The process of growth eventually produces a larger organism, but obviously more than a larger egg, spore, or bud. Parts of a developing system grow at different rates. Growth in all organisms is sooner or later differential (Thompson, 1942; Goss, 1964). In many animals and particularly in plants (in the shaping of leaves, flowers, etc.) differential growth is an important morphogenetic and repair process, although its role in the embryonic development of some animals has been doubted (Needham, 1964). The relative growth rates of different parts are expressed by the formula below (Huxley, 1932):

$$y = bx_k$$

or in logarithmic form:

$$\log y = \log b + k \log x$$

b = constant (value of x when $y = 1$)
k = constant (growth rate or slope of the line)
x = size (weight) of one part
y = size (weight) of another part

If $k = 1$, x and y grow at equal rates. If $k > 1$ or < 1 a

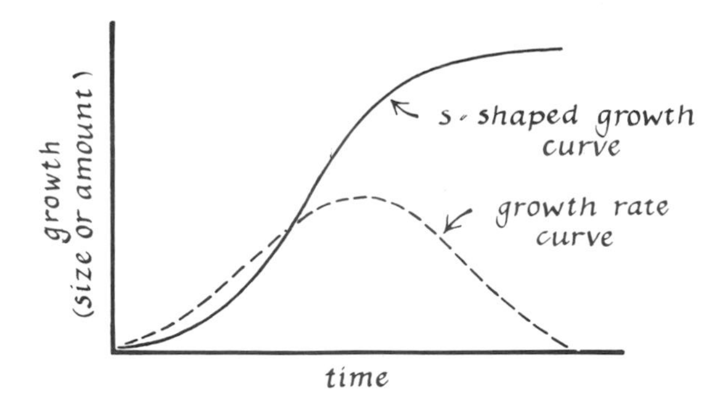

Figure 2-1. *The S-shaped growth curve and the growth rate curve (DeBeer, 1924; Astbury, 1945).*

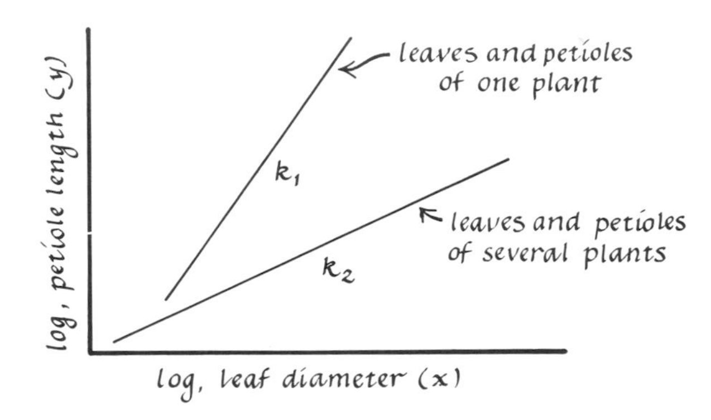

Figure 2-2. *Differential growth of the petiole and the leaf blade in* Nasturtium.

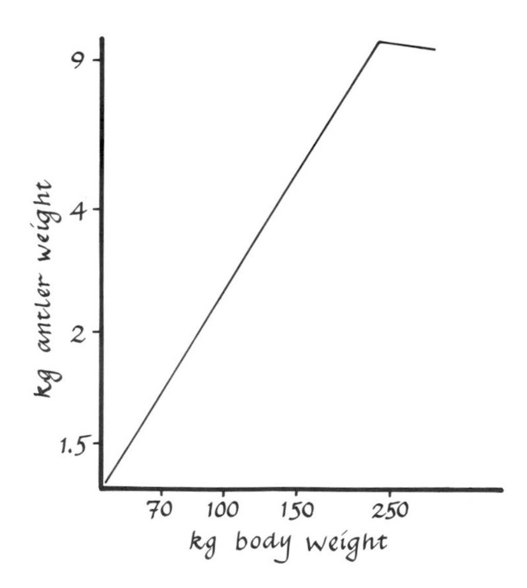

Figure 2-3. *Differential growth of the antlers of deer in proportion to the body weight (plotted logarithmically).*

change in shape or form occurs. The formula compares the rate of increase in size or weight of two different parts. In other words, different parts of an organism grow at different rates that show a constant relation to one another.

Differential growth plays a significant role in the development of form or shape in plants. Figure 2-2 compares growth of the petiole and leaf blade in *Nasturtium* (Pearsall, 1927). When the logarithm of the petiole length (y in Figure 2-2) is plotted against the logarithm of the blade diameter (x) a straight line is obtained. The slope of the line (k_1 or k_2) is a measure of the growth rate. The petiole grows at a consistently faster rate than does the leaf blade, but the different rates remain in constant proportion to one another, so the line is straight. Is this the result of an unequal distribution of growth-promoting substances (such as O_2, CO_2, food, hormones) and growth-inhibiting substances in the developing organism? Although a definite answer cannot be given for most specific examples of differential growth, it seems probable that growth, like some other activities of cells, is a response by the cells to their environment.

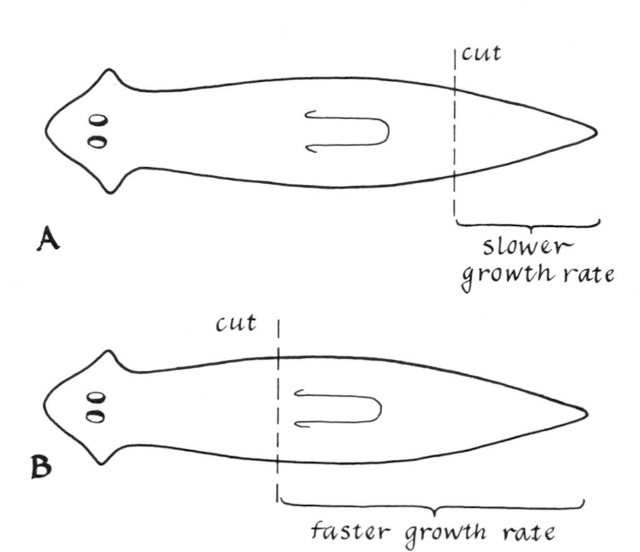

Figure 2-4. *Regeneration in* Planaria. *The larger the amount of tissue removed (compare B with A) the more rapid is the growth and replacement of the lost part.*

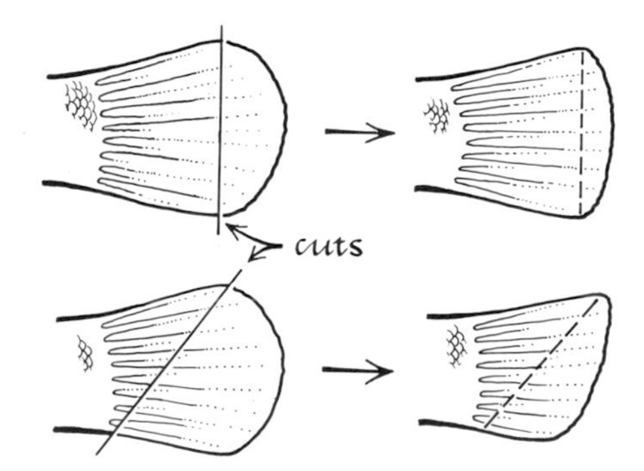

Figure 2-5. *Regeneration of the tail in the marine fish* Fundulus.

A well-known example of differential growth in animals is the growth of the antlers of deer in proportion to body weight (Huxley, 1932). Figure 2-3 shows the relatively rapid growth of the antlers in comparison with the slower body growth during regeneration of the antlers each season (Chapter 22). Also, growth of the antler stops (note top of line) each season when the antler weight is proportional to the k power of the body weight. The mechanism that controls the growth of the antlers so that their final size is proportional to the body weight is not known.

A similar phenomenon has been observed in the regeneration of planarians—increasing the amount of tissue removed increases the regeneration growth rate (Abeloos, 1932). This is illustrated in Figure 2-4. When a relatively large part of the body is removed, the rate of replacement is faster than when a smaller part is removed. Another example, described by Morgan (1901), is the regeneration of the tail in *Fundulus,* a marine fish. The fastest regeneration occurs in the region where the greatest amount of material has been removed (Figure 2-5).

The rate of repair (return to normal size for its age) in an explanted chick blastoderm increases with the

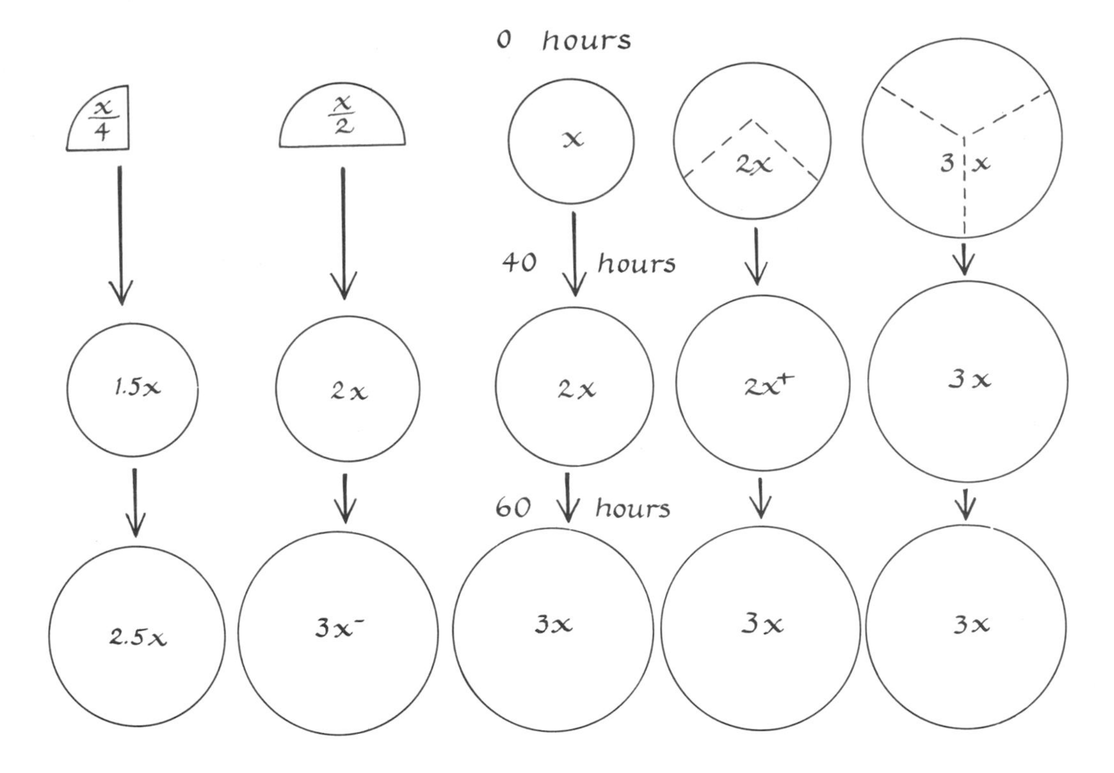

Figure 2-6. *Size of the chick blastoderm and growth rate. Influence of the original size (number of cells) of a group of explanted, unincubated chick blastoderm cells on the rate of increase in area of the group.* x *equals the size of a normal whole blastoderm. From* Introduction to Cell Differentiation *by Nelson T. Spratt, Jr., Copyright © 1964 by Reinhold Publishing Corporation, by permission of Van Nostrand Reinhold Company.*

amount of tissue removed (Figure 2-6), but if the blastoderm is made abnormally large for its age by fusing two or three blastoderms, the opposite effect is obtained—there is a depression in the growth rate. When the developmental age of the cells of this precociously large blastoderm becomes proportional to the number of cells characteristic of a normal single blastoderm of its age, normal growth rate is resumed (Spratt, 1964). This phenomenon illustrates the influence of cell number on the rate of cell division. The relationship occurs both in populations of unicellular organisms and in multicellular organisms (Chapter 16), but little is known about the mechanisms that control it.

Differential growth is an important morphogenetic mechanism in plants, but the direction of growth (the direction in which cell addition occurs) is equally important. One interesting example is the role of directional growth in transforming the ovary of a flower into a fruit. In spherical gourds, the axes of the mitotic spindles are oriented more or less equally in all directions and cell elongation is equal in all directions. In elongated gourds the spindle axes are mostly aligned in the direction of expansion of the fruit (see Figure 11-22) (Sinnott, 1939, 1960).

In summary, the differential growth formula fits many cases but it has only descriptive value and does not reveal the mechanism of growth control.

Differentiation

Cellular differentiation is cell specialization. Early in development cells are generally not specialized for a particular function and presumably are capable of carrying out various activities within the limits set by the genome of the species. Cell specialization is thus the selective use of only a portion of the cell's genetic repertory (Spratt, 1964). How this selection is accomplished is a major problem confronting the

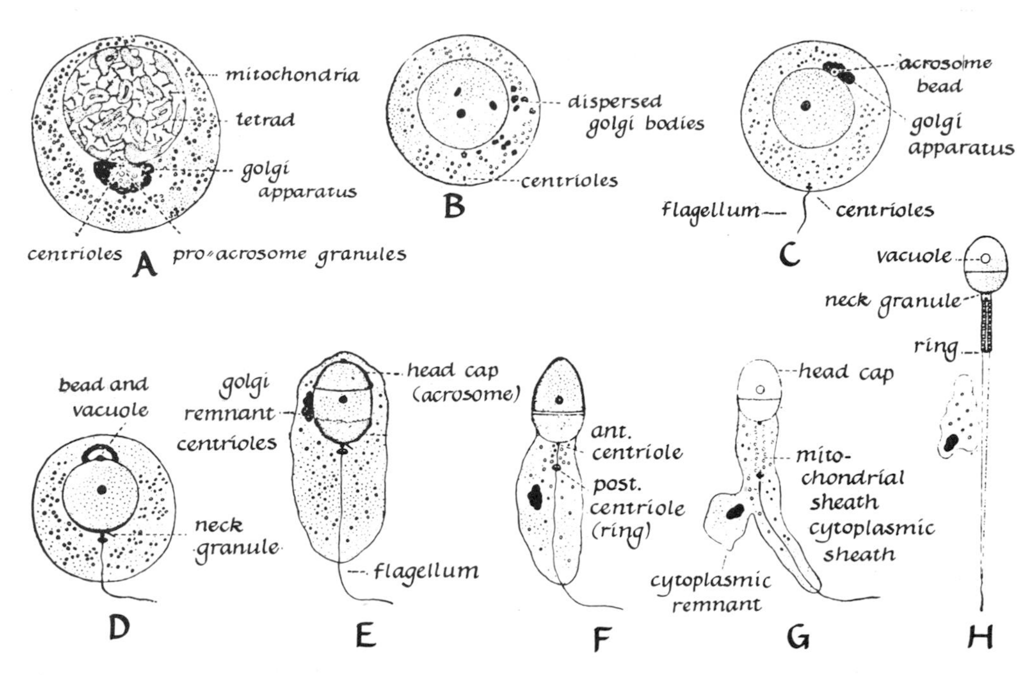

Figure 2-7. *Intracellular differentiation during development of the human spermatozoan. × 1000-2500. A, spermatocyte; B, spermatid; C-G, stages in transformation; H, mature spermatozoan. From Arey,* Developmental Anatomy, *7th edition. W. B. Saunders Company, Philadelphia, 1965.*

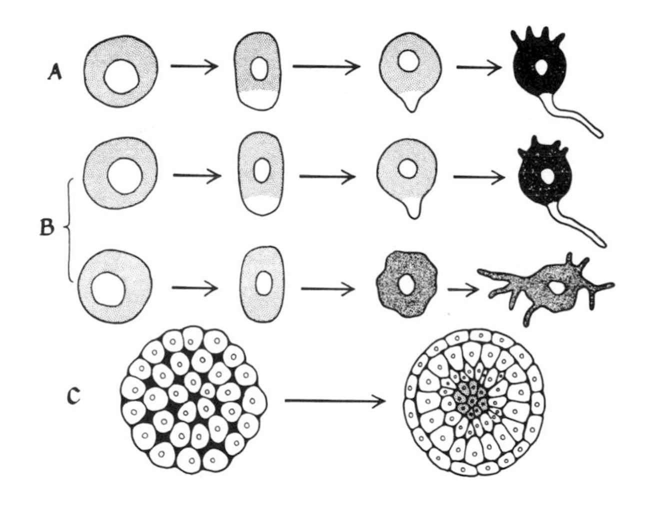

Figure 2-8. *A, intracellular differentiation; B, intercellular differentiation; C, intercellular differentiation in a cell population. From* Introduction to Cell Differentiation *by Nelson T. Spratt, Jr., Copyright © 1964 by Reinhold Publishing Corporation, by permission of Van Nostrand Reinhold Company.*

student of development. First we shall examine briefly two aspects of cellular differentiation.

Intracellular (Differentiation in Time)

Changes in the function or structure of a single cell with the passing of time constitute intracellular differentiation. Excellent examples are the formation of an egg or sperm from the morphologically unspecialized primordial germ cells of some animals and tissue cells of plants. Stages in the development of the human spermatozoan illustrate this type of cellular differentiation (Figure 2-7).

The final consequence of the process, the spermatazoan, is a highly differentiated cell. The sperm is specialized to fertilize an egg and can do nothing else. This process of differentiation in time occurs in all cells of multicellular organisms, particularly during reproduction and regeneration of lost parts.

Table 2-1. Kinds of cellular differentiation.

	Kinds of Changes	
	Undifferentiated Cell	*Differentiated Cell*
Function	generalized	specific
Shape	simple	complex
Internal structure	simple	complex
Size	more uniform	more diverse
Responsiveness	great	lessened
Motility	great	depressed
Mitotic activity	great	depressed
Quantity of cell products	small	large
Type of cell products	simple (nonspecific)	special (specific)
Metabolism	generalized	specialized
Microenvironment	simple	complex
Cell number	small	large
Tissue architecture	simple	complex
Ploidy level	haploid or diploid	polyploid (in plants)
General capacities	great (diverse)	small (restricted)

From *Introduction to Cell Differentiation* by Nelson T. Spratt, Jr., Copyright © 1964 by Reinhold Publishing Corporation, by permission of Van Nostrand Reinhold Company.

Intercellular (Differentiation in Space)

Intercellular differentiation is the process by which two or more cells become different from one another. Examples of this are: epidermis and nerve cell formation in a population of ectodermal cells; leaf, petal, and pistil formation in the cell population at the growing tip of a plant; vascular tissue formation within the stem of a plant; and blood cell formation within the bone marrow of animals.

The difference between intracellular and intercellular differentiation is diagramed in Figure 2-8. The two aspects are closely interrelated, however, and often there is no sharp distinction between them. Differentiation in time occurs in all kinds of organisms, but differentiation in space occurs in multicellular forms only.

Kinds of Differentiation

Cells specialize in many ways. A few of the properties of a cell that may be changed during differentiation are listed in Table 2-1.

From the table we can draw three general conclusions about the differentiation process. (1) It involves one, a few, or many changes in form (shape, size, internal structure) and activity (kinds of cell products, cell division rate, responsiveness, metabolism) of the cells. (2) In general, all changes are in the direction of limitation in expression of the capacities. (3) Association of cells in populations, as in multicellular organisms, makes possible a greater degree of specialization than would otherwise occur. Biochemical processes underlie and characterize both the cellular properties listed in Table 2-1 and the changes in these properties that occur during cellular specialization (Papaconstantinou, 1967).

A curious type of differentiation is the specialization of cells to die during the developmental process. Ordinarily one thinks of development as the building up, not the destruction, of new parts, but dead cells are indispensable to the functioning of many organisms. For example, the feathers of birds and the hairs of mammals are constructed out of dead cells. Besides this, cell death is an important process in development of plants and animals (reviews of Glücksman, 1951; Saunders, 1966). Especially in animals, both in embryos and adults, cell death eliminates tissues or cells that are needed only for short periods. In some patterns of continuous development in adults the addition of cells is balanced by the death and loss of cells. We shall encounter numerous examples of this in Chapter 4.

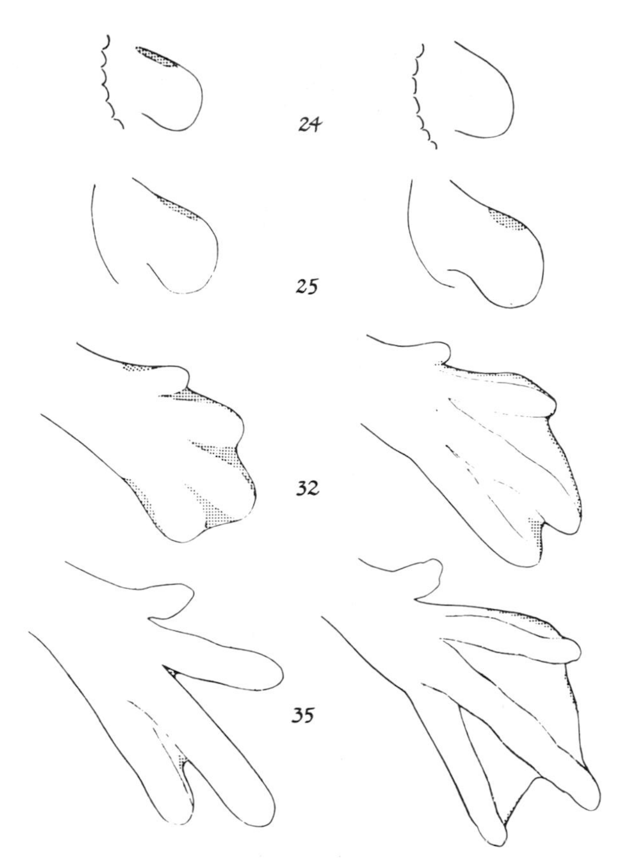

Figure 2-9. *The patterns of cell death in the developing leg of the chick (left) and duck (right) embryos. The numbers denote Hamburger-Hamilton stages of development. From Saunders and Fallon. Copyright 1966 by Academic Press.*

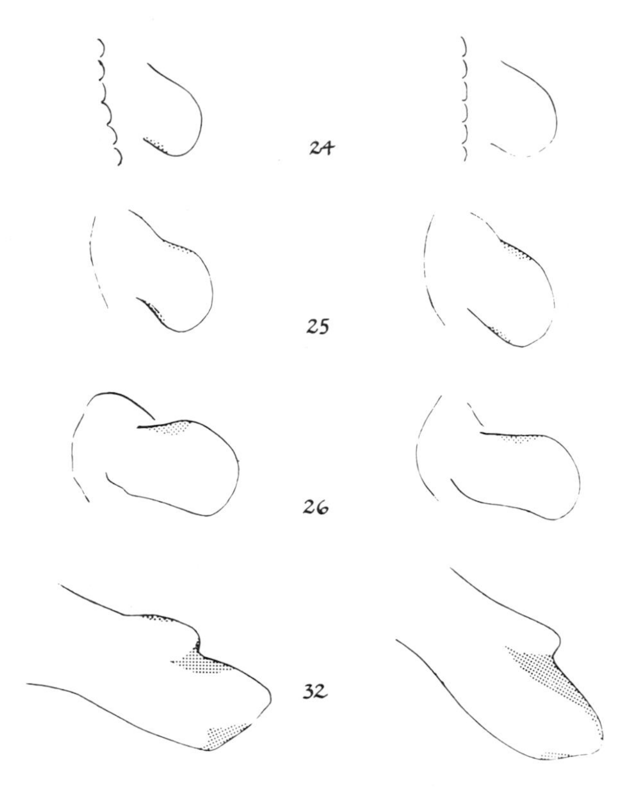

Figure 2-10. *Zones of cell death in the superficial mesoderm of wing buds of the chick (left) and duck (right) as revealed by vital staining with Nile Blue. The numbers indicate Hamburger-Hamilton stages. From Saunders and Fallon. Copyright 1966 by Academic Press.*

During embryonic and metamorphic development organs or tissues are lost (for example, the pronephric and mesonephric kidneys in mammals, the tail and gills of frog tadpoles, and various larval tissues in insects). Among many examples of the role of cell death in embryos is the sculpturing of the limb in the 3- to 9-day-old chick embryo (Saunders, 1966), illustrated in Figures 2-9 and 2-10.

Saunders cites evidence that cell death may be important in the separation of the digits in the mouse, rat, mole, and man, as well as in the chick. Cell death is important in the morphogenesis of the fruiting bodies of cellular slime molds (Chapter 6), and also occurs in the development of plant organisms. For example, cell death occurs in the woody and outer bark portion of tree trunks; in the growing root, forming the protective root cap; possibly in the formation of an abscission zone in the petiole of a leaf, allowing the leaf to be shed in autumn (Leopold, 1967); and in the development of the archegonium, the female reproductive organ in many kinds of plants (Figure 10-20). The leaves of the common house plants *Monstera* and *Philodendron* are pierced with naturally formed holes. Whether these are the result of cell death in local areas has not yet been studied. In general, both localized cell death and cell addition are normal developmental processes.

Cellular Interactions

As a multicellular organism develops from a reproductive unit, the number of its cells significantly increases. However, unlike the cells in a population of unicellular organisms, the cells are integrated to form a single organism. Each cell is under the influence of its neighboring or distant cells and influences its neighboring or distant cells. Because of this interaction, the cells are interdependent to varying degrees. Every cell behaves in the context of the cell population of which it is a part. This is shown in Figure 2-11. Because of these reciprocal influences, the cells of multicellular organisms, unlike those of unicellular organisms, become functionally specialized so that a "division of labor" is possible. Indeed, this sharing of activities is an important mechanism for functional integration of the population. It is clear that cellular interactions are involved in the development of all multicellular organisms, plant or animal (Chapter 21).

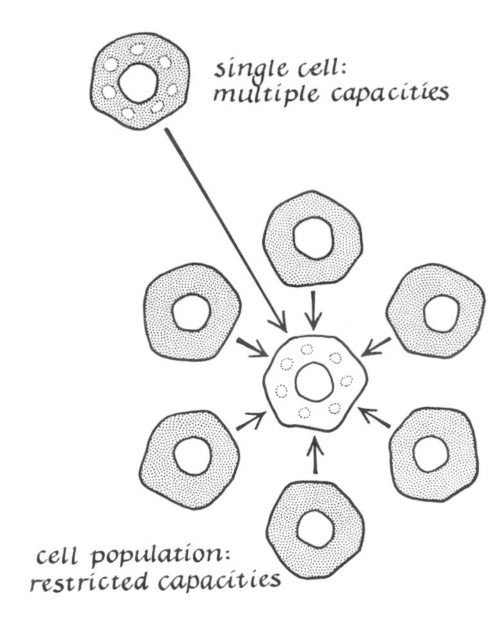

Figure 2-11. *Illustration of the restriction of a cell's capacities as a consequence of the influence of surrounding cells.*

Movement

Movement of individual cells and of tissues, particularly in developing animals, plays an important role in the shaping of the organism and its parts. We can recognize three general types of movement in development.

Tissue Movement

The formation of germ layers (primitive tissues such as ectoderm, mesoderm, and endoderm) in many animal embryos is partially the result of movement of whole areas in the surface to the interior. This type of movement is characteristic of gastrulation. Figure 2-12 diagrams a median longitudinal section of an amphibian embryo egg through an early and later stage of gastrulation. A consequence of the movement of the mesodermal area to the interior beneath the ectodermal area is the specialization of the ectoderm cells as nerve rather than skin cells.

Another dramatic type of tissue movement is the posterior movement of the primitive streak (a rodlike aggregation of cells) in bird embryos during the elongation of the embryo body in front of the streak. This movement, clearly visible in time-lapse movies of explanted embryos, is illustrated in Figure 2-13. (See also Chapter 12, Figures 33 and 35.)

Other types of morphogenetic movement are folding (formation of the neural tube, digestive tract), invagination (formation of the inner ear vesicle), and evagination (formation of the eye vesicles) during development of a vertebrate embryo. These movements are illustrated in Figure 2-14.

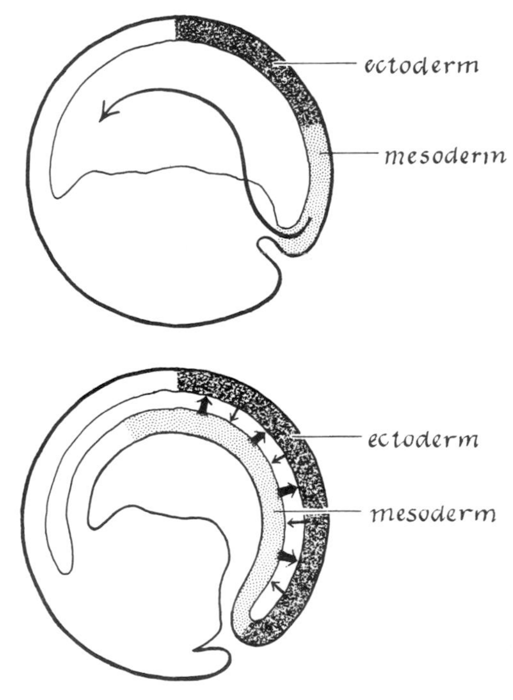

Figure 2-12. *The primary consequence of gastrulation movements in an amphibian embryo: interaction between ectodermal and mesodermal tissue layers. From* Introduction to Cell Differentiation *by Nelson T. Spratt, Jr., Copyright © 1964 by Reinhold Publishing Corporation, by permission of Van Nostrand Reinhold Company.*

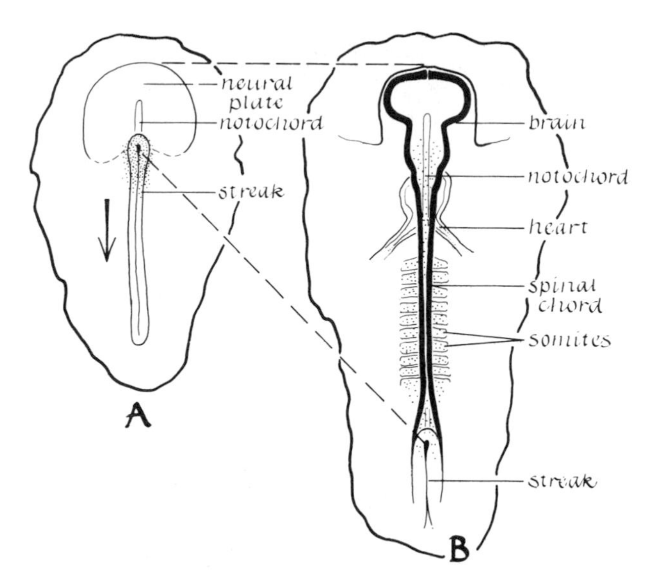

Figure 2-13. *Posterior movement of the primitive streak and formation of the embryo body axis of the chick embryo.*

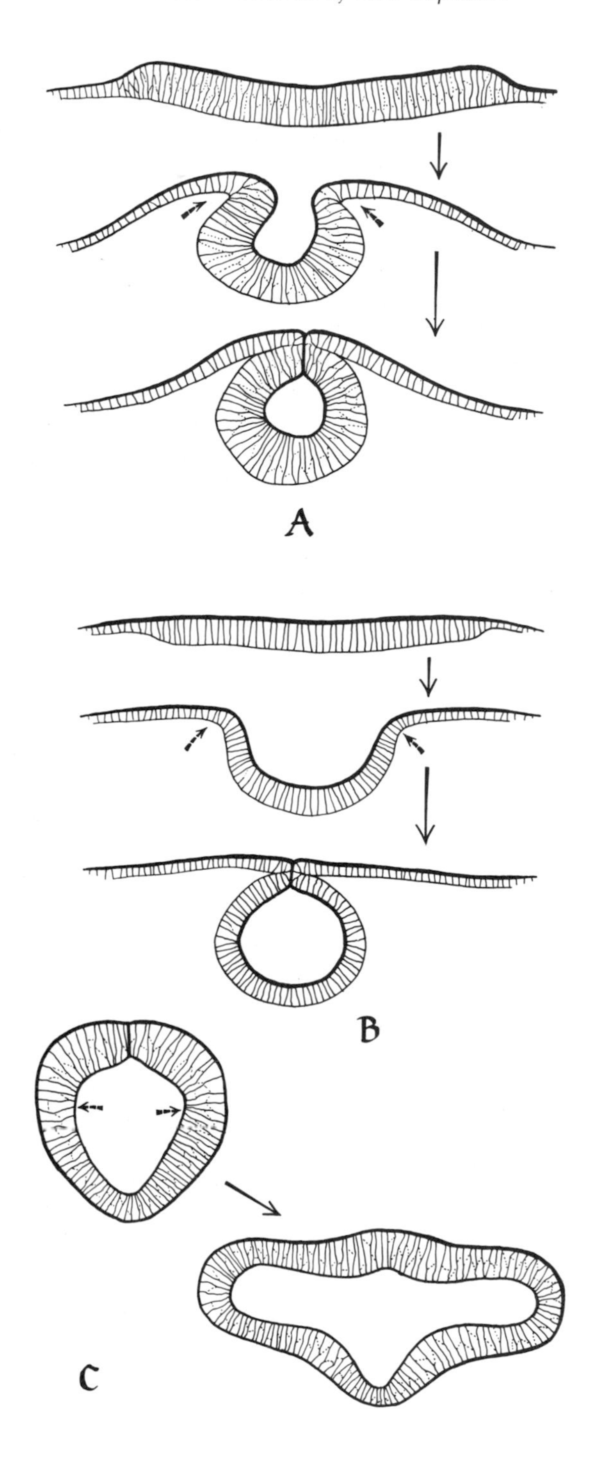

Figure 2-14. *Morphogenetic tissue movements involving folding (A), invagination (B), and evagination (C).*

A shows the rolling up of the flat neural plate to form the neural tube. B shows the invagination of the flat ear placode to form the inner ear vesicle (otocyst). C shows the evagination of the lateral sides of the neural tube in the brain region to form the eye vesicles.

Individual Cell Movement

Among the many examples of individual cell movements in animal embryos are the migration of neural crest cells from the dorsal part of the neural tube and the aggregation of these cells to form ganglia, cartilages, etc. (Figure 2-15A), the migration, aggregation, and fusion of myoblast cells to form the multinucleated myotubules of muscles; and the aggregation and sorting out of artificially dispersed mixtures of tissue cells (Chapter 8). In the life cycle of the cellular slime molds (Chapter 6), migration and aggregation of the individual amebae results in formation of a multicellular organism (Figure 2-15B).

Intracellular Movement

Intracellular movement of cytoplasmic regions in some animal eggs before or after fertilization is undoubtedly of significance to later development. This cytoplasmic movement, called "ooplasmic segregation," may establish an orderly pattern of distribution of cytoplasmic components—a type of intracellular differentiation. The kind of movement described for the egg of the marine worm, *Chaetopterus* (Costello, 1948) is diagramed in Figure 2-16. In this case the movements begin when the egg is shed into seawater. As illustrated in the figure, the ectoplasmic cap of the egg flows down and covers the endoplasmic area. The small endoplasmic area at the animal pole enlarges and the maturation spindle becomes attached there. Cytoplasmic streaming (directional movement) is begun by fertilization in some animal eggs (tunicate and amphibian). Intracellular cytoplasmic movements also occur in maturing and fertilized plant eggs. For example, in the maturing egg of the fern *Pteris aquilinum*, DNA moves outward from the nucleus into the cytoplasm. When this egg is fertilized, evaginations of the nuclear membrane (Chapter 13) appear to move out and apparently play a role in the "capture of the spermatozoan" (Bell, 1964). The cytoplasmic movements of the fertilized egg of the cotton plant (Jensen, 1964) are diagramed in Figure 2-17.

We shall examine the significance of the cytoplasmic structure of eggs in Chapter 13, and some additional examples of ooplasmic segregation will be described. For the present it is sufficient to note that cytoplasmic differentiation resulting from the movement of components probably is an important guideline for the later development of these eggs.

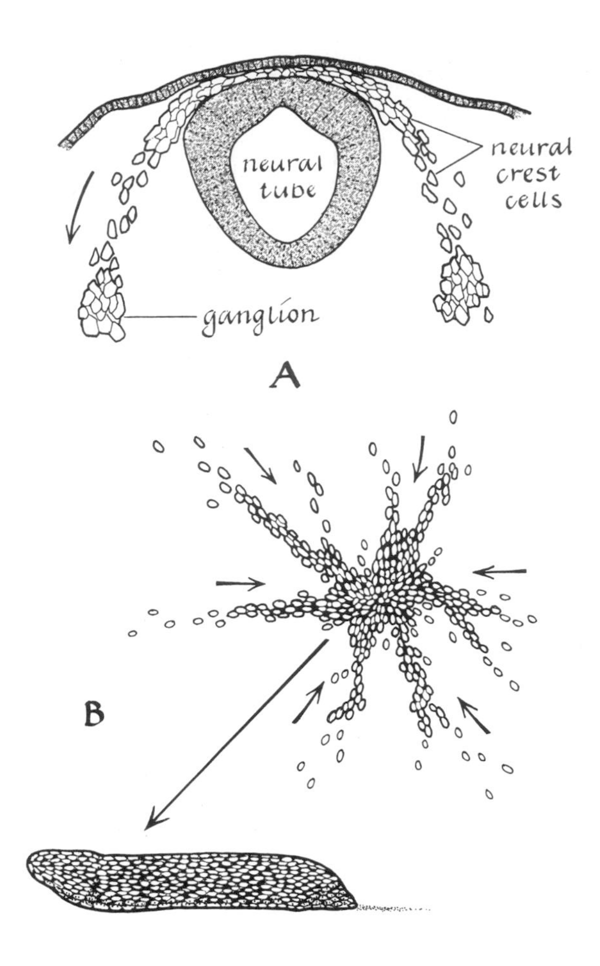

Figure 2-15. *A, individual cell movements and aggregation of cells to form ganglia; and B, aggregation of amebae to form a pseudoplasmodium (slug) in the cellular slime mold.*

Metabolism

The metabolic processes by which energy is made available for development underlie all of the other processes: growth, differentiation, cellular interactions, and movement. In addition, energy is needed for the maintenance of the structural organization and functional activity of each level of complexity. Changes in metabolism (respiration, sources of energy, activities of enzymes) occur during development and in some cases can be correlated with developmental events. Indeed, enzyme activity and functional differentiation are correlated (reviews of Moog, 1952; Boell, 1955). It seems probable that the same metabolic machinery furnishes energy for all the basic processes, including maintenance (Boell, 1955). There is no evidence that the sources of energy employed by developing animal organisms are significantly different from those of the fully formed adult.

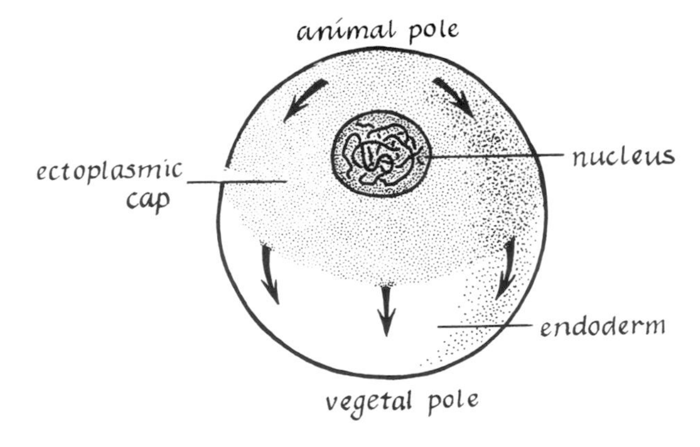

Figure 2-16. *Intracellular movement (ooplasmic segregation) of the ectoplasmic cap to cover the endoplasm, following fertilization of the egg of* Chaetopterus.

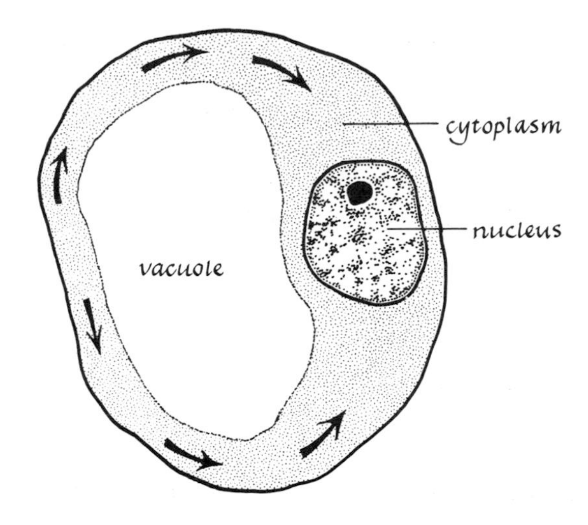

Figure 2-17. *Cytoplasmic movements in the fertilized egg (zygote) of a cotton plant.*

Conclusion

Probably no single basic process of development is exclusively characteristic of any stage of development. It is likely that the processes are in operation at most stages of development. Recognition of the basic processes in the context in which they occur is a first step in understanding development. Knowledge of the mechanisms that control these processes will be the second step. How the control mechanisms are coordinated to insure the orderly expression of inherited potential is the third step, the ultimate goal. Progress in these directions has been slow but sufficient to encourage continuing attempts to understand how developmental phenomena are controlled.

References

Abeloos, M. 1932. La Régénération et les Problèmes de la Morphogenèse. Gauthier-Villars, Paris.

Astbury, W. T. 1945. The forms of biological molecules. In Essays on Growth and Form, pp. 309–354. Oxford University Press, Oxford.

Bell, P. R. 1964. The membranes of the fern egg. In M. Locke, ed., Cellular Membranes in Development, pp. 221–231. Academic Press, New York.

Boell, E. J. 1955. Energy exchange and enzyme development during embryogenesis. In B. H. Willier, P. Weiss, and V. Hamburger, eds., Analysis of Development, pp. 520–555. W. B. Saunders, Philadelphia.

Bonner, J. T. 1952. Morphogenesis: An Essay on Development. Princeton University Press, Princeton, N.J.

Costello, D. P. 1948. Oöplasmic segregation in relation to differentiation. Ann. New York Acad. Sci. **49:**663–683.

DeBeer, G. R. 1924. Growth. Edward Arnold, London.

Glücksmann, A. 1951. Cell deaths in normal vertebrate ontogeny. Biol. Rev. Camb. Phil. Soc. **26:**59–86.

Goss, R. J. 1964. Adaptive Growth. Logos Press, London.

Huxley, J. S. 1932. Problems in Relative Growth. Methuen, London.

Jensen, W. A. 1964. Cell development during plant embryogenesis. In Meristems and Differentiation, pp. 179–202. Brookhaven Symp. Biol., Upton, New York.

Lenhoff, H. H., and W. F. Loomis. 1961. The Biology of Hydra. University of Miami Press, Coral Gables, Florida.

Leopold, A. C. 1967. The mechanism of foliar abscission. Symp. Soc. Exp. Biol. **21:**507–516.

Moog, F. 1952. The differentiation of enzymes in relation to functional activities of the developing embryo. Ann. New York Acad. Sci. **55:**57–66.

Morgan, T. H. 1901. Regeneration. Macmillan, New York.

Needham, A. E. 1964. The Growth Process in Animals. Pitman, London.

Needham, J. 1950. Biochemistry and Morphogenesis, 2nd ed. Cambridge University Press, Cambridge.

Papaconstantinou, J. 1967. Metabolic control of growth and differentiation in vertebrate embryos. In R. Weber, ed., The Biochemistry of Animal Development, pp. 57–113. Academic Press, New York.

Pearsall, W. H. 1927. On the relative sizes of growing plant organs. Ann. Bot. **41:**549–556.

Saunders, J. W. 1966. Death in embryonic systems. Science **154:**604–612.

Saunders, J. W., and J. F. Fallon. 1966. Cell death in morphogenesis. In M. Locke, ed., Major Problems in Developmental Biology, pp. 289–314. Academic Press, New York.

Sinnott, E. W. 1939. The cell-organ relationship in plant organization. Growth Symp. **3:**77–86.

Sinnott, E. W. 1960. Plant Morphogenesis. McGraw-Hill, New York.

Spratt, N. T., Jr. 1947. Regression and shortening of the primitive streak in the explanted chick blastoderm. J. Exp. Zool. **104:**69–100.

Spratt, N. T., Jr. 1964. Introduction to Cell Differentiation. Reinhold, New York.

Thompson, D. W. 1942. On Growth and Form. Cambridge University Press, Cambridge.

3 The Storage, Transmission, and Use of Genetic Information: Sequential and Differential Gene Action

The sequence of stages by which organisms develop is not fortuitous but specified by the genome of the species. In this respect, developmental biology can almost be equated with developmental genetics, the study of the phenotypic expression of inherited capacities. In molecular biology, the central problem of development is what kinds of molecular mechanisms and inherited chemical information provide the instructions for the sequence of stages, their characteristic features, and their component processes. What is the chemical nature of the hereditary material passed from generation to generation of an organism, and how is this information used to construct a new individual?

Earlier studies had suggested the role of a special group of organic chemicals—nucleic acids—in heredity, but not until the early 1950s did the studies of Pollister, Swift, and Alfert clearly implicate deoxyribonucleic acid (DNA) rather than the proteins also present in the chromosomes as the hereditary material of all organisms (except some viruses in which the hereditary molecules are ribonucleic acid, RNA). The structure of DNA was later deduced by Watson and Crick (1953) and Wilkins (1956) to be a linear sequence of nucleotides: adenylic, guanylic, cytidylic, and thymidylic acids.

The specific sequence of nucleotides is now known to be a code—a kind of language specifying the sequence of amino acids, the "building blocks" of polypeptides and proteins (see reviews of Sullivan, 1967; DeBusk, 1968). This discovery was important because the diverse types of cells in an organism are largely characterized by the kinds of enzymes and other proteins they contain. It is now known that the DNA in these cells specifies the synthesis of their enzymes and other proteins. The enzymes synthesized under control of the DNA determine the products found in cells. These cellular products, many of which are capable of self-assembly, determine the characteristics of cellular organelles such as microtubules and membranes. These organelles help control the shape and function of cells, as we shall see in Chapter 25. Development involves the emergence of order and the increasing complexity of structure and function. Small units of structure associate to form larger units, eventually whole cells and multicellular organisms. Thus, there is a structural and developmental continuity from DNA to the completely developed organism.

It is reasonably certain that all cells of a multicellular organism contain the same kind of inherited information, passed from a cell to its daughter cells during cell division. However, it is also established that specialized cells such as nerve, muscle, and skin cells synthesize different proteins. These differences cannot result solely from inherited information; they must be the result of different ways in which the information is used. In this chapter we shall first review our information about the role of the genetic code in the storage, transmission, and use of inherited information. Then, we will survey some evidence for the constancy of DNA and for sequential and differential gene action in development.

Control of Protein Synthesis

In describing how the genes control developmental processes and, specifically, how they control the synthesis of enzymes and other cell proteins, our first concern is with the method of replication of DNA during cell division so that both daughter cells receive the same genetic information. A second concern is the method by which the genetic instructions in DNA are copied by RNA. Finally, we examine the method by which RNA copies control the synthesis of proteins.

Replication of DNA

The chemical composition of DNA, the basic hereditary material, is an assembly of nucleotides—adenylic (A), thymidylic (T), guanylic (G), and cytidylic (C) acids—whose order in a DNA molecule is the basis of the genetic code (Sinsheimer, 1967). A linked set of deoxyribonucleotides is shown in Figure 3-1. Each nucleotide consists of a purine base (such as adenine and guanine) or a pyrimidine base (such as cytosine and thymine), the five-carbon sugar deoxyribose, and a phosphate group. The nucleotides are linked together by the phosphate groups. Polynucleotide chains of great length are formed in this way.

The four nucleotides in which the genetic message is written, A T G C, are used by the cell in groups of three, called codons. Codons, which may be compared to words, specify which amino acids are to be linked to form a protein. There are 64 codon groups but only 20 commonly used amino acids. Thus, there is usually more than one codon that specifies a particular amino acid. For example, the DNA codons CCA, CCG, CCT, and CCC all specify glycine. The codons for the amino acids are given in Figure 3-2 (Sinsheimer, 1967).

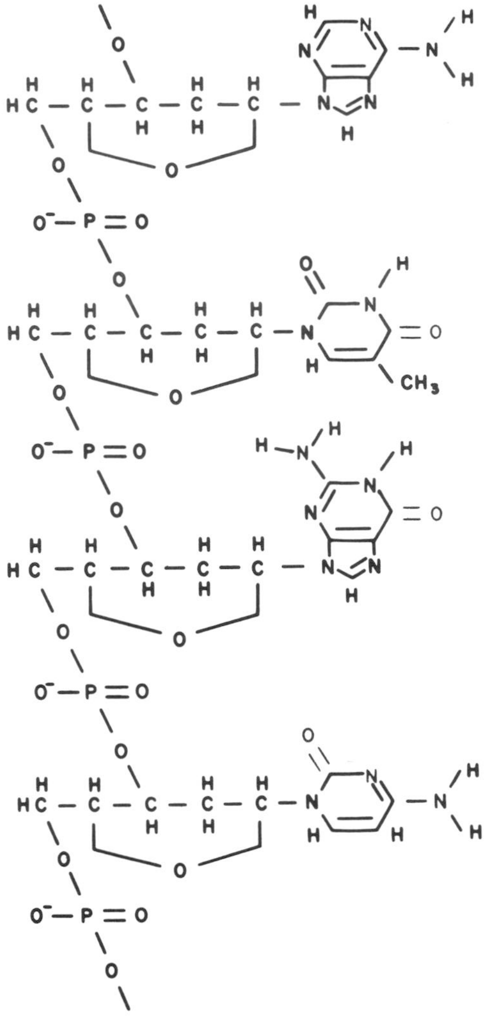

Figure 3-1. *A linked set of deoxyribonucleotides.*

The DNA codons were not determined directly but through experiments in which a complementary nucleic acid, RNA, was used as a template for the incorporation of amino acids into protein. Some DNA and RNA triplets do not code for any amino acid but mark the end of a sequence of triplets that code for a particular polypeptide or protein. Because of this device, the entire DNA strand is not read at once.

We have tentatively accepted the hypothesis that every cell of an organism has the complete and same genetic information (the same DNA code), which must be copied precisely every time a cell divides.

Codon	Amino acid	Codon	Amino acid	Codon	Amino acid	Codon	Amino acid	Codon	Amino acid	Codon	Amino acid	Codon	Amino acid	Codon	Amino acid
AAA	PHE	AGA	SER	ATA	TYR	ACA	CYS	TAA	ILEU	TGA	THR	TTA	ASPN	TCA	SER
AAG		AGG		ATG		ACG		TAG		TGG		TTG		TCG	
AAT	LEU	AGT		ATT	END?	ACT	TRY	TAT	MET	TGT		TTT	LYS	TCT	ARG
AAC		AGC		ATC		ACC		TAC		TGC		TTC		TCC	
GAA	LEU	GGA	PRO	GTA	HIS	GCA	ARG	CAA	VAL	CGA	ALA	CTA	ASP	CCA	GLY
GAG		GGG		GTG		GCG		CAG		CGG		CTG		CCG	
GAT		GGT		GTT	GLUN	GCT		CAT		CGT		CTT	GLU	CCT	
GAC		GGC		GTC		GCC		CAC		CGC		CTC		CCC	

Figure 3-2. *DNA codons and the amino acids they specify in protein synthesis. PHE = phenylalanine; LEU = leucine; SER = serine; TYR = tyrosine; CYS = cystine; ILEU = isoleucine; MET = methionine; THR = threonine; ASPN = asparagine; LYS = lysine; ARG = arginine; PRO = proline; HIS=histidine; GLUN=glutamine; VAL=valine; ALA= alanine; ASP=aspartic acid; GLY=glycine; GLU=glutamic acid; TRY=tryptophan.*

The cell uses a special device to do this. First, DNA synthesizes its complement from a pool of deoxyribonucleotides; A pairs with T and G with C. The complementary DNA then directs the synthesis of its complement, an exact copy of the original DNA, by the same kind of pairing. The complementary chain carries no information—it is nonsense—but is a device for replication of the sense chain (Figure 3-3).

Original DNA (Sense) Chain		*Synthesis of Complementary DNA*		*Synthesis of New DNA Chain*
	Pairing		Pairing	
A	⟷	T	⟷	A
G	⟷	C	⟷	G
G	Step 1	C	Step 2	G
T		A		T
C		G		C
T		A		T
C		G		C
A		T		A
A		T		A
G		C		G

Figure 3-3. *The method of replication of DNA.*

With rare exceptions, the DNA molecule is a double chain of nucleotides, the chain containing usable information (the sense chain) being paired with its exact complement. In cell division the sense chain is a template for synthesis of its complementary chain, and the complementary chain is a template for synthesis of a new sense chain, as is shown in Figure 3-4.

Transcription of DNA

DNA not only duplicates itself with every cell division, but also functions as a template on which a complementary RNA is synthesized under control of an RNA polymerase. Some of this RNA is a template for the synthesis of proteins. The process of RNA synthesis is called transcription and the process by which RNA controls protein synthesis is called translation. The sequence of processes by which DNA duplicates itself and directs the synthesis of proteins is summarized by the diagram below.

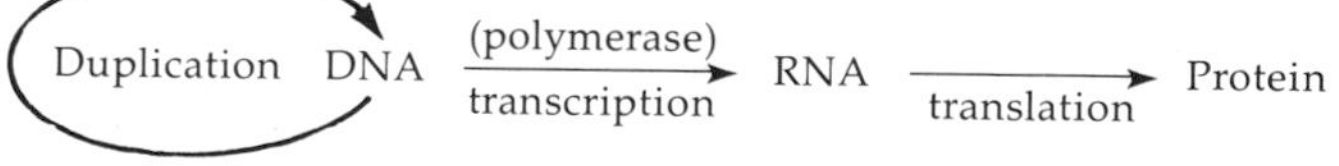

Duplication ↻ DNA —(polymerase) transcription→ RNA —translation→ Protein

Details of these processes have been worked out in microorganisms (see reviews of Atwood, 1965; DeBusk, 1968). In the transcription process, the

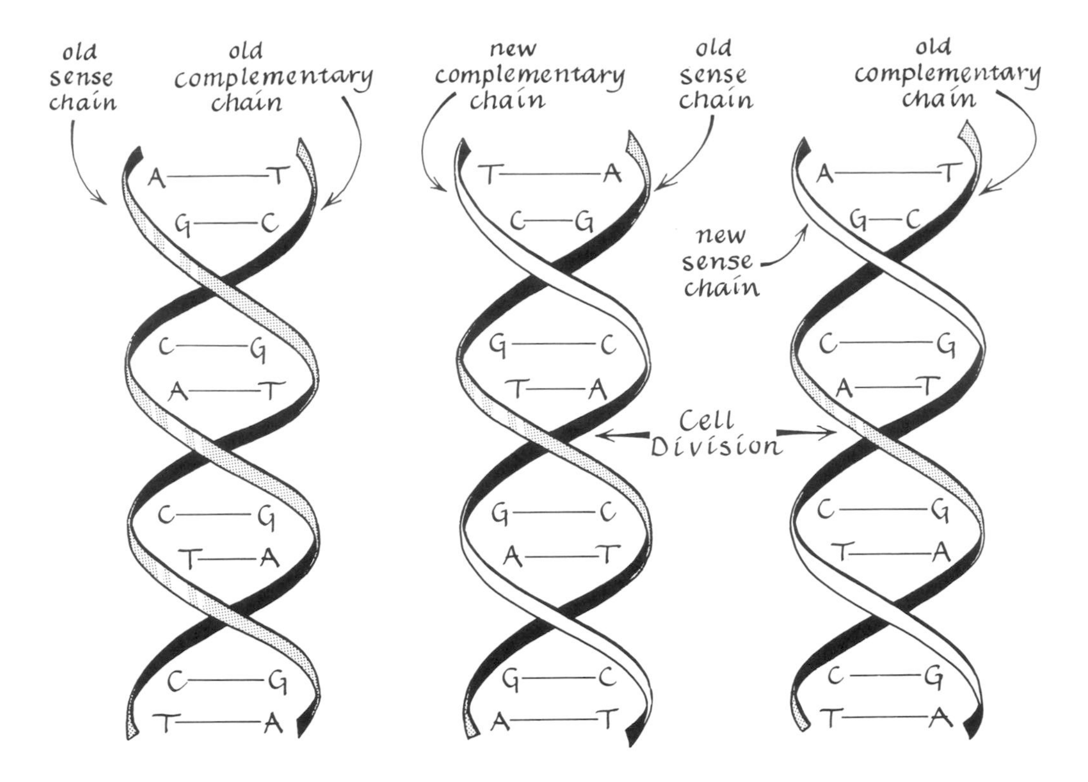

Figure 3-4. *The sense and complementary chains of DNA and the method of their replication.*

nucleotides of DNA—deoxyadenylic, thymidylic, deoxyguanylic, and deoxycytidylic acids—serve as a template to synthesize the complementary RNA—uridylic (U), adenylic (A), cytidylic (C), and guanylic (G) acids respectively. The nucleotide uridylic acid replaces thymidylic acid in RNA. Some of this RNA is a template for the sequential linking of the amino acids to form a protein.

According to the currently accepted sequence, first a DNA-like RNA, messenger RNA (mRNA), is transcribed. The mRNA leaves the nucleus, enters the cytoplasm, and associates with another kind of RNA in the ribosomes (rRNA). The rRNA is also a product of transcription of other parts of the DNA. Smaller RNA molecules called transfer RNA (tRNA) that are specific for each amino acid are transcribed from still other portions of the DNA, combine with the corresponding amino acids, and attach to the mRNA on the ribosome. Through the joint action of the three classes of RNA—mRNA, rRNA, and tRNA—the amino acids are linked in a specific sequence to form a polypeptide chain, the basic structural unit of a protein. The mRNA determines the amino acid sequence. Even a synthetic RNA such as polyuridylic acid can act as mRNA, specifying the synthesis of the polypeptide polyphenylalanine. This significant discovery of Nirenberg and Matthaei (1961) was a breakthrough in deciphering the genetic code (see Nirenberg, 1963). As one might expect from the fact that not all the inherited capacities of an organism are used at once, only about 5 to 10 percent of the information in the DNA of a chromosome is transcribed at any one time during development (McCarthy and Hoyer, 1964; Paul, 1967).

Translation of RNA

The diagram in Figure 3-5 illustrates the supposed mechanism by which the information in a specific RNA molecule is translated into a specific sequence of amino acids in a protein (Warner et al., 1962; Rich et al., 1963).

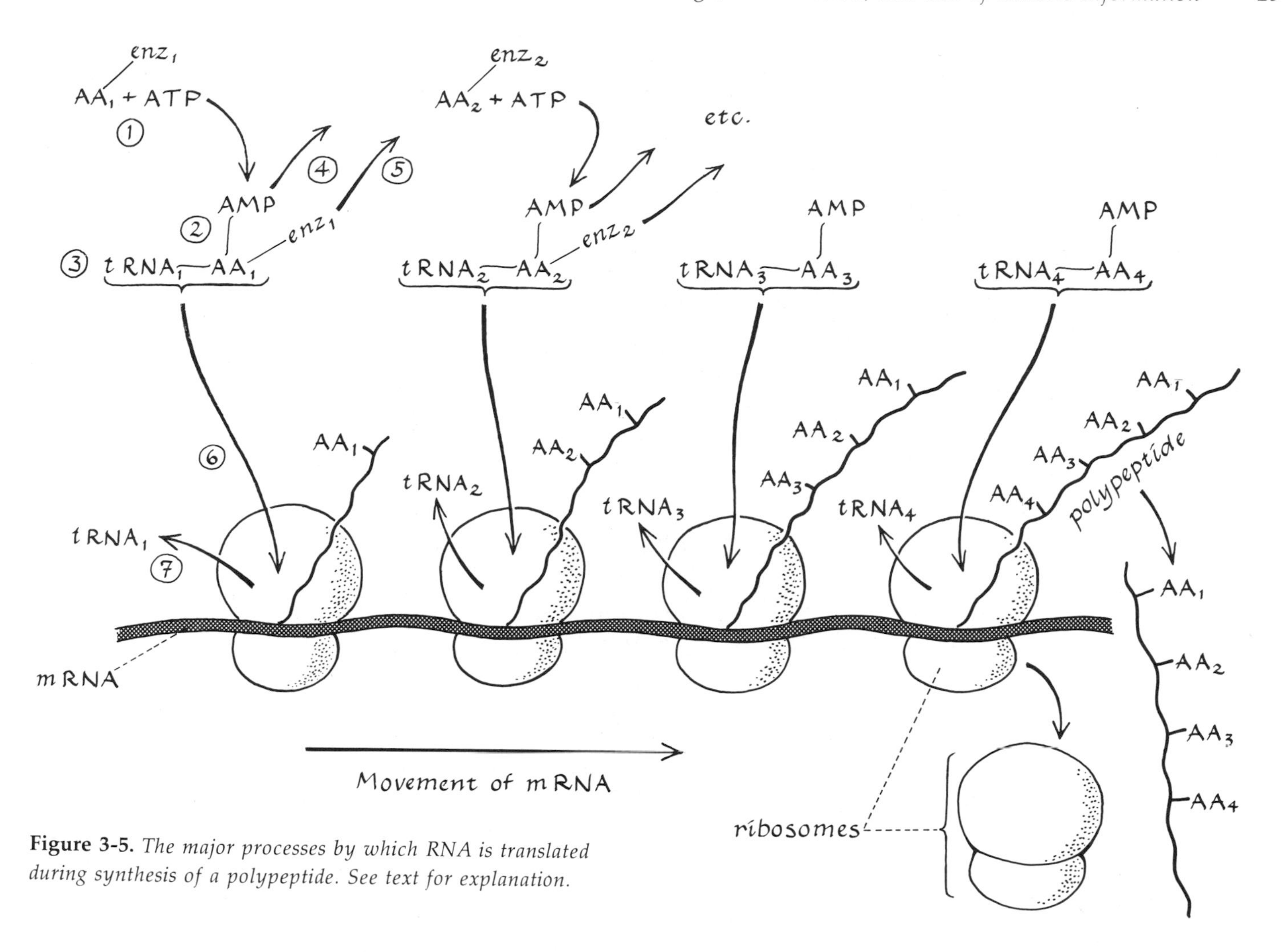

Figure 3-5. *The major processes by which RNA is translated during synthesis of a polypeptide. See text for explanation.*

As shown in the figure, each amino acid (AA) is "activated" by a specific enzyme and joined to adenosine triphosphate (ATP) before being assembled to form the polypeptide chain of a protein. In this process, two of the three phosphate groups of the ATP are released; the remaining adenosine monophosphate (AMP) becomes attached to the amino acid by an energy-rich bond, ≃. Finally the activator enzyme ~ amino acid attaches to a specific molecule of transfer RNA. In this process the molecule of AMP is discarded, and the energy-rich bond binds the amino acid to the transfer RNA molecule. The activating enzyme then moves away. The tRNA molecule mediates via its anticodon triplet in attaching the amino acid to its complementary code word in the mRNA strand and then is released. Another tRNA-AA complex enters the ribosome and, if specified by the codons in mRNA, is joined to the first AA by the energy-rich bond it received from the ATP. The energy-rich bonds become the peptide bonds by which the amino acids are linked to form a polypeptide. As the process continues, the mRNA probably moves through the ribosomes and the amino acid chain grows longer (see review of Loewy and Siekevitz, 1963). An increasingly long chain of amino acids constituting a polypeptide is thus built up rapidly, at the rate of about one amino acid per half second. The transfer RNA molecules are released and can be used again for further protein synthesis. The mRNA molecule can also be used again, but it eventually becomes degraded and new mRNA molecules must be synthesized.

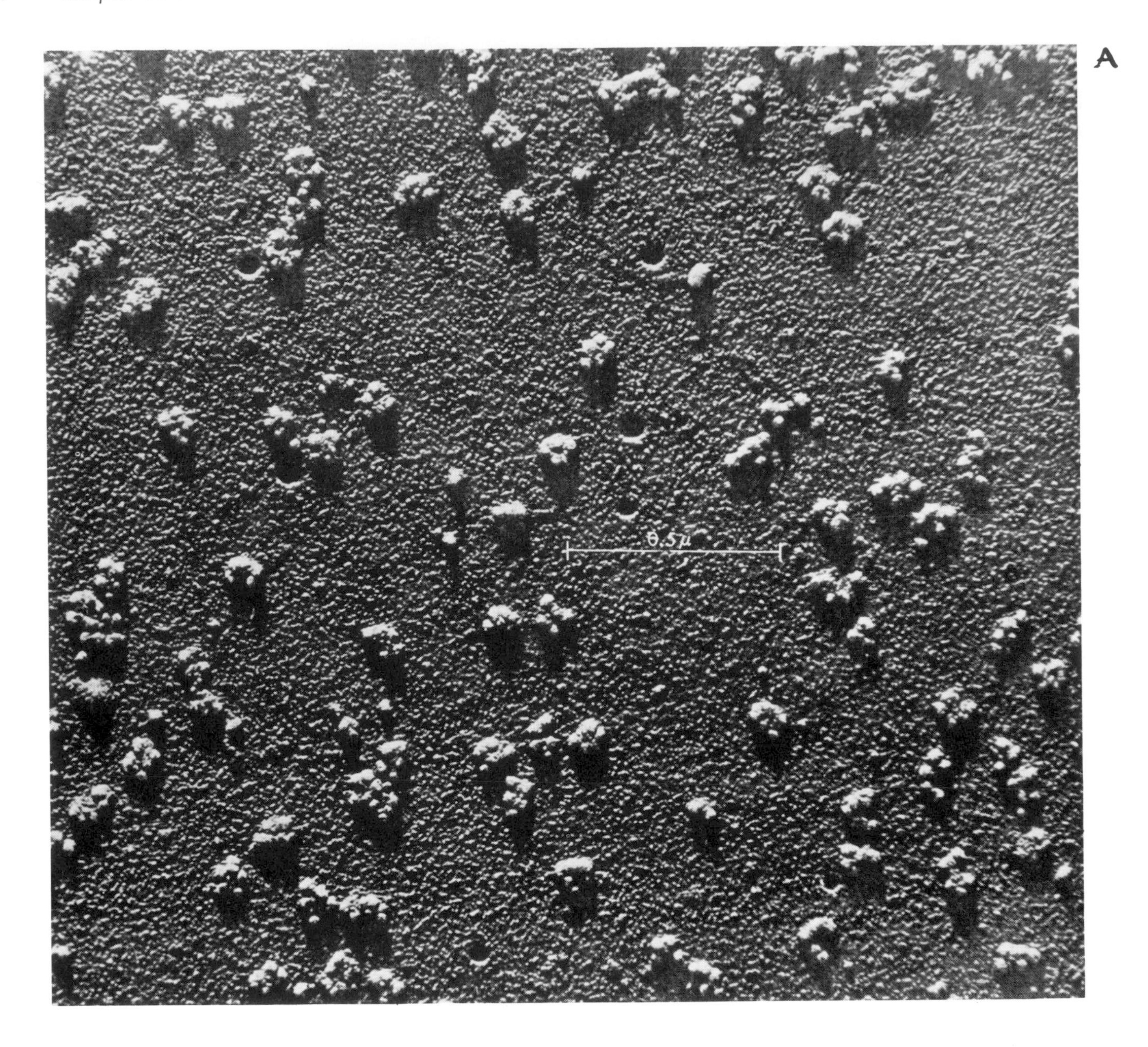

Studies of Rich et al. (1963) and Slater et al. (1963) with rabbit reticulocytes indicate that protein synthesis occurs on clusters of ribosomes (polysomes). The clusters, as revealed in electron micrographs, consist of four to six ribosomes held together by a thin strand (Figure 3-6A and B). Polysomes also occur in plant cells (Figure 3-6C). Since Nomura et al. (1960) had demonstrated that mRNA is associated with ribosomes in the bacterium *E. coli,* Rich and co-workers believe that the strand is a long mRNA molecule. Polysomes are found in many plant, animal, and microbial cells, and they appear to represent protein synthesis in action.

If every cell of an organism contains the same DNA, then different kinds of cells must result from differential use of the coded information. In terms of nucleic acids, cellular differentiation would be, in many instances, the result of synthesizing different kinds of mRNA molecules. In the following section we shall consider some evidence supporting this attractive hypothesis of tissue specificity of messenger RNA as well as that of DNA equivalence in diverse cell types.

Figure 3-6. *Electron micrographs of polysomes in animal (A and B) and plant (C) cells. A, from Warner et al.,* Science *138: 1399–1403. Copyright 1962 by the American Association for the Advancement of Science. B, from Slater et al., copyright 1963, by Academic Press. C., from Jensen and Park,* Cell Ultrastructure, *courtesy Dr. E. Newcomb and Dr. H. Bonnett, copyright 1967 by Wadsworth Publishing Co.*

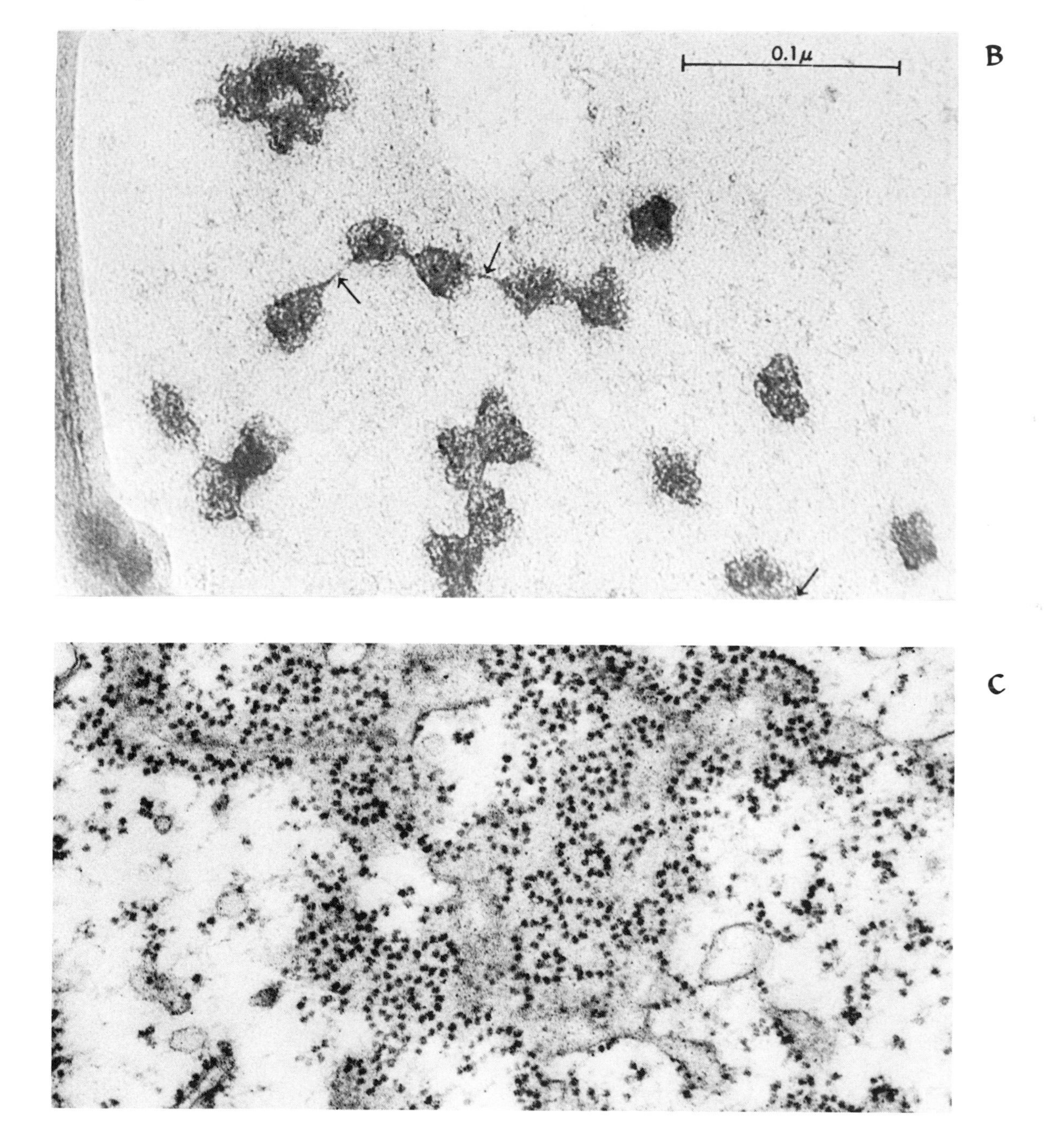

DNA Equivalence and Its Differential Transcription

Mirsky and Ris (1949) assayed various animal cells (erythrocytes, sperm and liver cells of chicken, carp, and frog) for their DNA content per nucleus. They found that in a given animal the amount of DNA per nucleus is constant, and that sperm cells have one-half the DNA content of the somatic cells. An exception to DNA constancy has been reported (Rosenkranz, 1965) in which the DNA of gastrula and adult tissues of the sea urchin *Arbacia* has a different base ratio, but this is unconfirmed.

DNA-DNA hybridization experiments provide a direct test for DNA equivalence in all cells of an organism (Nygaard and Hall, 1963; McCarthy and Hoyer, 1964). In the experiments of McCarthy and Hoyer, DNA was extracted from mouse embryo, mouse brain, liver, kidney, thymus, and spleen and depolymerized into the two single strands by heating. The separated strands were dispersed in an agar column (Figure 3-7A). When, for example, single-stranded liver ^{14}C-labeled DNA was allowed to diffuse through the agar column, labeled hybrid molecules were formed in the agar not only between the agar-imbedded liver DNA and the diffusing DNA but also between the diffusing-liver DNA and the kidney, thymus, and other DNAs (Figure 3-7C). The hybrid molecules are double stranded and result from the pairing of complementary nucleotides. When labeled kidney DNA was allowed to diffuse through the column, the binding of kidney with liver DNA also tended to be complete (10 percent error in the method); it was therefore concluded that the nucleotide sequences in the DNA of the two cell types is equivalent. A similar result was obtained when DNA from other tissues hybridized with liver DNA. Thus DNA from one kind of mouse cell hybridizes with DNAs from other types of mouse cells just as well as with DNA from the same cell type. However, when DNA from *Bacillus subtilis* was used, no hybridization occurred.

In other experiments of McCarthy and Hoyer, RNA from several mouse tissues (spleen, liver, kidney) was hybridized with mouse embryo DNA. For example, when liver RNA diffused through the agar column containing DNA, hybrids (DNA-RNA) were formed. Upon saturation, no more hybrid molecules were formed when more liver RNA was added. However, when kidney RNA was then added to the column, additional hybrid molecules were formed

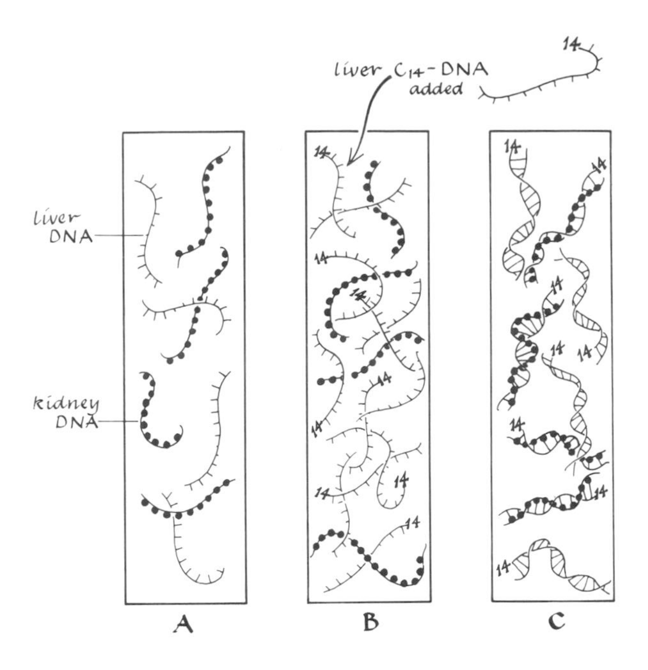

Figure 3-7. *Hybridization of mouse liver and kidney DNA.*

because the kidney RNA could form hybrids with complementary sequences in the liver DNA other than those used to form the RNA's characterizing liver cells. This means that kidney RNA has a different nucleotide sequence from liver RNA. In other words, different types of RNA molecules are transcribed by different cell types in the mouse.

Paul (1967) and Paul and Gilmour (1968) have reported additional evidence for differential transcription of DNA: in hybridization experiments 50 percent more DNA was transcribed in mammalian bone marrow than in thymus. A large common complement of DNA was transcribed in both tissues, but a large unique sequence was transcribed only in bone marrow. These and other results suggest organ-specific transcription and masking of DNA (see Ursprung and Smith, 1965). DNA is also transcribed differentially during cell differentiation in the shoot and root apices of plants. In roots a four- to five-fold increase in RNA per cell is accomplished by a change in the ratio of purine to pyrimidine bases (Heyes, 1963). This indicates that a change in the composition of RNA is involved.

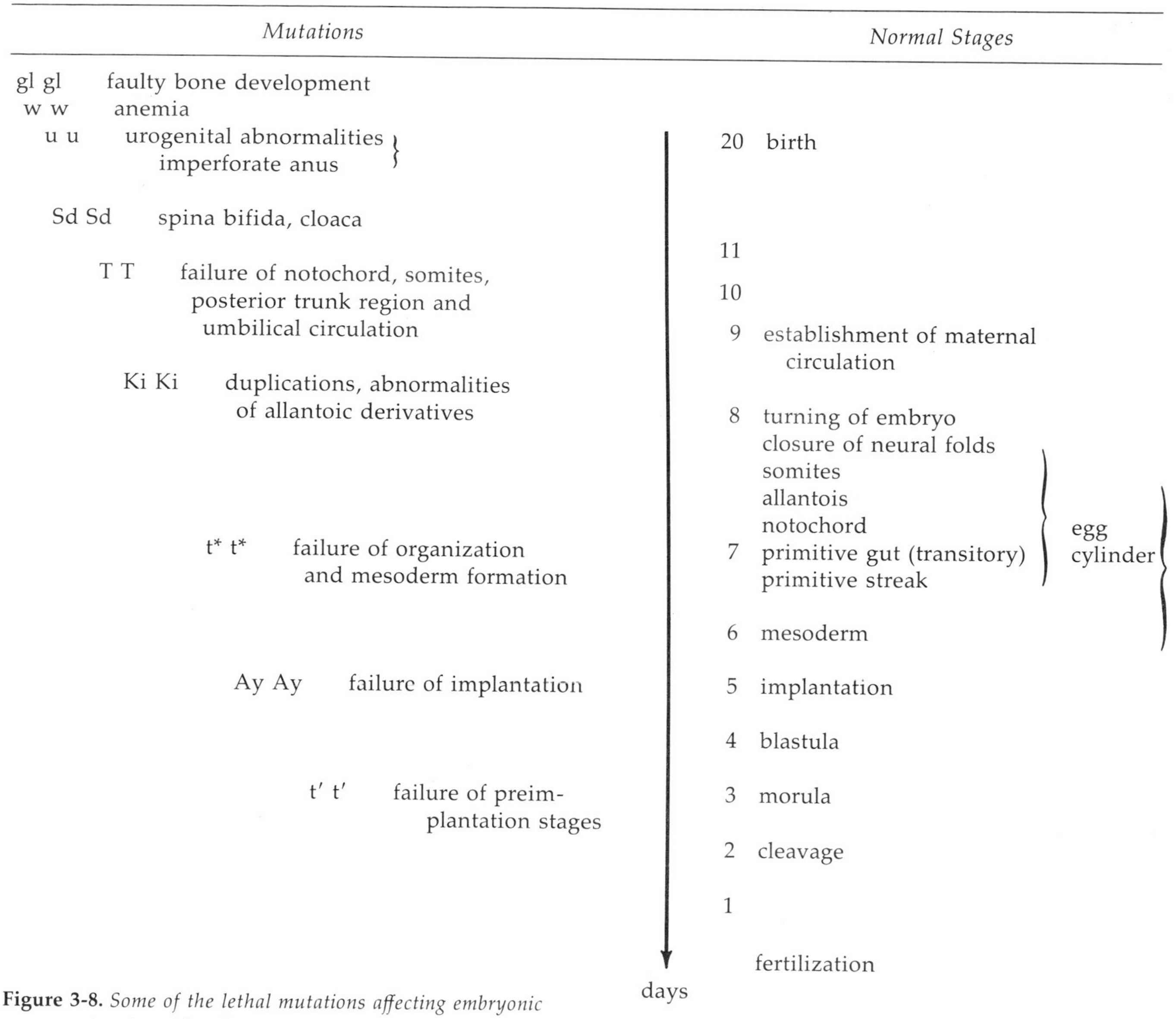

	Mutations	Days	*Normal Stages*
gl gl	faulty bone development		
w w	anemia		
u u	urogenital abnormalities, imperforate anus	20	birth
Sd Sd	spina bifida, cloaca		
		11	
T T	failure of notochord, somites, posterior trunk region and umbilical circulation	10	
		9	establishment of maternal circulation
Ki Ki	duplications, abnormalities of allantoic derivatives	8	turning of embryo
			closure of neural folds, somites, allantois, notochord (egg cylinder)
t* t*	failure of organization and mesoderm formation	7	primitive gut (transitory), primitive streak (egg cylinder)
		6	mesoderm (egg cylinder)
Ay Ay	failure of implantation	5	implantation
		4	blastula
t′ t′	failure of preimplantation stages	3	morula
		2	cleavage
		1	
			fertilization

Figure 3-8. *Some of the lethal mutations affecting embryonic processes in mice. After Dunn in Willier, Weiss and Hamburger,* Analysis of Development. *W. B. Saunders Company, Philadelphia, 1955.*

Sequential Gene Action

As noted above, the orderly, sequential expression of genes is required by the sequence of stages in development. Studies of mutant genes provide direct evidence for the action of different genes at different stages (Stern, 1955). Figure 3-8 shows some lethal mutations affecting embryonic processes in mice.

Note in the figure that mutant genes interfere with embryonic processes from the morula stage to birth. In rare instances very early gene action has been detected—for example, immediately after fertilization in *Drosophila* (Counce, 1956); in other instances genes act long after birth. An example of this is the action of genes located on the X chromosome in abnormal human males with the combination

XXY. In this case development is normal until puberty, when the tubules of the testis atrophy and the individual becomes eunuchoid. Sequential gene action may mean not only one gene acting after another, but also sequential action of groups of genes which direct a program of synthesis. The programs may correspond to the stages of development. The time of action of many genes controlling development is not known, either because mutations in these genes have not been detected or because the mutations are lethal. Directly observing the activities of normal (wild-type) genes would greatly facilitate our understanding of the relation between sequential gene action and the stages of development.

Denis (1966, 1968) used the method of DNA-RNA hybridization to approach this problem. He found that none of the genes active in the tadpole stage of *Xenopus laevis,* the South African toad, are active during cleavage, but an increasing proportion of these genes are active in gastrula and neurula stages. He also found that some genes active during gastrulation apparently are not active at later stages. A similar program has been found in other embryos, such as those of sea urchins. As we shall see in Chapter 17, a program of nucleic acid synthesis probably underlies all patterns of development.

It is generally accepted that all cells of a developing or adult organism contain the same kinds of genes in their nuclei but use them differently in response to varying influences outside the genome. However, numerous examples of quantitative differences in information in nuclei of the same organism exist. Some genes have more copies in some cells than in others (as is the case in many eggs), and some cells of an organism contain more copies of the entire genome than others. The latter occurs in germ line and somatic line cell nuclei of the parasitic round worm *Ascaris* and of some insects, in the nuclei of salivary gland cells and other cells of insect larvae, and in the differentiated cells of plants, which are often polyploid (Torrey, 1959). Differences in the quantity of information in cells do not cause this differentiation. Rather, the genome is probably responding to external influences, either by loss of copies of the whole genome (for example, chromatin diminution in *Ascaris;* see Chapter 20) or amplification of the entire genome (as in the giant salivary gland chromosomes of *Drosophila*). However, direct proof of the maintenance of a full genome in all kinds of differentiated animal cells has not been found, despite existing examples of the conservation of genetic potentiality (Chapter 20; see also Schultz, 1965; Ebert and Kaighn, 1966).

Some evidence supports the hypothesis that cell differentiation results from differential use of the same information and thus that different genes are active in different kinds of cells and at different stages of development.

Differential Gene Action

Study of Giant Chromosomes

Apparent evidence of differential gene action is provided by the giant *polytene* chromosomes in the salivary glands and other larval tissues of dipteran insects such as *Drosophila, Sciara,* and *Chironomus.* These chromosomes result from repeated replication of DNA without chromosome, nuclear, or cell division and have up to 1,024 times the haploid DNA value in *Drosophila* and 4,096 times the haploid DNA value in *Sciara.* The striking feature of polytene chromosomes is their banding pattern, in which the bands correspond to gene loci. An inactive locus appears as a sharp band, an active locus as a "puff" (Figure 3-9).

Puffing Patterns

Different tissues of larval insects exhibit different patterns of puffing. In one tissue a particular locus may be a sharp band, while in another the same locus may be puffed. This occurs also in different lobes of the salivary gland (Figure 3-9). Furthermore, developmental changes within a single tissue are paralleled by corresponding changes in the puffing pattern (Beerman, 1956, 1961, 1966). However, in any one tissue only 10 to 20 percent of the bands form a puff at one time or another during development (Clever, 1966). This suggests that only a small number of genes are active in different tissues and at different stages of development. Figure 3-10 diagrams this observation.

The correlation between puffing patterns in different tissues indicates that we are observing cellular differentiation at the chromosomal and gene level. There is evidence that the puffs in giant chromo-

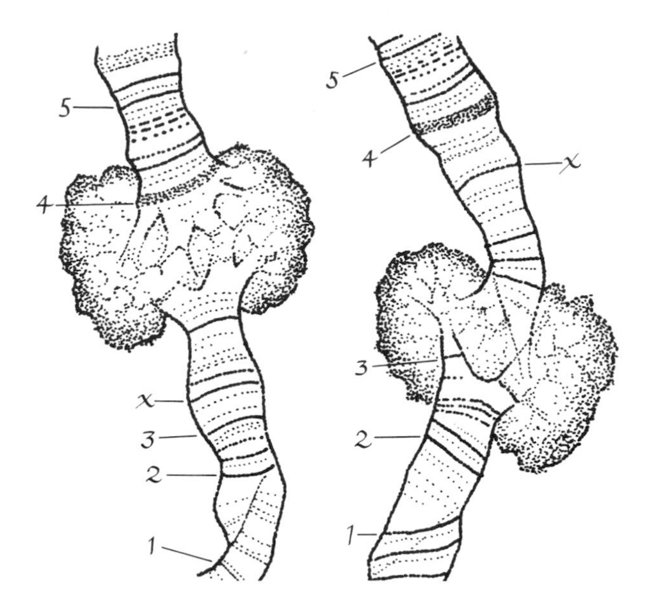

Figure 3-9. *The same chromosome region from different lobes of the salivary gland of* Irichocladius vitripennis *showing a strikingly different puffing pattern. The numbers indicate identified gene loci with respect to locus, x. Redrawn from Beerman (1952).*

somes are loci of the synthesis of RNA. In pulse-labeling experiments in which ³H-uridine (an RNA precursor) is injected into larvae, labeled RNA occurs almost exclusively in the puffs within 15 minutes. From two to twelve hours after injection, labeled RNA increases in the cytoplasm, but RNA in the puffs does not increase after two hours, indicating a transfer of RNA from the nucleus to the cytoplasm. Electron microscope studies on *Chironomus* larvae provide some evidence for the transfer of RNA-protein fibrils through pores in the nuclear membrane (Stevens and Swift, 1966). The RNA formed in the puffs is mRNA, which specifies the kind of protein to be synthesized in the cytoplasm (see reviews of Beerman, 1956, 1966; Clever, 1965; Pavan, 1965; Pelling, 1964). In addition to the puffs producing RNA, there are also puffs in other regions of the chromosome in which localized DNA replication occurs (Plaut and Nash, 1964; Swift, 1962). This was demonstrated, in the same sort of experiment as for RNA, by use of ³H-thymadine, a DNA precursor. The significance of localized DNA replication for cellular differentiation is not clearly understood but, as with RNA puffs, the patterns are not random. Further study of polytenic chromosomes—which

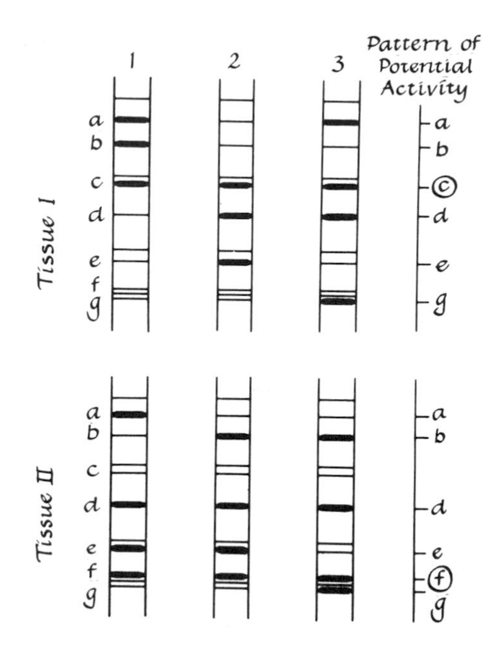

Figure 3-10. *Differences and similarities in the puffing patterns of different tissues* (I-II) *and different larvae* (1-3); a-g, *puff-forming loci. Note that only loci* c *and* f *form puffs peculiar to one of the two tissues, respectively. Except for these loci, the potential puffing activity of the two tissues is identical. From Clever, 1966, courtesy* American Zoologist.

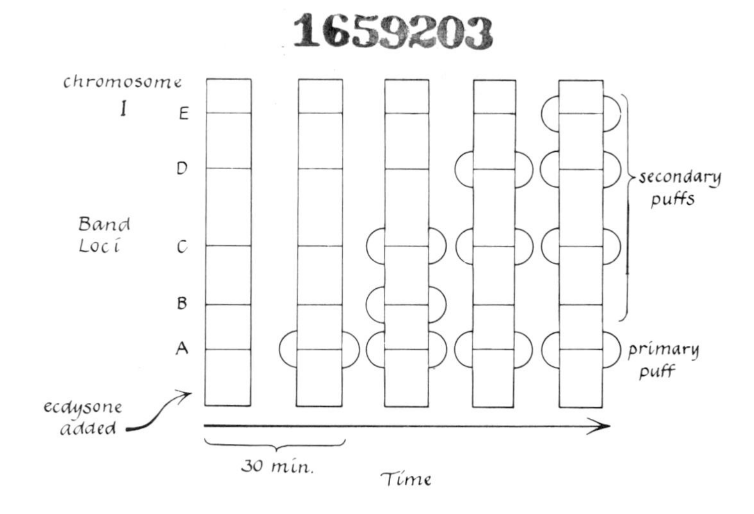

Figure 3-11. *A naturally occurring sequence of puffs in chromosome I of* Chironomous *larvae induced by injection of ecdysone. From the data of Clever et al. (1969).*

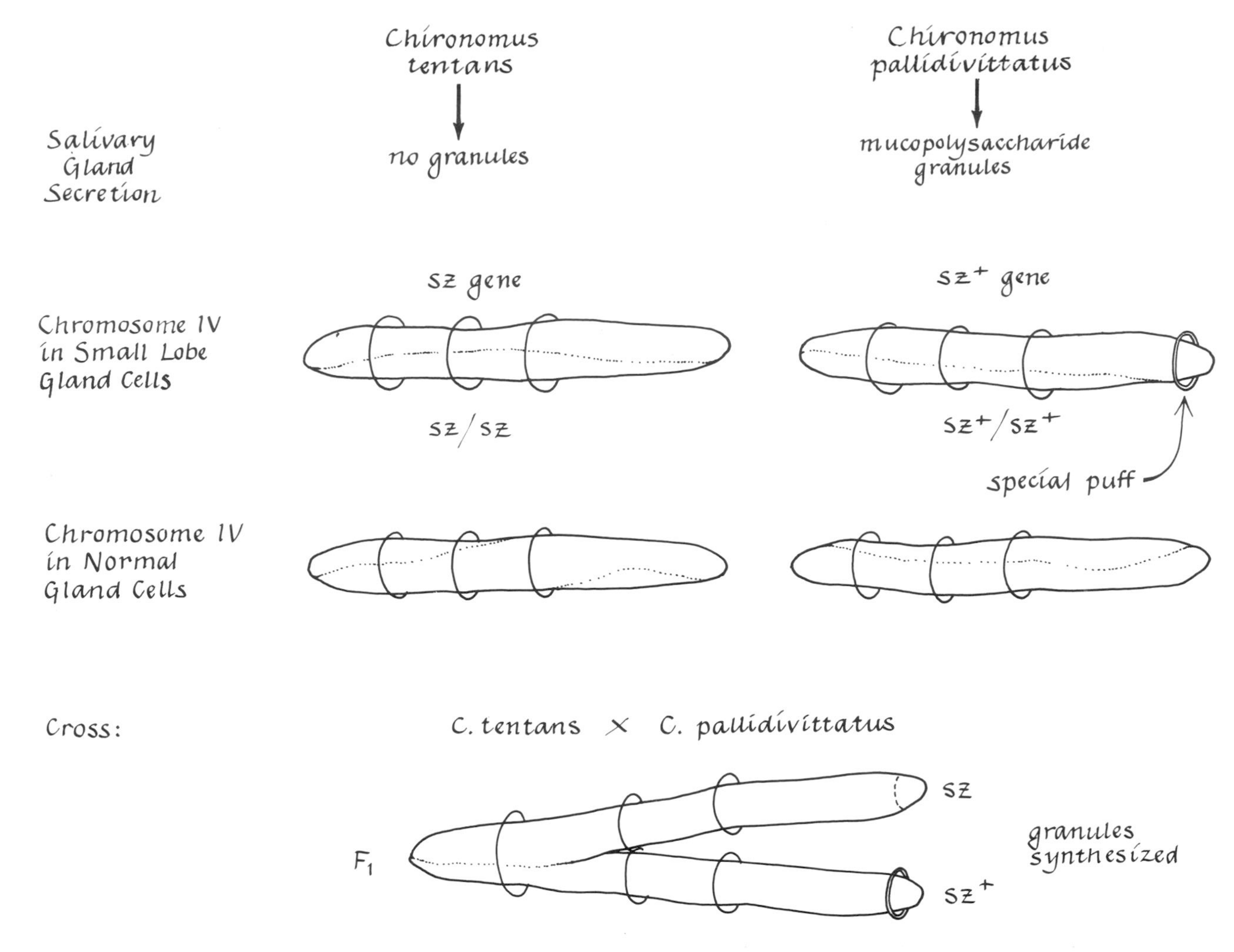

Figure 3-12. *Correlation between a specific puff locus and a specific gene product. From the data of Beerman (1966).*

exhibit visibly localized gene activity, unlike the smaller and more compact chromosomes found in most cells—will be necessary for a solution to this problem.

Environmental Modification of Puffing Patterns

Understanding the mechanisms that control puffing patterns is equivalent to understanding the control of gene action. Some information has been provided by use of ecdysone, the molting hormone of insects (Chapter 19). This hormone, in concentrations of 10^{-7} μg/mg larval weight injected into *Chironomus* larvae, induces a natural sequence of puffs in one chromosome (Clever et al., 1969). This is shown in Figure 3-11. Within one hour after the injection of ecdysone, a puff appears in another chromosome. Ecdysone induces a primary puff that may be an initiator locus for chains of secondary puffs (Beerman, 1963; Clever, 1964, 1966). Thus, the activity of some early reacting genes might release gene products that initiate the activity of other genes. If the activity of early reacting genes is blocked experimentally with the substance actinomycin-D, puffs that normally would have formed do not appear (Clever, 1966; Clever et al., 1969).

A control system in the nucleoplasm may also function in determining patterns of puffing in

Chironomus salivary gland chromosomes. The system is governed by the relative concentrations of Na^+ and K^+ ions. At an early stage of development, a high Na^+ to K^+ ratio induces certain loci to become active. Later a gradual change to a high K^+ to Na^+ ratio activates sequentially different sets of genes. Kroeger (1963) and Becker (1964) demonstrated this by altering the ionic composition of a culture medium in which isolated nuclei were placed. The extent to which ecdysone or other hormones or the composition of the microenvironment surrounding the genome control patterns of gene action remains to be discovered. Possibly, ecdysone controls the Na^+ to K^+ ratio through an unknown mechanism. This problem is similar to that of the mechanism of induction (Chapter 21).

Evidence for Puffing as a Visible Manifestation of Gene Activity

Beerman's experiments (1961, 1966) clearly indicate that a specific puff locus is directly correlated with a specific gene product. The experimenters crossed two species of *Chironomus.* In the species *pallidivittatus,* mucopolysaccharide granules are present in the cytoplasm of cells in the small lobe of the salivary gland. In *tentans* these granules are not present. In the hybrid, granules are present. The experiment is illustrated in Figure 3-12.

In the cells of the small lobe of *pallidivittatus,* chromosome IV exhibits a puff near its end. No puff is formed at the corresponding region of chromosome IV in *tentans.* Furthermore, the presence of the special puff is correlated with the presence of granules in the cytoplasm of the small lobe cells. Because the locus that puffs in *pallidivittatus* is a dominant one, sz^+ gene, the hybrid synthesizes granules. Beerman asks whether the recessive, *sz,* locus in *tentans* contains the same information as the sz^+ locus of *pallidivittatus.* Is use of this information blocked in *tentans?* Similar studies linking chromosomal puffs with cytoplasmic constituents have been reported by Laufer et al. (1964).

Study of Oocyte Chromosomes

Another visible manifestation of localized gene activity is found in the relatively long, "unwound"

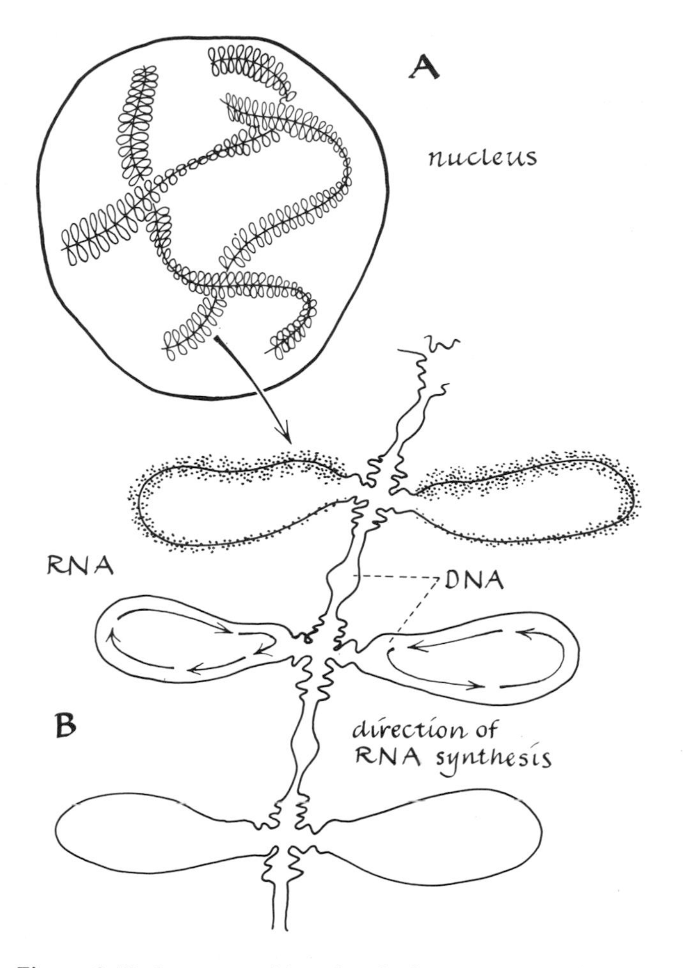

Figure 3-13. *Structure of lampbrush chromosomes in the nucleus of the oocyte of* Triturus. *Note that a chromosome contains two DNA strands and that RNA synthesis apparently starts at the base of a loop and travels around the loop (arrow in B). From the data of Gall (1963).*

chromosomes of oocyte nuclei in several animal species. The oocyte chromosomes of the salamander, *Triturus,* have been studied most often (Gall and Callan, 1962; Gall, 1963; Edstrom and Gall, 1963; Davidson et al., 1964). These chromosomes, called "lampbrush" chromosomes, have a fuzzy appearance caused by numerous loops of DNA strands extending out from the axis of the chromosome (Figure 3-13). As indicated in the figure, the DNA loops are covered with RNA. Radioautographic studies after injection of the animal with ^3H-uridine reveal that the loops are actively engaged in RNA synthesis. In *Triturus* there are about 20,000 loops in the oocyte nucleus. Gall and others suggest that the entire genome may function during oogenesis. This seems reasonable because, during oogenesis, the cytoplasmic information for later development is assembled. Note also in Figure 3-13 that the synthesis of RNA apparently starts at the base (one pole) of a loop and travels around the loop. This suggests a polarized use of information coded in DNA.

Conclusion

The discovery of the biochemical basis of heredity is particularly significant to students of heredity and development because it provides a means of understanding many aspects of inheritance and development in molecular terms. As we have seen, the hereditary material of all organisms consists of a linear assembly of nucleotides in groups of three, specifying which amino acids are to be linked to form polypeptides and proteins. The genetic language, DNA, is replicated with each cell division and transcribed into a complementary language, messenger RNA. In the cytoplasm, messenger RNA is translated, in conjunction with ribosomal and transfer RNA, into a specific amino acid sequence of the protein.

As there is good evidence that the diverse types of cells of a developing organism contain the same genetic instructions, it is obvious that the instructions are being selectively used. There is also evidence that the inherited instructions are used sequentially as development progresses from stage to stage. Thus, a major problem is the nature of the mechanisms that control differential transcription and translation of genetic information in different cells and the sequential or programmatic use of this information.

References

Atwood, K. C. 1965. Transcription and translation of genes. In M. Locke, ed., Reproduction: Molecular, Subcellular and Cellular, pp. 17–38. Academic Press, New York.

Becker, H. J. 1964. Die Genetischen Grundlagen der Zelldifferenzierung. Naturwiss. **51:**205–211; 230–235.

Beerman, W. 1956. Nuclear differentiation and functional morphology of chromosomes. Cold Spring Harbor Symp. Quant. Biol. **21:**217–232.

Beerman, W. 1961. Ein Balbiani-Ring als Locus einer Speicheldruesen-mutation. Chromosoma **12:**1–25.

Beerman, W. 1963. Cytological aspects of information transfer in cellular differentiation. Am. Zool. **3:**23–32.

Beerman, W. 1966. Differentiation at the level of the chromosomes. In Cell Differentiation and Morphogenesis, pp. 24–54. North Holland Publishing Co., Amsterdam.

Clever, U. 1964. Actinomycin and puromycin: effects on sequential gene activation by ecdysone. Science **146:**794–795.

Clever, U. 1965. Chromosomal changes associated with differentiation. In Genetic Control of Differentiation, pp. 242–253. Brookhaven Symp. Biol., Upton, N.Y.

Clever, U. 1966. Gene activity patterns and cellular differentiation. Am. Zool. **6:**33–41.

Clever, U., H. Bultmann, and J. M. Darrow. 1969. The immediacy of genomic control in polytenic cells. In E. W. Hanly, ed., Problems in Biology: RNA in Development, pp. 403–423. University of Utah Press, Salt Lake City.

Counce, S. J. 1956. Studies on female sterility genes in *Drosophila melanogaster:* I. The effects of the gene deep-orange on embryonic development. Ztschr. Ind. Abst. Vererb. **87:**443–461.

Davidson, E. H., W. C. Alfrey, and A. E. Mirsky. 1964. On the RNA synthesized during the lampbrush phase of amphibian oögenesis. Proc. Nat. Acad. Sci. U.S. **52:**501–508.

DeBusk, A. G. 1968. Molecular Genetics. Macmillan, New York.

Denis, H. 1966. Gene expression in amphibian development: II. Release of the genetic information in growing embryos. J. Mol. Biol. **22:**285–304.

Denis, H. 1968. Role of messenger ribonucleic acid in embryonic development. Adv. Morph. **7**:115–150.

Ebert, J. D., and M. E. Kaighn. 1966. The keys to change: factors regulating differentiation. In M. Locke, ed., Major Problems in Developmental Biology, pp. 29–84. Academic Press, New York.

Edstrom, J. E., and J. G. Gall. 1963. The base composition of ribonucleic acid in lampbrush chromosomes, nucleoli, nuclear sap, and cytoplasm of *Triturus* oöcytes. J. Cell Biol. **19**:279–284.

Gall, J. G. 1963. Chromosomes and cytodifferentiation. In M. Locke, ed., Cytodifferentiation and Macromolecular Synthesis, pp. 119–143. Academic Press, New York.

Gall, J. G., and H. G. Callan. 1962. H^3-uridine incorporation in lampbrush chromosomes. Proc. Nat. Acad. Sci. U.S. **48**:562–570.

Heyes, J. K. 1963. The role of nucleic acids in cell growth and differentiation. Symp. Soc. Exp. Biol. **17**:40–57.

Kroeger, H. 1963. Chemical nature of the system controlling gene activities in insect cells. Nature **200**:1234–1235.

Laufer, H., Y. Nakase, and J. Vanderberg. 1964. Developmental studies on the Dipteran salivary gland: I. The effect of Actinomycin-D on larval development, enzyme activity, and chromosomal differentiation in *Chironomus thummi.* Devel. Biol. **9**:367–384.

Loewy, A., and P. Siekevitz. 1963. Cell Structure and Function. Holt, Rinehart and Winston, New York.

McCarthy, B. J., and B. H. Hoyer. 1964. Identity of DNA and diversity of messenger RNA molecules in normal mouse tissues. Proc. Nat. Acad. Sci. U.S. **52**:915–922.

Mirsky, A. E., and H. Ris. 1949. Variable and constant components of chromosomes. Nature **163**:666–667.

Nirenberg, M. W. 1963. The genetic code: II. Sci. Amer. **208**:80–94.

Nirenberg, M. W., and J. H. Matthaei. 1961. The dependence of cell-free protein synthesis in *Escherichia coli* upon naturally occurring or synthetic polyribonucleotides. Proc. Nat. Acad. Sci. U.S. **47**:1588–1602.

Nomura, M., B. D. Hall, and S. Spiegelman. 1960. Characterization of RNA synthesized in *Escherichia coli* after bacteriophage T_2 infection. J. Mol. Biol. **2**:306–326.

Nygaard, A. P., and B. D. Hall. 1963. A method for the detection of RNA-DNA complexes. Biochem. Biophys. Res. Comm. **12**:98–104.

Paul, J. 1967. Masking of genes in cytodifferentiation and carcinogenesis. In A. V. S. DeReuck and J. Knight, eds., Cell Differentiation, pp. 198–207. Ciba Found. Symp. Little, Brown, Boston.

Paul, J. and R. S. Gilmour. 1968. Organ-specific restriction of transcription in mammalian chromatin. J. Mol. Biol. **34**:305–316.

Pavan, C. 1965. Nucleic acid metabolism in polytene chromosomes and the problem of differentiation. In Genetic Control of Differentiation, pp. 222–241. Brookhaven Symp. Biol., Upton, N.Y.

Pelling, C. 1964. Ribonukleinsaure-Synthose der Riesenchromosomen: Autoradiographische Untersuchungen an *Chironomus tentans.* Chromosoma **15**:71–122.

Plaut, W., and D. Nash. 1964. Localized DNA synthesis in polytene chromosomes and its implications. In M. Locke, ed., The Role of Chromosomes in Development, pp. 113–135. Academic Press, New York.

Rich, A., J. B. Warner, and H. M. Goodman. 1963. The structure and function of polyribosomes. Cold Spring Harbor Symp. Quant. Biol. **28**:269–285.

Rosenkranz, H. S. 1965. DNA and the embryonic development of the American sea urchin *Arbacia punctulata.* Biol. Bull. **129**:419 (abstr.)

Schultz, J. 1965. Genes, differentiation, and animal development. In Genetic Control of Differentiation, pp. 116–147. Brookhaven Symp. Biol., Upton, N. Y.

Sinsheimer, R. L. 1962. The structure of DNA and RNA. In J. M. Allen, ed., The Molecular Control of Cellular Activity, pp. 221–243. McGraw-Hill Book Company, New York.

Sinsheimer, R. L. 1967. The Book of Life. Addison-Wesley, Reading, Mass.

Slater, H. S., J. R. Warner, A. Rich, and C. E. Hall. 1963. The visualization of polyribosomal structure. J. Mol. Biol. **7**:652–657.

Stern, C. 1955. Gene action. In B. H. Willier, P. Weiss, and V. Hamburger, eds., Analysis of Development, pp. 151–169. W. B. Saunders, Philadelphia.

Stevens, B. J. and H. Swift. 1966. RNA transport from nucleus to cytoplasm in *Chironomus* salivary glands. J. Cell Biol. **31**:55–77.

Sullivan, N. 1967. The Message of the Genes. Basic Books, New York.

Swift, H. 1962. Nucleic acids and cell morphology in dipteran salivary glands. In J. M. Allen, ed., The Molecular Control of Cellular Activity, pp. 73–125. McGraw-Hill Book Company, New York.

Torrey, J. G. 1959. Experimental modification of development in the root. In D. Rudnick, ed., Cell, Organism, and Milieu, pp. 189–222. Ronald Press, New York.

Ursprung, H., and K. D. Smith. 1965. Differential gene activity at the biochemical level. In Genetic Control of Differentiation, pp. 1–13. Brookhaven Symp. Biol. Upton, N.Y.

Warner, J. R., A. Rich, and C. E. Hall. 1962. Electron microscope studies of ribosomal clusters synthesizing hemoglobin. Science **138**:1399–1403.

Watson, J. D., and F. H. C. Crick. 1953. The structure of DNA. Cold Spring Harbor Symp. Quant. Biol. **18**:123–131.

Wilkins, M. H. F. 1956. Physical studies of the molecular structure of deoxyribose nucleic acid and nucleoprotein. Cold Spring Harbor Symp. Quant. Biol. **21**:75–90.

Part Two:

Patterns of Development

4 *Patterns of Continuous Development*

Continuous developmental processes occur in all organisms from the time of their inception until death. Development does not cease with adulthood. Even aging is a manifestation of development that leads to senescence, the terminal stage. We begin our survey of patterns of development with adult organisms for three reasons. First, it is necessary to emphasize that development is a fundamental aspect of life. Continuous or periodic development maintains the integrity of the adult organism by counteracting the wear and tear to which all organisms are subject. The cessation of development is synonymous with the death of the organism. Second, the processes of continuous development are the most common patterns of development. Processes such as the continuous formation and destruction of blood and skin cells are part of everyday life, even though we may be oblivious to them. Other continuous developmental processes are growing grass, hair, and fingernails. Third, the mechanisms controlling continuous development in adults appear less complex than the mechanisms controlling either the replacement of a lost part (regeneration) or the formation of a new organism (reproduction). Although this assumption has not been demonstrated experimentally, the guidelines for continuous development seem to be present in the adult, eliminating the problem typical of many patterns of regeneration and reproduction of progressively creating guidelines for subsequent development.

Continuous development involves either the increase in size of an adult by the addition of cells, or the replacement of cells that are constantly being lost with no increase in size of the adult. An example of the former process is the continuous growth and increase in size of many plants. An extreme case is the continuous development of redwood trees for more than 2,000 years. Examples of the latter process are the balanced addition and loss of cells, the "turnover systems" in the skin, digestive tract, and blood of animals. The two types of continuous development are diagramed in Figure 4-1. (Goss, 1964). In Type IA, no special cells divide, but randomly distributed cells divide to increase the size of the population. In Type IB, special stem cells repeatedly divide, leaving one of the two cells produced a stem cell and increasing the size of the population. In Type II, the addition of cells by division is balanced by the death and loss of cells and the existing size of the organism is maintained.

Particularly in plants, continuous development involves an increase both in cell number and in cell size. Thus, an increase in size of fully formed organs in an animal and of the small but fully formed leaf of a plant can be the result solely of an increase in size of the cells.

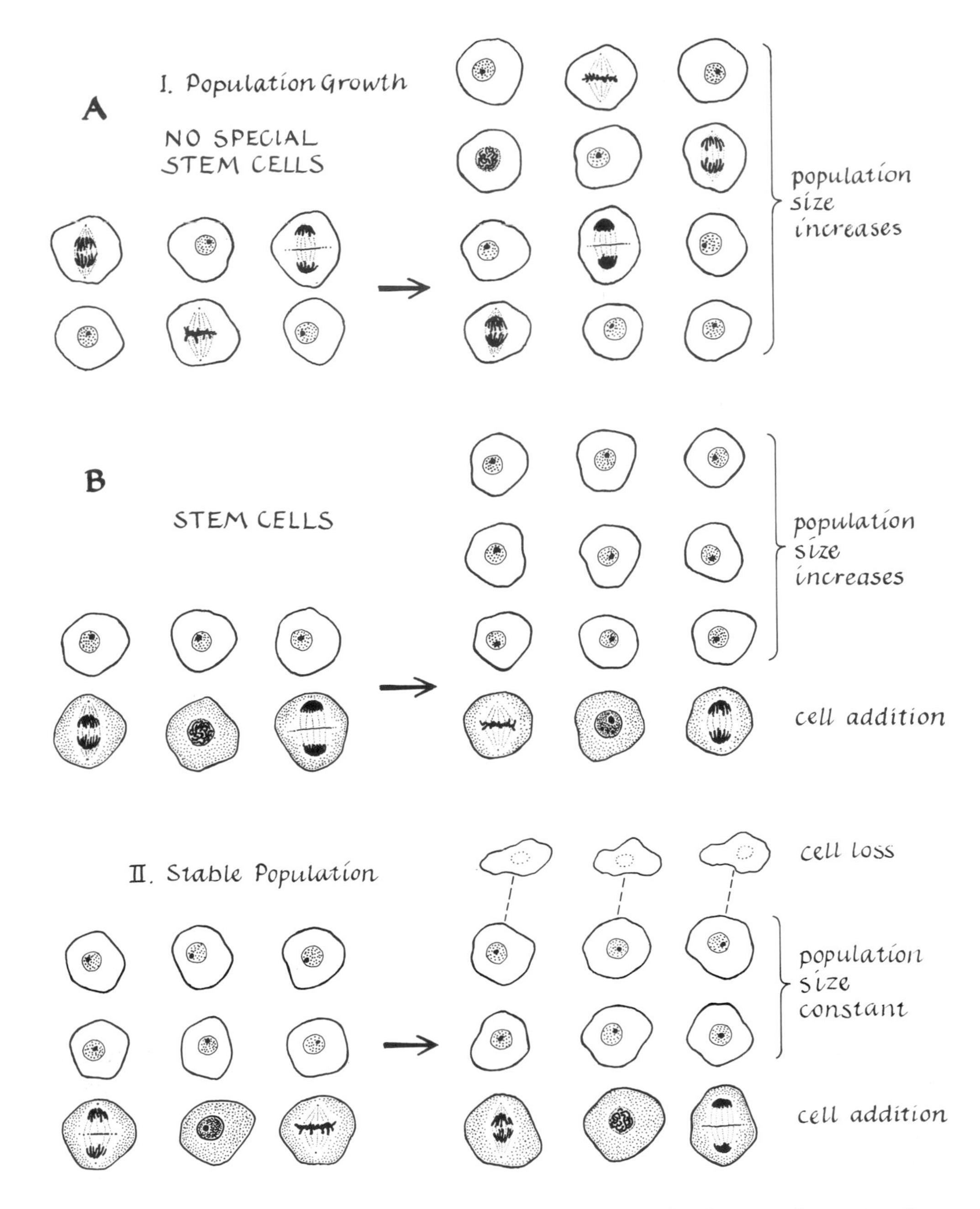

Figure 4-1. *Types of continuous development. See explanation in text.*

In this chapter we shall first examine some well-known examples of continuous or periodic development in generally distributed tissues, such as the cambium of adult plants and the skin of animals; we shall then look at examples of more localized developing tissues, for instance, the shoot apex of plants and the hair and feathers of animals.

Generally Distributed Developing Tissues

Vascular Cambium or Meristem (Woody Plants)

In a cross section of the stem of a woody, dicotyledonous plant the vascular tissue forms a cylinder with a ring of wood (xylem) inside and a ring of bark (phloem, pericycle, and cortex) outside. This layer of tissue, sometimes only one cell diameter thick, is the lateral or vascular cambium. The cambial cells, which are relatively thin-walled and densely protoplasmic, are capable of relatively rapid division. Tissue composed of cells with division capacities, called meristematic tissue, is found not only in plants but also in some animals. The number of dividing cells in the cambium can be tremendous, up to 500 million in a large tree. When the cambium is active its cells divide and produce xylem and phloem cells. Another type of lateral cambium located in the phloem region, called the cork cambium, produces cork cells but no xylem or phloem cells. This additional growing region can develop as a response of the stem to injury.

When cambial cells divide, the newly formed daughter cells inside the cambium differentiate into the various types of xylem or wood cells; those outside the cambium become the phloem, pericycle, and cortex cells. The cambial cells themselves remain unchanged and continue to proliferate daughter cells (Sinnott, 1946, 1960). These relations are illustrated in Figures 4-2, 4-3, and 4-4.

Through intermittent cell proliferation in the cambium, the stem increases in diameter. Cambial activity is seasonal and results in formation of the rings seen in a cross section of a tree trunk. A ring of new wood is formed each growing season, so the age of a tree can be determined by counting the number of rings in the trunk. Two important processes of development are illustrated by the activity of the lateral cambium—growth (cell division and enlargement) and differentiation (cell specialization).

Cell Division and Enlargement

The presence of a wall in plant cells seems to require a type of division unlike that in most animal cells. When the chromosomes separate in anaphase and reach the poles of the mitotic spindle, a fibrous structure, the phragmoplast, forms in the equatorial plane of the spindle. A cell plate (middle lamella) forms on the phragmoplast and expands from its edges until contact is made with the existing cell walls (Figure 4-5). Electron microscope studies show that the middle lamella is formed by fusion of vesicles pinched off the Golgi apparatus. The contents of the Golgi vesicles become the new middle lamella and the membranes surrounding them become the plasma membrane of the two daughter cells.

When the cambial cell divides longitudinally (Figure 4-5), one cell remains a cambial cell, but the other specializes as a nondividing xylem or phloem cell (Figure 4-2). A radial (transverse) division of the cambial cell will result in two cambial cells. Particularly in plants, the orientation of the mitotic spindle is important in determining the position of the two daughter cells, which determines their paths of specialization. The increase in size of one or both daughter cells that typically follows is an important type of growth in plants. The factors determining the position of the mitotic spindle are largely unknown, but clearly the direction of cell division and cell enlargement is important in the shape of plants.

Cell Specialization

Two extreme types of cellular differentiation occur in the mitotic descendants of a cambial cell: disintegration of the nucleus in the sieve tube cells of the phloem and cell death in the xylem cells. Phloem and xylem cells are illustrated in Figures 4-6 and 4-7. Electron microscope studies of the products of cambial cell differentiation reveal the ultrastructural changes involved, for example, increased synthesis of endoplasmic reticulum, increased vacuolation, and changes in the structure of mitochondria in the specializing products of a cambial cell (Esau, 1963). Another primary aspect of cellular differentiation in plants is the formation of specific cell products (see Chapter 23). For example, the formation of cell walls from primarily extracellular products is a significant aspect of plant cell specialization. The biosynthesis of diverse nonstructural products such as pigments and alkaloids is also an important type of cell differentiation in plants.

Much of the diversity in plant development is in the arrangement of cells to form various tissues and organs. Patterns of cellular associations are as significant as morphological specializations are. As

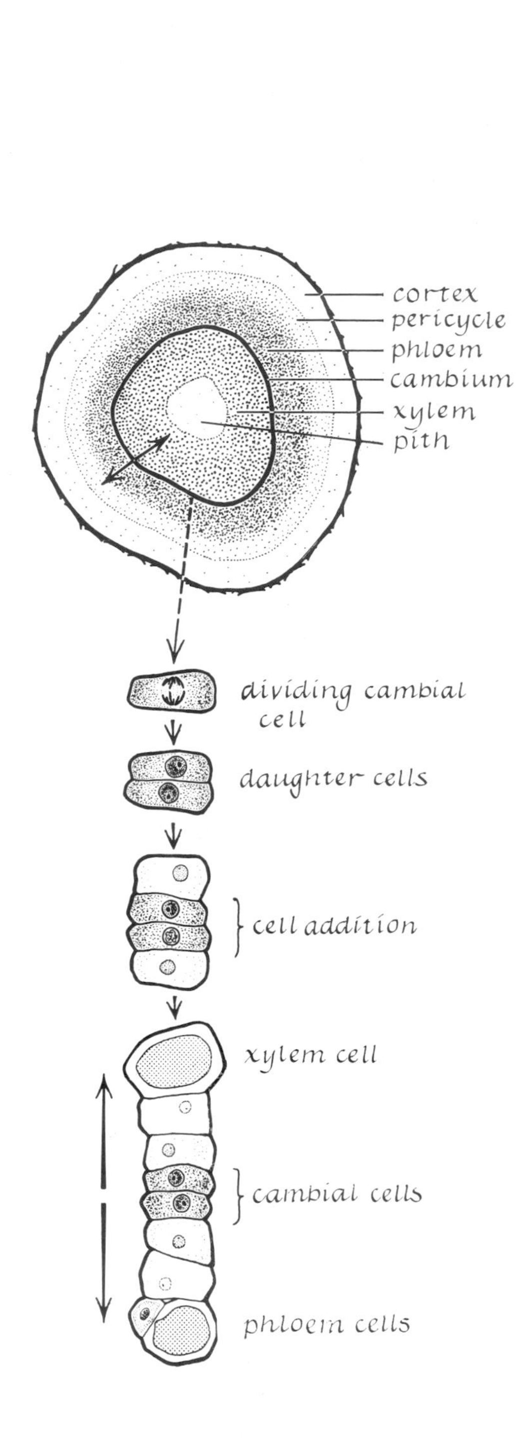

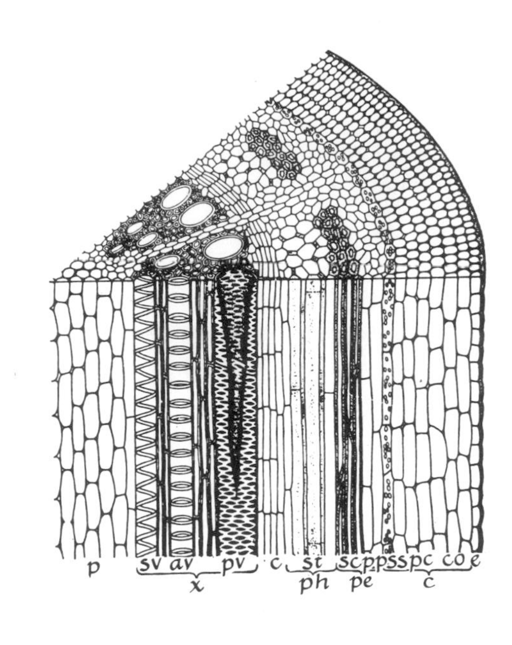

Figure 4-3. *Detail of the types of tissues produced by the cambium.* P, *pith;* x, *xylem;* c, *cambium;* ph, *phloem;* pe, *pericycle;* co, *cortex. From Brown,* The Plant Kingdom *(1935), reproduced by permission of Mrs. Brown.*

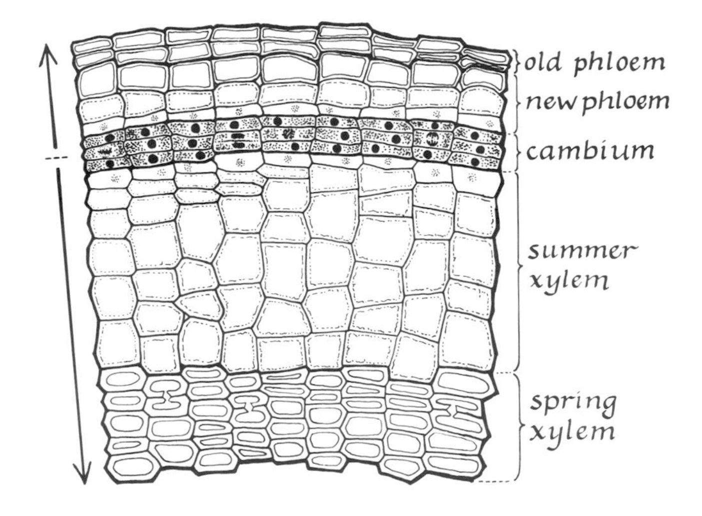

Figure 4-2. *Diagram of the relationship between cambial, xylem, and phloem cells in the stem of a plant. Note that the products of division of the cambial cells differentiate as xylem cells if they lie inside the cambial layer, but as phloem cells if they lie outside the cambial layer.*

Figure 4-4. *Cambium of a pine stem. The cambial cells are small and densely protoplasmic. The youngest xylem and phloem cells are also small but increase in size as they become older. Maturation of the xylem and phloem cells results in a significant increase in cell-wall synthesis.*

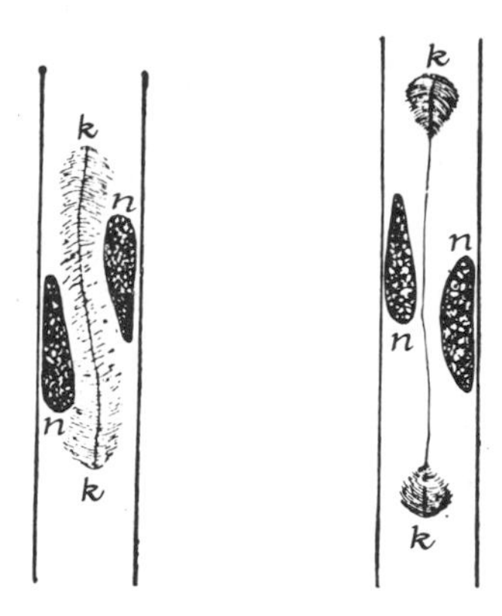

Figure 4-5. *Stages in division of a cambial cell showing the method of formation of the new cell wall. From* Botany: Principles and Problems *by Sinnott and Wilson. Copyright 1955 by McGraw-Hill Book Company. Used with permission of McGraw-Hill Book Company.*

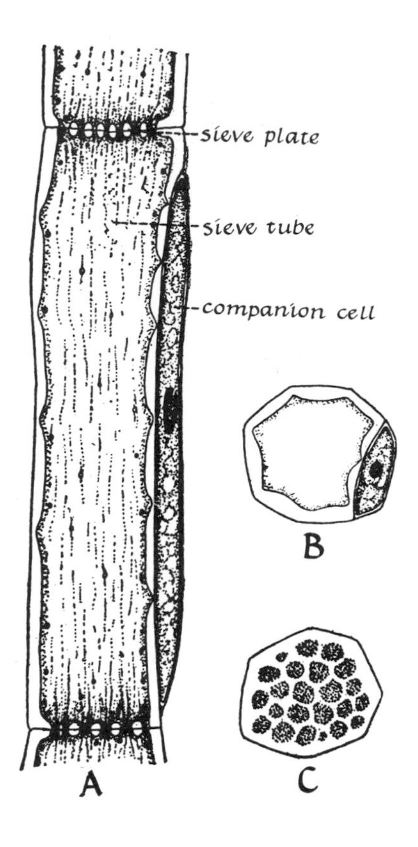

Figure 4-6. *Phloem structure in a herbaceous stem. A, longitudinal section through the sieve tube and companion cell of a squash plant. Note the connection of the sieve tube with adjacent sieve tubes via the sieve plate. B, transverse section through a sieve tube and companion cell. C, transverse section through a sieve plate. From* Botany: Principles and Problems *by Sinnott. Copyright 1963 by McGraw-Hill Book Company. Used with permission of McGraw-Hill Book Company.*

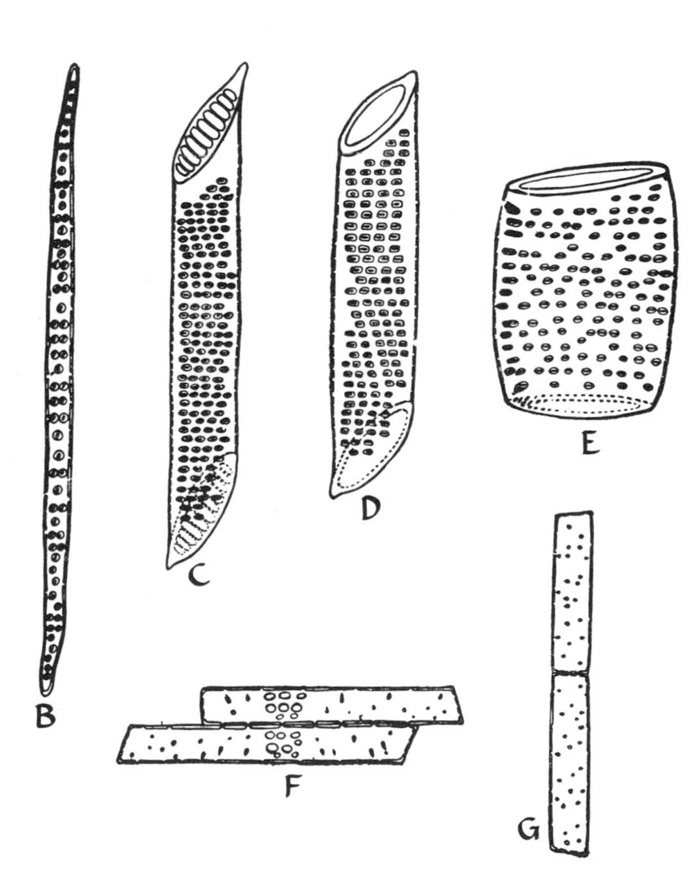

Figure 4-7. *Types of cells found in wood. A, fiber; B, tracheid; C, D, and E, types of vessel cells; F, ray cells; G, wood parenchyma cells. From* Botany: Principles and Problems *by Sinnott. Copyright 1963 by McGraw-Hill Book Company. Used with permission of McGraw-Hill Book Company.*

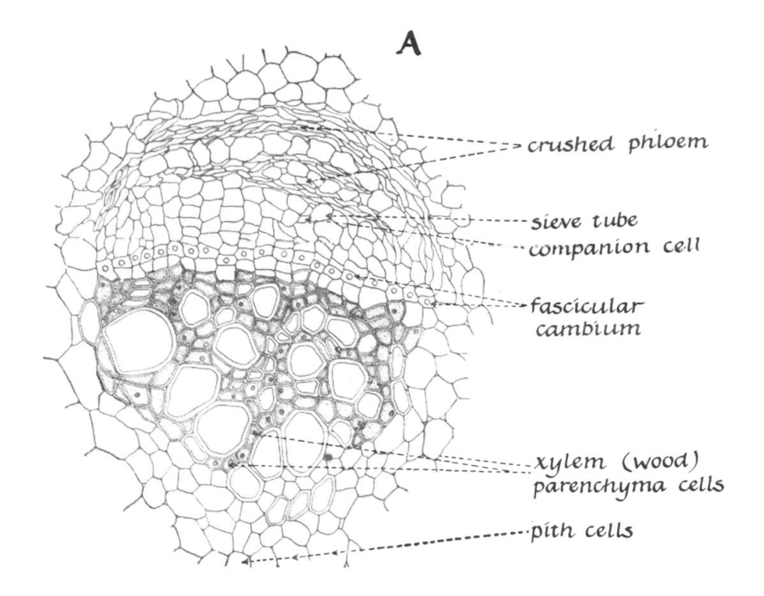

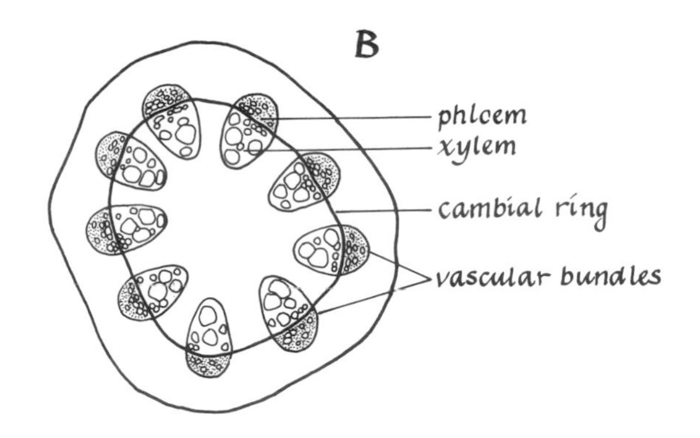

Figure 4-8. *A, vascular bundle, and B, arrangement of vascular bundles in a cross section of a young stem of* Aristolochia *(birthwort), a dicotyledon. A from Holman and Robbins, 1940, courtesy John Wiley and Sons, Inc.*

Sinnott (1960) notes, among the many problems that confront us are the understanding of mechanisms that control the proportion of xylem and phloem tissue, the patterns of tissues in the vascular bundles, and the arrangement of the bundles (see Chapters 23 and 26). These patterns in two groups of seed plants, dicotyledons and monocotyledons, are illustrated in Figures 4-8 and 4-9. Note that the various types of cells in a vascular bundle and the bundles themselves are arranged in a definite pattern in the stem. In dicotyledons and usually in monocotyledons these patterns originate in the growing tip, the shoot apex of the plant, the activity of which we shall describe below. In dicotyledonous plants the vascular bundles are arranged in a continuous or an interrupted ring, but in the stems of monocotyledonous plants they are scattered. The vascular bundles of dicotyledons increase in diameter as a result of cell division in the cambial layer, and the diameter of the stem significantly increases. The monocotyledons have no permanent cambium in the fully formed vascular bundles and, therefore, no comparable increase in the diameter of the stem. The stems of both kinds of plants increase in length by cell addition in the apical growing regions.

Mechanisms of Cell Differentiation

Why are the products of cambial cell division inside the cambium so different from those outside it? Also, what controls the patterned, nonrandom arrangement of the cell types? Two explanations are theoretically possible. One could suppose that the cambial cell systematically segregates cytoplasmic and nuclear determinants as it divides. Such a hypothesis is illustrated in Figure 4-10A. An alternative explanation is that the cells produced inside and outside by division of a cambial cell become different because the environment around the two cells is different (Figure 4-10B).

Unequal cell divisions resulting in daughter cells of different sizes or of different cytoplasmic constitution are common in plants (see Figure 4-37). An apparent example of unequal cell division involving segregation of cytoplasmic components is the differentiation of idioblast cells in *Ricinus,* the castor oil plant. In this plant an undifferentiated meristematic cell divides to form two unlike daughter cells, one containing tannin (a red pigment) and unsaturated fats, and the other lacking detectable amounts of

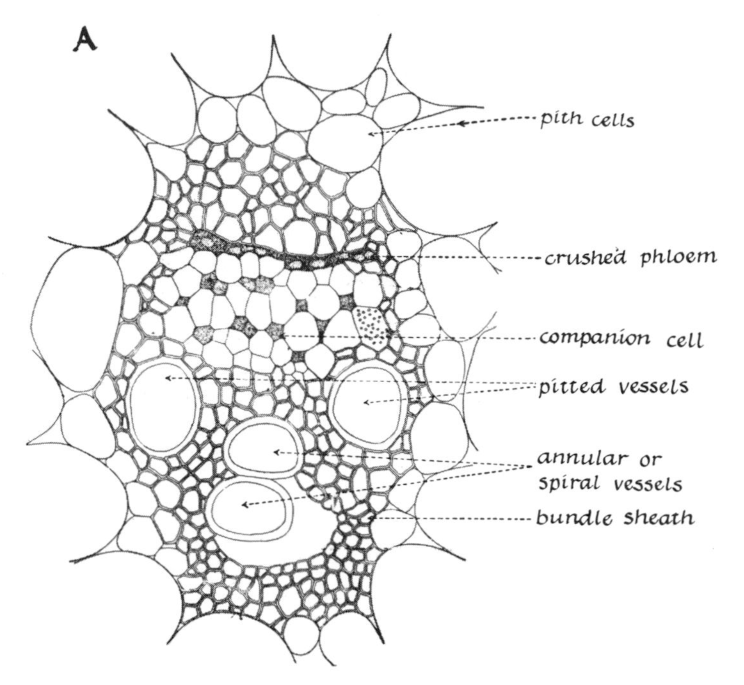

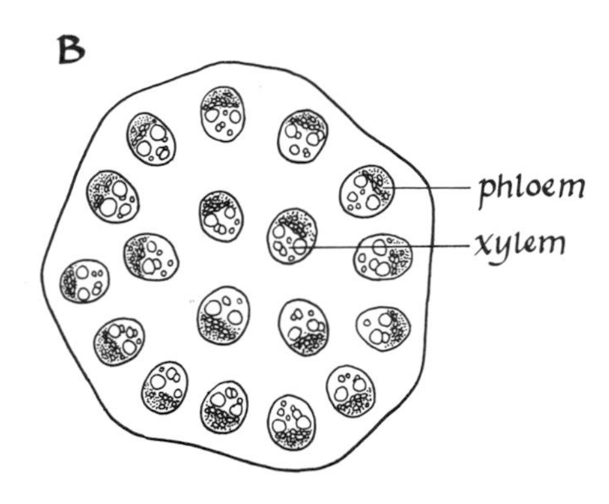

Figure 4-9. *A, vascular bundle, and B, arrangement of vascular bundles in a cross section of the stem of* Zea mays *(corn), a monocotyledon. A from Holman and Robbins, 1940, courtesy John Wiley and Sons, Inc.*

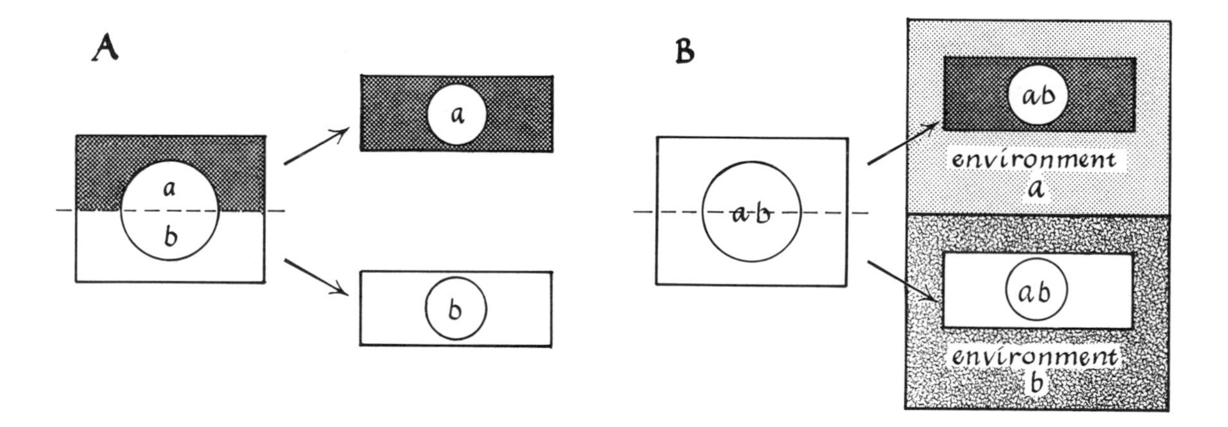

Figure 4-10. *Two alternative methods of cellular differentiation.*

these substances (Figure 4-11). It is not clear, however, from this observation of Bloch (1948) that an actual segregation of cytoplasmic components takes place. The cells may have been induced to form tannin because of their position.

Most evidence indicates that a cytoplasmic segregation mechanism is unlikely in the differentiation of cambial cells, although it undoubtedly occurs in other kinds of cellular differentiation, for example, in the early development of plant (Chapter 10) and some animal embryos (Chapters 13 and 15). Segregation of nuclear determinants, although theoretically a good way of ensuring orderly cell differentiation, remains to be discovered in developing plants and animals. Experimental analysis of the differentiation of lateral cambial cell products is lacking, but studies of similar continuously developing systems, such as the apical cambium in the growing shoot tip, suggest a different mechanism of control.

Many studies (reviewed by Sinnott, 1960; Wardlaw, 1965; Torrey, 1967; and others) indicate that supracellular influences may control the specialization of plant cells and the orderly relationship of these cell types to one another (see Figures 4-8 and 4-9 and Sinnott and Bloch, 1939). Thus cells inside the cambium are surrounded by an environment of composition different from that surrounding cells outside the cambium (Figure 4-12). As the cambial cells are specialized for division, the problem is what keeps them undifferentiated. As we noted above, one daughter cell resulting from division of a cambial stem cell remains a cambial stem cell. Perhaps the environment of the cambial stem cells prevents their differentiation. This explanation seems possible because the cambial cell ring tends to remain in about the same position relative to the outer surface as the stem increases in diameter. The phloem does not continue to increase in thickness as the xylem does, as the cells in the outer layer of the phloem are constantly dying and being sloughed off, a process superficially like the production and loss of skin cells in animals.

Studies of developing plants and animals continue to reinforce the view that a cell's microenvironment determines its developmental behavior. The history of experimental embryology demonstrates that the position of a cell plays an important role in its specialization. Interaction between the genome and the environment is the core of most modern theories of development. However, cytoplasmic segregation mechanisms are not inconsequential, as evidence

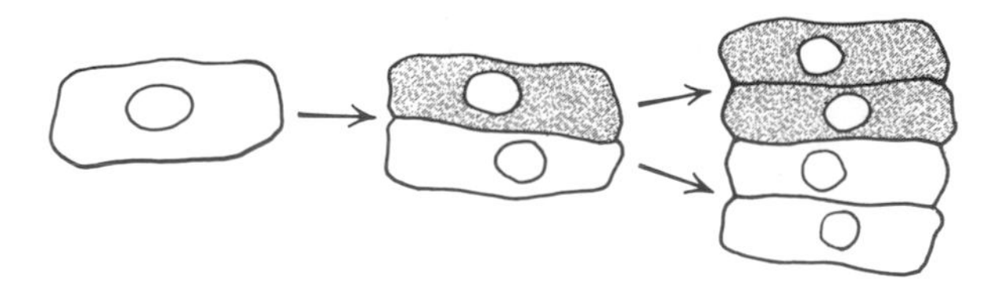

Figure 4-11. *Apparent segregation of cytoplasmic components in the differentiation of idioblast cells in the castor oil plant* Ricinus.

does exist for cytoplasmic segregation in the development of plants and animals. A more general discussion of the mechanisms of cellular differentiation will be given in Chapter 16.

Stratum Germinativum

The stratum germinativum in the skin of animals is a continuously proliferating tissue, comparable in this respect to the cork cambium of plants. The difference between the lateral cambium and the germinative layer is that all products of germinative layer cells are displaced toward the surface of the skin (outward) by the addition of new cells below them (Figure 4-13). As the cells are pushed outward, they undergo progressive changes including the loss of the capacity to divide. Just outside the germinative layer, the *stratum Malpighii,* which is several cell diameters thick, the cells synthesize keratohyaline granules. Further out, the cells replace the granules with a fluid substance called eleidin, and near the surface the cells flatten and their nuclei degenerate. Finally the cells become completely cornified and filled with the protein keratin. At the surface of the skin, the cells are dead and are continually being lost or washed off (Bloom and Fawcett, 1962). In this turnover system cell loss balances cell addition.

As in the cambial system, two basic processes of development—cell proliferation and cell specialization—are involved in skin formation. Similar questions arise in understanding the control of these processes. What prevents the germinative cells from becoming keratinized, or why do they continue cell division, remaining stem cells throughout the life of the organism? Perhaps their constant environment induces them to continue division. Would a single germinative cell, if isolated and grown in culture on a nutrient medium, continue to divide or would it become specialized in response to the changed environment?

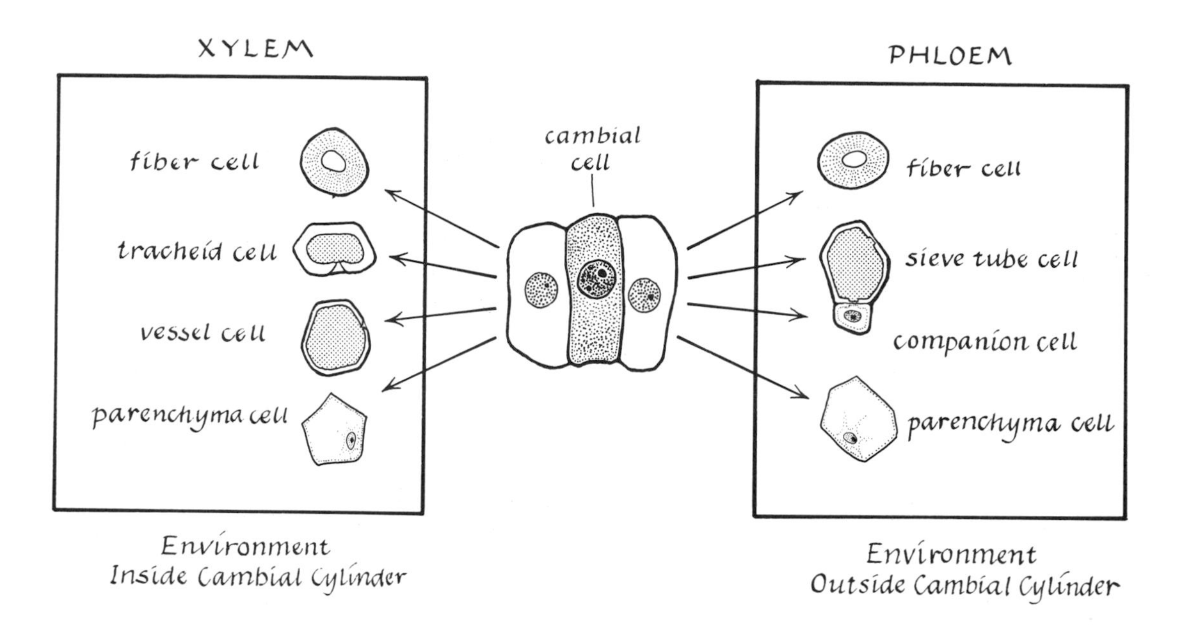

Figure 4-12. *Diagram of the possible influence of the microenvironment inside and outside the cambial cylinder on differentiation of cambial daughter cells.*

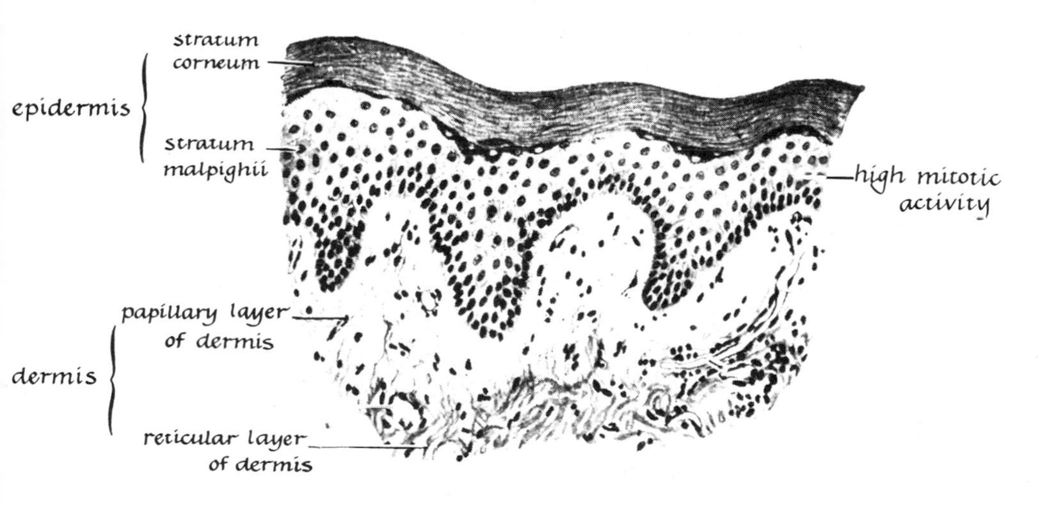

Figure 4-13. *Section through the skin of the human shoulder × 125. Note that relatively great mitotic activity occurs in the stratum malpighii, the germinative layer of the epidermis. From Alexander A. Maximov in Bloom and Fawcett:* A Textbook of Histology, *8th edition. W. B. Saunders Company, Philadelphia, 1962.*

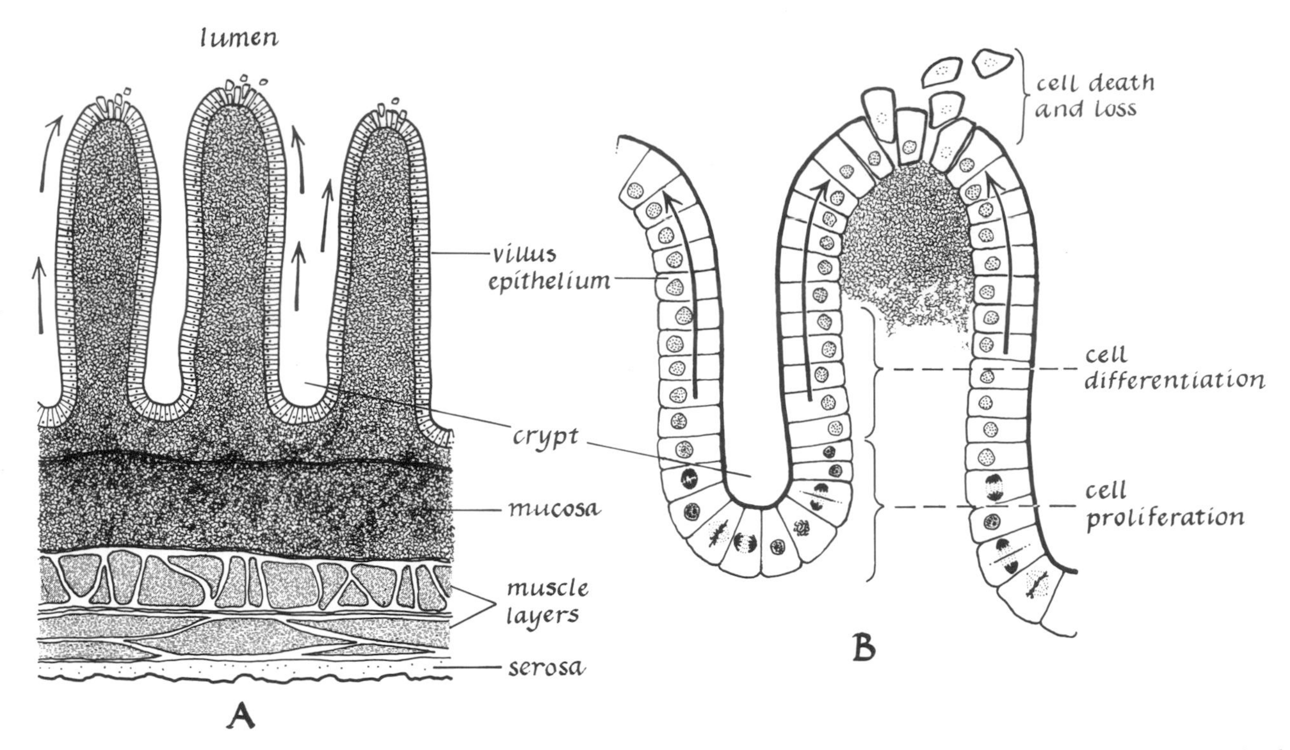

Figure 4-14. *Continuous cell proliferation and loss in the intestinal epithelium of the human.*

The skin is a complex organ consisting of the epidermis at the base of which the germinative cells are located, and the dermis, just below the germinative layer (Figure 4-13). Experiments demonstrate that processes of cellular interaction are involved in skin development. If the epidermis of chick embryo skin is separated from the dermis and explanted in culture, cell proliferation in the germinative layer quickly ceases. Within about 18 hours after the dermis is replaced, proliferation and differentiation of the explanted skin resumes. The dermal cells or their products are an all-important property of the environment that promotes cell division in the germinative layer (Wessels, 1962, 1967). The influence of one type of tissue on another is an example of induction. Examples of this type of cellular interaction and some suggested mechanisms will be found in Chapter 21.

Intestinal and Esophageal Epithelium

Another generally distributed developing system is found in the intestinal wall of adult animals. Between and at the bases of the villi are the crypts of Lieberkühn, in which relatively unspecialized dividing cells are located. Apparently all the differentiated epithelial cells covering the villi—columnar cells, goblet cells, argentaffin cells—are produced by the cells at the base of the villi (Bloom and Fawcett, 1962). This process is illustrated in Figure 4-14. As the cells are produced, they move up the surface of the villi, become differentiated, and eventually are cast off at the tips. In this way, the epithelial cell population of the villi is replaced approximately every 3 to 7 days (Questler and Sherman, 1959; Leblond, 1965).

Continuous formation of stratified, squamous epithelium also occurs in the esophagus. Studies of Leblond et al. (1959) and Messier and Leblond (1960) indicate that mitotic activity depends on the position of the cells in the basal layer of the digestive tract. Cells that are pushed out of the basal layer cease to divide, eventually die, and are sloughed off.

Some unsolved problems concerning continuously developing systems are: what keeps such turnover systems in balance and what happens when cell loss exceeds cell replacement? The latter may be a manifestation of an aging process.

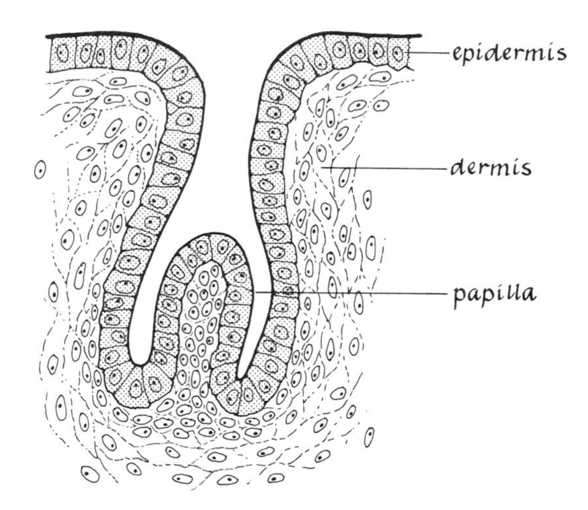

Figure 4-15. *A section through a follicle in the skin of a vertebrate. The papilla is the primordium of a hair, feather, or scale.*

Localized Developing Tissues and Structures

In adult animals and plants, many localized tissues are either continuously or intermittently developing. We shall consider briefly a few of these, but first we will examine the two systems that have been most thoroughly studied.

Hairs, Feathers, Scales

Individual hairs on our heads are continuously growing in length for long periods of time. However, the hair of animals with seasonal molting and of man as well, is shed and replaced by new hair. We do not usually notice this because the shedding and replacement of hairs are continuous and overlapping in time. Feathers, which are also epidermal derivatives, usually grow to maximum size, are molted, and then are replaced by new feathers. They are more complex in structure than hair, so their development is also more complex. Nevertheless the basic development of hair, feathers, and scales is similar.

Hairs, feathers, and scales grow out of follicles, invaginations of the epidermis into the dermal layer. At the bottom of each follicle is a structure called a papilla, a group of dermal cells. The papilla is covered by epidermis continuous with the lining of the follicle and the surface of the skin (Figure 4-15).

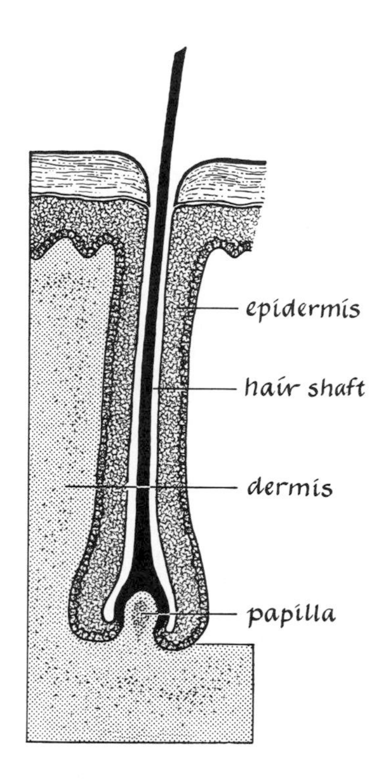

Figure 4-16. *A section through a hair follicle.*

The epidermis covering the papilla contains many rapidly dividing cells. As these cells divide, some remain near the papilla and continue to divide, but others are pushed upward or outward, cease dividing, and begin to differentiate. The formation of contour feathers in a juvenile bird follows shedding of the down feathers formed during embryonic development. Epidermal cells of the papilla multiply to form a thick ring of cells known as the collar. Proliferation of cells from the collar produces an epidermal cylinder surrounding a core of pulp derived from the dermal cells of the papilla (Figure 4-17D). The barbs of the feather develop from the inner cells of the cylinder as a series of parallel ridges (Figure 4-17A). The shaft forms along the dorsal side of the cylinder (Figure 4-17B and C) and, as the barb ridges increase in length, they shift to a more dorsal position and ultimately fuse with the shaft. Thus, the tip of the feather is the oldest part, more basal parts being formed by continuous cell proliferation in the collar (Rawles, 1964). This contrasts with the method of growth of other structures such as the shoot or root apices in plants, which grow from the tip rather than from the base.

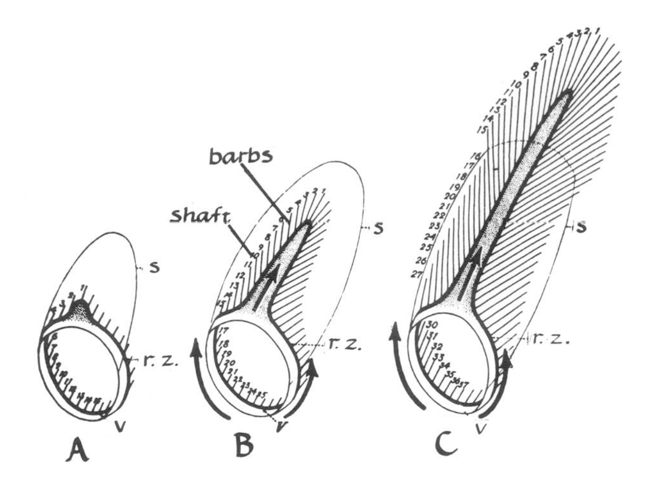

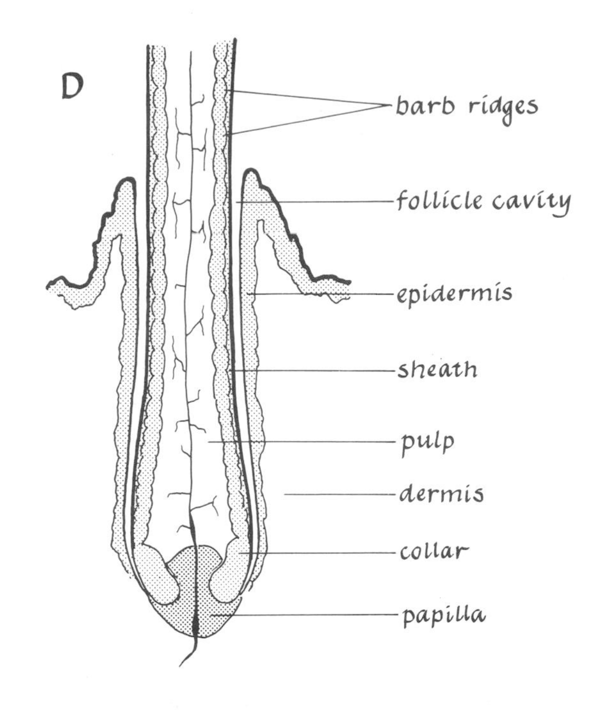

Figure 4-17. *The development of a feather. A, B, and C, stages in development of the shaft and barbs by concrescence of the cells of the collar, a ring of relatively unspecialized cells. The collar lies at the base of the follicle and surrounds the base of the papilla, D. The arrows indicate the concrescence of the collar cells and the elongation of the shaft. Modified from Domm, Gustafson, and Juhn in* Sex and Internal Secretions, *© 1939, The Williams and Wilkins Co., Baltimore, Md. 21202, U.S.A. Reproduced by permission.*

The relationship of the dividing epidermal cells to the papilla in hair, feathers, and scales is different (compare Figures 4-16 and 4-17) but in all cases the same basic processes of growth, cell differentiation, and cellular interaction occur. In hair, five distinct types of epidermal cells are formed; in feathers probably more. Keratin synthesis is the most obvious aspect of cellular differentiation in all these epidermal derivatives. Cell death is the final stage, as the completed hair, feather, or scale is a dead structure.

Different kinds of hairs and feathers develop on different body regions of the same mammal or bird. Experimental studies of Rawles (1964) have shown that the type of epidermal derivative—feather, scale, or beak—that develops in the chick embryo is controlled by the kind of underlying dermis (Chapter 21). This is another example of the role of induction in development. Evidently, the dermis of the papilla controls the type of hair or feather development throughout life (Lillie and Wang, 1941; Juhn, 1952). The presence of the dermal papilla is necessary for continued formation of the hair and feathers.

In most mammals and birds the hair or feathers are colored. Feathers especially have many diverse, genetically determined patterns (spotted, speckled, or striped) that are colored by the class of pigments known as melanins. There are species specific and body-region specific patterns of coloration in both hairs and feathers. The melanocytes, cells that synthesize the pigment, arise from the neural crest of the embryo and later migrate into the developing feathers and skin. The pattern seems to result from an interaction between the cells of the hair or feather and the melanocytes (see Rawles, 1948; Cohen, 1966 for reviews, and Chapter 26). The melanocytes in the feather collar release pigment granules that are picked up by the barb cells as the feather elongates (Strong, 1902).

Teeth

In animals such as rodents, the teeth are continuously developing but are worn down by the chewing habits of these animals. In many fish such as the

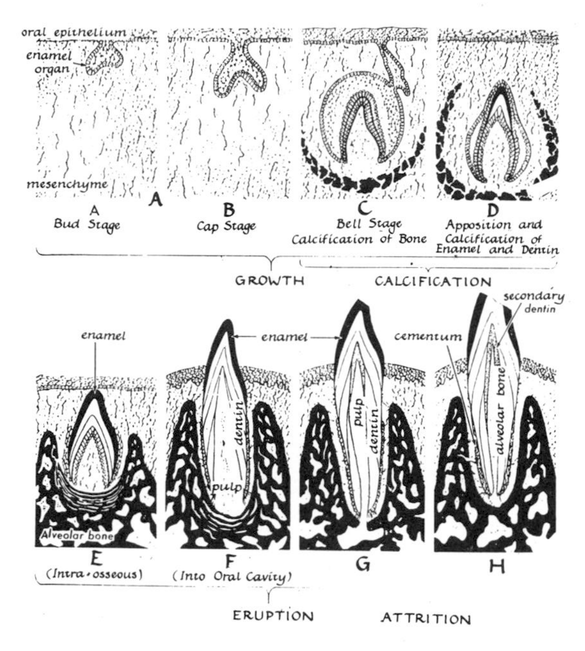

Figure 4-18. *The development of a human deciduous incisor. Enamel and bone are drawn in black. Redrawn from Schour and Massler in Bloom and Fawcett:* A Textbook of Histology, *8th edition. W. B. Saunders Company, Philadelphia, 1962.*

carp, which has pharyngeal teeth of endodermal origin, the teeth are deciduous but are continually replaced throughout the life of the fish. Man has two sets of teeth, the deciduous teeth of childhood, which begin to erupt about four months after birth, and the permanent teeth, which appear between the sixth and thirteenth years. The general features of development of a human tooth are illustrated in Figure 4-18. An especially interesting feature of tooth development is the synthesis of enamel, which covers the portion of the tooth extending above the oral epithelium, and dentin, which surrounds the pulp cavity. Dentin resembles bone in its structure and chemical composition; enamel is made almost entirely of calcium salts in the form of apatite crystals. Both enamel and dentin are products of cell activity. In this case, the process of cell specialization depends on cell function, that is, synthesis of extracellular products rather than a change in cell morphology. Nevertheless, the products influence the morphology of the cells that produced them. Other examples of this kind of cell specialization and the role it plays in development will be summarized in Chapter 23.

What controls the number of teeth typical for each species of animal? Why does man have only two sets of teeth but some fishes have many? Is the development of the enamel organ (Figure 4-18A) from the ectoderm induced by the underlying mesoderm? Is this development one of numerous examples of interaction between mesoderm and ectoderm in development of ectodermal derivatives? More intensive research is necessary to provide answers to such questions.

Apical Meristems

Adult plants and some adult animals engage in continuous development through the activity of apical growing and differentiating regions. The term apical meristem describes such localized growing regions in both plants and animals (Hyman, 1940). Apical growing regions are developmental centers exhibiting apical dominance (Chapter 19), that is, control by these regions of the development of adjacent regions or of the remainder of the organism. Apical dominance can result from more rapid cell division or greater production of hormonelike substances that inhibit cell activity in subapical regions. The shoot apex is an excellent example of apical dominance, but we shall also study other examples.

Shoot Apex

The shoot apex is a continuously or seasonally developing group of cells at the growing tip of the stem and its branches. It is relatively small—a few hundred microns in diameter in most flowering plants but up to several millimeters in some ferns. The apex is continuously producing from 1,000 to 500,000 new cells for formation of leaves and for elongation of the stem. The patterned arrangement of the vascular bundles in the stem (Figures 4-8 and 4-9) is determined by the position and orientation of cell division just below the promeristem region (Figure 4-19). Leaves are formed in an orderly sequence just below the promeristem. The leaves arise as leaf primordia, radially symmetrical protrusions, which may form opposite one another (as shown in Figure 4-19), in whorls, alternately, or in various spiral arrangements, depending upon the species of plant. Arrangement of leaves on the stem is called phyllotaxy.

Developmental processes in the shoot apices of many kinds of plants have been studied extensively

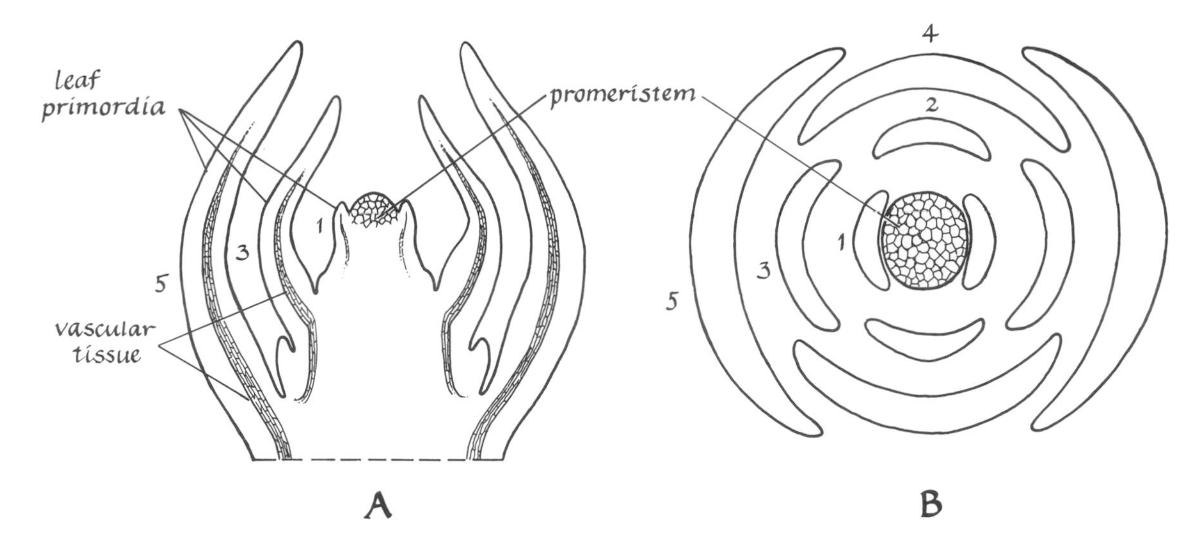

Figure 4-19. *The organization of a dicotyledonous shoot apex in which oppositely arranged leaves are formed. The numbers denote the order in which leaf primordia are formed.*

(see reviews of Wardlaw 1965, 1968; Wetmore, 1956; Sinnott, 1960; Gifford, 1963; Torrey 1966, 1967; and others). The shoot apices of all kinds of multicellular plants have the same basic organization (cellular architecture) and functional activities (Wardlaw, 1957, 1965, 1968), although in shape they can be flat, dome-shaped, or thin and elongated, extending a considerable distance above the youngest leaf primordia as in some monocotyledons. The apex is usually radially symmetrical, and in principle is a geometric system composed of the zones shown in Figure 4-20. The distal region consists of either a single cell (in most ferns) or a group of initial or stem cells (in seed plants). The region below this consists of a layer or layers of meristematic cells in which localized growth centers arise. Still further from the tip of the apex is a region in which leaf primordia develop from the localized growth centers and cellular differentiation begins. In the subapical region the shoot widens and elongates, the leaf primordia enlarge, and vascular differentiation continues. At the base of the apical meristem is a region of tissue maturation. Cell division occurs throughout the apex: in the promeristem, in regions in which a leaf primordium will arise, and between the primordia and the young leaves themselves. The latter cell divisions, accompanied by cell elongation, space the leaves and elongate the stem.

Vascular plants have three different types of cellular arrangement in the promeristem region (Torrey, 1967). In some ferns, a single apical cell repeatedly divides and, through further divisions of its progeny, produces all other cells of the shoot (Figure 4-25). In gymnosperms (for example, conifers) a small group of cells at the tip of the apex divides both parallel (periclinally) and perpendicular (anticlinally) to the

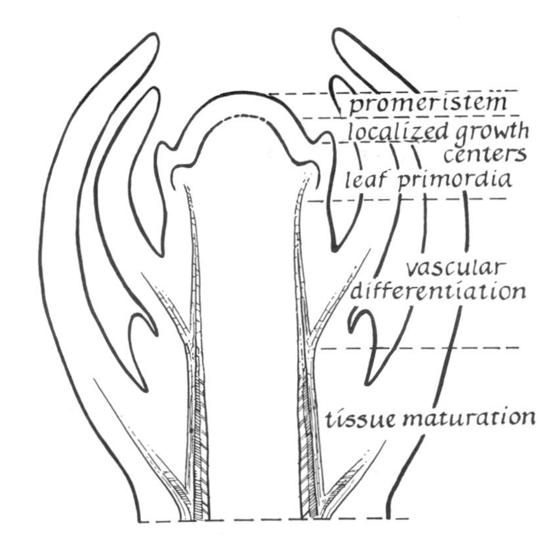

Figure 4-20. *Zones of developmental activity in the typical shoot apex. Based on the data of Wardlaw (1957).*

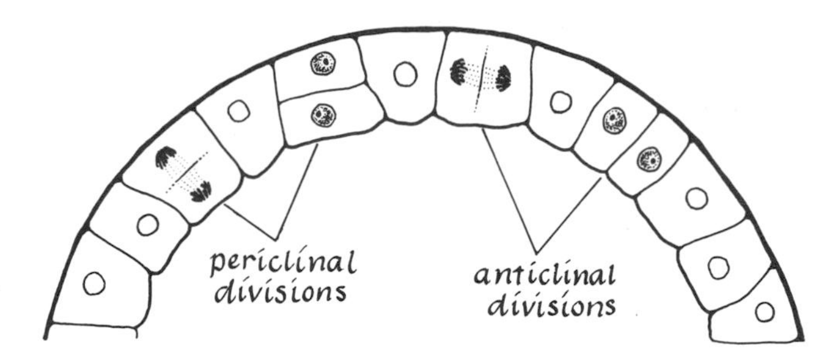

Figure 4-21. *Periclinal and anticlinal divisions in the outer tunica of the shoot apex of a gymnosperm.*

surface of the apex (Figure 4-21). In angiosperms the promeristem typically consists of discrete layers of cells: two tunica cell layers cover the surface of the apex and a variable number of corpus cell layers lie under the tunica. The basic arrangement of cells and cell layers in an angiosperm apex is shown in Figure 4-22. The outer and inner tunica layers increase in area by anticlinal divisions (the new cell wall is perpendicular to the surface). The outer tunica develops into the epidermis of the stem and leaves. The corpus cells divide in random orientation but mostly anticlinally, creating the major bulk of the stem and inner part of the leaf tissues. Where the leaf primordia will form are localized regions of active cell division in the corpus. In general, the tunica cells divide less actively than the corpus cells. However, when groups of cells are isolated and cultured in vitro, the tunica cells of some plants seem to have greater mitotic activity than the corpus cells (Wetmore and Garrison, 1966).

Even a superficial study of the shoot apex reveals that the basic processes of cell division, cell enlargement, and cell specialization that occur in and adjacent to the apex are similar to those found in many developing systems. The prime feature of the shoot apex is its continuous or periodic growth. Although the general arrangement of cells in the apex is maintained for long periods of time, its cellular composition constantly changes except in cases (for example, ferns) where one or a few cells function as permanent stem cells. The growth activity and pattern in the apex of angiosperms has been demonstrated by marking the tunica cells of the promeristem of *Lupinus* (lupine) with powdered charcoal or by destruction of a few cells (Soma and Ball, 1963). A marked cell in the geometric center of the apex or a single

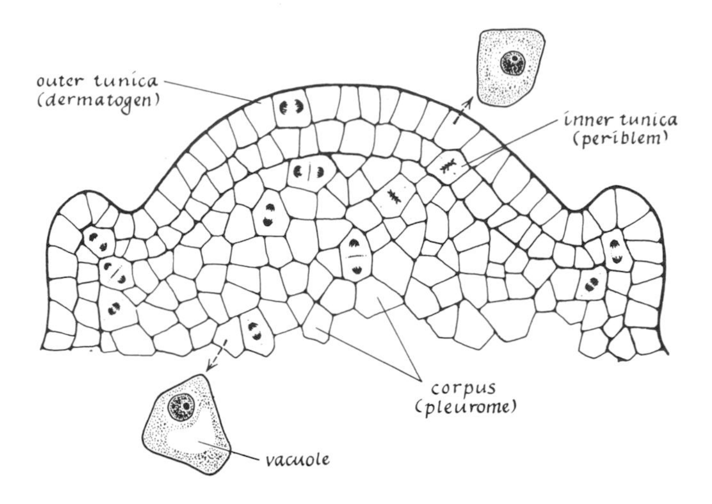

Figure 4-22. *The basic pattern of cells and cell layers in the angiosperm apex.*

punctured cell moves in a radial direction, showing that the most rapidly dividing cells are in or close to the center or summit of the apex. However, histochemical studies indicate that maximal DNA synthesis occurs in the localized growth region just below the summit in some plants (Buvat, 1955, and many others), but also occurs in or very close to the summit (Gifford, 1960). In *Lupinus* the marked cells do not move outward in a radially symmetrical pattern but appear to follow a spiral course. This is not unexpected, because the leaves of *Lupinus* are arranged in a spiral around the stem (spiral phyllotaxis). The major problem presented by the growing apex is the nature of the information that directs the growth pattern. This information results, for example, in spiral, opposite and intermittent, or alternate and intermittent arrangement of the leaves on the stems of different plants. The patterns are genetically controlled, but how the information is used to control the arrangement of leaf primordia remains an unsolved problem despite intensive study. We shall examine this problem later.

Cellular differentiation to form the vascular tissues of the stem and leaves begins in the localized growth centers below the apex. Below each leaf primordium some cells develop as tracheal cells, others as parenchyma cells. The tracheal cells (a vascular tissue) are arranged in rows separated from one another by parenchyma cells. A single row of tracheal cells and its relation to the base of a leaf primordium is shown in Figure 4-23.

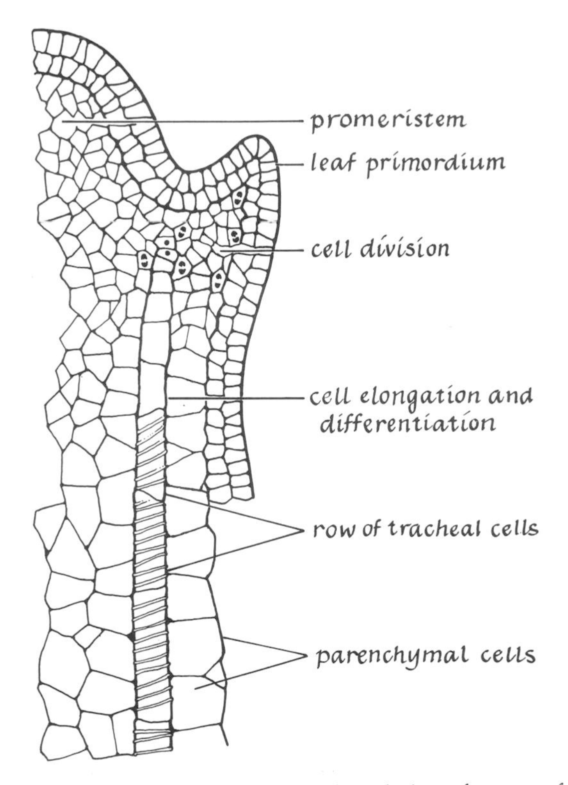

Figure 4-23. *Diagram illustrating the relation of a row of tracheal cells to the base of a leaf primordium.*

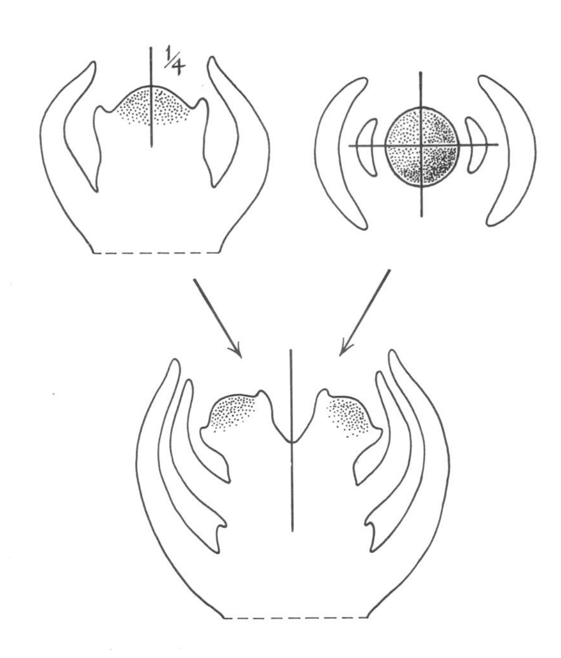

Figure 4-24. *Diagram illustrating regenerative capacities of the shoot apex of an angiosperm. Each ¼ of the apex forms a complete shoot apical system.*

Addition of cells to an existing row of tracheal cells occurs progressively within the meristem at the base of the leaf primordium as the stem elongates. Why do these cells become new tracheal cells whereas adjacent cells produced by the meristem become parenchyma cells? One might suppose that the position of the cells controls their differentiation, but would the meristem produce new tracheal cells in the absence of existing tracheal cells? Apparently not, for in higher plants an isolated apex that does not include the youngest leaf primordium and part of the tracheal elements will not continue its normal activities (Ball, 1946). For example, if the promeristem of a potato plant is severed by a cut from the subjacent stem but left seated on the stem, the growth of the apex will stop. This suggests the importance of cytoplasmic continuity for further activity between the specializing stem tissues and the apex (Sussex, 1963). This dependence of the shoot apex on products of its prior activity is, however, not found in all shoot apical systems, as we shall presently see.

How does a row of differentiated tracheal cells control the differentiation of the as yet uncommitted cell directly above it? Sussex and Clutter (1968) suggest that an induction mechanism may be involved, but no direct evidence exists for the transfer of molecules from a tracheal cell to the cell directly above it. If induction involves a transfer of molecules, this transfer must be polarized, because cells adjacent to but not directly above a newly formed tracheal cell became parenchyma, not tracheal cells. The hypothesis of polarized induction is attractive because the tracheal cells are connected to one another by plasmodesmata, the cytoplasmic strands connecting cells during and after cell plate formation in mitosis.

The Shoot Apex as a Target for Experimentation

In the sections that follow, an outline of experimental approaches to understanding the functioning of the shoot apex will be presented. Only by experimentation can most of the properties and developmental capacities of the shoot apex be revealed. The promeristem itself does not transform into stem, leaves, and other plant parts, but continually moves away from these products of its activity. Nevertheless, the apical center is necessary for continued formation of the plant shoot because its removal stops elongation of the stem below it.

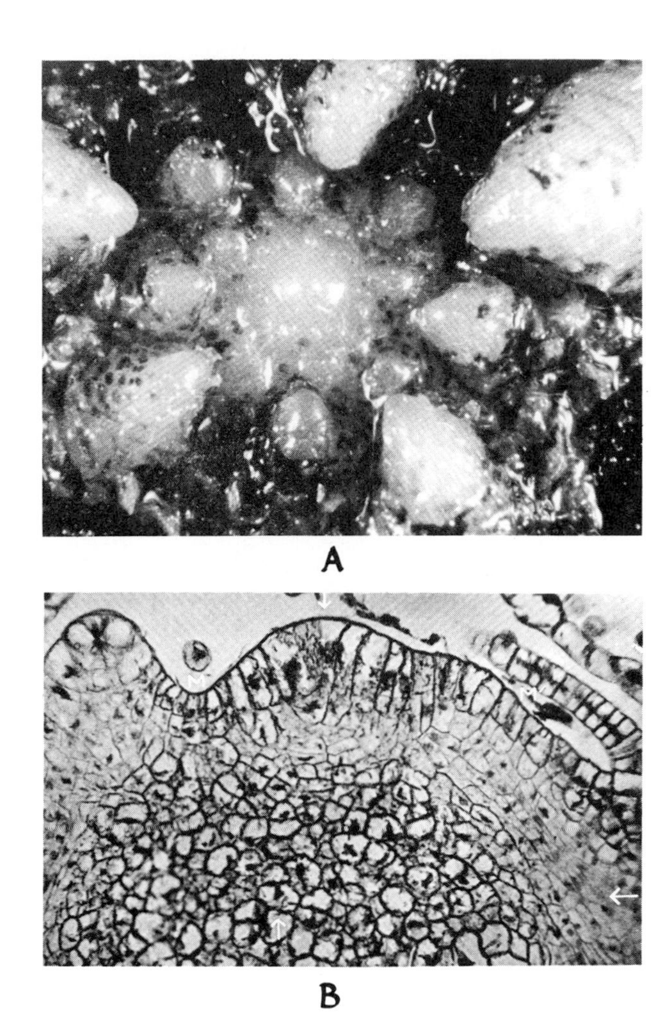

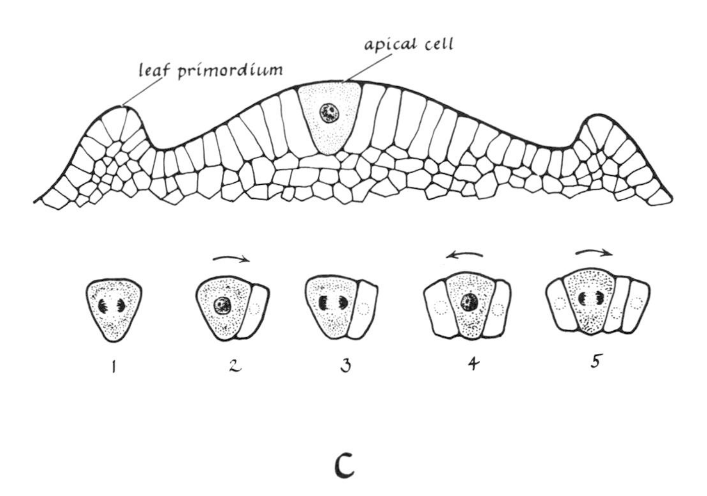

Figure 4-25. *Organization of the shoot apex in ferns. A, photograph from above, of the living apex in* Dryopteris dilatata, *showing the apical meristem in the center surrounded by enlarging leaf primordia (×20). B, longitudinal section through the apex;* ac, *apical cell;* m-m′, *prism-shaped cells of the apical meristem;* iv, *incipient vascular tissue;* p, *pith;* c, *cortex. C, diagram of how the single apical cell functions as a stem cell for other cells of the apex. A and B from Wardlaw:* Organisation and Evolution in Plants *(1965), by permission of Longman Group Limited, London, and Drs. Wardlaw and Cutter.*

Regenerative capacities. In angiosperms, extirpation or destruction of a few promeristem cells does not block the continuous leaf- and stem-forming activities of the shoot apex (Ball, 1948; review of Sussex, 1963). Subdivision of the promeristem region of *Lupinus albus* by vertical cuts into halves or quarters results in the formation of new apical systems in each fractional part (Ball, 1948) (Figure 4-24).

These experiments demonstrate that the shoot apex of angiosperms, particularly the promeristem region, as a developmental field (Chapter 26), is a population of unspecialized rapidly dividing cells, any part of which (perhaps within minimum size limits) is capable of functioning like the whole system (Ball, 1946). A suggested explanation of the continuous developmental activity of the shoot apex is that by always moving away from the products of its activity, the apex maintains a constant, unchanging relationship with its environment. The cells produced by the shoot apex progressively occupy microenvironments of changing composition. Possibly, environmental influences direct the specialization of the cells. The significance of a developmental field may be in the progressive establishment of a microenvironmental pattern, consisting of a quantitatively graded distribution of cell products, to which cells respond differentially according to their positions. We shall discuss this more fully when we consider general mechanisms of development in Chapters 16–22.

Autonomy and organization of the shoot apex in ferns. The promeristem exhibits the impressive autonomy of ferns. When isolated by excision and placed on a nutrient medium (salts and sucrose), the promeristem will continue to form new leaves and stem and eventually a whole plant (Wetmore, 1953;

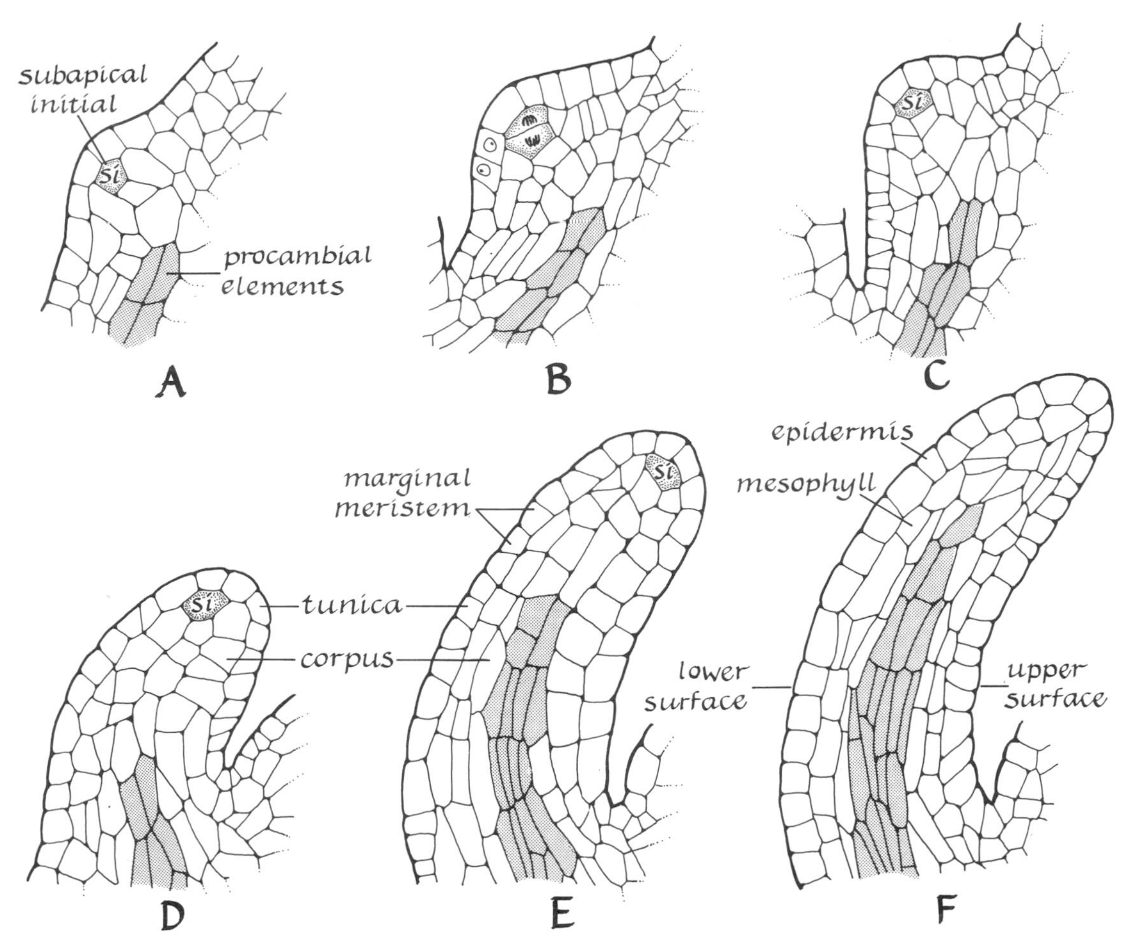

Figure 4-26. *Early stages in development of the leaf of* Linum. *Modified from Girolani, 1954. Copyright Monumental Printing Co. Reproduced by permission.*

Wardlaw, 1965, 1968; review of Torrey, 1967). Ferns (for example, *Dryopteris*) have a single large apical cell that functions as a stem cell and thus as the "mother cell" of all other cells of the shoot (Figure 4-25). Killing this single cell ends all further apical growth (Wardlaw, 1949). The tetrahedral apical cell divides alternately and anticlinally on its inner sides. This is partially shown in Figure 4-25C.

Leaf-forming activities: the spacing and arrangement of leaves. The formation of leaves, a major activity of the shoot apex, presents two problems: how the leaf primordia develop into mature leaves and why they form in an orderly sequence and a specific phyllotactic pattern. First we shall examine how leaves develop.

Early stages in development of the leaf primordium of a flax plant *(Linum)* are illustrated in Figure 4-26. The leaf primordium originates in the division of a single cell, the subapical initial *(si)* in the inner tunica layer (Figure 4-26A and B). This is followed by division of surrounding tunica cells and later by division in the adjacent corpus region. These divisions form the petiole and the basal portion of the midrib of what will become the mature leaf (Figure 4-26D, E, and F). The midrib initially increases in length by apical growth—continued division of the subapical cell—and later by intercallary growth. The lamina or blade of the leaf begins to form as a result of divisions in a marginal meristem that develops along the lateral edges of the leaf primordium (Figure 4-26F). At this stage the young leaf is only a fraction

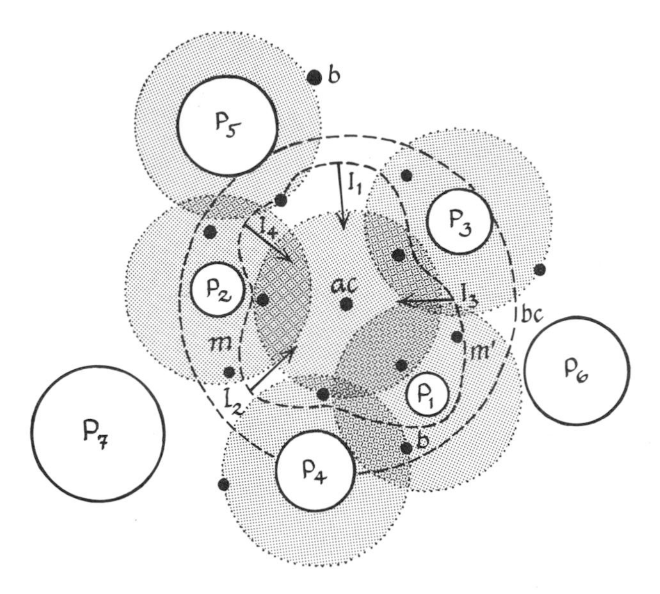

Figure 4-27. *Representation of the large apex of* Dryopteris *from above. The physiological fields of the top cycle of leaf primordia* (P_1-P_5) *and of the apical cell group* (ac) *are indicated by stippled circles.* P_1 *denotes the youngest visible primordium,* P_5 *the oldest.* I_1-I_4, *order and position (arrows) in which the next leaf primordia will be formed.* I_2 *denotes the primordium that will form after* I_1 *begins to transform into* P_1. *From Wardlaw* Organisation and Evolution in Plants *(1965), by permission of Longman Group Limited, London, and the author.*

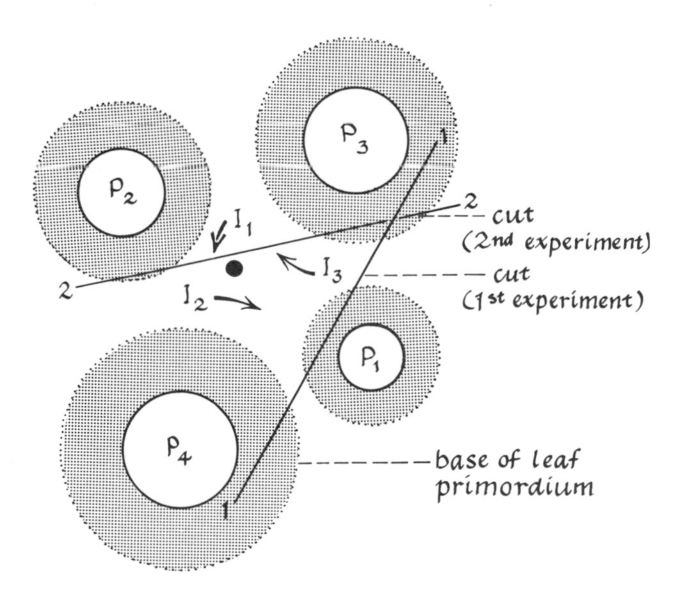

Figure 4-28. *Diagram of operations on the apex of* Dryopteris.

of a millimeter in length, yet the basic architecture of the adult leaf is already present. Later development is mainly by overall but differential intercallary growth (Foster, 1936).

Much of our knowledge about leaf development, especially what cells of the primordium form particular parts of the mature leaf, comes from the study of chimaeras, which are produced by grafting a bud or stem of one species onto the stem of another. In some chimaeras the tunica layers are of one genotype (for example, white cells) and the corpus of another (green cells). Thus the different genotypes are cell markers by which the fate of a cell can be traced. The role of chimaeras in the development of color patterns in leaves is discussed in Chapter 26.

As we have seen, the arrangement of leaves on the stem is determined by the positions of the leaf primordia on the shoot apex. What determines the positions, the exact spacing, and the sequence of leaf and bud primordia? This problem has been studied extensively (reviewed by Wardlaw, 1965; Wetmore and Garrison, 1966; Torrey, 1967; and others). A standardized notation (see Figure 4-27) used by botanists to describe the dynamic morphology of the shoot apex makes it possible to map both the positions of the leaf primordia and the prospective positions of leaf primordia yet to be formed (arrows in Figure 4-27). Figure 4-27 also shows the fields surrounding each leaf primordium. No other primordium can be initiated within these fields. Note the spiral pattern of the position of the primordia. One theory having some experimental support that explains the mathematical precision of leaf initiation proposes that a new primordium will arise in the "next available space." This space is in the part of the apex least occupied by already formed primordia, relatively far from the newest primordium and closer to older primordia (for example, P_3 and P_5 in Figure 4-27), but still as far as possible from any leaf primordia. The "next available space" theory, proposed by Snow and Snow (1948), has been discussed and tested by Wardlaw (1949, 1965). When the youngest visible primordium in the fern *Dryopteris* was isolated by a vertical tangential cut (P_1 in Figure 4-28), a displacement of the second incipient primordium (I_2) toward the wound usually resulted. The operation eliminated possible inhibitory influence of P_1 and increased the space available to the developing primordium (I_2). In a second experiment when I_1 was isolated, the prospective position of I_3 was displaced toward the wound. Unusually large leaf primordia in ferns may be produced by isolating the primordia from adjacent ones by cuts.

The results of Wardlaw's experiments can be interpreted as a demonstration that a leaf primordium may form wherever there is space for it. Removal of an existing primordium or any part of the promeristem region would allow more expansion than normal to occur in the apex, with shifting of the system in that direction. The apical system is normally a continuously expanding one. However, Wardlaw (1949, 1965, 1968) has suggested that each primordium is surrounded by a "physiological field" that presumably inhibits the initiation of another primordium within the area (Figure 4-27). Thus, the production of an unusually large primordium by isolating it from adjacent ones has been interpreted as reducing the inhibitory influence of the surrounding primordia, but no evidence exists for or against the presence of inhibitory substances.

An alternative hypothesis is that the increased size of the isolated primordium is a consequence of its having more room in which to grow. The shoot apex is a balanced system. Any interference that upsets the balance might lead to a tendency of the system to restore the balance. The stability of wholeness in organisms in contrast to that of fractional parts is a well-established principle in biology. The apical system is probably under tension because surgical cuts into it did not heal but widened out. The apex tends to expand, but this expansion is probably controlled by architectural restraints built into the system so that expansion is directed upward and outward with the two directions of expansion coordinated. In this way, there is always a next available space for leaf development.

The leaves of plants arise as separate localized and discontinuous structures. However, postulation of the presence of fields around each primordium does not explain the discontinuity, for one still must explain the spacing of the fields. Possibly, the localized centers of cell division that underlie each leaf primordium arise only where the concentration of metabolites or indole acetic acid (a naturally occurring plant hormone commonly abbreviated IAA) is at the threshold of response of the cells. IAA, one of several plant growth hormones called auxins, is synthesized primarily in the growing regions of plants, and its synthesis is maximal in the promeristem region of the shoot apex (Chapter 19). This results in a graded

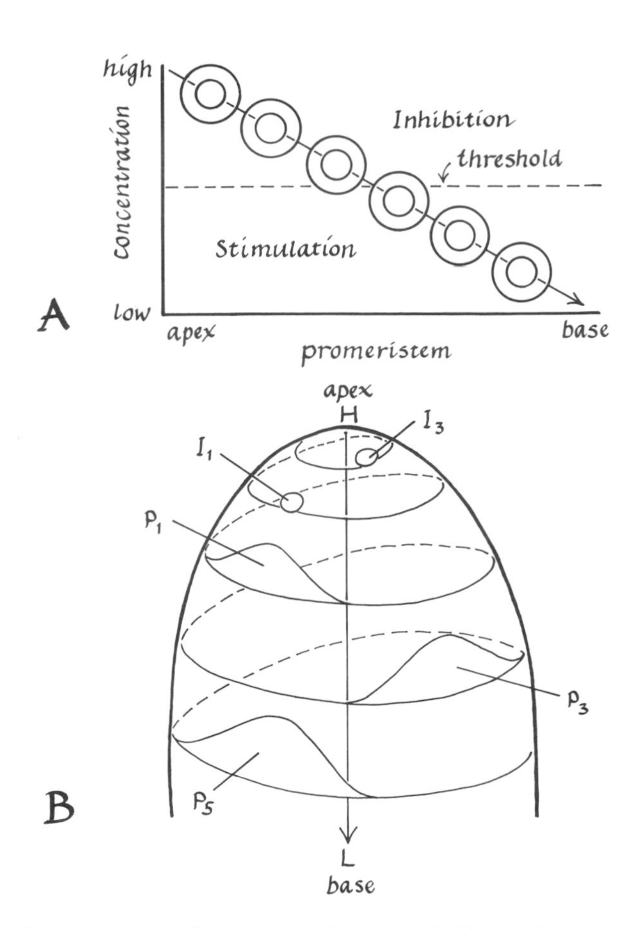

Figure 4-29. *A, the concept of the threshold, and B, its possible role in the spacing of leaf primordia in the shoot apex.*

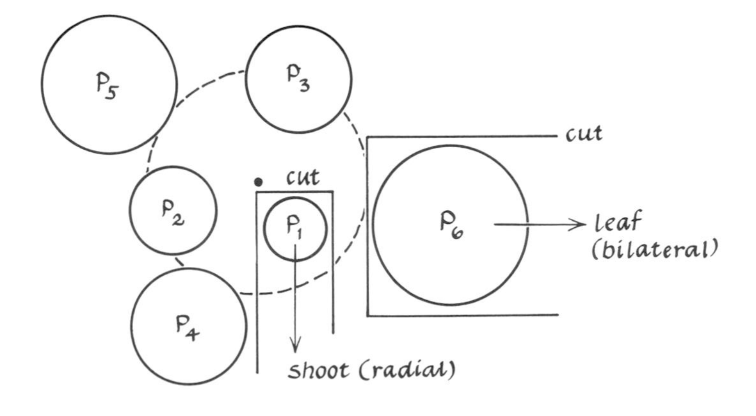

Figure 4-30. *Diagrams of experiments in which leaf primordia of different ages of* Osmunda *were isolated and grown in culture.*

distribution of auxin concentration, high at the apex and decreasing toward the base (Figure 4-29).

Cells contain or are surrounded by different auxin concentrations relative to their distance from the tip of the apex. The threshold, the responsiveness of the cell, may be reached either by increasing the concentration of a stimulating substance or by decreasing the concentration of an inhibiting substance; or the same substance may be inhibiting at a high but stimulating at a lower concentration (Figure 4-29). Auxins are known to have this effect on plant growth (Went and Thimann, 1937; Leopold, 1964; Zeevaart, 1966; and others). We shall return to the role of hormones in development in Chapter 19.

It is conceivable that initiation of a leaf primordium might occur only at a specific auxin concentration. This would occur when a group of cells occupies a definite distance from the apex of the promeristem. Cells in positions where auxin concentration would be higher would not respond by increased cell division activity until they were displaced further from the tip of the apex where auxin concentration would no longer be inhibitory.

We have engaged above in pure speculation, which can only be useful as a target for experimental attack. Nevertheless, the problem of discontinuity in an originally continuous system is common to the spacing of leaf primordia and to metameric structures in animals—segments in some worms, somites in vertebrate embryos, and spacing of hairs, feathers, bristles, and stomata—which we shall examine in Chapter 26.

Bilateral symmetry of leaves. The leaf primordia of the apical meristem are arranged in a radially or spirally symmetrical pattern. Mature leaves have different upper and lower surfaces, and one half is the mirror image of the other half, in most leaves. Thus leaves are bilaterally symmetrical, but the leaf primordia are radially symmetrical. Wardlaw (1965) suggests that faster growth on the outside of the primordium converts it into the flattened structure of the leaf. But what directs this differential growth in the leaf primordium? Is it influenced by the remainder of the apex? The experiment (Steeves, 1966) illustrated in Figure 4-30 attempts to answer this question. In the experiment, young and older leaf primordia of the fern *Osmunda* were excised and grown in culture on a medium containing sucrose plus mineral salts. The systematic isolation and culture of progressively older leaf primordia produced the results summarized in Table 4-1. Steeves' results illustrate an important principle in development, that the pattern of development of a structure is progressively "determined"—the primordium progressively attains its independence from the whole and is able to continue its program of normal development. Before the process of determination is complete, an isolated part may develop in an abnormal way; for example, a leaf primordium may develop into a shoot. This phenomenon illustrates the principle of regulation. Thus, in the experiment some cultured leaf primordia developed into whole plants.

Isolated Primordium	*Structure Formed in Vitro*	
	Leaves	*Shoots*
P_1	2	7
P_2	2	12
P_3	4	10
P_6	12	8
P_7	16	4
P_8	17	1
P_{10}	20	0

Table 4-1.
Results obtained by Steeves (1966) when increasingly older leaf primordia of *Osmunda* were isolated and grown in culture, illustrating the principles of regulation and progressive determination.

In earlier experiments on the determination of leaf primordia in the fern *(Dryopteris)* Wardlaw (1949) isolated an I_1 primordium by a vertical tangential cut and found that a shoot developed instead of a leaf. Cutter (1956) found that a primordium as old as P_3 in *Dryopteris* would still develop into a shoot when isolated by vertical cuts.

It thus appears that the bilaterally symmetrical pattern of development of a leaf primordium depends on its continuity with the remainder of the shoot apex. Could the induction or selection of the leaf-type program of development be a consequence of the position of the primordium in a gradient of some substance produced in the promeristem apex or in the stem below? It may be significant that the incipient vascular strands below the leaf primordium must be severed for the leaf primordium to develop into a shoot (Wardlaw and Cutter, 1956).

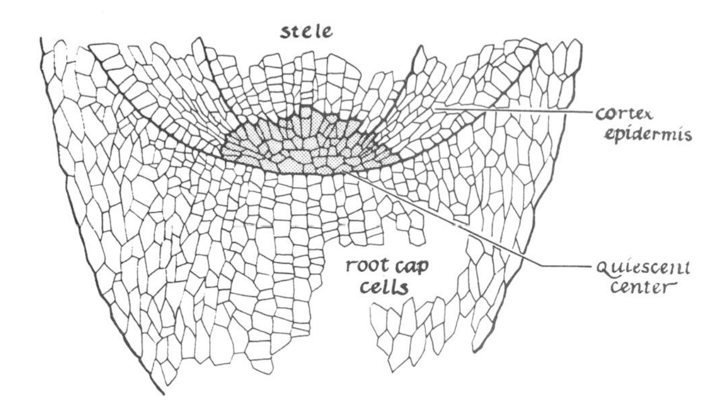

Figure 4-31. *Median section of the root apex of corn.*

Flower-forming activity. Under certain conditions (change in photoperiod, temperature, moisture) the shoot apex ceases its leaf-forming activity and transforms into a flower. This interesting and complex process will be examined after we have discussed the process of sexual reproduction in plants (Chapter 11).

The Root Apex

Like the shoot apex, the root apex is a continuously developing region, exhibiting much the same basic processes of cell division, cell specialization, and cell enlargement (Torrey, 1963, 1967). The cell architecture

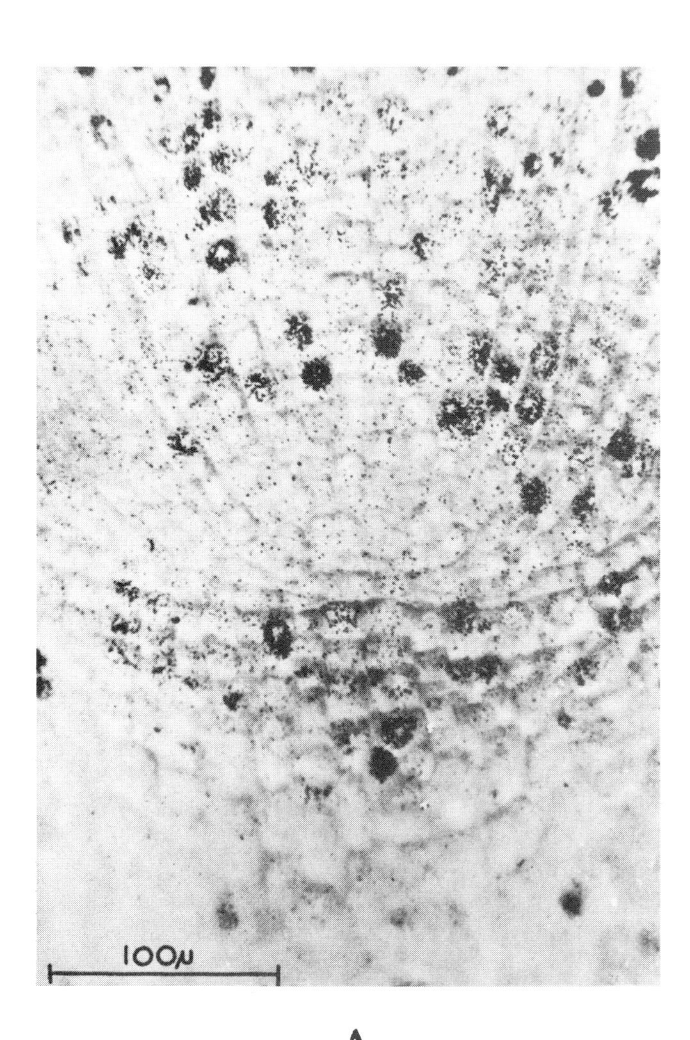

A

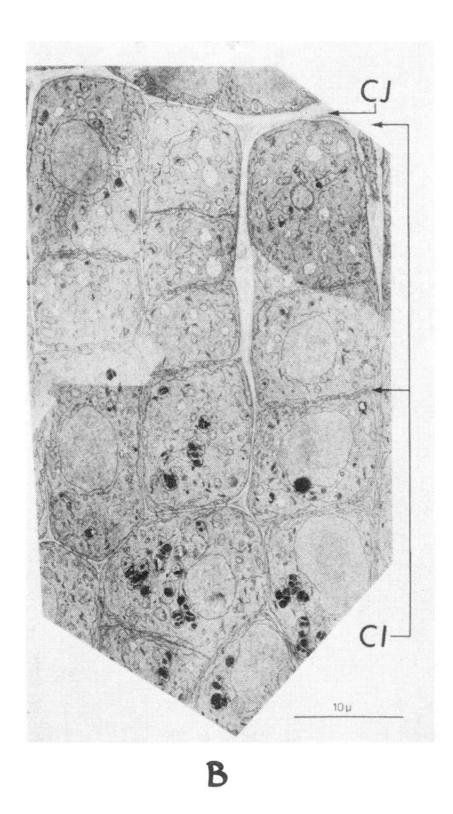

B

Figure 4-32. *A, radioautograph of a median section of the root of corn fed for one day with ^{3}H-thymidine. Note the relative absence of silver grains over the cells of the quiescent center. From Clowes:* Apical Meristems. *Copyright 1961 by Blackwell Scientific Publications Ltd., Oxford. B, electron micrograph mosaic of meristematic cells of the root cap of* Zea mays *(corn).* cj, *boundary between quiescent center and cap initials.* ci, *cap initial cells. From Clowes and Juniper:* Plant Cells. *Copyright 1968 by Blackwell Scientific Publications, Ltd., Oxford.*

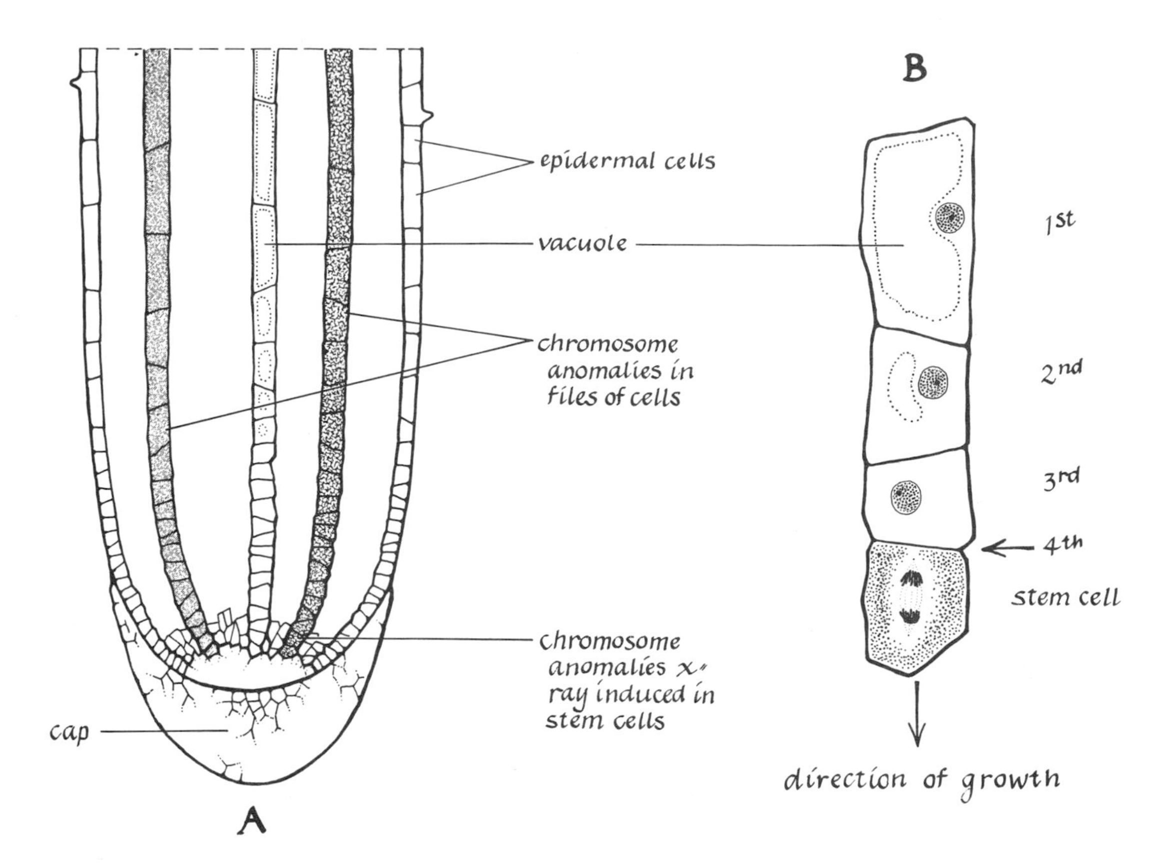

Figure 4-33. *A, the transmission of x-ray induced chromosome anomalies in meristematic cells to the descendants of these cells; B, sequential division products of a stem cell of the root.*

of the typical root apex of corn is illustrated in Figures 4-31 and 4-32. Particularly significant is the presence of a quiescent center in which cell division only infrequently occurs. The mitotic cycle in the quiescent center may be up to 15 times as long as that of the surrounding cells in the meristem region. Note the assimilation of ^{3}H-thymidine by cells surrounding but not in the quiescent center (Figure 4-32). The quiescent center cells are more resistant to x-radiation and high concentrations of tritiated thymidine than other cells (Clowes, 1963).

Probably a relatively small number of meristematic cells surrounding the quiescent center function as stem cells, giving rise in an orderly fashion to all other cells of the root. X-ray induced chromosome anomalies in some cells of the meristem region of the root (Figure 4-33A) are transmitted to the descendants of these cells (reviewed by Torrey, 1967).

An increase in size and vacuolation of the cells derived by division of the meristematic cells accompanies cell differentiation. As the stem cells divide and their products of division elongate, they are pushed downward through the soil, maintaining in this way a constant relationship to the root cap (Figure 4-33B).

A careful analysis of root growth has been done by Goodwin and Stepka (1945) and Erickson and Goddard (1951). The mitotic index pattern in the root of *Zea mays* (corn) is illustrated in Figure 4-34. No cell division was found beyond 2.5 mm from the tip. Erickson and Goddard found that the region in which maximum cell elongation occurred (3 to 5 mm from tip) had no dividing cells. In other words, the meristem of the root tip is more or less sharply localized, as is obvious from the study of mitosis in onion root tips. More detailed study (Jensen and Kavaljian, 1958, 1963) has revealed a complex pattern of cell division within the meristematic region of the onion root tip (Figure 4-35).

Cellular elongation and specialization occur above the meristem region in the root. In some roots, cellular differentiation involves an increase in the number of chromosome sets in the nucleus, that is, an increase in ploidy level (Clowes, 1961). A possible significance of polyploidy in the differentiated cells of plants will be discussed in Chapter 16.

Epidermal cells of the root divide unequally into a large and a smaller cell. The smaller cell, called a trichoblast, develops a root hair, a cytoplasmic extension of the cell (Figure 4-36). The trichoblast arises from the protoplasmically more dense region of the parent cell (Bunning, 1952; Jensen, 1963). Studies by Avers (1958), who used histochemical methods, indicate that the trichoblast is richer in cytochrome oxidase than the larger cell. This unequal cell division may be an example of cytoplasmic segregation (Figure 4-36C).

Initial aspects of cellular differentiation in the root tip include in addition to vacuole formation and protein synthesis, a manyfold increase in area of the endoplasmic reticulum, increase in the amount and size of Golgi bodies per cell, changes in the ultrastructure of mitochondria, and an increase in their number per cell in the root cap. These changes in cells of the stele and cap are illustrated in Figure 4-37.

In the root, as in the stem, the primary vascular tissues are arranged in a definite pattern. Typically, in the root of dicotyledons the xylem tissue occurs in longitudinal strands that radiate from the cross-sectional center and alternate with the phloem tissue (Figure 4-38). Cross sections through the developing root of a pea plant illustrate the typical triarch pattern, blocked out within 0.5 mm of the root tip. This pattern is determined by the site and orientation of cell divisions immediately above the root apex (Torrey, 1963). As in the stem, the pattern of vascular tissues in the root is probably determined by supracellular influences that control the pattern of distribution and orientation of cell divisions in the promeristem and adjacent region.

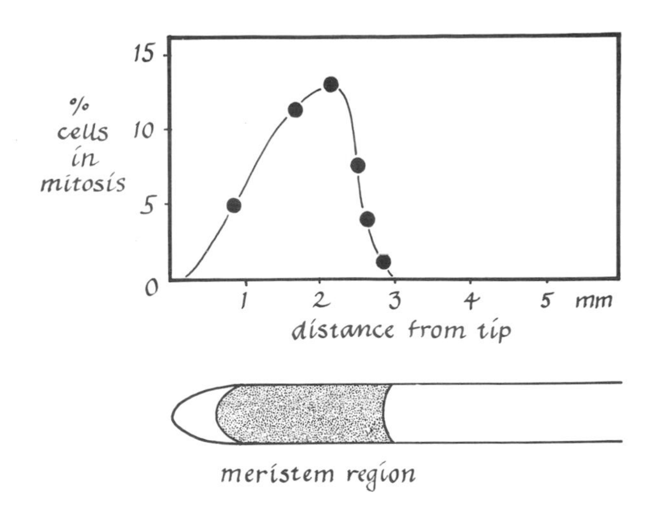

Figure 4-34. *Mitotic index pattern in the root of corn.*

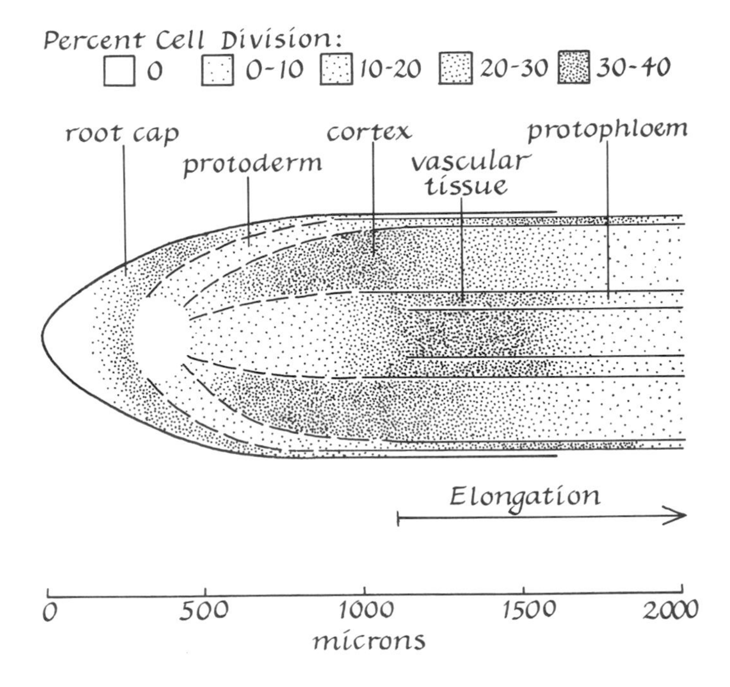

Figure 4-35. *Distribution of mitotic figures in the root, showing the abundance of divisions in the apical meristem region. From Jensen and Kavaljian, 1958. Copyright Monumental Printing Co. Reprinted by permission.*

Intercallary Meristems

In monocotyledonous plants such as corn or grasses, elongation of the stem occurs not only at

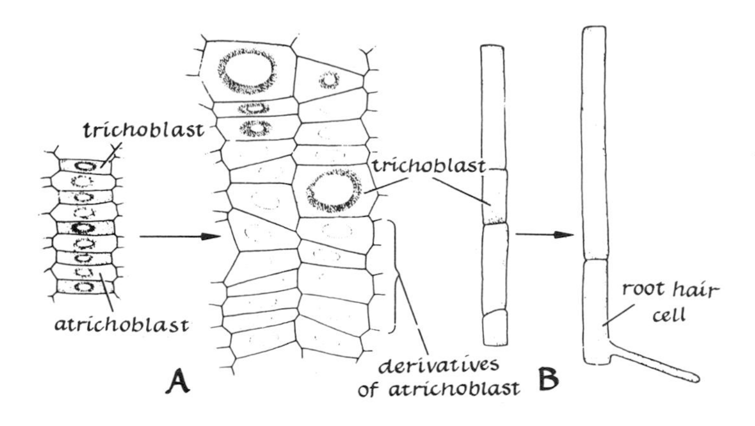

Figure 4-36. *Differentiation of root hairs. A, early stage in differentiation of trichoblast and atrichoblast cells. Each trichoblast forms a single root hair cell. B, later stages in development of a root hair cell of* Phleum *(timothy). C, unequal division of a parent cell resulting in formation of a root hair cell rich in cytochrome oxidase. A and B reprinted with permission of The Macmillan Company from* Development in Flowering Plants *by Torrey. Copyright © John G. Torrey, 1967.*

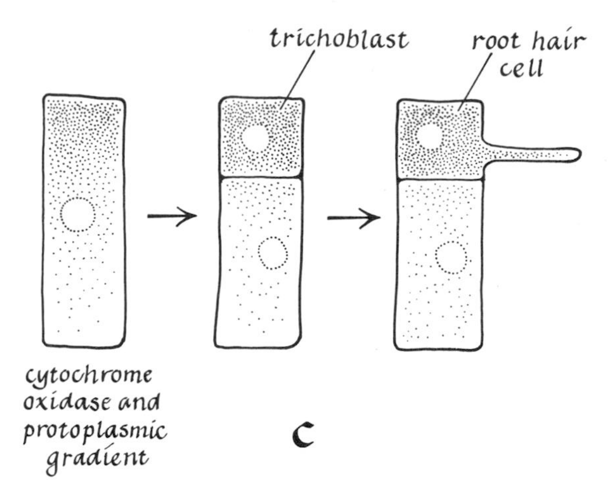

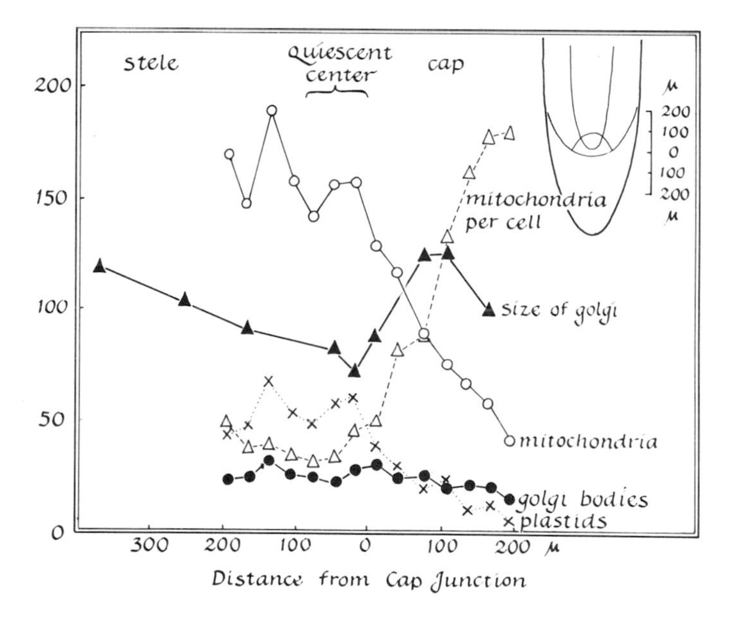

Figure 4-37. *Cellular differentiation in the root tip. Note the difference in numbers of mitochondria, Golgi bodies, and plastids per unit volume of cytoplasm in the stele, quiescent center, and cap. The insert shows the region of the root from which the measurements were taken. From Clowes and Juniper, 1964. Courtesy Clarendon Press, Oxford.*

the tip but also in nodes spaced at intervals along the stem. As a result of cell proliferation in the nodes and cell elongation in the internodal regions the stem continues to grow in length after growth at the apex has ceased. Leaves are formed a considerable distance below the apex of the stem in a nodal region. When grass is cut, the more distal intercallary meristems are removed but the basal ones continue to produce leaves and contribute to elongation of the stem.

Animal Buds

Many colonial hydroids *(Cnidaria)* develop their arborescent patterns through activity of localized apical or intercallary growth centers. Three main types of growth centers comparable to meristems in plants occur. In some hydroids (Figure 4-39A) the meristem is at the base of the hydranth, a hydralike structure that has a hypostome (mouth region) surrounded by tentacles. This meristem continuously produces successive ramifying branches and new hydranths. The whole colony originates from the meristem of the terminal hydranth.

The meristem becomes subdivided by continuous branching, so that with an increased number of hydranths comes an increased number of meristems. Apparently, the activity of these meristems is not interrupted. In another type of meristem (Figure 4-39B), growth and differentiation are discontinuous. The meristem produces a hydranth, then stops activity, and at a later time resumes activity and produces another hydranth. The third type (Figure

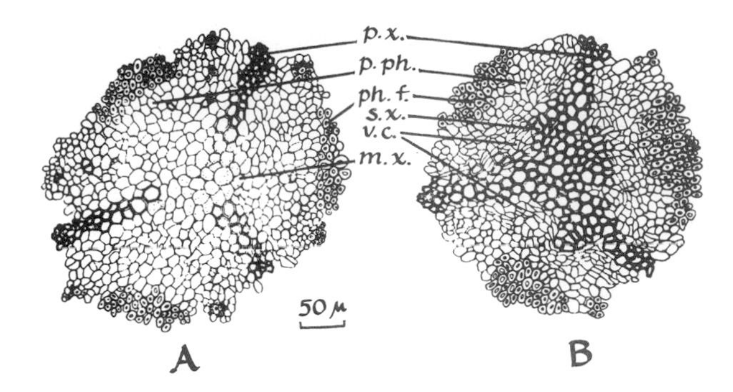

Figure 4-38. *Diagrams representing cell outlines of the central cylinder region of pea roots: A, at the level of primary vascular tissue differentiation; B, at the level of the initiation of the vascular cambium.* Px, *primary xylem;* pp, *primary phloem;* pf, *primary phloem fibers;* sx, *secondary xylem;* vc, *vascular cambium;* mx, *immature primary xylem. From Torrey, 1963. Courtesy Cambridge University Press.*

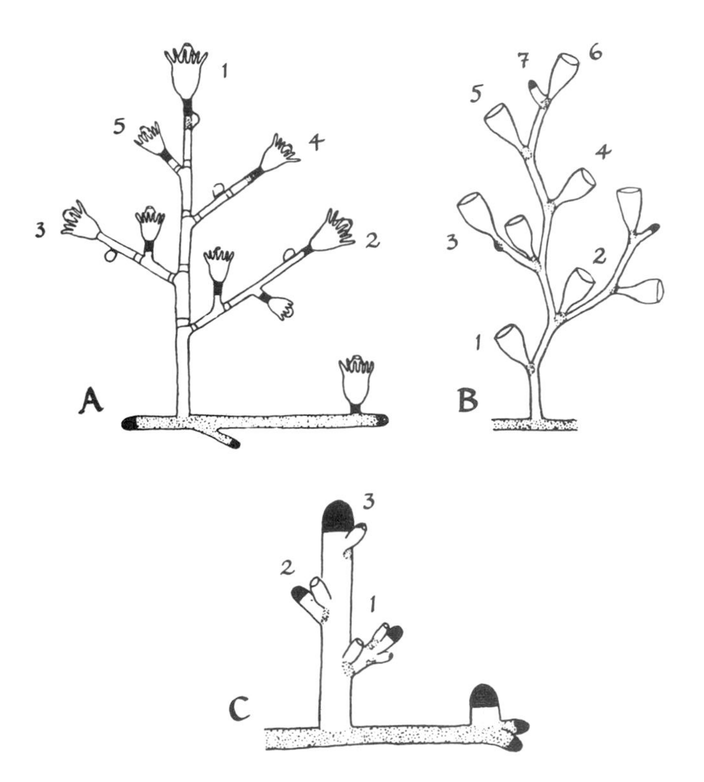

Figure 4-39. *Growth patterns in different hydroids. A, monopodial; B, sympodial; C, apical growth. Meristematic zones* m *are colored black. From John Tyler Bonner,* Morphogenesis: An Essay on Development, *Copyright* © *1952 by Princeton University Press. Reprinted by permission of Princeton University Press.*

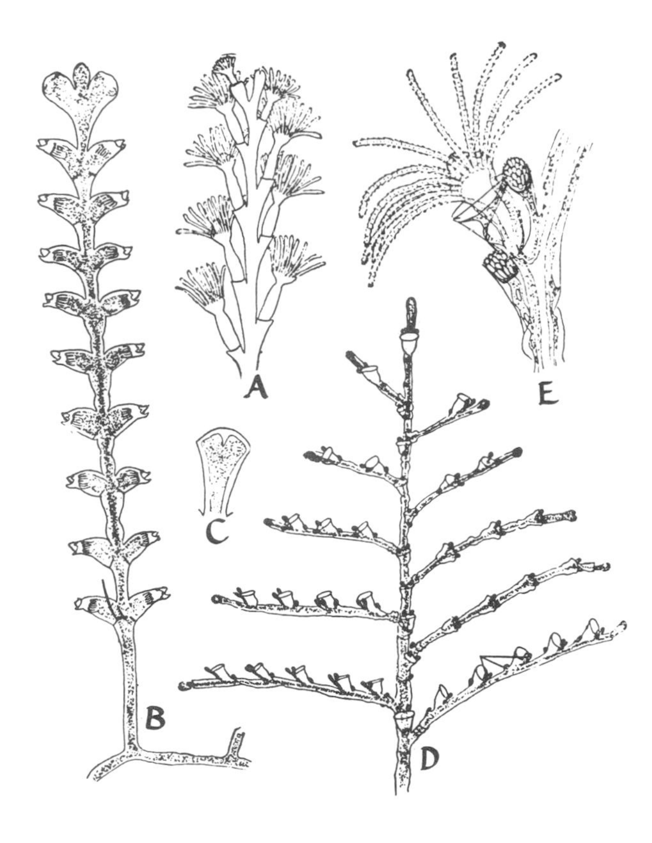

Figure 4-40. *"Phyllotactic" arrangement of buds and hydranths in colonial hydroids. A, sympodial growth; B, dichotomous branching with apical meristem; C, growing point of* Sertularia; *D, apical meristems in* Plumularia; *E, feeding hydranth (polyp) of* Plumularia. *From* The Invertebrates, Volume 1, Protozoa Through Ctenophora, *by Hyman. Copyright 1940 by McGraw-Hill Book Company. Used with permission of McGraw-Hill Book Company.*

4-39C), has a true apical meristem that produces buds which form both lateral hydranths and meristem buds. The buds form lateral branches on which the process of hydranth and meristematic bud formation is repeated. The result is a complex colony (Hyman, 1940).

In many colonial hydroids, the buds and hydranths develop at definite positions on the stalk, thus simulating a type of phyllotaxis (Komai, 1951). Three types of growth patterns in colonial hydroids are shown in Figure 4-40.

Continuous growth by bud formation also occurs in some tunicates (sea squirts) and results in a colony

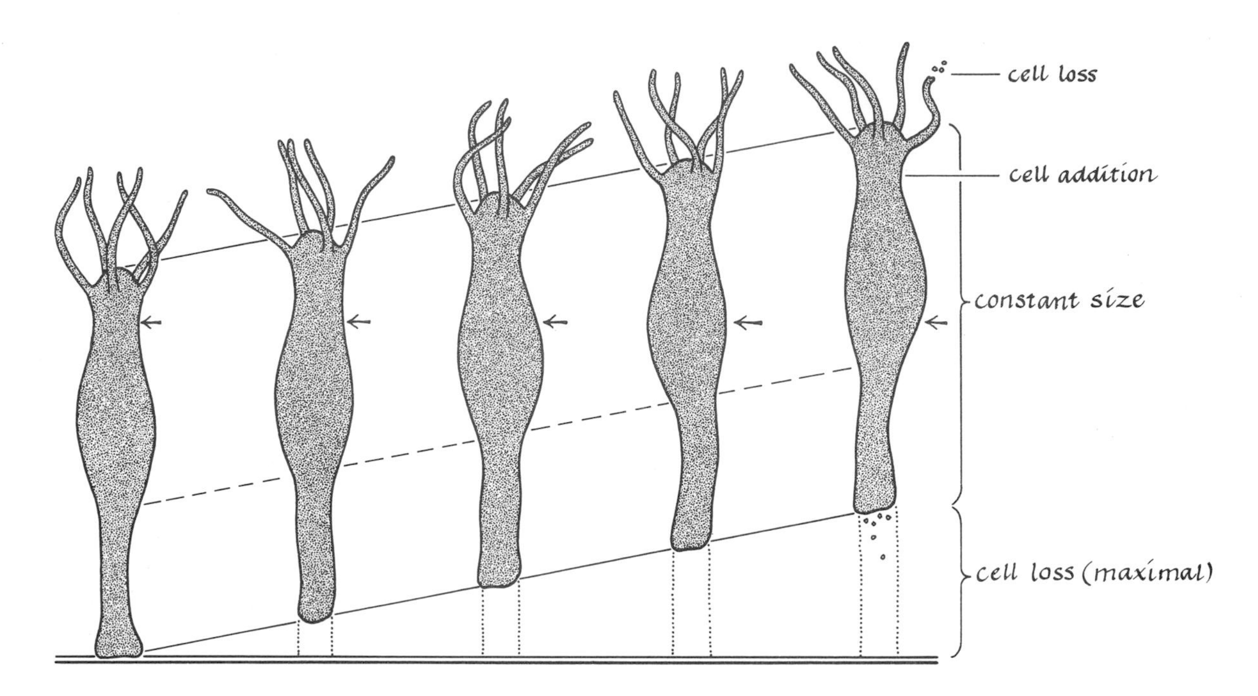

Figure 4-41. *Diagram illustrating the turnover system in* Hydra. *The arrows denote the position of a group of marked cells at successive time intervals, showing that growth occurs just below the tentacles.*

of connected individuals (Chapter 9). In animal bud formation and development the basic processes of cell division, elongation, or change in cell shape and cell specialization occur. Meristematic localized growth in animals is not limited to colonial hydroids or tunicates but occurs in the growth of bones, horns, hair, feathers, and skin. Although the term meristem is not usually used to refer to localized or generally distributed growing regions of animals, the similarity between these regions in plants and animals cannot be denied.

The Whole Organism as a Turnover System

Hydra is a particularly interesting example of a continuously developing organism. The region just below the hypostome functions like a localized meristem of a plant (Brien and Reniers-Decoen, 1949; Burnett, 1962; Brien, 1968; Campbell, 1968). Other hydroids such as *Tubularia* (a marine hydralike organism) exhibit continuous development of a similar type. In the turnover system of the hydra, the loss of cells at the tentacle tips and at the base of the stalk is balanced by cell addition below the hypostome (Figure 4-41). A continuous production of cells flows outward from the meristematic region, but the hydra does not increase in size. The cell population of hydra is renewed about every fifteen days. In some respects it can be said that hydra, not its component cells, is immortal.

Blood Cell Formation

The various specialized blood cells in vertebrates are continuously formed from stem cells in the marrow of the long bones and from lymphoblasts in the lymph glands (Figure 4-42). The stem cells usually remain in the bone marrow but their products enter the circulating blood by slipping through the walls of the venous sinusoids (Bloom and Fawcett, 1962; Maloney and Patt, 1969). However, stem cells may also leave the bone marrow and circulate in the blood (Hodgson et al., 1968). Furthermore, it is generally accepted that stem cells can emigrate from the spleen and "seed" hemopoiesis in the bone marrow

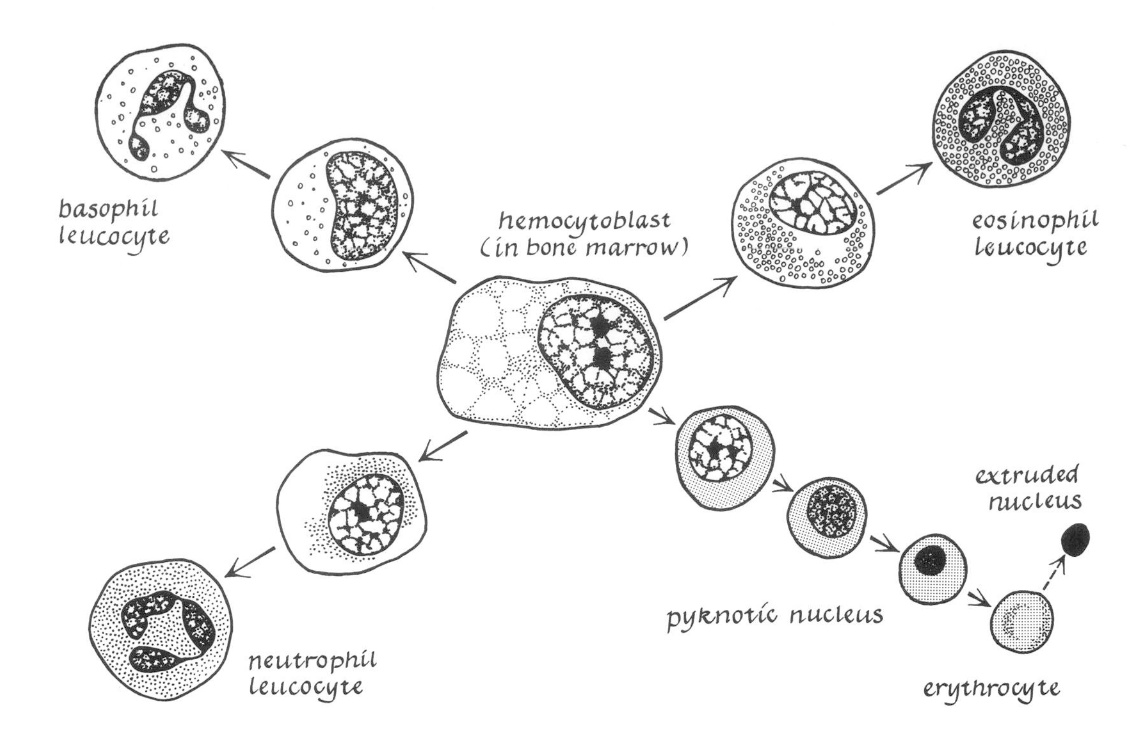

Figure 4-42. *Formation of various blood cell types from a stem cell in the bone marrow of the human adult.*

(Kretchmar, 1969). The diagram in Figure 4-43 summarizes some aspects of the unitarian theory of hemopoiesis.

Approximately 2 to 3 × 10^{11} blood cells are produced and destroyed daily in the human adult (Galbraith et al., 1965; Lajtha, 1967). It would be difficult to find a more rapid turnover system involving complex processes of cell differentiation. The stem cell precursor of the erythrocytes has not been definitely identified (Marks and Kovach, 1966), but studies of Cartwright et al. (1964) indicate that the erythroblasts and myelocytes (Figure 4-43) are the only cells that incorporate ^{3}H-thymidine into DNA. Perhaps these cells, not the hemocytoblasts, are the pleuripotent stem cells.

Mechanisms that control the differentiation of blood cells are not known, but studies in which presumed stem cells of the mouse were implanted in bone marrow and spleen suggest that the stroma tissue of the spleen and marrow (that is, the microenvironment) directs the differentiation of the cells (Wolf and Trentin, 1968). The proportions of the various cell types appear to be controlled by a feedback mechanism (Feldman and Bleiberg, 1967). Other evidence indicates that the hormone erythropoietin is the direct regulator of red blood cell formation in mammals and the primary inductor of erythroid differentiation (Goldwasser, 1966).

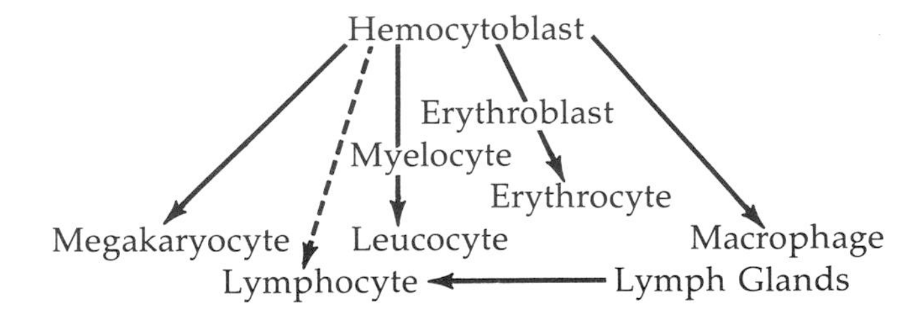

Figure 4-43. *Illustration of the unitarian theory of hemopoiesis in the bone marrow.*

Gametogenesis

The male primate and the rooster are species in which spermatozoa develop continuously from

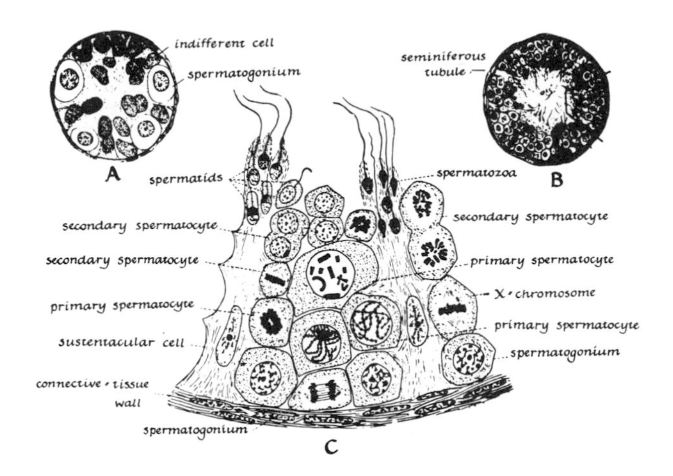

Figure 4-44. *Spermatogenesis in the human testis tubules. A, section through tubule of a newborn (×400); B, section through tubule of adult (×115); C, stages in formation of spermatozoa in area outlined in B. From Arey:* Developmental Anatomy, *7th edition. W. B. Saunders Company, Philadelphia, 1965.*

Figure 4-45. *Uterine and ovarian cycles, correlated with changes in level of the sex hormone estrogen in the human female. Modified from Arey:* Developmental Anatomy, *7th edition. W. B. Saunders Company, Philadelphia, 1965.*

spermatogonia from the time of sexual maturity until death. Stages in formation of spermatozoa in the human testis are illustrated in Figure 4-44.

In females, egg development from oogonia may occur seasonally, periodically, or continuously. All these cases are examples of the differentiation of highly specialized types of cells, eggs and sperm. The sperm cell is so specialized that it can do nothing but fertilize an egg. An egg, when fertilized or artificially activated, can only develop into a whole organism—it cannot bypass embryonic development and develop directly into special cell types, such as nerve cells or heart cells. Later, we shall examine in more detail the special properties of eggs, sperm, and other reproductive units. There is no evidence that oogonia, spermatogonia, or even primordial germ cells can develop into anything other than gametes. However, the type of gamete (whether egg or sperm), regardless of the genetic sex of the germ cell, can be determined by the male or female hormonal composition of its environment (see Chapter 19). This again is an example of the principle of environmental influences.

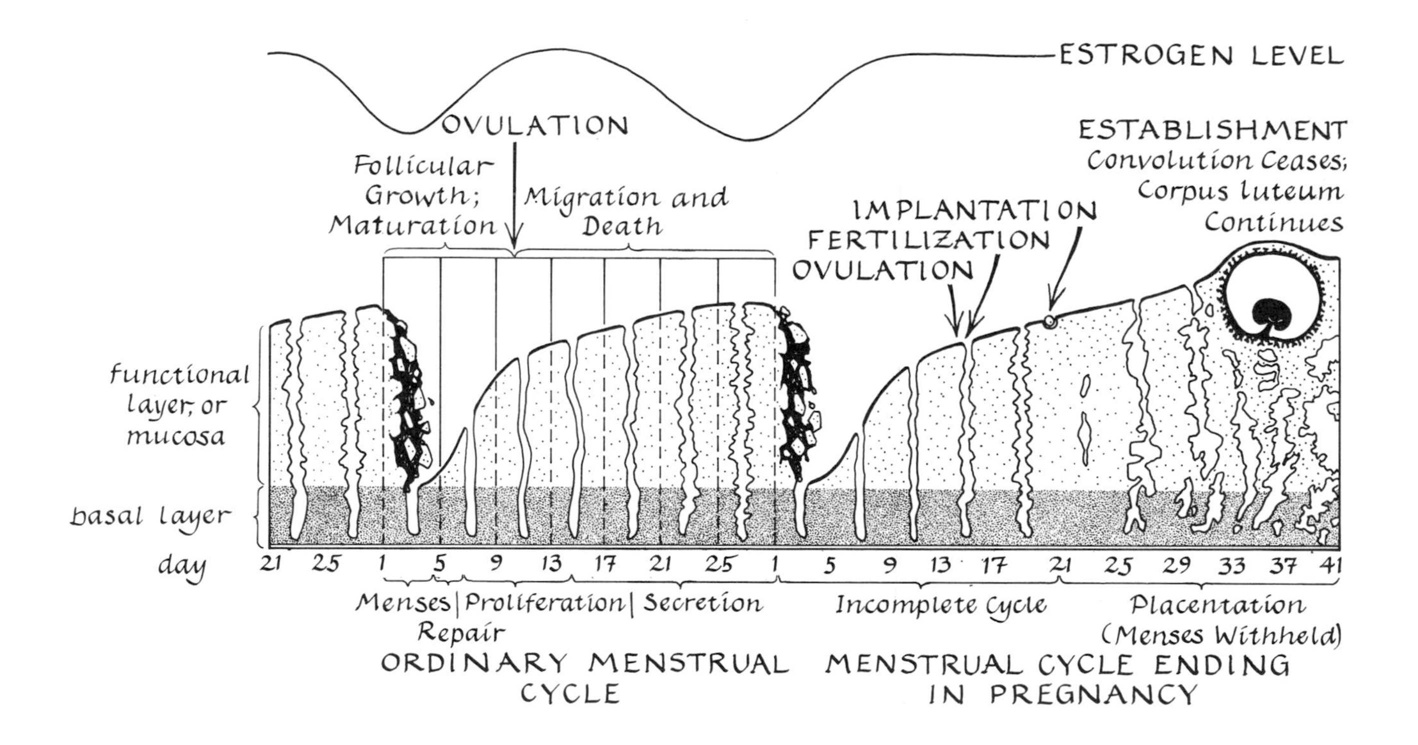

Uterine Endometrium

Regression or breakdown, repair, growth, differentiation, followed by regression or breakdown, and so forth, of the lining of the uterus in mammals occurs periodically but continuously from sexual maturity until menopause or its equivalent. At the cellular and tissue levels these developmental events are truly dramatic. The events are controlled by the cyclic rise and fall of the hormone concentration in the blood of the female (Arey, 1965). Cyclic changes in the uterus, correlated with changes in the level of estrogen, are illustrated in Figure 4-45.

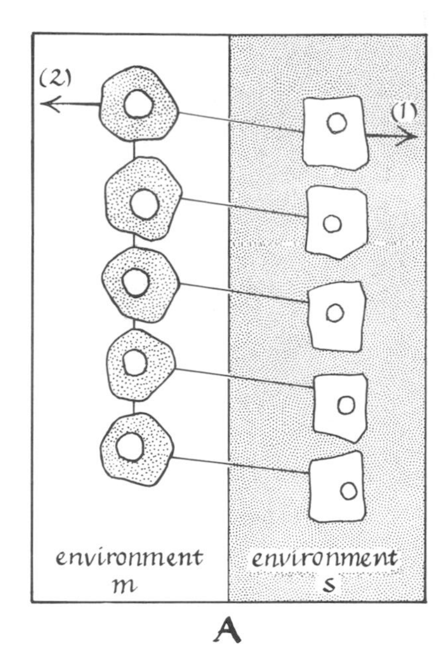

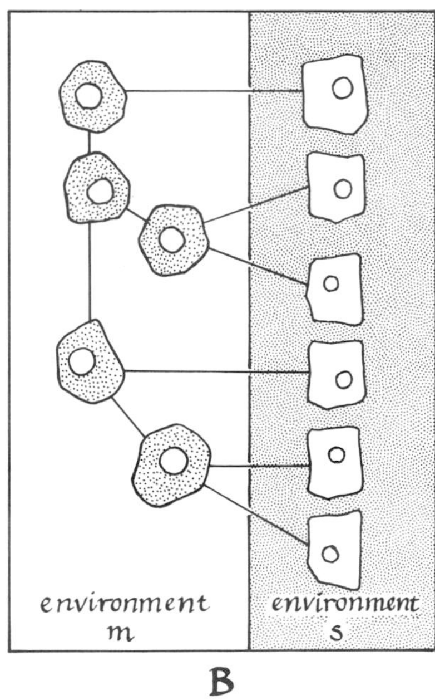

Figure 4-46. *Diagrams of the distinction between A, a permanent stem cell population, and B, a temporary stem cell population. Environment M favors cell division and maintenance of a stem cell population. Environment S favors cell specialization. (1), daughter cell moves into an area favoring specialization. (2), stem cells and meristem region grow away from the daughter cell.*

The Phenomenon of Aging

The rate of continuous development in organisms eventually diminishes, and the process of aging becomes overt. With the possible exception of organisms like hydra (and even this is debated—Strehler, 1962), organisms appear to be genetically programmed to age, and sooner or later the terminal process of senescence sets in (Sinex, 1966; VonHahn, 1966). Senescence is usually defined as a decrease in survival capacity that generally begins in the post-reproductive period. Among the many manifestations of senescence at the cell level in animals are the accumulation of cell products. These include the addition of pigments and lipids in the cells, an increase in calcium and cholesterol content of the cells, and a decrease in cell permeability (Strehler, 1962). Extracellular products, such as collagen in animals and woody tissues in plants, also increase. However, these accumulations have not been demonstrated to be the cause of aging; they may be by-products of a more fundamental change. Aging in plants is generally held to be the final stage in the program of orderly development (Carr and Pate, 1967). Not only does the whole organism age but some of its organs can age at a more rapid rate. This is obvious in the aging and loss of leaves in plants (Woolhouse, 1967).

None of the several theories of the cause of aging—including genetic and developmental theories, which space does not permit us to discuss (see Strehler, 1962; Curtis, 1966)—is alone adequate to explain the phenomenon. Students of the problem agree that aging is a complex process which results from many contributing factors, including increased susceptibility to disease and a decrease in reparative capacities. This aspect of development certainly merits more study than it has received.

Conclusions

The many patterns of continuous development in adult organisms have in common the presence of pre-existing guidelines. The structural and physiological organization of the adult constitutes a basic guideline for further developmental activities. In patterns of reproduction, which we shall consider later, some of the guidelines for development must be progressively formed during development.

Another feature common to all patterns of continuous development is the presence of unspecialized cells that are the source of newly specialized cells. In some instances, permanently unspecialized cells are produced. In other patterns, there is a complete turnover of cells and thus no permanently unspecialized cells. The distinction between these two kinds of unspecialized cells is illustrated in Figure 4-46. Figure 4-46A shows a permanent stem cell that divides in such a way that one daughter cell remains in the meristematic region as a stem cell and the other cell either moves away from the site of its formation into an environment favoring cell specialization or is left behind, as the stem cell and meristematic region move away from it (as in the case of a shoot or root apical stem cell). The latter daughter cell which does not remain in the region favoring cell division becomes specialized. In Figure 4-46B no permanent stem cells are shown, because both division products of a temporary stem cell may move out of the meristematic region into an environment favoring cell specialization.

Some stem cells (for example, the single apical cell in some ferns) are clearly irreplaceable, because their removal stops all addition of new cells. The properties of stem cells are generally a consequence of their positions in the organism. For example, studies of the continuous formation of stratified squamous epithelium in the esophagus of the rat indicate that mitotic activity is a consequence of the position of the cells in the basal layer of the gut. Even in some shoot apical systems there seems to be no permanent stem cell population. It may be tentatively concluded that, in many continuously developing systems, any cell which happens to occupy a position favoring cell multiplication will function as a stem cell as long as it remains in that position (Figure 4-46B). What is constant in many patterns of continuous development is a structured, patterned, and orderly microenvironment, not the cell population, which is continually "flowing through" this environment. We shall consider the development of this environment in Chapter 22.

References

Arey, L. B. 1965. Developmental Anatomy, 7th ed. W. B. Saunders, Philadelphia.

Avers, C. J. 1958. Histochemical localization of enzyme activity in the root epidermis of *Phleum pratense*. Am. J. Bot. **45**:609–613.

Ball, E. 1946. Development in culture of stem tips and subjacent regions of *Tropaeolum majus* L. and of *Lupinus albus* L. Am. J. Bot. **33**:301–318.

Ball, E. 1948. Differentiation in the primary shoots of *Lupinus albus* and of *Tropaeolum majus* L. Symp. Soc. Exp. Biol. **2**:246–262.

Bloch, R. 1948. The development of the secretory cells of *Ricinus* and the problem of cellular differentiation. Growth **12**:271–284.

Bloom, W., and D. W. Fawcett. 1962. Histology. W. B. Saunders, Philadelphia.

Brien, P. 1968. Blastogenesis and morphogenesis. Adv. Morph. **7**:151–203.

Brien, P., and M. Reniers-Decoen. 1949. La croissance, la blastogénèse, l'ovogénèse chez *Hydra fusca* (Pallas). Bull. Biol. France-Belg. **83**:293–386.

Bünning, E. 1952. Morphogenesis in plants. Surv. Biol. Prog. **2**:105–140.

Burnett, A. L. 1962. The maintenance of form in *Hydra*. In D. Rudnick, ed., Regeneration, pp. 27–52. Ronald Press, New York.

Buvat, R. 1955. Le méristèm apical de la tige. Année. Biol. **31**:596–656.

Campbell, R. D. 1968. Cell behavior and morphogenesis in hydroids. In M. M. Sigel, ed., Differentiation and Defense Mechanisms in Higher Organisms, pp. 22–32. Waverly Press, Baltimore.

Carr, D. J., and J. S. Pate. 1967. Aging in the whole plant. Symp. Soc. Exp. Biol. **21**:559–599.

Cartwright, G. E., J. W. Athens, and M. M. Wintrobe. 1964. The kinetics of granulopoiesis in normal man. Blood **24**:780–803.

Clowes, F. A. L. 1961. Apical Meristems. Blackwell, Oxford.

Clowes, F. A. L. 1963. The quiescent center in meristems and its behavior after irradiation. In Meristems and Differentiation, pp. 46–58. Brookhaven Symp. Biol., Upton, N.Y.

Clowes, F. A. L., and B. R. Juniper. 1964. The fine structure of the quiescent centre and neighboring tissues in root meristems. J. Exp. Bot. **15**:622–630.

Cohen, J. 1966. Feathers and patterns. Adv. Morph. **5**:1–38.

Curtis, H. J. 1966. A composite theory of aging. Gerontol. **6**:143–149.

Cutter, E. G. 1956. Experimental and analytical studies of pteridophytes: XXXVI. Further experiments on the developmental potentialities of leaf primordia in *Dryopteris cristata* Druce. Ann. Bot. N. S. **21**:343–372.

Erickson, R. O., and D. R. Goddard. 1951. An analysis of root growth in cellular and biochemical terms. Growth **15**(suppl.): 89–116.

Esau, K. 1963. Ultrastructure of differentiated cells in higher plants. Am. J. Bot. **50**:495–506.

Feldman, M., and I. Bleiberg. 1967. Studies on the feedback regulation of hemopoiesis. In A. V. S. DeReuck and J. Knight, eds., Cell Differentiation, pp. 79–92. Ciba Found. Symp. Little, Brown, Boston.

Foster, A. S. 1936. Leaf differentiation in angiosperms. Bot. Rev. **2**:349–372.

Galbraith, P. R., L. S. Valberg, and M. Brown. 1965. Patterns of granulocyte kinetics in health, infection, and carcinoma. Blood **25**:683–692.

Gifford, E. M., Jr. 1960. Incorporation of tritiated thymidine into nuclei of shoot apical meristems. Science **131**:360.

Gifford, E. M., Jr. 1963. Developmental studies of vegetative and floral meristems. In Meristems and Differentiation, pp. 126–137. Brookhaven Symp. Biol., Upton, N.Y.

Girolani, G. 1954. Leaf histogenesis in *Linum usitatissimum*. Am. J. Bot. **41**:264-273.

Goldwasser, E. 1966. Biochemical control of erythroid cell development. In A. Moscona and A. Monroy, eds., Current Topics in Developmental Biology, pp. 173–211. Academic Press, New York.

Goodwin, R. H., and W. Stepka. 1945. Growth and differentiation in the root tip of *Phleum pratense*. Am. J. Bot. **32**:36–46.

Goss, R. J. 1964. Adaptive Growth. Logos Press, London.

Hodgson, G., E. Guzmán, and C. Herrera. 1968. Characterization of the stem cell population of phenylhydrazine-treated rodents. In Symposium on the Effect of Radiation on Cellular Proliferation and Differentiation. Monaco.

Hyman, L. H. 1940. The Invertebrates: Protozoa through Ctenophora. McGraw-Hill Book Company, New York.

Jensen, W. A. 1963. Specialization of the plant cell. In D. Mazia and A. Tyler, eds., General Physiology of Cell Specialization, pp. 53–60. McGraw-Hill Book Company, New York.

Jensen, W. A., and L. G. Kavaljian. 1958. An analysis of cell morphology and the periodicity of division in the root tip of *Allium cepa*. Am. J. Bot. **45**:365–372.

Jensen, W. A., and L. G. Kavaljian. 1963. Plant Biology Today. Wadsworth Publishing Company, Inc., Belmont, Calif.

Juhn, M. 1952. Functional persistence of embryonic determination in feathers and late developmental stages in spurs. Ann. New York Acad. Sci. **55**:133–141.

Komai, T. 1951. Phyllotaxis-like arrangement of organs and zooids in some medusae. Am. Nat. **85**:75–76.

Kretchmar, A. L. 1969. Comment on cover. Exp. Hemat. **19**:V-VI. Oak Ridge National Laboratory, Oak Ridge, Tenn.

Lajtha, L. G. 1967. Proliferation kinetics of steady state cell populations. In Control of Cellular Growth in Adult Organisms, pp. 97–105. Academic Press, New York.

Leblond, C. P. 1965. The time dimension in history. Am. J. Anat. **116**:1–27.

Leblond, C. P., B. Messier, and B. Kopriwer. 1959. Thymidine-H^3 as a tool for the investigation of the renewal of cell populations. Lab. Invest. **8**:296–308.

Leopold, A. C. 1964. Plant Growth and Development. McGraw-Hill Book Company, New York.

Lillie, F. R., and H. Wang. 1941. Physiology of development of the feather: V. Experimental morphogenesis. Physiol. Zool. **14**:103–133.

Maloney, M. A., and H. M. Patt. 1969. Origin of repopulating cells after localized bone marrow depletion. Science **165**:71–73.

Marks, P. A., and J. S. Kovach. 1966. Development of mammalian erythroid cells. In A. Moscona and A. Monroy, eds., Current Topics in Developmental Biology, pp. 213–252. Academic Press, New York.

Messier, B., and C. P. Leblond. 1960. Cell proliferation and migration as revealed by radioautography after injection of thymidine-H^3 into male rats and mice. Am. J. Anat. **106**:247–285.

Questler, H., and F. G. Sherman. 1959. Cell population kinetics in the intestinal epithelium of the mouse. Exp. Cell Res. **17**:420–438.

Rawles, M. E. 1948. Origin of melanophores and their role in development of color patterns in vertebrates. Physiol. Revs. **28**:383–408.

Rawles, M. E. 1964. Tissue interactions in the morphogenesis of the feather. In A. G. Lyne and B. F. Short, eds., Biology of the Skin and Hair Growth, pp. 105–128. Proc. Symp. Canberra, Australia. Elsevier, New York.

Sinex, E. M. 1966. Genetic mechanisms of aging. J. Gerontol. **21**:340–346.

Sinnott, E. W. 1946. Botany, Principles and Problems. McGraw-Hill Book Company, New York.

Sinnott, E. W. 1960. Plant Morphogenesis. McGraw-Hill Book Company, New York.

Sinnott, E. W., and R. Bloch. 1939. Changes in intercellular relationships during the growth and differentiation of living plant tissues. Am. J. Bot. **26**:625–634.

Snow, M., and R. Snow. 1948. On the determination of leaves. Symp. Soc. Exp. Biol. **2**:263–275.

Soma, K., and E. Ball. 1963. Studies of the surface growth of the shoot apex of *Lupinus albus*. In Meristems and Differentiation, pp. 13–45. Brookhaven Symp. Biol. Upton, New York.

Steeves, T. 1966. On the determination of leaf primordia in ferns. In E. Cutter, ed., Trends in Plant Morphogenesis, pp. 200–219. Wiley, New York.

Strehler, B. L. 1962. Time, Cells, and Aging. Academic Press, New York.

Strong, R. M. 1902. The development of color in the definitive feather. In Contributions from the Zoological Laboratory of the Museum of Comparative Zoology at Harvard College, Vol. XL, pp. 147–185. Harvard University Press, Cambridge, Mass.

Sussex, I. M. 1963. The permanence of meristems: developmental organizers or reactors to exogenous stimuli? In Meristems and Differentiation, pp. 1–12. Brookhaven Symp. Biol. Upton, N. Y.

Sussex, I. M., and M. E. Clutter. 1968. Differentiation in tissues, free cells, and reaggregated plant cells. In M. M. Sigal, ed., Differentiation and Defense Mechanisms in Lower Organisms, pp. 3–12. Waverly Press, Baltimore.

Torrey, J. G. 1963. Cellular patterns in developing roots. Symp. Soc. Exp. Biol. **17**:285–314.

Torrey, J. G. 1966. The initiation of organized development in plants. Adv. Morph. **5**:39–91.

Torrey, J. G. 1967. Development in Flowering Plants. Macmillan, New York.

VonHahn, H. P. 1966. A model of regulatory aging of the cell at the gene level. J. Gerontol. **21**:291–294.

Wardlaw, C. W. 1949. Experiments on organogenesis in ferns. Growth **13**(suppl.):93–113.

Wardlaw, C. W. 1957. On the organisation and reactivity of the shoot apex in vascular plants. Am. J. Bot. **44**:176–185.

Wardlaw, C. W. 1965. Organization and Evolution in Plants. Longman, London.

Wardlaw, C. W. 1968. Morphogenesis in Plants. Methuen, London.

Wardlaw, C. W., and E. G. Cutter. 1956. Experimental and analytical studies of pteridophytes: XXXI. The effect of shallow incisions on organogenesis in *Dryopteris aristata* Druce. Am. Bot. N. S. **20**:39–56.

Went, F. W., and K. V. Thimann. 1937. Phytohormones, Macmillan, New York.

Wessels, N. K. 1962. Tissue interactions during skin histodifferentiation. Devel. Biol. **4**:87–107.

Wessels, N. K. 1967. Differentiation of epidermis and epidermal derivatives. New England Jour. Med. **277**:21–33.

Wetmore, R. H. 1953. The use of "in vitro" cultures in the investigation of growth and differentiation in vascular plants. Abnormal and Pathological Plant Growth, pp. 22–40. Brookhaven Symp. Biol., Upton, N.Y.

Wetmore, R. H. 1956. Growth and development of the shoot system of plants. In D. Rudnick, ed., Cellular Mechanisms in Differentiation and Growth, pp. 173–190. Princeton University Press, Princeton, N.J.

Wetmore, R. H., and R. Garrison. 1966. The morphological ontogeny of the leafy shoot. In E. Cutter, ed., Trends in Plant Morphogenesis, pp. 187–199. Wiley, New York.

Wolf, N. S., and J. J. Trentin. 1968. Hemopoietic colony studies: V. Effect of hemopoietic organ stroma on differentiation of pleuripotent stem cells. J. Exp. Med. **127**:205–214.

Woolhouse, H. W. 1967. The nature of senescence in plants. Symp. Soc. Exp. Biol. **21**:179–213.

Zeevaart, J. A. 1966. Hormonal regulation of plant development. In Cell Differentiation and Morphogenesis, pp. 144–179. Wiley, New York.

5 *Regeneration*

Patterns of Regeneration

Besides being able to maintain integrity or increase in size, most adult organisms have the capacity in varying degrees to repair injuries to themselves. This capacity for repair or reconstitution is often called "regeneration." Strictly speaking, to regenerate means to reform something that has already been generated or formed, which is different from forming something that did not exist before. However, such a strict definition excludes many important repair processes and gives a misleading view of an important developmental phenomenon. Regeneration is usually defined by a zoologist as the process of replacing a lost part—that is, regrowth or reformation in an adult of a part lost through accident. Typically, only the part lost is replaced, as in the regeneration of a new hand following amputation of the original hand of a salamander. This kind of regeneration is obviously different from the continuous physiological regeneration that occurs in turnover systems. Another kind of regeneration, usually termed wound-healing, replaces a lost part by a different kind of tissue, such as replacement of torn skin by scar tissue (Needham, 1952; Seilern-Aspang and Kratochwil, 1965; Schilling, 1968). In still another kind of regeneration more than the lost part is replaced (typically the case in plants) or it is replaced by a different part, for example, the lost eye stalk of a crab by an antenna.

The term regeneration is customarily applied to repair processes in adult organisms. However, a similar process occurs in early developmental stages of many organisms. In these cases, not only can the young stage replace a lost part, but a separated part often can reconstitute a whole organism. The capacity of an experimentally isolated part to form a whole organism is usually called *regulation.* Morgan (1901) includes regulation as a type of regeneration, but the replacement of a lost part and formation of a whole organism from a part are not exactly the same process (see Chapter 15).

Regeneration involving replacement of a lost part and nothing more or different from this part is relatively rare in adult plants, for usually much more than the lost part is formed following injury to the plant. In some respects, the process in plants is artificial induction of asexual reproduction, in which the plant, after removal of a part, forms a whole new plant. The general distinction between repair processes in animals and in plants is illustrated in Figure 5-1. Note that the missing part of the leaf is not replaced, but a whole new plant develops from the cut surface, or more commonly from the petiole (stem) of the isolated leaf or a fragment of it. Not all kinds of leaves have regenerative capacity but stem pieces of many kinds of plants can regenerate a whole plant. In the planarian, only the missing part is replaced. Figure 5-1 also shows the result of development by an isolated single cell of an animal and a plant. In the animal, a single cell placed in a nutrient culture medium may divide repeatedly to form a population of cells with little or no overall organization, but in plants a single phloem cell in culture may divide to form a cell population from which a complete plant can develop (Steward, 1963).

For our purposes it is only necessary to keep in mind the general differences between regeneration in plants and in animals and to be aware that exceptions occur. In some plant systems, such as the leaf primordium of a fern (Steeves, 1962, 1966), cap regeneration in *Acetabularia* (Figure 5-4), and fruiting bodies of some fungi (Goebel, 1908), only the missing part is replaced. Conversely, in some animals (coelenterates and tunicates) more than the missing part is replaced, and occasionally a whole new organism can be formed (Berrill, 1961).

The basic component processes—growth, cell differentiation, movement, and cellular interactions—occur in regeneration, but in most cases the origin of the cells replacing a lost part or forming a new organism is not the same as the origin of the cells that formed the original part during reproductive development. In contrast to reproductive development, regeneration is a relatively direct reconstruction of the lost part, in most cases not involving a long series of stages (Abeloos, 1932). Nevertheless, repair processes in regeneration probably involve the same interaction between the inherited information in the genome and the composition of the microenvironment as does continuous development.

Many of the numerous patterns of regeneration have been studied extensively, especially limb regeneration in vertebrates. Most of the earlier literature is cited in the reviews of Hay (1966), and Schmidt (1968), to mention a few. The following examples illustrate this aspect of developmental biology.

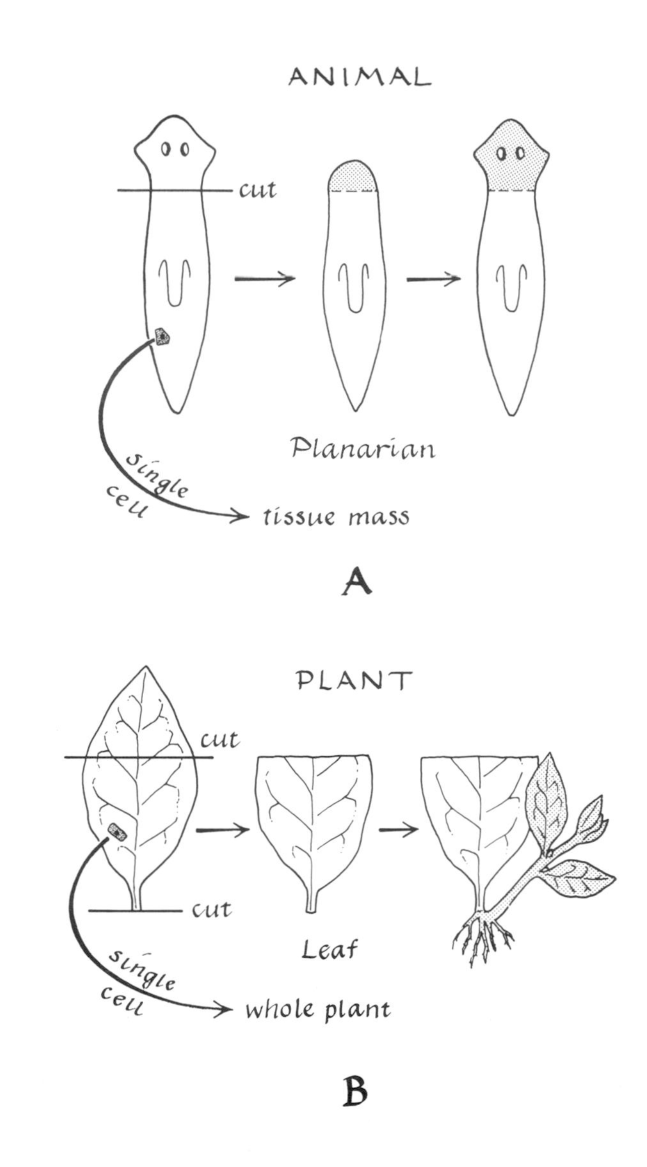

Figure 5-1. *Comparison of regeneration in animals (A) and plants (B).*

Regeneration in Plants

In general, plants possess far greater repair capacities than animals. Even a single, isolated tissue cell from an adult plant can reconstitute the whole plant under appropriate conditions of culture in vitro (reviews by Steward, 1963, Steward et al., 1966 and 1969).

Colonial Algae

In colonial algae such as *Gonium*, in which the colony consists of 4 to 16 cells according to the species, *Pandorina* (4 to 32 cells), *Eudorina* (16 to 64 cells), *Pleordorina* (32 to 128 cells), and *Volvox* (500 to 50,000 cells), an isolated single cell is capable of dividing and reforming a new colony (Bock, 1926; Brown, 1935; Smith, 1933). However, only the reproductive cells of *Pleodorina* and *Volvox* are capable of forming a new colony. On the other hand, one or more cells removed from a colony are not replaced. Thus, colonial algae exhibit the type of regeneration similar to asexual reproduction, but are not capable of replacing the missing part of a colony. Note in Figure 5-2 that two cells removed from a colony of *Gonium*

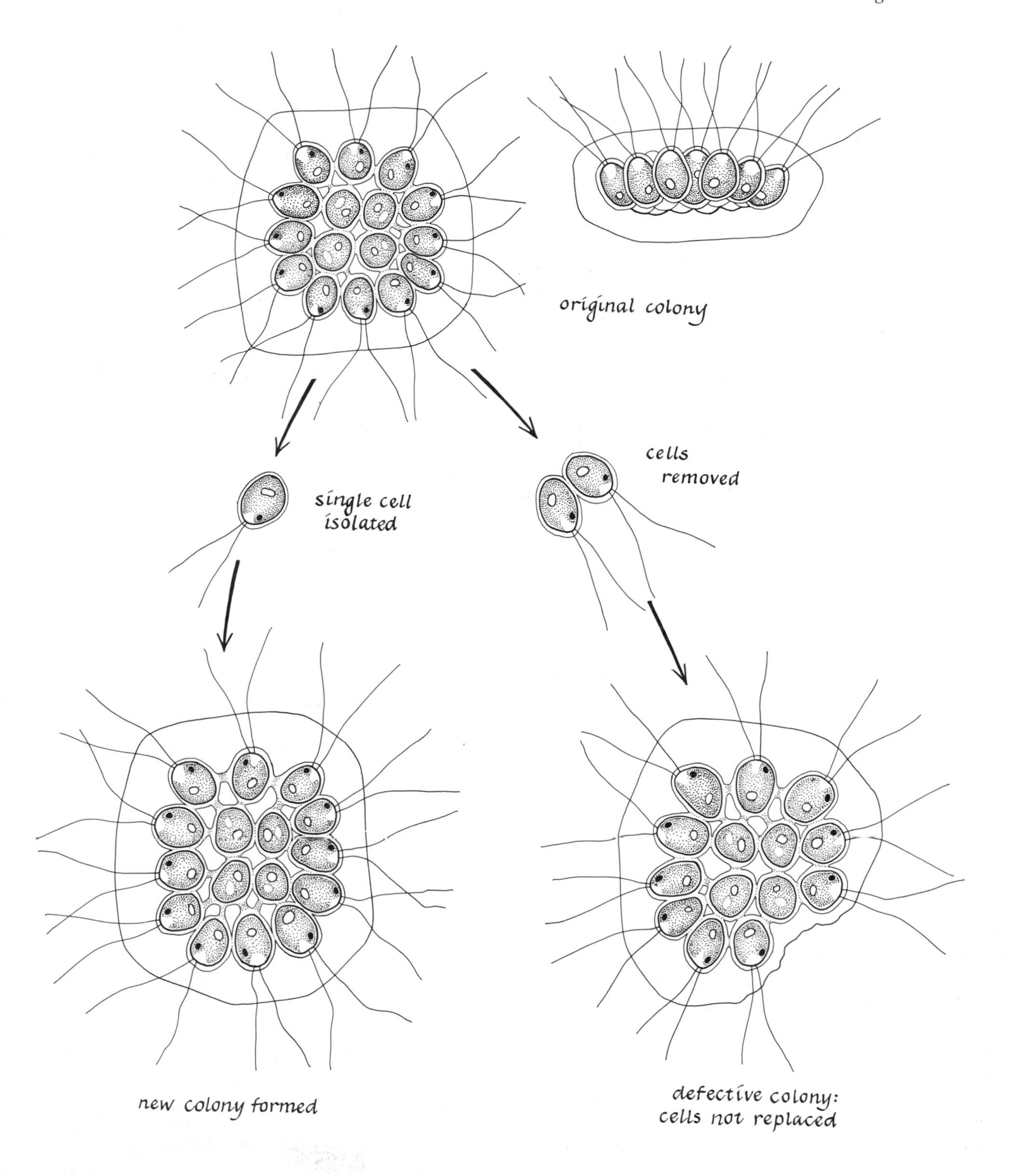

Figure 5-2. *Type of regeneration characteristic of* Gonium *and other colonial algae.*

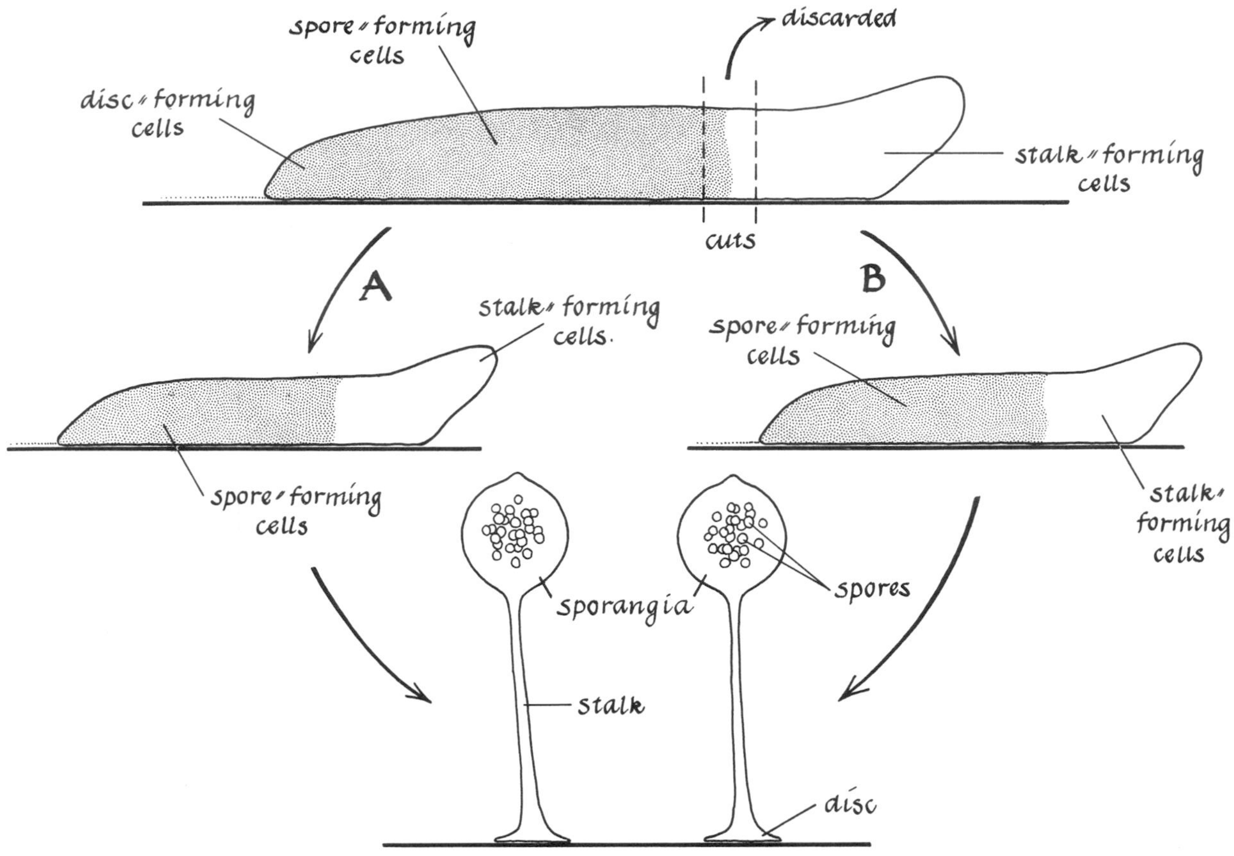

Figure 5-3. *Regeneration in the pseudoplasmodium (slug) of a cellular slime mold. A, transformation of prospective spore-forming cells into stalk-forming cells. B, transformation of prospective stalk-forming cells into spore- and disc-forming cells.*

are not replaced, but a single isolated cell can form a new colony of 16 cells.

Are the guidelines that direct the development of an isolated single cell into a new colony progressively formed, and do these guidelines control the pattern of limited cell division? Or is the system genetically programmed for a definite number of cell divisions? Such a program may be the reason why the colony is unable to replace a few missing cells. In this respect the system has mosaic qualities similar to those exhibited by some animal embryos. Isolation of a single cell results in a change in its environment, and this may be sufficient to reactivate the genetic program. However, knowledge of the actual control mechanism must await further experimentation.

Slime Molds (Myxomycetes)

The migrating *pseudoplasmodium,* a multicellular stage in the life cycle of the cellular slime molds (see Chapter 6) is capable of a type of regeneration. When the pseudoplasmodium or slug, which is about 0.1 to 2 mm long and consists of hundreds or thousands of individual amebae, is cut into two pieces, each piece reforms the missing part by reorganization of the remainder without cellular multiplication (Figure 5-3). As indicated in the figure, there are three different kinds of cells (or amebae): disc-forming cells, spore-forming cells, and stalk-forming cells. A significant feature of this pattern of regeneration is the formation of two complete slugs, each about one-half normal size but containing the normal

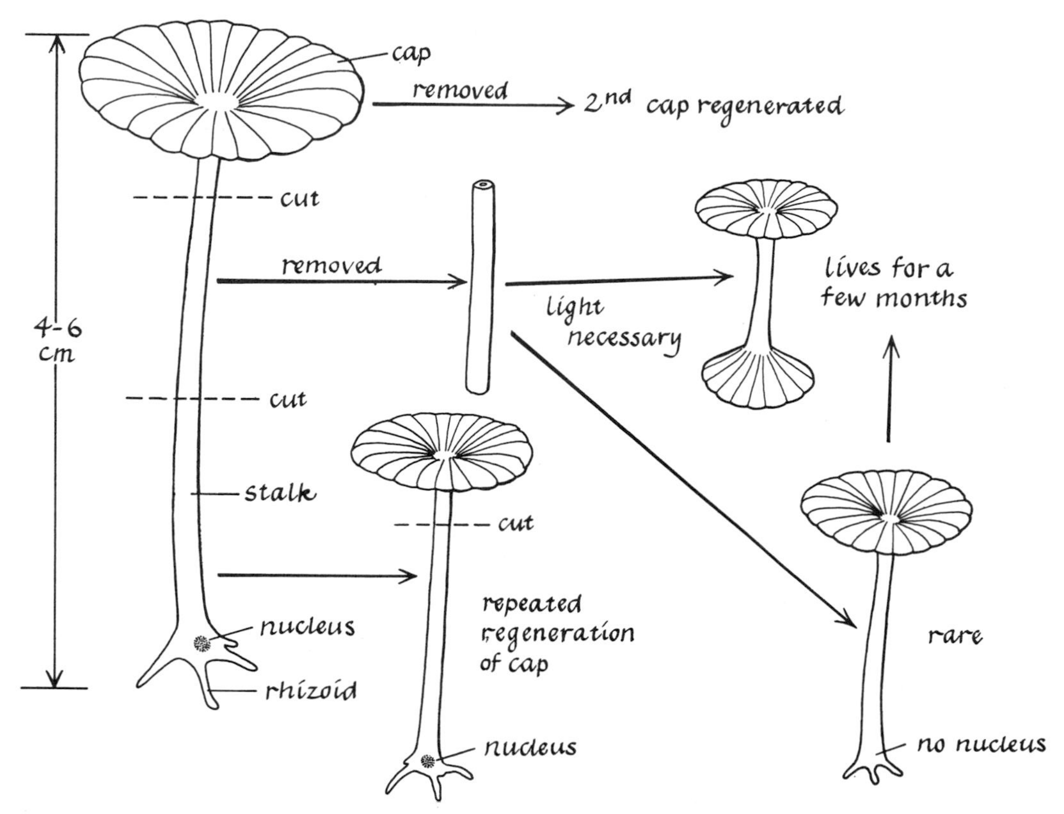

Figure 5-4. *Regeneration in* Acetabularia. *An isolated cap plus a small piece of attached stalk will regenerate another cap; an isolated piece of the stalk regenerates a cap at each end and lives for a few months. In both cases regeneration occurs in the absence of the nucleus (located in the rhizoid).*

proportion of disc, spore, and stalk amebae in the fruiting body it forms later (Bonner, 1952, 1967). The reorganization thus involves transformation of some prospective spore-forming cells into stalk-forming cells (Figure 3A) and transformation of some prospective stalk-forming cells into spore- and disc-forming cells (Figure 3B). The significance of these transformations will be discussed in Chapter 6.

Other myxomycetes are not cellular, but are multinucleate cytoplasmic masses that move slowly like a large ameba. Under appropriate conditions of light, moisture, and exhaustion of the bacterial food supply, the whole cytoplasmic mass (called a plasmodium) transforms into fruiting bodies (see Chapters 10 and 23). The plasmodium probably cannot reconstitute a missing part, but pieces of the plasmodium can continue the life cycle and produce fruiting bodies.

A Single Cell Alga

Acetabularia, sometimes called "the mermaid's wine glass," is a particularly interesting coenocytic (multinuclear) green alga, because at one stage in its life cycle it has only one large nucleus, located in one of the *rhizoids* (root-like structures) (Figure 5-4). Because of this feature and because the uninucleate form is relatively large, *Acetabularia* has been the object of many experiments (Hammerling, 1966). The uninucleate stage is capable of striking regeneration. Even a part lacking the nucleus is capable of regenerating a new cap because of the presence of stable mRNA molecules (Chapter 17). This regeneration is shown in Figure 5-4.

The capacity of a piece of the stalk to regenerate one or two caps increases with the increase in length

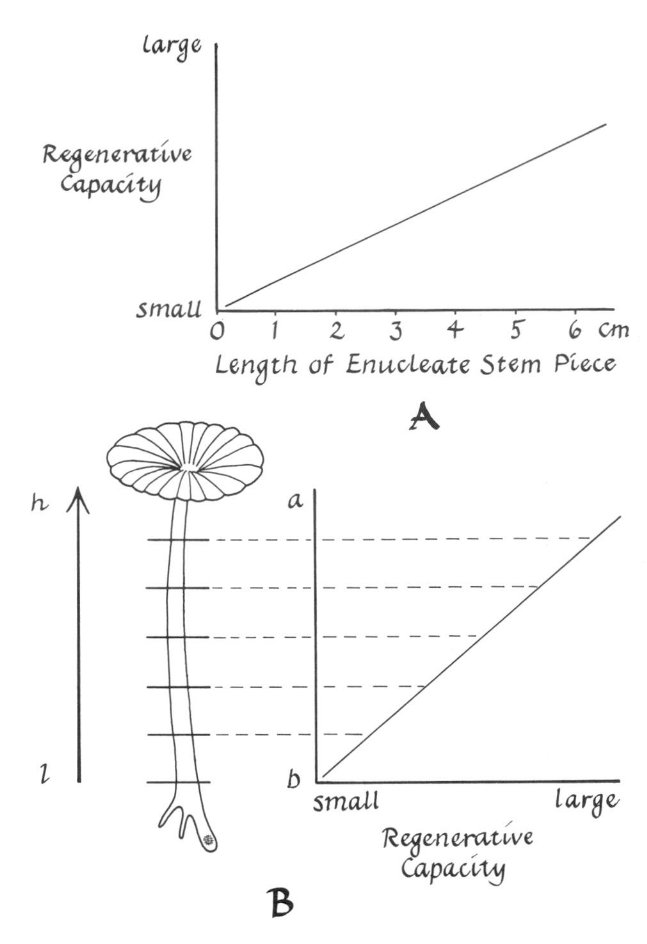

Figure 5-5. *Graphs of the increase in regenerative capacity of* Acetabularia. *A, when increasing the length of the enucleated stem piece, and B, in equal-length stem pieces isolated progressively closer to the cap.*

of the piece (Figure 5-5A). The regenerative capacity of stalk pieces of the same length is greater in those taken from the apical region (near the cap) than in those near the rhizoid (Hammerling, 1966). This is illustrated in Figure 5-5B. These results suggest that there is a gradient in the distribution of a cap-forming substance, with the highest concentration in the stem region just below the cap. DNA and RNA net synthesis is extensive in the enucleate pieces, presumably a consequence of the presence of chloroplasts that are relatively autonomous of the nucleus (Brachet, 1968) (see Chapter 28).

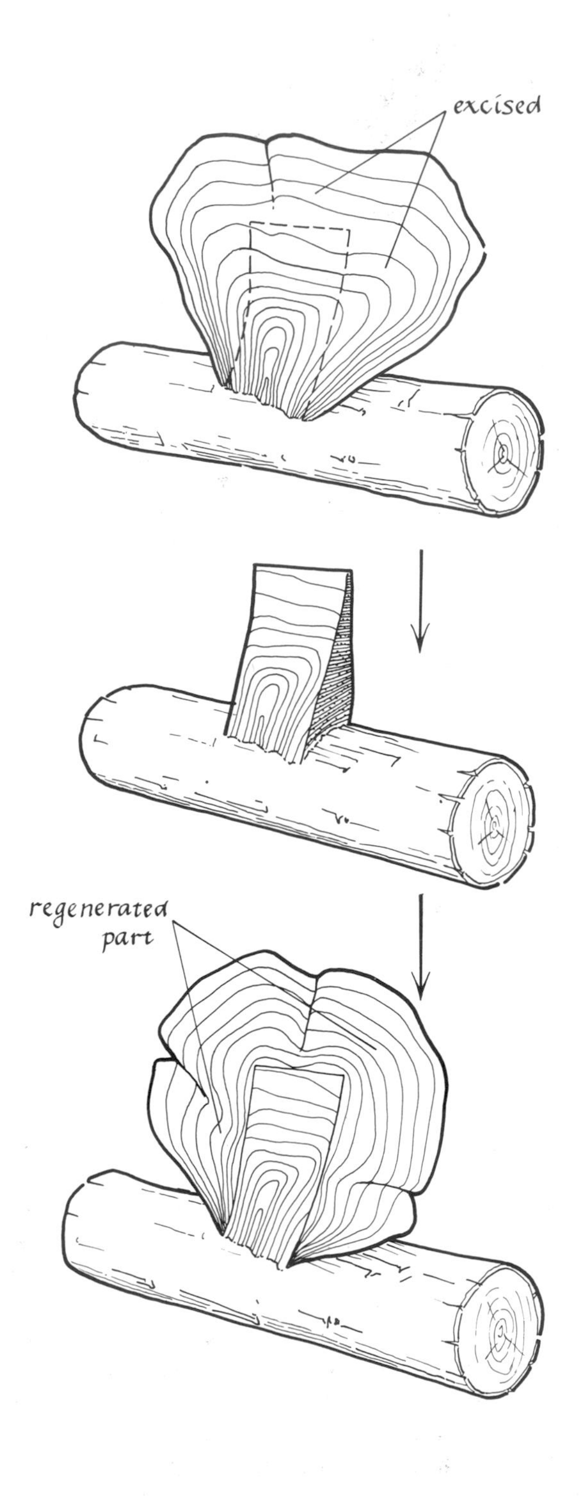

Figure 5-6. *Regeneration by the fruiting body of* Stereum *of excised peripheral parts.*

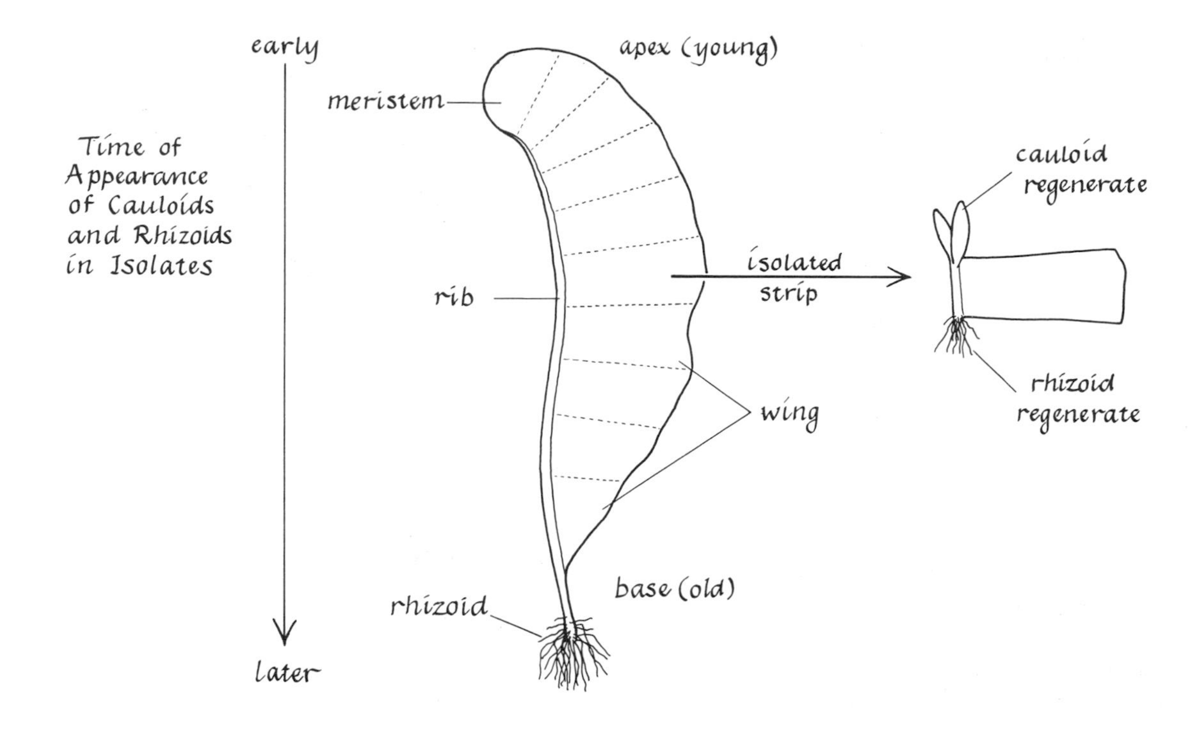

Figure 5-7. *Apical-basal differences in the rate of regeneration of cauloids by isolated pieces of* Riella.

A Fungus

Regeneration involving replacement of the lost part has been described by Brefeld (1876) in *Coprinus*, a mushroom. The multinucleated zygospore (formed by sexual fusion of hyphae in the mycelium) germinates by pushing out a tube. If this tube is cut off a new tube regenerates from the zygospore, not from the cut surface of the amputated tube, and develops by formation of filaments into the mushroom, which is the fruiting body of the fungus.

The fruiting body of the Basidiomycete fungus, *Stereum hirsutum,* is capable of replacing the lost part (Goebel, 1908). This is illustrated in Figure 5-6. The fruiting body of this fungus is not formed suddenly like that of many other mushrooms, but grows perennially and continuously which may underlie its capacity for regeneration of the type shown. The rapidly formed fruiting bodies of other mushrooms do not replace lost parts, but an excised part at early stages can reconstitute the whole fruiting body.

Liverworts

Regeneration in multicellular plants involves *dedifferentiation,* that is, changes in differentiated cells leading to formation of less specialized cells. The dedifferentiated cells then redifferentiate to form more specialized plant tissue. However, as we have noted before, more than the part lost is formed. Dedifferentiation in the liverwort *Riella helicophylla* involves a number of changes in the specialized cells: (1) increase in nucleolar size (increase in rRNA synthesis); (2) increase in arginine-rich proteins; (3) starch synthesis; (4) nuclear migration to the center of the cell; (5) chloroplast division or budding; and (6) mitosis (Stange, 1964). *Riella* has a relatively simple organization (Figure 5-7) consisting of a multicellular rib and a unistratose wing. The meristem is located at the apex and along the edge of the wing of the plant (the youngest part), rhizoids at the base (the oldest part). Pieces of the plant are capable of regenerating rhizoids from the basal cut surface and

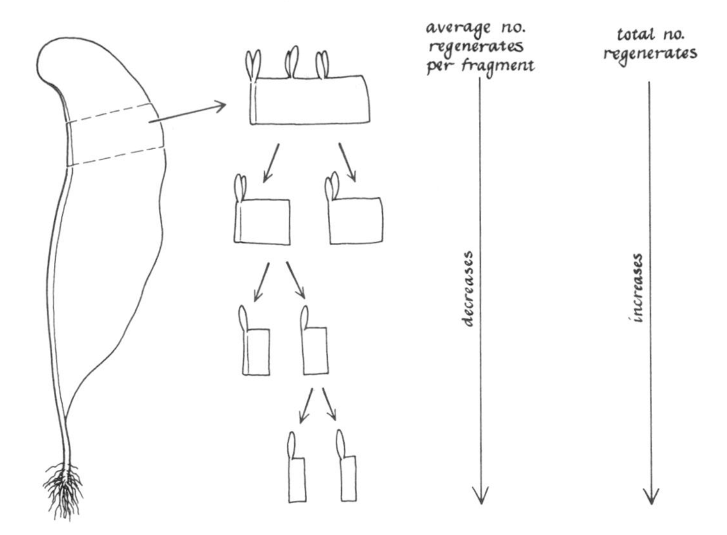

Figure 5-8. *Relationship between the size of an isolated piece of* Riella *and the number of cauloids regenerated.*

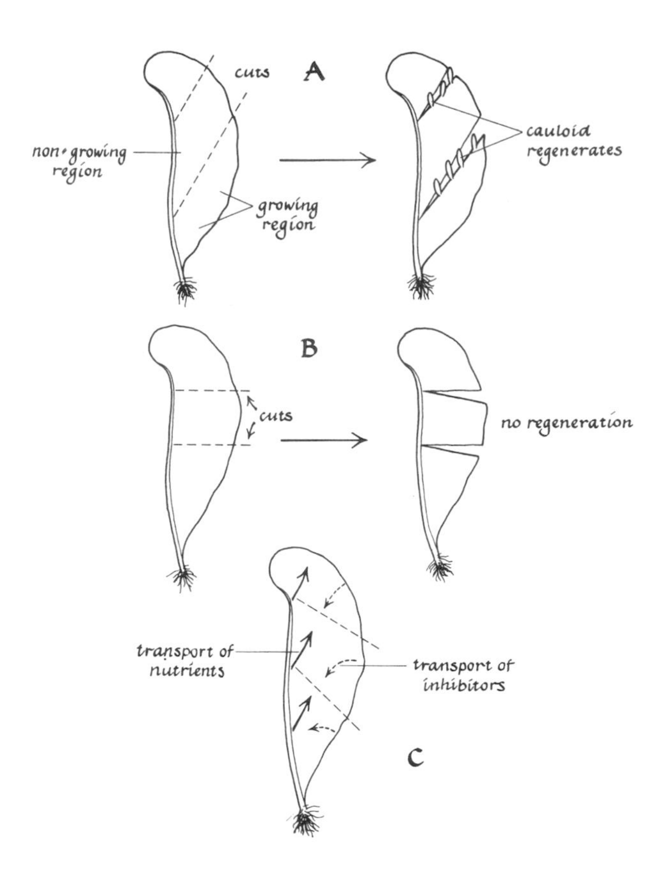

Figure 5-9. *A and B, diagrams illustrating the necessity of interrupting the continuity between growing and nongrowing regions to induce regeneration in* Riella. *In A and C, cuts through the blade in the positions shown block the transport of nutrients from the rib to the meristem near the edge of the blade and prevent the inhibition of growth in regions closer to the rib by auxins produced in the meristem.*

cauloids (bladelike structures) from the apical cut surface. Thus the apical-basal polarity of the whole plant is maintained in the pieces. By examining further patterns of regeneration we shall see that maintenance of polarity, the difference between one end and the other of an organism, is an almost universal principle of regeneration. This polarity is probably caused by a gradient in metabolic activity or in concentration of hormones.

The time of appearance of cauloids and rhizoids after isolation of a segment of the plant is dependent on the age of the cells. Dedifferentiative changes begin earlier in pieces from the apical region of the blade. There is also a relationship between the size of the isolated piece and the number of regenerates (cauloids and rhizoids) forming at the cut surfaces. The larger the area of the fragment, the higher is the number of regenerates. However, the total number of regenerates can be increased by further subdivision of an isolated fragment. These relations are diagramed in Figure 5-8.

Theoretically, subdivision of a fragment down to single cells should result in a number of regenerates equal to the number of cells in the fragment. This result has been approached in several cases. In *Riella*, interruption of the continuity between growing and nongrowing regions of the wing is a necessary stimulus for regeneration (Figure 5-9). Stange (1964) suggests two explanations for the effect of interrupting the normal "directional relations between the cells." First, if the transport of building materials (nutrients) from nongrowing cells near the rib to meristematic cells at the blade edge is interrupted by a cut in the proper plane, the nutrients accumulate in the nongrowing cells, providing them with substrates for cell division and enlargement (Figure 5-9C). This result suggests that the cells compete for nutrients. Second, interruption isolates the nongrowing regions from inhibition by auxins (growth hormones) synthesized mainly in the meristem regions (Figure 5-9C). There seems to be no good evidence for deciding between these alternatives. However, nontoxic concentrations of synthetic auxin applied to fern or moss gametophytes do not inhibit regeneration. Presence of the apical meristem in pieces of moss and fern prothallia seems to inhibit the formation of new meristems but not the regeneration of rhizoids.

Ferns

In the fern *Pteris vittata*, single cells of the gametophyte have been isolated by pricking and killing surrounding cells with a fine glass needle (Figure 5-10A). The isolated cells divide and may give rise to a whole new *prothallium* (gametophyte generation) (see Chapters 10 and 14). Cells in different parts of the prothallium were isolated in this way (Ito, 1962). Ito demonstrated the inverse relationship between the age of the isolated cell and the time required to form the first protonema cell (Figure 5-10B).

In the experiment illustrated in Figure 5-10, alternating cells from the base to the apex of the prothallium were isolated. At the base of the prothallium (older cells) only 1 to 6 days were required to form the first protonema cells from the isolated cell. At the apex, where younger cells are formed by the apical meristem, 21 days were required to form the first cell from the isolated cell. However, after cells from the apical region had begun to divide (Figure 5-10B), they formed a larger number of cells than those isolated from the base. In other words, isolated younger cells require more time to reach a state in which they can engage in mitosis but, when they reach that stage, they divide more rapidly than older cells.

While cells in the apical region were actively dividing, probably because of relatively higher concentrations of hormones produced by cells of the meristem, cells near the base had ceased dividing. Thus, when apical region cells are isolated their mitotic activity is depressed, probably because of their removal from the source of growth-stimulating hormone. The division of cells isolated from the basal region increases, possibly because the cells are removed from the inhibiting hormones produced by the apical cells. Why cells near the apex are stimulated but those near the base are inhibited by the same hormones might be explained by their difference in age. However, the cells differ in other respects as well, so one might expect them to respond differently.

As we have already seen in Chapter 4, when older leaf primordia of ferns are removed from the shoot apex and grown in culture, they may develop into whole leaves. Even parts of such a primordium of *Osmunda cinnamomea* can form whole leaves with bilateral symmetry (Steeves, 1966). As indicated in Figure 5-11A, division of the primordium into right and left halves results in reconstitution of a whole primordium by each half. Still more interesting is the development of pieces 2 and 3 shown in Figure 5-11B, which must form new apical meristems to become whole primordia capable of forming whole leaves. Pieces of sporophyte tissue of the fern *Phlebodium* (Ward, 1963) and of gametophyte tissue of the moss *Splachnum* (von Maltzahn and MacNutt, 1960) can regenerate into whole plants. In Chapter 14 we shall consider aspects of regeneration in mosses and ferns in relation to the problem of alternation of haploid and diploid generations.

As noted in the case of the liverworts, regeneration in fern gametophytes also depends on disruption of the relationship between the dominant mitotically active and the less mitotically active regions. This is illustrated in Figure 5-12 for the fern prothallium (Meyer, 1953). Similarly, in some moss plants a half leaf left attached to the stem of the gametophyte does not regenerate, but if the leaf is detached, regeneration occurs (review of Stange, 1964).

Higher Plants

Fragments of many kinds of flowering plants are capable of reforming a whole plant in which the original polarity is maintained. Thus, leaves and buds develop from the apical end of the piece and roots from the basal end. The formation of adventitious roots on stem or leaf cuttings is a significant kind of regeneration in higher plants. Such adventitious roots also develop from the stems of English ivy and on the runners of a strawberry plant (Figure 9-30). Although adventitious buds can develop on stem and leaf cuttings, buds capable of producing shoots rarely develop on roots. The nature of the stimulus that induces the formation of adventitious roots and buds is unclear, but plant hormones produced in the leaves or buds of the cutting are evidently involved (Chapter 19).

Pieces of flowering plants, even as small as a single cell (Figure 5-13), may form whole plants under appropriate environmental or culture conditions (see

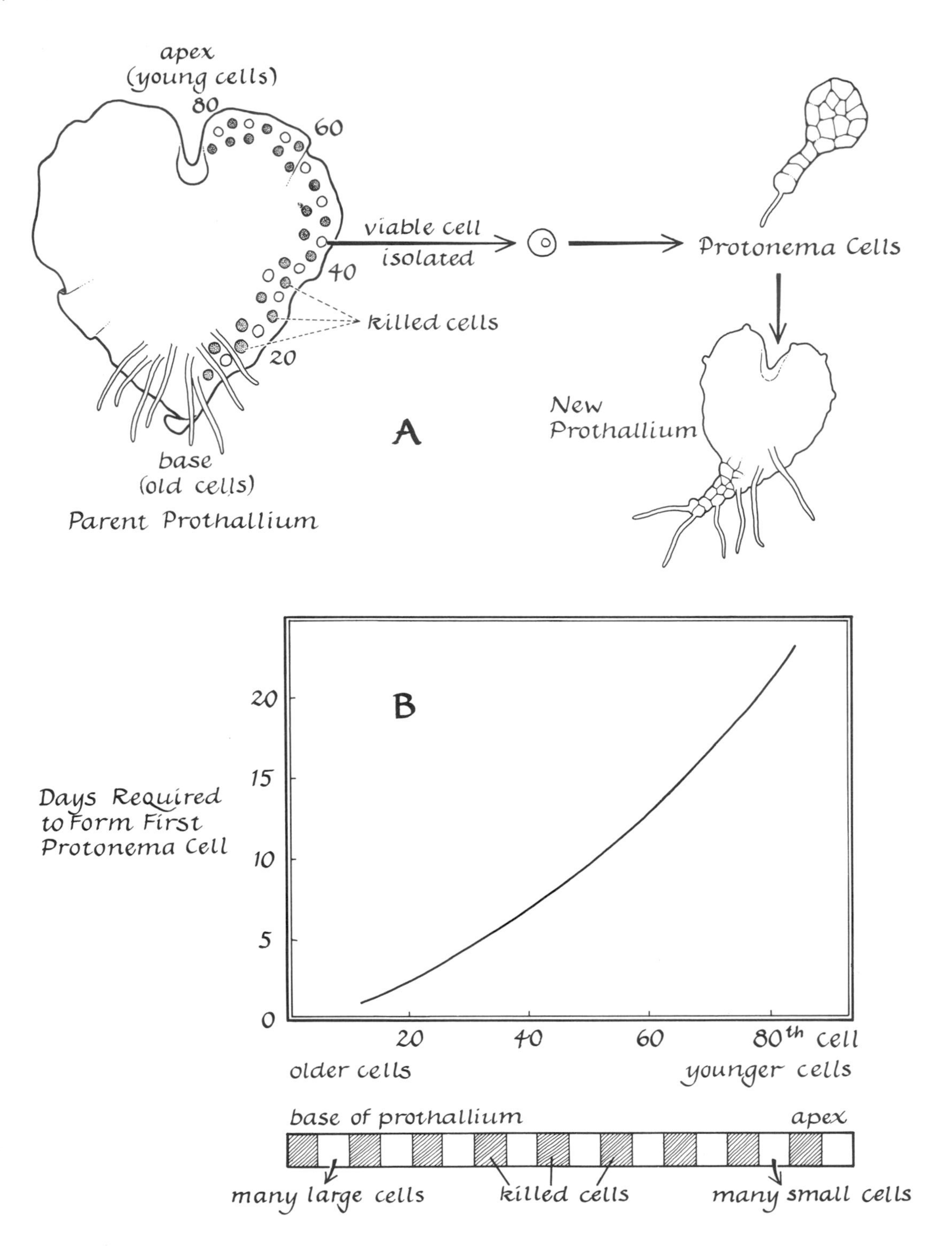

Figure 5-10. *A, formation of a new prothallium by isolated cells of a parent prothallium. B, graph of the inverse relationship between the age of an isolated cell and the length of time required to form the first protonema cell.*

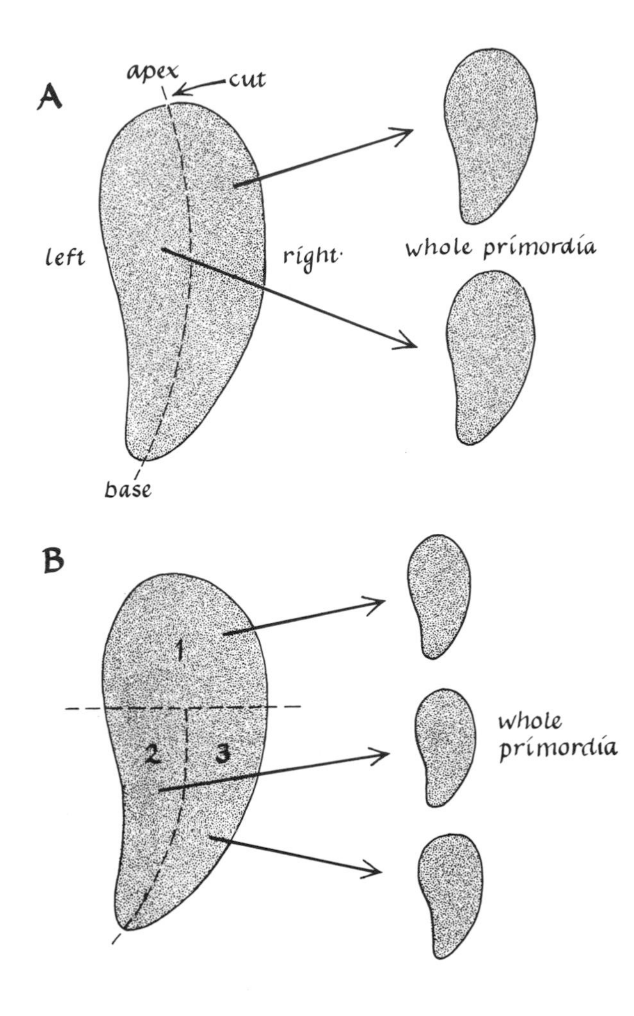

Figure 5-11. *Formation of whole leaf primordia by isolated fractions of a cultured leaf primordium of* Osmunda. *A, right and left halves isolated; B, apical third (1) and right and left halves (2 and 3) of the basal two-thirds isolated.*

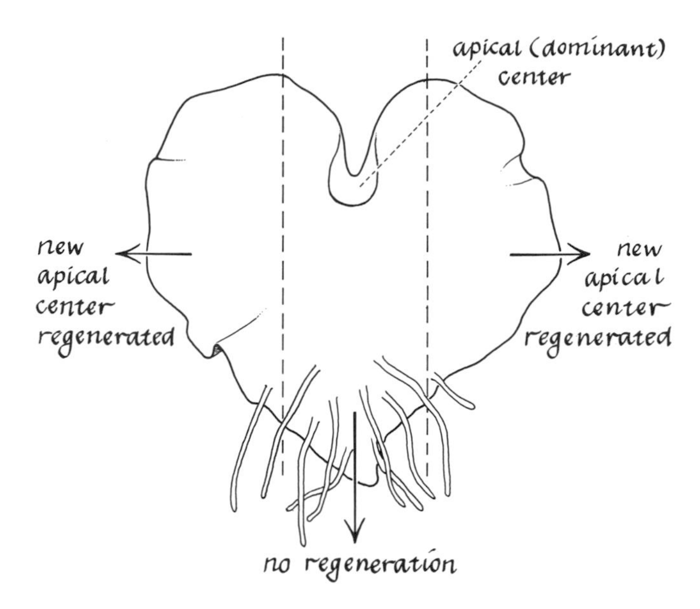

Figure 5-12. *To induce regeneration in a fern prothallium, it is necessary to disrupt the continuity between growing and nongrowing regions.*

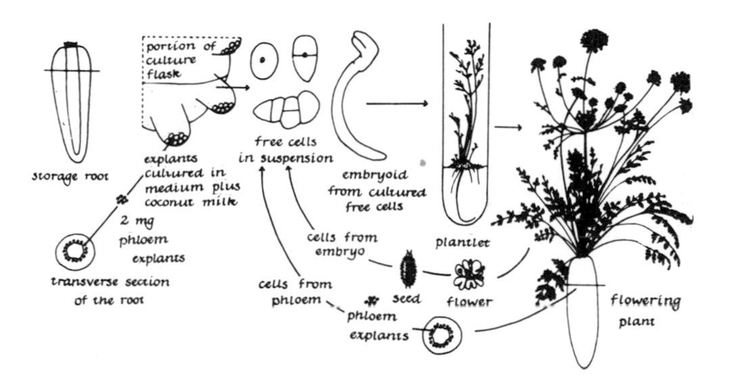

Figure 5-13. *Formation of a whole plant by free cells of* Daucus carota *(carrot), cultured in vitro. Redrawn from Steward et al. Copyright 1966 by Academic Press.*

reviews of Sinnott, 1960; Steward, 1963; Steward et al., 1966, 1969). This capacity of part of a plant to form a new plant is basic to the use of cuttings for propagation in agriculture and horticulture.

Typically, when isolated tissues of adult plants are grown in vitro on nutrient culture media they undergo overt dedifferentiation and form a callus, a mass of unspecialized cells. Under appropriate environmental and nutrient conditions, cells of the callus will redifferentiate to form plant organs and even a whole plant body.

Regeneration in Animals

Regenerative capacity varies greatly in the animal kingdom, with both simpler organized animals such as Cnidaria and more complex animals such as amphibians exhibiting this capacity. The following section contains a few well-known examples of animal regeneration.

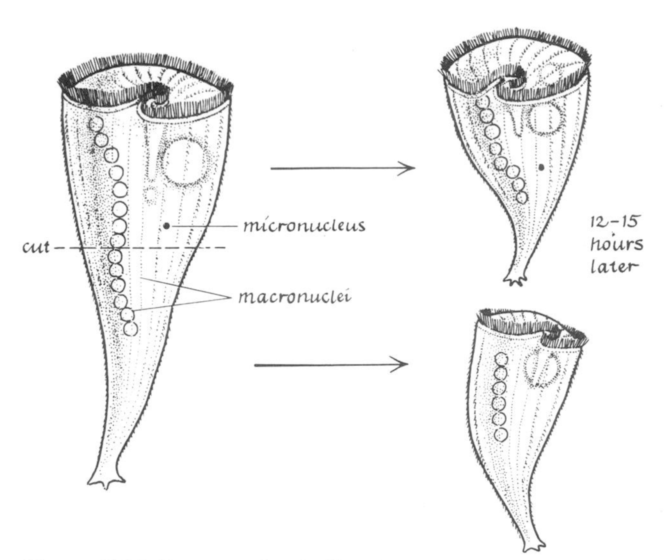

Figure 5-14. *Regeneration in* Stentor.

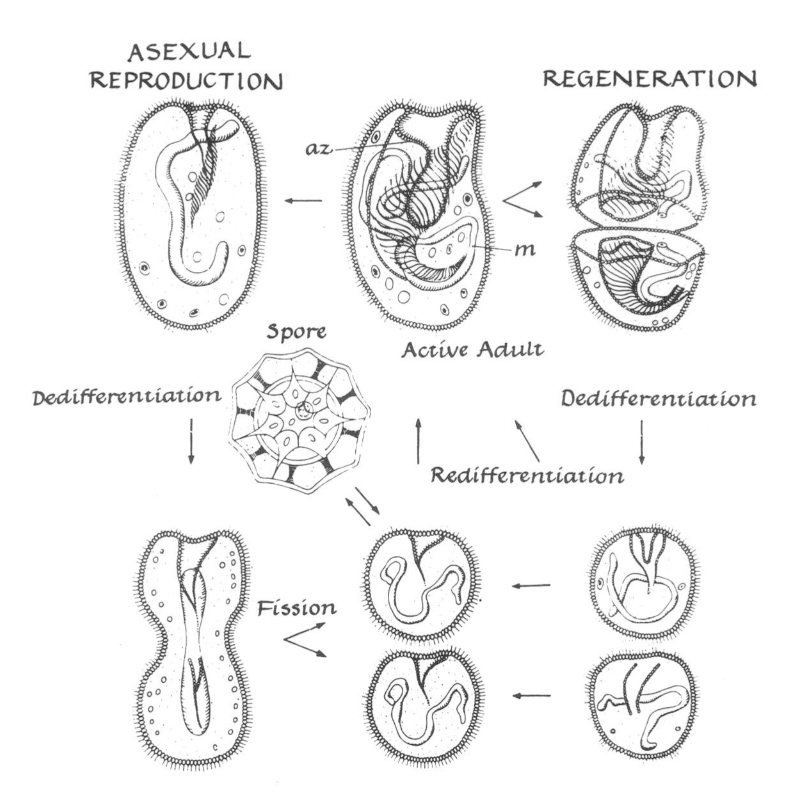

Figure 5-15. *Asexual reproduction and regeneration in* Bursaria. *Note that dedifferentiation and redifferentiation of cytoplasmic structures occur during both processes. From Lund,* J. Exp. Zool., *Volume 24, p. 21, 1907. Courtesy Wistar Press.*

Protozoa

Among the protozoa, the ciliates exhibit striking regenerative capacities after loss of part of the unicellular body (reviews of Weisz, 1954; Tartar, 1941, 1961, 1962). Halves of *Stentor coeruleus* (Figure 5-14) or pieces as small as a thousandth of the animal can regenerate a whole organism, provided one lobe of the macronucleus and part of the cortex are present (Weisz, 1948). A micronucleus (Figure 5-14) is not necessary for regeneration.

Enucleation of a regenerating piece of *Stentor* at an early stage in the process halts regeneration sooner than enucleation at a later stage, a result which suggests that products formed in sufficient quantity under control of the macronucleus make further regeneration possible.

As regeneration in a unicellular organism such as *Stentor* involves not only rearrangement of structures but also formation of new structures (organelles) by division of pre-existing structures, it is not surprising that substances which interfere with protein synthesis (ribonuclease, puromycin) or block DNA transcription (actinomycin-D) prevent regeneration (reviewed by Hay, 1966; Tartar, 1967). *Stentor* and other ciliates reproduce asexually by transverse fission, a process involving dedifferentiation of organelles, as described by Lund (1907) in *Bursaria*. As illustrated in Figure 5-15, regeneration and asexual reproduction are similar (Chapter 9).

Porifera (Parazoa)

Sponges display poor capacities to replace a missing part (Needham, 1952). In contrast, the ability of dispersed sponge cells to reaggregate and reconstitute a whole sponge is remarkable (Wilson, 1907; reviewed by Berrill, 1961) (see Chapter 8).

Cnidaria (Coelenterates)

Tubularia (hydrozoan). It has long been known that the hydranth of *Tubularia*, a marine hydrozoan, (Figure 5-16A) drops off under unfavorable environmental conditions (such as too high a temperature) and then is replaced by a new one (Dalyell, 1814). Elongation of the stem then occurs by mitosis in the zone of growth at the base of the hydranth as in *Hydra* (see Brien and Reniers-Decoen, 1949). Normal stages in

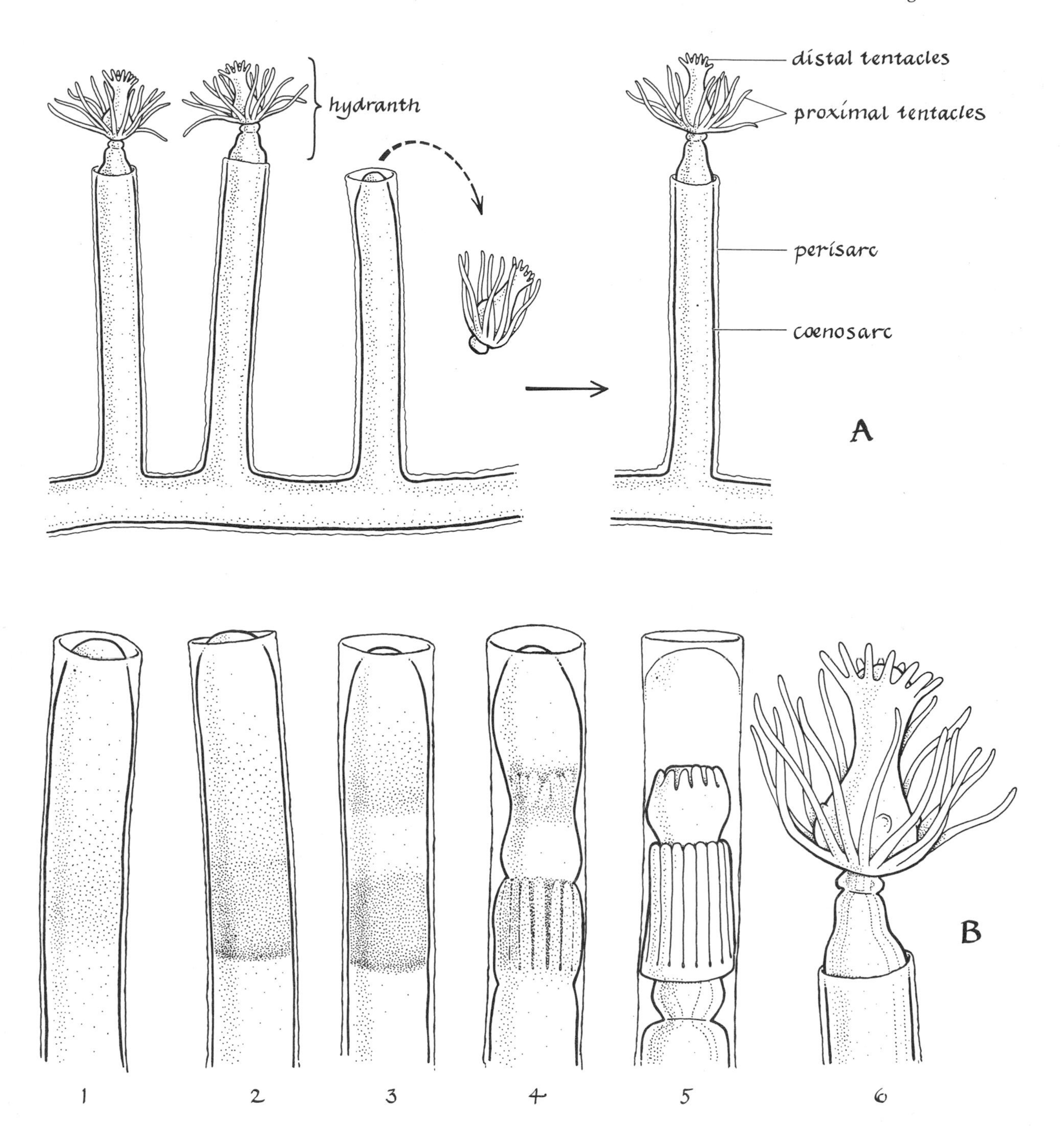

Figure 5-16. *A, colony of* Tubularia, *showing loss and regeneration of a hydranth. B, normal stages in hydranth regeneration.*

regeneration of a hydranth are illustrated in Figure 5-16B (see review of Berrill, 1961). Emergence of the new hydranth occurs within 2 to 3 days.

In stem pieces about 5 to 10 mm long the hydranth is usually formed at the distal end. In longer pieces of stem (15 mm ±) and shorter pieces (2mm ±) a hydranth may develop at both ends (Barth, 1940, 1955; Steinberg, 1954; Berrill, 1961). If a piece of the perisarc, the tube of chitin secreted by the animal (Figure 5-16A), is removed without injury to the underlying tissue, one or two hydranths may develop in that region (Zwilling, 1939). Thus cutting into the coenosarc, the underlying tissue, is not a necessary stimulus for regeneration (Figure 5-17). Exposure of the coenosarc to oxygen in the sea water appears to be the stimulus. Zwilling also found that the rate of regeneration increased significantly with increase in the area of exposed coenosarc.

Regeneration in hydroids seems to involve the rearrangement of already specialized cells, a process called morphallaxis (Morgan, 1901), and the differentiation of neoblasts, which are unspecialized, reserve cells sometimes called interstitial cells (Tardent, 1963). Time-lapse movies show that, in some species of *Tubularia*, tissue moves toward the distal end of a stem piece. This presumably results in an increased cell population density at the distal end and a depletion of cells at the proximal end (Steinberg, 1955), which may be one reason the new hydranth forms at the distal end. *Tubularia* and other hydroids exhibit a gradient in regenerative capacity. For example, in pieces of the same length, those taken from the more distal (younger) regions regenerate a new hydranth sooner than those from more proximal (older) parts of the stem (Steinberg, 1954). This is illustrated in Figure 5-18.

Hydra. Apparently the first study of regeneration in *Hydra* was reported by Trembly in 1744. He turned *Hydras* inside out and found that the animals reestablished their normal cellular architecture. He also demonstrated that small pieces of *Hydra*, except for the tentacles, regenerate a perfect organism. In 1897, Peebles found that pieces of *Hydra* cut out successively closer to the base regenerate at a progressively slower rate. She also found that the smallest piece capable of regenerating an oval cone and at least one tentacle was a sphere of tissue about .17 mm in diameter. More recently it has been found that isolated,

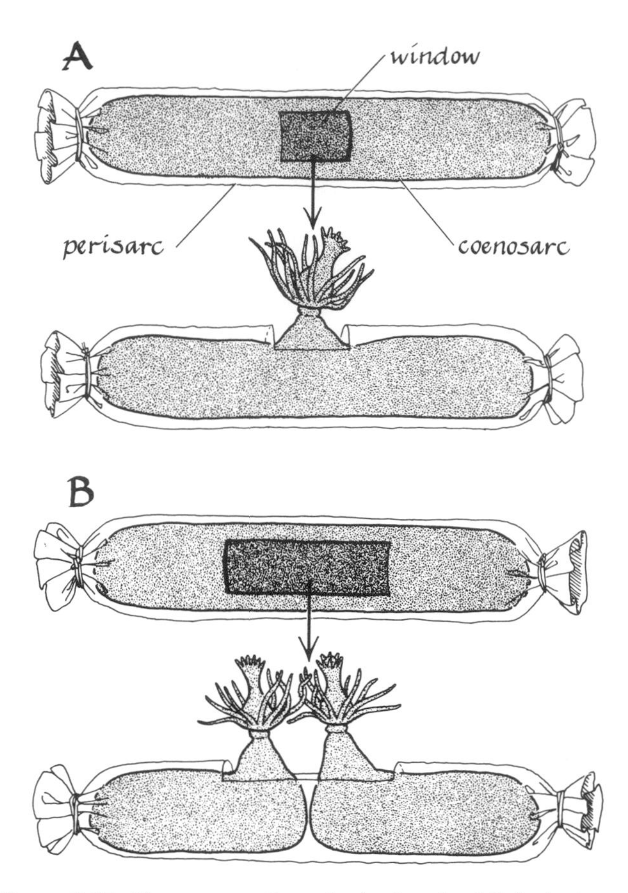

Figure 5-17. *The regeneration of a hydranth of* Tubularia; *cutting a window in the perisarc exposes the coenosarc to seawater.*

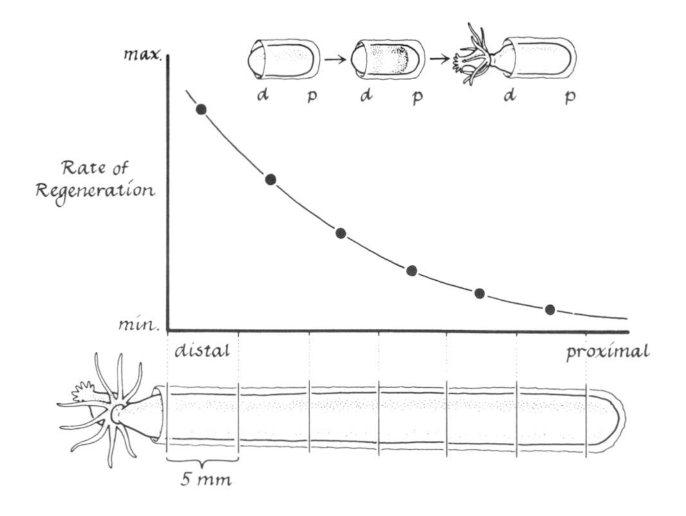

Figure 5-18. *The rate of regeneration of a hydranth by distal and more proximal pieces of the* Tubularia *stem.*

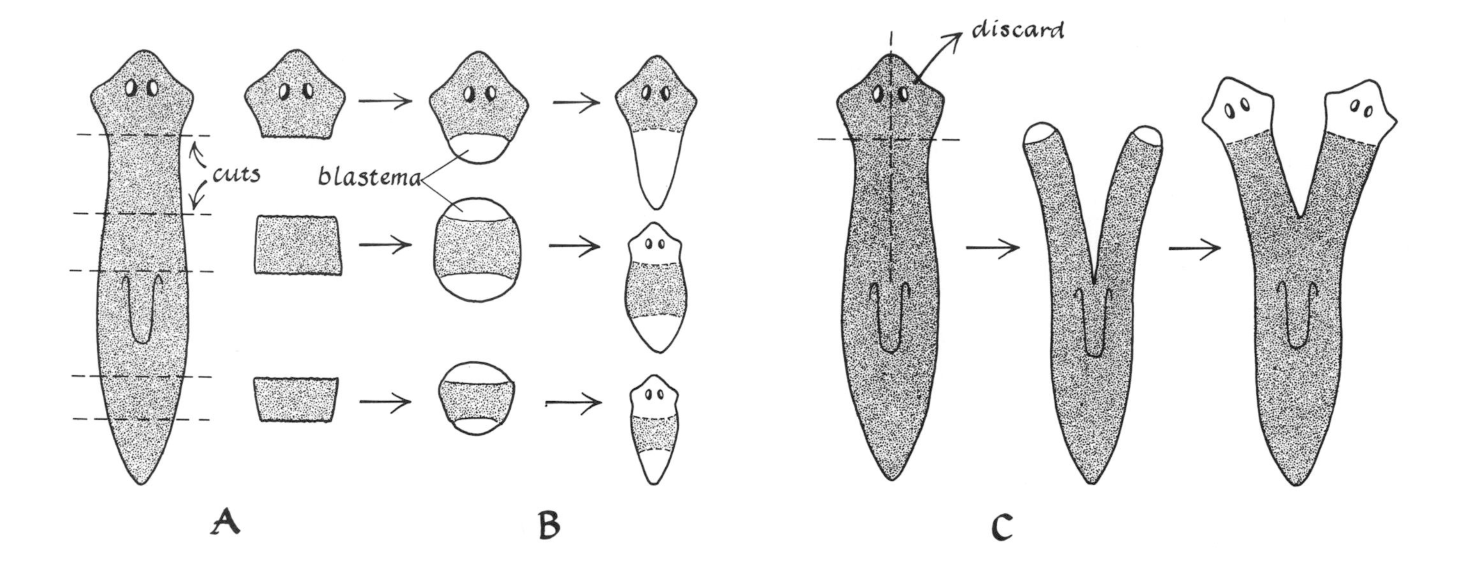

Figure 5-19. *Head and tail regeneration in* Planaria. *A, regeneration of body and tail by isolated head and regeneration of head and tail by isolated middle and posterior body parts. B, regeneration of two heads following removal of the head and partial longitudinal splitting of the body.*

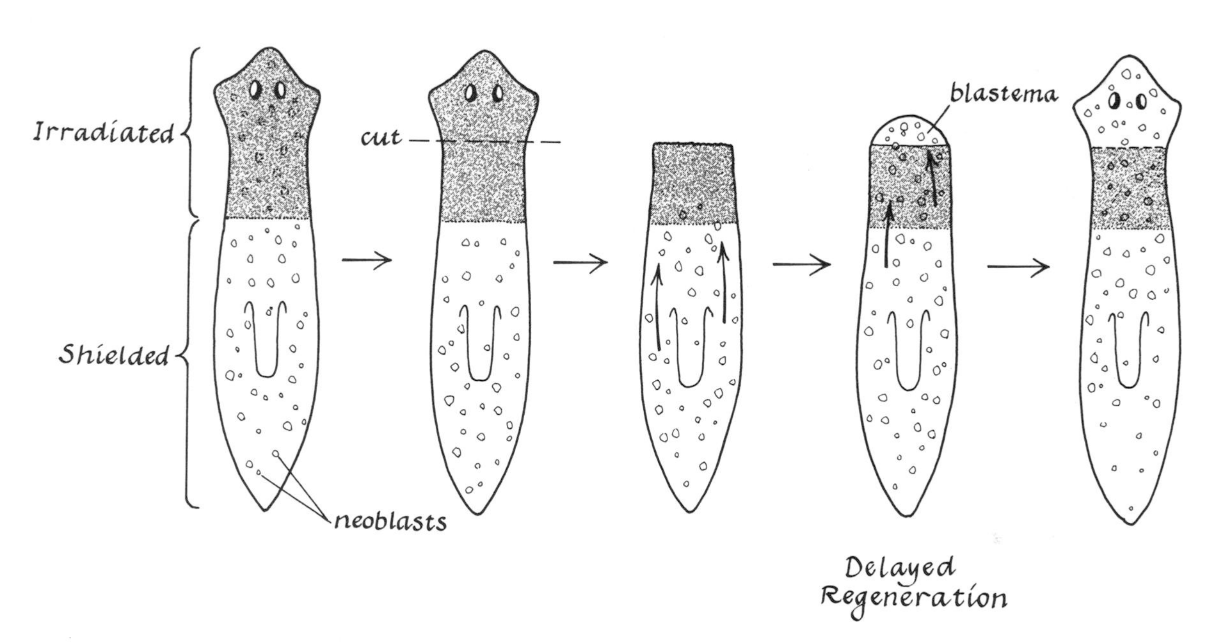

Figure 5-20. *Role of neoblasts in regeneration of lost parts, revealed by the experiments of Wolff (1962).*

pure epidermis or endodermis fragments can regenerate a whole *Hydra* (Lowell and Burnett, 1969).

Flatworms

The ability of planarians to regenerate heads, tails, and other body parts has been known at least since 1814, when Dalyell wrote that planarians are "almost to be called immortal under the edge of the knife." Since then, over 500 papers describing experiments on planarians have been published. In his book, *Regeneration*, Morgan (1901) describes his classical experiments on regeneration. Some of his experiments are illustrated in Figure 5-19A and B. If the head is removed by a transverse cut and the body is partially split by a longitudinal cut (Figure 5-19C), a *blastema* (a collection of unspecialized cells) may form on both anterior cut surfaces, and two heads rather than one will regenerate (Child, 1941). Each partial half tends to reconstitute a whole animal. An excellent review of earlier studies of regeneration in flatworms is found in Child's book, *Patterns and Problems of Development* (1941).

Evidence explaining the origin of the blastema is still unsatisfactory (Hay, 1966). According to Lender (1962) and Stephan-Dubois (1965), neoblasts distributed throughout the body migrate to the cut surface and form the blastema. Evidence for the role of neoblasts comes from irradiation experiments such as the one illustrated in Figure 5-20.
Irradiation selectively kills the neoblasts and reduces their number in the unshielded part of the planarian. After removal of the head, only more posteriorly located neoblasts are available for blastema formation and more time is required for their migration through the irradiated portion. Removal of part of the body of a planarian appears to stimulate mitosis of the neoblasts, further implicating them in blastema formation (Stephan-Dubois, 1965).

In some planarians, morphallaxis (rearrangement of remaining cells) rather than addition of cells at the cut surface is the primary mode of regeneration (Morgan, 1901). This is illustrated in Figure 5-21. Note that the isolated piece undergoes form changes involving elongation of the pigment stripes.

In *Planaria*, as in *Tubularia*, there is a gradient in the rate of regeneration of the head and the size of the reformed head (Child, 1941; Brønsted, 1955). See Figure 5-22.

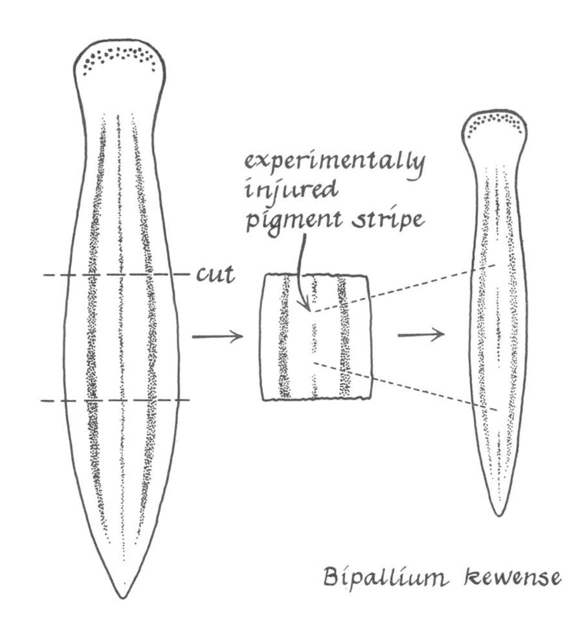

Figure 5-21. *Morphallaxis during regeneration in the land planarian* Bipalium kewense.

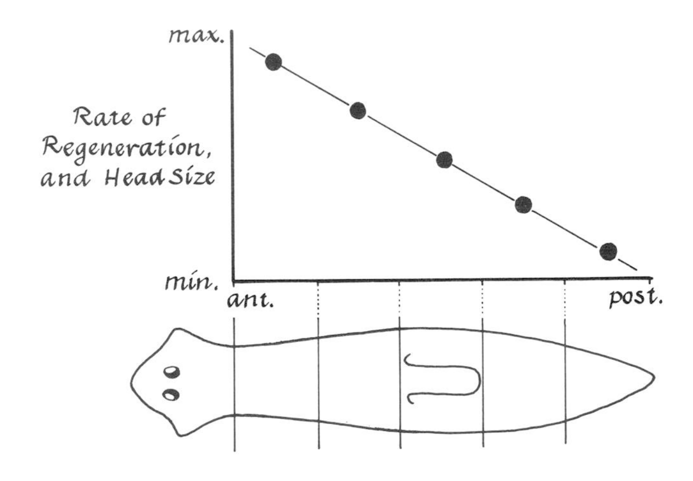

Figure 5-22. *The rate of regeneration of the head and head size by isolated anterior and posterior body parts of* Planaria.

Biologists have been speculating for a long time about the morphological and physiological bases of polar gradients (Harrison, 1898; Child, 1941; Steinberg, 1955; Rose, 1957; Tardent, 1963; Flickinger, 1967). Child suggested that axial gradients of metabolic activity existed. Tardent demonstrated a distal-proximal gradient in the number of interstitial cells

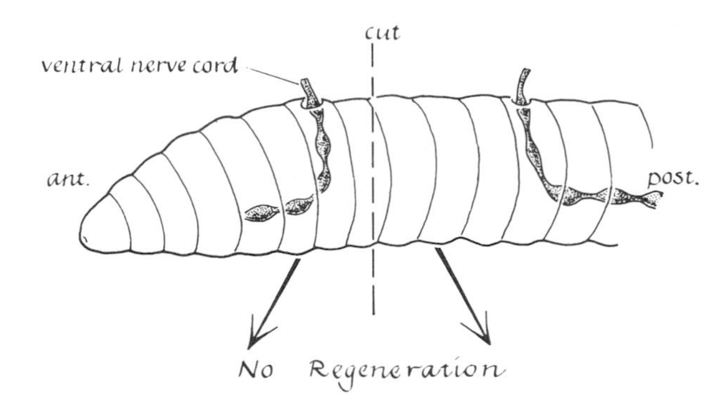

Figure 5-23. *Experiment demonstrating the importance of the ventral nerve cord in regeneration of annelids.*

in *Tubularia* stems. Lender (1965) reported a head-to-tail gradient of neoblasts in planarians. The graded differences in the intact organism underlie the graded differences in the regeneration capacity of isolated parts. In other words, the regenerative guidelines are already present in the parts at the time of their isolation.

Annelids and Nemerteans

In general, annelids and nemerteans exhibit spectacular regenerative powers. In the nemertean *Lineus socialis,* the only limit to regeneration is that the length of a fragment must exceed its width and must contain a piece of ventral nerve cord. In *Chaetopterus variopedatus* and some other polychaete annelids, a single segment, even a segment lacking gonads, can reconstitute a complete worm. Nevertheless, some annelids have an anterior-posterior gradient in regenerative capacity (Berrill, 1961). According to Herlant-Meewis (1964), regeneration in some annelids depends on migration of neoblasts to the cut surface unless such cells are already present there. In general, however, new parts are formed from old parts, germ layer for germ layer.

The importance of the ventral nerve cord for regeneration in annelids has long been known (Morgan, 1901, 1902; Avel, 1947, cited by Herlant-Meewis, 1964). Cutting and deviation of the nerve cord (as shown in Figure 5-23) prevents both posterior regeneration of the anterior piece and anterior regeneration of the posterior piece. In some annelids, tail regeneration depends on the presence of the brain for the first few days after transection. Hyperactivity of the neurosecretory cells in the brain following cutting suggests the role of a brain hormone in promoting regeneration (Herlant-Meewis, 1964; Clark, 1969). Removal of the brain at any stage in regeneration of posterior segments halts the further formation of segments (Golding, 1967).

Arthropods

Regeneration of appendages in larval insects and crustacea has been reviewed by Needham (1965). A particularly interesting type of regeneration, *autotomy,* the spontaneous casting off of an appendage followed by regeneration, is found in crustacea. For example, if the claw of a crab is seized or crushed, the leg breaks off spontaneously, by sudden contraction of the extensor muscle, at a preformed plane across the second leg joint. A new claw regenerates at the next molt (Bliss, 1960). Occasionally the lost part is replaced by a different but homologous part, for example, an antenna instead of an eyestalk (Morgan, 1901; Needham, 1952). The origin of the regenerative cells in autotomy remains to be demonstrated.

Echinoderms

Starfish and sea cucumbers are capable of extensive regeneration. A whole animal can form from one arm and part of the central disc in starfish (*Asteroids*) and brittle stars (*Ophiuroids*). The sea cucumber (a *Holothuroid*) can regenerate its internal organs after they have been eviscerated as a consequence of rough handling or some other irritation. Rough handling of a starfish may also cause autotomy by spontaneous constriction and casting off of an arm (Hyman, 1955). The source of regeneration cells is not clear (Anderson, 1965).

Tunicates (Sea Squirts)

The remarkable regenerative properties of this group of protochordates are excellently reviewed by Berrill (1961). The anatomy of a tunicate is shown in Figure 5-24.
A piece of the septum, epicardium, or atrial epithelium can regenerate the entire tunicate body. Other tissues such as epidermis, pharynx, and intestine are capable only of regenerating the same tissue.

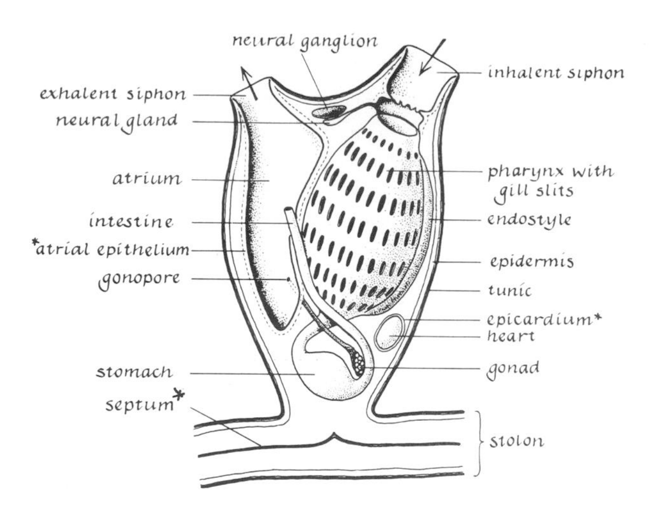

Figure 5-24. *Anatomy of a typical tunicate. (Starred body parts are capable of regenerating an entire body.)*

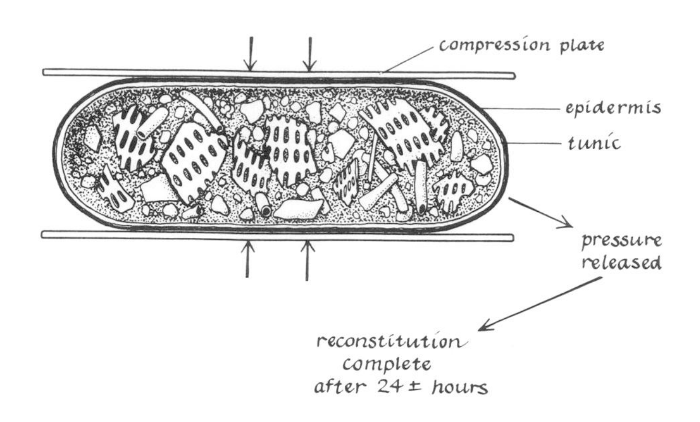

Figure 5-25. *Disruption of a young tunicate zooid by pressure, and reassembly of tissue fragments to reconstitute the normal anatomy.*

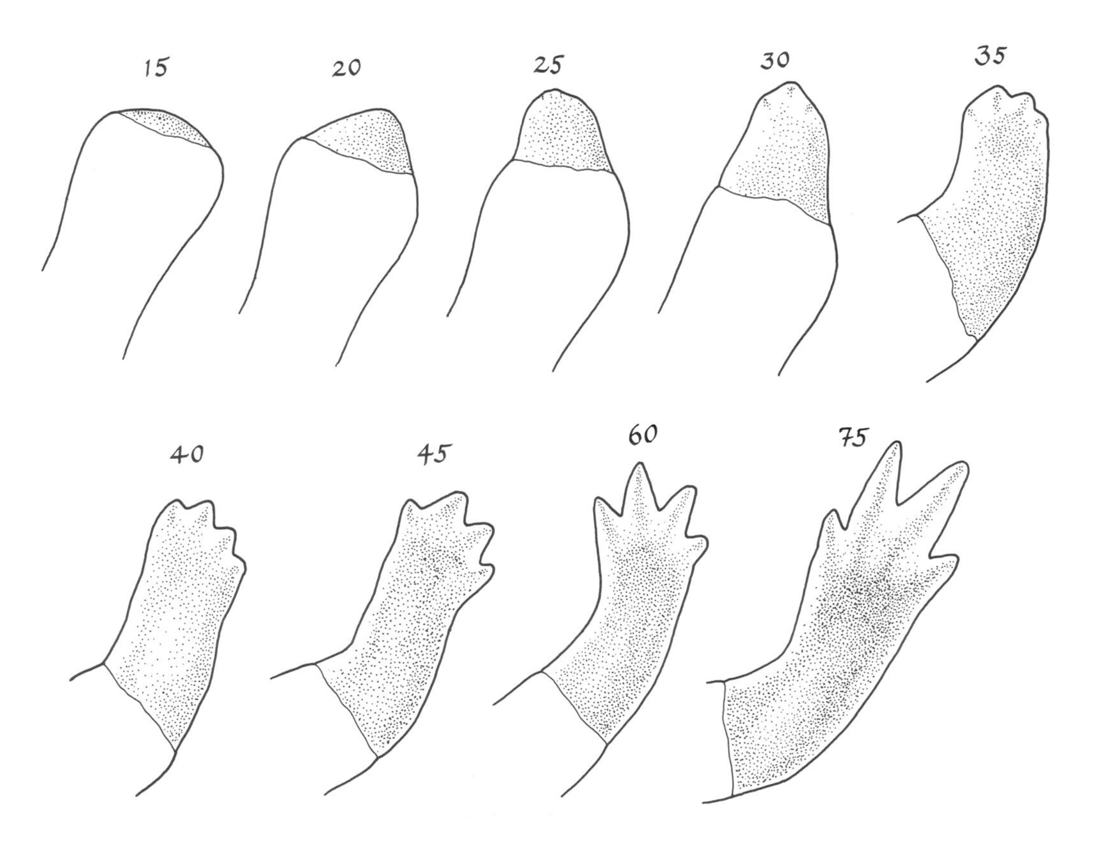

Figure 5-26. *Stages in regeneration of the amputated forelimb of an adult* Triturus. *The numbers refer to the number of days following amputation.*

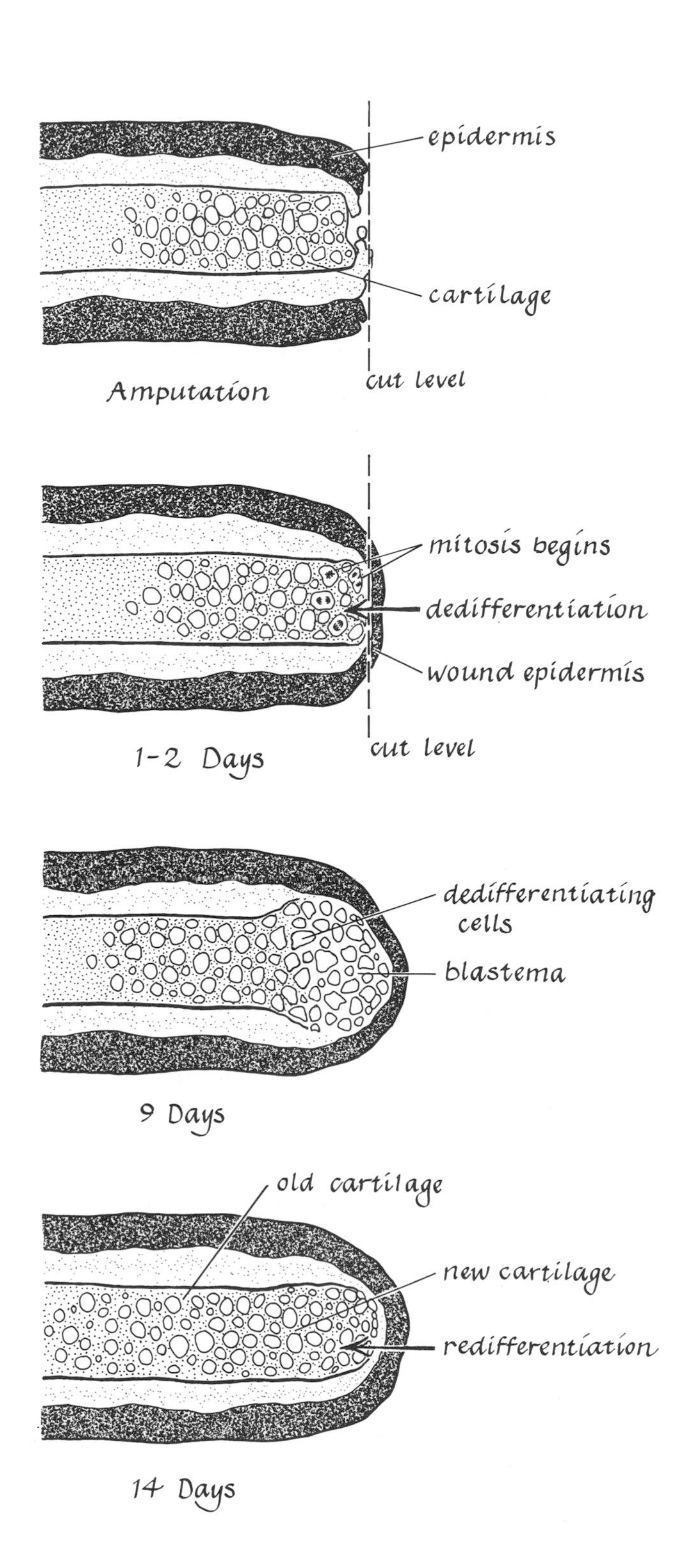

Figure 5-27. *Stages in regeneration of the finger of* Ambystoma.

When a young individual tunicate (a zooid) is carefully compressed between plates, the internal organs are completely disrupted and break into fragments, but the epidermis and test (or tunic) remains intact (Figure 5-25). If the pressure is then released fragments of all the organs reassemble into their organ systems and the adult anatomy is reconstituted after about twenty-four hours. The guidelines controlling this assembly are not known. In a somewhat similar experiment Scott and Schuh (1963) inserted minced tissue of the tunicate *Amaroecium* into a common tunic and found by radioautographic methods that like tissues reaggregated with like tissues. Similar phenomena in the reaggregation of dispersed tissue cells of vertebrates are discussed in Chapter 8. Tunicates subjected to trauma may eviscerate, after which new viscera regenerate from the atrial epithelium.

Vertebrates

Among the vertebrates, salamanders possess the most remarkable regenerative capacities. Other vertebrates exhibit some regenerative capacities—regeneration of the fin, scales, gills, and even the brain (Chapter 24) in some fishes; tail and hindlimb in frog larvae; tail by a process of autotomy in lizards; liver in mammals (reviewed by Hay, 1966; Boss, 1969). Recently, Mizell (1968) has described regeneration of portions of the foot and toes in the newborn opossum. We shall primarily study regeneration in the urodele amphibians (salamanders).

Limb regeneration. Stages in the regeneration of the amputated forelimb of the adult urodele *Triturus viridescens* are illustrated in Figure 5-26.

A more detailed analysis of the events in regeneration is given by the diagrams of finger regeneration in a living *Ambystoma* larva (Figure 5-27, modified from Hay, 1966). Following amputation (Figure 5-27A), epidermal cells at the edges of the wound migrate over the cut surface. However, the dermis and basement membrane remain at the edges of the wound (Figure 5-27B). During the next ten days histolysis results in the release of cartilage cells from their matrix and an accumulation of mesenchymal-like cells. Studies of Thornton (1938, 1968), Hay (1966), and others suggest that the blastema cells (Figure 5-27C) are derived by a process of overt dedifferentiation of cells (for example, muscle and cartilage

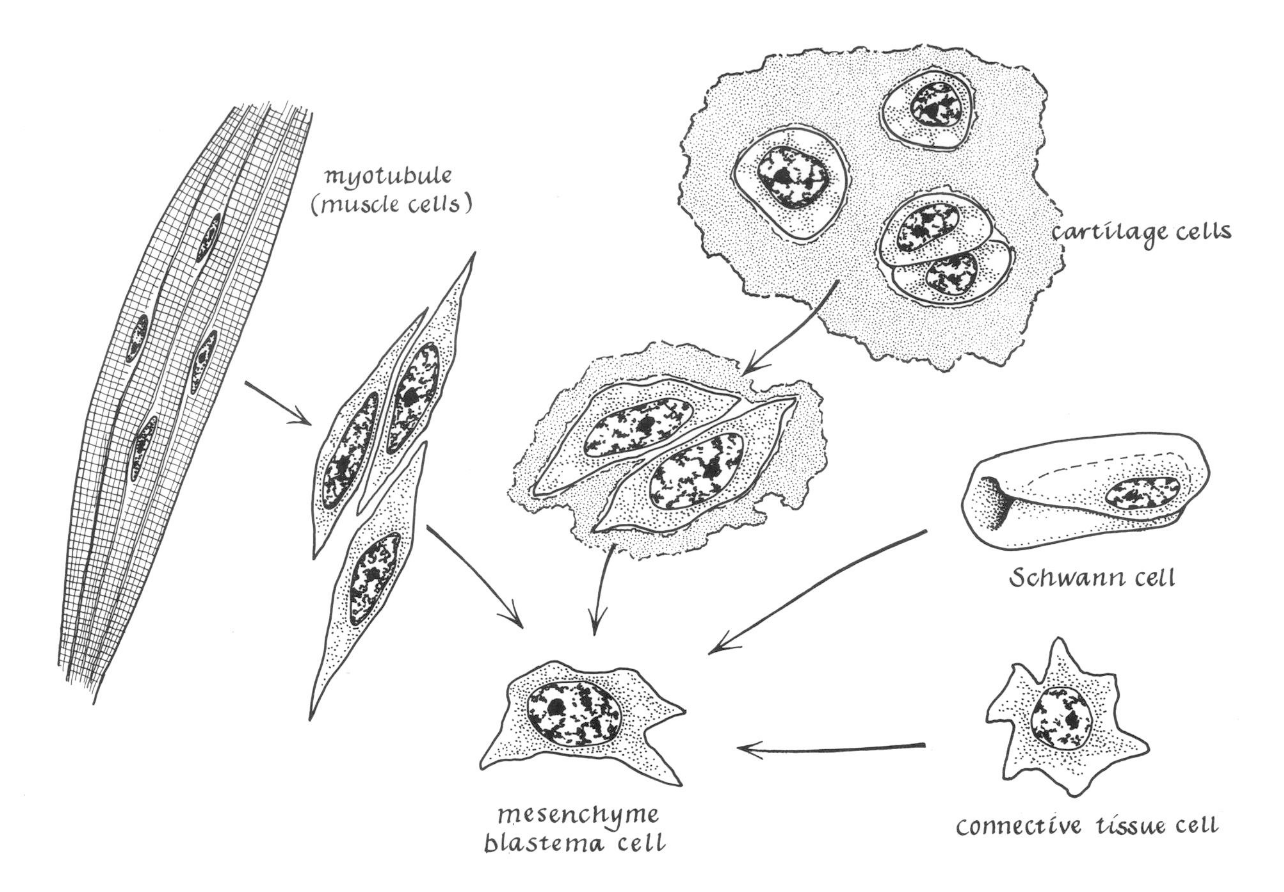

Figure 5-28. *Origin of mesenchymal cells of the blastema in amphibian limb regeneration.*

cells) at and near the cut surface. Evidence for this comes from studies in which a labeled DNA precursor, thymidine, was applied to the limb stump ten days after amputation. The labeled thymidine was incorporated into the DNA of the cells, for when radioautographs were made five days later, the label was found in blastema cells (review of Flickinger, 1967). The origin of the mesenchymal cells of the blastema is illustrated in Figure 5-28.

The mesenchymal cells of the blastema then divide and redifferentiate into muscle and cartilage cells, as illustrated in Figure 5-29 (Hay, 1966). Whether redifferentiation of the diversely derived mesenchymal cells is along the original path of dedifferentiation is a major problem that we shall presently discuss.

The local origin of the cells that participate in regeneration of amphibian limbs has been known since Hertwig (1927) performed the experiment illustrated in Figure 5-30. A haploid limb was grafted to a diploid salamander and, after healing, the graft limb was amputated through the upper arm. Cells of the regenerated limb were all haploid. Newer studies with chimaeric limbs have confirmed these studies (Anton, 1965).

Are the blastema cells completely uncommitted (totipotent) in respect to their redifferentiation into tissues of the new limb? Some evidence suggests totipotency, some does not. If the humerus bone in the upper arm is removed from a salamander limb and the limb is amputated through the upper arm, the part of the humerus distal to the level of the cut and all the more distal skeletal structures are regenerated (Fritsch, 1911; Weiss, 1925) (Figure 5-38). The new bone is thus formed from blastema cells whose normal fate was not bone formation. Evidence shows that the blastema may be a collection of dividing, unspecialized cells (a developmental *field*), because

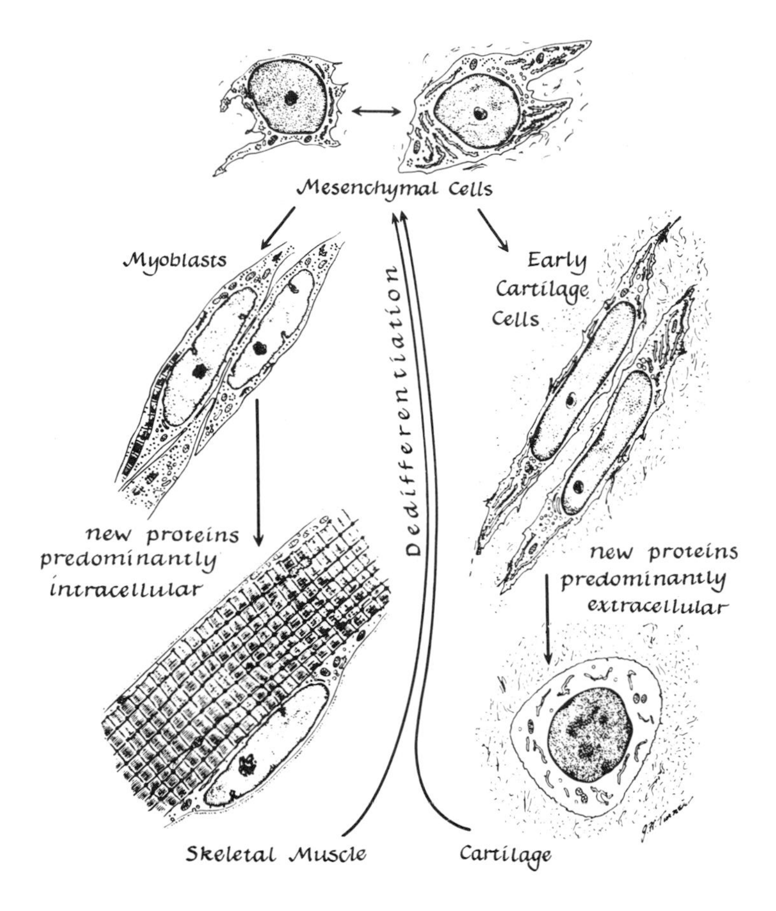

Figure 5-29. *Diagram summarizing the changes in fine structure that occur in the mesenchymal cells of the blastema during regeneration. (From Elizabeth Hay, "Cytological Studies of Dedifferentiation and Differentiation in Regenerating Amphibian Limbs," in* Regeneration, *edited by Dorothea Rudnick. Copyright © 1962 by the Ronald Press Company, New York.)*

parts of a blastema may form a whole limb and two blastemas fused together may form a single limb (Weiss, 1925).

It has been claimed that the blastema is totipotent and uncommitted, for a limb blastema of the salamander *Triton cristatus* grafted to the tail of a larval stage formed a tail, and a tail blastema grafted to a limb stump formed a limb (Guyénot, 1927). However, these experiments cannot be accepted as conclusive, because it is not clear what the actual cellular source of the regenerate was. The grafted blastema may have degenerated, and a new blastema formed by the host animal may have participated in the process. Labeling of the graft or host cells or use of haploid and diploid animals will be necessary to determine the source of the regenerate. To demonstrate the totipotency of labeled blastema cells, cells of a blastema formed after cutting the intestine of the adult newt *Diemictylus viridescens* were labeled with ^{3}H-thymidine and implanted into a 14-day limb blastema (Oberpriller, 1967). In some regenerates labeled cartilage cells were found. Of course, no cartilage is normally found in the intestine. The blastema thus appears to be totipotent.

Nevertheless, there is evidence indicating that the blastema is not simply a collection of indifferent and equipotential cells. Faber (1965), grafted young blastemas to foreign locations on the host animal and found that their differentiation was not altered. In earlier experiments of Mettetal (1939), a blastema developing on a limb amputated through the proximal part of the lower arm of *Triton cristatus* was grafted to the head or back of a host animal. The youngest blastemas formed only digits of the hand; an older blastema formed digits and the carpus; and a still older blastema formed digits, carpus, and the missing part of the lower arm. It is especially interesting that the progressive increase in potency of the transplanted blastema with increase in its age is the same as the sequence of embryonic development of the limb—distal to proximal (Faber, 1965). The studies of Stocum (1968) provide further evidence that the blastema is not a collection of equipotential cells, even in the early cone and pallett stages (Figure 5-26), in which the blastema has a cytologically and overtly homogeneous cell population. When Stocum grafted separated distal, proximal, anterior, and posterior halves of cone and pallett stage limb blastemas of larval *Ambystoma maculatum* to the dorsal fin of the same animal, he observed that only parts of the limb skeleton regenerated (Figure 5-31). These experiments indicate that the blastema has mosaic properties (Chapter 15) as early as the cone stage. Stocum also explanted blastemas at this stage in vitro and found that the isolated blastemas were capable of differentiating precartilage and striated muscle cells. This demonstrates that the cone and later stage blastema is a self-differentiating and thus partly autonomous system. (Recall the autonomy of the apical meristem, described in Chapter 4.)

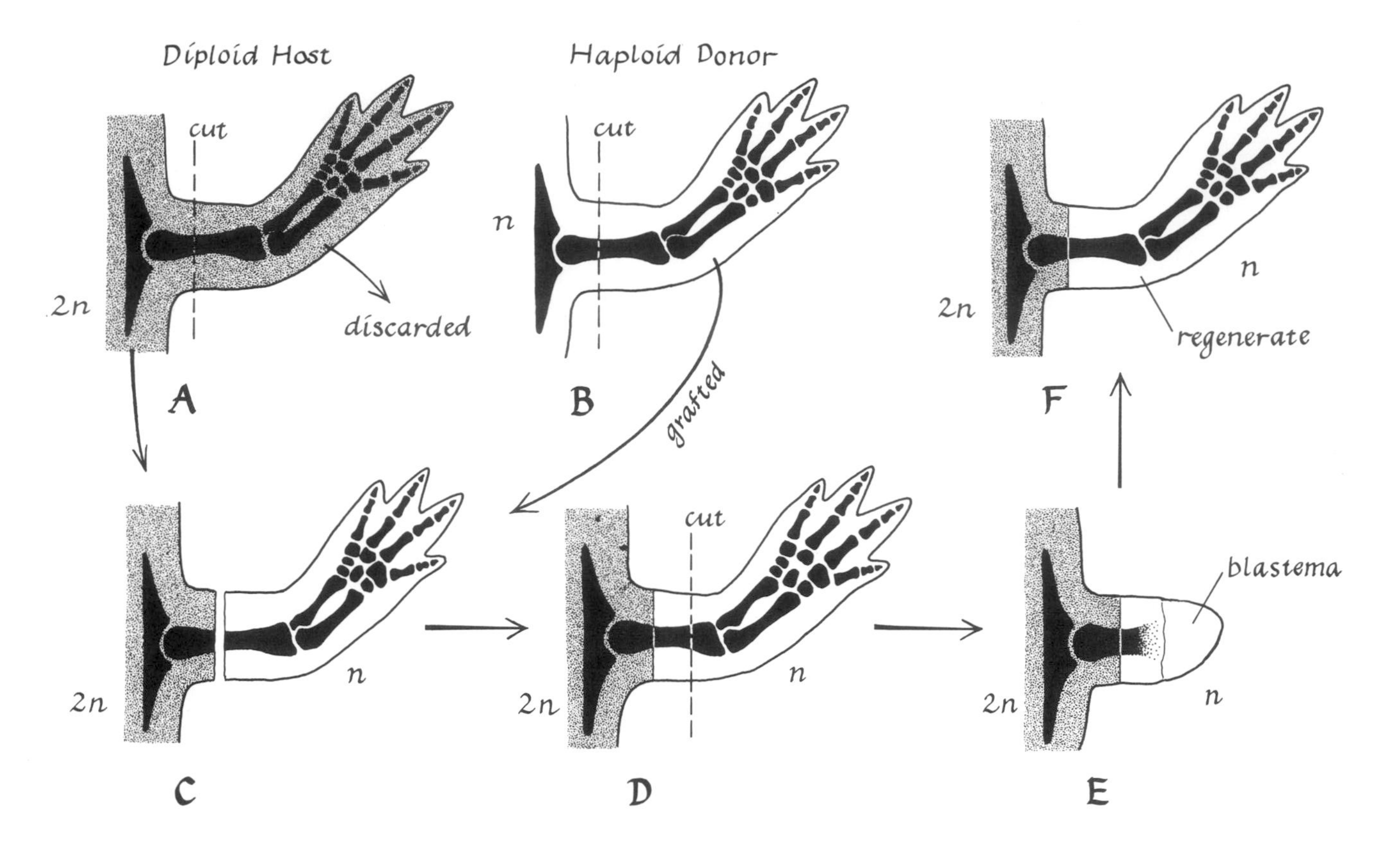

Figure 5-30. *Experiment demonstrating the local origin of blastema cells in amphibian limb regeneration.*

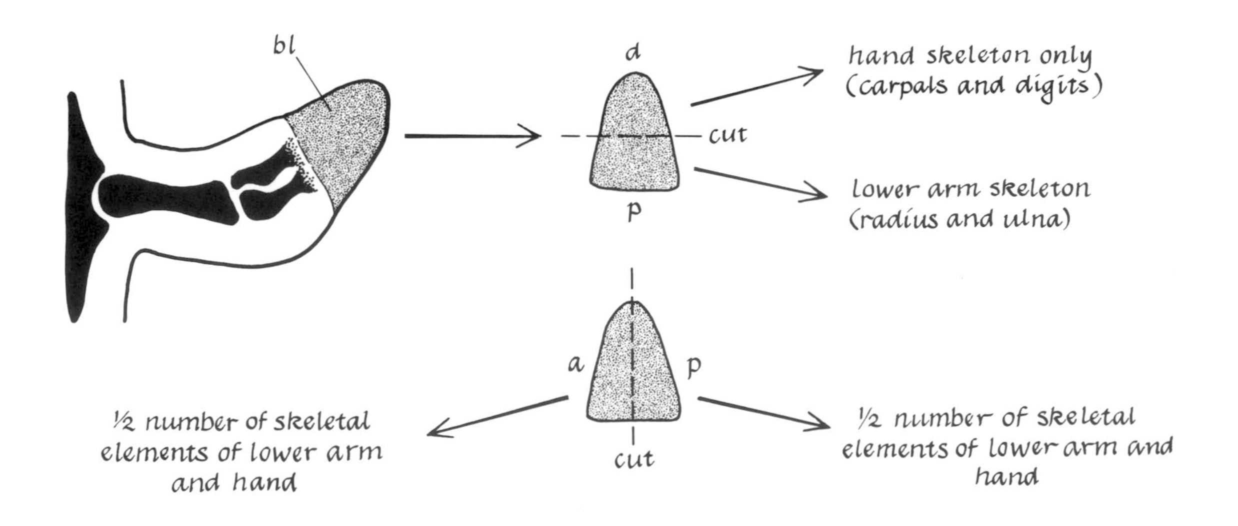

Figure 5-31. *Experiments indicating the mosaic properties of the blastema of larval* Ambystoma maculatum.

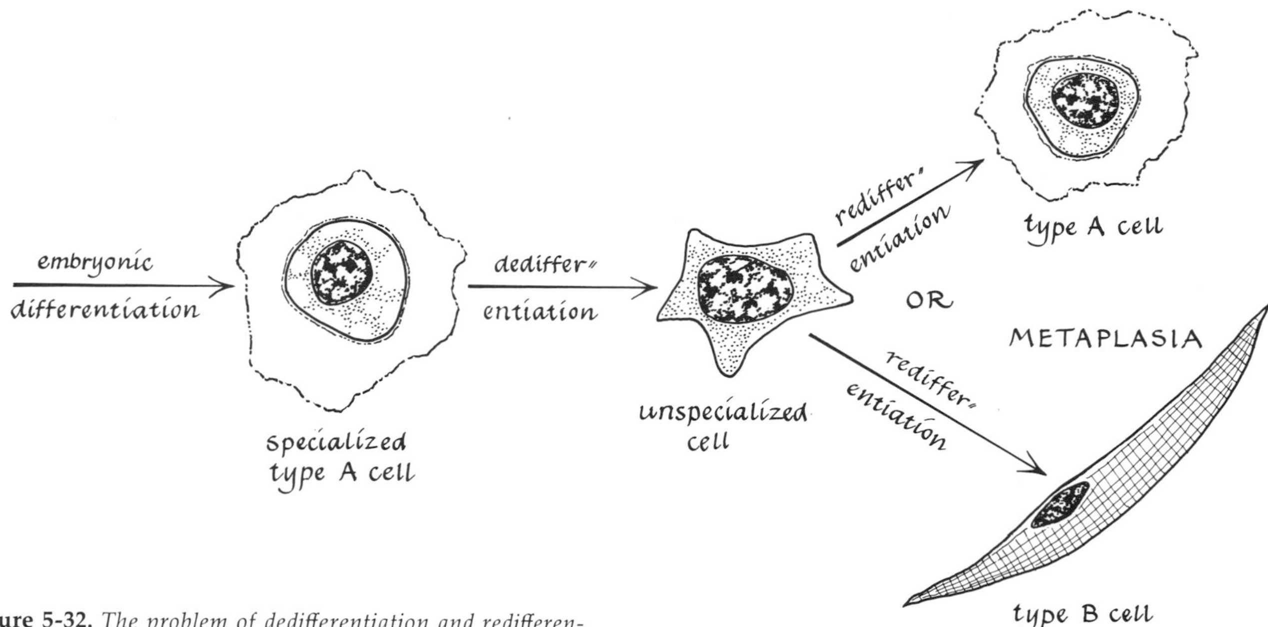

Figure 5-32. *The problem of dedifferentiation and redifferentiation in amphibian limb regeneration.*

As we shall see later in this chapter, relatively specialized cells in *Hydra* and *Planaria* evidently dedifferentiate and then redifferentiate along new pathways. Similarly, there is evidence for despecialization and respecialization of cells in an amphibian limb undergoing regeneration (Hay, 1966, 1968; Haynes et al., 1968). However, respecialization in vertebrate regeneration has not been proved to occur along a path different from that of the original specialization. The problem is illustrated in Figure 5-32. Dedifferentiation can result in an apparently unspecialized cell. However, many investigators believe that true dedifferentiation occurs only when redifferentiation into a different type of cell (metaplasia) takes place (Figure 5-32, Type B cell).

Cartilage cells undergo dedifferentiation, as attested by the ultrastructural changes carefully followed by Hay (1965, 1966). Schmidt (1968) doubts that any change in the program of gene action occurs during the redifferentiation of limb blastema cells of urodele amphibians. Steen's studies (1968) of limb regeneration in the axolotl *(Siredon mexicanum)* indicate that ^{3}H-thymidine labeled cartilage cells dedifferentiate and redifferentiate into cartilage cells. This cell type thus remained stable through the entire process of regeneration (Figure 5-32). Labeled muscle contributed cells to cartilage but the labeled cells in the cartilage may not have been derived from labeled muscle cells. Connective tissue cells in the muscle may have transformed into cartilage cells. Discovering whether metaplasia occurs in regeneration will depend on more precise methods of following cells, either by genetic or other labeling methods.

In experiments in which a Chimaeric limb (*Triton cristatus* mesoderm and *Triton vulgaris* ectoderm) was amputated, no evidence for transformation of cell type was found (Anton, 1965; see also review of Thornton, 1968). Anton's experiment is diagrammed in Figure 5-33. The Chimaeric limb was produced by grafting ectoderm from a neurula or gastrula of *T. vulgaris* over the limb forming area of a *T. cristatus* embryo of the same age after removal of host ectoderm in this region. Cells forming mesodermal elements of the regenerate were derived from the mesodermal cells of the host, and the epidermis of the regenerate was formed by the ectodermal cells of the donor. Ectodermal cells were not transformed into mesodermal cells or vice versa. Hay and Fishman (1961) found no evidence that *Triturus* ectodermal cells (whose nuclei were labeled radioactively) contributed to the blastema. However, Rose and Rose (1965) reported, on the basis of similar

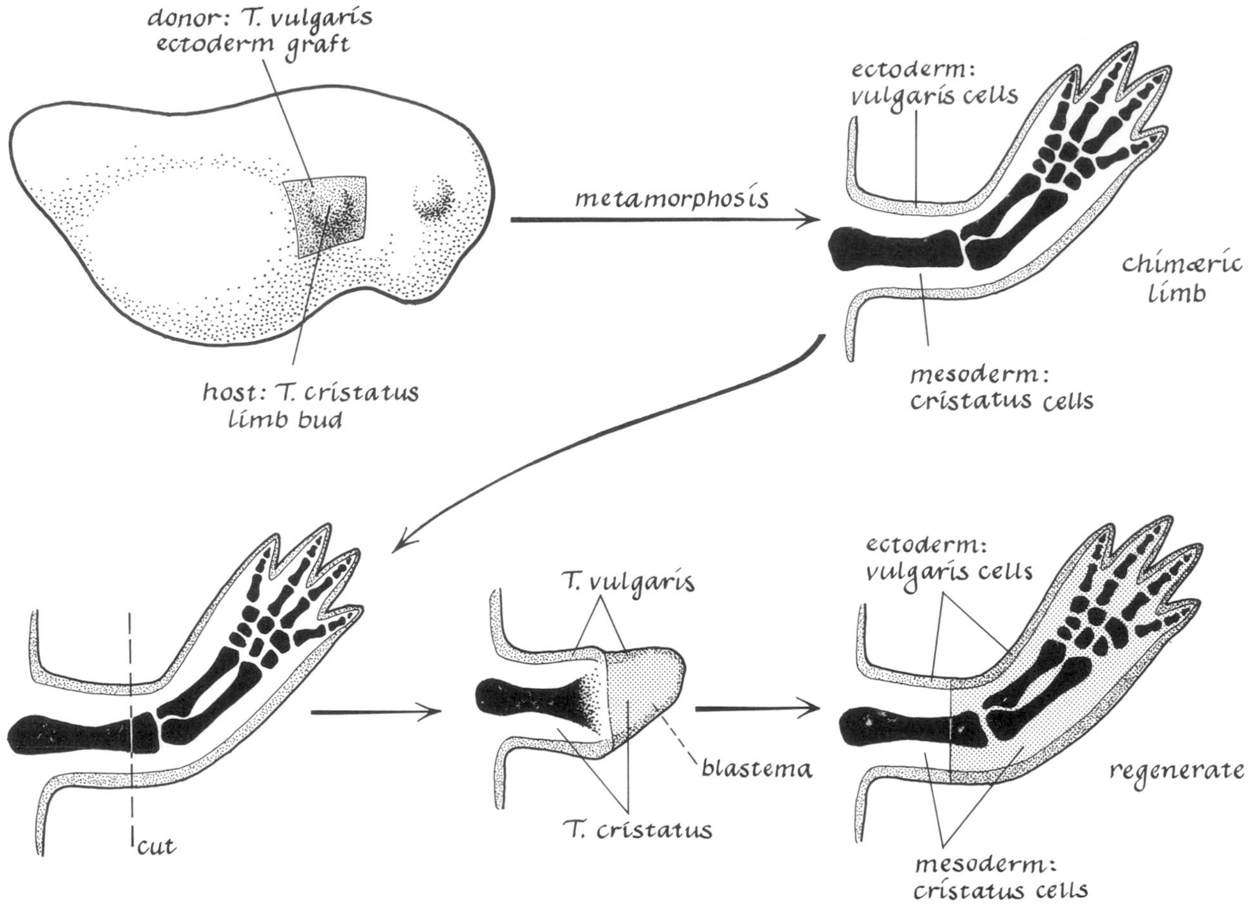

Figure 5-33. *Regeneration in a chimaeric limb* (Triton cristatus *mesoderm and* Triton vulgaris *ectoderm).*

experiments, that radioactively labeled cells of ectodermal origin were found in the blastema. The problem remains unsolved, but most evidence from studies on amphibians suggests that metaplasia does not occur. However, Rose and Shostak (1968) reported that vitally stained gastrodermal cells in the planarian *Phagocata gracilis* dedifferentiated into neoblasts and redifferentiated into parenchymal cells. This appears to be an example of metaplasia, but more reliable methods of labeling the gastrodermal cells are necessary before safe conclusions can be made.

To complicate the problem of whether metaplasia occurs in regenerating limbs is the possible presence of relatively unspecialized cells, such as connective tissue or other mesenchymal cells. An example of dedifferentiation of intestinal wall cells in the adult newt and redifferentiation of these into cartilage cells was cited earlier, but this result cannot be accepted as proof of dedifferentiation of a specialized intestinal cell followed by redifferentiation in a new direction (Oberpriller makes no such claim). Some cells in the blastema formed by the cut intestine may not have become specialized, and these may be the cells that respond to the limb blastema environment. To prove that specialized cells can undergo metaplasia during vertebrate regeneration, a genetically labeled clone of dedifferentiated cells must be implanted into a limb blastema. If labeled cartilage or bone cells are found, then it can be said with reasonable confidence that metaplasia has occurred (Figure 5-32).

The adult frog is not capable of completely regenerating an excised limb, but some regeneration of limb parts can be stimulated by trauma of the cut

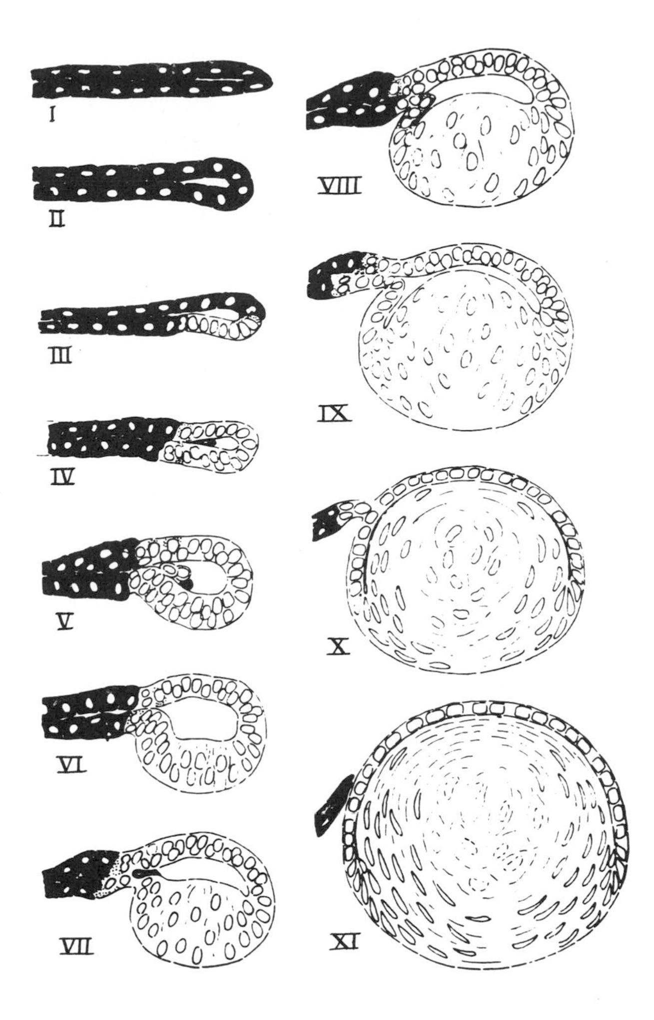

Figure 5-34. *Stages in Wolffian lens regeneration in adult* Triturus viridescens. *The Roman numerals are stages in regeneration. The pigmented iris cells are indicated by black cytoplasm; depigmented regenerate cells by white cytoplasm. From Yamada. Copyright 1967 by Academic Press.*

surface (Polezhayev, 1946), by treatment of the limb stump with strong salt solutions (Rose, 1944), by transplantation of adrenal glands to the jaws (Schotté and Wilbur, 1958; Schotté, 1961), and by increasing the nerve supply to the cut surface (Singer, 1952). Later in this chapter we shall examine the role of nerves in limb regeneration.

Trauma induces activity of proteolytic enzymes leading to autolysis and dedifferentiation. This is followed by an increase in dipeptidase activity, synthesis, growth, and differentiation (Urbani, 1965). However, in all these cases of artificially induced regeneration the outgrowths are abnormal (Goode, 1967).

Retina regeneration. In many urodeles the neural retina degenerates when the eye is transplanted from one eye socket to another or when the optic nerve is cut. Then cell division occurs in the pigmented layer of the retina and some cells migrate inward, toward the cavity of the eye. The inner cells lose their melanin pigment granules and form the new neural retina by continued cell division. The cells in the outer region of the pigmented layer regain the structure and function of retinal pigment cells. The transformation of pigmented cells into nerve cells is a clear example of overt dedifferentiation followed by redifferentiation along a new path (Stone, 1950; Reyer, 1962). A new optic nerve grows from the neurons of the new retina into the visual center of the brain, and normal vision in the eye is restored about 77 days after the operation (Chapter 14).

Lens regeneration. After removal of the lens from a salamander of the family *Salamadridae*, a new lens regenerates from the dorsal part of the iris. This process is called Wolffian lens regeneration, after Wolff (1895), one of the first embryologists to study it. As in neural retinal regeneration from the pigmented retina, depigmentation of the iris cells precedes their differentiation into the highly specialized lens fiber cells (Reyer, 1954, 1962; Stone, 1950). Studies of the cellular and subcellular processes in Wolffian lens regeneration conducted by Yamada (1967) and others reveal the sequence of changes induced by removal of the original lens. These changes include depigmentation of the iris cells, DNA synthesis and multiplication of the depigmented cells, cessation of DNA synthesis, polysome formation (Chapter 3), and synthesis of γ-crystallins as the lens fibers appear. Thus lens regeneration also shows overt dedifferentiation. Some stages in Wolffian lens regeneration are indicated in Figure 5-34.

Wolffian lens regeneration does not occur in the amphibian *Xenopus laevis*, the South African toad. After removal of the original lens, a new lens regenerates by invagination of the corneal epithelium—a process similar to that occurring in lens formation during the embryonic development of amphibians and other vertebrates (Freeman, 1962; Overton, 1965).

Principles and Control Mechanisms of Regeneration

Maintenance of Polarity

The maintenance of the original polarity of an organism is a feature common to all patterns of regeneration. Regenerative processes tend to restore the original apico-basal or anterior-posterior differences and symmetry in animals and plants. The following examples illustrate this principle.

Plants

Polarity of regeneration in plants is well established (reviews of Bloch, 1943, and Sinnott, 1960). When the apico-basal axis of the liverwort *Riella helicophylla* is bisected, rhizoids develop from the cut edge of the apical half and cauloids develop from the cut edge of the basal half, restoring the original polarity (see Figures 5-7 and 5-8). Polarity of regeneration in the formation of roots and shoots from pieces of the stem and branches of willow trees was studied long ago (Vöchting, 1877; Pfeffer, 1900-1906). Although capable of being reversed, the polarity of most plants is remarkably stable and largely independent of environmental influences, especially in young twigs. However, an abnormal environmental influence brought about by immersing the cut apical end of a stem in water will change the pattern of regeneration, resulting in root formation at the apical end and shoot formation at the basal end. Examples of maintenance of polarity in regenerating pieces of stems or branches and of reversal of polarity are shown in Figure 5-35. Shoots develop from the apical end of a piece of willow stem suspended in the normal position in moist air, and roots arise from the basal end. When the stem is suspended with the basal end up, roots continue to develop from the basal end and shoots from the apical end, thus maintaining polarity. Roots normally form from the anatomically basal end, but polarity can be reversed by immersing the apical end of a stem piece in water or soil. This procedure causes poor development of the regenerates but, when the pieces are later turned right side up, vigorous development of shoots from the apical end and roots from the basal end results, indicating that the environmentally induced reversal

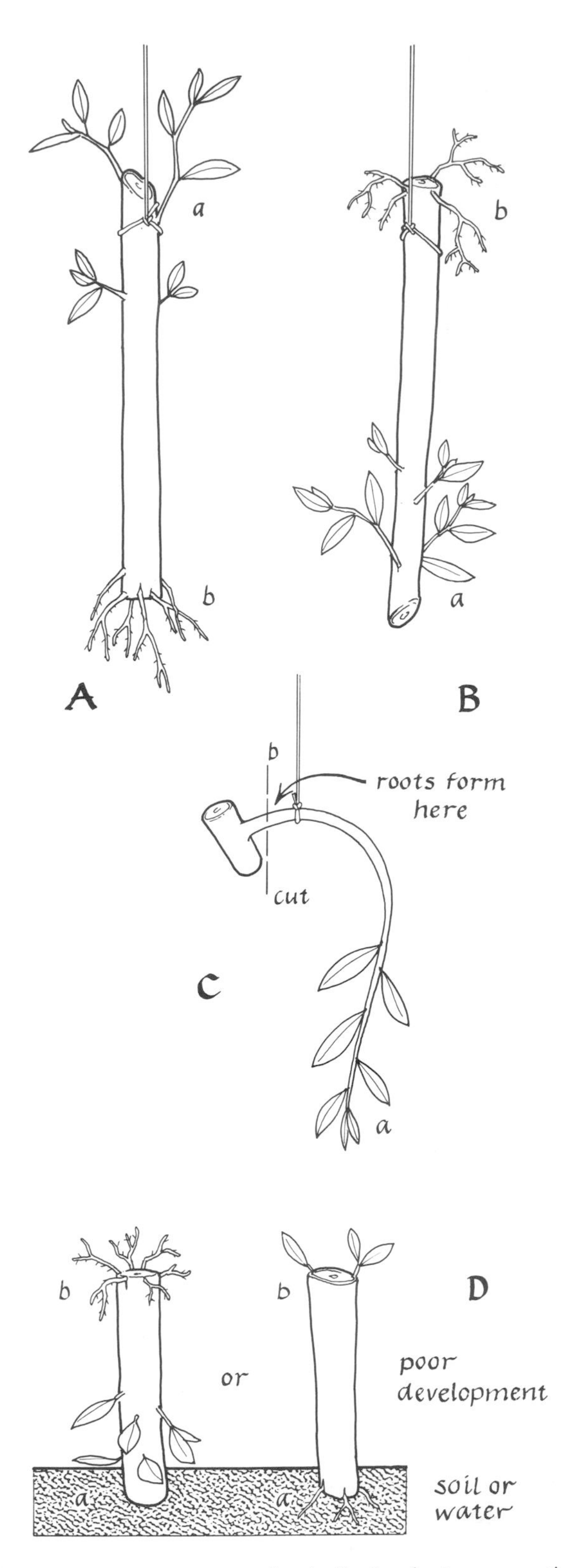

Figure 5-35. *Maintenance of polarity in plant regeneration.*

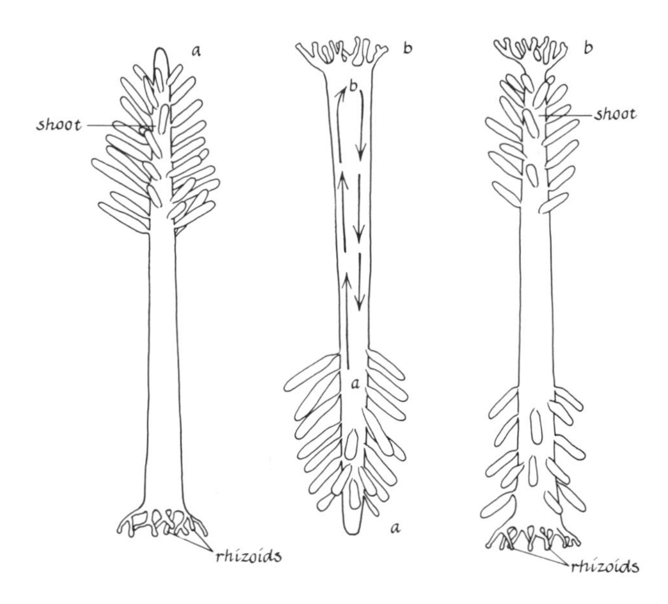

Figure 5-36. *Apparent reversal of polarity in* Bryopsis. *Arrows denote the rotation of the cytoplasm in a plant held in an inverted position.*

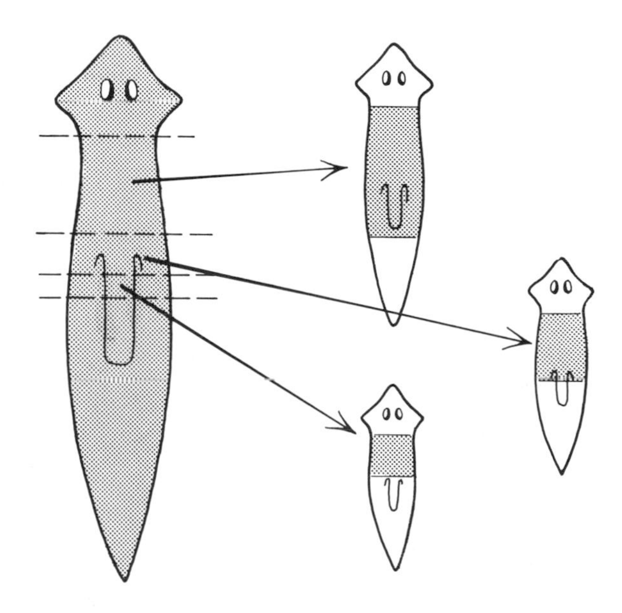

Figure 5-37. *Maintenance of polarity in small regenerating pieces of* Planaria.

of polarity is only temporary. Vöchting (1892) also discovered that grafts of part of one plant to another are usually successful only if the polarity of the host and grafted piece coincide.

The polar organization of a plant has a physiological basis in the movement of hormones and other substances, and this polarity may extend even to single cells in multicellular plants (Chapter 19). The substances do not simply diffuse through the plant but are transported in a particular direction that apparently is controlled by a structural polarity in the cells. Bünning's (1953) suggestion that establishment of a polar axis in the cell is basic to cell differentiation in plants may explain why it is so difficult to reverse the polarity of regenerating plant parts or organs. Electrical polarities are found in many organs and in the plant body as a whole. However, the significance of these electrical potentials in relation to the polarity of regeneration remains obscure at the present time (Sinnott, 1960).

An experimental reversal of polarity in an intact alga *Bryopsis* was reported by Steinecke (1925). This feather-shaped marine alga has no cell walls separating the nuclei, and polar organization can be reversed completely if the plant is inverted (Figure 5-36). A new shoot develops at the basal end and new holdfasts appear at the apical shoot end. Steinecke stained apical and basal regions with vital (nontoxic) dyes and found that an apical to basal rotation of the cytoplasm occurred. Further experiments indicated that light, not gravity, is responsible for the rotation in the inverted plant. Thus, not a reversal but a rotation of the polar organization occurred in the cytoplasm.

Animals

Maintenance of polarity in animal regeneration is also well established but, in contrast to plants, experimental reversal of polarity in hydroids and flatworms has been easily accomplished. In protozoa such as *Stentor*, polarity seems to be rigidly maintained in regeneration even of very small parts (Weisz, 1954). In hydroids, *Tubularia* and *Hydra*, the hydranth is regenerated at the distal end of medium length stem pieces. In flatworms the anterior-posterior polarity even of small pieces is retained (Morgan, 1901; Child, 1941; Flickinger, 1967). This is illustrated in Figure 5-37.

In most studies of limb and tail regeneration in urodeles, polarity relations are maintained. In some instances, amputation is followed by proximally directed regression from the cut surface but this usually does not involve the whole limb. When regeneration occurs, only parts distal to the unregressed portion of the limb are formed (Hay, 1966). Thus it seems to be a general principle of regeneration in animals that only parts distal to the cut surface are regenerated. When removal of the humerus from the limb of a salamander is followed by amputation through the middle level of the upper arm, only the part of the humerus distal to the amputation level is regenerated (Figure 5-38).

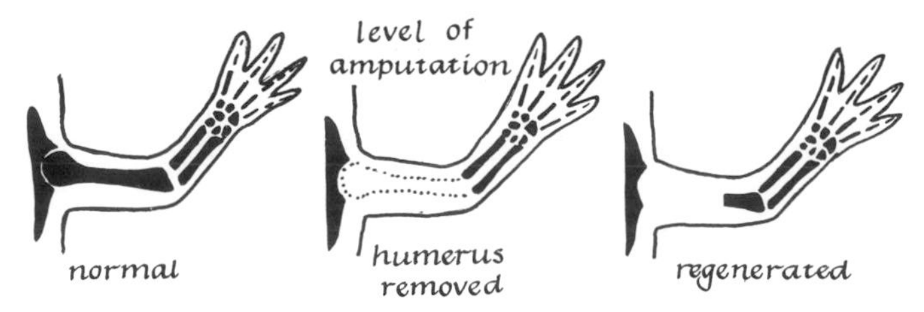

Figure 5-38. *Regeneration of a forelimb in a newt from which the humerus had been previously removed. After Weiss in Balinsky:* An Introduction of Embryology, *3rd edition. W. B. Saunders Company, Philadelphia, 1970.*

Experimental Reversal of Polarity

The normal polarity of regeneration in animals can be experimentally reversed. For example, in *Tubularia* the regeneration of a hydranth by a medium length piece of stem (10 to 15 mm long) normally occurs at the distal end. A hydranth can be forced to form at the proximal end by inserting the distal end into the sand (Morgan, 1903) or by ligating the stem piece just after it is cut out (Steinberg, 1955, and others). This is illustrated in Figure 5-39. The reason ligation of the stem enables regeneration to occur at the proximal end will be explained in the section on *Mechanism of Dominance in Animals.*

Polarity of hydranth regeneration in the hydroid *Obelia* can be reversed by placing a piece of cut stem in an electric field of sufficient strength (Lund, 1925, 1947). With a potential difference of 10 mv, distal regeneration of the hydranth was suppressed in a piece 10 mm long when the distal end faced the cathode but not when it faced the anode. In studies of pieces of *Tubularia* stems through which an electric current is passing, Rose (1963, 1967) suggests that positively charged particles produced by a distally developing hydranth primordium migrate to the cathode end of the stem and inhibit regeneration there.

Reversal of polarity in *Hydra,* in which a hydranth can form in the proximal end of a fragment, has been induced by treatment of the fragments with extracts from homogenized hydras (review of Goss, 1969). The extract contains a substance called a stimulator (not yet identified) produced in the hypostome region that diffuses proximally. The graded distribution of this substance is believed to constitute the basis of polarity in *Hydra.*

Reversal of polarity in regenerating pieces of the planarian *Dugesia* has been achieved by imbedding the pieces in agar and exposing them to a direct cur-

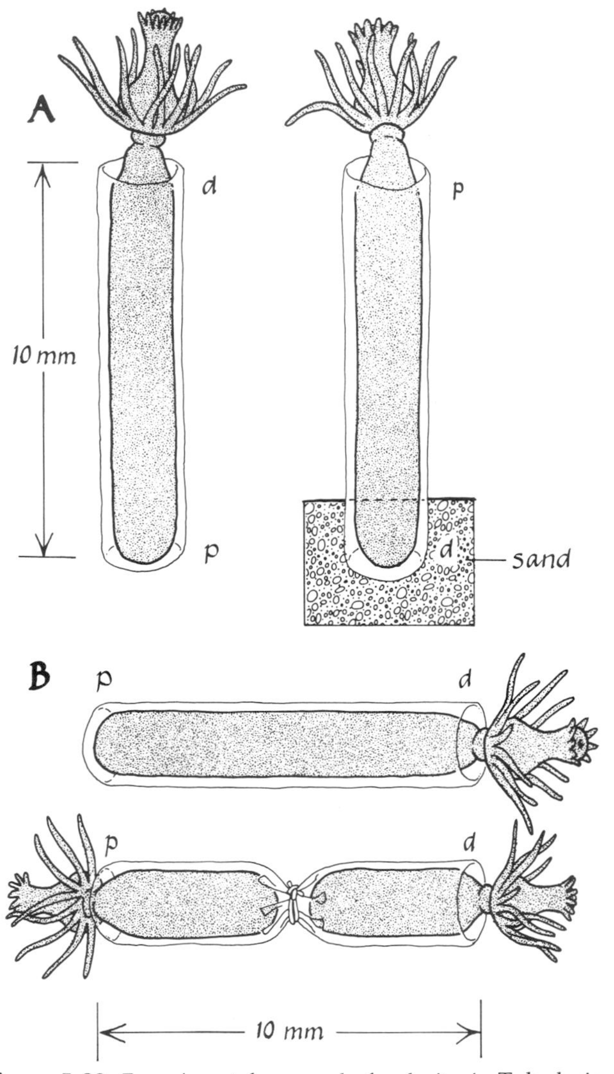

Figure 5-39. *Experimental reversal of polarity in* Tubularia: *A, by insertion of the distal end of a stem piece in sand; B, by ligating the stem piece immediately after its isolation. In B, hydranths form at both the distal and proximal ends.*

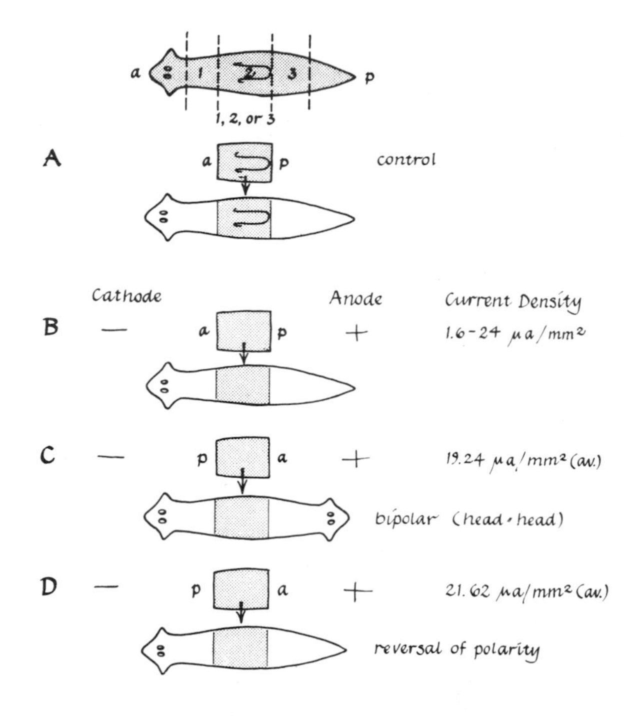

Figure 5-40. *Reversal of polarity in regenerating pieces of the planarian* Dugesia *by exposure to direct current flow.*

rent flow for five days (Marsh and Beams, 1952). Some of these experiments are diagramed in Figure 5-40.

Normal polarity of regeneration (Figure 5-40B) is maintained at all current strengths (densities) used, provided the anterior cut surface faces the cathode. When it faces the anode at a current density of 19.24 $\mu a/mm^2$, regeneration of the head occurs at both the anterior and posterior cut surfaces. At the highest density used, when the anterior end faces the anode, head regeneration at that end is inhibited.

In the examples of reversal of polarity just cited, the electric current presumably causes an electrophoresis (differential movement of substances in accord with the charge they carry) in the regenerating system. If this differential movement actually occurs, a redistribution of inhibitors or other regeneration controlling agents and, thus, a reversal of polarity may result.

The polarity of regeneration in a planarian may also have an underlying head-to-tail gradient in metabolic activity. Thus, when the anterior cut surface is immersed for a day or so in a solution of antimetabolites such as colcemide or chloramphenicol, head regeneration can occur on the posterior cut surface with the result that a bipolar, head-to-head worm is formed (Flickinger and Coward, 1962). As the antimetabolites also inhibit DNA and RNA synthesis and cell division, they probably act by obliterating an anterior-posterior gradient in mitotic activity, normally high in the head and lower in the tail.

An example of reversal of polarity in limb regeneration in the urodele has been described by Dent (1954) and by Butler (1955). The experiment and the result are illustrated in Figure 5-41.

As indicated in Figure 5-41B the hand is amputated, and in C, the stump is inserted in a pocket cut in the side of the salamander. After the stump is healed in and supplied with blood vessels and nerves, the limb is amputated through the upper arm region (Figure 5-41D). In E, a blastema forms on the cut surface of the limb, now attached to the body only by its distal end. In F, a forearm and hand regenerate, with distal structures formed on the proximal part of the upper arm. This result indicates that a reversal of polarity has occurred in the arm permitting a hand rather than a whole new animal to form.

The experiments cited indicate that the polarity of regeneration in animals, unlike that in plants, can be experimentally and permanently reversed.

Origin of Special Regeneration Cells

Of all the sources of cells for regeneration, the participation of special regeneration cells (interstitial cells in hydroids and neoblasts in flatworms) has been most critically examined. Aside from evidence for the existence and participation of special regeneration cells in hydroids (Tardent, 1954; 1963), planarians (Stephan-Dubois, 1965), and annelids (Herlant-Meewis, 1964), the main point of contention is the origin of these cells. Are permanent cells specialized for regeneration, or might such cells be derived by dedifferentiation from relatively specialized tissue cells? In several kinds of organisms regeneration cells have been found to arise from tissue cells. Some experiments indicating this will now be considered.

Brien and Reniers-Decoen (1955) claim, on the basis of experiments on *Hydra* in which x-rays selectively destroy or block division of interstitial cells,

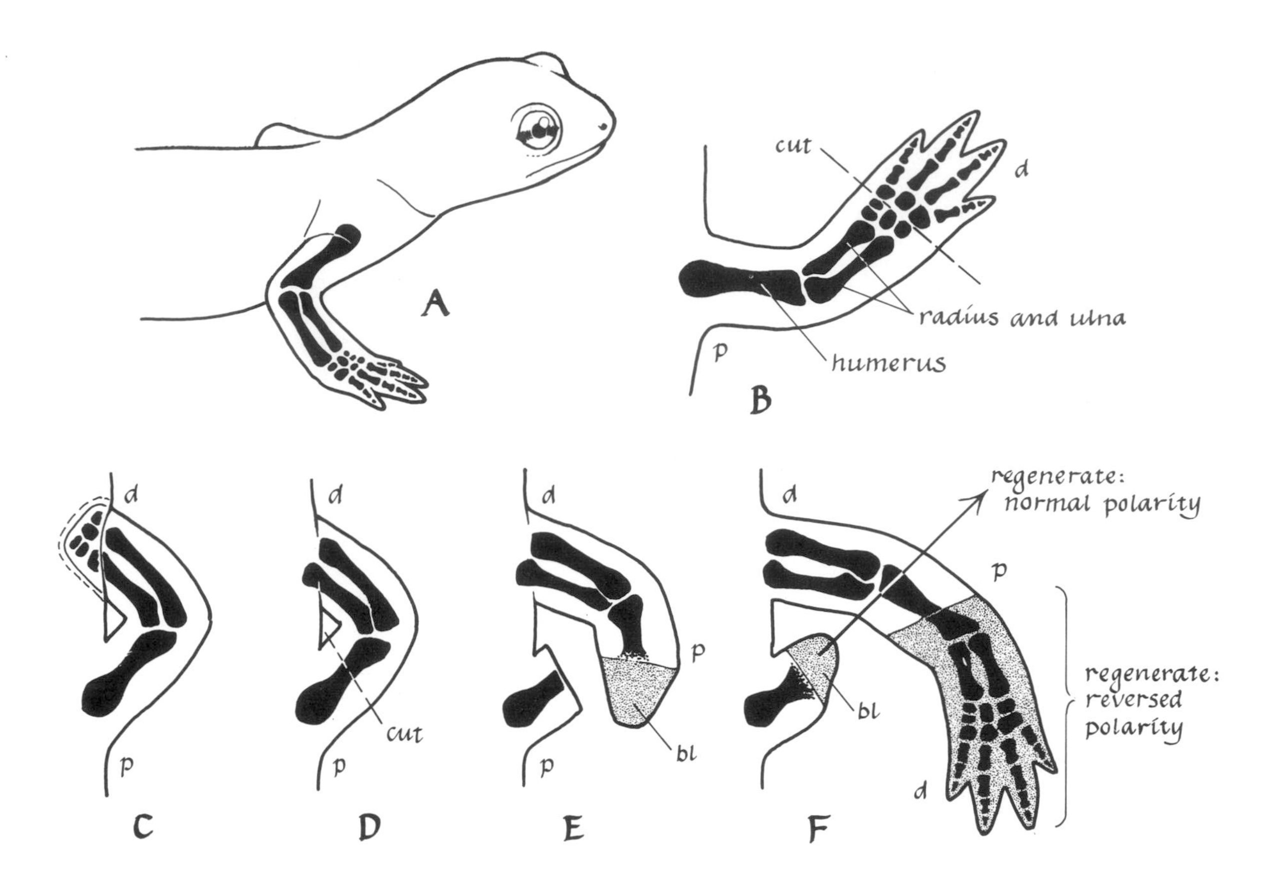

Figure 5-41. *Reversal of polarity by reversal of the proximal-distal axis of the forelimb of the urodele. See text for further description.*

that the interstitial cells (I-cells) are of secondary importance in regeneration. Perhaps more convincing evidence for the dispensability of I-cells is the demonstration by Haynes et al. (1968) that small fragments of the endodermis in *Hydra viridis* (a species lacking I-cells in the endodermis) may regenerate whole animals when cultured in a medium containing sodium, calcium, potassium, and magnesium salts. These investigators also have reported that I-cells, which are not normally present in the basal disc of *Hydra*, appear after severe trauma (temperature shock and physical abrasion) to the disc. I-cells appear to arise from the dedifferentiation of gland cells in the endodermis. The dedifferentiated gland cells can redifferentiate into cnidoblasts, nerve cells, or gametes under environmental conditions consisting of the proper anions and cations (Burnett, 1968).

In studies of the jellyfish *Aurelia*, Steinberg (1963) obtained good regeneration from isolated ectodermal fragments selected from regions containing no I-cells. Following the wounding of a flatworm, the rate of detachment of gland cells from the intestine accelerates and these cells apparently transform into neoblasts (Woodruff and Burnett, 1965).

Injection of blood cells of the colonial tunicate *Perophora viridis* into x-rayed, nonbudding colonies results in resumption of budding (Freeman, 1964). This result suggests that blood cells may function like stem cells or neoblasts.

Although the existence of a permanent population of special regeneration cells has not been satisfactorily proved, it appears that such cells are not ab-

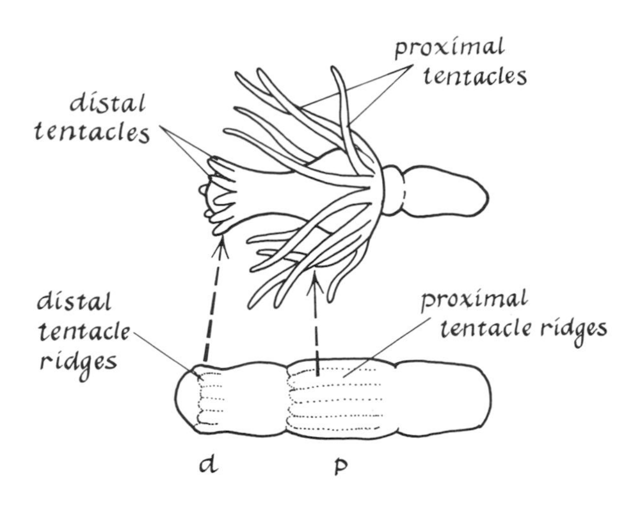

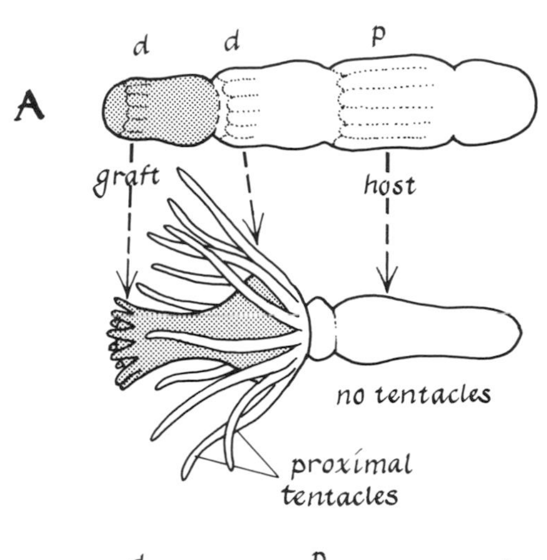

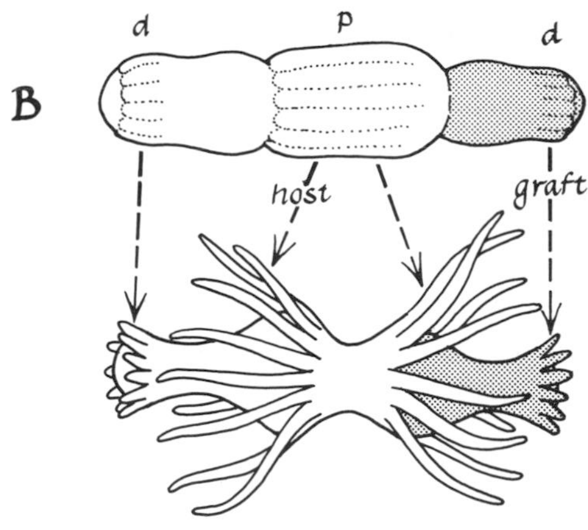

Figure 5-42. *Distal dominance in* Tubularia *regeneration.*

solutely necessary for regeneration in cnidarians. The evidence from studies of flatworms is less conclusive.

Control Mechanisms in Regeneration

In general, the control mechanisms for regeneration in animals appear to be inherited, but the control mechanisms in plants arise epigenetically in most instances. In Chapter 22 we shall discuss in detail how progressively formed guidelines might arise. For the present, we shall be concerned only with the operation of guidelines present either in the body of the organism from which a part has been removed or from the beginning in the isolated part undergoing regeneration.

Apical and Distal Dominance

In both lower and higher plants, the shoot apex dominates the growth of more basally located lateral buds. This apical dominance prevents regeneration of pieces of a fern prothallium containing the apical center. As we shall see in Chapter 19, auxins—plant hormones produced in relatively high concentration by apical centers—appear to be the mediators of this dominance. As we saw earlier in this chapter, animals have a different form of dominance. When a structure regenerates, the part is almost always distal to the site of amputation. However, the most distal region is first replaced, followed by more proximal parts. The order of replacement is thus distal to proximal. In the case of limb regeneration in urodeles, an apical proliferation center is first established and through its activity proximal cells are added (Faber, 1965). For example, after amputation of a limb through the upper arm near the shoulder, the order of replacement is: fingers, carpus of the hand, wrist, lower arm, elbow, and upper arm. Differentiation in the blastema begins at the distal end and progresses proximally. The distal region not only develops first but prevents the formation of distal structures proximal to it. Rose's (1963, 1964) grafting experiments in *Tubularia* clearly demonstrate this (see Figure 5-42). Any piece of *Tubularia* stem above a minimum size can form a hydranth, but it never forms that part which has already started to form anterior to it. In general, distal regenerating parts are the most developmentally active, as are the apical growing regions in both plants and animals. Apical developmental centers such as shoot and root apices in plants and regeneration blastemas in animals tend to be dominant regions, controlling the activities of proximal or basal regions. Thus, in Figure 5-42A the graft of a distal part into the distal end of a larger piece transforms what would have been distal tentacles in the host into proximal tentacles, and suppresses formation of a second set of proximal tentacles. In Figure 5-42B, doubling of the proximal tentacle region and development of two sets of distal tentacle ridges forms two hydranths facing opposite directions.

From the experiments cited, we can conclude that both apical dominance and the distal to proximal order of replacing lost parts are additional principles of regeneration.

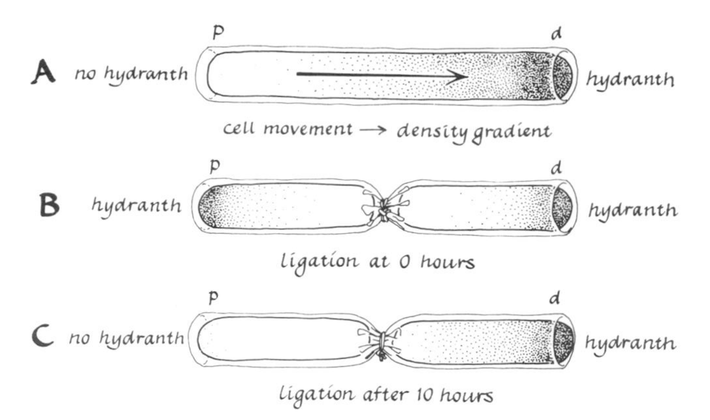

Figure 5-43. *Cell population density as a mechanism of dominance in* Tubularia. *Arrows denote the direction of cell movement as revealed in time-lapse motion pictures of pieces of the stem.*

The Mechanism of Dominance in Animals

Rose (1957, 1963, 1967) has suggested that failure of hydranths to form at the proximal end of a 5 to 10 mm long piece of *Tubularia* stem is the result of inhibitor substances produced by regenerating distal primordia. The inhibitor substances seem to be positively charged particles that move in a polarized (proximal) direction (Rose, 1967). Other studies (Akin and Akin, 1967) indicate that a regeneration inhibitor (extracted from homogenates of regenerating stems) is a protein or polypeptide. In studies on the hydroid *Corymorpha palma,* an extract from homogenized hydranths inhibits hydranth regeneration in stem pieces of the same or different species. The inhibitor has been identified as a nematocyst toxin containing tetramethylammonium (Davis, 1967). Much more work must be done to prove that a specific inhibitor substance is a general mechanism of dominance.

Another mechanism of dominance, that dominant regions have a greater cell population density than surrounding or adjacent regions, has been proposed by Steinberg (1954, 1955). In *Tubularia,* regeneration of a hydranth occurs wherever a cut is in the perisarc, provided there is a sufficient cell population, regardless of proximal-distal polarity. Time-lapse motion pictures of *Tubularia* stem pieces reveal a proximal to distal movement of tissue (Figure 5-43A). If a 10 mm long piece of stem is ligated immediately after its isolation, regeneration occurs at both ends (Figure 5-43B) because the cells in the proximal section migrate to the cut surface by reversing their polarity of movement. If ligated ten hours after isolation, only distal regeneration occurs (Figure 5-43C) because in the ten-hour interval cells have moved out of the proximal region and the cell population is of insufficient density to support regeneration.

Nevertheless, the studies on *Corymorpha* imply that inhibitors are involved and that the cell movements are incidental to the inhibitors. However, it is not clear why the inhibitors presumably produced in higher concentration by regions of high cell population density (the distal end of a stem piece) prevent proximal rather than distal regeneration.

Further evidence that inhibitors are involved in the control of regeneration is revealed in recent studies showing that regeneration of the basal disc in *Hydra viridis* can be inhibited by grafting another basal disc to the *Hydra* (MacWilliams and Kafatos, 1968). Thus, a structure already formed can inhibit the development of another structure of the same type. This phenomenon is strikingly illustrated by grafting heads and tails to different body levels of planarians (review of Brønsted, 1955). Results of these experiments are illustrated in Figure 5-44. The mechanism of this type of inhibition is not known with certainty, but some evidence of specific, homologous inhibitor substances in the regeneration of planarians has been presented (review of Lender, 1965). Ziller-Sengel (1965) reports that extracts of the pharynx will inhibit pharyngeal regeneration. Lender cites cases of inhibition by head extracts of brain regeneration. When tail extracts were used as a control no inhibitory effect was observed.

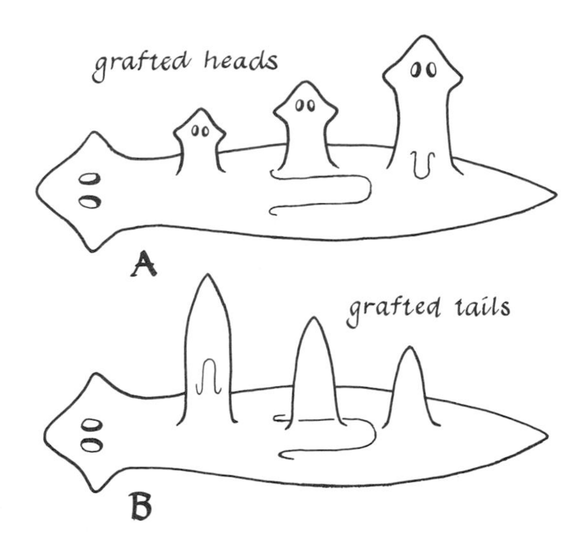

Figure 5-44. *Heads and tails grafted to different body regions of planaria. A, maximal growth of the head occurs when it is grafted more posteriorly. B, maximal growth of the tail occurs when it is grafted more anteriorly.*

Granted that specific inhibitor substances play a role in the control of regeneration and growth processes, the identity of the substances has yet to be discovered.

The Role of Hormones

In 1926, Schotté reported that limb regeneration does not occur in adult newts that have been hypophysectomized. However, larval stages of these animals are able to regenerate limbs or tails after removal of the hypophysis (Tassava, et al., 1968). The hypophyseal hormone, ACTH (adrenocorticotrophic hormone), seems to participate in regeneration processes in adults but not in larval *Ambystoma.* As observed in the section on limb regeneration, implantation of adrenal glands into young hypophysectomized frogs *(Rana clamitans)* induced some degree of limb regeneration following amputation. Tail regeneration in the lizard *Anolis curolinensis* is virtually abolished by hypophysectomy (Licht and Howe, 1969).

Thyroid hormones inhibit limb regeneration in amphibians but hypothyroidism seems to increase the rate of regeneration (review of Schmidt, 1968). Much work must be done before we can gain a clear picture of the role of hormones in controlling regeneration (see Schotté, 1961; Rose, 1964).

The Role of Nerves

Some degree of regeneration can be stimulated in adult frogs, lizards, and the newborn opossum (which normally do not regenerate limb parts) by experimentally increasing the nerve supply to the region of amputation (Schotté and Wilbur, 1958; Simpson, 1961; Mizell, 1968). Although urodele limbs that have developed embryologically without a nerve supply can regenerate without nerves (Yntema, 1959), nerves must be present at early stages for regeneration in adult urodeles (Schotté and Butler, 1941; Singer, 1952). Nerves at the cut surface are also necessary for regeneration not only in earthworms as noted above but also in many invertebrates (see review of Kiortsis and Trampusch, 1965).

The mechanism of action of the nerves in regeneration is not known, but Singer (1960, 1965) has suggested that the agent stimulating regeneration may be more concentrated in nerves. Granules resembling the neurosecretory granules of many invertebrates have been observed in nerve processes of the newt *Triturus.* These granules may be involved in transport of the regeneration stimulating agent (Lentz, 1967).

It has been reported that denervation of the newt limb results in a decrease in the synthesis of DNA, RNA, and protein by cells of the blastema (Dresden, 1969). The quantity, not the quality, of nerve fibers or neural substance at the amputation surface controls the regeneration of the limb in urodeles. About ⅓ to ½ the number of nerve fibers normally present at the cut surface is the threshold number below which no regeneration occurs (Singer, 1965). More recent studies (Rzehak and Singer, 1966) of *Rana sylvatica* and *Xenopus laevis* indicate that the threshold for regeneration may not be exclusively a function of the number of nerve fibers present. If the diameter of the nerve fibers is considered, the threshold can be expressed as the amount of neuroplasm available at the amputation surface. Even on this basis in *Triturus* and *Ambystoma,* species that have good limb regeneration capacities, the amount of neuroplasm is far greater than in anurans, which have neither the number of nerve fibers nor the quantity of neuroplasm occurring in urodeles.

Influence of the Old Part on the Regenerate

There are many examples of the influence of the organism on the regeneration of a missing part. In plants, the half leaf of a moss gametophyte does not regenerate if left attached to the parent plant (Stange, 1964). In algae, bryophytes, and ferns, parts having an apical cell or meristem generally regenerate only rhizoids. In plants, the region from which a part is isolated influences the regeneration of that part. Leaf cuttings from the flowering part of many kinds of plants (for example, *Begonia*) regenerate from the petiole and produce flowers at once. Those taken from a nonflowering plant produce only leaves (Goebel, 1898). The formation of flowers by *Achimenes,* a tropical plant, from a leaf cutting is shown in Figure 5-45. Goebel and Sachs suggested that a "flower-forming stuff" was present in the leaves of the flowering part of the plant. We know now that "florigens" (flower-inducing substances) do exist in plants (Chapter 19).

The age of an organism also influences the pattern of regeneration of an isolated part. Isolated pieces of an adult sporophyte or gametophyte of a moss hybrid *(Physcomitrium-Funaria)* form only a protonema (filamentous structure). In young sporophytes, however, apically and basally isolated regions produce different regeneration products (Bauer, 1963). This is illustrated in Figure 5-46.

A particularly interesting example of control of regeneration by existing structures is found in the biradial ctenophore *Vallicula multiformis* (Freeman, 1967). In this organism "half-animals" were produced by a transverse cut through the apical organ. Such animals do not regenerate the missing set of tentacles or a canal system on the cut side, but do regenerate the missing half of the apical organ (Figure 5-47A).When a half-animal is cut into three parts (Figure 5-47B), the part containing the apical organ regenerates as a half-animal, other parts as whole animals. When whole animals are cut into three parts, all parts regenerate into whole animals. When the apical organ of a half-animal is grafted into a whole animal, it functions like the organ of a whole animal in that a part containing it regenerates as a whole animal. This result indicates that the apical organ of a half-animal is not a functionally whole apical organ but becomes whole in the whole animal host. Thus the apical organ of the half-animal cannot "direct" the regeneration of a whole animal when

Figure 5-45. *Formation of flowers by a leaf cutting from flowering* Achimenes Haageana, *a hybrid tropical plant. At the basal end of the severed leaf stalk an adventitious shoot bearing flowers has developed. From* Organography of Plants *by K. Goebel (Clarendon Press, Oxford).*

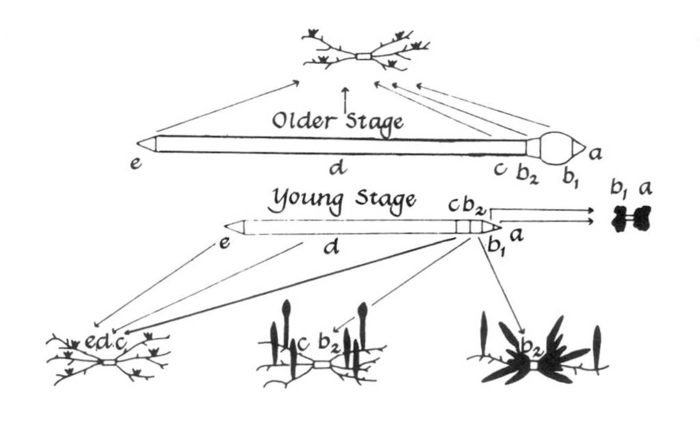

Figure 5-46. *Regeneration products of the different zones of sporangia in the hybrid moss* Physcomitrium piriforme × Funaria hygrometrica *at different ages. A, apical cell and adjacent cells. B1, meristem producing the capsule. B2, meristem contributing to the seta. C, extension zone. D, fully differentiated part of the seta. E, foot of sporangium. From* J. Linn. Soc. (Bot.) *58, 1963. Published by permission of the Linnean Society of London.*

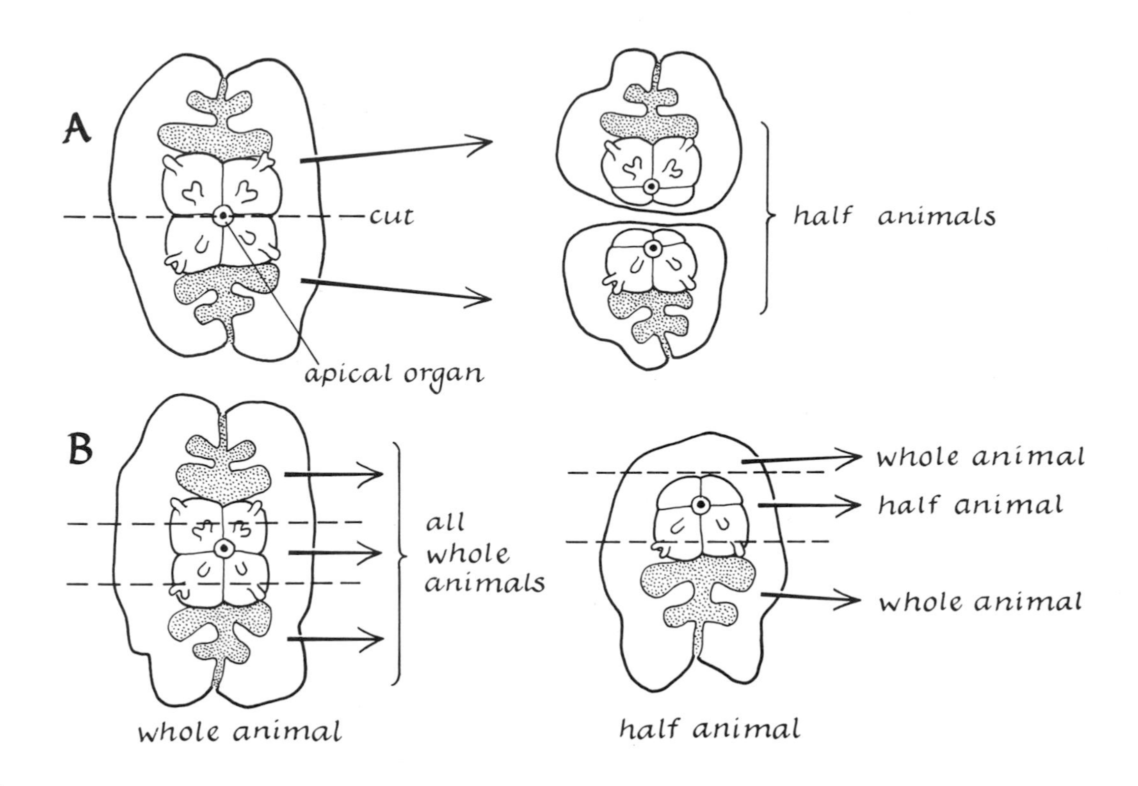

Figure 5-47. *Control of regeneration by existing structures in the old part of the ctenophore* Vallicula multiformis.

resident in the isolated middle piece of the half-animal. When a half-animal having the grafted apical organ of a whole animal is tested for regeneration, all pieces regenerate as whole animals. Perhaps some interaction between the functionally half-apical organ and the whole animal pattern restores functional wholeness in the organ.

Conclusion

In our survey of some patterns of regeneration in diverse organisms we have seen that in plants more than the part lost or removed is usually replaced. Rarely is only the lost part replaced. Instead, regeneration in plants is more like asexual reproduction than like regeneration in animals. This is clearly the case when a single, isolated cell of the adult plant is capable of dividing and building up a cell population to form a complete new individual plant. Nothing like this has been observed in any isolated animal cell.

In most kinds of regeneration in animals only the missing part is reformed, but in a few exceptional cases a fragment of the adult may reconstitute a whole new organism. In both plants and animals, regeneration is usually preceded by dedifferentiation of adult cells and involves both a rearrangement and a respecialization of cells or a reorganization within the cytoplasm of unicellular organisms. In some cases the missing part is replaced by rearrangement of the remaining portion of the organism, but in most cases of regeneration cells are added to form a callus (in plants) or a blastema (in animals).

The basic mechanisms controlling regeneration at the cellular and molecular levels are probably the same as those that controlled the original development of the organism and its parts. However, the stages through which a lost part is replaced or through which an isolated part reconstitutes a whole organism are not the same as those leading to their original development. Regeneration is directed toward restoration of the symmetry and pattern of the whole organism or generation of symmetry in an isolated fragment. Apparently the stimulus for regeneration is an interruption in the unity of the organism. The result of programmed interaction between the genome and its environment is the attainment of the stable state that characterizes the adult organism. The stable internal microenvironment of maturity includes a marked decrease in programmed gene action of the specialized cells. Although purely speculative, it is suggested that when the stable state is disturbed, the gene action program is again set in motion and regeneration occurs.

One way in which the stable state may be disturbed is by a change in the microenvironment surrounding the cells. When part of an organism is cut off, the environment of the cells at the cut surface is immediately changed. This sets in motion cellular activities and transformations commensurate with the properties of the surface cells. The environment of cells below the surface of the wound is also changed. How far from the site of the trauma the change in environment extends is a challenging problem. It is probably a definite distance in each type of regeneration system. Although the obvious reconstitution processes may be localized, the processes are probably initiated in some organisms by changes in cells (for example, meristematic cells, ganglion or brain cells, hormone-producing cells) far from the cut surface.

An almost universal principle in regeneration is the maintenance of the polarity and symmetry of the organism. Probably the first regenerative response is the establishment of a developmental field with polarized and symmetrical properties. This field, potentially capable of replacing the missing part or of developing into a whole organism, can serve both as a preformed and an epigenetically formed guideline for the development of the regenerate. Such a guideline is a microenvironmental system to which the cells respond according to their positions. This results in selective and sequential gene action.

In general, the region of the missing part to be replaced first is always the most distal or apical region, whether or not cell addition is primarily localized there. Reconstitution then proceeds in an apical to basal or distal to proximal direction under control of the distal and dominant region.

References

Abeloos, M. 1932. La régénération et les problèmes de la morphogenèse. Gauthier-Villars, Paris.

Akin, G. C., and J. R. Akin. 1967. The effect of trypsin on regeneration inhibitors in *Tubularia*. Biol. Bull. **133:**82–89.

Anderson, J. M. 1965. Studies on visceral regeneration in sea-stars: II. Regeneration of pyloric caeca in *Asteriidae,* with notes on the source of cells in regenerating organs. Biol. Bull **128:**1–23.

Anton, H. J. 1965. The origin of blastema cells and protein synthesis during forelimb regeneration in *Triturus.* In V. Kiortsis and H. A. L. Trampusch, eds., Regeneration in Animals and Related Problems, pp. 377–395. North Holland Publishing Co., Amsterdam.

Barth, L. G. 1940. The process of regeneration in hydroids. Biol. Bull. **15:**405–420.

Barth, L. G. 1955. Regeneration: invertebrates. In B. H. Willier, P. A. Weiss, and H. Hamburger, eds., Analysis of Development, pp. 664–673. W. B. Saunders, Philadelphia.

Bauer, L. 1963. On the physiology of sporangium differentiation in mosses. J. Linn. Soc. Bot. **58:**343–351.

Berrill, N. J. 1961. Growth, Development and Pattern. W. H. Freeman, San Francisco.

Bliss, D. E. 1960. Autotomy and regeneration. In T. Waterman, ed., The Physiology of Crustacea, Vol. 1, pp. 561–585. Academic Press, New York.

Bloch, R. 1943. Polarity in plants. Bot. Rev. **9:**261–310.

Bock, F. 1926. Experimentelle Utersuchungen an Koloniebildenden Volvocaceen. Arch. Protist. **56:**321–356.

Bonner, J. T. 1952. Morphogenesis: An Essay on Development. Princeton University Press, Princeton, N.J.

Bonner, J. T. 1967. The Cellular Slime Molds, 2nd ed. Princeton University Press, Princeton, N.J.

Brachet, J. 1968. Synthesis of macromolecules and morphogenesis in *Acetabularia*. In A. Moscona and A. Monroy, eds., Current Topics in Developmental Biology, pp. 1–36. Academic Press, New York.

Brefeld, O. 1876. Die Entwicklungs-geschichte der Basidiomyceten. Bot. Zeit. **34:**48–62.

Brien, P., and M. Reniers-Decoen. 1949. La croissance, la blastogénèse, l'ovogénèse chez *Hydra fusca* (Pallas). Bull. Biol. France-Belg. **83:**293–386.

Brien, P., and M. Reniers-Decoen. 1955. La signification des cellules interstitiells des hydres d'eau douce et le problème de la rēserve embryonnaire. Bull. Biol. France-Belg. **89:**258–325.

Brønsted, H. V. 1955. Planarian regeneration. Biol. Rev. Camb. **30:**65–126.

Brown, W. H. 1935. The Plant Kingdom. Ginn and Co., New York.

Bünning, E. 1953. Entwicklungs-und Bewegungs physiologie der Pflanze. Springer Verlag, Berlin.

Burnett, A. L. 1968. The acquisition, maintenance, and lability of the differentiated state in *hydra*. In H. Ursprung, ed., Stability of the Differentiated State, pp. 109–127. Springer Verlag, New York.

Butler, E. G. 1955. Regeneration of the urodele forelimb after reversal of its proximo-distal axis. J. Morph. **96:**265–281.

Child, C. M. 1941. Patterns and Problems of Development. University of Chicago Press, Chicago.

Clark, R. 1969. Endocrine influences in annelids. Gen. Comp. Endocrinol. **2**(suppl.):572–581.

Dalyell, J. G. 1814. Observations on some interesting phenomena in animal physiology exhibited by several species of planaria. A. Balfour, Edinburgh.

Davis, L. V. 1967. The source and identity of a regeneration-inhibiting factor in hydroid polyps. J. Exp. Zool. **164:**187–194.

Dent, J. N. 1954. A study of regeneration emanating from limb transplants with reversed proximo-distal polarity in the adult newt. Anat. Rec. **118:**841–856.

Dresden, M. H. 1969. Denervation effects on newt limb regeneration: DNA, RNA, and protein synthesis. Devel. Biol. **19:**311–320.

Faber, J. 1965. Autonomous morphogenetic activities of the amphibian regeneration blastema. In V. Kiortsis and H. A. L. Trampusch, eds., Regeneration in Animals and Related Problems. North Holland Publishing Co., Amsterdam.

Flickinger, R. A. 1967. Biochemical aspects of regeneration. In R. Weber, ed., Biochemistry of Animal Development, pp. 303–337. Academic Press, New York.

Flickinger, R. A., and S. J. Coward. 1962. The induction of cephalic differentiation in regenerating *Dugesia dorotocephala* in the presence of the normal head and in unwounded tails. Devel. Biol. **5:**179–204.

Freeman, G. 1962. Lens regeneration from the cornea in *Xenopus laevis*. J. Exp. Zool. **154:**39–65.

Freeman, G. N. 1964. The role of blood cells in the process of asexual reproduction in the tunicate *Perophora viridis*. J. Exp. Zool. **156:**157–184.

Freeman, G. N. 1967. Studies on regeneration in the creeping ctenophore *Vallicula multiformis*. J. Morph. **123:**71–84.

Fritsch, C. 1911. Experimentelle Studien über Regenerationsvorgänge des Gliedmassenskeletts der Amphibien. Zool. Jahrb. Abt. Allgem. Zool. Physiol. Tiere, **30:**377–472.

Goebel, K. 1898. Organographie der Pflanzen: I. Allgemeine Organographie. Jena (trans., 1900; Clarendon Press, Oxford).

Goebel, K. 1908. Einleitung in die Experimentelle Morphologie der Pflanzen. Teubner, Leipzig.

Golding, D. W. 1967. Neurosecretion and regeneration: I. Regeneration and the role of the supraesophageal ganglion. Gen. Comp. Endocrinol. **8:**348–355.

Goode, R. P. 1967. The regeneration of limbs in adult anurans. J. Emb. Exp. Morph. **18:**259–267.

Goss, R. J. 1969. Principles of Regeneration. Academic Press, New York.

Guyénot, E. 1927. La problème morphogénétique dans la régénération des Urodèles determination et potentialites des régénérats. Rev. Suisse Zool. **34:**127–154.

Hammerling, J. 1966. Nucleo-cytoplasmic relationships in the development of *Acetabularia.* In R. A. Flickinger, ed., Developmental Biology, pp. 23–47. Wm. C. Brown, Dubuque, Iowa.

Harrison, R. G. 1898. The growth and regeneration of the tail of the frog larva studied with the aid of Born's method of grafting. Arch. Entw. Mech. **7:**430–485.

Hay, E. D. 1962. Cytological studies of dedifferentiation and differentiation in regenerating amphibian limbs. In D. Rudnick, ed., Regeneration, pp. 177–210. Ronald Press, New York.

Hay, E. D. 1965. Limb development and regeneration. In R. L. DeHaan and H. Ursprung, eds., Organogenesis, pp. 315–336. Holt, Rinehart and Winston, New York.

Hay, E. D. 1966. Regeneration. Holt, Rinehart and Winston, New York.

Hay, E. D. 1968. Dedifferentiation and metaplasia in vertebrate and invertebrate regeneration. In H. Ursprung, ed., Stability of the Differentiated State, pp. 85–108. Springer Verlag, New York.

Hay, E. D., and D. A. Fishman. 1961. Origin of the blastema in regenerating limbs of the newt *Triturus viridescens.* Devel. Biol. **3:**26–59.

Haynes, J. F., A. L. Burnett, and L. E. Davis. 1968. The influence of inorganic ions on the differentiation of gastrodermal cells in *Hydra.* In M. M. Sigel, ed., Differentiation and Defense Mechanisms in Lower Organisms, pp. 49–56. Waverly Press, Baltimore, Md.

Herlant-Meewis, H. 1964. Regeneration in annelids. Adv. Morph. **4:**155–215.

Hertwig, G. 1927. Beiträge zum Determinations-und Regenerations-problem mittels der Transplantation haploidkerniger Zellen. Arch. Ent. Mech. **111:**292–316.

Hyman, L. H. 1955. The Invertebrates: Echinodermata. The Coelomate Bilateria. McGraw-Hill Book Company, New York.

Ito, M. 1962. Studies on the differentiation of fern gametophytes: I. Regeneration of single cells isolated from cordate gametophytes of *Pteris vittata.* Bot. Mag. Tokyo. **75:**19–27.

Kiortsis, V., and H. A. L. Trampusch, eds. 1965. Regeneration in Animals and Related Problems. North Holland Publishing Co., Amsterdam.

Lender, T. 1962. Factors in morphogenesis of regenerating fresh water planaria. Adv. Morph. **2:**305–331.

Lender, T. 1965. La régénération des planaires. In V. Kiortsis and H. A. L. Trampusch, eds., Regeneration in Animals and Related Problems, pp. 95–111. North Holland Publishing Co., Amsterdam.

Lentz, T. L. 1967. Fine structure of nerves in the regenerating limb of the newt *Triturus.* Am. J. Anat. **121:**647–670.

Licht, P., and N. R. Howe. 1969. Hormonal dependence of tail regeneration in the lizard *Anolis carolinensis.* J. Exp. Zool. **171:**75–83.

Lowell, R. D., and A. L. Burnett. 1969. Regeneration of complete hydra from isolated epidermal explants. Biol. Bull. **137:**312–320.

Lund, E. J. 1907. Reversibility of morphogenetic processes in *Bursaria.* J. Exp. Zool. **24:**1–34.

Lund, E. J. 1925. Experimental control of organic polarity by the electric current: V. The nature of the control of organic polarity by the electric current. J. Exp. Zool. **41:**155–190.

Lund, E. J. 1947. Bioelectric Fields and Growth. University of Texas Press, Austin.

MacWilliams, H. K., and F. C. Kafatos. 1968. *Hydra viridis:* inhibition by basal disk of basal disk differentiation. Science **159:**1246–1247.

Maltzahn, K. E. von, and M. M. MacNutt. 1960. Intracellular configuration and dedifferentiation in *Splachnum ampullaceum.* L. Nature **187:**344–345.

Marsh, G., and H. W. Beams. 1952. Electrical control of morphogenesis in regenerating *Dugesia tigrina:* I. Relation of axial polarity to field strength. In R. A. Flickinger, ed., Developmental Biology, pp. 60–77. Wm. C. Brown, Dubuque, Iowa (1966).

Mettetal, C. 1939. La régénération des membres chez la salamandre et le triton. Arch. Anat. Histol. Embryol. **28:**1–214.

Meyer, D. E. 1953. Über den Verhalten einzelner isolierter prothalliumzellen und dessen bedeutung für Korrelation und Regeneration. Planta **41:**642–647.

Mizell, M. 1968. Limb regeneration induction in the newborn opossum. Science **161:**283-286.

Morgan, T. H. 1901. Regeneration. Macmillan, New York.

Morgan, T. H. 1902. Experimental studies of the internal factors of regeneration in the earthworm. Arch. Entw. Mech. **14:**562–591.

Morgan, T. H. 1903. Some factors in the regeneration of *Tubularia.* Arch. Entw. Mech. **16:**125–154.

Needham, A. E. 1952. Regeneration and Wound-Healing. Wiley, New York.

Oberpriller, J. 1967. A radioautographic analysis of the potency of blastemal cells in the adult newt *Diemictylus viridescens.* Growth **31:**251–296.

Overton, J. 1965. Changes in cell fine structure during lens regeneration in *Xenopus laevis.* J. Cell Biol. **24:**211–222.

Peebles, F. M. 1897. Experimental studies on *Hydra.* Arch. Entw. Mech. **5:**794–819.

Pfeffer, W. 1900–1906. The Physiology of Plants, 2nd ed. Clarendon Press, Oxford.

Polezhayev, L. W. 1946. The loss and restoration of regenerative capacity in the limbs of tailless amphibia. Biol. Rev. Camb. **21:**141–147.

Reyer, R. W. 1954. Regeneration of the lens in the amphibian eye. Quart. Rev. Biol. **29:**1–46.

Reyer, R. W. 1962. Regeneration in the amphibian eye. In D. Rudnick, ed., Regeneration, pp. 211–265. Ronald Press, New York.

Rose, F. C., and S. M. Rose. 1965. The role of normal epidermis in recovery of regenerative ability in x-rayed limbs of *Triturus.* Growth **29:**361–393.

Rose, F. C., and S. Shostak. 1968. The transformation of gastrodermal cells to neoblasts in regenerating *Phagocata gracilis* (Leidy). Exp. Cell Res. **50:**553–561.

Rose, S. M. 1944. Methods of initiating limb regeneration in adult anura. J. Exp. Zool. **95:**149–170.

Rose, S. M. 1957. Cellular interactions during differentiation. Biol. Rev. Camb. **32:**351–382.

Rose, S. M. 1963. Polarized control of regional structure in *Tubularia.* Devel. Biol. **7:**488–501.

Rose, S. M. 1964. Regeneration. In J. A. Moore, ed., Physiology of the Amphibia, pp. 545–622. Academic Press, New York.

Rose, S. M. 1967. Polarized inhibitory control of regional differentiation during regeneration in *Tubularia:* III. The effects of grafts across seawater-agar bridges in electric fields. Growth **31:**149–164.

Rzehak, K., and M. Singer. 1966. Limb regeneration and nerve fiber in *Rana sylvatica* and *Xenopus laevis.* J. Exp. Zool. **162:**15–22.

Schilling, J. A. 1968. Wound healing. Physiol. Revs. **48:**374–423.

Schmidt, A. J. 1968. Cellular Biology of Vertebrate Regeneration and Repair. University of Chicago Press, Chicago.

Schotté, O. E. 1926. Hypophysectomie et régénération chez les Batraciens urodèles. C. R. Soc. Phys. Hist. Nat. (Geneva) **43:**67–72.

Schotté, O. E. 1961. Systemic factors in initiation of regenerative processes in limbs of larval and adult amphibians. In D. Rudnick, ed., Synthesis of Molecular and Cellular Structure, pp. 161–192. Ronald Press, New York.

Schotté, O. E., and E. G. Butler. 1941. Morphological effects of denervation and amputation of limbs in urodele larvae. J. Exp. Zool. **87**:279–322.

Schotté, O. E., and J. F. Wilber. 1958. Effects of adrenal transplants upon forelimb regeneration in normal and in hypophysectomized frogs. J. Embryol. Exp. Morph. **6**:247–261.

Scott, F. M., and J. E. Schuh. 1963. Intraspecific reaggregation in *Amaroecium constellatum* labeled with tritiated thymidine. Acta Embryol. Morphol. Exp. **6**:39–54.

Seilern-Aspang, F., and K. Kratochwil. 1965. Relation between regeneration and tumor growth. In V. Kiortsis and H. A. L. Trampusch, eds., Regeneration in Animals and Related Problems, pp. 452–473. North Holland Publishing Co., Amsterdam.

Simpson, S. B., Jr. 1961. Induction of limb regeneration in the lizard *Lygosoma laterale* by augmentation of the nerve supply. Proc. Soc. Exp. Biol. Med. **107**:108–111.

Singer, M. 1952. The influence of the nerve in regeneration of the amphibian extremity. Quart. Rev. Biol. **27**:119–200.

Singer, M. 1960. Nervous mechanisms in the regeneration of body parts in vertebrates. In D. Rudnick, ed., Developing Cell Systems and their Control, pp. 115–133. Ronald Press, New York.

Singer, M. 1965. A theory of the trophic nervous control of amphibian limb regeneration, including a reevaluation of quantitative nerve requirements. In V. Kiortsis and H. A. L. Trampusch, eds., Regeneration in Animals and Related Problems, pp. 20–32. North Holland Publishing Co., Amsterdam.

Sinnott, E. W. 1960. Plant Morphogenesis. McGraw-Hill Book Company, New York.

Smith, G. M. 1933. Freshwater Algae of the United States. McGraw-Hill Book Company, New York.

Stange, L. 1964. Regeneration in lower plants. Adv. Morph. **4**:111–153.

Steen, T. P. 1968. Stability of chondrocyte differentiation and contribution of muscle to cartilage during limb regeneration in the axolotl *(Siredon mexicanum)*. J. Exp. Zool. **167**:49–78.

Steeves, T. A. 1962. Morphogenesis in isolated fern leaves. In D. Rudnick, ed., Regeneration, pp. 177–151. Ronald Press, New York.

Steeves, T. A. 1966. On the determination of leaf primordia in ferns. In E. Cutter, ed., Trends in Plant Morphogenesis, pp. 200–219. Wiley, New York.

Steinberg, M. S. 1954. Studies on the mechanism of physiological dominance in *Tubularia*. J. Exp. Zool. **127**:1–26.

Steinberg, M. S. 1955. Cell movement, rate of regeneration, and the axial gradient in *Tubularia*. Biol. Bull. **108**:219–234.

Steinberg, S. N. 1963. The regeneration of whole polyps from ectodermal fragments of scyphistoma larvae of *Aurelia aurita*. Biol. Bull. **124**:337–343.

Steinecke, F. 1925. Über Polarität von Bryopsis. Bot. Arch. **12**:97–118.

Stephan-Dubois, F. 1965. Les neoblasts dans la régénération chez les Planaires. In V. Kiortsis and H. A. L. Trampusch, eds., Regeneration in Animals and Related Problems, pp. 112–130. North Holland Publishing Co., Amsterdam.

Steward, F. C. 1963. Growth and organization in free cell cultures. Meristems and Differentiation. Brookhaven Symp. Biol., pp. 73–88. Upton, N.Y.

Steward, F. C., A. E. Kent, and M. O. Mapes. 1966. The culture of free plant cells and its significance for embryology and morphogenesis. In A. Moscona and A. Monroy, eds., Current Topics in Developmental Biology, pp. 113–154. Academic Press, New York.

Steward, F. C., M. O. Mapes, and P. V. Ammirato. 1969. Growth and morphogenesis in tissue and free cell culture. In F. C. Steward, ed., Plant Physiology, Vol. V. Academic Press, New York.

Stocum, D. L. 1968. The urodele limb blastema: a self-differentiating system: I. Differentiation *in vitro*. II. Morphogenesis and differentiation of autografted whole and fractional blastemas. Devel. Biol. **18**:441–480.

Stone, L. S. 1950. Neural retina degeneration followed by regeneration from surviving retinal pigment cells in grafted adult salamander eyes. Anat. Rec. **106:**89–109.

Tardent, P. 1954. Axiale verteilungs-gradienten der Interstitiellen Zellen bei Hydra und Tubularia und ihre bedeutung für die Regeneration. Arch. Entw. Mech. **146:**593–649.

Tardent, P. 1963. Regeneration in the hydrozoa. Biol. Rev. Camb. **38:**293–333.

Tartar, V. 1941. Intracellular patterns: facts and principles concerning patterns exhibited in the morphogenesis and regeneration of ciliate protozoa. Growth **5**(suppl.): 21–40.

Tartar, V. 1961. The Biology of *Stentor.* Pergamon Press, New York.

Tartar, V. 1962. Morphogenesis in *Stentor.* Adv. Morph. **2:**1–26.

Tartar, V. 1967. Morphogenesis in protozoa. In T. T. Chen, ed., Research in Protozoology, pp. 1–116. Pergamon Press, New York.

Tassava, R. A., F. J. Chlapowski, and C. S. Thornton. 1968. Limb regeneration in *Ambystoma* larvae during and after treatment with adult pituitary hormones. J. Exp. Zool. **167:**157–164.

Teir, H., and T. Rytoma, eds. 1967. Control of Cellular Growth in Adult Organisms. Academic Press, New York.

Thornton, C. S. 1938. The histogenesis of muscle in the regenerating forelimbs of larval *Amblystoma punctatum.* J. Morph. **62:**17–47.

Thornton, C. S. 1968. Amphibian limb regeneration. Adv. Morph. **7:**205–249.

Trembley, A. 1744. Mémoires pour servir à l'histoire d'un genre de polypes d'eau douce, à bras en forme de cornes. Leide.

Urbani, E. 1965. Proteolytic enzymes in regeneration. In V. Kiortsis and H. A. L. Trampusch, eds., Regeneration in Animals and Related Problems, pp. 39–55. North Holland Publishing Co., Amsterdam.

Vöchting, H. 1877. Ueber Teilbarkeit im Pflanzenreich und die Wirkung innerer und aüsserer kräfte auf Organbildung an Pflanzentheilen. Arch. Physiol. (Pfluger) **15:**153–190.

Vöchting, H. 1892. Ueber Transplantation am Pflanzenkörper. Untersuchungen zur Physiologie und Pathologie. Laupp'schen Buchhandlung, Tubingen.

Ward, M. 1963. Developmental patterns of adventitious sporophytes in *Phlebodium aureum.* J. Linn. Soc. Lond. Bot. **58:**377–380.

Weiss, P. 1925. Unabhängigkeit der Extremität-enregeneration von Skolett (bei *Triton cristatus*). Arch. Ent. Mech. **104:**359–394.

Weisz, P. B. 1948. Time, polarity, size, and nuclear content in the regeneration of *Stentor coeruleus.* J. Exp. Zool. **107:**269–287.

Weisz, P. B. 1954. Morphogenesis in protozoa. Quart. Rev. Biol. **29:**207–229.

Wilson, H. V. 1907. On some phenomena of coalescence and regeneration in sponges. J. Exp. Zool. **5:**245–258.

Wolff, E. 1962. Recent researches on the regeneration of planaria. In D. Rudnick, ed., Regeneration, pp. 53–84. Ronald Press, New York.

Wolff, G. 1895. Entwickelungsphysiologische studien. Arch. Entw. Mech. **1:**380–390.

Woodruff, L. S., and A. L. Burnett. 1965. The origin of the blastemal cells in *Dugesia tigrina.* Exp. Cell Res. **38:**295–305.

Yamada, T. 1967. Cellular and subcellular events in Wolffian lens regeneration. In A. Moscona and A. Monroy, eds., Current Topics in Developmental Biology, Vol. 2, pp. 247–283. Academic Press, New York.

Yntema, C. L. 1959. Blastema formation in sparsely innervated and aneurogenic forelimbs of *Amblystoma.* J. Exp. Zool. **142:**423–439.

Ziller-Sengel, C. 1965. Inhibition de la régénération du pharynx chez les planaires. In V. Kiortsis and H. A. L. Trampusch, eds., Regeneration in Animals and Related Problems, pp. 193–201. North Holland Publishing Co., Amsterdam.

Zwilling, E. 1939. The effect of removal of the perisarc on regeneration in *Tubularia crocea.* Biol. Bull. **76:**90–103.

6 *Cellular Associations: Cellular Slime Molds*

In the development of multicellular organisms from unicellular and multicellular reproductive units, patterns of cell association emerge. As the number of cells increases, the cells specialize and arrange themselves in orderly patterns constituting tissues, organs, and whole organisms. The association of cells to form structural units may be the result of oriented cell divisions, especially in plants, in which the dividing cells remain together. Cellular associations may also be caused by movement and aggregation of cells.

In this chapter and the two following ones we shall examine the development of cellular associations and the capacity of experimentally dispersed tissue and organ cells to reaggregate and reconstitute the original tissue or organ architecture.

The cellular slime molds, Acrasieae, exhibit clearly and dramatically the transformation from a unicellular to a multicellular state. These organisms live on decaying organic matter in damp forests, but they can easily be brought into the laboratory and cultured on media that support the growth of bacteria, their natural food. Among a number of species, *Dictyostelium discoideum* has been studied intensively (Raper, 1935; Bonner, 1952, 1959, 1967; Shaffer, 1962; Gregg, 1964; Sussman, 1956, 1967; and others).

Life Cycle

At one stage in the life cycle of a slime mold the individual organism is a single ameboid cell about 3 to 5 mm in diameter. If a single ameba is isolated and placed on an appropriate culture medium, it feeds on bacteria and divides. Then, the daughter amebae divide until a large clonal population of amebae is formed. When the population reaches a critical density the amebae begin to aggregate by streaming toward one or more points. The amebae do not move individually toward the center of the aggregation but collect into moving streams, much like tributaries of a river. Eventually one or more aggregates of 1,000 to 2,000 amebae is formed (Figure 6-1). The aggregate, which is now a multicellular organism, is a sausage-shaped body 1 to 2 mm long (Figures 6-1, 6-2, and 6-3). In this aggregation, the amebae do not fuse. They retain their individuality but function collectively in a coordinated manner to move the entire mass over the culture substratum. The migrating collection of cells (called a pseudoplasmodium or slug) moves about for a length of time that depends on environmental conditions, then stops and raises itself into the air to form a fruiting body. This process is called culmination. The fruiting

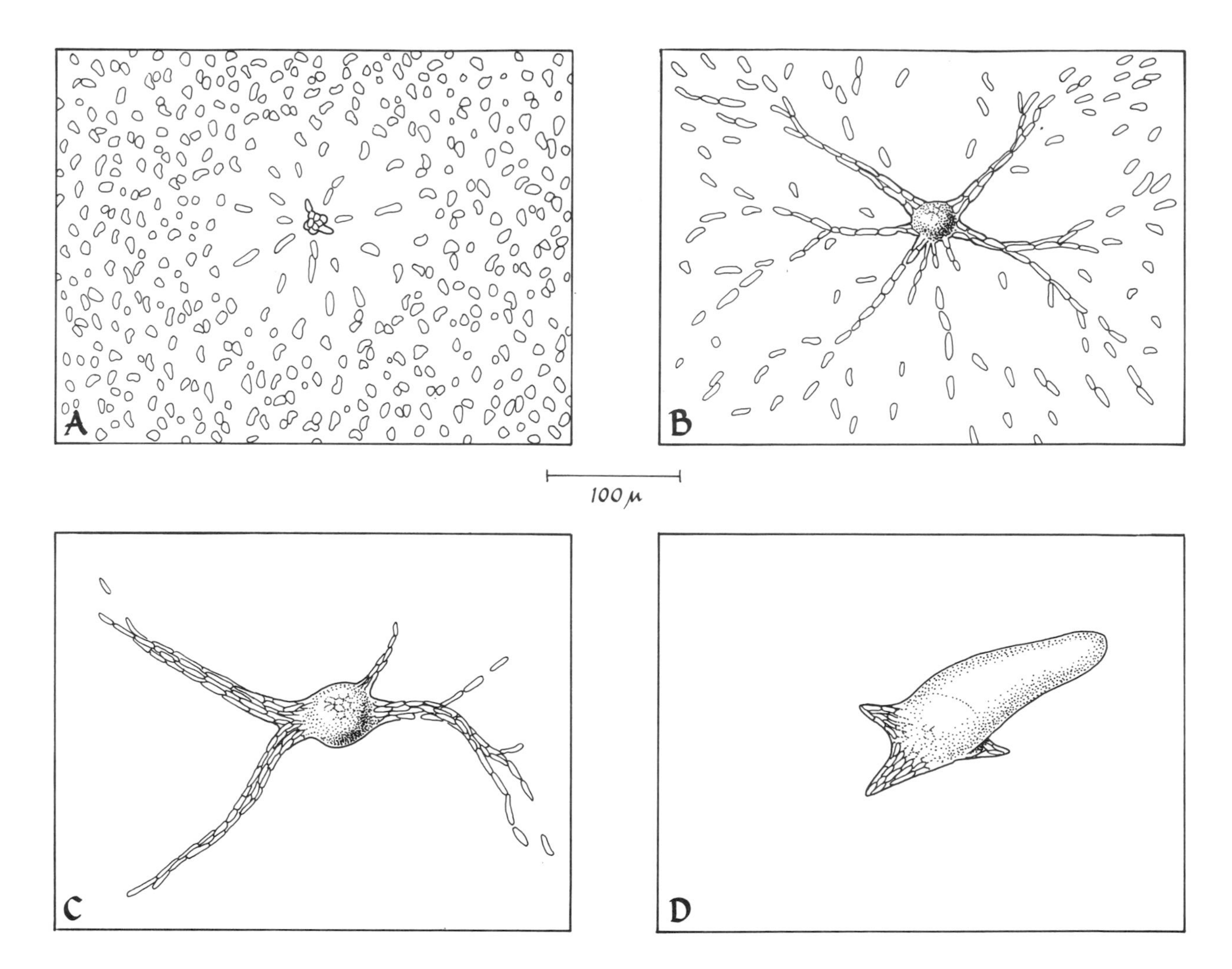

body consists of a ball of spore cells on top of a long thin stalk. The spore cells eventually drop from the fruiting body; under appropriate conditions of humidity, each spore germinates and releases a single ameba. The processes of cell division, aggregation, and so forth, then repeat. The life cycle is diagramed in Figures 6-2A and B. We now examine in more detail the stages of the life cycle.

Growth (Vegetative) Phase

During the growth stage the amebae divide about once every three hours. The amebae are equipotential in that any one can start a clonal population that goes through the complete life cycle (Bonner, 1967). Satisfactory growth of the amebae has been obtained on a medium containing salts, glucose, proteose peptone, and yeast extract (Sussman and Sussman, 1967), but maximal growth seems to depend on the presence of living or dead bacteria.

Just before the next stage, aggregation, a number of changes occur in the amebae. The staining properties (appearance of metachromatic granules in the cytoplasm) change, food vacuoles disappear, the cell and nucleus decrease in size, and the cell changes shape. Apparently induced by the depletion of the bacteria supply (starvation), these changes can be considered intracellular differentiation. Whether intercellular differentiation also occurs is debatable. Note that a food supply is not necessary for the developmental transformations of the life cycle, as a

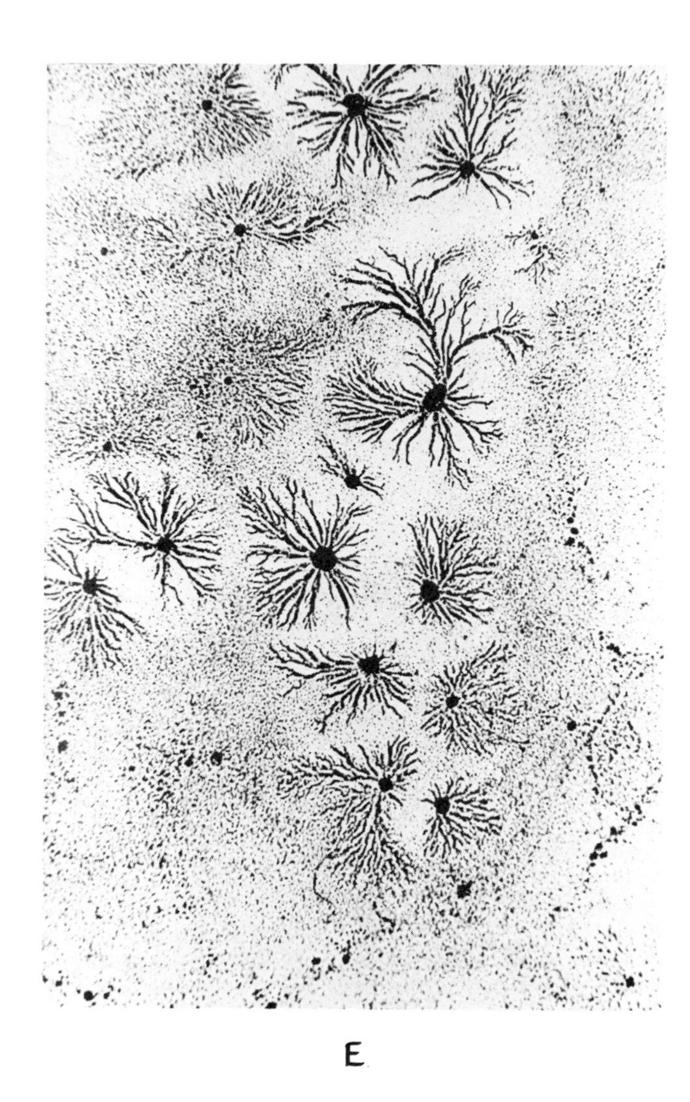

Figure 6-1. *Four stages in the aggregation of the slime mold* Dictyostelium *taking place under water on the bottom of a glass dish. A, the beginning of aggregation. B and C, successive stages of aggregation, showing thickening of the streams and enlargement of the center. D, the final pseudoplasmodium. From J. T. Bonner,* Morphogenesis: An Essay on Development *(copyright 1952 by Princeton University Press). Reprinted by permission of Princeton University Press. E, aggregation in* Dictyostelium discoideum *showing many aggregates. From J. T. Bonner,* The Cellular Slime Molds *(2nd rev. edition, copyright © 1967 by Princeton University Press). Photograph by K. B. Raper. Reprinted by permission of Princeton University Press.*

high concentration of amebae may aggregate and form fruiting bodies, and the spores may germinate and repeat the cycle on a nonnutrient agar medium.

Aggregation Phase

Many workers have noted that depletion of the bacterial food supply will stimulate the aggregation process (Bonner, 1967). Indeed, aggregation can be delayed indefinitely by replenishment of the food supply provided the population is not allowed to continue increasing in density. The most important stimulus for aggregation is apparently an increase in population density, as an increased food supply does not prevent aggregation in a dense population.

Aggregation patterns vary greatly depending on environmental conditions, such as whether the amebae are under water. Time-lapse motion pictures reveal that the amebae move toward a center of aggregation not smoothly but in pulses. Centers of aggregation may begin to form, disperse, and then form in another location. Polarity of the aggregating amebae is well known. If the center toward which an ameba is moving is cut out and placed behind the ameba, the ameba will not back up but will turn around and move toward the new location of the center. As noted above, the amebae move in streams toward the center (Figure 6-1). A single ameba will move into a stream of amebae, others will join the stream, and all will eventually reach the center of aggregation.

During the aggregation phase, few if any cell divisions occur. No clear evidence has been presented for any kind of sexual fusion process, although this has been claimed by some workers.

Migration Phase

In the center where the inflowing streams of amebae merge, a cone-shaped aggregate begins to form (Figure 6-2). When nearly all the amebae have moved into the center, the cone-shaped aggregate bends over and becomes the multicellular pseudoplasmodium—"pseudo" because cell fusion to form a syncytium (multinucleated protoplast) does not occur, as it does in the "true" slime molds. The pseudoplasmodium then leaves the site of its formation and glides over the surface of the substratum. The slug is encased in a slime sheath secreted by the

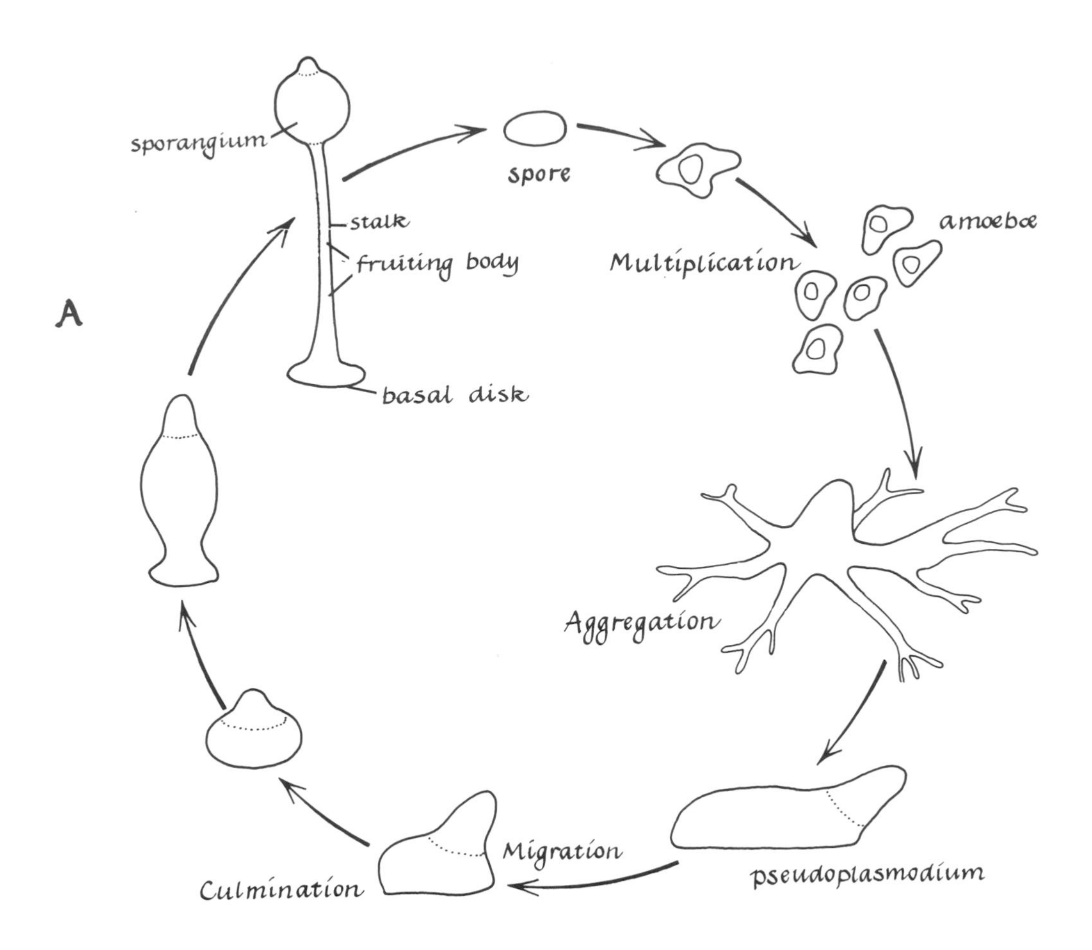

amebae, which collapses behind the slug as it moves (Figure 6-3).

The slug has definite anterior and posterior ends. Although all the amebae in a slug may have been derived from a single parent ameba (the slug is a clone of cells), the slug exhibits cell differentiation. Thus, the anterior cells (prospective stalk-forming cells of the fruiting body) are larger than the posterior cells (prospective spore-forming cells). When the slug is stained with the dye haematoxylon, the posterior cells stain more darkly than the anterior cells. Staining with the vital dyes (neutral red or nile blue sulfate) at late stages of migration results in a densely stained anterior part and a blanched posterior part. The line of demarcation between the two cell types is sharp. Staining with the periodic acid-Schiff reagent for polysaccharides shows that the division line between spore-forming and stalk-forming cells is sharply delineated. The stalk-forming cells also have greater alkaline phosphatase activity (Bonner, 1967) (Figure 6-4).

Culmination Phase

After a period of migration, the slug stops moving and the anterior end turns upward (Figure 6-2A and B). This is followed by an upward movement of the mass in which spore cells are pulled away from the substratum and lifted up by the formation and elongation of the stalk (Figure 6-2A and B)—a dramatic phenomenon when viewed in time-lapse motion pictures. The stalk-forming amebae secrete cellulose and eventually die. The only amebae in the slug to survive and reproduce the next generation are the spore-forming amebae.

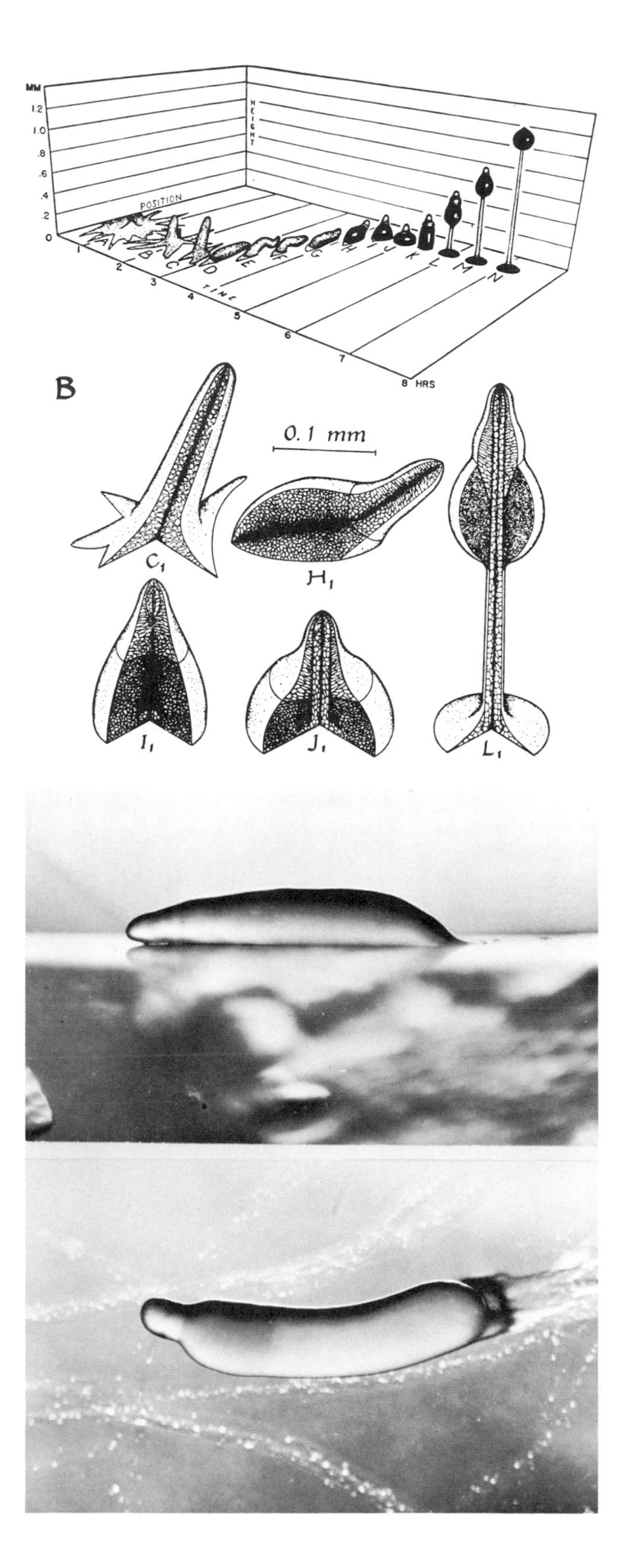

Figure 6-2. *A, life cycle of the cellular slime mold* Dictyostelium. *B, morphogenesis of* Dictyostelium *represented in a three-dimensional graph, with drawings of the cellular structure at stages corresponding to those indicated in the graph. A, from Kerr, N.,* Principles of Development, *1967, Dubuque, Iowa. Wm. C. Brown Company, Publishers. B, from J. T. Bonner,* Morphogenesis: An Essay on Development *(copyright 1952 by Princeton University Press). Reprinted by permission of Princeton University Press.*

Biochemical Events in the Cycle

Studies by Sussman and Sussman (1956–1969) have added significantly to our knowledge about the biochemical processes that underlie the slime mold life cycle. A few aspects of biosynthesis and their correspondence in time with morphological changes are shown in Figure 6-5.

An acid mucopolysaccharide (MP) begins synthesis in small quantity during migration, reaches a maximum value (about 1 to 2 percent of the total dry weight) at the end of culmination, and is found only in the spores of the fruiting body (White and Sussman, 1963). This polysaccharide is not present in the amebae in the vegetative stage or in the stalk and basal disc cells. The enzyme responsible for the incorporation of galactose into uridine diphosphate galactose polysaccharide transferase (UDPGT in Figure 6-5) is absent from vegetative amebae but appears during transformation of the aggregate into a slug, about one hour before mucopolysaccharide can be detected. The enzyme reaches peak activity at about 21 hours after the onset of aggregation and about three hours later is released from the cells into the medium. This release appears to be a mechanism for eliminating an unnecessary enzyme (Sussman, 1967). Both the accumulation and release of the enzyme are sensitive to the coincident inhibition of protein synthesis (by cycloheximide) and the prior inhibition of RNA synthesis (by actinomycin). Another enzyme, UDP glucose pyrophosphorylase

Figure 6-3. *Migrating pseudoplasmodium of* Dictyostelium discoideum. *Above, side view; below, view from above. From J. T. Bonner,* The Cellular Slime Molds *(2nd rev. edition, copyright © 1967 by Princeton University Press). Photographs by D. R. Francis. Reprinted by permission of Princeton University Press.*

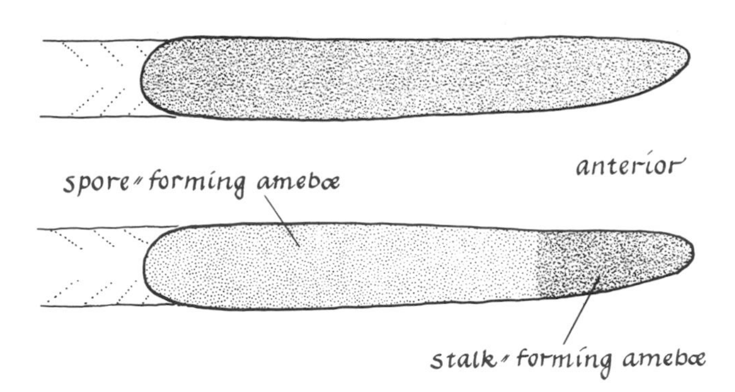

Figure 6-4. *Drawing showing how a vitally stained migrating cell mass of uniform coloration will change to one with a deeply stained tip and lighter posterior portion.*

Figure 6-5. *The program of biosynthesis in the life cycle of* Dictyostelium. *Modified from N. Kerr, 1967, and based on the data of Sussman and Sussman, 1956–1969.*

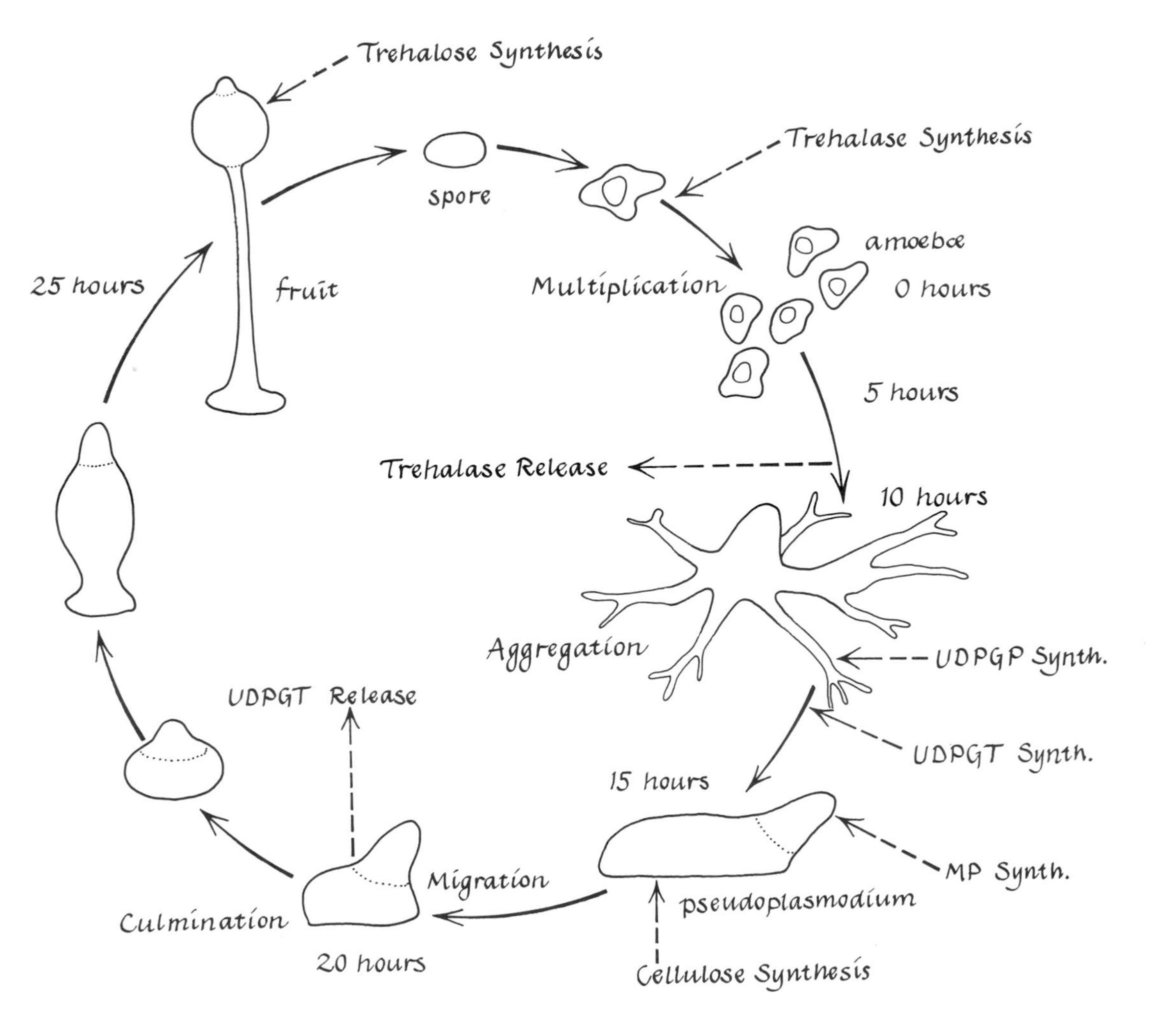

(UDPGP), which is involved in the synthesis of UDP-galactose, begins to be synthesized rapidly at about 12 hours, reaches a peak at 20 hours, and then declines (Sussman and Sussman, 1969). Finally, the sugar trehalose is synthesized in the spores. The enzyme trehalase, which degrades the trehalose, is not present in the spores but exhibits high activity in the amebae after spore germination (Ceccarini, 1967).

The correspondence of biochemical and developmental processes in the life cycle is summarized on an approximate time scale in Figure 6-5. In other studies, immunoelectrophoretic methods have been used to show the synthesis of new antigens and the disappearance of others that were originally present in the life cycle (Gregg, 1961, 1964).

The biochemical studies of specific syntheses reveal the programmatic action of the genome in directing the stages of development (Sussman and Sussman, 1969). Identification of gene products is one thing; discovery of the mechanism that regulates the sequential synthesis of these products is quite another. The problem for students of development is what controls the operation of the genetic program.

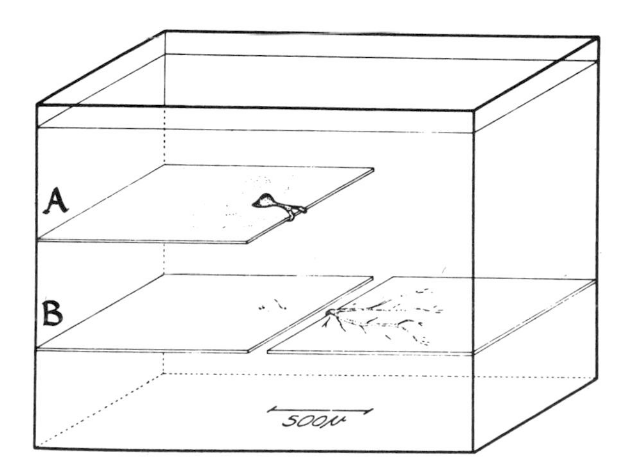

Figure 6-6. *Two experiments revealing chemotaxis in the aggregation of cellular slime mold amebae. A, the amebae are attracted from the lower surface around the edge of the coverslip to a center on the upper surface; B, the amebae are attracted to the center on the left hand coverslip across a gap. From J. T. Bonner,* Morphogenesis: An Essay on Development *(copyright 1952 by Princeton University Press). Reprinted by permission of Princeton University Press.*

Mechanism of Aggregation

In 1942 Runyon demonstrated that chemotaxis, the response of an organism to a gradient in concentration of a chemical substance, is part of the aggregation process. When amebae were placed on two sides of a cellophane membrane, aggregation patterns on the two sides were perfectly aligned one above the other, suggesting that substances produced by one group of aggregating amebae diffuse through the cellophane pores and stimulate aggregation in the other group. Bonner clearly demonstrated that the chemotactic agent was a freely diffusible substance. He named the substance acrasin (see also Shaffer, 1953). Two experiments demonstrating its existence are illustrated in Figure 6-6.

The chemical nature of acrasin was not known until 1968 when it was found to be 3′, 5′—adenosine monophosphate, commonly abbreviated cyclic AMP. This compound was shown to be degraded by the enzyme phosphodiesterase (Konijn et al., 1968). Cyclic AMP is also present in bacteria *(Escherichia coli)* and mammals. Apparently this is the reason why aggregating amebae will stream toward clumps of *E. coli* (Bonner, 1967).

One should not conclude that the aggregation process is simply the streaming of amebae up a single concentration gradient to the point of highest concentration (Shaffer, 1956; Bonner, 1967). Mechanical factors are also involved. Because acrasin causes sensitive amebae to become sticky and to secrete acrasin, they adhere to one another and streams of amebae are pulled to the center almost as a unit. The direction of movement within a stream appears to be mechanically controlled. Shaffer demonstrated that the amebae in a stream release as much acrasin as those in the center. Thus there is no evidence for an overall gradient of acrasin concentration. Instead there seems to be a relay system of small, temporary gradients, beginning at the aggregation center and spreading outward.

The problem of what initiates the aggregation process still exists. Workers in the field generally hold that some cells in the population of dividing amebae begin to secrete acrasin before others. Time-lapse movies of *Dictyostelium minutum* show that an aggregation center originates from a single cell called a "founder cell." The founder amebae differ from other amebae in the population by being nonmotile and having few pseudopodia (Gerisch, 1968). As cloning does not diminish the capacity for center formation, the founder cells must arise from non-founder cells. A founder cell is essentially a differentiated cell, but its differentiation is reversible; it can turn into a motile cell again. Early stages in center formation in *D. minutum* are illustrated in Figure 6-7, which is based on the studies of Gerisch.

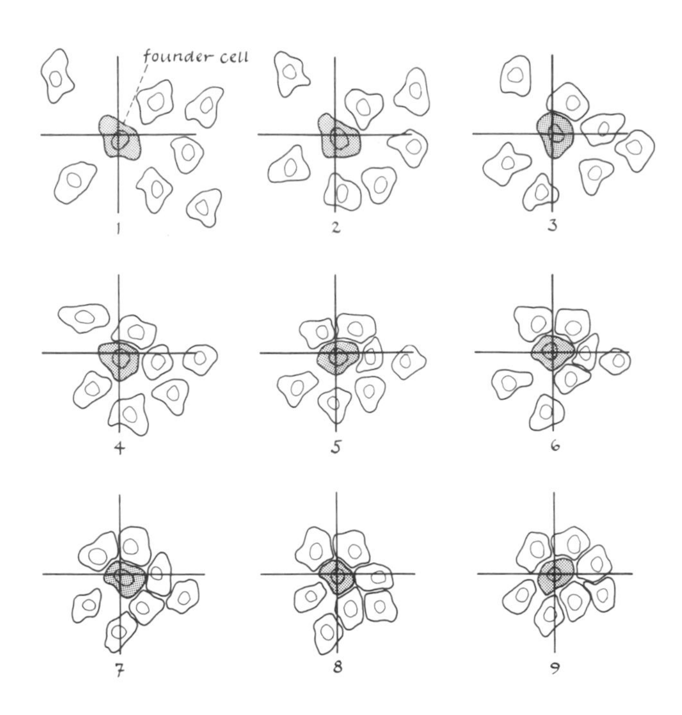

Figure 6-7. *Early stages in center formation in* D. minutum. *The founder cell is shown in black.*

In *D. discoideum*, similar founder cells presumably occupy the center. Because acrasin is rapidly destroyed by a specific phosphodiesterase produced by the amebae, a steep concentration gradient results. Amebae near the "center initiating cell" move toward and aggregate around it. These cells then begin to secrete acrasin, and more outlying cells orient in the new gradient and move toward the center. How cyclic AMP attracts the amebae is a problem yet to be solved.

Another problem is what controls the size and number of aggregates arising in a population of amebae. Studies of Sussman (1956) have shown that each species or mutant has a cell population density at which the number of fruiting bodies will be maximal. For example, in *D. discoideum* the density is 200 cells/mm^2 (Figure 6-8), in *D. purpureum* it is 100 cells/mm^2, and in the "bushy" mutant of *D. discoideum* it is 350 cells/mm^2. The number of amebae in the slug and fruiting body is relatively constant for each species: two thousand one hundred cells in *D. discoideum*, twenty-four cells in the "fruity" mutant of this species. However, when the total number of vegetative amebae in the population is less than the number that usually forms the slug but the population density is optimal, smaller slugs and fruiting bodies may form. Fractions of a migrating slug may form very small fruiting bodies (see Chapter 5). These results mean that a particular cell density is optimal. Below this optimum, there may be only enough cells to form a smaller than maximum number of centers, or they may be so dispersed that no centers form (density below 80 cells/mm^2). For reasons not clearly understood, crowding above the optimum inhibits center formation (Figure 6-8).

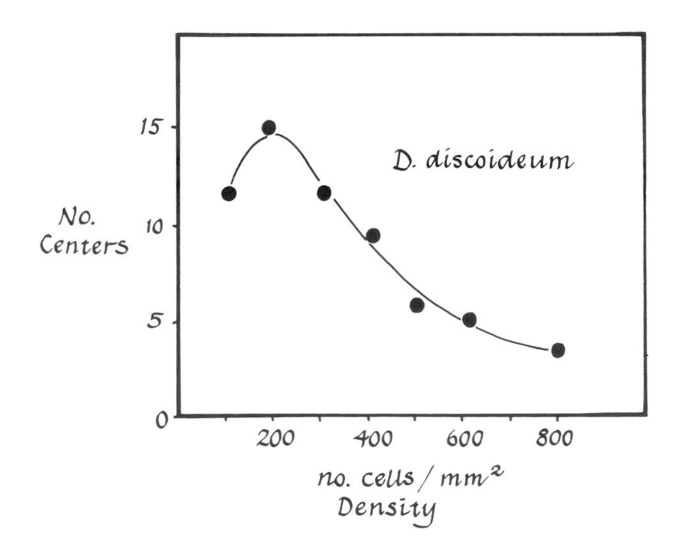

Figure 6-8. *The relationship between the population density of amebae and the number of centers formed.*

At one time it was thought that specific genetically different cells in the population (for example, one cell of every 2,100 cells in *D. discoideum*) initiated aggregation and that their number would control the number of aggregation centers or fruiting bodies. However, any ameba taken from the growth phase of the cycle can produce a cell population in which

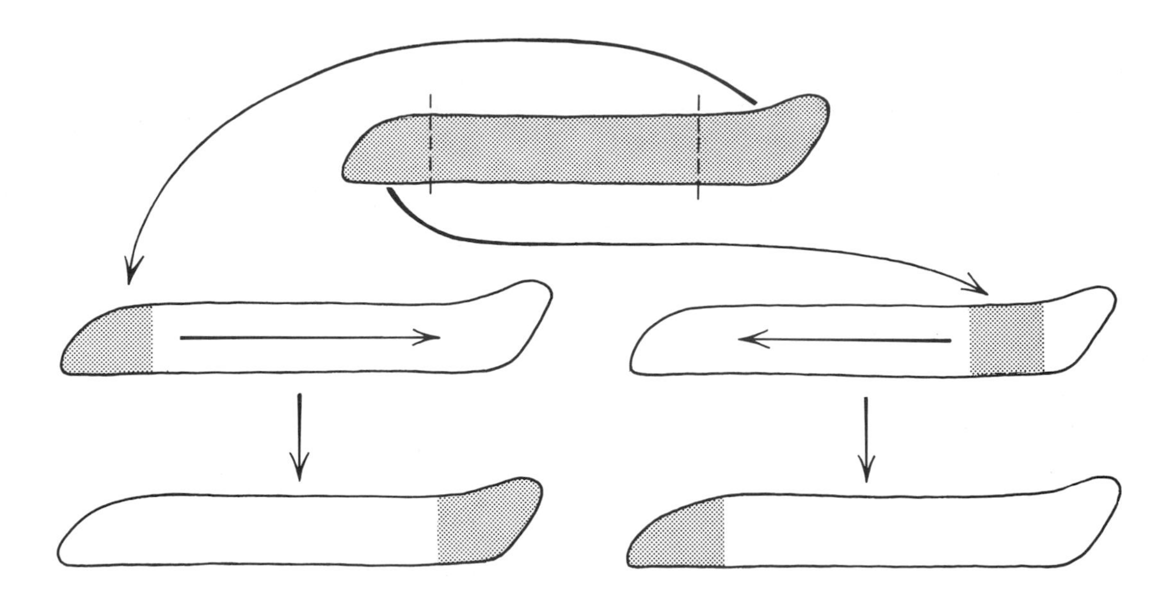

Figure 6-9. *Change in position of vitally stained amebae in the pseudoplasmodium of* Dictyostelium.

aggregation centers form. On the basis of more recent studies (Bonner, 1967), it appears that aggregation is not dependent on a definite ratio of initiator cells to other cells in the population. Thus, for drops of water containing 200 *D. discoideum* amebae, aggregation occurred in all drops. Even in drops containing only 100 cells, aggregation occurred in about 75 percent of the drops. Furthermore, there is evidence for a center inhibitor substance that in some unknown way mediates the spacing of the centers in a large population (Bonner, 1967).

Mechanism of Movement

The rate of movement of the slug is about 0.3 to 2.0 mm per hour at 20° C (Bonner, 1967). The mechanism of movement is not clearly understood, but it is known that the slime sheath does not move and that all the amebae are active. However, Bonner has shown that all the amebae do not move at the same rate. When nile blue vitally stained amebae from the anterior end of a slug were grafted to the posterior end of a colorless slug, the stained cells moved to the anterior end within two or three hours and remained there during prolonged migration (Figure 6-9).

Conversely, when a group of colored posterior cells was grafted to the anterior half of a colorless slug, these cells were slowly displaced to a posterior position, where they remained. According to Bonner, the experiment shows only that a relative change in position of the cells occurred, not that intrinsic differences in rate of movement were the primary cause of the shift. Still many studies have shown that amebae do not randomly distribute themselves during migration (Shaffer, 1964).

The anterior end of the slug produces more acrasin than the remainder. This was shown by separating the tip from the remainder and allowing both pieces to compete in a field of sensitive amebae. More amebae aggregated around the tip than around the remainder.

Mechanism of Culmination

The most important environmental factor initiating culmination is a decrease in humidity. Culmination, which is striking in *D. discoideum*, has been carefully studied by Bonner (1959, 1967). Initial stages in stalk formation are shown in Figure 6-10, and major features of the entire process are diagramed in Figure 6-11. After the slug stops moving (Figure 6-11A), the anteriormost stalk-forming cells invaginate as a group through the spore-forming cell mass until they reach the substratum where the basal disc cells are located (Figure 6-11B and C). Until the fruiting body is completely formed, no stalk-forming cell remains for long at the apex (the original anterior end of the slug). Stalk-forming cells are constantly moving to the top of the part of the stalk formed by invagination (Figure 6-11D). Here, the stalk-forming cells stop moving and pile on top of one another, contributing to elongation of the stalk (Figure 6-11E and F). The diagrams of Figure 6-11 are only approximate, because many features of the process are not yet clear.

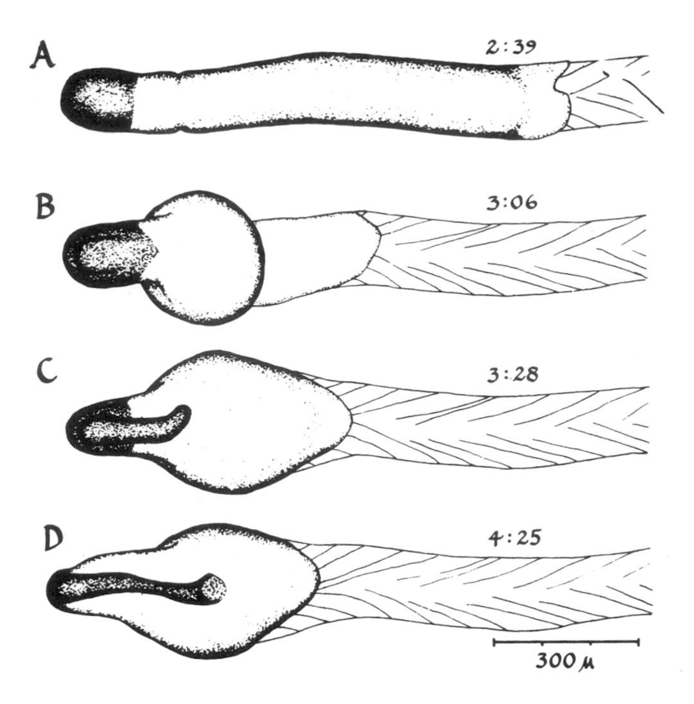

Figure 6-10. *Camera lucida drawings A-D (surface view), showing how the stalk is first formed at the tip and then pushed downward through the pre-spore cells during culmination. The dark tip was obtained by grafting the tip of a vitally stained slug to a decapitated unstained slug. From J. T. Bonner,* The Cellular Slime Molds *(2nd rev. edition, copyright © 1967 by Princeton University Press). Reprinted by permission of Princeton University Press.*

Mechanism of Differentiation

As we have seen, any ameba (except those trapped in the stalk) has the capacity at any stage in the life cycle to start a colony when isolated. Yet we have also noted that two readily distinguishable types of cells exist in the slug stage: stalk-forming and spore-forming cells. The origin of these two cell types, particularly in a clonal population of amebae, is the

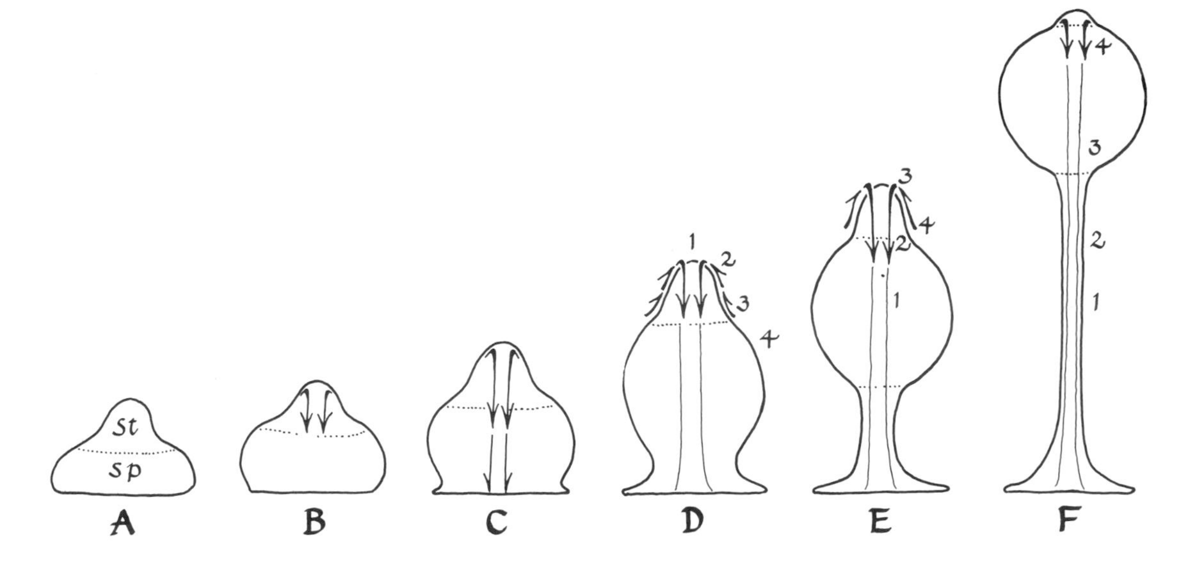

Figure 6-11. *Major features of the culmination process.*

problem confronting us. The fact that isolated spore-forming or stalk-forming cells can produce both spores and stalk cells attests to the equipotentiality of the cell types. Raper (1940, 1941) demonstrated that each type of cell can transform into the other.

When the migrating slug was cut into four parts, each part formed a fruiting body, with both stalk and spore cells even in pieces that consisted exclusively of spore-forming or stalk-forming cells (Figure 6-12). When migration of the parts continues for about 24 hours before culmination, the proportion of stalk and spore cells in the fruiting body approximates that of normal development. A period of time is required for the transformation process to restore the normal proportion of spore cells to stalk cells. (This result has been confirmed by Bonner and others.) The mechanism that regulates the proportion of spore cells to stalk cells is a basic problem in this area of developmental biology.

It is known that the pseudoplasmodium is polarized and that the anterior tip (highest level of acrasin production) dominates the remainder (Raper, 1940, 1941). This is illustrated in Figure 6-13. Two anterior tips grafted to the sides of a migrating slug pulled the slug into three pieces, each of which reorganized to form three migrating slugs. The basis of polarity in the slug is not known, but Shaffer (1964) has suggested that it may reside in the polarity of individual amebae or possibly in the distribution of relatively fast and slow moving amebae. He has presented evidence that the amebae line up end to end during aggregation, and he surmises that this relation is maintained in the slug. Thus, cutting a slug in half does not change the polar organization—it simply reduces the extent. He suggested that differences between anterior and posterior ends are quantitatively graded. If this is true, then the most anterior amebae in a posterior half would be different from the most posterior amebae in that half, just as anterior and posterior amebae differ in the whole slug. One may postulate an intercellular, anterior to posterior microenvironmental gradient, possibly in acrasin concentration. Further, the amebae would respond differently according to their position in this gradient, becoming either spore-forming or stalk-forming cells when their thresholds of response are reached. However, polarity and the final differentiation of stalk-forming and spore-forming amebae in the slug and fruiting body could result from the fact that, during culmination, the anterior end of the slug is exposed first to the liquid-air interface. For example, when agglomerates of amebae in liquid culture are

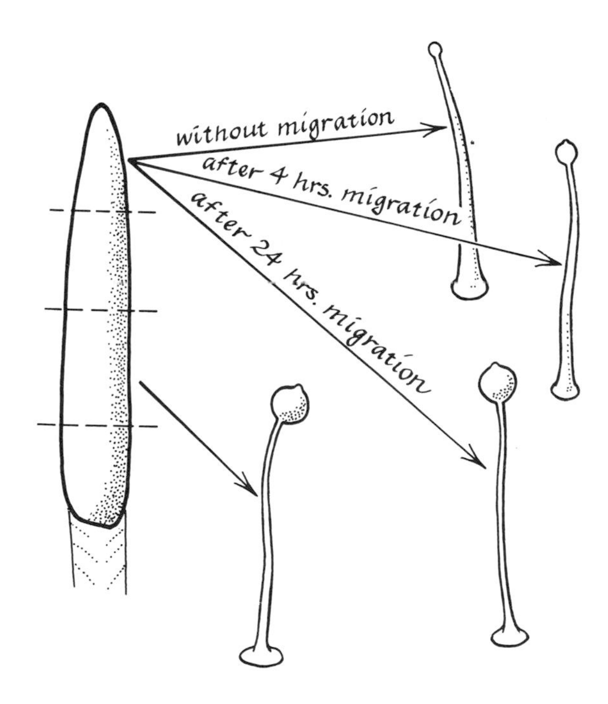

Figure 6-12. *Transformation of pre-spore into pre-stalk forming amebae and vice versa in isolated pieces of a pseudopodium.*

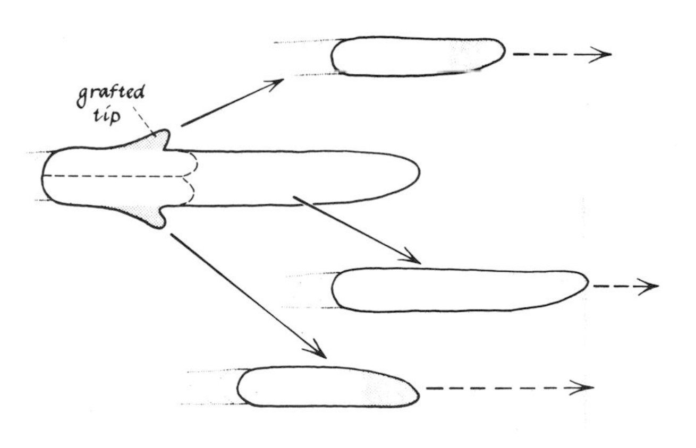

Figure 6-13. *Experiment illustrating dominance of the anterior tip of the slug over the remainder. Two tips grafted to the sides of a slug pulled the slug into three pieces, each of which formed a slug.*

inserted into a water film between two interfaces, double slugs with opposed polarities can develop (Gerisch, 1968). This is shown in Figure 6-14.

Bonner (1967) has presented evidence for the redistribution of amebae between aggregation and migration. This observation supports the idea that the two cell types develop during the migration phase, not before aggregation, but the problem is by no means solved.

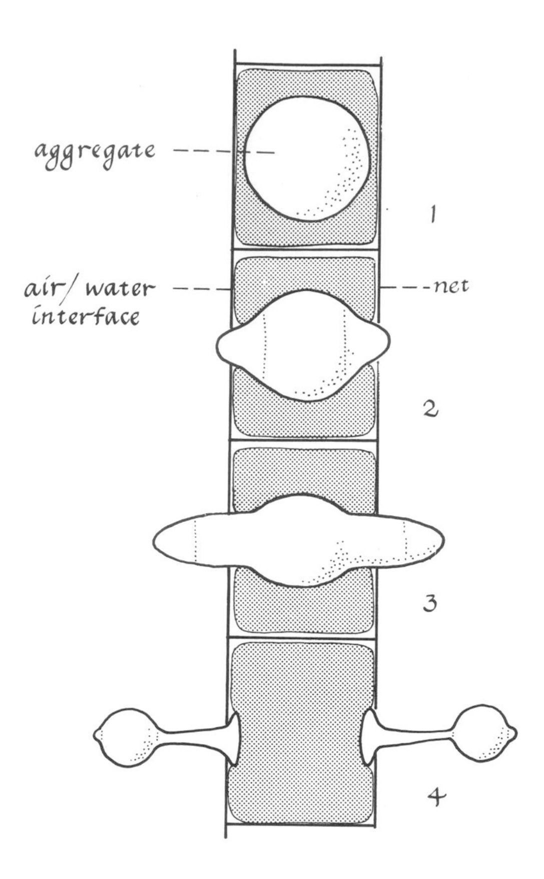

Figure 6-14. *Influence of the liquid-air interface on polarity and differentiation in an aggregate of amebae. 1-4, stages in culmination.*

Sorting Out of Species and Strains

When spores or amebae of different species are mixed, they sort themselves out to form separate fruiting bodies. Two species may form a common aggregate but then form separate fruiting bodies (Raper and Thom, 1941). Thorough mixing of the migrating cell masses of two species results in formation of a single fruiting body in which spores of the two species remain genetically stable.

When two strains of *D. mucoroides* were mixed at the aggregation or migration stage, variable results were obtained, from the formation of separate fruiting bodies to the formation of a single fruiting body with separate spore masses.

In an experiment reported by Kahn (1964), a mutant of *D. purpureum* that was incapable of aggregation was mixed with wild type amebae. The aggregateless cells entered the fruiting body even when in great excess of the wild type. Remarkably, all the spores produced wild type progeny. Presumably, the wild type cells "cured" the mutant cells of their inability to aggregate, a remarkable result that must be confirmed.

In general, mixed cells of different genotypes in the cellular slime molds sort themselves out. In Chapter 8, we shall encounter a similar phenomenon when cells of different species in sponges are mixed, but a very different phenomenon when dispersed tissue cells of vertebrates are mixed.

Conclusion

Attainment of the multicellular state by cellular aggregation, so clearly illustrated by cellular slime molds, is not an exclusive property of these organisms. However, the method of becoming multicellular in these organisms has been most extensively studied. These studies have raised many questions, such as: why does aggregation occur in the first place? Some type of synergism may be involved that, as Bonner has suggested, takes the place of a sexual process yet to be found in the cellular slime molds. What is the mechanism of regulation? In the slime molds, at least, the capacity of a part to reconstitute the whole is not complicated by cell division. The restoration of proportion between spore-forming and stalk-forming cells in fragments of a pseudoplasmodium means changes in the fates of the cells. These changes are presumably controlled by supracellular guidelines established by prior cellular activity. The nature of these guidelines is not known, but one thing seems clear—the guidelines are progressively established during the aggregation process and are capable of being reconstituted in fragments.

References

Bonner, J. T. 1947. Evidence for the formation of cell aggregates by chemotaxis in the development of the slime mold *Dictyostelium discoideum.* J. Exp. Zool. **106:**1–26.

Bonner, J. T. 1952. The pattern of differentiation in amoeboid slime molds. Am. Nat. **86:**79–89.

Bonner, J. T. 1959. Differentiation in social amoebae. In From Cell to Organism: Readings from Scientific American, pp. 7–14. W. H. Freeman, San Francisco.

Bonner, J. T. 1967. The Cellular Slime Molds, 2nd ed. Princeton University Press, Princeton, N.J.

Ceccarini, C. 1967. The biochemical relationship between trehalase and trehalose during growth and differentiation in *Dictyostelium discoideum.* Bioch. Biophys. Acta. **148:**114–124.

Gerisch, G. 1968. Cell aggregation and differentiation in *Dictyostelium.* In A. Moscona and A. Monroy, eds., Current Topics in Developmental Biology, pp. 157–197. Academic Press, New York.

Gregg, J. H. 1961. An immunoelectrophoretic study of the slime mold *Dictyostelium discoideum.* Devel. Biol. **3:**757–766.

Gregg, J. H. 1964. Developmental processes in cellular slime molds. Physiol. Revs. Camb. **44:**631–656.

Kahn, A. J. 1964. Some aspects of cell interaction in the development of the slime mold *Dictyostelium purpureum.* Devel. Biol. **9:**1–19.

Konijn, T. M., D. S. Barkley, Y. Y. Chang, and J. T. Bonner. 1968. Cyclic AMP: a naturally occurring acrasin in the cellular slime molds. Am. Nat. **102:**225–233.

Raper, K. B. 1935. *Dictyostelium discoideum,* a new species of slime mold from decaying forest leaves. J. Agric. Res. **50:**135–147.

Raper, K. B. 1940. Pseudoplasmodium formation and organization in *Dictyostelium discoideum.* J. Elisha Mitchell Sci. Soc. **56:**241–282.

Raper, K. B. 1941. Developmental patterns in simple slime molds. Growth **5**(Suppl.): 41–76.

Raper, K. B., and C. Thom. 1941. Interspecific mixtures in the Dictyosteliaceae. Am. J. Bot. **28:**69–78.

Runyon, E. H. 1942. Aggregation of separate cells of *Dictyostelium* to form a multicellular body. Collecting Net **17:**88.

Shaffer, B. M. 1953. Aggregation in cellular slime molds: in vitro isolation of acrasin. Nature **171:**975.

Shaffer, B. M. 1956. Acrasin, the chemotactic agent in cellular slime molds. J. Exp. Biol. **33:**645–657.

Shaffer, B. M. 1962. The acrasina. Adv. Morph. **2:**109–182.

Shaffer, B. M. 1964. Intracellular movement and locomotion of cellular slime mold amebae. In R. D. Allen and N. Kamiya, eds., Primitive Motile Systems in Cell Biology, pp. 387–405. Academic Press, New York.

Sussman, M. 1956. The biology of the cellular slime molds. Ann. Rev. Microbiol. **10:**21–50.

Sussman, M. 1967. Evidence for temporal and quantitative control of genetic transcription and translation during slime mold development. Fed. Proc. **26:**77–83.

Sussman, M., and R. Sussman. 1969. Patterns of RNA synthesis and of enzyme accumulation and disappearance during cellular slime mold cytodifferentiation. Symp. Soc. Gen. Microbiol. **19:**403–435.

Sussman, R., and M. Sussman. 1967. Cultivation of *Dictyostelium discoideum* in axenic medium. Biochem. Biophys. Res. Comm. **29:**53–55.

White, G. J., and M. Sussman. 1963. Polysaccharides involved in slime mold development: II. Water-soluble acid mucopolysaccharide(s). Biochem. Biophys. Acta. **74:**179–187.

7 Cellular Associations: Colonial Organisms

In studying the cellular slime molds, we saw how an aggregation of individual amebae became a multicellular organism. A far more common way of becoming multicellular is by repetitive cell division. A survey of the increasingly complex cellular associations, from the simplest nonsymmetrical collections of algae and bacterial cells to the more orderly and symmetrical associations in colonial organisms, illustrates many basic problems and principles of reproductive development in multicellular organisms.

The transformation from a unicellular to a multicellular state involves far more than controlled increase in size and cell number. Basic problems to be solved, besides discovering the mechanism that controls the size of the organism and its parts, are what controls shape (symmetry) and, in many cases, what controls cellular specialization. In this chapter we shall survey some patterns of cell association and specialization that arise in the development of relatively simple organisms. Cellular associations, particularly in plants but also in animals, play a very important role in development, no less important than the specialization of individual cells. Wardlaw (1965) has called attention to the similarity in structure of alga cells and higher plant cells of protozoa and animal tissue cells. The cells of multicellular organisms are not more complex than unicellular organisms. The complexity of the organism arises from cellular associations.

We shall first examine a few examples of cellular association resulting from cell division in colonial algae, fungi, and colonial protozoa. Following this, we shall look at some remarkable aggregations in bacteria and algae, and finally we shall examine associations of individual organisms so closely unified as to form a "superorganism," more than a colony of organisms.

Cell Associations Resulting from Cell Division

Cell Colonies in Algae

In many colonial organisms of the group of green algae, the unit of structure is a flagellated cell such as *Chlamydomonas* (Brown, 1935). This basic type of cell is illustrated in Figure 7-1. *Chlamydomonas* has two flagella, a cup-shaped chloroplast, a reddish stigma (eye spot), and a pyrenoid (a starch- and protein-containing organelle).

Irregular Associations

During the life cycle of some species of *Chlamydomonas,* the unicellular organism secretes a jelly-like coating and, after resorption of its flagella, divides into a variable number of cells that remain within the jelly coat (Figure 7-2). The cells are arranged in several subgroups of 2 to 8 cells, in no regular arrangement. Another organism, *Palmella* (closely related to *Chlamydomonas*), does not have a flagellated stage and exists permanently in the jelly. This primitive type of cell association illustrates the important integrative role of cell products (jelly, in this case) in holding otherwise independent cells together. In some algae related to *Palmella,* the colony has a specific shape although the cells are irregularly arranged within it. The jelly determines the shape of the colony. (The role of cell products in the development of shape is so important that we shall devote Chapter 23 to it.) Whether there is any integration of the cells beyond this mechanical role of the jelly is not known.

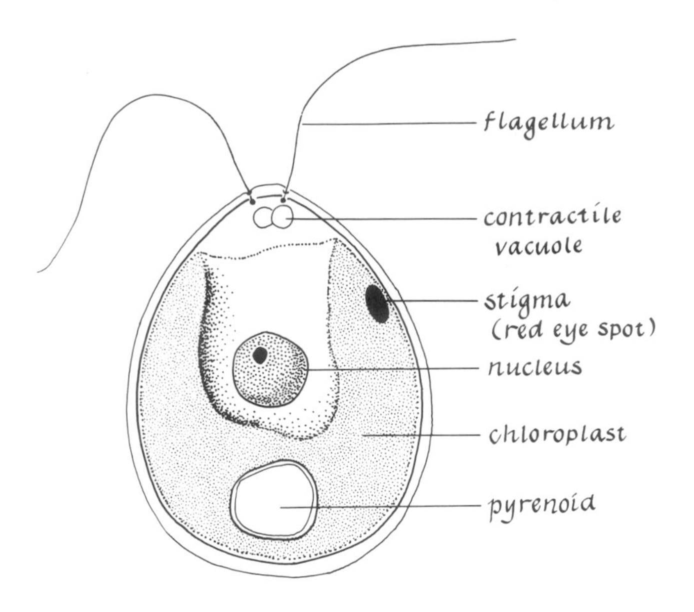

Figure 7-1. *The basic structure of* Chlamydomonas.

Radially Symmetrical Associations

In *Volvocales* (colonial green algae related to *Volvox*), the various species can be arranged in a series of progressively more complex associations of *Chlamydomonas*-like cells. The associations are all radially symmetrical in two or three planes. The series illustrates not only an increase in the number of cells in the colony but an increasing "division of labor" among the cells. In this arrangement, we can see the beginnings of cell specialization. In these colonies, the cells are so closely integrated that the group functions as a single individual. They have developed from a single reproductive cell and thus present us with problems of the mechanisms of integration perhaps similar to those operating in multicellular plants and animals.

Gonium. Perhaps the simplest radial association of cells is *Gonium sociale*—four cells held together by jellylike cell products. In other species such as *Gonium pectorale,* the assembly consists of 16 cells arranged in an almost square, flat plate (Harper, 1912; Brown, 1935; Fritsch, 1935; Scagel et al., 1966). As in *Gonium sociale,* the cells are held together by a gelatinous cell product (Figure 7-3).

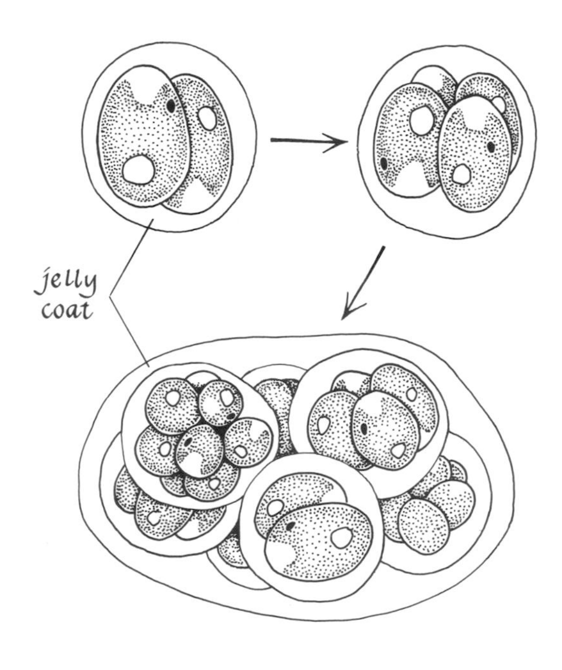

Figure 7-2. *Asexual reproduction in* Chlamydomonas. *Formation of the palmella stage.*

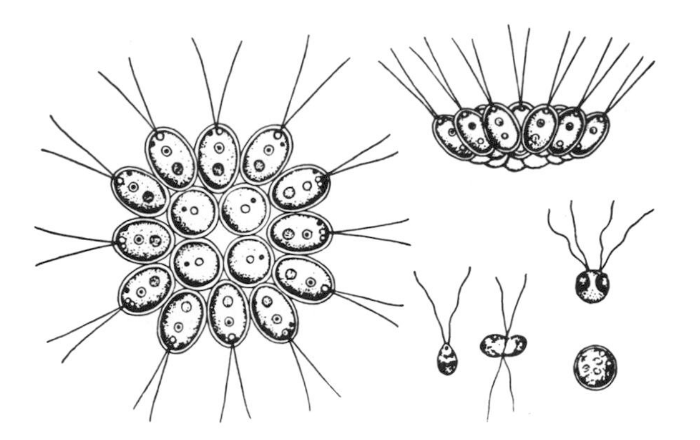

Figure 7-3. *Top and side view of a colony, gamete, and fusion of gametes to form a zygote in* Gonium pectorale. *From Brown:* The Plant Kingdom, *1935. Reproduced by permission of Mrs. Brown.*

Figure 7-4. *Cell lineage map of* Gonium pectorale.

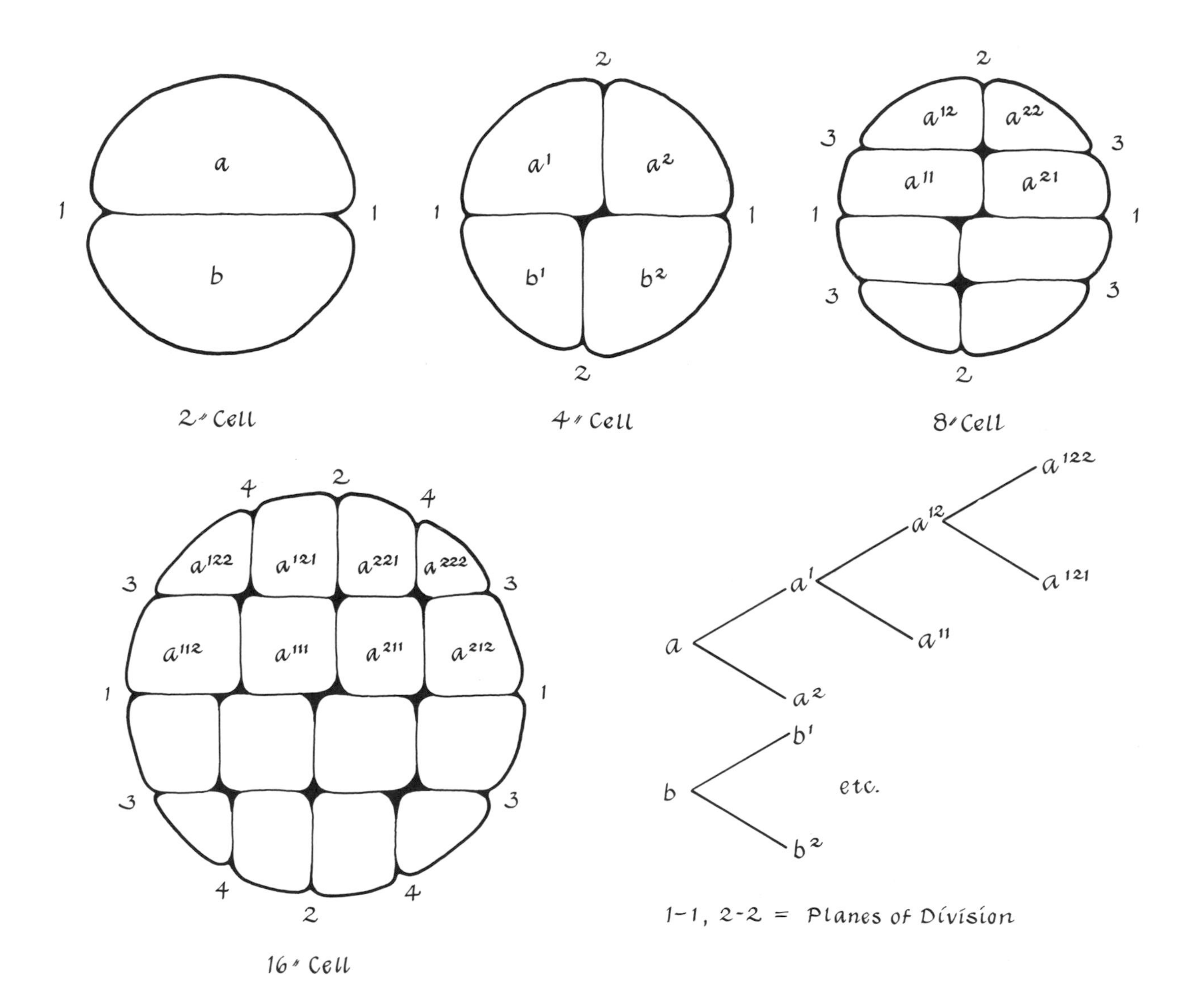

All cells of *Gonium* are alike in that each is capable of forming a new colony when isolated (Chapter 5). Under specific conditions, the cells escape from the colony and function as isogametes (morphologically similar gametes that fuse sexually). The orderly arrangement of cells in *G. pectorale* (4 central and 12 peripheral cells) results from an orderly program of changes in the orientation of the mitotic spindles and thus in the planes of cell division (Hartman, 1924; Gerisch, 1959). The pattern of cell division is so regular that a map of cell lineage showing the origin of cells at progressive stages of the colony (Figure 7-4), can be constructed. The notation generally used to indicate the lineage of a cell is shown in the figure. The colony-forming cell divides into two cells, designated *a* and *b*. Both *a* and *b* divide again to give a^1, a^2, b^1, and b^2 at the four-cell stage. To form the eight-cell stage, a^1 divides to form a^{11} and a^{12}, and so forth. A cell lineage map demonstrates the principle of orderly development, the 1 to 1 relationship between parts at an early stage and parts at a later stage of development. As we shall see again and again, programs of normal development are clearly deterministic. However, programs may be modified experimentally and perhaps changed. Note in Figure 7-4 that the pattern of cell division planes (1-1 to 4-4) is not radially symmetrical. The spindles are all in one plane, but they alternate in direction in a definite pattern: the second plane of division is at right angles to the first, the third is parallel to the first, and the fourth is parallel to the second. The mechanisms controlling this orderly pattern of division are unknown. Similarly, nothing is known about the mechanisms controlling the orderly pattern of cleavage in animal eggs. The final symmetry of the adult colony is eventually achieved by a clockwise rotation of the cells. We know nothing about the underlying mechanism of these cell movements.

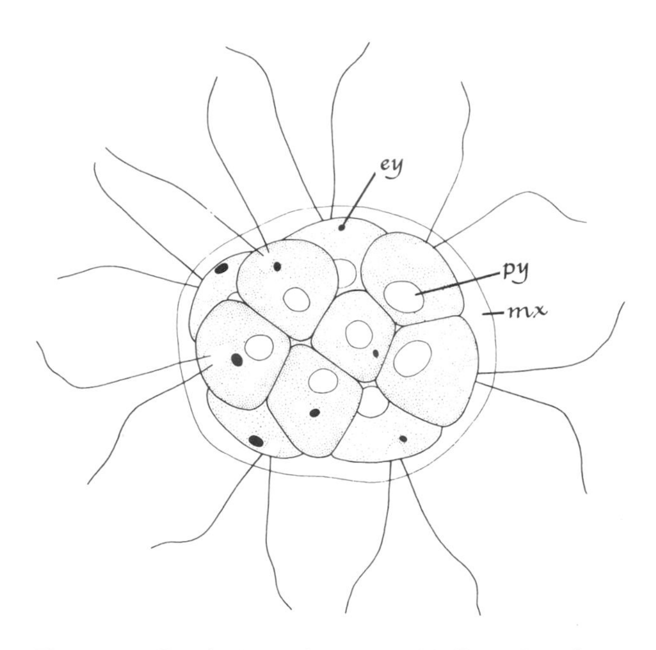

Figure 7-5. Pandorina *colony,* × *2,000. From Scagel et al.,1966. Courtesy of Wadsworth Publishing Company, Inc.*

Pandorina. Unlike *Gonium,* this colony of 16 cells (rarely 8 or 32) is radially symmetrical in three planes, the cells being packed closely together into a polarized sphere (Figure 7-5). Some cells of the colony function as motile heterogametes (large and small gametes), but large gametes don't always fuse with small gametes and vice versa (Scagel et al., 1966). Formation of such gametes is a primitive type of *sexual differentiation.* Why some cells function as gametes and others do not is unknown. However, all cells of the colony are able to form a new colony by asexual reproduction.

Eudorina. Each of the 32 cells (16 or 64 in some species) of this colony may be capable of forming a new colony, but in some species the four anterior cells seem to be purely vegetative (Fritsch, 1952). The cells of the colony are arranged to form a hollow sphere and are held together by a gelatinous matrix (Figure 7-6). Heterogametes are formed by separate colonies that can be considered as male and female (Brown, 1935). It also has been reported that some species of *Eudorina* are *oogamous,* with a small yellow-green sperm and a large spherical green egg (Scagel et al., 1966). Why some colonies form the smaller (male) gametes and others the larger (female) gametes is not known.

Pleodorina. The development of this radially symmetrical and polarized colony of 32 to 128 cells (depending on the species) has been studied by Gerisch (1959). The cells are held together by a gelatinous matrix that prevents them from touching one another.

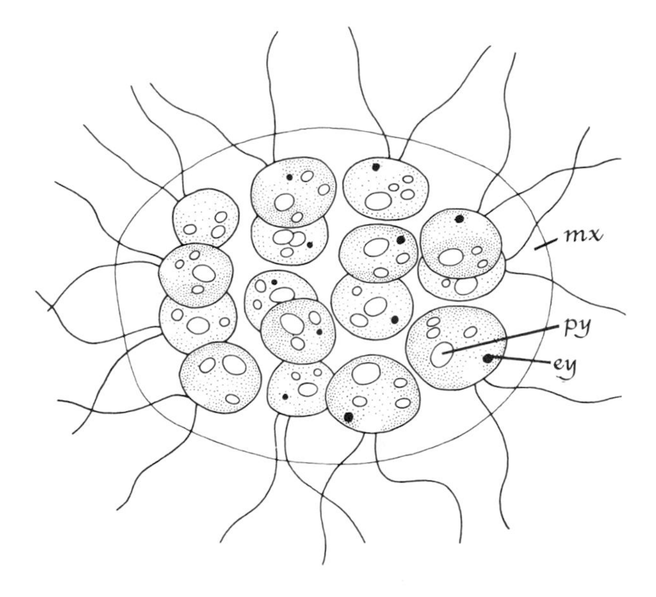

Figure 7-6. *Colony of* Eudorina, *× 1,075. Scagel et al., 1966. Courtesy of Wadsworth Publishing Company, Inc.*

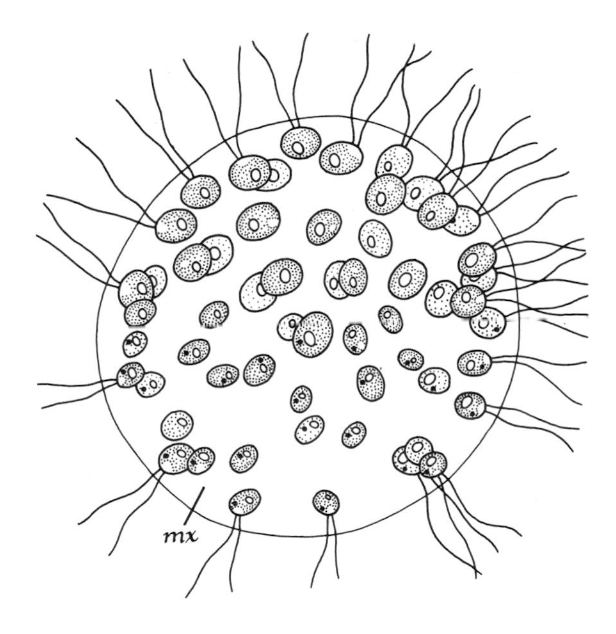

Figure 7-7. *Morphological diversity in a colony of* Pleodorina californica. *Large arrow indicates the polarity of the colony. Small arrows at bottom indicate the direction of the colony's rotation. Scagel et al., 1966. Courtesy of Wadsworth Publishing Company, Inc.*

Unlike the colonial organisms previously discussed, with possible exceptions in *Eudorina,* some cells differentiate as somatic cells (not capable of reproduction), others as generative cells. In one species, *P. californica,* the somatic cells are located at the anterior pole as the colony swims, and the generative cells constitute the remainder of the colony (Figure 7-7). The generative cells are larger and more numerous than the somatic cells, which possess a stigma (Figure 7-9). The ratio of somatic to generative cells is about 3 to 5 according to Gerisch (1959). As the colony swims, it rotates in a clockwise direction around the polar axis.

The pattern of cell division, as in *Gonium,* is regular, and Gerisch has constructed a cell lineage map. Development through the sixteen-cell stage is illustrated in Figure 7-8. The division pattern is a clockwise spiral, somewhat like the spiral cleavage of annelid and mollusc eggs. The prospective somatic cells are centrally located at the 16-cell stage; the generative cells are peripheral. Cell displacement then occurs, so that the somatic cells move to one pole and the generative to the other.

When generative cells are experimentally isolated, they are capable of producing new colonies asexually by division. The somatic cells produce incomplete, atypical colonies (Gerisch, 1959).

The mature colony is apparently integrated by cytoplasmic strands connecting the cells with one another (Fritsch, 1935). According to Gerisch, the cells are arranged in an orderly pattern with respect to the plane of flagella insertion, the plane of flagella vibration, and the plane of cell division (Figure 7-9).

After experimental disruption of the 8-cell stage (Figure 7-8) with a fine glass needle so that some prospective somatic cells lie at the edges of the dividing mass, the cell pattern in the completed colony is very abnormal (Gerisch, 1959). Such a colony may swim in circles or may have somatic cells at the generative pole or too few somatic cells. This system thus has distinct *mosaic* properties; that is, it is not capable of repairing gross defects and is similar to cleaving eggs of mulluscs, annelids, and tunicates (see Chapters 12 and 15).

Volvox. Among radially symmetrical cell associations in algae, the maximum in size and degree of cell specialization is attained in *Volvox.* The colony is a sphere growing to about 0.5 mm in diameter.

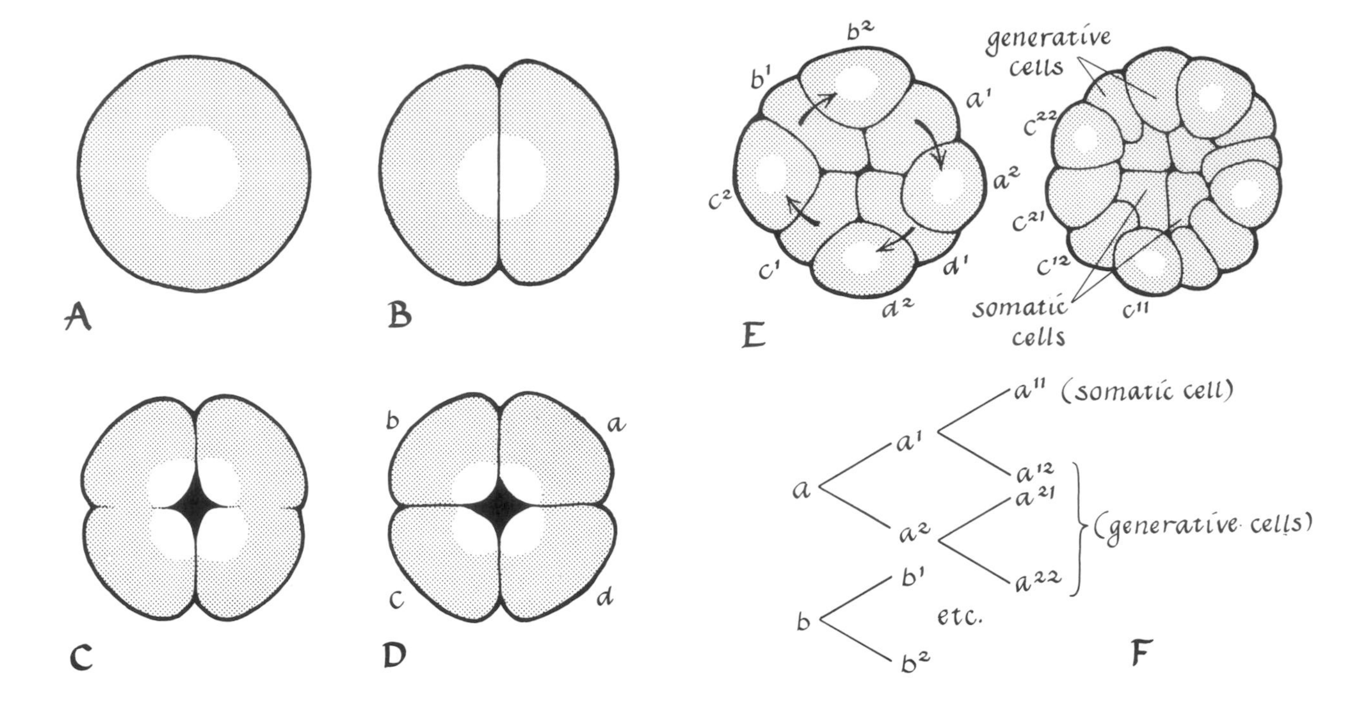

Figure 7-8. *Development through the 16-cell stage of* Gonium. *A-D, first and second divisions; E, later divisions resulting in formation of the 16-cell stage; F, diagram of lineage of somatic and generative cells.*

The wall of the sphere is one cell thick, built up by a pattern of cell division (Pocock, 1933) in which the spindle axes are mostly parallel to the surface of the developing colony (Figure 7-10).

In different species, the gonidium (a relatively large colony-forming cell) divides a definite number of times to form the total cell number characteristic of the species. The gonidium seems programmed to divide a limited number of times, and this potentiality is inherited. The same phenomenon is found in many mosaic patterns of animal egg development. Is this block to continued cell division also programmed in the genome or is it induced by external influences? Present evidence implies that the large gonidial cell is specialized (like eggs and buds) for repetitive but limited growth.

The sphere is polarized, having an anterior half of vegetative cells and a posterior half in which many of the 500 to 50,000 cells of the colony (in different species) are reproductive. Some reproductive cells are specialized as colony mother cells capable, by repeated division, of producing daughter colonies. Other reproductive cells are specialized as male initial cells (producing sperm) or larger female initial cells (developing into eggs). The colony swims in a polarized direction, anterior end forward, and ro-

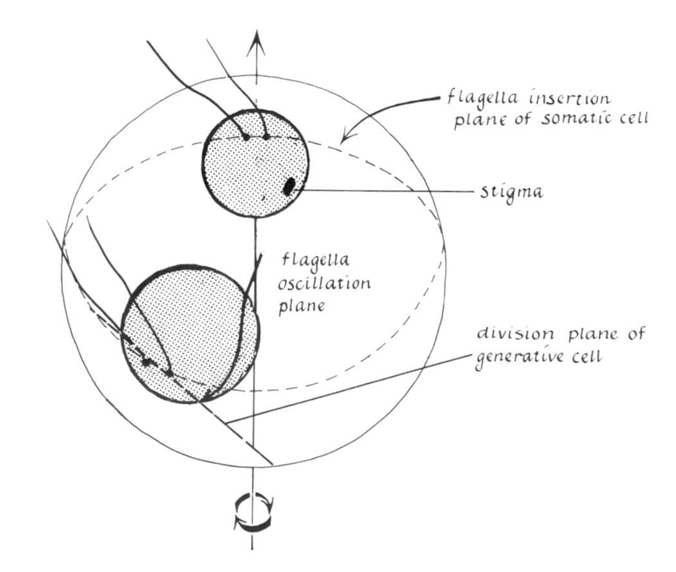

Figure 7-9. *Diagram of the orderly arrangement of cells in* Pleodorina *with respect to the planes of flagella insertion, flagella vibration, and cell division.*

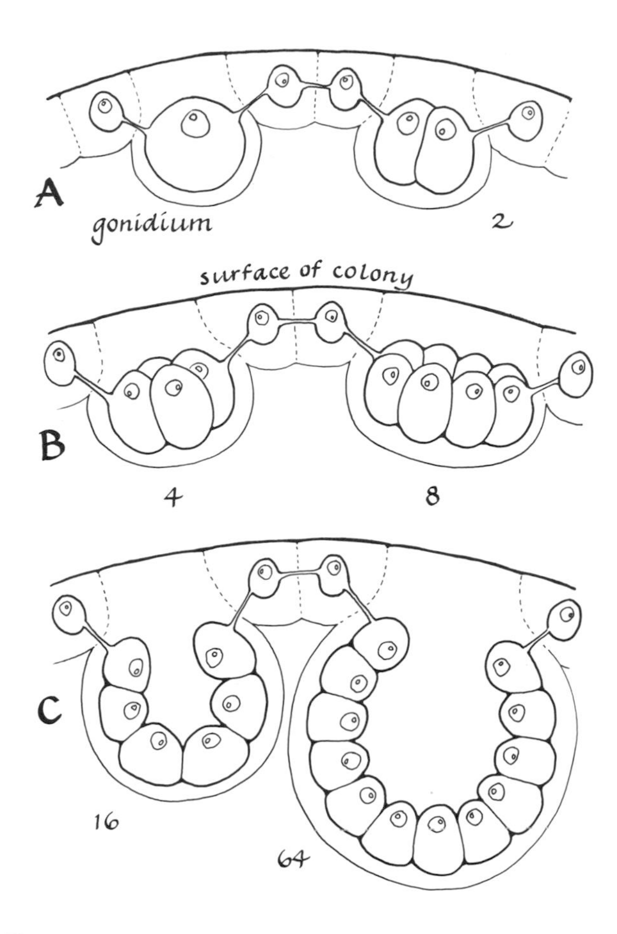

Figure 7-10. *Stages in development of the gonidium into a colony in* Volvox.

tates about its axis in either a clockwise or counterclockwise direction (Janet, 1923; Bower, 1930). In some species, most cells of a colony are connected by cytoplasmic strands, plasmodesma. These connections may be the basis for the coordinated beating of the biflagellate cells, which makes possible the polarized and unified movement of the colony as a single individual (Figure 7-11).

Of special interest is the problem of the mechanism by which the polarized pattern of cell specialization is attained in development of both the gonidium cell and the egg cell (Berrill, 1961). From the detailed studies of Janet (1923) it appears that the cells acquire special properties from their position in the aggregate mass, that is, from the composition of the local environment. Berrill (1961) suggests that this environment is graded, which may be the basis of the graded pattern in distribution of specialized cell types. Experimental studies of *Volvox* development, which are almost nonexistent, would help in clarifying the roles played by environment and cytoplasmic inheritance in development from either a gonidium or an egg. In any case, a striking aspect of development of *Volvox* and all the radially symmetrical colonial algae is their orderly geometric pattern of cell division. One can thus pinpoint at the gonidial cell stage the polarity of the future colony and probably also the fate of different parts of the gonidial cell (Figure 7-12).

Filamentous Associations

Quite a different geometry of cell association is the arrangement of cells in tandem to form long filaments. Cell division of terminal cells or cells along the length of the filament permits an almost unlimited increase in filament length. In spherically symmetrical or radially symmetrical cellular associations (such as discs), the surface-to-volume ratio decreases with an increase in the size (radius) of the aggregate. Optimal relationships between the organism and its external environment cannot be maintained in an indefinitely enlarging sphere or disc. Inside-to-outside or peripheral-to-central relationships change drastically with an increase in size of such aggregates. There are also purely mechanical limits to increase in the size of spheres or discs. In a single filament (a cylinder), the surface-to-volume ratio decreases with an increase in the length of the filament but not as rapidly as in a sphere or disc, because the diameter of the filament remains approximately constant as cells are added. The large size of many higher plants is probably possible only because of terminal or intercallary growth of special apical meristem cells at one or both ends of a filament. The simplest filament arrangement is found in the alga *Geminella,* in which cells are held together by a gelatinous cell product (Figure 7-13). Note that the mitotic spindles are all in the same plane, so that the filament elongates but does not increase in diameter (Smith 1933).

Another type of filamentous growth occurs in the alga *Prasinocladus.* In this case, the filament is also a cell product. Stages in development are shown in Figure 7-14. After the flagellated stage attaches to a substratum it secretes a new membrane around it-

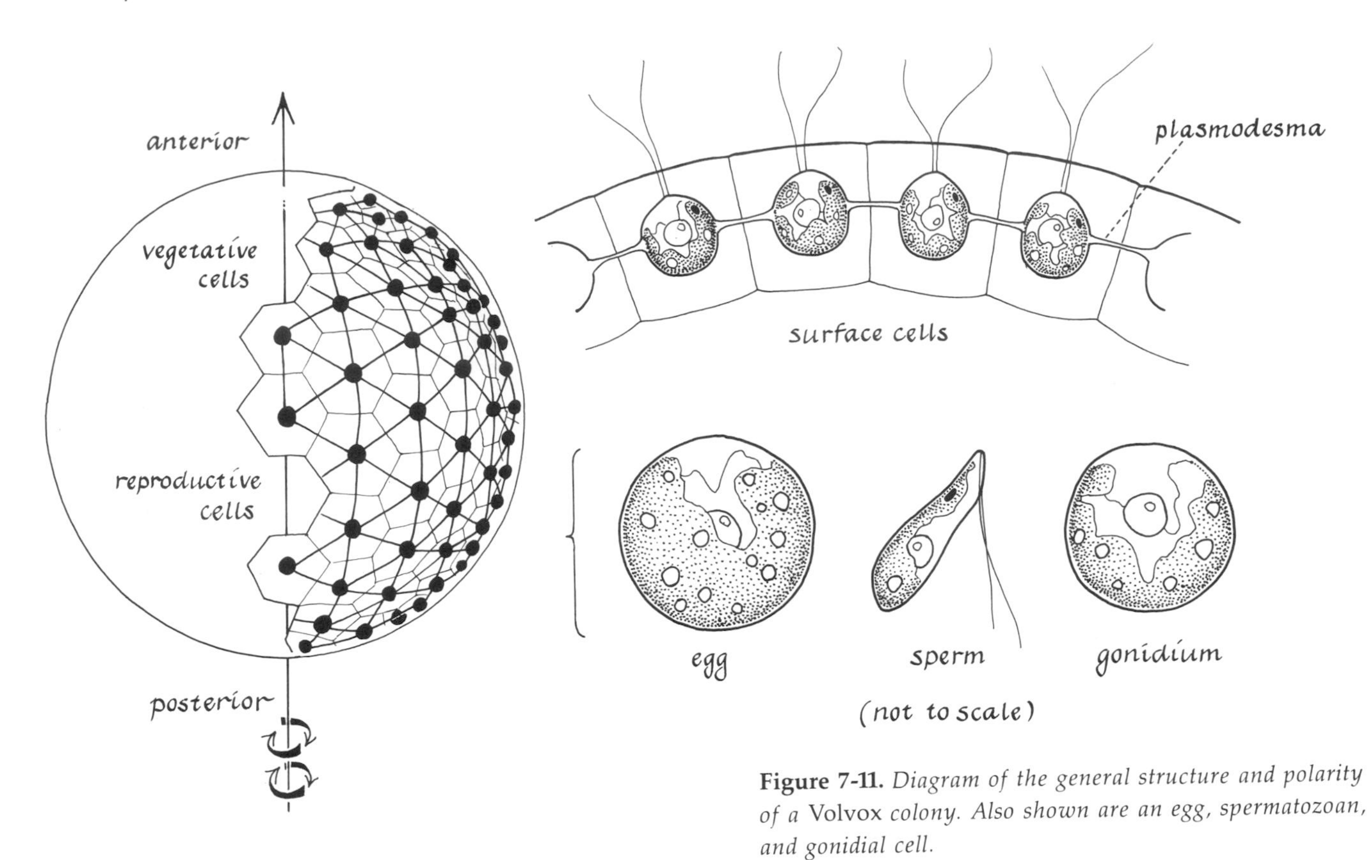

Figure 7-11. *Diagram of the general structure and polarity of a* Volvox *colony. Also shown are an egg, spermatozoan, and gonidial cell.*

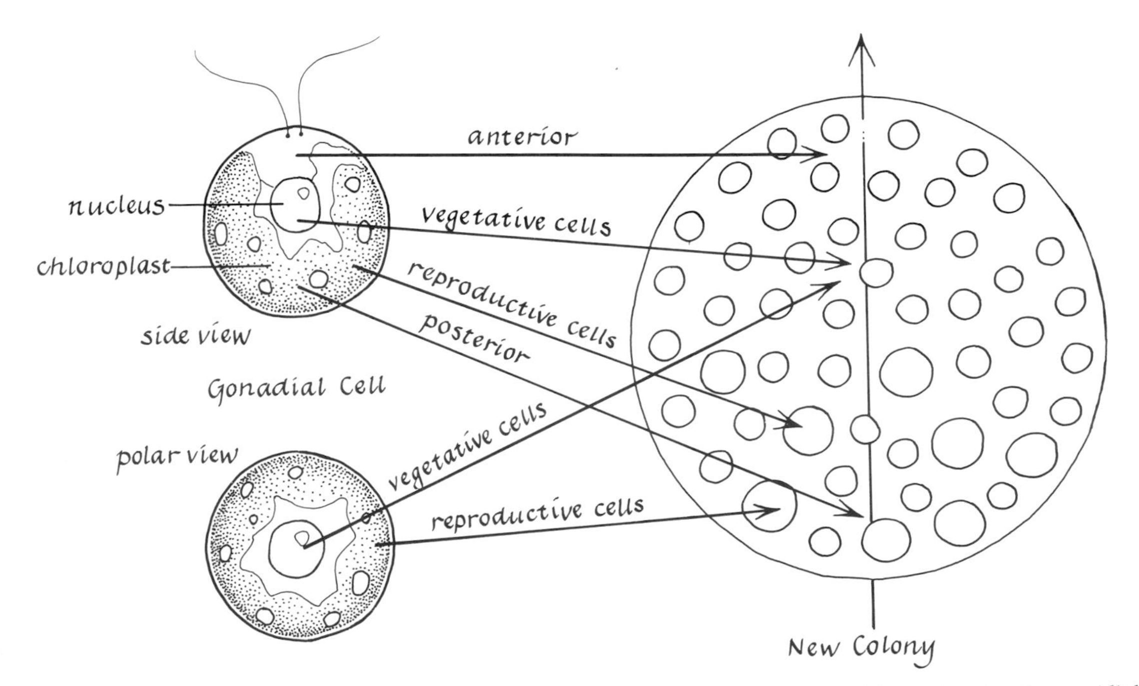

Figure 7-12. *The normal fate of parts of a gonidial cell.*

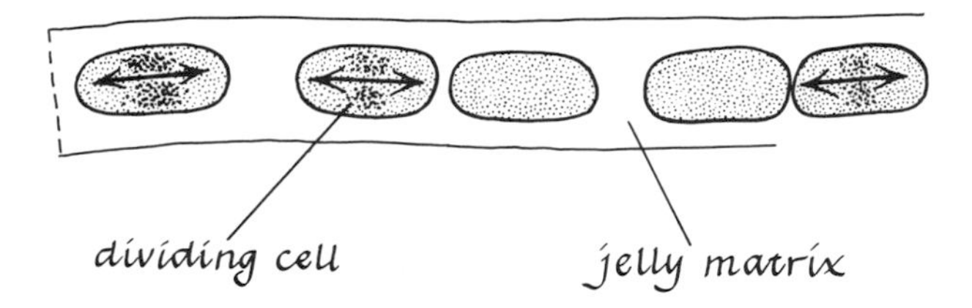

Figure 7-13. *Arrangement of cells of* Geminella *to form a filament.*

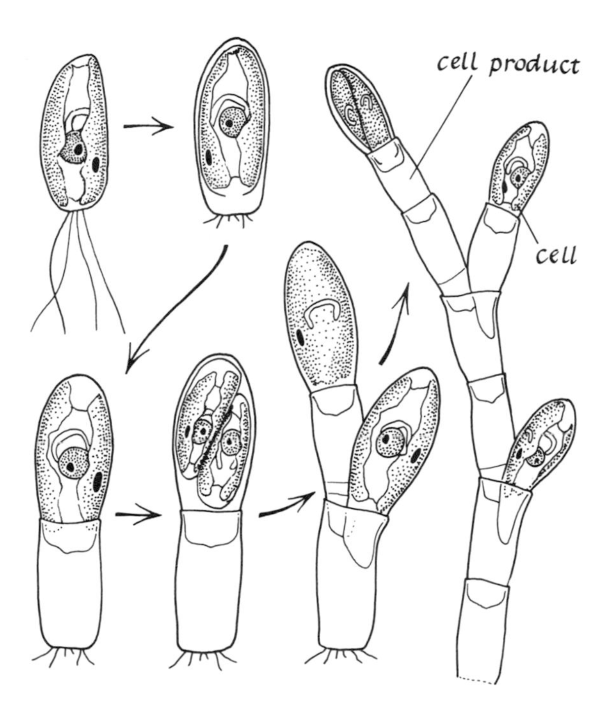

Figure 7-14. *Formation of a filament by secretion of cell products in* Prasinocladus.

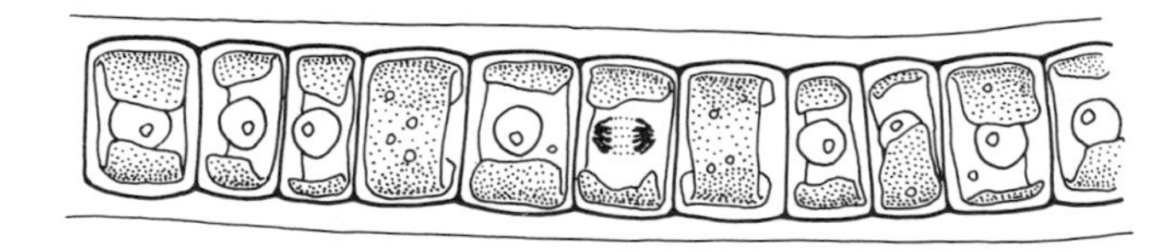

Figure 7-15. *Filamentous cell association in* Ulothrix.

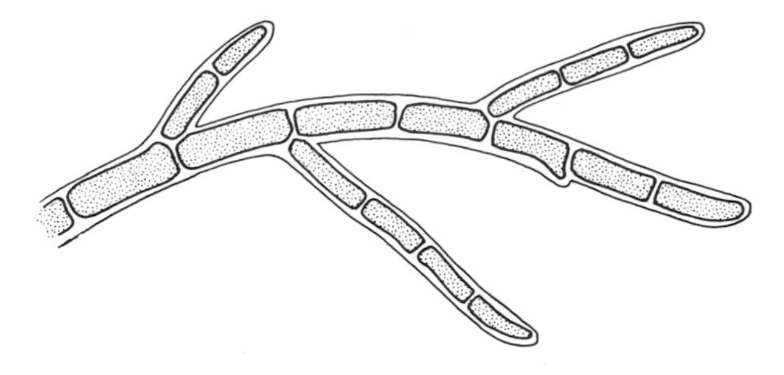

Figure 7-16. *Branching of the filament in* Cladophora.

self. The original apical wall of the flagellate ruptures and the protoplast moves upward. These activities are repeated, and the original protoplast is pushed upward on an increasingly long filament made of cell products. When the protoplast divides, branches are formed (Brown, 1935). Thus, rhythmical cell membrane synthesis elongates the filament.

A close, cell-to-cell type of filament (without a matrix) is found in *Stigeoclonium* and *Ulothrix* (Brown, 1935). Here the filament elongates by cell division within the filament (Figure 7-15).

A more complex filamentous pattern of development occurs when the filament branches. For reasons not known, when specific mitotic spindles are oriented at an angle with the long axis of the filament, a branch is formed as in *Cladophora* (Figure 7-16). The pattern of branching is species specific. In some species, the branches remain single; in others the branches branch again.

Filament elongation may be achieved not only by division of almost any cell but also by basal or apical growth. For example, the blue-green alga *Rivularia* elongates by cell addition just behind a specialized cell (a heterocyst), whereas another blue-green alga, *Scytonema*, elongates by division of a terminal cell (Figure 7-17).

The filament as a unit of structure. Filamentous cell association permits, by such processes as fasciculation (linear aggregation of filaments), branching, and almost limitless elongation, more complex structures than are possible in spherical cell associations. Indeed, the filament can be considered the basic structural unit of all plants. This idea was discussed long ago by Haberlandt (1914) and more recently by Fritsch (1952).

One example of the filamentous structure of a multicellular plant is the green alga *Coleochaete* (Figure 7-18). The body of this plant is constructed of radiating filaments connected by lateral branches. The result is a flat disc of cells that expands outward by cell division in the marginal meristem. The disc is often one cell diameter thick, but in some species it is two or three cell layers thick. In some species the radiating filaments are very clear (Fritsch, 1952).

Increasing complexity in the architecture of the red and brown algae seems to have developed by the fasciculation of filaments (Fritsch, 1952; Dixon, 1966). Some examples of fasciculation are illustrated in Figure 7-19. Note that apical and intercallary meristem cells present in the algae play an important role in determining the shapes of these plants (Taylor, 1937; Sinnott, 1960).

Association of filaments may occur by branching and fasciculation of the branches with the main stem, or an increase in thickness of a filament may occur by longitudinal septation.

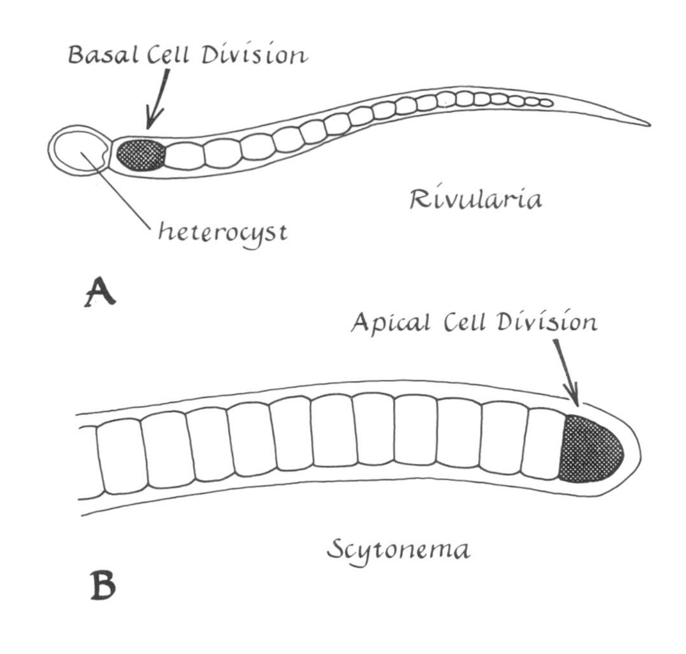

Figure 7-17. *Elongation of a filament by basal cell division in* Rivularia, *and by apical cell division in* Scytonema.

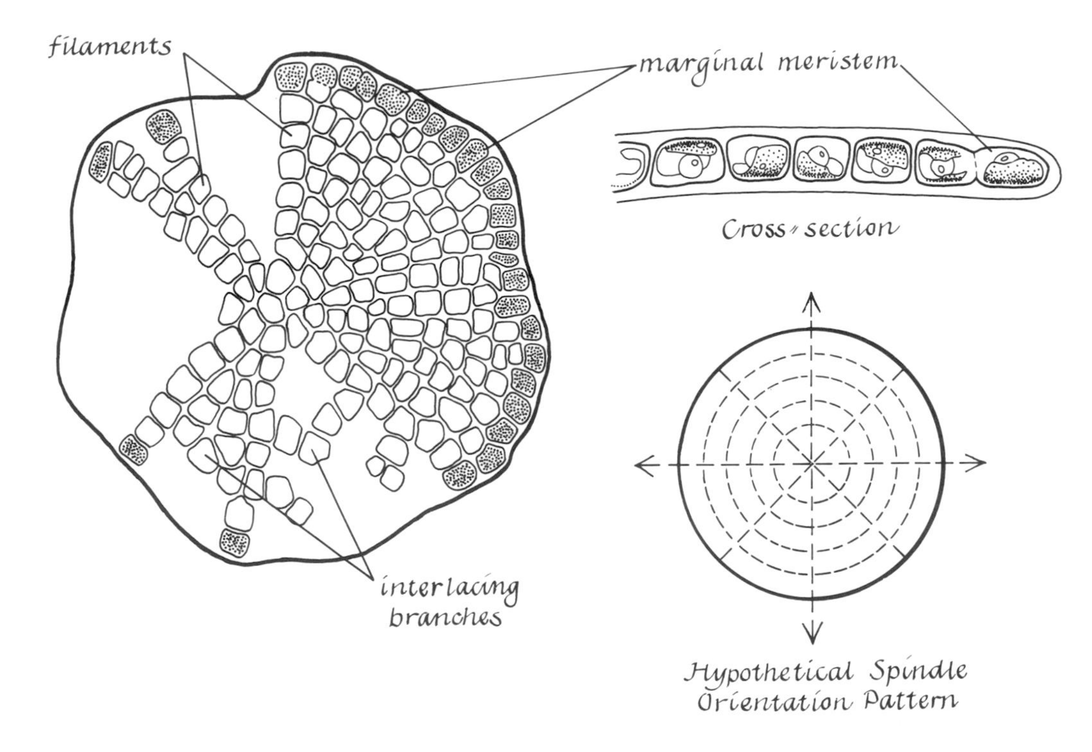

Figure 7-18. *Diagram of the filamentous structure of* Coleochaete.

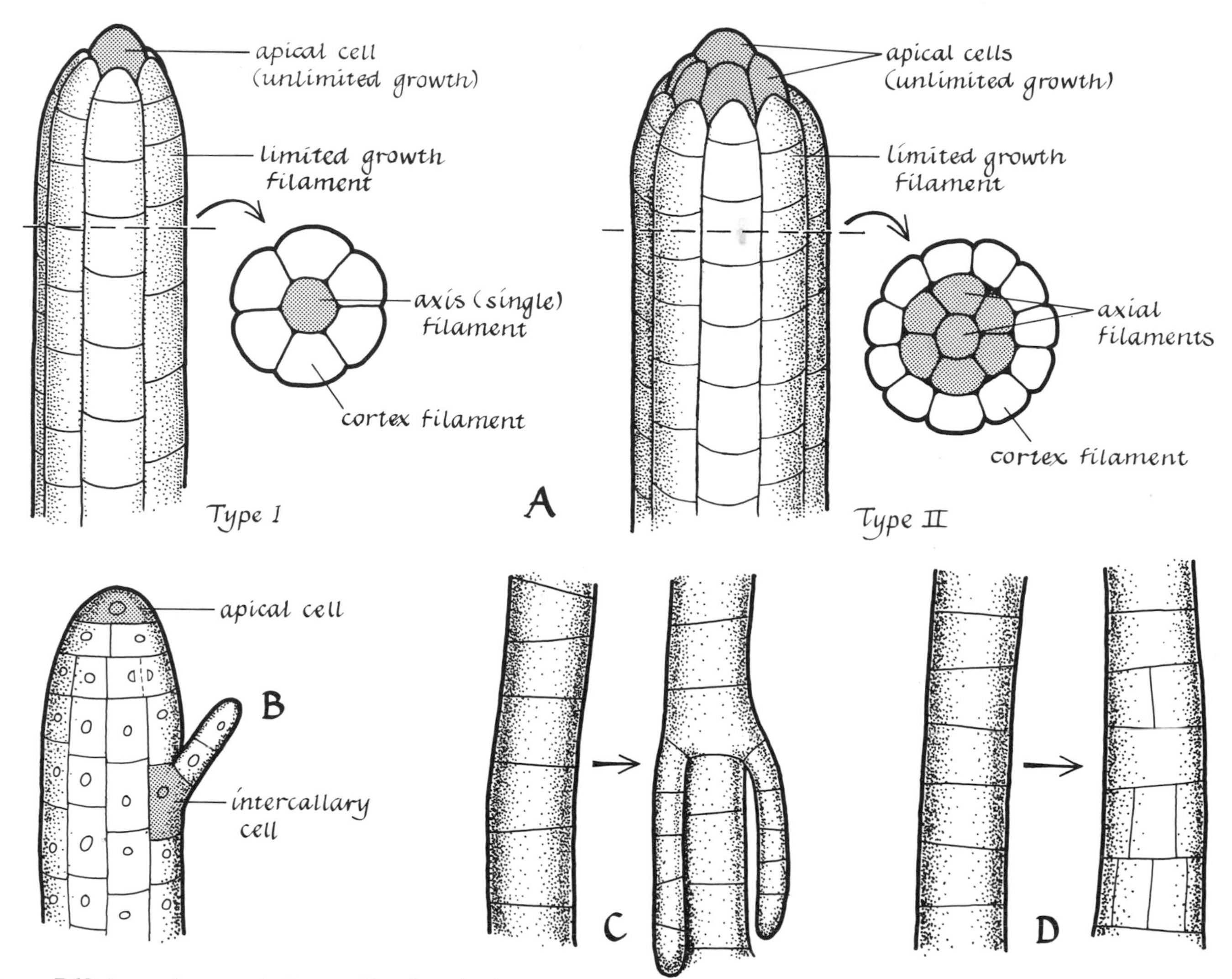

Figure 7-19. *Increasing complexity resulting from fasciculation of filaments in red and brown algae. A, Type I: single axial filament; Type II: several axial filaments. B, apical and intercallary meristem cells. C, fasciculation of branch with main stem. D, increase in thickness by longitudinal separation.*

The filamentous unit structure of mushroom fruiting bodies is familiar to anyone who has dissected them under a microscope. Development of the fruiting body from the fusion of underground hyphae (filaments) of the mycelium (a network of filaments) is a process of fasciculation of hyphal threads, as the fruiting body emerges almost overnight after a rainy spell (Bonner et al., 1956; Hawker, 1965). This is illustrated in Figure 7-20.

In the complex fruiting bodies of the larger fungi, differentiation of the hyphae into skeletal, generative, and binding hyphae adds to the complexity of the cellular association (Corner, 1950).

The mechanisms that guide filament associations are unknown. Long ago Gurwitsch (1922, 1927) proposed a "field theory"—a set of environmental conditions guiding cell behavior. We shall consider this idea in Chapter 26, noting here that hormones are probably important constituents of the internal environment of the developing mushroom. Filament associations in brown algae may lead to very large, even treelike structures. Even in higher plants, the

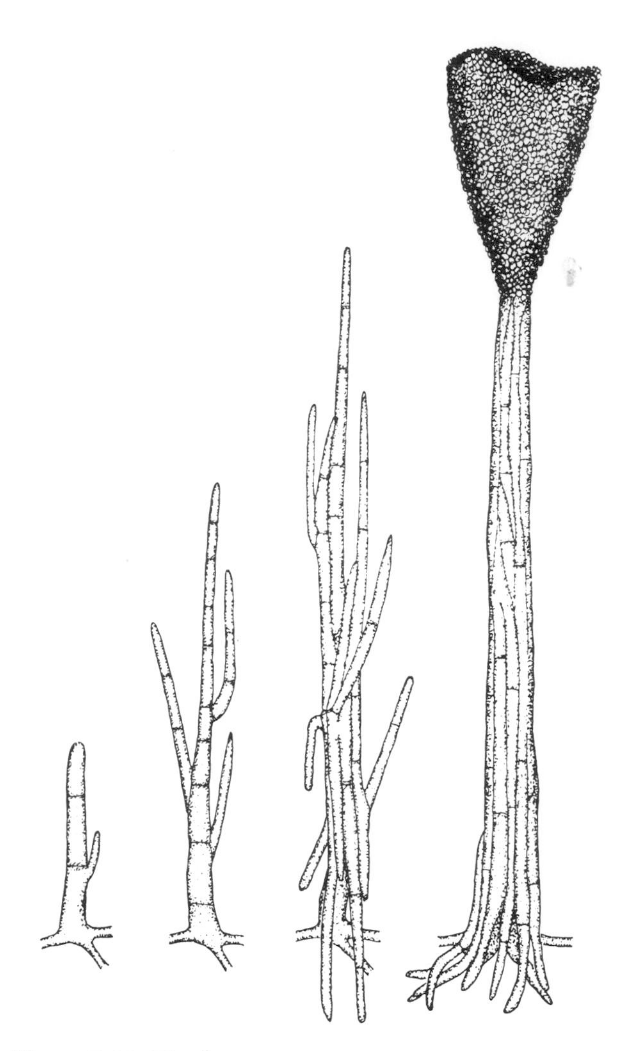

Figure 7-20. *The filamentous structure of the fruiting body in a mushroom of the coremium type. From J. T. Bonner,* Morphogenesis: An Essay on Development *(copyright 1952 by Princeton University Press). Reprinted by permission of Princeton University Press.*

arrangement of cells in filaments (usually called files of cells) is evident in the root and shoot (Chapter 4). Both the cells of a file and the cells of primitive algal filaments are descended from one mother cell in the plant.

Spindle orientation and colony shape. In cell associations resulting from repeated division of a single cell, the shape of the cell mass will be determined (in the absence of cell movements) by the orientation of the mitotic spindles—and the planes of new cell formation (Figure 7-21). The result may be a flat disc, a sphere, a filament, a cylinder, or even a cube.

When the spindle axes are all in one plane but alternately at right angles, the result will be a flat plate of cells. This occurs in the blue-green algae *Merismopedia* and *Tetrapedia* (Brown, 1935) and in the green alga *Monostroma* (Figure 7-21). When the spindle axes are at right angles to one another in two planes, a two-layered structure will result, as in *Ulva*. The colony is square in outline when it consists of 4, 16, 64 cells, and so forth. When the spindle axes are at right angles to one another in three planes the result is a cube-shaped colony (Figure 7-22).

The mechanisms controlling the shape of cell colonies arising by division (that is, those controlling orderly mitotic spindle orientation) remain undiscovered. Harper (1908, 1918) suggested that cell addition tends to occur in the direction of compression. The axes of the mitotic spindles would thus lie in the plane of maximum compression (Figure 7-23). Harper described the phenomenon as "the pressure-contact stimulus determination of growth direction." It is known that pressure applied to groups of explanted chick embryo cells stimulates proliferation of these cells (Spratt, 1958; Spratt and Haas, 1960). The growth of the femur of a 10-day chick embryo is enhanced if the explanted bone is forced to push a weight (Haaland, 1968). However, we know almost nothing about the influence of physical forces on cell behavior.

According to another idea about the orientation of mitotic spindles ("Sachs' Rule," cited by Bonner, 1952), the axis of the spindle is in the same direction as the long axis of the cell, the plane of division thus being at right angles to this axis. Exceptions are not uncommon; for example, the pattern of cell division in the vascular cambium (Chapter 4). Particularly in radially symmetrical colonies, division is usually synchronous. All conditions being equal, cells of the same age reach the division stage at the same time. We shall see examples of this in the cleavage of plant and animal eggs.

In addition to the orientation of cell division, the direction in which individual cells enlarge is important in shaping the plant. Initially the cells produced by division of a parent cell are more or less cuboid, but they commonly elongate in one direction. This elongation is largely a function of cell-wall formation, and the plane in which the cell wall is built is crucial. If the plane is parallel to the surface of the stem, the

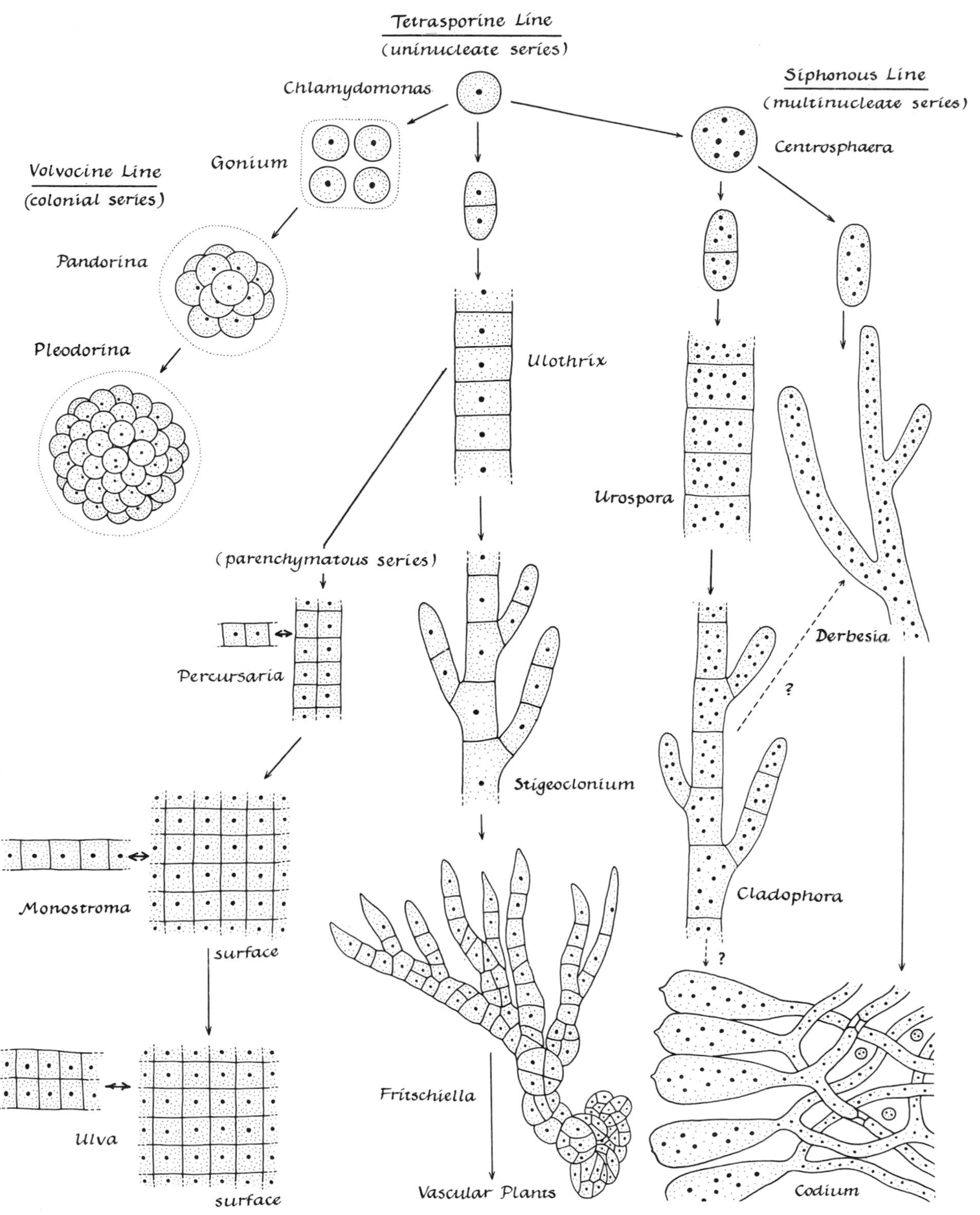

Figure 7-21. *Colony shapes resulting from different planes of cell and nuclear divisions in green algae. From Scagel et al., 1966. Courtesy Wadsworth Publishing Company, Inc.*

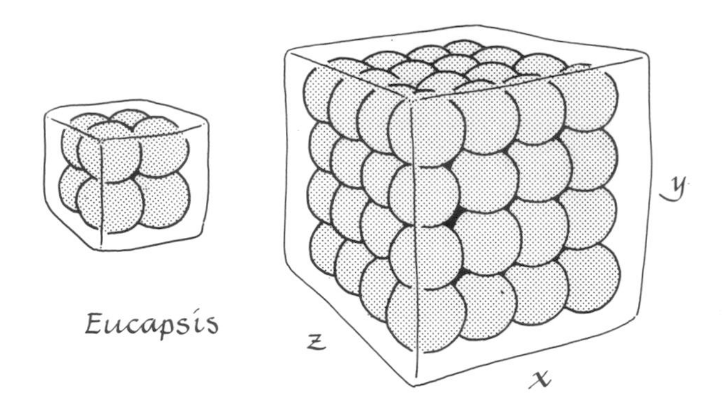

Figure 7-22. *Cube-shaped colony of* Eucapsis.

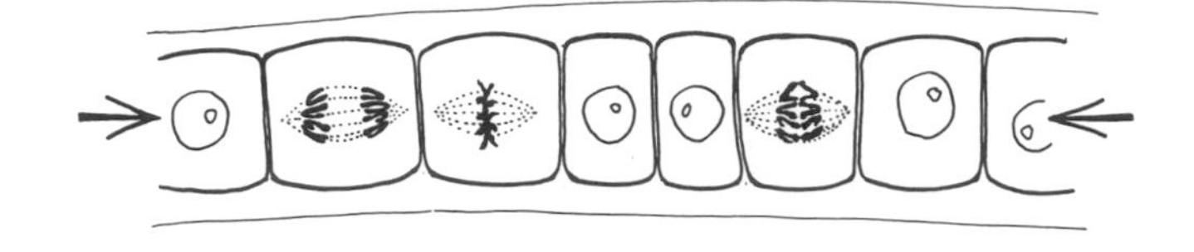

Figure 7-23. *Cell addition in the plane of maximum compression.*

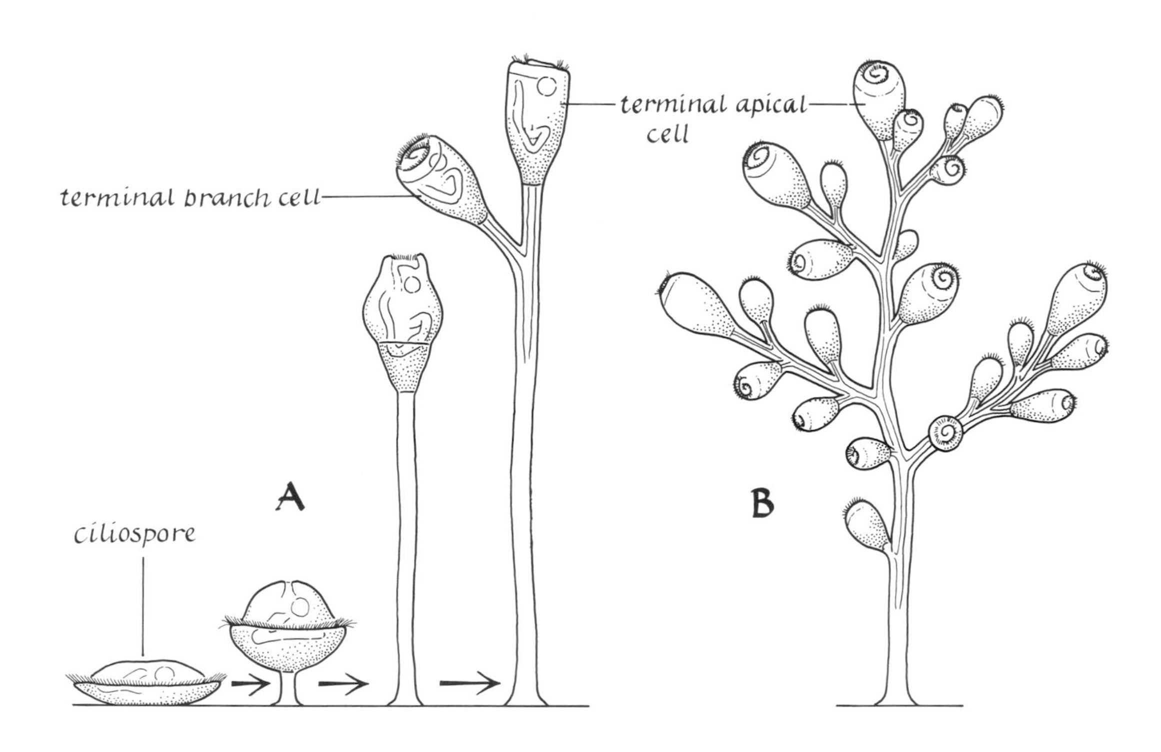

Figure 7-24. *A, development of* Zoothamnium alternans *from a ciliospore; B, alternately branched colony of many individuals.*

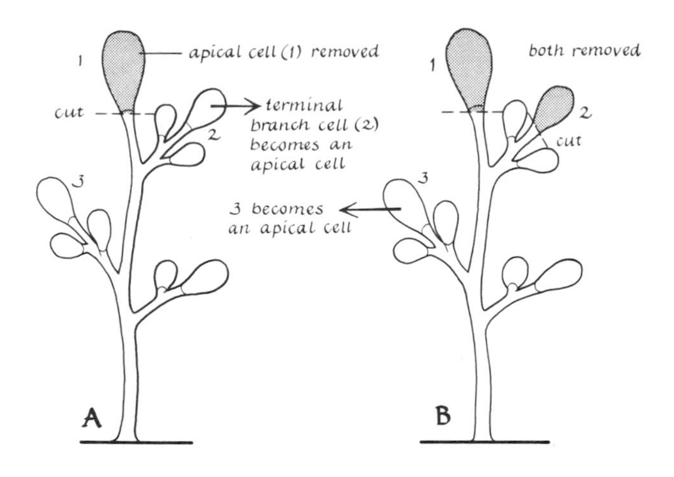

Figure 7-25. *Apical dominance in a colony of* Zoothamnium. *A, removal of the apical cell (1) results in a terminal branch cell becoming the new apical cell; B, removal of the apical cell and terminal branch cell (2) results in terminal branch cell (3) becoming an apical cell of the colony.*

stem may increase in girth; if the plane is in a cross section, increase in length may occur. The microtubules, ultrastructural elements widely distributed in the cytoplasm of both plant and animal cells (Chapter 25), appear to play a role in controlling the direction of cell-wall formation (Newcomb, 1969). We shall examine cell-wall formation in more detail in Chapter 23.

Cell Colonies in Protozoa

An excellent example of animal cell association resulting from cell division is found in the colonial protozoan *Zoothamnium* (Fauré-Fremiet, 1930). At one stage in its life cycle, the animal is a ciliospore (Chapter 12). After swimming about for a period of time, the ciliospore comes to rest on the substratum, secretes an adhesive at its lower surface, and raises itself up on a stalk that is a cell product (Figure 7-24A).

Division of the protoplast then occurs, giving rise to two cells. One becomes a terminal cell, the other a terminal branch-forming cell. The process is repeated to produce an alternately branched colony of many "individuals" (Figure 7-24B). The branching pattern is a result of division of the apical cell into another apical cell and a terminal branch cell. If the apical cell is removed, the nearest terminal branch cell transforms into an apical cell and continues the upward growth of the colony (Summers, 1938, 1941). Summers' experiments are shown in Figure 7-25.

The apical cell functions like a stem cell and exhibits apical dominance, but the mechanism of this is unknown. Recall (Chapter 4) that similar dominance

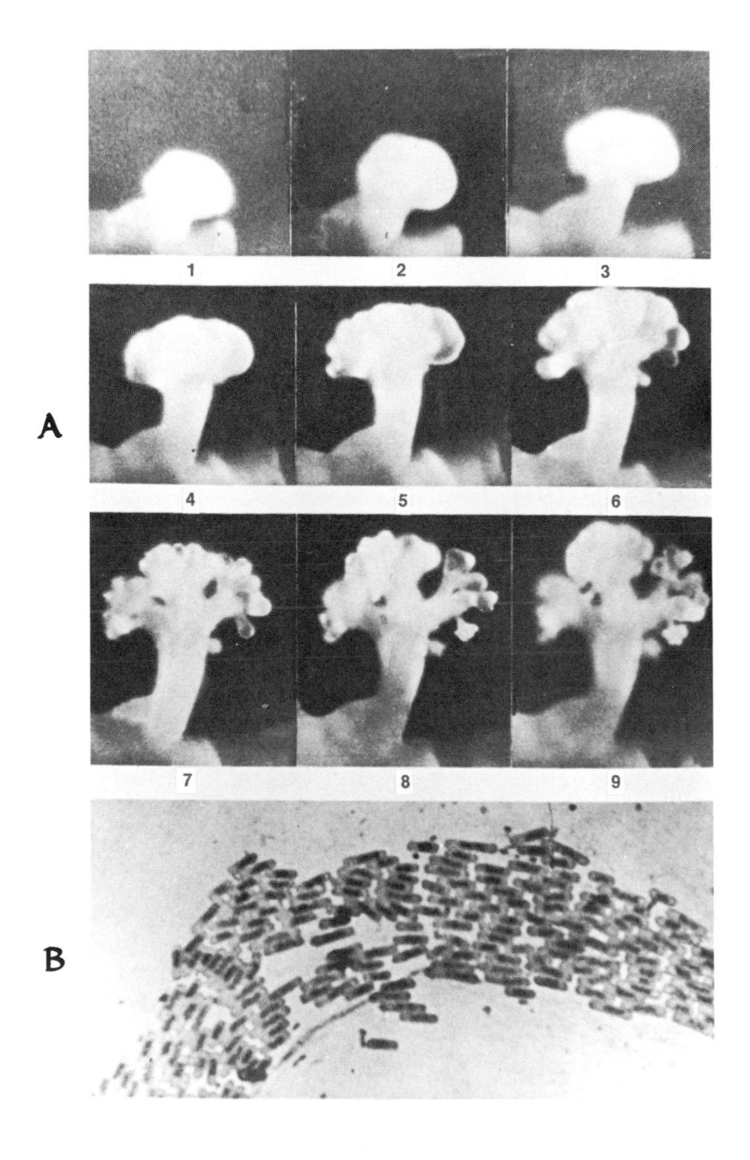

Figure 7-26. *A, stages in fruiting-body formation at approximately hour intervals in the myxobacterium* Chondromyces crocatus. *B, the bacterial rods are streaming toward an aggregation center. Photographs by E. A. Wheaton. From J. T. Bonner,* Morphogenesis: An Essay on Development *(copyright 1952 by Princeton University Press). Reprinted by permission of Princeton University Press.*

of the shoot apex in plants over more basally located lateral buds (potential stem apices) is better understood. The hormone auxin (indole acetic acid) produced in higher concentration by the shoot apex inhibits activity in the lateral buds in some plants (Chapter 19). Recall that apical dominance is also exhibited by the cellular slime mold slug; acrasin (cycle AMP) may mediate this dominance (Chapter 6).

Cell Associations Resulting from Cell Aggregation

Associations of cells by the movement (ameboid, ciliate, or flagellate) of individual cells toward one another to form and increase the size of an aggregate occur throughout the animal kingdom, in lower plants, and in microorganisms. Many beautiful and puzzling aggregation processes in bacteria, algae, and fungae have received little attention by biologists (Bonner, 1952). A few examples are illustrated below.

Myxobacteria

What appears to be but probably is not a simple cellular aggregation is found in the myxobacteria or slime bacteria (Thaxter, 1892; Stanier, 1942; reviews of Bonner, 1952, and Quinlan and Raper, 1959). In the species *Chondromyces crocatus*, motile rod-shaped cells aggregate to form a multicellular mass encased in the gelatinous product of the cells. As we shall see in Chapter 23 the product might be involved in formation of the fruiting body. The clump increases in size by cell division and by fusion of separate clumps. Suddenly the movement of the mass stops and the process of forming a fruiting body begins (Figure 7-26).

Studies of the developing fruiting body show that the bacteria lie in the uppermost portion, and all point in their direction of movement (Figure 7-27). The stalk is composed largely of gelatinous material. Finally, large groups of bacteria become encased in cysts (spores) that rupture and release the bacteria on germination.

In the mechanism directing the movements of individual bacteria, the cells seem to be guided by their contact with the slime (Stanier, 1942). Unlike the cellular slime molds, the cells do not seem to undergo chemotaxis during aggregation (see Chapter 6).

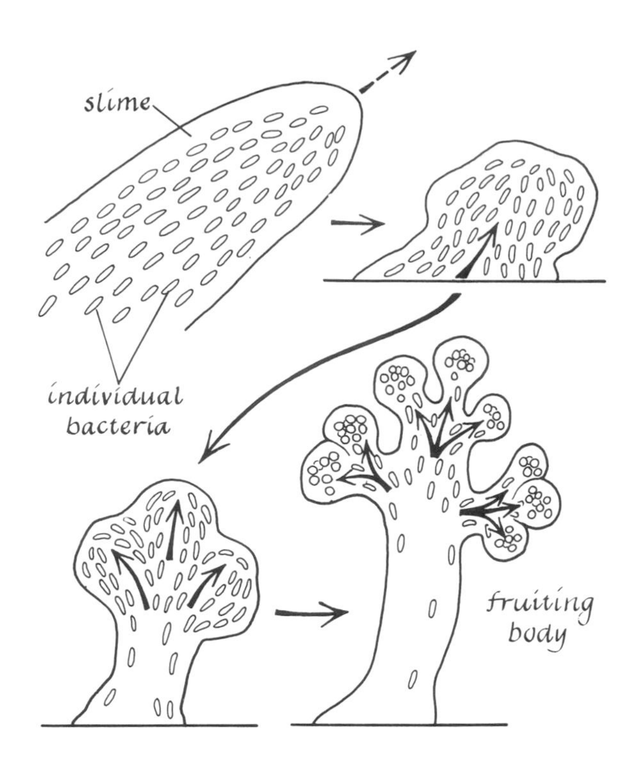

Figure 7-27. *Diagrams of the positions of the bacterial rods during fruiting-body formation.*

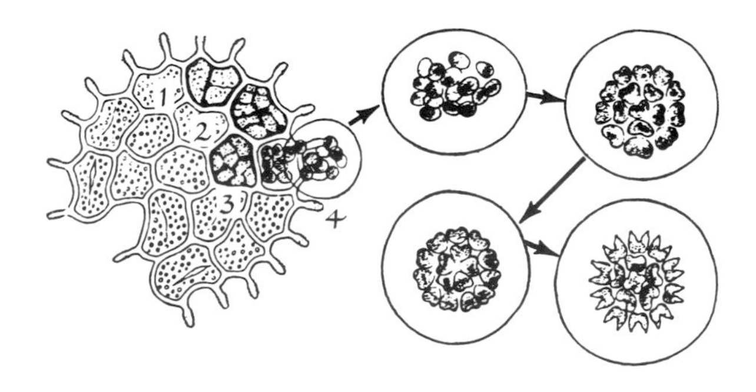

Figure 7-28. *Asexual development in* Pediastrum. *1-4, successive steps in the cleavage and emergence of a daughter colony. Also shown are stages in the aggregation of the zoospores to form a new colony. From Brown,* The Plant Kingdom, *1935. Reproduced by permission of Mrs. Brown.*

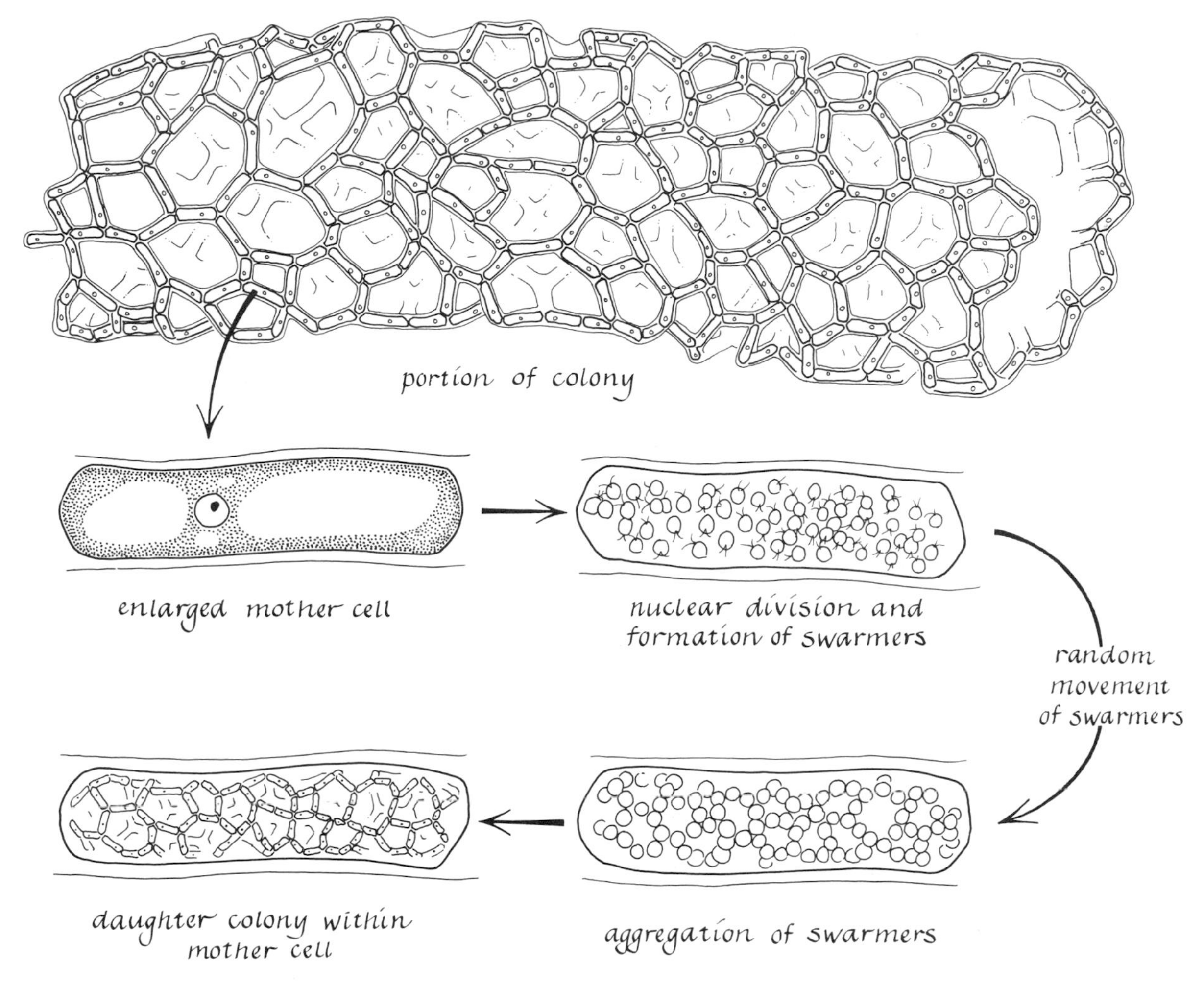

Figure 7-29. *Formation of a daughter colony in* Hydrodictyon *by aggregation of swarmers to form a net within a colony mother cell.*

Green Algae

Pediastrum

In this example of cell association, the colony-forming mother cell divides into between 4 and 64 cells. The daughter colony, encased in a vesicle, is released from the parent colony. Each biflagellate cell of the new colony swims about violently for three or four minutes, then aggregates to form the radically symmetrical flat disc of the mature colony (Smith, 1916). For reasons not entirely clear, irregular colonies are also produced. In some respects, the aggregation seems similar to the fitting together of geometrically regular but different pieces of a jigsaw puzzle. The aggregation takes only about 10 minutes (Figure 7-28). Why a disc is formed instead of the more likely sphere is unknown.

Hydrodictyon

A similar method of colony formation by aggregation of flagellated cells occurs in *Hydrodictyon reticulatum*, the water net. The mature colony is a cylindrical, sausage-shaped, netlike arrangement of individual cells; the colony may reach a length of 2 to 3 feet (Figure 7-29). On the average, there are six cells per mesh (range from 3 to 8). The colony mother cell (any cell of the net) divides many times within its cell wall to produce thousands of motile biflagellate cells called swarmers or zoospores. After a period of random movement within the wall of the mother cell, these cells stop moving, lose their flagella, and plaster themselves against the wall of the mother cell, forming a minute net. In this way, the wall of the mother cell is a mold for formation of the new net (Harper, 1908). The spherically symmetrical motile cells, probably in response to a pressure-contact stimulus, elongate to form the cell type of the mature net.

Fungi

In the *Phycomycetes* (water molds), fusion of hyphae (coenocytic, that is, multinucleated filaments) to form the mycelial mat is a type of aggregation that forms a colony. However, in this colony there are no individual cells but a multinucleated fabric of cytoplasmic strands. According to Park and Robinson (1966), positive chemotropism (bending of the tips of the hyphae toward one another) seems to be involved in fusion of the tips to form the mycelium. Similar fusion of hyphal threads occurs in Ascomycetes ("sac" fungi such as *Penicillium*, source of the antibiotic penicillin). Fusion may occur between two different genotypes resulting in the formation of a heterocaryon, a cytoplasmic mass containing genetically different nuclei.

Artificial Cell Aggregations in Higher Plants

Many more studies have been done on the capacity of artificially dispersed animal tissue cells to reaggregate than on plant tissue cells. In one experiment testing this capacity in dispersed hypocotyl (young root) cells of the *Eucalyptus* tree, the dispersed root cells had to be pushed together by packing them in cellophane dialysis bags with a centrifuge (Sussex and Clutter, 1968). In the artificially formed aggregates from single cells, the cells are not connected by plasmodesma (cytoplasmic bridges). No xylem cell differentiation occurs even after auxin (IAA) is added to the culture medium, even though IAA stimulates xylem formation in tissue cultures of plant callus cells (Chapter 22).

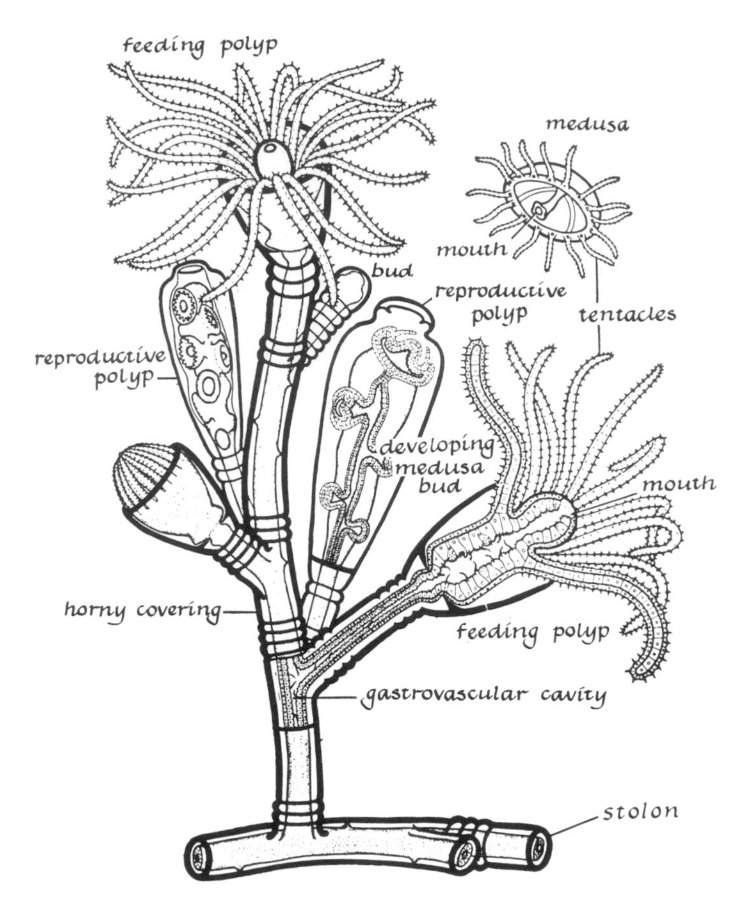

Figure 7-30. *Diagram of a colony of* Obelia. *From Ralph Buchsbaum,* Animals Without Backbones, *copyright University of Chicago Press, 1948.*

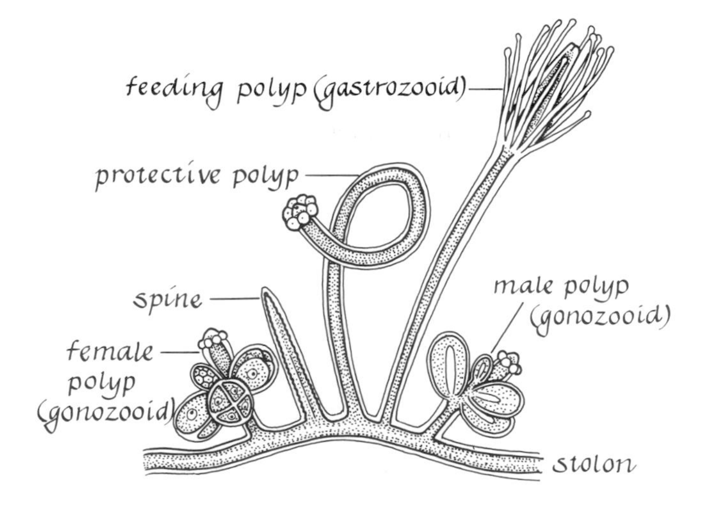

Figure 7-31. *Division of labor in a colony of* Hydractinia: *an example of polymorphism.*

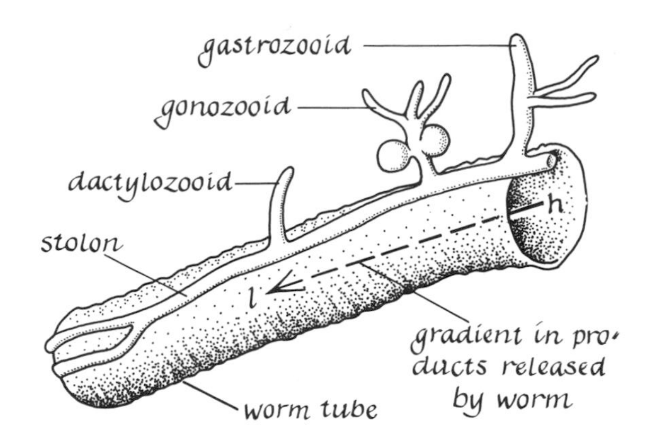

Figure 7-32. *Influence of position (microenvironment) on the differentiation of polymorphic zooids of* Proboscidactyla *growing on the tube formed by a marine annelid.*

When dispersal is incomplete and small groups of cells are cultured, plasmodesma are present. If IAA is added to the medium containing the small cell groups, xylem (tracheal cell) differentiation occurs inside the cell group. The experiments suggest that interconnection of cells via the plasmodesma is necessary for tracheal cell differentiation. Neither the single cells nor the artificially reaggregated single cells were capable of xylem differentiation. The plasmodesma seem to be necessary for xylem formation.

As noted in Chapter 5, cultured single cells of some plants can divide and produce a callus, from which a whole plant may develop. Although cell division can occur in some free cells, the rate of division appears to be higher in cell aggregates. Thus, cellular associations are necessary for repeated and continuous cell division (Street and Henshaw, 1963). The free root cells probably do not undergo any differentiation, but they do show senescent changes.

Colonies of Individual Multicellular Organisms

Cellular associations in which the colony is an integrated collection of multicellular individuals are common in the hydroid coelenterates (Hyman, 1940). Individuals of such colonies exhibit varying degrees of independence and specialization. For example, *Obelia* has feeding individuals and reproductive individuals (Figure 7-30).

A more complex "division of labor" among individuals of a colony is found in *Hydractinia*, another colonial hydroid (Figure 7-31).

Functional and structural specialization of individuals of a colony is known as polymorphism. A particularly challenging problem in polymorphic coelenterates is the nature of the mechanism that controls development of the diverse zooids. Spiral (protective) zooids, polyps of *Podocoryne* that grow on the shell of a hermit crab, are located at the rim of the shell only if the shell is inhabited. Cultures grown on glass slides do not develop such zooids (Braverman, 1960). Some evidence (Loomis, 1961) indicates that the partial pressure of carbon dioxide (pCO_2) in cultures of *Hydractinia* may stimulate spiral zooid development.

The hydroid *Proboscidactyla flavicirrata* grows on the tube formed by a marine annelid worm. The colony is axially patterned with stolons running parallel to the tube and with gastrozooids, gonozooids, and dactylozooids located in sequence from the rim of the tube back to the stolon region (Campbell, 1968). Differentiation of the polymorphic zooids in this sequence is probably induced by their positions relative to the open end of the tube. The concentration of substances released by the worm would be higher at the rim of the tube than farther down its length, and the type of zooid formed could be ascribed to its position in a gradient of these substances (Figure 7-32). As the worm increases the length of the tube by adding to its rim, each gastrozooid and its associated gonozooids move forward individually, but all are connected by the anastomosis (fusion) of the stolons.

Still more elaborate polymorphism develops in *Physalia*, the Portuguese man-of-war. Feeding, protective, feeler, male and female individuals are attached to one bright blue balloon-shaped individual that functions only as a float for the colony.

Colonial Associations without a Morphological Basis

In concluding this chapter, we should note that some close associations of cells and individual organisms seem to have no structure or matrix uniting them. In the colonial diatom *Nitschia putrida* the individuals are separate but may form a swarm that moves as a unit (Figure 7-33). When the swarm encounters an obstruction, it may split and each part flow around the object, or it may divide permanently into two separate swarms (Wagner, 1934).

The swarming of *Sciara* larvae and ants (Wheeler, 1911), the schooling of fish, and the flocking of birds are well-known examples of organism associations sufficiently close that the group often behaves as a single unit (reviewed by Emerson, 1939, and Bonner, 1952). In ant, bee, and some other insect colonies, there is a functional division of labor among the members. In honeybee colonies, the division of labor constitutes a model of organism differentiation similar to cellular differentiation. A colony consists of a queen, sterile females (workers), and males (drones). Some of the workers collect honey and pollen, others build new combs, and still others specialize in parental duties, caring for the brood. This differentiation is a developmental process related to the age of the individual. Each worker engages in these activities in successive periods of her life. First, the young worker sweeps and cleans the cells for about three days, then she begins to feed the older larvae. About three days later, she feeds the younger larvae. At the age of ten days, the worker stops feeding the larvae and begins new activities, including the building of cells. On the twentieth day, the worker becomes a guard and inspects every bee arriving at the entrance of the hive. Finally she becomes a forager, flying out to collect honey and pollen (Tinbergen, 1970).

Conclusion

Our survey of colonial organisms illustrates many diverse types of cellular association. In all types, the development of the completed pattern of cell arrangement seems to be achieved by a pattern of cell division or cellular aggregation, either independent of or influenced by the surrounding environment. The guidelines directing the development of these associations are partly preformed and partly epigenetic. In radially symmetrical associations, cell-division capacity seems limited primarily by the genotype, not by influences of the extracellular environment. This is indicated by the fact that removal of part of the aggregate does not induce resumption of cell-division activity (see Chapter 5). The arrangement of cells in geometric order is also a result of built-in guidelines that control the orientation of the mitotic spindles. We know practically nothing about the mechanism controlling programmatic use of the inherited information.

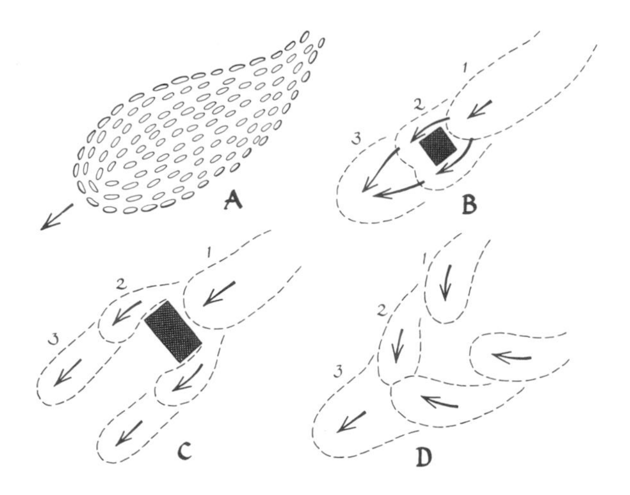

Figure 7-33. *A, swarm of the colonial diatom* Nitschia. *B, swarm flowing around an obstruction; C, division of a swarm as it flows around an obstruction; D, fusion of two swarms.*

In filamentous cellular associations, almost unlimited growth is possible, not only because surface-volume relations do not change as drastically as in radial or spherical associations but also because terminal or basal growth is possible. Removal of part of a filament not involving the apical or basal region may be followed by resumption of cell-division activity. Directional growth of a filament is also the result of oriented cell division, but is probably influenced by the surrounding environment as well. The filamentous arrangement of cells permits more architectural complexity than a spherical one, as the filaments can become associated in higher orders of complexity than would be possible with spheres, in the same way that more complex designs can be made with matches than with an equal number of marbles. The association of filaments seems to be influenced greatly by guidelines progressively formed as a consequence of cellular interactions.

Finally, the significance of cellular associations is the possibility of division of labor among the cells, probably resulting from the development of a complex internal microenvironment. Such an environment is possible in close associations of individual cells and in close associations of multicellular individuals (such as hydroids) and may lead to the structural specialization of individuals (polymorphism) or to the functional specialization of individuals (schools of fish, flocks of birds, and honeybee colonies).

Once cellular association and cell specialization have occurred, an association tends to be remarkably stable. As we shall see in the next chapter, many patterns of cellular association are reconstituted after artificial disruption and dispersal of the separate cells.

References

Berrill, N. J. 1961. Growth, Development and Pattern. W. H. Freeman, San Francisco.

Bertalanffy, L. von, and J. H. Woodger. 1933. Modern Theories of Development. Oxford University Press, Oxford.

Bonner, J. T. 1952. Morphogenesis: An Essay on Development. Princeton University Press, Princeton, N.J.

Bonner, J. T., K. K. Kane, and R. H. Levey. 1956. Studies on the mechanics of growth in the common mushroom, *Agaricus campestris.* Mycologia **48:**13–19.

Bower, F. O. 1930. Size and Form in Plants. Macmillan, London.

Braverman, M. H. 1960. Differentiation and commensalism in *Podocoryne carnea.* Am. Midland Nat. **63:**223–225.

Brown, W. H. 1935. The Plant Kingdom. Ginn and Co., Boston.

Campbell, R. D. 1968. Colony growth and pattern in the two-tentacled hydroid, *Proboscidactyla flavicirrata.* Biol. Bull. **135:**96–104.

Corner, E. J. H. 1950. Clavaria and Allied Genera. Clarendon Press, Oxford.

Dixon, P. S. 1966. On the form of the thallus in the Florideophyceae. In E. Cutter, ed., Trends in Plant Morphogenesis, pp. 45–63. Wiley, New York.

Emerson, A. E. 1939. Social coordination and the superorganism. Amer. Midland Nat. **21:**181–206.

Fauré-Fremiet. 1930. Growth and differentiation of the colonies of *Zoothamnium alternans* (Clap. and Lachm.). Biol. Bull. **58:**28–51.

Fritsch, F. E. 1935. The Structure and Reproduction of the Algae. Macmillan, New York.

Fritsch, F. E. 1952. The evolution of a differentiated plant: a study in cell differentiation. Proc. Linn. Soc. Lond. **163:**218–233.

Gerisch, G. 1959. Die Zelldifferenzierung bei *Pleodorina californica* (Shaw) und die Organisation der Phytomonadinenkolonien. Arch. Protist. **104:**292–358.

Gurwitsch, A. 1922. Über den Begriff des embryonalen Fieldes. Arch. Entw. Mech. **51:**383–415.

Gurwitsch, A. 1927. Wieterbildung und Verallgemeinerung des Fieldbegriffes. Arch. Entw. Mech. **112:**433–454.

Haaland, J. E. 1968. Interaction of ionic environment and stress on a bone forming system. (Doctoral thesis, University of Minnesota).

Haberlandt, G. 1914. Physiological Plant Anatomy (trans. 4th ed. by M. Drummond). Macmillan, London.

Harper, R. A. 1908. The organization of certain coenobic plants. Bull. Univ. Wisconsin, no. 207.

Harper, R. A. 1912. The structure and development of the colony in *Gonium.* Trans. Am. Micros. Soc. **31:**65–83.

Harper, R. A. 1918. The evolution of cell types and contact and pressure responses in *Pediastrum.* Mem. Torrey Bot. Club **17:**210–240.

Hartman, M. 1924. Untersuchungen über die Morphologie und Physiologie des Formwechsels der Phytomonadinen (Volvocales): IV. Über die Veränderung der Koloniebildung von *Eudorina elegans* und *Gonium pectorale* unter dem Einfluss ausserer Bedingungen. Arch. Protist. **49:**375–395.

Hawker, L. E. 1965. The physiology of development in fungi. In Encyclopedia of Plant Physiology, Vol. 15, pp. 716–757. Springer Verlag, Berlin.

Hyman, L. H. 1940. The Invertebrates: Protozoa through Ctenophora. McGraw-Hill Book Company, New York.

Janet, C. 1923. Le Volvox. Mém. 3. Paris.

Loomis, W. F. 1961. Feedback factors affecting sexual differentiation in *Hydra littoralis.* In H. Lenhoff and W. F. Loomis, eds., The Biology of Hydra and Some Other Coelenterates, pp. 337–362. University of Miami Press, Coral Gables, Fla.

Newcomb, E. H. 1969. Plant microtubules. Ann. Rev. Plant Physiol. **29:**253–288.

Park, D., and P. M. Robinson. 1966. Aspects of hyphal morphogenesis in fungi. In E. Cutter, ed., Trends in Plant Morphogenesis, pp. 27–44. Wiley, New York.

Pocock, M. A. 1933. *Volvox* in South Africa. Ann. S. Africa Museum **16:**523–646.

Quinlan, M. S., and K. B. Raper. 1959. Development of the Myxobacteria. In Encyclopedia of Plant Physiology, Vol. 15, pp. 596–611. Springer Verlag, Berlin.

Scagel, R. F., R. J. Bandoni, G. E. Rouse, W. B. Schofield, J. R. Stein, and I. M. C. Taylor. 1966. An Evolutionary Survey of the Plant Kingdom. Wadsworth Publishing Company, Inc., Belmont, Calif.

Sinnott, E. W. 1960. Plant Morphogenesis. McGraw-Hill Book Company, New York.

Smith, G. M. 1916. Cytological studies in the Protococcales: II. Cell structures and zoospore formation in *Pediastrum boryanum* (Turp.) Ann. Bot. **30:**467–479.

Smith, G. M. 1933. Freshwater Algae of the United States. McGraw-Hill Book Company, New York.

Spratt, N. T., Jr. 1958. Analysis of the organizer center in the early chick embryo: IV. Some differential enzyme activities of node center cells. J. Exp. Zool. **138:**51–80.

Spratt, N. T., Jr., and H. Haas. 1960. Importance of morphogenetic movements in the lower surface of the young chick blastoderm. J. Exp. Zool. **144:**257–275.

Stanier, R. 1942. A note on elasticotaxis in myxobacteria. J. Bact. **44:**405–412.

Street, H. E., and G. G. Henshaw. 1963. Cell division and differentiation in suspension cultures of higher plant cells. Symp. Soc. Exp. Biol. **17:**234–256.

Summers, F. M. 1938. Form regulation in *Zoothamnium alternans.* Biol. Bull. **74:**130–154.

Summers, F. M. 1941. The protozoa in connection with morphogenetic problems. In G. N. Calkins and F. M. Summers, eds., Protozoa in Biological Research, pp. 772–817. Columbia University Press, New York.

Sussex, I. M., and M. E. Clutter. 1968. Differentiation in tissues, free cells, and reaggregated plant cells. In M. M. Sigel, ed., Differentiat on and Defense Mechanisms in Lower Organisms, pp. 3–12. Waverly Press, Baltimore, Md.

Taylor, W. R. 1937. Marine Algae. University of Michigan Press, Ann Arbor.

Thaxter, R. 1892. On the myxobacteriaceae, a new order of schizomycetes. Bot. Gaz. **17:**389–406.

Tinbergen, N. 1970. The growth of social organizations. In C. E. Johnson, ed., Contemporary Readings in Behavior, pp. 2–18. McGraw-Hill Book Company, New York.

Wagner, J. 1934. Beiträge zur Kenntnis der *Nitschia putrida,* Benecke, insbesondere ihrer Bewegung. Arch. Protist. **82:**86–113.

Wardlaw, C. W. 1965. Organization and Evolution in Plants. Longman, London.

Wheeler, W. M. 1911. The ant-colony as an organism. J. Morphol. **22:**307–325.

8 *Cellular Associations: Invertebrates and Vertebrates*

In the development of most animals and plants, cellular associations are established by cellular proliferation in which the cells remain in one group. In some instances, however, cellular association results from the movement and aggregation of cells—for example, the aggregation of apparently unspecialized cells to form the embryo body in one tunicate and in annual fishes. Cellular aggregation also occurs in the development of some tissues, structures, or organs of many invertebrates and vertebrates. The formation of ganglia (collections of nerve cell bodies) in vertebrates is the result of the migration and aggregation of neural crest cells (cells originating in the top of the neural folds [Chapter 2] during formation of the neural tube in the embryo). Other examples are found in the development of blood vessels (particularly veins), the skeleton of echinoderm larvae, and skeletal muscles of vertebrates (in which separate myoblast cells aggregate and fuse to form the multinucleate myotubules). Apparently, specialized cells of the same or different histotypes do not move and aggregate to form tissues, organs, or organ systems in normal embryonic development of invertebrates or vertebrates. In regenerative development of hydroids, however, specialized cells do migrate and aggregate to form tissues or organs in morphallaxis (Chapter 5). Further, specialized, stable cells of older embryos, after experimental dispersal to the single cell state, can reassemble and reconstitute tissues.

The primary purpose of this chapter is to show how experimentally dispersed tissue cells of both invertebrates and vertebrates are capable of reassembling and reconstituting the tissue or organ from which they were derived. Studies of cell behavior under these artificial but controlled conditions have yielded information about cellular relationships that would be difficult to discover in the intact organism. The studies also reveal that differentiated cells which have reached a recognizable level of morphological specialization are very stable, even in isolation. It is remarkable that these cells are able to reassemble, sort out, and reconstruct their original tissues from suspensions of mixed types of cells (Chapter 16). Mechanisms controlling this process are only partially known, but clearly the individuality of cell type is maintained.

Invertebrates

Sponge Cell Reaggregation

The association of cells in sponges is so loose that they have been described as animals below the tissue grade of organization (Hyman, 1940). However, several different types of cells are recognizable in sponges. These include epithelial pinacocytes, flagellated choanocytes (collar cells), collencytes, scleroblasts (skeleton-forming cells), and archeocytes (unspecialized cells). The archeocytes are relatively large, basophyllic cells capable of forming all the other types of cells (reviews of Hyman, 1940; Berrill, 1961). These cell types are illustrated in Figure 8-1.

Studies of the reaggregation process in sponges began with Wilson (1907), who devised a method for dispersing the cells of mature sponges. When whole sponges were placed in a bag of fine bolting silk cloth and the bag was squeezed into sea water, separate cells and cell clusters were obtained. The cells settled to the bottom of the container and in a day or two reaggregated to begin formation of small sponges. As can be seen in time-lapse movies, the single cells move about like amebae at random and coalesce by chance with one another. As more cells coalesce and join groups of already coalesced cells, netlike aggregates are formed. In aggregates of about 2,000 cells, cellular arrangement typical of the adult species is attained after five or six days (Galtsoff, 1925). In cell mixtures of the red sponge *Microciona prolifera* and the yellow sponge *Cliona,* the two species sorted out and aggregated separately. As tissue cell types in sponges are of doubtful stability, some cellular transformation rather than sorting out may be involved in the reconstitution.

In studies of the details of the reaggregation process in the freshwater sponge *Ephydatia,* Brien (1937) found that the epidermis cells of the aggregate form a peripheral epithelium enclosing and isolating the mesenchyme, archeocyte, choanocyte, and other cell types from the external environment. Next, the central region of the mass undergoes histolysis (cell death) to form the lacunae of the exhalent system of the sponge. The cellular debris so produced is ingested by the archeocytes. Finally, choanocytes in the peripheral region line the developing chambers (Figure 8-2). During the reconstitution process, many cells, probably the archeocytes, are dividing. Probably both sorting out of cell types and formation of new cell types by archeocytes are involved in reconstitution of freshwater sponges.

Both the proportions of cell types and the genotypes of the cells are important in the reaggregation process. If the ratio of choanocytes to archeocytes is relatively large, the archeocytes ingest the choanocytes and precipitate involution (degeneration) to the extent that no new sponge develops (Brien, 1937). An involuting aggregate will not fuse with a non-

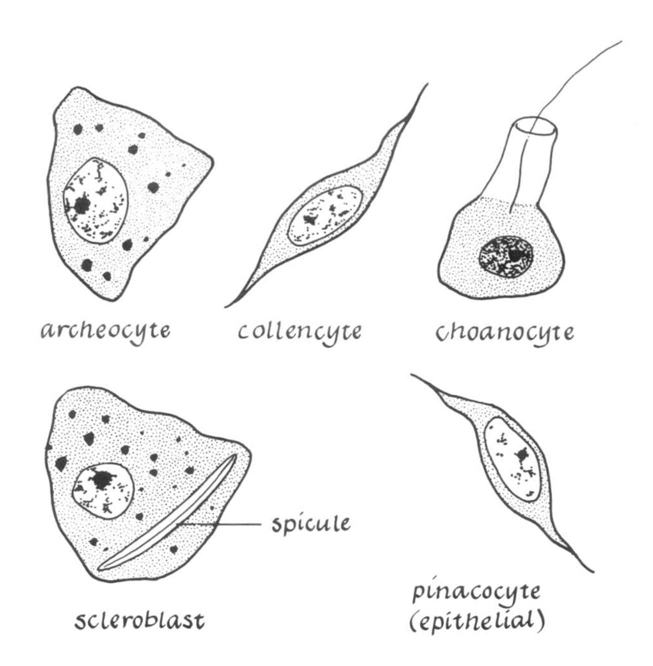

Figure 8-1. *Sponge cell types.*

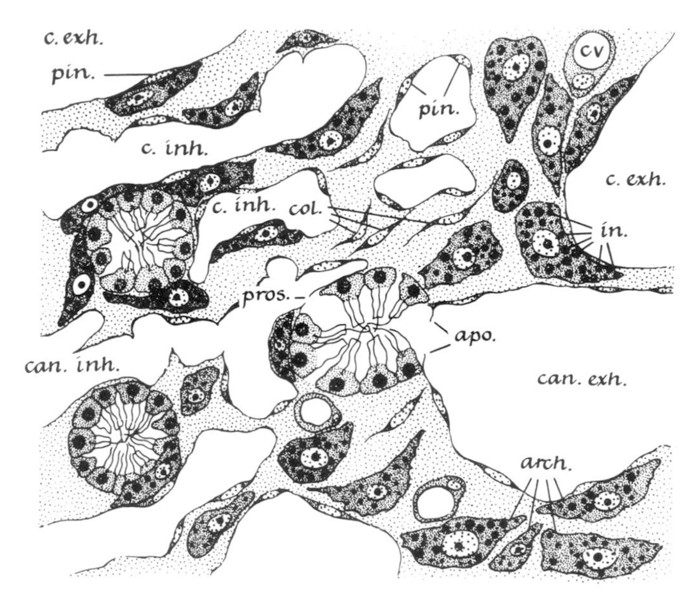

Figure 8-2. *Late stage in differentiation in a reaggregate of cells of the freshwater sponge* Ephydatia. [Cinh., *inhalent canal;* C exh., *exhalent canal;* Pin., *pinacocyte;* Arch., *archeocyte;* Col., *collencyte;* Pros., *prosopyle;* Apo., *apopyle;* Cv., *contractile vacuole.*] *From* Growth, Development, and Pattern *by N. J. Berrill. W. H. Freeman and Company. Copyright © 1961.*

involuting aggregate but will fuse with another involuting aggregate. As already noted, reaggregation and sorting are usually species specific. Mixtures of *Microciona* and *Cliona* or of *Microciona* and *Haliclona* initially form a single aggregate, but later the two species sort out to form separate sponges (Galtsoff, 1925; Spiegel, 1954; Humphreys, 1967; Moscona, 1968a).

Studies by Humphreys (1963, 1967) indicate that calcium and magnesium ions and a cell surface substance are necessary for reaggregation. Humphreys believes that the divalent ions are necessary for the adhesion of the cell surface macromolecules. Thus, washing *Microciona* and *Haliclona* cells in seawater containing no calcium or magnesium removes the surface molecules and prevents aggregation. Addition of the supernatant from this washing of *Microciona* cells to washed, dispersed *Microciona* cells at 5 C permits aggregation to occur. At 22 C the washed cells can resynthesize the surface molecules. Similar results have been reported by Moscona (1968a, 1968b). The surface molecules appear to be species specific, as the supernatant from washed *Microciona* cells does not enhance aggregation when added to washed *Haliclona* cells (Humphreys, 1967; Moscona, 1968b).

Humphreys and Moscona are attempting to identify the macromolecules. They appear to be glycoprotein particles. Electron microscopy of the supernatant reveals particles about 20–25 Å in diameter (Moscona, 1968b). Humphreys (1967) found larger particles, possibly clusters of the smaller ones. No doubt the molecular architecture (the stereochemical structure) of the cell surface plays the important role in cellular adhesions and associations, not only in sponges but perhaps in all organisms.

Dissociation and Reaggregation of Molluscan Cells

Reaggregation of dispersed ovotestis cells of the snail *Helix pomatia* has been reported by Farris (1968). The most complete dissociation of cells was obtained in a solution of trypsin, galactose, and trehalose. Unlike the process necessary with chick and mammal tissue cells, calcium- and magnesium-free solutions need not be used with the snail ovotestis. Reaggregation was accomplished by centrifuging for 10 minutes to remove the dispersal solution and was followed by reconstitution of histotypically organized tissue in which the cell types (nurse cells, gametocytes, sperm, eggs, epithelial cells, and so forth) were in the same relation as those in normal ovotestis. These results demonstrate the capacity of dispersed, specialized tissue cells and gametes of a complex adult organ to reconstitute the original tissue architecture.

Dissociation and Reaggregation of Echinoderm Embryo Cells

Cells of sea urchin embryos at the blastula and gastrula stages can be dispersed by washing in calcium-free seawater followed by mechanical agitation. Isolated single cells in a suspension containing cells from many embryos undergo self-reaggregation in stationary dishes, but centrifuging or rotation of the suspension results in an increase in the number and size of the aggregates. A brief description of the reaggregation process in *Paracentrotus lividus* and other species studied by Giudice and Mutolo (1970) follows.

Stages in normal development and in the reaggregation and development of the reaggregates of these cells are illustrated in Figure 8-3.

As shown in Figure 8-3B, a blastocoel forms in the reaggregate between 9 and 20 hours after aggregation begins, and at approximately the same time an archenteron (primitive gut) forms. The archenteron, however, is not formed by invagination of the vegetal side of the blastula (as in normal development) (Figure 8-3A, *m*) but apparently by cell aggregation and formation of a cavity within the aggregate. After three days, many reaggregates develop into remarkably normal pluteus stage larvae with skeletal rods (Figure 8-3B). Significantly, in reaggregation separate embryos form from a suspension containing dissociated cells of many blastulae. Both small and large aggregates are formed, but typically only those approximating the size of a normal blastula develop into plutei. However, giant plutei with extra arms and archenterons, thus appearing polyembryonic, are also obtained. No larvae are formed in reaggregates of cells from late gastrula stages (Figure 8-3A, *m* and *n*). The dissociated cells appear to be of stable type, as they do not lose their animal (ectodermal) and vegetal (endodermal-mesodermal) characteristics. Another significant aspect of reaggregation is the failure of normal changes in metabolism (such as the increase in alkaline phosphatase activity) to occur until the morphology normal for these changes is attained in the aggregates. In suspensions containing cells from two species of sea urchins, cell adhesion was found to be species specific, but within particular age limits cell adhesion was not stage specific. Complex problems regarding the mechanisms that direct reaggregation await further investigation.

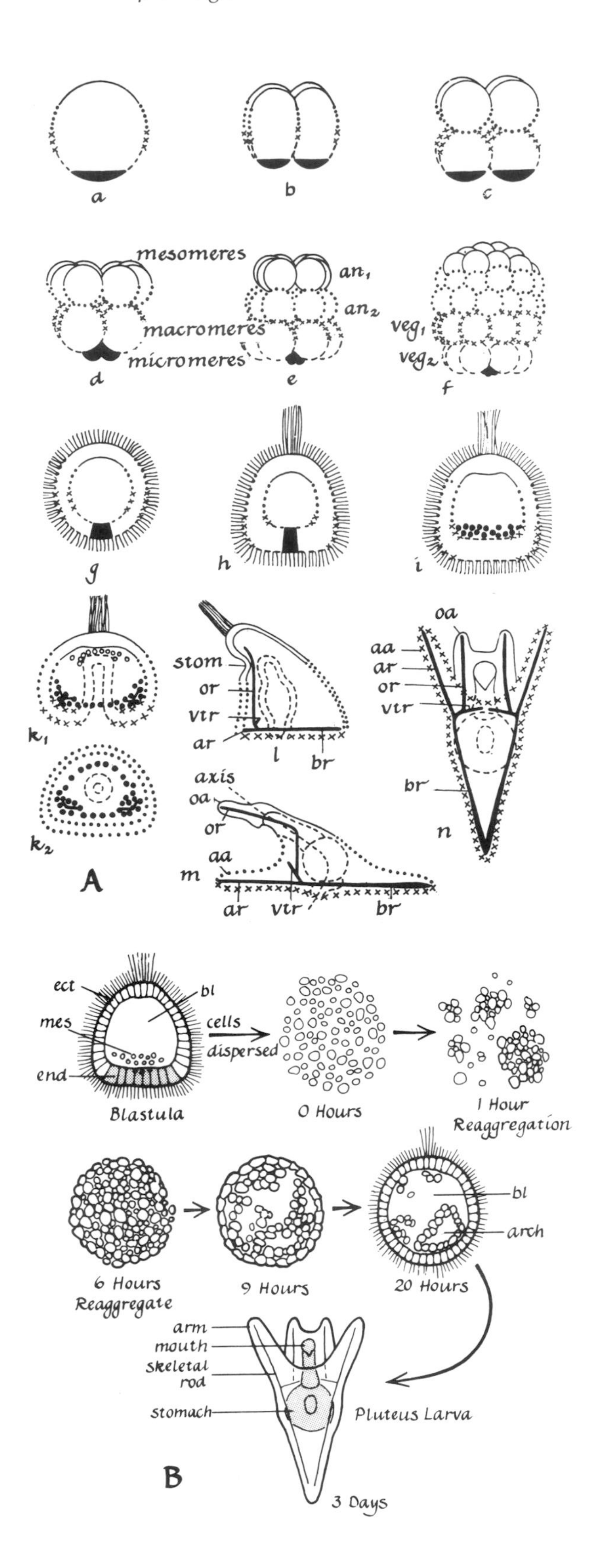

Figure 8-3. *A, normal stages in development of the sea urchin egg of* Paracentrotus *into the pluteus larva. B, diagrams of reaggregation of dispersed blastula cells in* Paracentrotus lividus *and reconstitution of the pluteus larva. A, from Horstadius, 1939,* Biological Reviews *Cambridge, Volume 14. Copyright Cambridge University Press.*

Cellular Aggregation in the Tunicate Salpa

Cellular aggregation has been described as occurring in the normal early development of the *Salpa* embryo. This was first described by Brooks (1893) and is discussed by Berrill (1961). Early cleavage of the egg is quickly masked by the inward migration of follicle cells which remain attached to the extruded egg. The result is one of the most amazing developmental performances in the animal kingdom. The follicle cells arrange themselves into a crude modeling of the form of the future tunicate, almost a framework around which the rapidly multiplying embryo cells aggregate. Eventually, the follicle cells disintegrate, leaving the aggregated embryo cells in the new form of the embryo. The embryo cells seem at least partially guided in their assembly by the follicle cell arrangement. Clearly, any preformed structure or polarity of the *Salpa* egg before cleavage must have little significance as a guideline for development. The embryo cells aggregate with definite polarities and symmetries before the cells specialize. Presumably, the positions of the cells determine their later differentiation, but this has not been experimentally demonstrated.

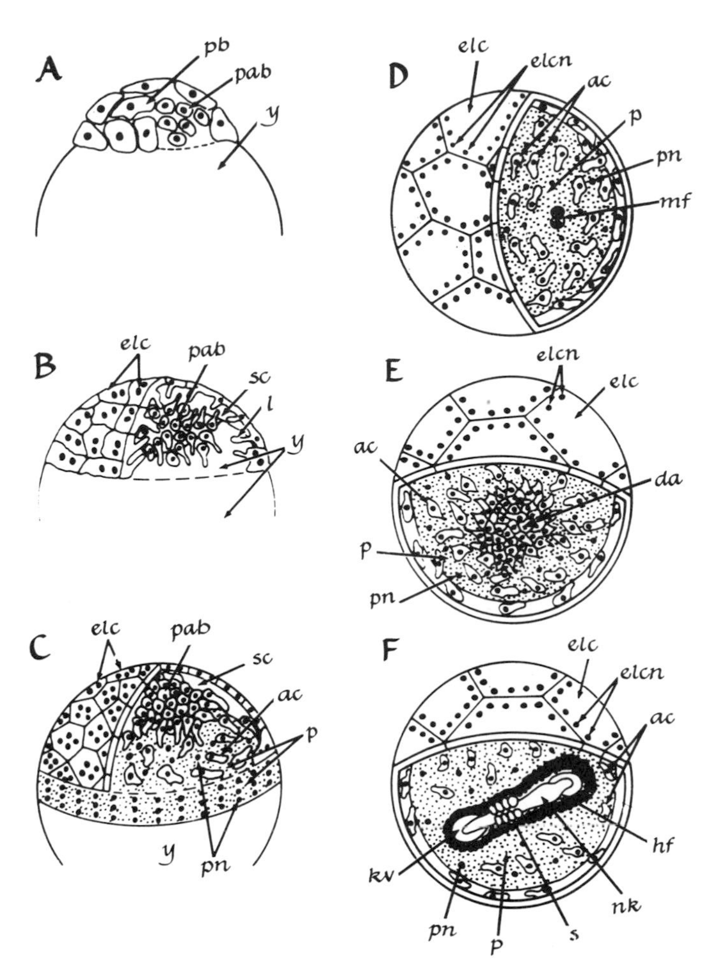

Figure 8-4. *Diagrams of dispersal and reaggregation of cells during normal embryonic development of the annual fish* Austrofundulus myersi. *From Wourms, 1967. Copyright © 1967 by the Thomas Y. Crowell Company.*

Reaggregation of Drosophila Embryo Cells

Studies reported by Fox et al. (1968) reveal the capacity of dispersed cells in four-hour-old embryos of the fruitfly *Drosophila* to reaggregate. The cell suspensions contain small cells and larger, yolk, cells. The small cells divide at a logarithmic rate, and the yolk cells divide less frequently. Spontaneous aggregation of yolk cells was observed, but aggregation of both yolk and small cells was facilitated by placing the suspension on a shaking machine (a technique also used to speed up aggregation of sponge and vertebrate cells). After about thirty minutes on the shaking machine a number of aggregates were formed. Some simple patterns of cellular association and differentiation were observed, for example, the formation of rosette-like cell clusters, compact spheres, and tubule-like structures. As single cells remaining in suspension undergo no apparent changes, cellular interactions may be necessary for cellular differentiation (see Chapter 21). Cells remaining in suspension for one to ten days gradually lose their ability to reaggregate. Assuming that these cells are still viable, the adhesive surface molecules may have washed off during the period of suspension. It is also possible that packing the cells together creates an intercellular environment influencing their behavior. Such an environment would not be present around isolated cells.

Reaggregation of Moth Fat-Body Cells

When studying reaggregation of fat-body cells in saturnid moths, Walters (1969) found that a nondialyzable, heat-labile substance in the blood plasma of these moths was necessary for cellular adhesion and reaggregation. Calcium or magnesium ions were also required. No aggregation took place in physiological saline.

Vertebrates

Cellular Dispersal and Reaggregation in Annual Fishes

An unusual pattern of normal development has been discovered in "annual fishes." These are cyprinodontid fishes found in isolated water holes or puddles that dry up seasonally. The fish population survives the dry season only in the form of eggs buried in the mud. After beginning development, the eggs may undergo three resting periods (diapauses) during development. Our special interest in them is that their early development is unlike that of other teleosts (bony fishes). Dispersion and reaggregation are shown in Figure 8-4.

According to Wourms (1967), early development is characterized by the dispersal and reaggregation of the cells. This begins with the dispersal of the ameboid cells (Figure 8-4*b* and *c*) lying between an outer (extra-embryonic) enveloping cell layer and an inner syncytial (periblast) layer. The ameboid cells remain dispersed for a day or two and increase by mitosis. Later, ameboid cells reaggregate to form the embryo body (Figure 8-4*e* and *f*). Apparently, the basic body form (head, trunk somites, and tail) results from aggregation of unspecialized cells followed by differentiation of the cells according to their positions.

What guides the reaggregation of the cells and later controls their differentiation is not known, but the periblast on which the reaggregation occurs probably plays an important role. The process is similar to the aggregation of slime mold amebae except that, in the fish, a preformed substrate may constitute the basic guideline. However, more experiments are necessary before we can draw any conclusions.

Reaggregation of Amphibian Embryo Cells

The capacity of dispersed cells at the neurula stage (Figure 8-5) in a number of species of amphibians has been demonstrated by Feldman (1955) and Townes and Holtfreter (1955). The ectoderm of an *Ambystoma* neurula in its vitelline membrane was dispersed by treatment with KOH at *p*H 9.8, and the neurula was returned to *p*H 8. The dispersed ectoderm cells then reaggregated and covered the partly denuded embryo. The embryo then developed in an almost normal way (Townes and Holtfreter, 1955).

Cell Combinations

In many experiments, Townes and Holtfreter dispersed cells (at the neurula stage) of the three germ layers (ectoderm, mesoderm, endoderm) and of structures such as the neural plate (Figure 8-6). Then, they mixed suspensions of these cells in various combinations and observed the pattern of reaggregation. Their method and the general type of result is shown in Figure 8-6.

A mixture of dispersed neural plate and epidermis cells first reaggregate at random, the cell types being intermingled. After about ten hours, the cells sort out, so that the epidermis cells occupy only the surface of the mass and the neural plate cells remain inside (Figure 8-6). Other combinations (mesoderm and endoderm; epidermis and mesoderm; neural plate, archenteron roof [axial mesoderm], and epidermis) are shown in Figure 8-7.

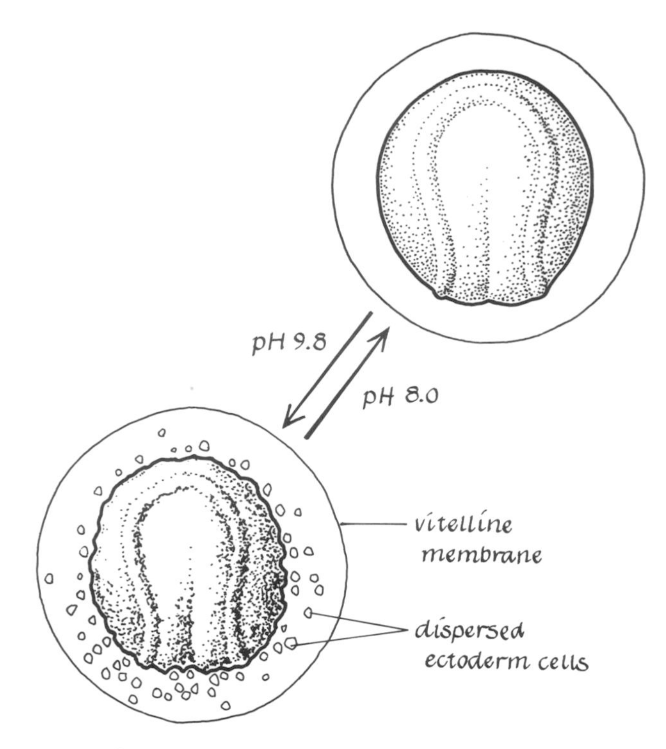

Figure 8-5. *Dispersal and reaggregation of ectoderm cells of the neurula of* Ambystoma.

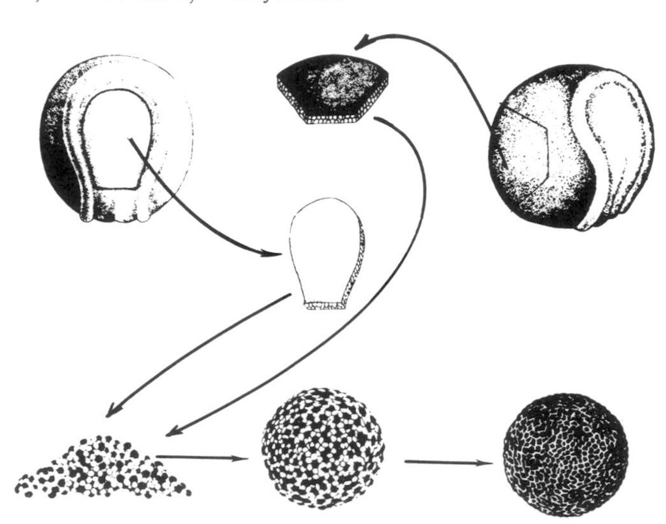

Figure 8-6. *Diagram of disaggregation, reaggregation and segregation in mixtures of epidermis and neural plate cells of* Ambystoma. *Epidermal cells are indicated in black. In segregation, the surface of the explant is covered with epidermis cells. From Townes and Holtfreter,* J. Exp. Zool., *Volume 128, p. 61, 1955. Courtesy Wistar Press.*

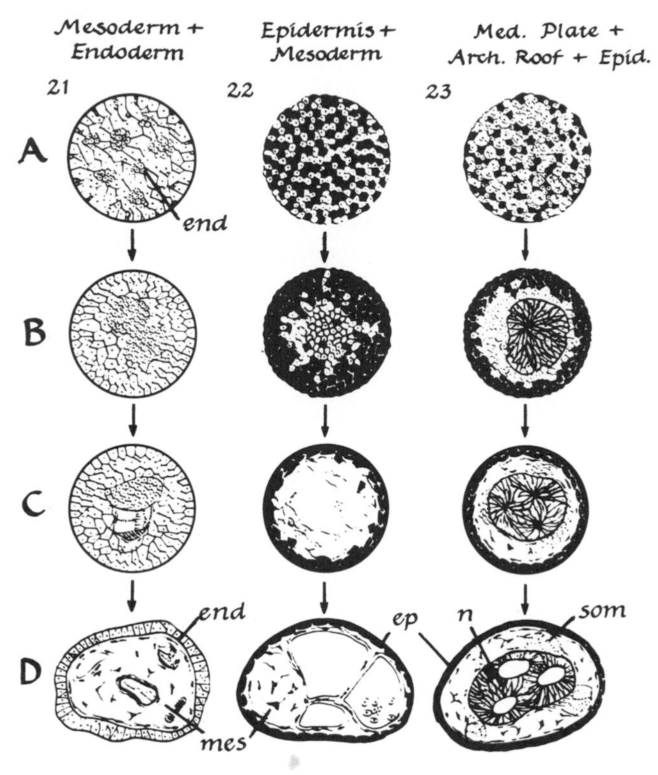

Figure 8-7. *Combining dissociated mesoderm and endoderm cells results in centripetal migration of the mesoderm cells. In combinations of dispersed epidermis and mesoderm cells, the epidermis cells move outward and the mesoderm cells move inward to form mesenchyme, coelemic cavities, and blood cells. In combinations of dispersed medullary (neural) plate, archenteron roof, and epidermis cells, sorting out results in formation of centrally located neural tissue surrounded by somites and epidermis covering the explant. From Townes and Holtfreter,* J. Exp. Zool., *Volume 128, p. 83, 1955. Courtesy Wistar Press.*

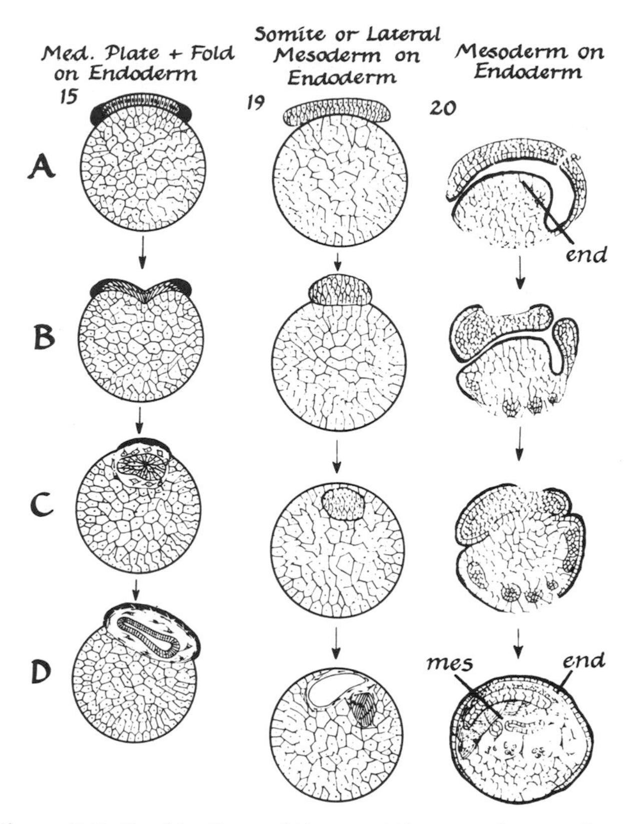

Figure 8-8. *Combinations of tissues at the neurula stage in* Ambystoma. *Combining neural fold and neural plate with endoderm results in partial invagination of the neural plate. Combining somite or lateral mesoderm with trunk endoderm results in incorporation of the mesoderm into the endoderm. Removal of the ectoderm from a neurula results in movement of the denuded mesoderm into the endoderm and spread of endoderm over the denuded embryo. From Townes and Holtfreter,* J. Exp. Zool., *Volume 128, pp. 74-79, 1955. Courtesy Wistar Press.*

Feldman (1955), studying reaggregation of *Triturus* cells, obtained similar results. No interconversion of cells occurred; therefore the reaggregating cells remain stable as to type. Unlike what would probably be the case with embryo dispersed cells at cleavage and blastula stages, the final position of neurula cells is in accord with their origins. Recall that the sorting out process is not so clear in the reconstitution of sponges from dispersed cells.

Tissue Combinations

Townes and Holtfreter (1955) have also tested the capacity of tissue fragments in artificial combinations to reestablish the normal tissue relations (Figure 8-8).

The tissue fragments move about until a more stable (normal) relationship is attained. In general, there are two kinds of movement: (1) inward movement (for example, neural plate or mesoderm in combination with epidermis or endoderm cells), and (2) outward movements and spreading (endoderm spreading over denuded mesoderm).

According to Townes and Holtfreter, each cell type has inherent migration tendencies and the stratification of reconstituted tissue layers corresponds to the normal germ layer relationship (Figure 8-7). Cells of like histotype adhere to one another when they meet. It seems probable that "self-marking" molecules on the cell surfaces may be progressively acquired as an aspect of cell differentiation (Burnett, 1956; Picken, 1960, for discussion). Spiegel (1954) has presented evidence for the presence of species-specific antigens on the surface of frog and *Triton* gastrulae. Binding the presumed antigens of frog cells with frog antisera inhibits the aggregation of frog ectoderm and endoderm cells but not that of *Triturus* gastrula cells. The role of the cell surface will be discussed more fully later in this chapter.

Reaggregation of Dispersed Chick Embryo Cells

Moscona (1952) discovered that trypsin dissociated tissues of 5 to 7 day incubated chick embryos, allowing the study of cellular associations in chick embryo cells. In some experiments trypsin or EDTA (ethylene diamine tetracetic acid) was used to disperse the tissues. EDTA preferentially chelates (binds) calcium ions. Using this method, Moscona (1956) and Trinkaus and Groves (1955) discovered that several types of tissue cells (neural retina, kidney, cartilage-forming, muscle-forming cells) after dispersal and removal of the dispersing medium, would reaggregate and reconstitute the tissues from which they originated. Furthermore, after mixing these dispersed cells, random reaggregation and then tissue-specific sorting occurs. The investigators found that reaggregation was facilitated by shaking the cell suspensions. Trinkaus and Lentz (1964) directly observed the sorting process in living cultures of chick retinal pigment cells (black) and chick heart ventricle cells (unpigmented). Figure 8-9 diagrams the result of sorting in a mixture of liver, pigmented retina, and chondroblast cells. The cartilage-forming cells occupy the innermost position, surrounded by pigmented retinal cells, which are surrounded by liver cells. Different tissue types begin to take definite positions in the aggregate during the sorting out process. Complete organs with remarkably normal tissue architecture may form within aggregates from the single cell dispersal of whole chick embryos (Weiss and Taylor, 1960).

Studies of reaggregation in 5 to 7 day old chick embryo cells reveal that the cells remain stable as to histotype. At this late stage, most cells are already specialized, and no transformation from one cell type to another occurs during the sorting out process. Furthermore, the final positions of the cells in the aggregate are in accord with their origin. During the early development of the chick embryo, however, the positions of cells determine to a great extent their specialization as distinct histotypes.

The surface of the cerebellum of the brain plays a role in controlling the orientation of dispersed and reaggregating neuroblasts (neuron-forming cells), as illustrated by the following experiment. Minced tissues of the cerebellum and optic tectum (visual center in the brain) were mixed and grafted to the chorio-allantoic membrane (extra-embryonic membrane) of a host chick embryo (Stefanelli et al., 1961). The orientation of the developing neurons is normal if an intact surface of the cerebellum is part of the graft, but not if this surface is missing. These results suggest that the arrangement of neurons in the brain depends primarily on surrounding structural features.

Reaggregation of Dispersed Adult Mammal Cells

Cells in tissue cultures of adult rabbit adrenal, aorta, heart, and cornea were dispersed with trypsin, and cell suspensions of each organ were placed in millipore filter diffusion chambers. The chambers were then implanted subcutaneously in serologically compatible litter mates. Histological examination of the reaggregated cells after two to three weeks revealed remarkable reconstitution of the tissue architecture of the organs. For example, cortex and medulla were present in the adrenal reaggregate, and heart tissue surrounding a chamber-like cavity developed in the heart reaggregate (Wada and Pollak, 1969).

Reaggregation of Mixtures of Dispersed Chick and Mouse Embryo Cells

Mixtures of 7 day old chick and 15 day old mouse embryo liver cells reaggregate to form a chimaeric liver with interspersed mouse and chick cells (Moscona, 1960). Mouse and chick chondrogenic cells also reaggregate to form a single cartilage mass bounded by a common matrix. In other experiments, embryonic chick and mouse chondroblasts and chick liver cells were dispersed and mixed together. After sorting, a single chimaeric cartilage mass formed separate from the reconstituted chick liver tissue

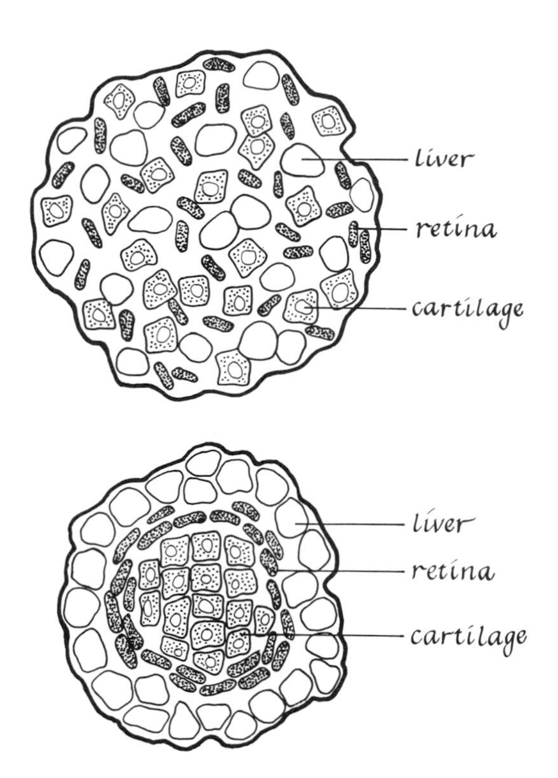

Figure 8-9. *Diagram of the sorting in a mixture of dispersed liver, pigmented retina, and chondroblast cells of a 5- to 7-day incubated chick embryo.*

(Moscona, 1957). Thus, cells of the same histotype adhere to one another to form chimaeric tissue, whereas cells of the same genotype (chick cartilage and liver) remain separate. Clearly, both the chick and mouse cells remain stable in respect to histotype. However, one cannot generalize that cells of the same histotype but different genotype never sort out by the species of origin. Burdick and Steinberg (1969) report that chick and mouse heart ventricle cells in mixed aggregates separate after about two days. Unpublished studies of Okada (cited by Burdick and Steinberg) show that mouse and chick embryonic kidney cells also sort out by genotype.

Even more complex cellular associations have been found (Garber and Moscona, 1964). Dissociated mouse and chick embryo skin cells reaggregate to form clusters of dermal cells covered by epidermis in which chimaeric follicles developed. However, if the dispersed mouse embryo skin cells are grown in monolayer culture for about ten days and then reaggregated by centrifuging, their capacity to reconstitute the tissue structure of the skin is lost (Moscona, 1960). Presumably, the cells are either injured by this treatment or the surface substances necessary for sorting out are lost and not replaced.

Mechanism of Sorting Out of Cells

Several hypotheses have been advanced to explain the mechanisms of cellular adhesion and sorting of vertebrate embryo cells in mixed aggregates (Townes and Holtfreter, 1955; Curtis, 1961, 1962; Steinberg, 1958, 1962, 1963, 1964; Moscona, 1960). In mixtures of two tissue types, sorting eventually leads to the reconstitution of one tissue peripheral to the other. For example, chondrogenic cells are surrounded by mesonephros (kidney) (Trinkaus and Groves, 1955); chondrogenic cells are surrounded by liver (Moscona, 1957); heart is surrounded by liver (Steinberg, 1963).

The surface structure of cells probably plays an important role in the cell aggregation, sorting out, and cohesion of cells of like histotype. Cell surfaces are a mosaic of large and functionally discrete macromolecular assemblies (Revel and Ito, 1967; Wallach, 1967). Steinberg (1958) suggests that the distribution of ions on the cell surface may play a part in the reaggregation of cells. The macromolecular structure of the cell surface in sponges seems to be important in aggregation (Moscona, 1968a, 1968b). The surface charge on cells may also be important, as cells with high negative charge are generally nonadhesive (Ambrose, 1967).

The role of the extra-cellular matrix of cell products in cellular adhesions is suggested by some studies of Moscona (1960). These products, which seem to bind the cells together, have been isolated from suspensions of chick embryo cells. They are digested by trypsin. Their addition to suspensions of cells rotating on a shaker forms larger aggregates than in control cultures. According to Moscona (1961, 1968a, 1968b), the matrix material is cell-type specific, only enhancing the aggregation of cells like those from which the substance was originally isolated. Electron microscopy of the cellular exudate indicates the presence of particles about 20 Å in diameter, the same order of magnitude as the particles in the supernatant of dispersed sponge cells. The presence of an intercellular matrix (which Overton, 1969, found to be fibrillar in reaggregates of chick embryo cells) cannot be doubted. Further studies on reaggregating 8-day-old chick liver and cartilage cells indicate that these cells form different products which presumably play a role in the sorting out process. Mucopolysaccharides appear to be released by the cells (Khan and Overton, 1969). When cells are dispersed, some of the surface macromolecules may leave the surface and become components of the intercellular matrix. The decreased capacity to aggregate of mouse skin cells grown as monolayers may result from the loss of surface macromolecules. Whether the intercellular matrix also plays a role in directing cell movements during sorting out has not been proved (Weiss, 1945; Moscona, 1960).

Townes and Holtfreter (1955) postulated active and directed cell migration as the cause of sorting out in mixtures of amphibian cells. Possibly the cells respond to radial chemical gradients established by diffusion of metabolites. This would result in one centrally located tissue surrounded by another, as diagramed in Figure 8-10.

Curtis (1961) has suggested that dissociation of *Xenopus* gastrula cells may induce physical changes in the cell surfaces from which various types of cells recover after variable periods of time. The final result would still be a cell mass of one type in the center of the aggregate surrounded by cells of the other type, as shown in Figure 8-10.

A third hypothesis to explain the distribution of two cell types following sorting out has been advanced by Steinberg (1962, 1963, 1964). He assumes that the forces of cohesion between cells of one histotype are specifically different from those between cells of another histotype. For example, the cohesion between cartilage-forming cells is greater than that between heart ventricle cells. The result of sorting out through differential adhesiveness is illustrated in Figure 8-10.

Depending on the strength and balance of cohesive forces between like cells and of adhesive forces between unlike cells, theoretically there could be: (A) no sorting out, (B) complete segregation and separation of the two cell types, or (C) the surrounding of one cell type by the other (Steinberg, 1963). The three possibilities are illustrated in Figure 8-11. Possibility 3 was obtained in nine of eleven different mixtures of two types of dispersed cells from 5 to 7 day incubated chick embryos. Some of Steinberg's results are illustrated in Figure 8-12.

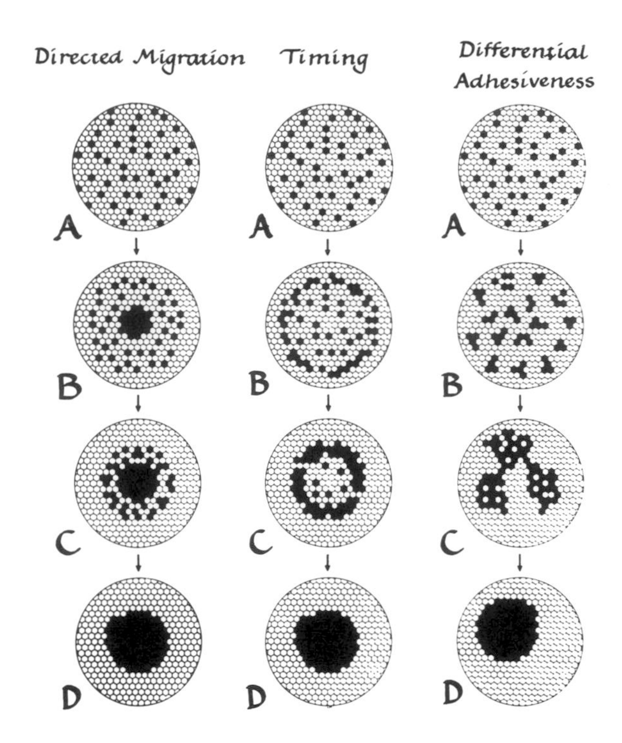

Figure 8-10. *The time course of segregation as it would appear if brought about by directed cell migration, by timing, or through differential cellular adhesion. From Steinberg, in* Cellular Membranes in Development. *Copyright 1964 by Academic Press.*

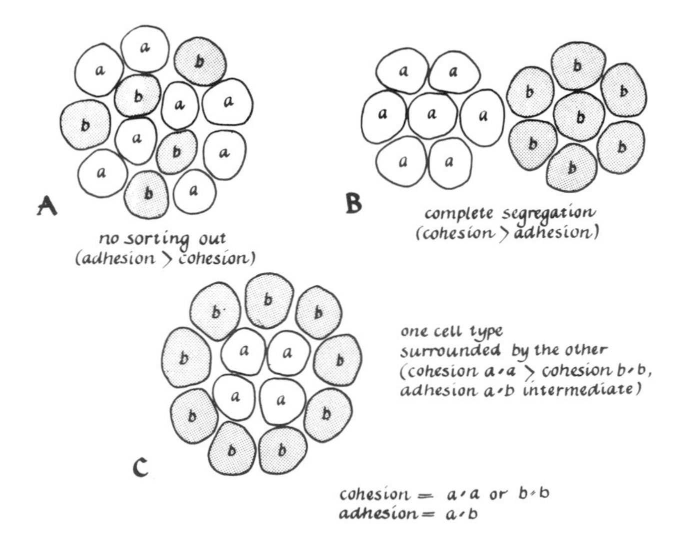

Figure 8-11. *Three possible relations following sorting out of two cell types in an aggregate.*

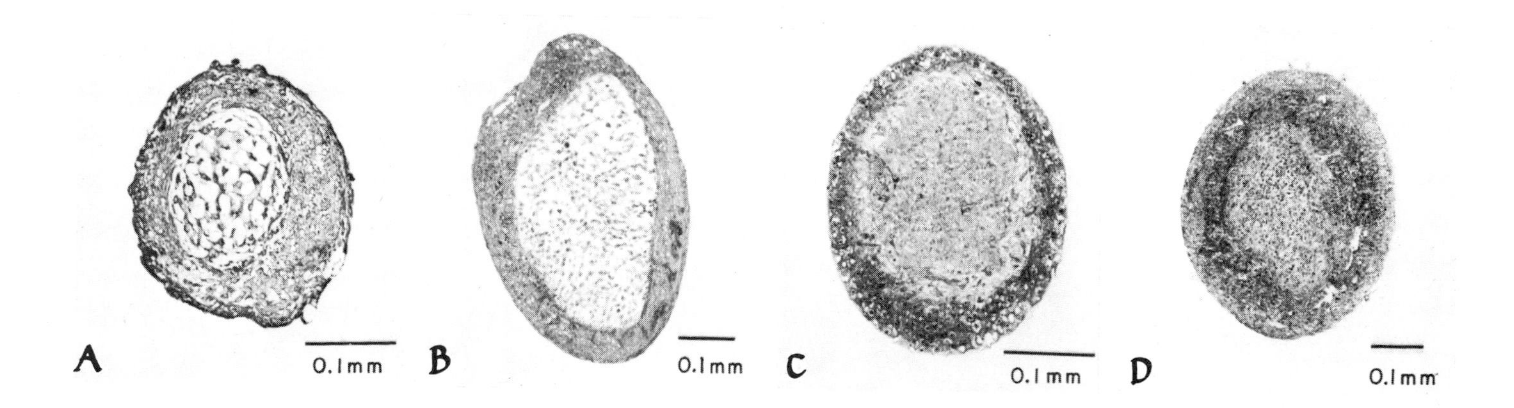

Figure 8-12. *A, section through an aggregate formed by dissociated 4-day limb-bud chondrogenic cells and 5-day heart ventricle cells of the chick embryo. Cartilage is surrounded by heart tissue. B, section through an explant formed by a fragment of chondrogenic 4-day limb-bud fused with a fragment of 5-day heart ventricle. The heart tissue surrounds the chondrogenic cells. C, section through an aggregate formed by dissociated 5-day heart ventricle cells and 5-day liver cells. The liver tissue envelops the heart tissue. D, section through an aggregate formed by dissociated 4-day limb-bud chondrogenic cells and 5-day liver cells. The liver tissue envelops the chondrogenic tissue. From Steinberg,* Science *(August 1963), 141:401–408. Copyright 1963 by the American Association for the Advancement of Science. Photographs courtesy of Dr. Steinberg.*

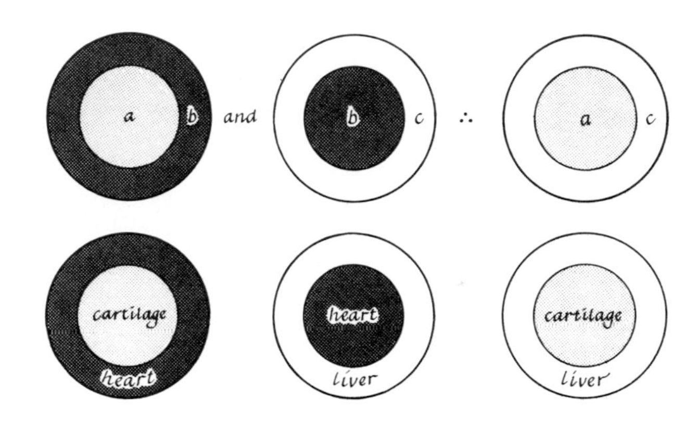

Figure 8-14. *Diagrams of the general rule that less cohesive cells, b, will envelop more cohesive cells, a. Thus cartilage is enveloped by heart, heart by liver, and cartilage by liver.*

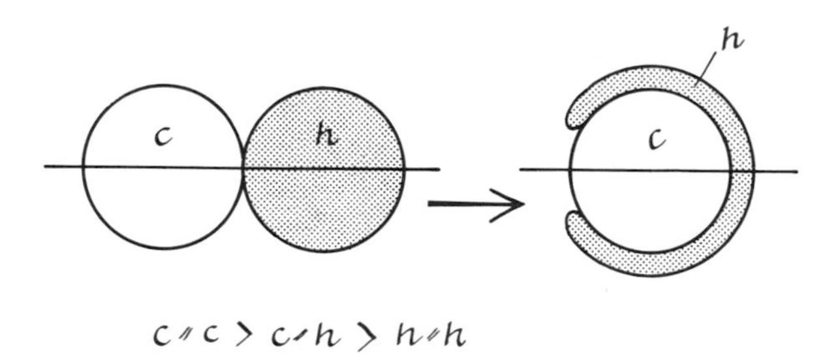

Figure 8-13. *Diagram of the shish-kebob method for fusing fragments of embryonic tissues.*

The characteristic distribution of two dispersed tissue cell types when sorting is complete has also been found when two fragments of tissue (four-day chick cartilage and five-day chick heart) are fused by impaling them shish-kebab fashion on a fine glass thread (Figure 8-13). In reaggregates of dispersed chondrogenic and heart cells, the chondrogenic cells are surrounded by heart cells, that is, C-C > H-H but C-H is intermediate (Figure 8-12A). As indicated in Figure 8-13, the same arrangement is obtained with intact tissue fragments by the spreading of the heart cells over the cartilage-forming cells. More recent studies of Steinberg (1970) suggest that differential adhesion operates both in reassembled cell types resulting from sorting out in mixed aggregates and in the spreading of one tissue mass over another.

From studying fifteen possible combinations of two cell types among six different tissues, Steinberg found that the following rule is obeyed: in any population of two types of mutually adhesive, motile cells, the less cohesive will envelope the more cohesive when sorting out is completed. Therefore, if cell type A is covered by cell type B and B by cell type C, then A will be covered by C. This prediction holds true, as illustrated in Figure 8-14. In several combinations of different tissues of chick embryos, Steinberg (1970) found the hierarchy of tissues for an internal position to be as illustrated in Figure 8-15. The diagram is a hypothetical composite deduced from the observation that each tissue indicated, except liver, is surrounded by the tissue external to it. As Steinberg has clearly demonstrated, intercellular products can influence the configuration of the multitissue aggregate.

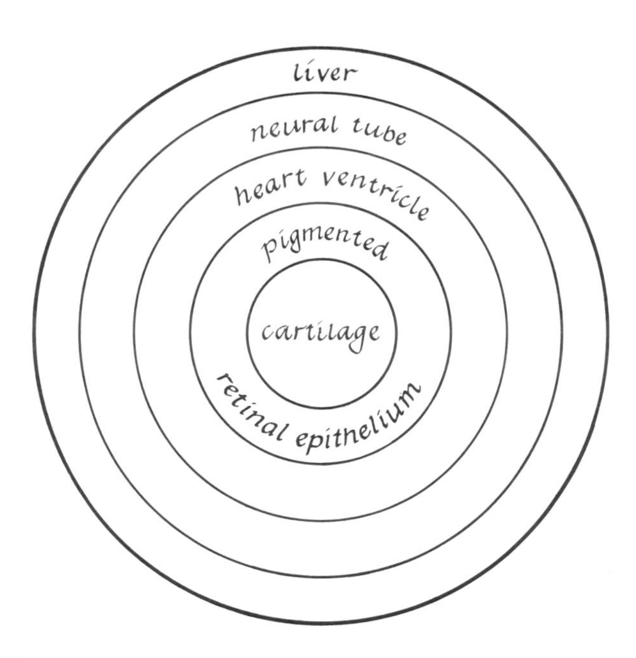

Figure 8-15. *Diagram of the hierarchy of tissues for an internal position in an aggregate.*

Conclusion

The embryo cells used in the studies of reaggregation and sorting out of experimentally dispersed cells are so specialized that their further differentiation is always in accord with their origin. (The reconstitution of sponges from dispersed cells may be a partial exception.) Unspecialized cells in intact younger embryos differentiate in accord with their position in the embryo. By contrast, the positions of specialized cells in mixed aggregates are a function of their nature, that is, determined by their specific properties. When the dispersed cells are those normally adjacent to one another in the embryo, their positions after sorting out in a mixed reaggregate closely simulate those within the embryo.

Although sorting out is a rarely used mechanism in normal embryonic development, studies such as those described here provide important information about cell-to-cell relationships not otherwise discernible in intact embryos. Probably, differences in adhesiveness reflecting the macromolecular architecture of the cell surfaces are responsible for the various types of spreading of one tissue over another or of the penetration of one tissue into another as in gastrulation. Cellular relationships are an important aspect of development, and only through studies of dispersed and reaggregating tissue cells can some of these relationships be discerned.

References

Ambrose, E. J. 1967. Possible mechanisms of the transfer of information between small groups of cells. In A. V. S. DeReuck and J. Knight, eds., Cell Differentiation, pp. 101–115. Little, Brown, Boston.

Berrill, N. J. 1961. Growth, Development, and Pattern. W. H. Freeman, San Francisco.

Brien, P. 1937. La réorganisation de l'eponge après dissociation par filtration et phénomènes d'involution chez *Ephydatia fluviatilis.* Arch. Biol. **48:**185–268.

Brooks, W. K. 1893. The genus *Salpa.* Johns Hopkins Biol. Lab. Mem. **2:**1–396.

Burdick, M. L., and M. S. Steinberg. 1969. Embryonic cell adhesiveness: do species differences exist among warm-blooded vertebrates? Proc. Nat. Acad. Sci. U.S. **63:**1169–1173.

Burnett, F. M. 1956. Enzyme, Antigen, and Virus. Cambridge University Press, Cambridge.

Curtis, A. S. G. 1961. Timing mechanism in specific adhesion of cells. Exp. Cell Res. Suppl. **8:**107–122.

Curtis, A. S. G. 1962. Cell contact and adhesion. Biol. Rev. Camb. **37:**82–129.

Farris, V. K. 1968. Molluscan cells: dissociation and reaggregation. Science **160:**1245–1246.

Feldman, M. 1955. Dissociation and reaggregation of embryonic cells of *Triturus alpestris*. J. Emb. Exp. Morphol. **3:**251–255.

Fox, A. S., M. Horikawa, and L. L. Ling. 1968. The use of *Drosophila* cell cultures in studies of differentiation. In M. M. Sigel, ed., Differentiation and Defense Mechanisms in Lower Organisms, pp. 65–84. Waverly Press, Baltimore, Md.

Galtsoff, P. S. 1925. Regeneration after dissociation (an experimental study of sponges): II. Histogenesis of *Microciona prolifera*. J. Exp. Zool. **42:**223–255.

Garber, B., and A. A. Moscona. 1964. Aggregation in vitro of dissociated cells: I. Reconstruction of skin on the chorioallantoic membrane from suspensions of embryonic chick and mouse skin cells. J. Exp. Zool. **155:**179–202.

Giudice, G., and V. Mutolo. 1970. Reaggregation of dissociated cells of sea urchin embryos. Adv. Morph. **8:**115–158.

Horstadius, S. 1939. The mechanics of sea urchin development studied by operative methods. Biol. Revs. Camb. **14:**132–179.

Humphreys, T. 1963. Chemical dissolution and *in vitro* reconstruction of sponge cell adhesions. I. Isolation and functional demonstration of the components involved. Devel. Biol. **8:**27–43.

Humphreys, T. 1967. The cell surface and specific cell aggregation. In B. D. Davis and L. Warren, eds., The Specificity of Cell Surfaces, pp. 195–210. Prentice-Hall, Englewood Cliffs, N.J.

Hyman, L. H. 1940. The Invertebrates: Protozoa through Ctenophora. McGraw-Hill Book Company, New York.

Khan, T., and J. Overton. 1969. Staining of intercellular material in reaggregating chick liver and cartilage cells. J. Exp. Zool. **171:**161–173.

Moscona, A. A. 1952. Cell suspensions from organ rudiments of chick embryo. Exp. Cell Res. **3:**535–539.

Moscona, A. A. 1956. Development of heterotypic combinations of dissociated embryonic chick cells. Proc. Soc. Exp. Biol. Med. **92:**410–416.

Moscona, A. A. 1957. The development in vitro of chimaeric aggregates of dissociated embryonic chick and mouse cells. Proc. Nat. Acad. Sci. U.S. **43:**184–194.

Moscona, A. A. 1960. Patterns and mechanisms of tissue reconstruction from dissociated cells. In D. Rudnick, ed., Developing Cell Systems and Their Control, pp. 45–70. Ronald Press, New York.

Moscona, A. A. 1968a. Aggregation of sponge cells: cell-linking macromolecules and their role in the formation of multicellular systems. In M. M. Sigel, ed., Differentiation and Defense Mechanisms in Lower Organisms, pp. 13–21. Waverly Press, Baltimore, Md.

Moscona, A. A. 1968b. Cell aggregation: properties of specific cell-ligands and their role in the formation of multicellular systems. Devel. Biol. **18:**250–277.

Overton, J. 1969. A fibrillar intercellular material between reaggregating embryonic chick cells. J. Cell Biol. **40:**136–143.

Picken, L. 1960. The Organization of Cells. Clarendon Press, Oxford.

Revel, J. P., and S. Ito. 1967. The surface components of cells. In B. D. Davis and L. Warren, eds., The Specificity of Cell Surfaces, pp. 211–234. Prentice-Hall, Englewood Cliffs, N.J.

Spiegel, M. 1954. The role of specific surface antigens in cell adhesion. Biol. Bull. **107:**130–155.

Stefanelli, A., A. M. Zacchei, and V. Ceccherini. 1961. Reconstituzione retiniche in vitro dopo disagregazione dell'abozzo oculare di embrione di pollo. Acta Emb. Morphol. Exptl. **4:**47–55.

Steinberg, M. S. 1958. On the chemical bonds between animal cells: a mechanism for type-specific association. Am. Nat. **92:**65–82.

Steinberg, M. S. 1962. On the mechanism of tissue reconstruction by dissociated cells: I. Population kinetics, differential adhesiveness, and the absence of direct migration. Proc. Nat. Acad. Sci. U.S. **48:**1577–1582.

Steinberg, M. S. 1963. Reconstruction of tissues by dissociated cells. Science **141:**401–408.

Steinberg, M. S. 1964. The problem of adhesive selectivity in cellular interactions. In M. Locke, ed., Cellular Membranes in Development, pp. 321–366. Academic Press, New York.

Steinberg, M. S. 1970. Does differential adhesion govern self-assembly processes in histogenesis? equilibrium configurations and emergence of a hierarchy among populations of embryonic cells. J. Exp. Zool. **173:**395–433.

Townes, P., and J. Holtfreter. 1955. Directed movements and selective adhesion of embryonic amphibian cells. J. Exp. Zool. **128:**53–120.

Trinkaus, J. P., and P. W. Groves. 1955. Differentiation in culture of mixed aggregates of dissociated tissue cells. Proc. Nat. Acad. Sci. U.S. **41:**787–795.

Trinkaus, J. P., and J. P. Lentz. 1964. Direct observation of type-specific segregation in mixed cell aggregates. Devel. Biol. **9:**115–136.

Wada, A., and D. J. Pollak. 1969. In vitro organ reconstruction from cells cultured in vitro. Cell and Tissue Kinetics **2:**67–74.

Wallach, D. F. H. 1967. Isolation of plasma membranes of animal cells. In B. D. Davis and L. Warren, The Specificity of Cell Surfaces, pp. 129–163. Prentice-Hall, Englewood Cliffs, N.J.

Walter, D. R. 1969. Reaggregation of insect cells in vitro: I. Adhesive properties of dissociated fat-body cells from developing saturnid moths. Biol. Bull. **137:**217–227.

Weiss, P. 1945. Experiments on cell and axon orientation in vitro: the role of colloidal exudates in tissue organization. J. Exp. Zool. **100:**353–386.

Weiss, P., and A. C. Taylor. 1960. Reconstitution of complete organs from single cell suspensions of chick embryos in advanced stages of differentiation. Proc. Nat. Acad. Sci. U.S. **46:**1177–1185.

Wilson, H. V. 1907. On some phenomena of coalescence and regeneration in sponges. J. Exp. Zool. **5:**245–258.

Wourms, J. P. 1967. Annual fishes. In F. H. Wilt and N. K. Wessells, eds., Methods of Developmental Biology, pp. 123–137. T. Y. Crowell, New York.

9 *Patterns of Asexual Reproduction*

Reproduction is the process by which an organism produces a new individual of the same species. The new individual may be similar to its parent, or it may be morphologically different. There are many ways in which organisms reproduce but only two basic categories: asexual and sexual reproduction. Some organisms reproduce by asexual methods exclusively, others by exclusively sexual methods, and still others by both asexual and sexual methods. In many organisms that can reproduce by both sexual and asexual methods, the type of reproductive pattern is controlled by environmental influences. The role of the external environment in selecting patterns of reproduction will be discussed in Chapter 22.

In sexual reproduction, the formation of a new individual usually requires two sexually different parents, or at least two different kinds of sex cells. The latter are produced either by separate parents described as belonging to sexually dimorphic species, or by a single parent called a hermaphrodite. In any case, sexual reproduction involves genetic recombination. Asexual reproduction, on the other hand, is accomplished by a single parent and involves no fusion or exchange of genetic material. In this and the following three chapters we shall examine patterns of reproduction and the problems of understanding the mechanisms involved.

In our survey of some methods of becoming multicellular (Chapters 6, 7, and 8), we noted the great diversity made possible by the various kinds of cellular associations in multicellular organisms. We have also observed that an increase in the number of associated cells makes possible higher orders of complexity, such as the specialization of cells in the multicellular organism. Reproduction in multicellular organisms is basically an opposite process: that of becoming smaller rather than larger—in many instances even becoming unicellular. The processes of becoming smaller or unicellular have received far less attention by biologists than the development of the reproductive structure into the new adult. As emphasized by Picken (1960): "The process of becoming unicellular, of producing a unicellular phase, which all metazoa regularly undergo, is as remarkable as that of becoming multicellular, but the implications of the former process have in part been lost sight of." The reproductive process requires that all the genetic information necessary to direct development of the new individual must be packaged in a reproductive unit. Although in many patterns of reproduction the process is essentially packaging information (in genes) in the nucleus of the cell, the necessary information is not located exclusively in the nuclear genes. The problem of the biologist is understanding the mechanisms used by each organism to set aside a potentially independent part of its body, either a part of a cell, a whole cell, or a group of cells. Different mechanisms are used by different organisms, and some organisms employ two or more mechanisms of reproduction, including both sexual and asexual devices. In this chapter, we shall survey and attempt to classify patterns of asexual reproduction. Methods of asexual reproduction vary from simple fragmentation of the adult to formation of specialized units, either multicellular (buds or gemmules) or unicellular (spores).

Fragmentation

In this method of reproduction, the new individual develops from a detached piece of the parent organism. This is the only method of reproduction of some brown algae (Phaeophyta), such as some free-floating species of *Sargassum* found in the Sargasso Sea. It is also a common method in blue-green algae such as *Ocillatoria*. In this filamentous alga, breaks occur where cells have died, as shown in Figure 9-1 (Brown, 1935).

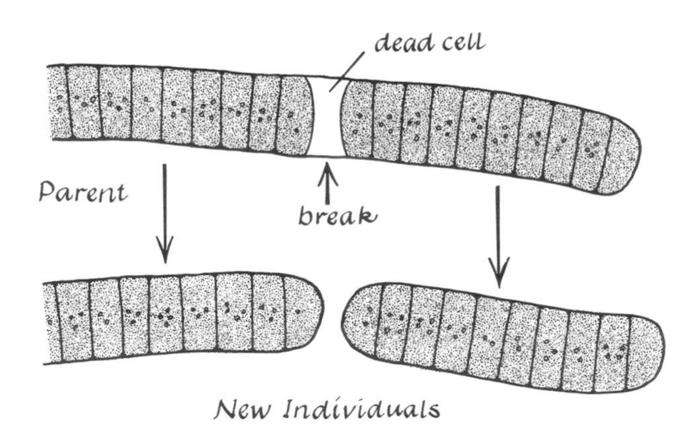

Figure 9-1. *Asexual reproduction by fragmentation in* Oscillatoria.

Fragmentation also occurs in animals. Among many examples are adult nemertean and polychaete annelid worms. As we saw in Chapter 5, minute pieces of nemerteans resulting from experimentally induced fragmentation can develop into whole worms, providing the length of a piece exceeds its width (review of Berrill, 1961).

In patterns of fragmentation, the entire mother organism is actually a unified collection of reproductive units. In annelids, the units may be single body segments. Some worms have morphologically visible breaking points corresponding to constrictions of the alimentary canal. Guidelines for formation of a new individual by fragmentation are mainly preformed in the fragment, but the process of reconstitution, similar to regeneration, is clearly a progressive process involving cell proliferation and cell differentiation. The symmetry and polarity of the new individual is directly inherited and is the same as that of the fragment and the mother organism. Development from fragments is thus direct, without the intervention of the complex stages that characterize the development of sexual reproductive units such as eggs. However, as we shall see in Chapter 13, some eggs possess guidelines directly inherited from the mother.

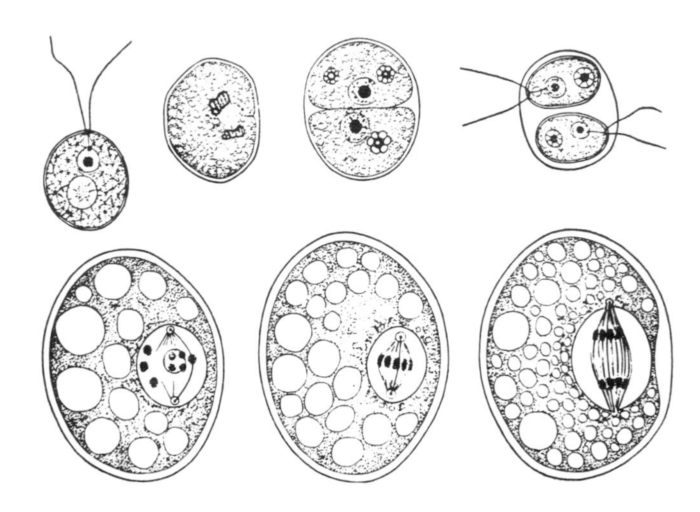

Figure 9-2. *Fission in* Chlamydomonas nasuta. *From Brown:* The Plant Kingdom, *1935. Reproduced by permission of Mrs. Brown.*

Fission

This method of reproduction, similar to fragmentation except that it is more orderly, is used by bacteria, protozoa, algae, coelenterates, nemerteans, annelids, echinoderms, and tunicates. Fission involves both the division of an organism by mitosis into two parts and the division or replication of its internal structures. In many cases, the morphology of the two parts is extensively reorganized. A few examples of fission are described below.

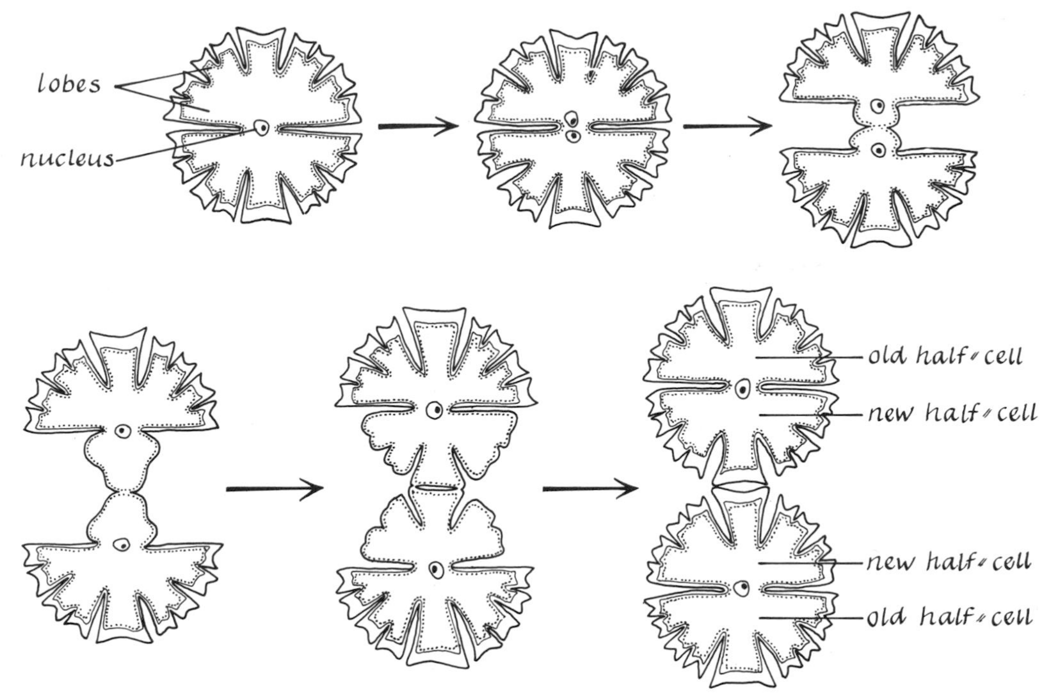

Figure 9-3. *Fission in* Micrasterias.

Chlamydomonas

In *Chlamydomonas* and other green algae, nuclear division precedes division of the parent cell (cytokinesis) and also of cytoplasmic organelles such as the stigma (eye spot), basal bodies of the flagella, and the pyrenoid (starch center). Fission in *Chlamydomonas nasuta* is shown in Figure 9-2. The two new individuals produced by division are halves of the parent cell, yet they are also whole new organisms capable of further reproduction by fission. Thus, in terms of reproductive potential each *Chlamydomonas* represents many individuals. It would be interesting to calculate how many generations must occur before the descendants of a single parent cell would contain no nongenomic molecules originally present in the parent.

Micrasterias

One striking example of the direct influence of the parent cell on the properties of its daughter cells is revealed by the study of fission in green algae of the group called desmids, particularly in the genus *Micrasterias* ("little star"). The pattern of division in this desmid is illustrated in Figure 9-3. In this biradially symmetrical alga, two new halves of the daughter cells are formed between the two old halves of the parent cell. The result is that in every *Micrasterias* individual, one-half of the body is younger than the other half. It would be difficult to find a better example of the continuity between parent and offspring (Figure 9-4). The lineage 1, 2, 3, 4 illustrates direct inheritance of one-half of the parent. The lineage 1, 2, 5, 6 illustrates indirect inheritance through the new half-cell formed by the parent. The new half-cell is essentially a unicellular reproductive unit capable of forming a complete *Micrasterias* by fission.

Not only is continuity obvious from the study of normal dividing *Micrasterias*, but it is also demonstrated by the reproduction of experimentally produced abnormal cells. Centrifuging the algae just before fission begins produces defective cells (Waris and Kallio, 1964). Among the abnormal types found are enucleate cells, binucleate cells, and diploid cells (most are normally haploid). Cold (shock) treatment at the telophase of division may result in uniradiate, aradiate, or more complex types of symmetry. Some examples of experimentally produced abnormal cells are shown in Figure 9-5, and the method of reproduction of two abnormal cells is illustrated in Figure 9-6. Note that the parent cell (Type I) produces an abnormal cell like itself plus two normal cells.

According to Waris and Kallio (1964), the abnormal form of *Micrasterias,* whether uniradiate, aradiate, or more complex, is transmitted through as many as one thousand generations without change in symmetry. These investigators suggested that the basis of symmetry is independent of the nucleus. However, the degree of complexity of the lobulation is influenced by the presence of the nucleus. Evidence for cytoplasmic control of symmetry or asymmetry is found in such observations as: (1) a defect in one half-cell is always on the same side of the cell through many generations; (2) the aradiate form (Figure 9-5) is stable (irreversible); and (3) the daughter cell formed by an enucleate cell has the same type of symmetry as the parent cell. Although Waddington (1962, 1963) could not detect any cytoplasmic fibrillar pattern in electromicrographs of *Micrasterias,* some radial pattern can be seen in living cells. Perhaps the pattern is not preserved in the electron microscopic preparations. What appear to be heritable nuclear mutations have also been produced by gamma irradiation of *M. Torregi* (Waris and Rouhiainen, 1970).

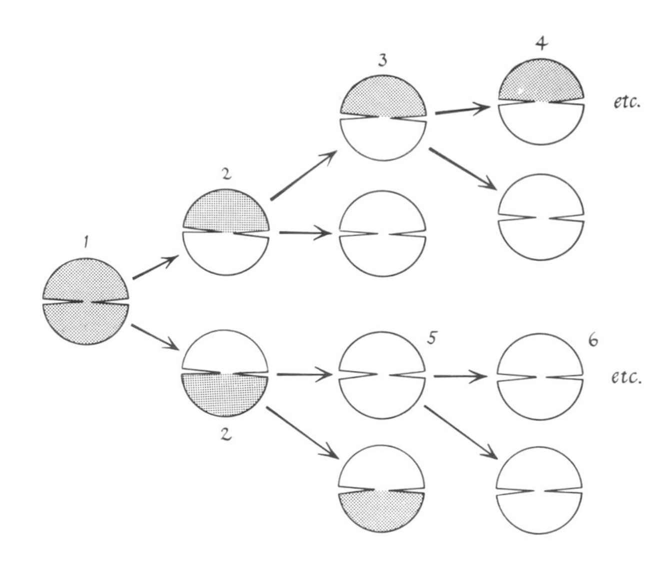

Figure 9-4. *Cell lineage and continuity between parent* Micrasterias *(1) and offspring (2-6) derived by fission. The shaded part shows direct inheritance of one-half of the parent.*

Figure 9-5. Micrasterias *cells. Upper: biradiate haploid, biradiate diploid, uniradiate haploid, aradiate haploid. Lower: enucleate cells—aradiate, uniradiate, biradiate, and triradiate. From Waris and Kallio in* Advances in Morphogenesis, *Volume 4. Copyright 1964 by Academic Press.*

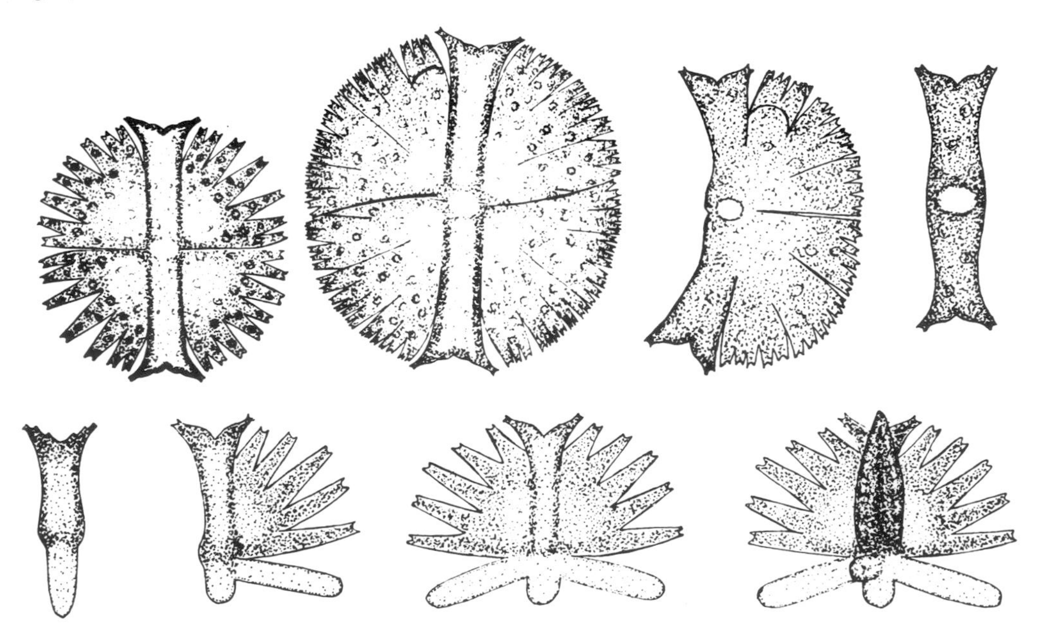

Stentor

The orderly process of transverse fission in *Stentor* has been extensively studied by Johnson (1893) and reviewed by Tartar (1961). Stages in its fission are illustrated in Figure 9-7. Extensive reorganization of cytoplasmic organelles and synthesis of new ones is involved in fission of *Stentor,* as it is in protozoa generally. The changes are almost identical to those occurring in the regeneration of one-half a *Stentor* (Chapter 5). A preformed fission plane is morphologically present in *Stentor* (Tartar, 1968), but according to Tartar, cutting the *Stentor* in half above or below this plane does not prevent fission.

Fission also occurs in complex multicellular organisms such as adult Sipunculid worms (Rice, 1970) and echinoderms (*Asteroidea, Ophiuroidea,* and *Holothuroidea*). Starfishes can divide in half through the disc along a preformed line that avoids the ossicles and leaves the arms intact (Hyman, 1955). The wound closes over and new arms are regenerated. In one species of starfish which has four madreporites (openings to the water vascular system), two anal openings, and eight arms, the disc splits to leave four arms, two madreporites, and one anus in each half. The parent in this species is essentially double. Sea cucumbers *(Holothuroidea)* undergo fission by transverse constriction, twisting, and pulling apart.

Zooid Formation

This process of reproduction is similar to fission but can be distinguished from it, because chains of two to seven individuals (zooids) are formed before the separation process. A classic example of zooid formation found in *Planaria* (Child, 1941) is illustrated in Figure 9-8. The plane of fission develops in the region just posterior to the pharynx and eyes, and a new pharynx develops before separation of the zooid from the parent worm. In other species of planarians (for example, *Dugesia paramensis*) the fission plane is anterior to the pharynx, and a new pharynx develops in the anterior half after separation. In still other planarians, the parent worm suddenly constricts or fragments, and the pieces form the missing organs.

In some flatworms such as *Stenostomum incaudatum* (Child, 1902, 1941), chains of about seven individuals are formed. The order of the division planes is

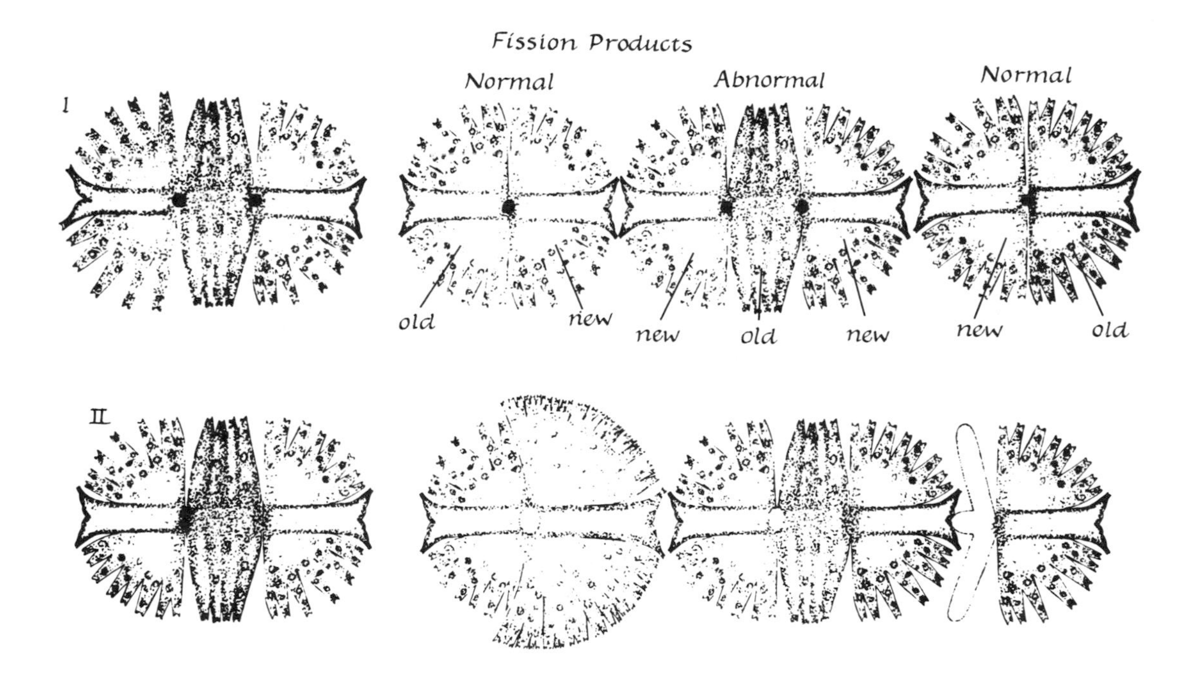

Figure 9-6. *Two complex types of abnormal* Micrasterias *cells and their method of replication. From Waris and Kallio in* Advances in Morphogenesis, *Volume 4. Copyright 1964 by Academic Press.*

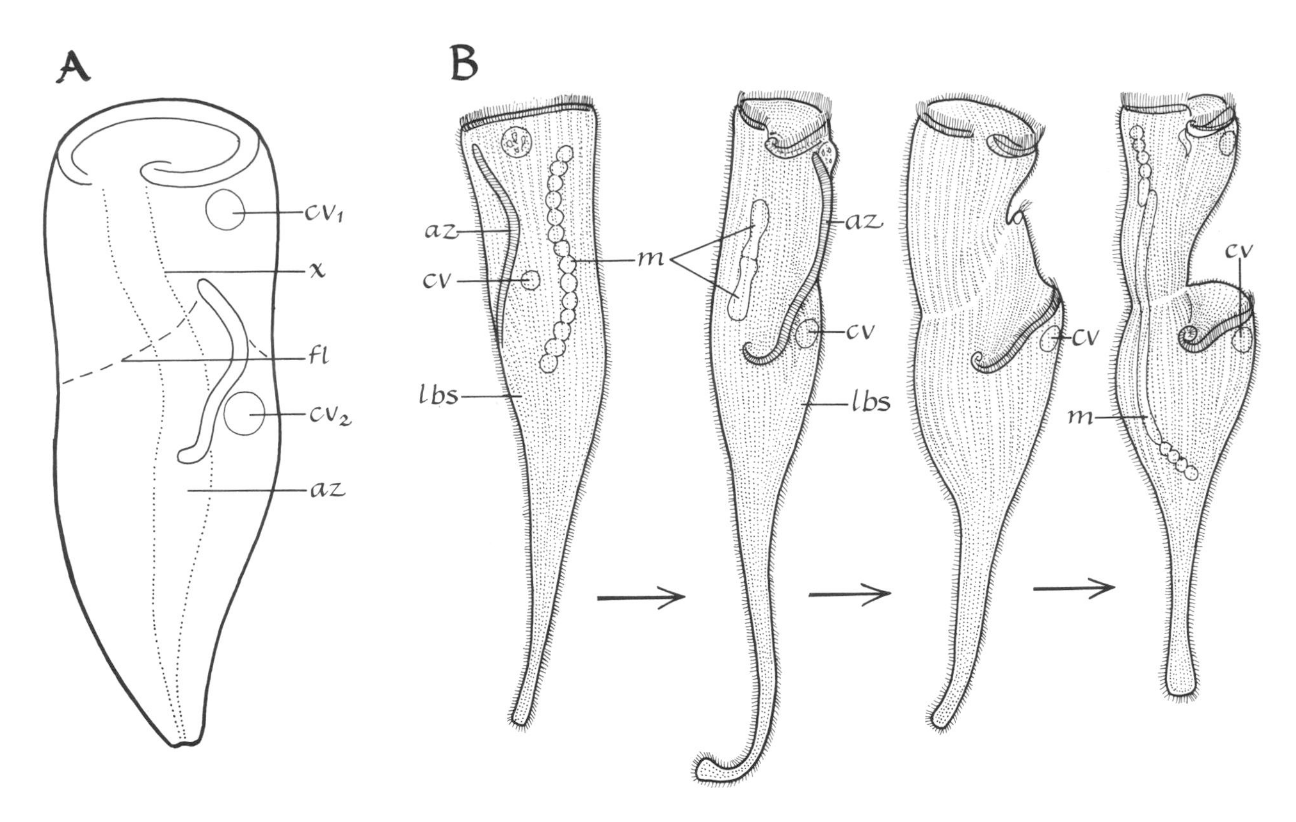

Figure 9-7. *A, diagram of* Stentor. A, *course of fission line,* fl; x, *path of cytoplasmic stripes;* cv1 *and* cv2, *old and new contrictile vacuoles, respectively.* B, *stages in fission,* m, *macronucleus;* cv, *new contrictile vacuole;* az, *new adoral zone. After Johnson and others.*

indicated in Figure 9-9. Note that the divisions are alternately anterior and posterior to the plane of the first division and that the new individuals are formed from the middle of the body of the original parent. Recall that new *Micrasterias* individuals are formed in a similar position. Study of the exact positions of the division planes indicates that the new fissions occur closer to new than to old heads. An analysis by Sonneborn (1930) revealed that the anterior region (including the original head) and the posterior region (including the tail) tend to remain unchanged, except that they age and lose fission capacity. The middle third of the worm grows and produces the fission zones (Berrill, 1961).

In some oligochaete annelids (for example, *Nais*), chains of zooids are produced by a fission zone near the middle of the worm which is a region of active cell division. In some polychaetes (such as *Autolytus*), a zooid (also called a stolon head) always develops on the anterior half of segment 14, specialized for this type of asexual reproduction. Stolon heads are formed before fission occurs. Removal of all the segments anterior to number 13 results not only in regeneration of a head from segment 13, but also

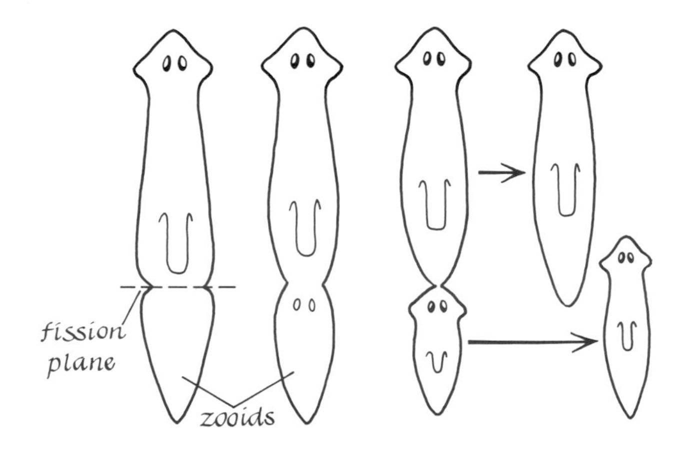

Figure 9-8. *Zooid formation and fission in* Planaria.

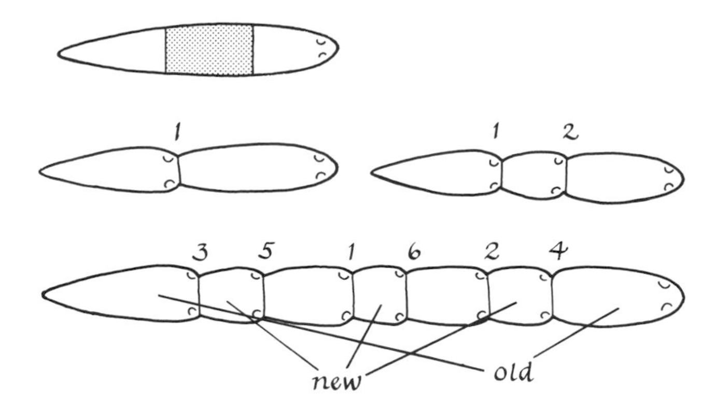

Figure 9-9. *Formation of zooids in* Stenostomum. *1-6, order of division planes. The shaded region indicates the part of the parent worm from which new individuals arise.*

formation of a stolon head from number 14 (Okada, 1929). In other polychaetes, chains of stolon heads may develop (Berrill, 1961).

A similar kind of zooid formation, usually described as strobilation, is found in tunicates and cnidarians, in the latter particularly in the group called Scyphomedusae (oceanic jellyfishes). As shown in Figure 9-10, zooids are progressively formed from the distal to the proximal part of the polyp (the hydroid form). The mature zooids are released in the order shown in the figure (1 to 15) and develop as ephyrae. Apical zooids, which are larger, form ephyrae with up to twelve lobes; basal zooids produce ephyrae with as few as four lobes (Berrill, 1961). Each ephyra develops into a medusa that reproduces sexually by formation of eggs and sperm when mature. The fertilized egg develops into a planula larva, which becomes a polyp.

In fission and zooid formation of annelids and flatworms and in strobilation of coelenterates and tunicates, the common method of interrupting the integrity of the organism to isolate a part for reproduction is the ingrowth of epidermis in a region of rapid cell

Figure 9-10. *Strobilation in the jellyfish* Aurelia.

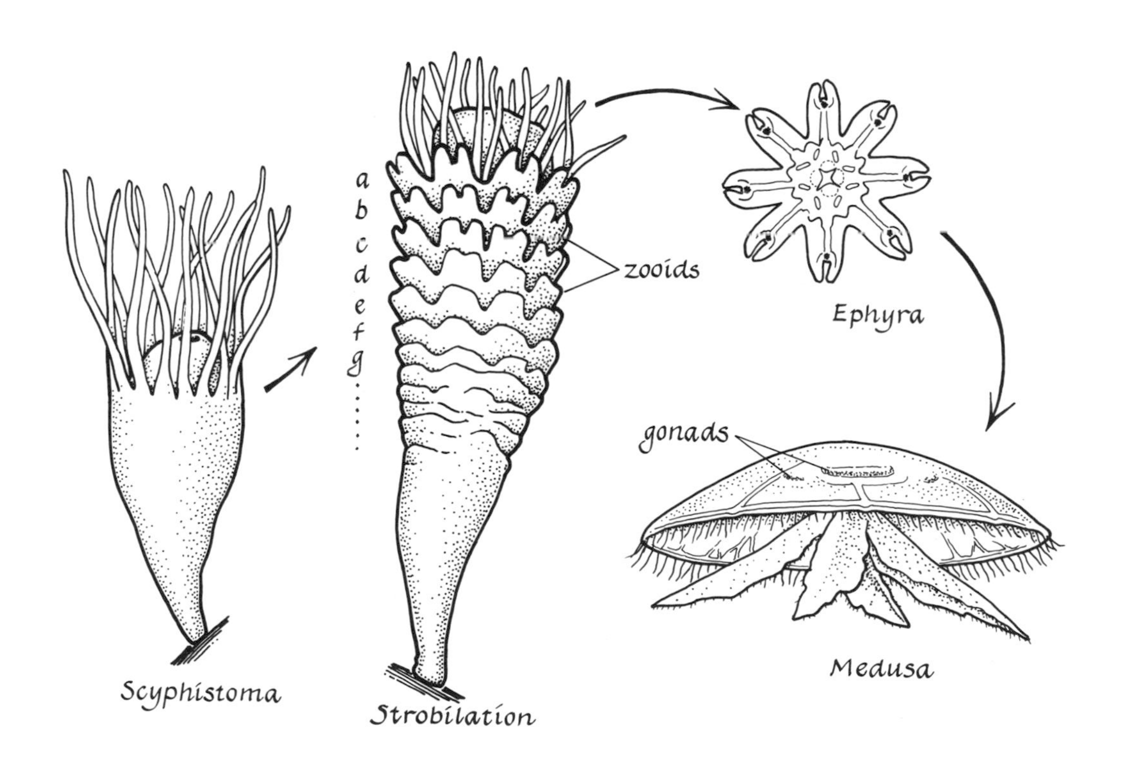

proliferation. In general, isolation of a part from the parent organism precedes (initiates?) reconstruction of the part into a whole individual. In some respects, this is different from the formation of specialized reproductive units (spores, buds, gemmules) that later become isolated from the parent organism. However, isolation of the prospective spore- or bud-forming cells from the parent may also initiate the specialization process (Berrill, 1961).

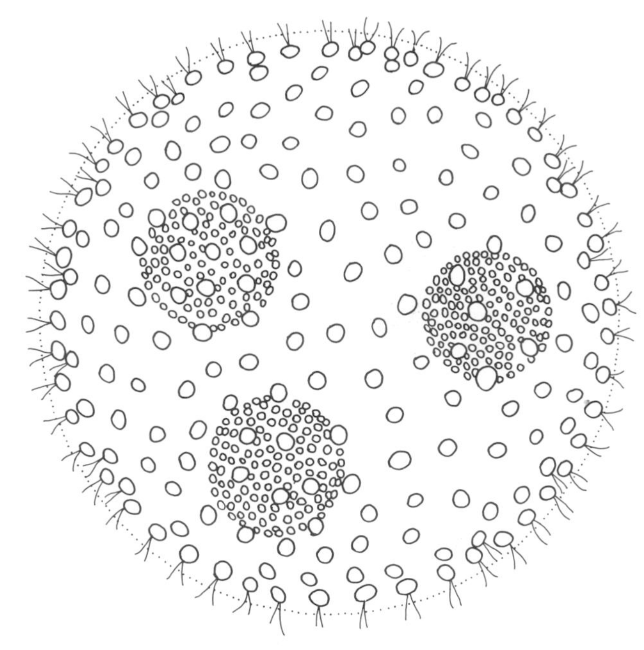

Figure 9-11. *Diagram of a* Volvox *colony, containing asexually produced daughter colonies derived from gonidial cells.*

Bud Formation (Blastogenesis)

Another common method of asexual reproduction is the formation of buds that become detached from the parent organism and function as specialized reproductive units. Some buds are unicellular, but most buds in both plants and animals are multicellular, consisting of several kinds of specialized cells arranged in a definite pattern. Their development is more epigenetic than that of fragments, fission products, or zooids; that is, the guidelines as a whole are progressively formed rather than preformed. There is such great diversity in the formation and development of buds in plants and animals that we can examine only a few representative examples.

Unicellular Buds

Unicellular buds are formed by bacteria, some algae, and the yeast *Saccharomyces.* Gonidial cells of *Volvox* (Chapter 7) and conidiospores produced by several kinds of fungi (Figure 7-35) are similar in function to unicellular buds. An adult *Volvox* colony containing daughter colonies that have developed from gonidial cells is illustrated in Figure 9-11. The gonidium divides to produce a cup-shaped structure that soon becomes a sphere, with only a small opening at the pole where it is attached to the parent colony. Flagella formed by the cells of the now multicellular bud are directed inward; but, by a remarkable process called inversion, the sphere turns inside

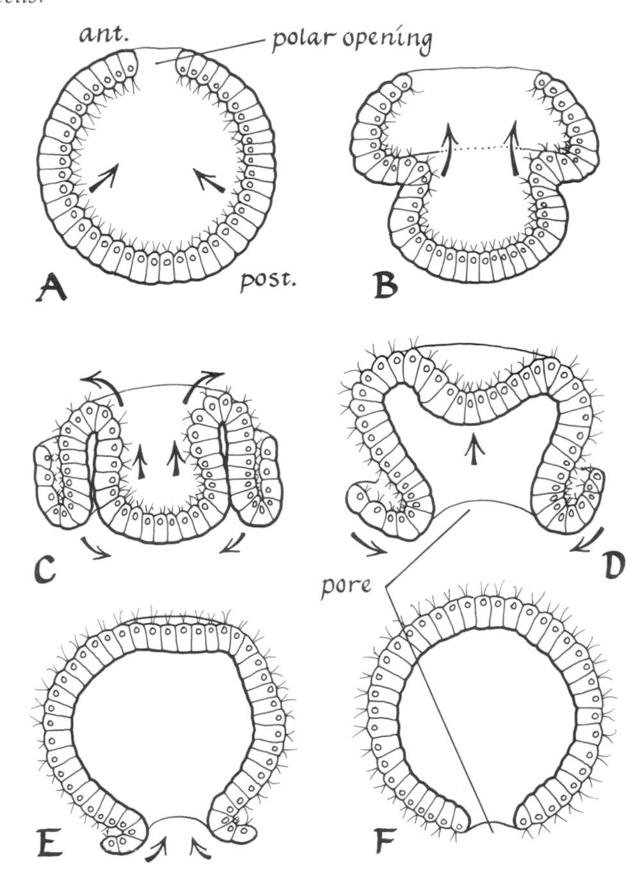

Figure 9-12. *The process of inversion in the development of a daughter colony of* Volvox. *A-F, stages in the process of inversion.*

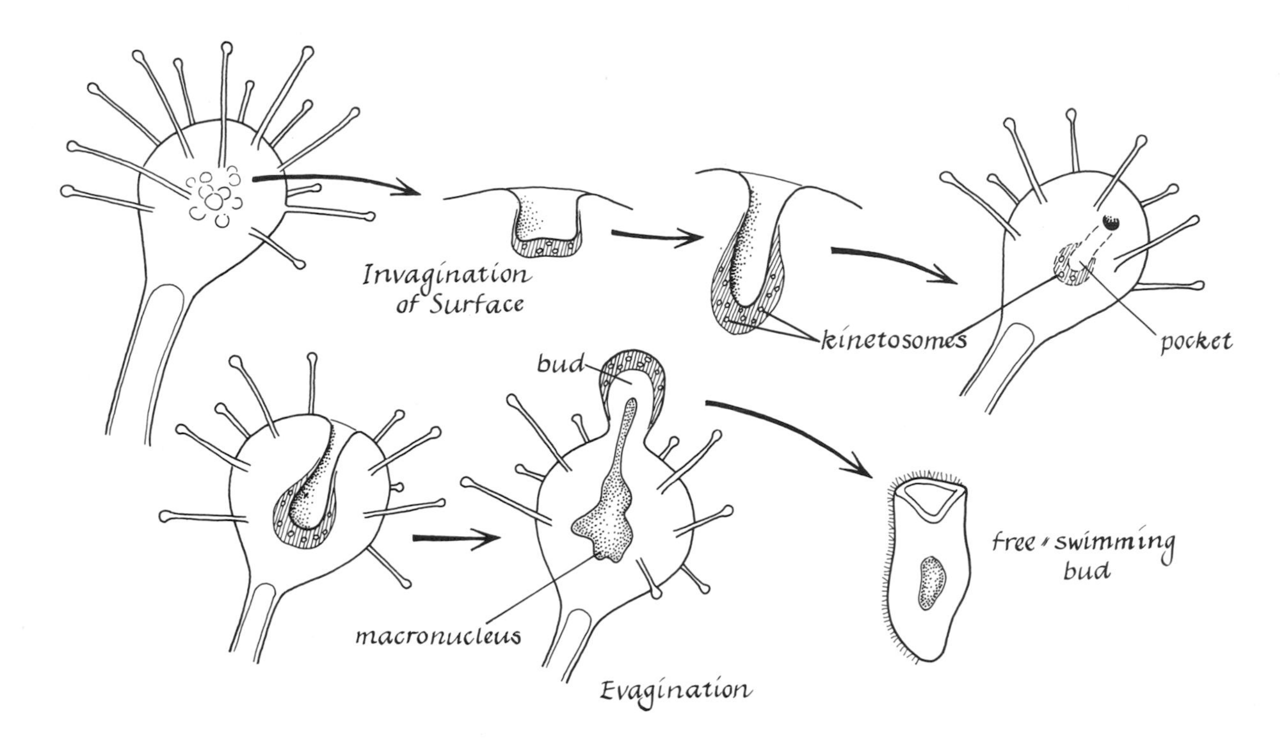

Figure 9-13. *The method of formation of a unicellular bud in the suctorian* Discophyra.

out through the polar opening. The result is a young colony with a structure like that of the parent. The process is illustrated in Figure 9-12. An equatorial constriction partially separates the posterior half of the sphere from the anterior; then the contracted smaller posterior half invaginates into the anterior half forming a temporarily two-layered vesicle with the original opening in the anterior half and a pore opposite this in the original equatorial region. As this continues, the expanding posterior half pulls down the anterior half and rolls it around the pore opening as the anterior opening dilates greatly. In this way the anterior half is turned inside out. One cannot help comparing this inversion process with gastrulation in animal embryos (Chapter 12). Both involve invagination and involution (rolling in) of cells around the rim of the pocket formed by invagination.

Among animals, unicellular buds are formed by suctorian, sporozoan, and some other protozoa. The suctoria reproduce by endogenously and exogenously formed buds. Endogenous budding in *Discophyra* is illustrated in Figure 9-13. During invagination, the area of the organism surface greatly increases. The result is a pocket extending almost completely from one side of the protozoan to the other, accompanied by a concentration of kinetosomes (cytoplasmic organelles) at the bottom of the invaginated pocket. Next, an evagination of the pocket, accompanied by a pushing out and inclusion of a lobe of the macronucleus, results in the completely formed bud, which is ciliated. The bud eventually separates from the parent and develops into a new adult suctorian (Guilcher, 1950). Other kinds of endogenous budding and an example of exogenous budding in suctoria are illustrated in Figure 9-14. In these examples of budding, the macronucleus of the parent produces one or more buds of its material, which is included in the developing endogenous or exogenous bud.

In some protozoa (Figure 9-15), chains of buds are produced in a manner similar to the formation of

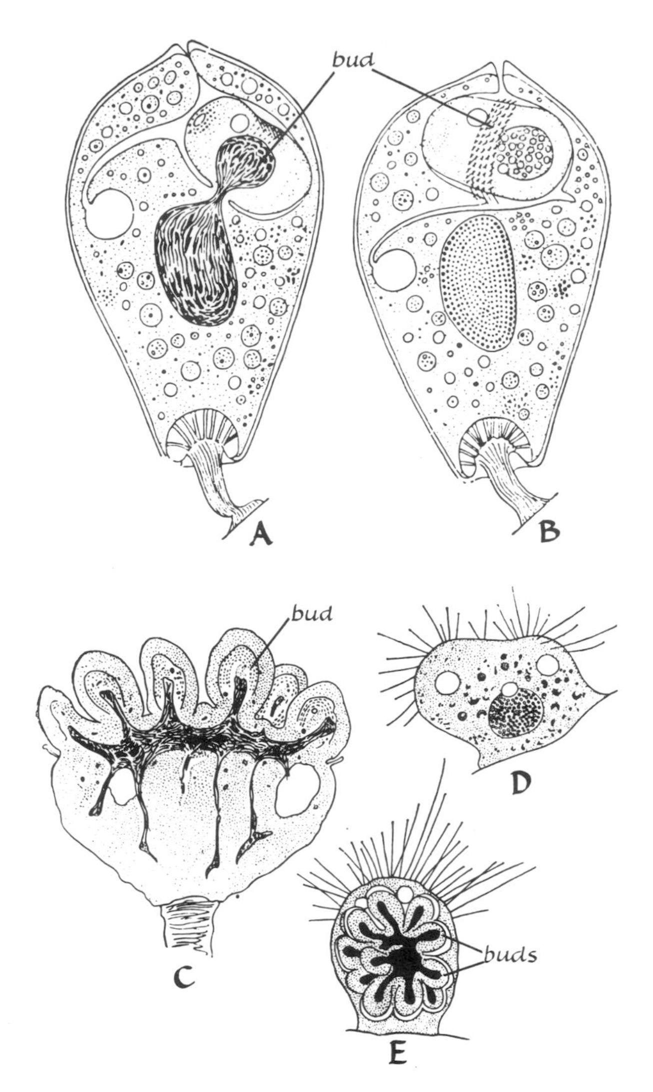

Figure 9-14. *A and B, Endogenous budding in the suctorian* Tokophyra cyclopum; *C, exogenous budding in the suctorian* Ephelota gemmipara; *D and E, endogenous budding in* Trichophyra salparum. *From* General Protozoology *by V. A. Dogiel (Clarendon Press, Oxford).*

chains of zooids in *Stenostomum* (Figure 9-9). Note that after the first bud, additional buds are formed between the first bud and the parent.

Multicellular Buds

Reproduction by multicellular buds occurs in both plants and animals. A well-known example in plants is the formation of latent buds, at notches in the

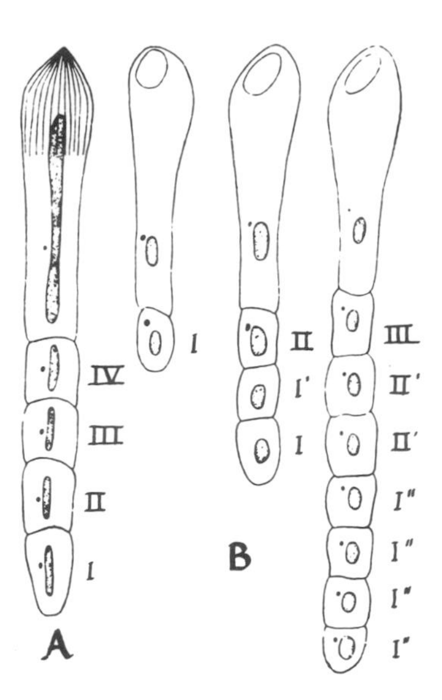

Figure 9-15. *Two types of linear, unicellular budding in astomatous ciliates: A,* Radiophyra *type; B,* Haptophyra *type. Order of bud formation is denoted by Roman numerals, order of division of buds denoted by primes. From* General Protozoology *by V. A. Dogiel (Clarendon Press, Oxford).*

fleshy leaves of *Bryophyllum* (Figure 9-16A), which become detached from the plant and fall on moist ground. The structure of a *Bryophyllum* bud is illustrated in Figure 9-16B. One or more buds at the edge of an isolated leaf develop into complete plantlings.

Other plants, such as some species of lilies, have specialized axillary buds that detach, drop to the ground, and develop into new plants. A more familiar example of reproduction by budding in plants is the development of new plants from the "eyes" or buds of the potato.

Reproduction by multicellular buds occurs widely throughout the animal kingdom, particularly in coelenterates and ascidians. Buds in *Hydra* and other hydroid coelenterates arise as protrusions of the body wall in a definite budding zone (Figure 9-17B)—a region of localized and accelerated cell division functionally similar to the meristematic region of a plant. Both ectoderm and endoderm in *Hydra* participate in bud formation (Brien, 1950). In the green hydra *H. viridis,* which contains algae (chlorellae) in its endoderm cells, an increase in the number of chlorellae is one of the first indications of budding. An early stage of bud formation in *H. viridis* is shown in Figure 9-17A. A bud is initiated as a

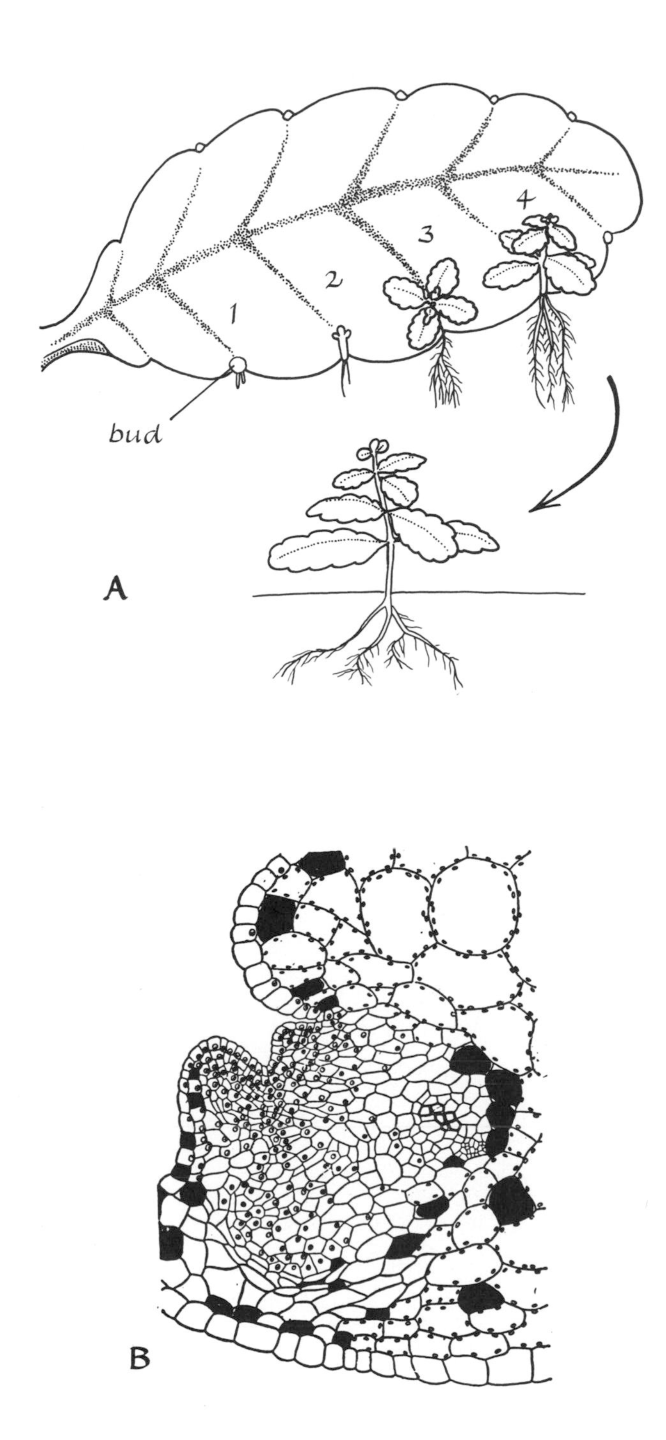

Figure 9-16. *A, development of a bud on the leaf of* Bryophyllum; *B, section through a notch in the leaf of* Bryophyllum *showing a "foliar embryo" with two leaf primordia and a root primordium buried in the tissue. B, from* Botany: Principles and Problems *by Sinnott and Wilson. Copyright 1955 by McGraw-Hill Book Company. Used with permission of McGraw-Hill Book Company.*

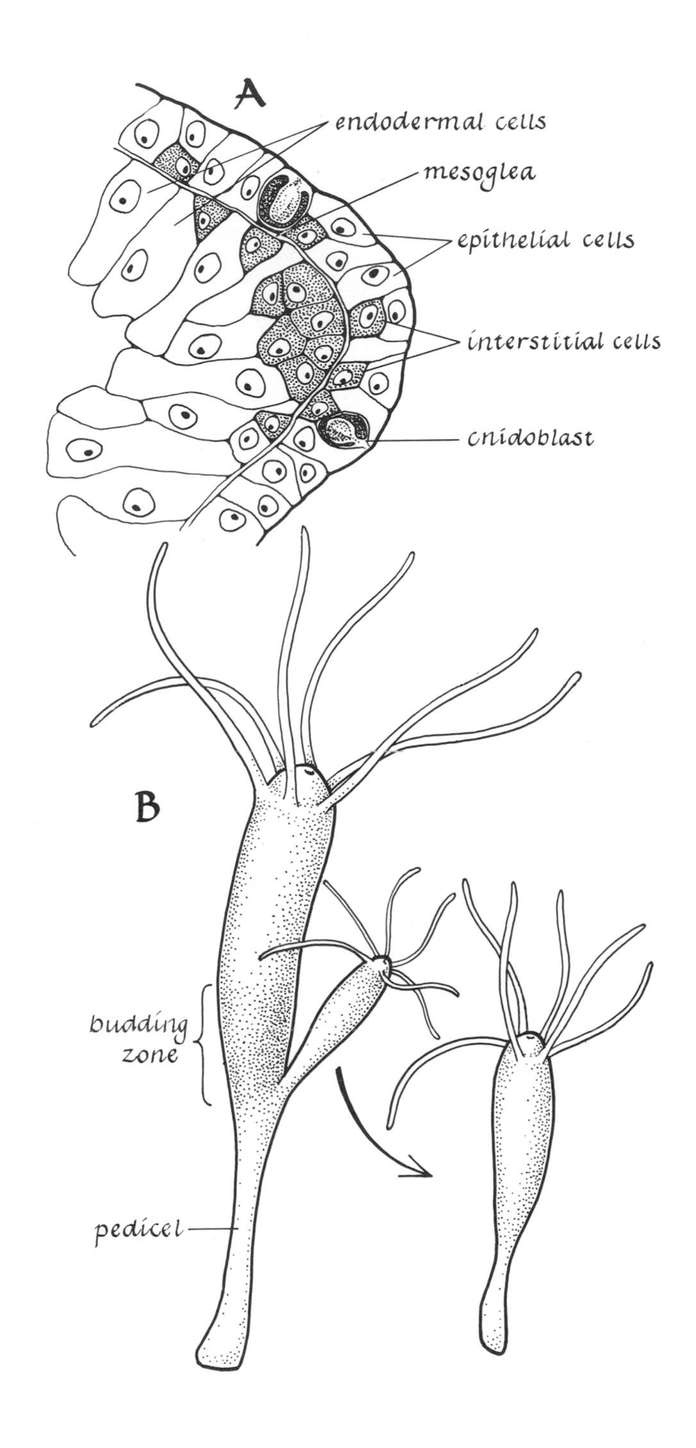

Figure 9-17. *A, section through a young bud of* Hydra. *B, budding zone and separation of a bud in* Hydra.

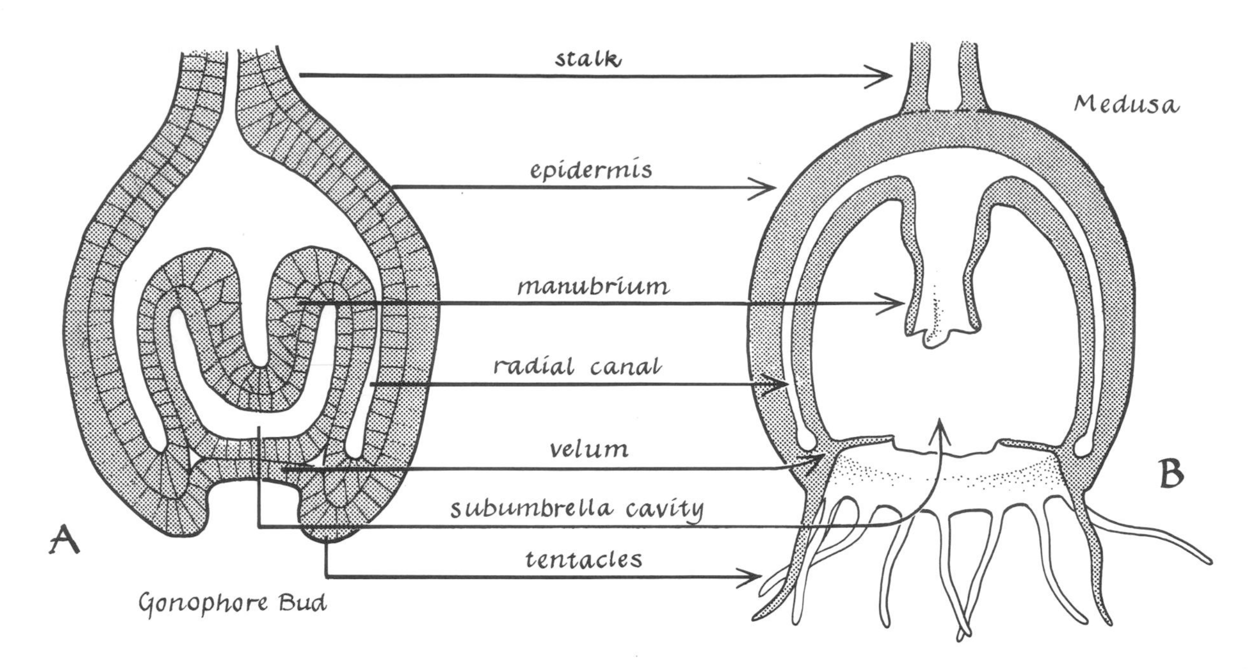

Figure 9-18. *Diagram of the relation between parts of a developing gonophore bud and parts of the medusa into which it develops.*

secondary growth center between the endodermal and epithelial layers of the body wall. Specialized cells such as cnidoblasts which form the highly complex nematocyst (stinging organelle in cnidarians), are already present in the ectoderm of the bud. The interstitial cells also develop into cnidoblasts and other types of cells. Eventually, the bud develops a mouth and set of tentacles, constricts at its base, and is freed as a new individual (Figure 9-17B).

The formation of buds that develop into new hydroid individuals on the hydroid phase of the organism is common in coelenterates. However, some coelenterates form buds on the hydroid phase that develop into medusae, the jellyfish phase of the animal. These buds develop inside special branches called gonophores (Figure 7-30). Development of buds on a gonophore is a complicated process, involving an ectodermal invagination, to form an entocodon around which the structures of the medusa (manubrium, radial canals, umbrella) develop (Goette, 1907; Hyman, 1940). The relationship between parts of the developing bud and structures of the completed medusa is shown in Figure 9-18.

The entocodon, which is hollow, expands and forms the lining of the subumbrellar cavity of the medusa. It is believed that the entocodon functions as a morphological guideline, directing the shaping of the medusa.

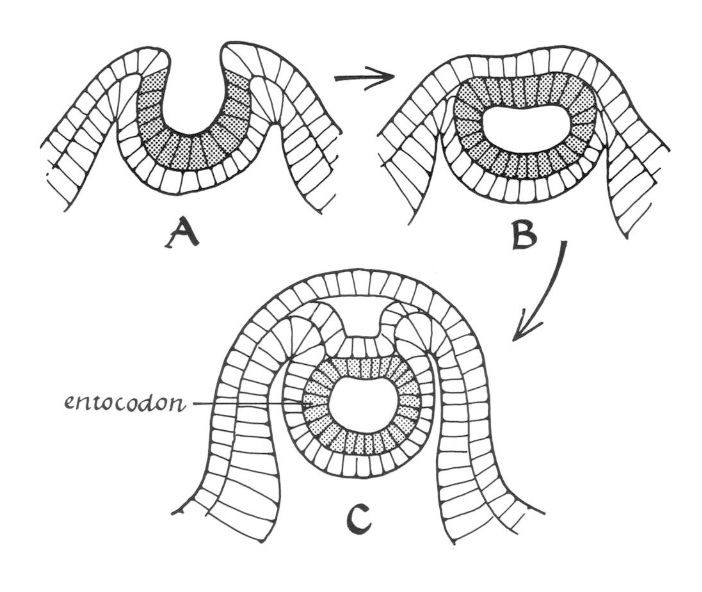

Figure 9-19. *Formation of the entocodon by invagination in a medusa bud.*

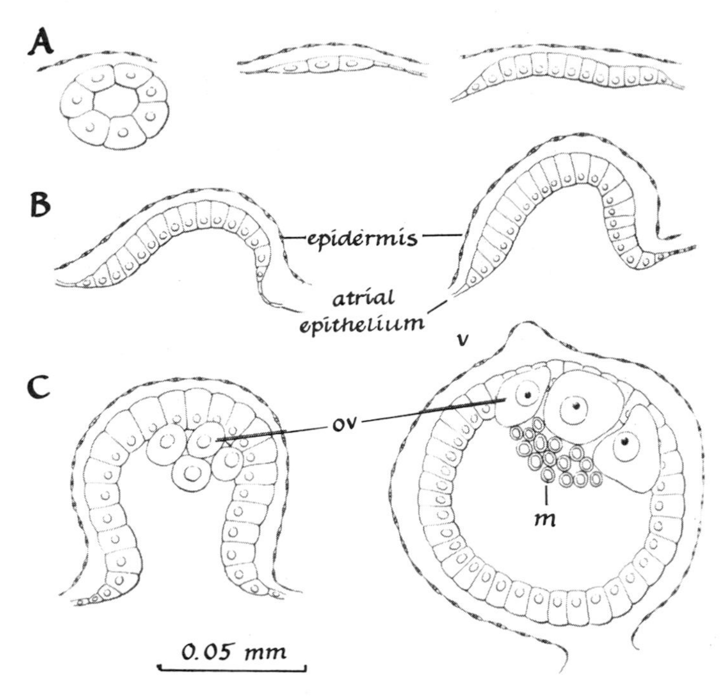

Figure 9-20. *Development of a bud of the tunicate* Botryllus. *A, surface and side view of the bud disc. B, arching of disc to form a hemisphere. C, closing of hemisphere and segregation of ova and spermatogonia. From* Growth, Development, and Pattern *by N. J. Berrill. W. H. Freeman and Company. Copyright © 1961.*

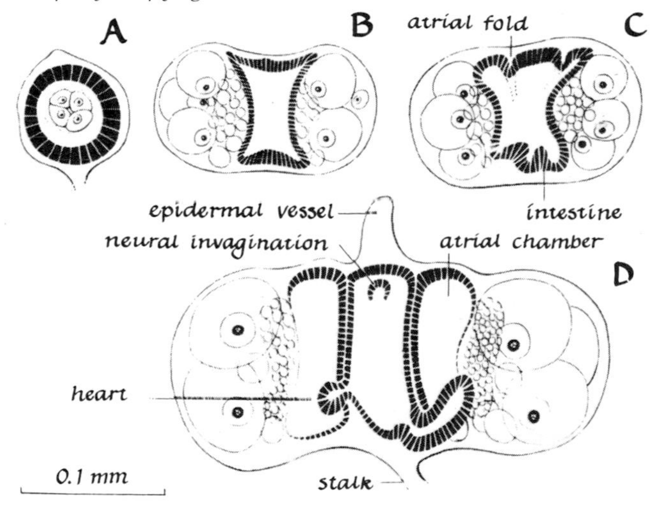

Figure 9-21. *Folding and modeling processes in development of a* Botryllus *bud. A-B, growth of hermaphrodite gonads; C, subdivision of inner vesicle into central and lateral compartments; D, evagination of central chamber to form intestine, heart, and neural organ. From* Growth, Development, and Pattern *by N. J. Berrill. W. H. Freeman and Company. Copyright © 1961.*

Other kinds of buds develop in some coelenterates (see Chapter 14), including buds that form on medusae and develop into more medusae. Early stages in formation of such buds, including the method of entocodon formation, are shown in Figure 9-19.

Budding is a common method of reproduction in bryozoa and ascidians (Berrill, 1961). The formation and early development of an ascidian bud is illustrated in Figure 9-20. The bud arises from a bud disc, a group of atrial epithelial cells below the epidermis on the side of the adult tunicate. The bud disc evaginates and forms a closed sphere. During the closing, ova and spermatogonia differentiate from the atrial epithelium. At this stage, a remarkable modeling process begins, which forms the organs of the adult by folding of the atrial epithelium. Some features of the folding and modeling process are illustrated in Figure 9-21.

Control mechanisms for this kind of direct morphogenesis are completely unknown. Why the evaginations occur in particular regions of the atrial epithelial sac and to what extent, if any, the parent organism furnishes guidelines is a challenge for students of development. However, some autonomy in bud development has been observed in the differentiation of the sex cells. Berrill (1961) has reported the relationship between the number of cells in a bud disc and the later development of the gonads (Figure 9-22). When the bud disc is small (less than 60 cells) no gonads develop in the new individual. With 60 to 120 cells in the disc, a testis but no ovary develops. With more than 120 cells, both an ovary and a testis form. This is a striking example of the relationship between cell number and the complexity of cell differentiation. Further examples of this common aspect of development will be described in Chapter 22 in relation to the role of the internal environment of a cellular aggregate.

Some ascidians reproduce asexually by formation of buds on the rootlike stolons. The stolon is a slender tube, consisting of ectodermal and endodermal layers with scattered mesodermal tissue between them, that grows from a special region at the posterior end of the endostyle of the parent organism (Figure 9-23).

In some ascidians, buds develop at spaced intervals along the stolon and form zooids that remain connected by the stolon. This occurs in *Perophora viridis*, in which (as noted in Chapter 5), the blood cells are necessary for zooid formation. In other tunicates, the stolon gradually becomes transversely subdivided into a chain of adjacent buds, usually called strobili, by constrictions in the ectodermal layer (Figure 9-23). Each bud develops into a new zooid that may or may not become separated from its neighbors. The most distal bud in the chain is the oldest, new buds being formed progressively near the attachment of the stolon to the parent. This process is fundamentally a multiple and serial type of fission.

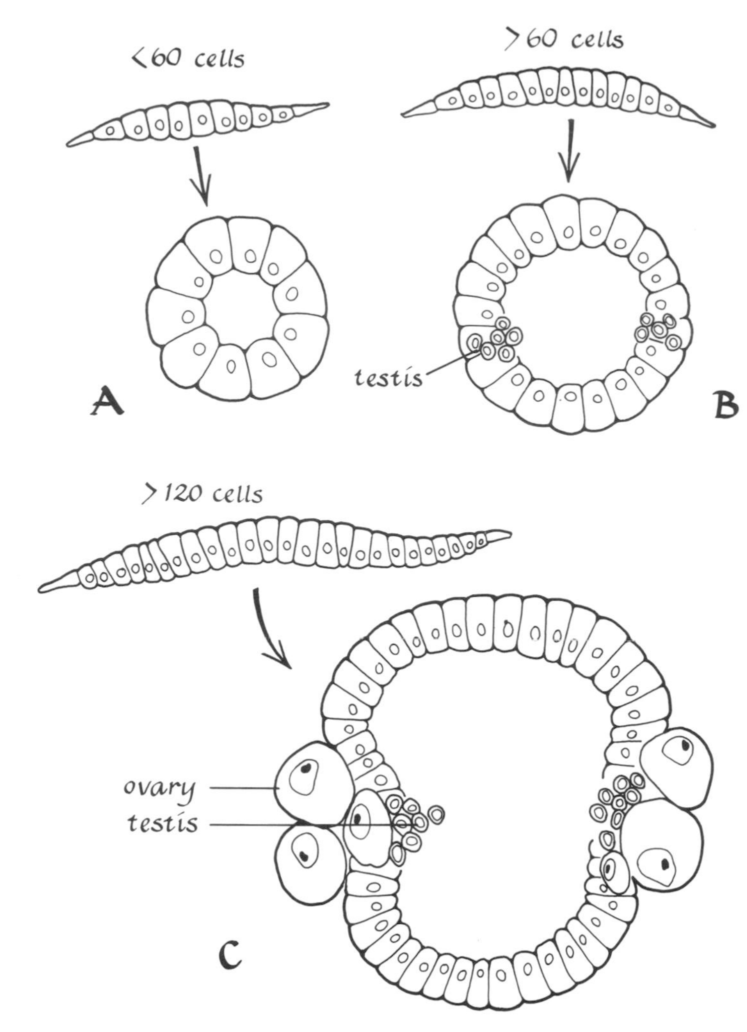

Figure 9-22. *Diagram of the relation between number of cells in a tunicate bud disc and later development of gonads.*

Embryo Buds

Polyembryony, the formation of two or more embryos from a single fertilized egg, occurs in both plants and animals. The extra embryos arise by what is essentially a budding process. Although budding in these cases occurs in a fertilized egg (zygote), the extra individuals, which usually bud off the original or primary embryo, are not the direct result of a sexual process.

In the bryozoan *Crisia*, as many as 100 secondary embryos may be budded off the primary embryo (Berrill, 1961). Secondary embryo formation is illustrated in Figure 9-24. The fertilized egg of *Crisia*, unlike most zygotes, both begins its development within a follicle and is multinucleated, as the result of nuclear division unaccompanied by cytoplasmic division. The zygote develops into a primary embryo on which exogenous buds soon form. The buds detach and become secondary embryos.

Polyembryony also occurs in some parasitic insects such as *Platygaster verualis* (Leiby and Hill, 1924). Four embryos, sometimes more, develop from a single egg. The first division of the zygote nucleus results in formation of a segmentation nucleus (which, together with surrounding cytoplasm, will divide to form the numerous embryos) and a polar nucleus (which also divides but does not participate in embryo formation), as shown in Figure 9-25. This formation of an embryonic and an extra-embryonic nucleus at the first division parallels a similar phenomenon common in the first division of a plant zygote, the formation of an embryonic and an extra-embryonic cell (see Chapter 10). In another parasitic insect *Litomastix*, as many as 2,000 embryos are formed by a single egg (Silvestri, 1936). In many eggs that exhibit polyembryonic development, first an amorphous mass of dividing cells is formed, then

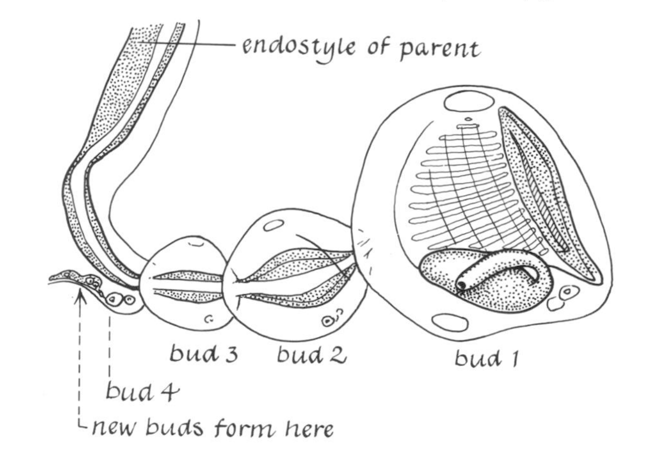

Figure 9-23. *Stolonic budding in a tunicate,* Pyrosoma.

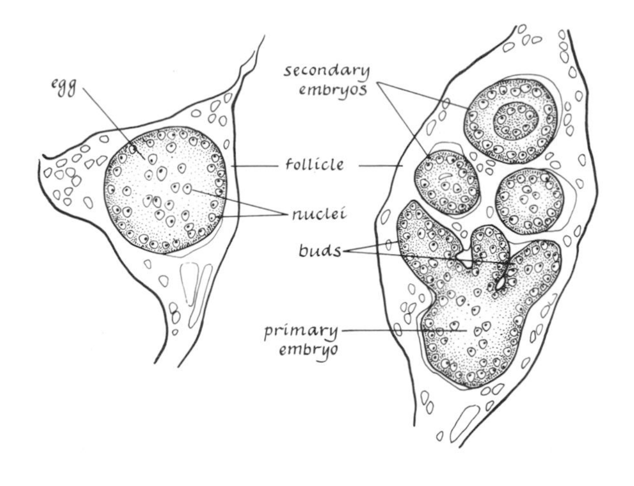

Figure 9-24. *Embryo budding in the bryozoan* Crisia.

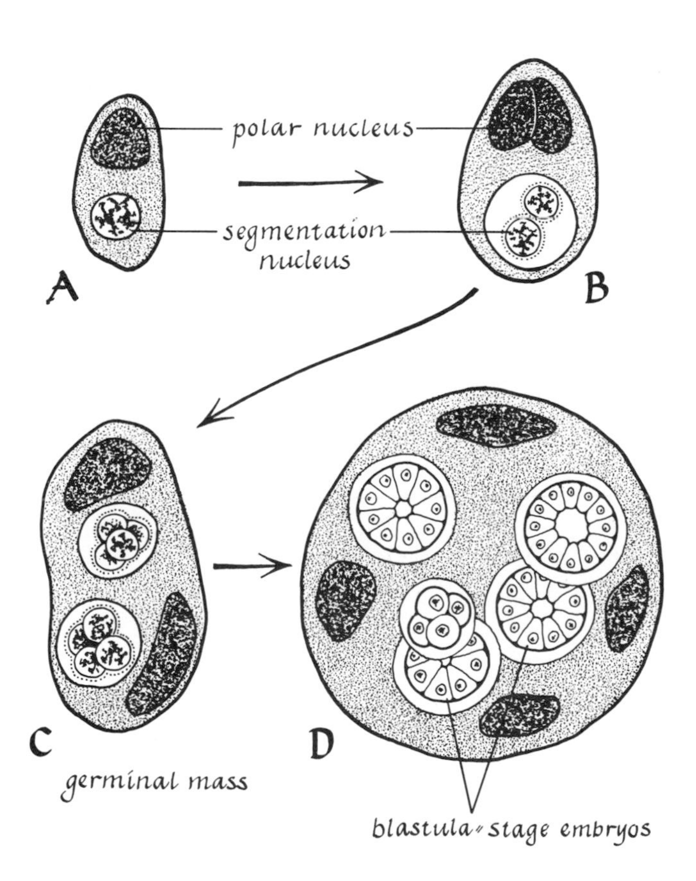

Figure 9-25. *Embryo budding in the parasitic insect* Platygaster.

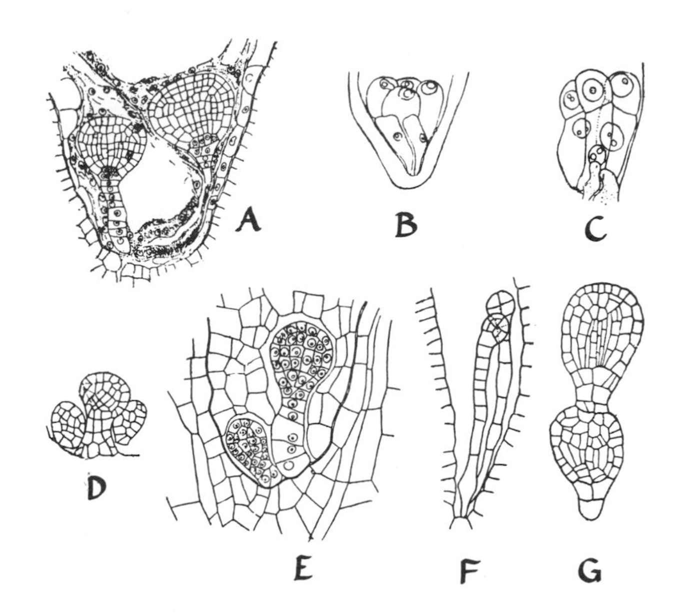

Figure 9-26. *Embryo budding in several species of angiosperms. From Wardlaw:* Embryogenesis in Plants, *1955. Courtesy Methuen and Co., Ltd.*

separate embryos arise by divisiòn or budding of the mass.

Embryo budding also occurs in the plant kingdom, commonly in gymnosperms and sporadically in angiosperms. As many as four embryos can develop during cleavage of a pine zygote, but usually only one of them develops to maturity (Wardlaw, 1955). Among the angiosperms, in orchids a filamentous proembryo can branch to yield several embryos, or small proembryo buds can develop into separate embryos. Embryos can also develop from buds formed on callus tissue derived from embryos cultured in vitro (see Chapter 11). Some examples of polyembryony in angiosperms are shown in Figure 9-26.

Polyembryony in vertebrates, usually called twinning or multiple twinning, can be considered a type of asexual reproduction by budding. Twins or multiple embryos rarely develop from the separation of cells at the two or four cell stage, as is popularly believed (Newman, 1940; Arey, 1965). Formation of separate embryos usually occurs at a much later stage and proceeds from multicellular embryo-forming centers called primitive streaks in birds and mammals, embryonic shields in fishes, and organizer centers in amphibians. For example, in the Texas armadillo, quadruplets are normally formed from one egg. Two primitive streaks arise at the blastocyst stage and each buds off another center, as is shown in Figure 9-27 (Patterson, 1913).

Most evidence regarding human twinning indicates that identical twins arise from division of the inner cell mass of the blastocyst derived from a single fertilized egg, as diagramed in Figure 9-28. Because they are formed within a single blastocyst, identical twins develop within a common chorionic sac and have a common placenta (Streeter, 1919; Newman, 1940; Arey, 1965). The Dionne quintuplets are believed to have arisen by a budding process (Arey, 1965).

Experimental polyembryony has been accomplished in several kinds of animals by surgically dividing the cleaving egg or much later stages into two or more parts. The embryo in ducks and chickens is a disc of several thousand cells at the time the egg is laid. This collection of cells is called a blastoderm. When it is divided into more parts with a fine glass needle, two or more embryo bodies may develop (Lutz, 1949; Spratt and Haas, 1960), as shown in Figure 9-29 and Figure 15-6.

The embryo body develops from the primitive streak, which is located at the posterior pole of the blastoderm (Figure 12-31). If the blastoderm is cut into pieces (as in Figure 9-29), a primitive streak and later an embryo body can form in each piece. The marginal zone, a boundary between the inner and outer areas, functions like a meristem and produces the extra primitive streaks by a process similar to budding.

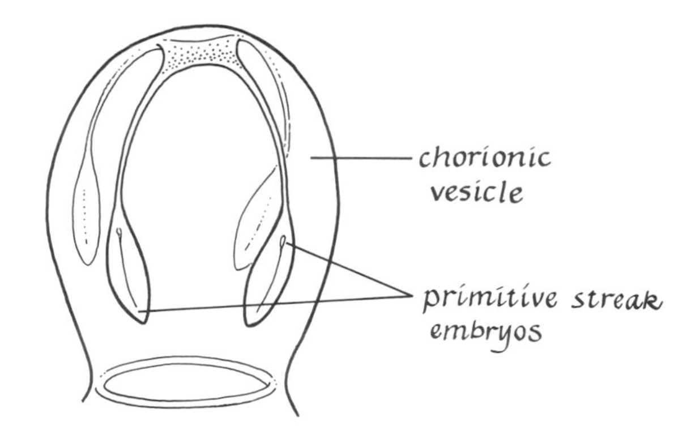

Figure 9-27. *Polyembryony in the Texas armadillo.*

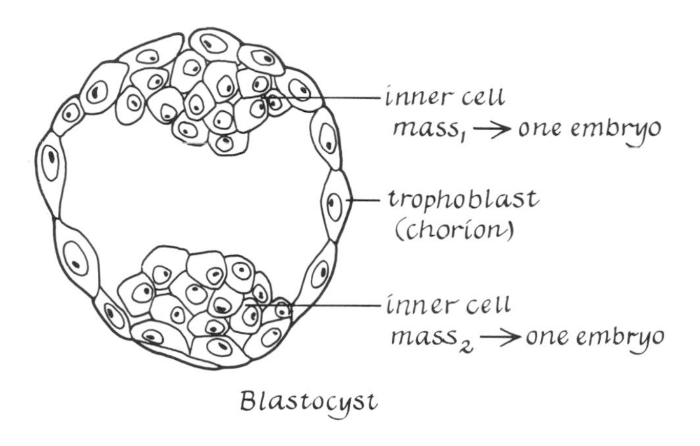

Figure 9-28. *Diagram of a human blastocyst, showing two inner cell masses that later develop into identical twins.*

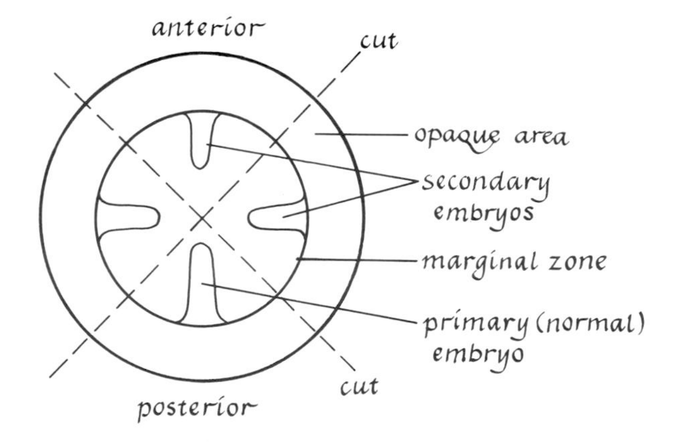

Figure 9-29. *Experimental polyembryony in the chick blastoderm.*

Stolons, Rhizomes, Bulbs

The formation of buds on the stolons of ascidians was described in the section on bud formation. When the individuals developing from the buds remain attached to the stolons, spreading colonies are formed as the stolons grow out. Similarly, many plants spread over wide areas by the outgrowth of stolons, which form buds when they touch the soil. In other plants, underground rootlike rhizomes, on which buds are formed at definite intervals, grow out. In both cases, the buds develop into new individual plants. For example, a single grass plant can extend its area by means of stolons until it forms a patch of turf. Plants like the strawberry produce stolons from which separate plants develop (Figure 9-30). Plants

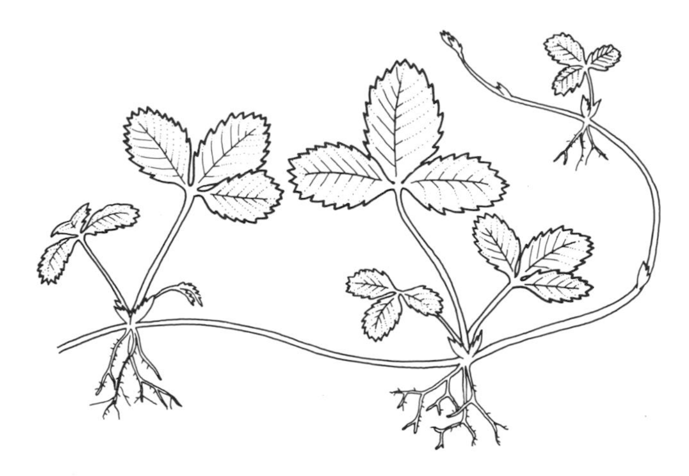

Figure 9-30. *Asexual reproduction by formation of buds on a stolon in the strawberry plant.*

that produce bulbs, like the onion and tulip, reproduce asexually by producing bulblets, which detach from the parent bulb and become new plants.

Gemmules, Statoblasts, and Gemmae

Gemmules, statoblasts, and gemmae are multicellular reproductive units similar to buds, but unlike most buds detach from the parent and remain dormant for a period of time before germinating to form the new individual. They are adaptations that tide the species over periods unfavorable to survival of the adult.

Gemmules, which are produced by freshwater and some marine sponges, consist of a mass of archeocytes covered by a chitinous envelope of columnar

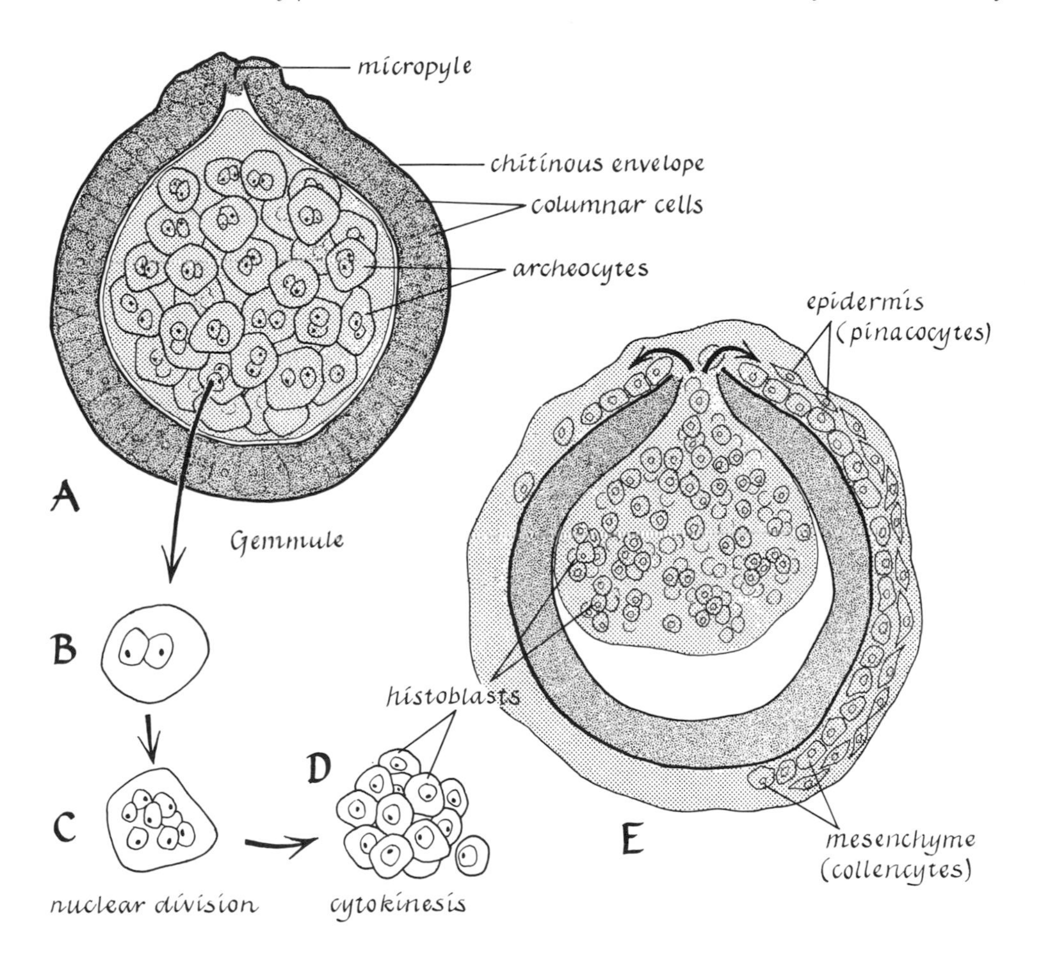

Figure 9-31. *Diagram of events in the germination of a gemmule of* Spongilla.

cells studded with spicules (Hyman, 1940; Berrill, 1961; Rasmont, 1962). When conditions, such as adequate water and warm temperatures, are favorable to survival of the adult, the gemmule germinates. The structure of a gemmule of the freshwater sponge *Spongilla* and the general features of the germination process are illustrated in Figure 9-31.

Germination begins with the division of the nuclei of the binucleate archeocytes (Figure 9-31B and C). The multinucleated archeocytes then undergo cytokinesis to form groups of uninucleate histoblasts (Figure 9-31D). The histoblasts move out of the gemmule through its opening, the micropyle, and spread over the outer surface of the gemmule (Figure 9-31E). The outermost histoblasts become pinacocytes of the epidermis of the new sponge, and the inner cells become collencytes (Chapter 8). The specialization of the histoblasts is influenced by the microenvironment, thus is a consequence of their positions, but when a new sponge is reaggregated from dispersed cells (Chapter 8), the position of the cells is largely a function of their prior specialization (Berrill, 1961).

Like gemmules, statoblasts are adaptations that enable the organism to survive unfavorable winter conditions in which the adult disintegrates. They are produced by freshwater *Bryozoa* of the ectoproct group, and consist of an internal mass of peritoneal cells surrounded by a group of epidermal cells and covered by a protective shell. Statoblasts develop as budlike structures on the funiculus of the adult (Figure 9-32A). The formation of a statoblast and an early stage in its germination are shown in Figure 9-32B.

Gemmae are typically multicellular units of considerable complexity, produced by liverworts, some mosses, and the "living fossil" plants, Lcyopods. Gemmae have highly specific structures and have been used for the identification of species. In liverworts, gemmae develop on apical leaf lobes at the apex of the shoot or in cuplike structures on the gametophyte (Figure 9-33A). Stages in formation of the gemma of a liverwort are shown in Figure 9-33B. The gemma develops on a stalk, from which it breaks off when mature, and the gametophyte develops when the gemma germinates from two meristematic regions (Figure 9-33A). In mosses, gemmae develop singly or in clusters on various parts of the gametophyte, but their position is constant in any one species. The gemmae of liverworts and mosses are

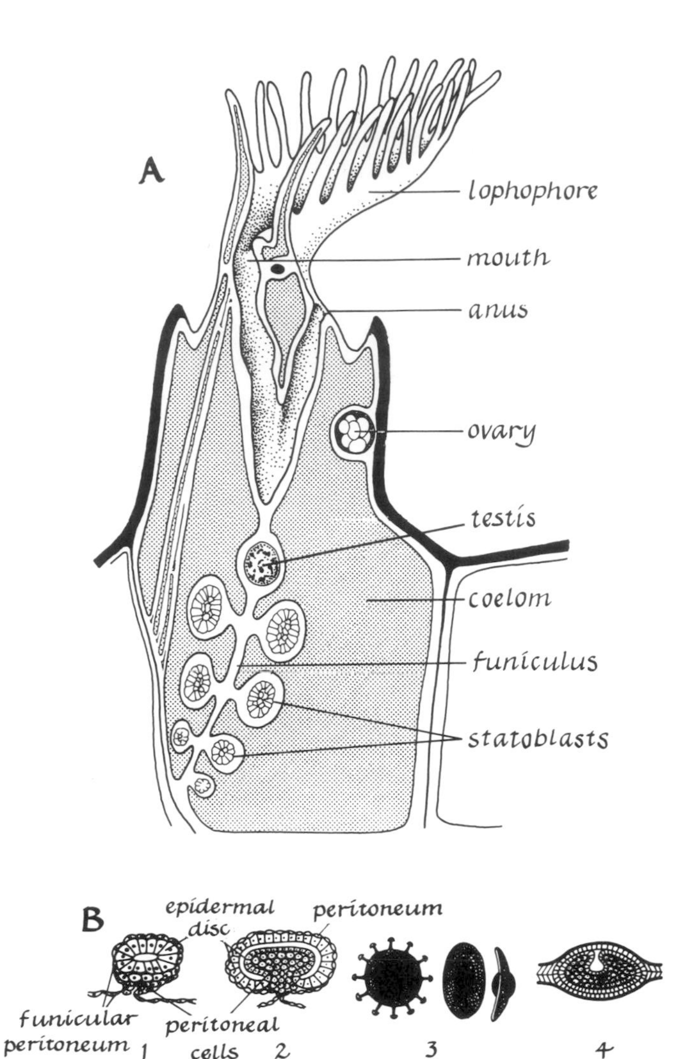

Figure 9-32. *A, formation of statoblasts; B, development of a statoblast: 1-2, statoblast developing within the funiculus of the parent Bryozoan; 3, mature statoblast shells; 4, early stage in germination. B, from* The Science of Zoology *by Weisz. Copyright 1966 by McGraw-Hill Book Company. Used with permission of McGraw-Hill Book Company.*

haploid structures (gemmules and statoblasts are diploid). However, in *Lycopodium* the gemmae are diploid, being formed on the sporophyte generation.

Spores

Unfortunately the term spore is often used loosely. Spores differ in their origin, structure, motility, and

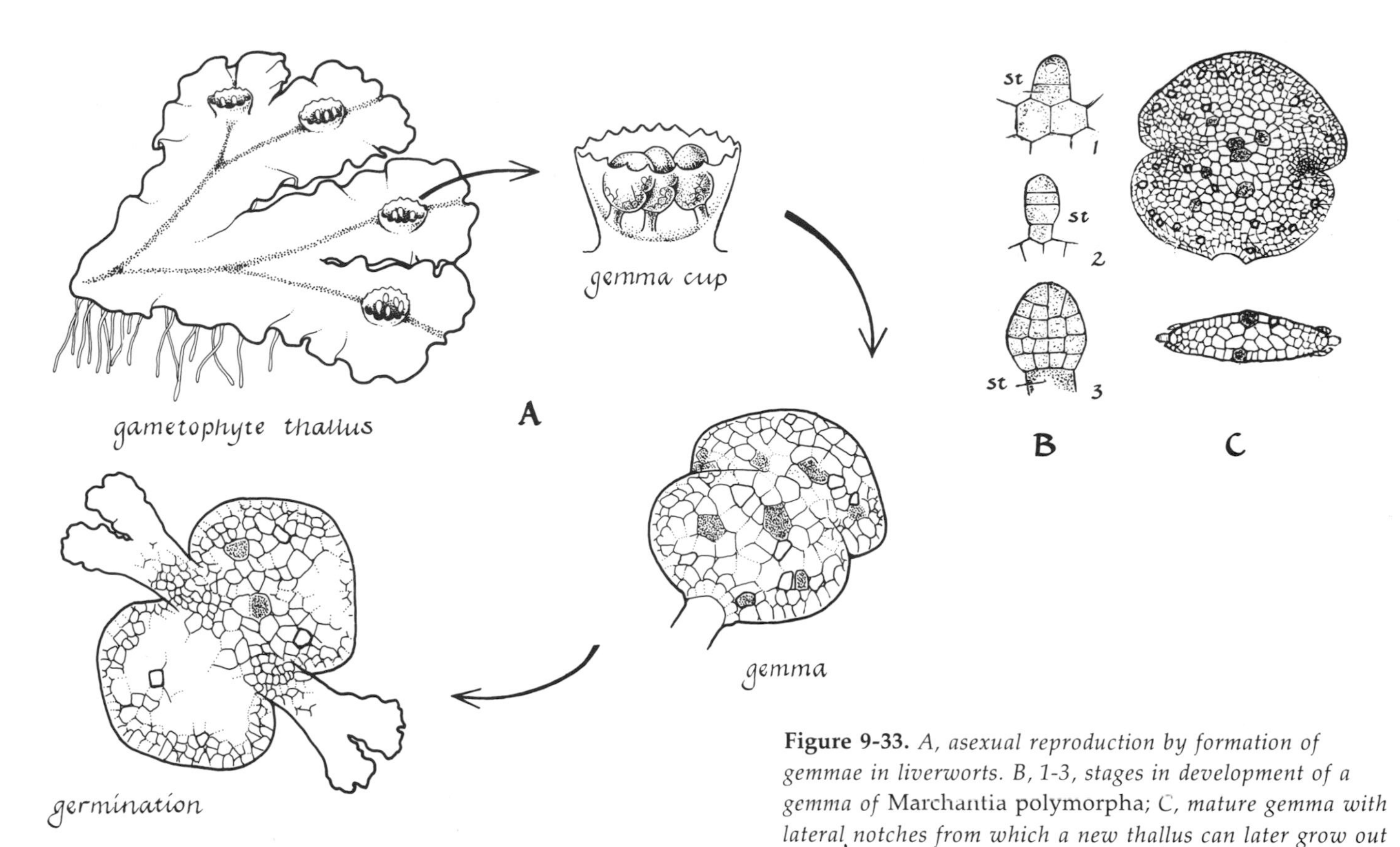

Figure 9-33. *A, asexual reproduction by formation of gemmae in liverworts. B, 1-3, stages in development of a gemma of* Marchantia polymorpha; *C, mature gemma with lateral notches from which a new thallus can later grow out and a transverse section at the level of the notches. B and C, from* Organography of Plants *by K. Goebel, 1900 (Clarendon Press, Oxford).*

type of organism they produce on germination. In general, spores are unicellular, uninucleate reproductive units. They are formed by many kinds of plants, from bacteria to flowering plants. In animals, spores are produced by some protozoans such as the Sporozoa, Foraminifera, and Radiolaria. Typically, spores are adaptations that enable the species to survive unfavorable periods. The spores of mosses, ferns, and some fungi have a thick protective coat or cell wall. The spores of algae and some fungi have a thin wall. Both haploid and diploid spores occur, but in this chapter we shall discuss only the haploid type. Diploid spores such as zygospores are actually fertilized eggs (zygotes) and are thus resting and protected stages in sexual reproduction (which will be discussed in Chapter 10).

Motile spores, called zoospores or planospores, are produced by many water molds and algae. Some of these are illustrated in Figure 9-34. Zoospores of the water mold *Saprolegnia* are products of mitosis in the haploid sporangium. They bear two flagella, one called a tinsel-type flagellum because of its lateral hairlike projections. The zoospores are released by the sporangium, encyst, and eventually germinate to form the parent type plant which is coenocytic (multinucleated and without walls separating the nuclei). In the green alga *Ulothrix* (Figure 9-34B) a cell in the haploid filament divides into a group of 4 to 16 cells, which become zoospores bearing four flagella. The zoospores escape from the cells in which they were formed, swim about, eventually settle down on a substratum, lose their flagella, elongate, and by transverse division develop into a new filament.

Many kinds of aplanospores (nonmotile spores) are found throughout the plant kingdom but among

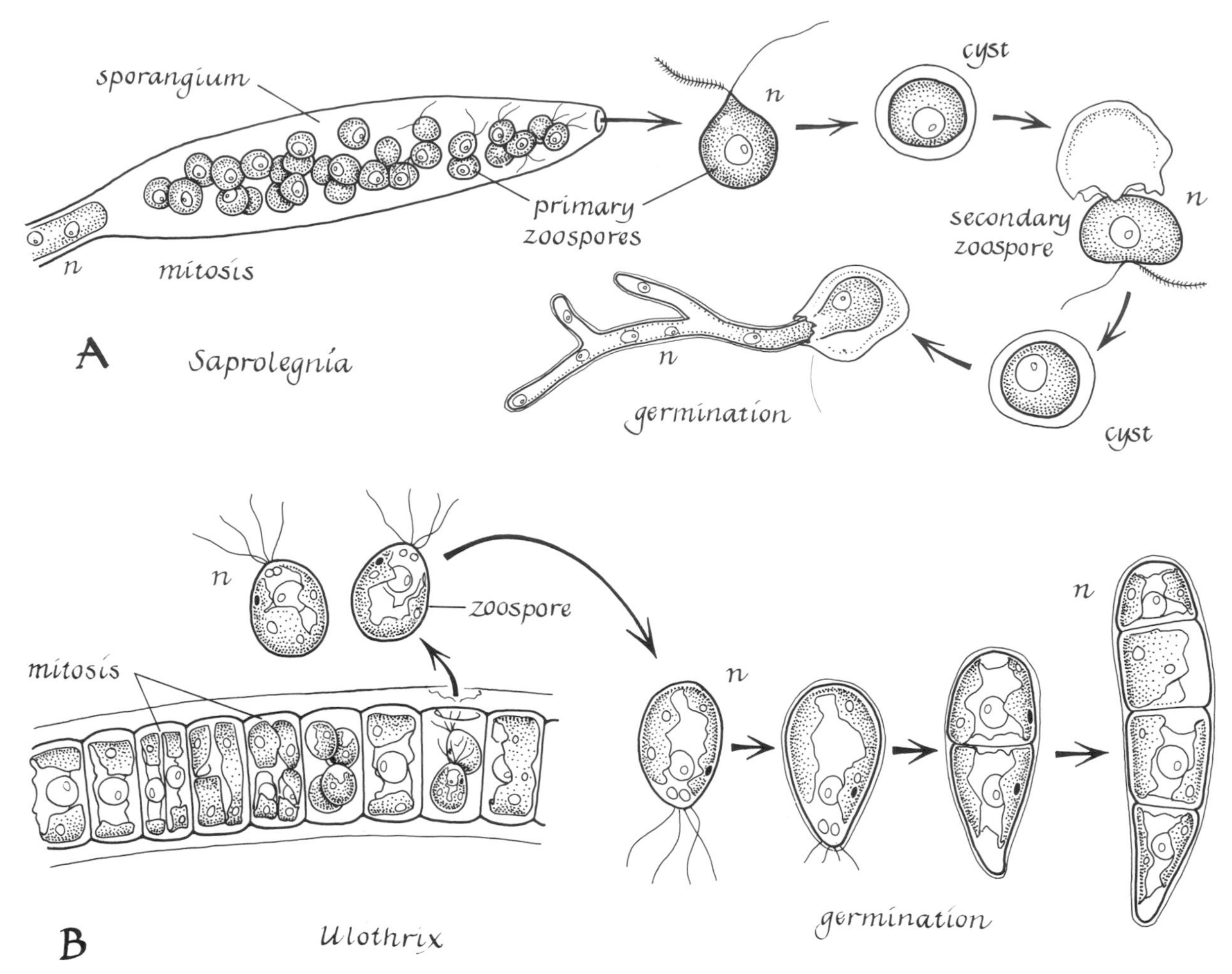

Figure 9-34. *Formation and germination of zoospores: A, in* Saprolegnia; *B, in* Ulothrix.

the animals they are found only in protozoa. Some examples of nonmotile spores are illustrated in Figures 9-35 and 9-36.

Conidiospores (Figure 9-35A and B) are produced in chains by the green and yellow mold *Aspergillus*, commonly found on leather, and by the blue mold *Penicillium*, found on lemons, jelly, and other organic material. Ascospores (Figure 9-36C and D) are produced in groups of eight inside a structure called an ascus. This spore is characteristic of sac fungi, Ascomycetes. Aplanospores are also produced by the higher groups of plants including liverworts, mosses, ferns, gymnosperms, and angiosperms. In the latter two groups, two kinds of spores are formed: microspores (pollen grains, which develop into the male gametophyte) and megaspores, which develop into the female gametophyte. The significance of these spores will be described in Chapter 11.

Mitotically and Meiotically Derived Spores

Many species of plants alternate a haploid, gamete producing generation with a diploid, spore producing generation. An example of the alternation of generations in the liverwort *Riccia* is diagramed in Figure 9-37. Aspects of this alternation, in which plants of the two generations are often morphologically different, will be discussed in Chapters 10, 11, and 14. In this chapter our purpose is to note the relationship between spore formation and the ploidy

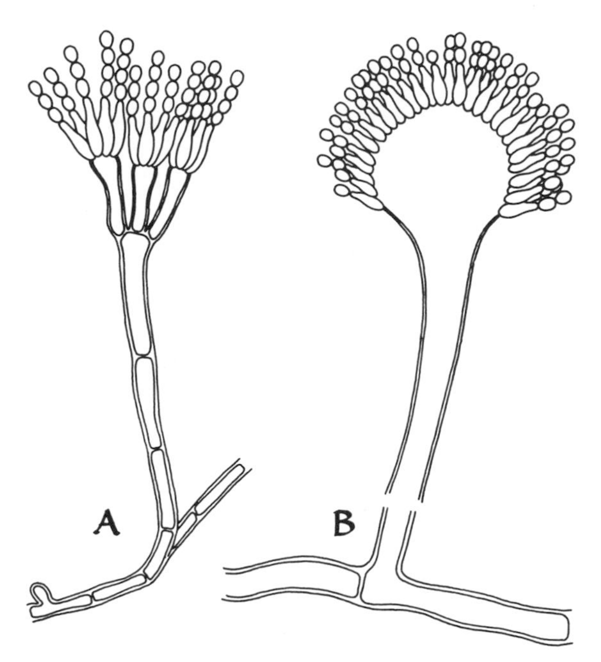

Figure 9-35. *Conidiophores and conidia: A,* Penicillium, *× 1,250; B,* Aspergillus, *× 1,000. From Scagel et al. Copyright © 1966 by Wadsworth Publishing Company, Inc.*

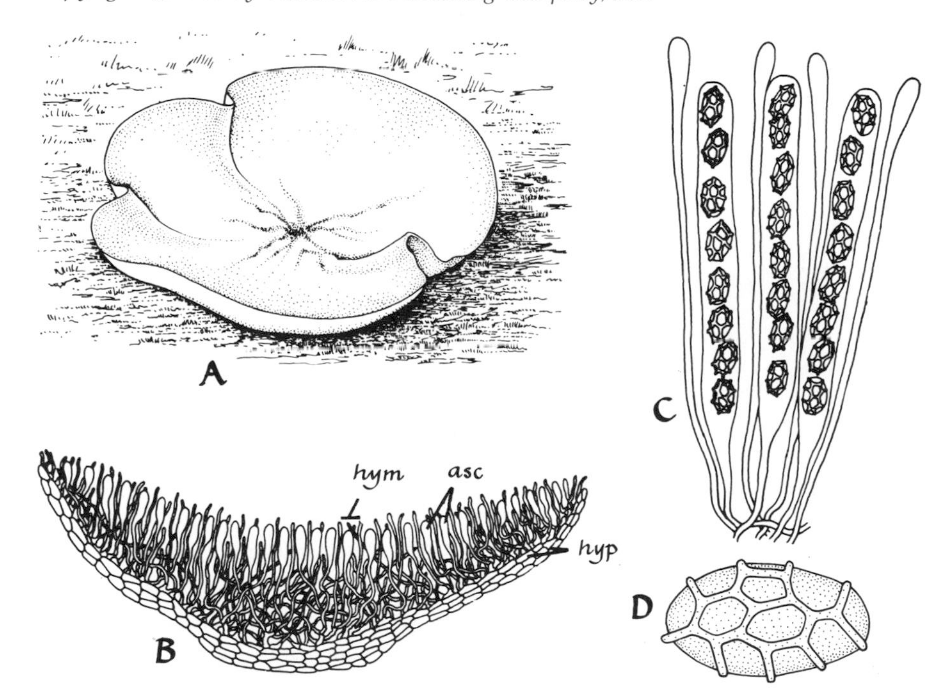

Figure 9-36. *A, habit sketch of the fruiting body of the ascomycete* Aleuria aurantia*; B, diagram of section through fruiting body; C, asci containing ascospores; D, single ascospore. From Scagel et al. Copyright © by Wadsworth Publishing Company, Inc.*

level of the plant, in other words, the relationship of sporogenesis to mitosis and meiosis.

Spores formed by haploid plants, such as the zoospores of *Saprolegnia* and *Ulothrix* (Figure 9-34) and the conidiospores of *Penicillium* and *Aspergillus* (Figure 9-35) are products of mitosis and are called mitospores. Spores formed by diploid plants are either mitospores or meiospores. Some spores are direct products of the two meiotic divisions—as in liverworts such as *Riccia* (Figure 9-37), in which the four spores remain together as tetrads. Others are indirect products of the meiotic divisions in which a mitotic division follows meiosis and leads to formation of eight spores. The ascospores of *Aleuria* (Figure 9-36) and many other ascomycetes are examples of the latter. However, in the yeast *Saccharomyces,* also an ascomycete, only four ascospores are formed. These spores which are either direct or indirect products of meiosis are called meiospores.

Spores formed by diploid plants are not always haploid; that is, they are neither direct nor indirect products of meiosis. For example, the diploid generation of *Allomyces,* a water mold, produces both diploid zoospores (which develop into plants like the parent) and meiotically derived haploid zoospores (which develop into haploid, gamete producing plants).

In meiospores, however, genetic segregation occurs during formation and results in diversity among the haploid plants that develop from the germinating spores. A striking example is the segregation for sex in the liverwort *Sphaerocarpus donnelli,* in which two spores of the tetrad develop into male and two into female gametophytes (Allen, 1919). We shall discuss the significance of meiosis more fully in the next chapter.

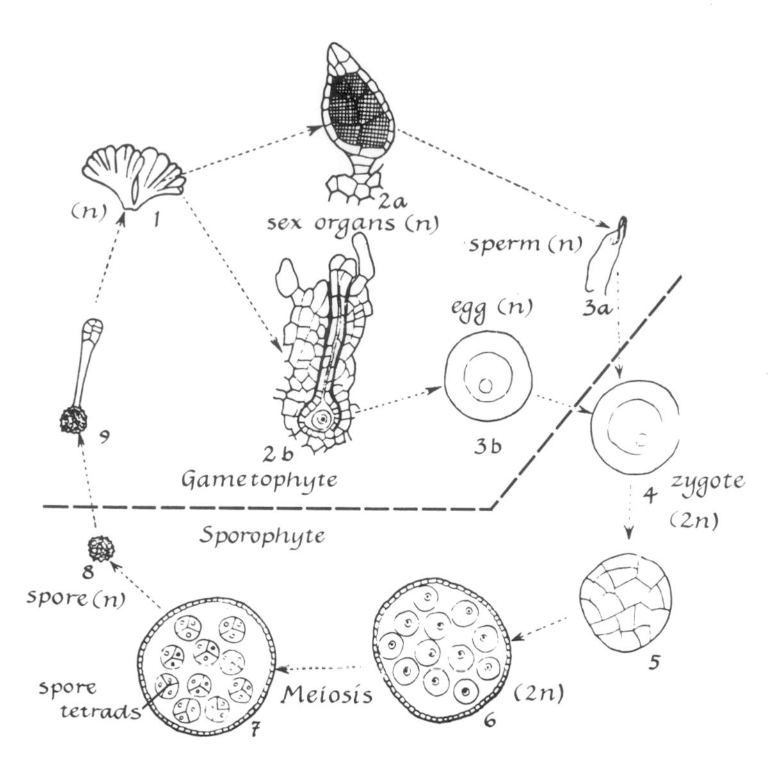

Figure 9-37. *Life cycle of the liverwort* Riccia, *illustrating the alternation of a spore-producing with a gamete-producing generation. The sex organs (gametes and spores) are greatly enlarged out of proportion to the size of the mature gametophyte shown in 1. From Holman and Robbins, 1940. Courtesy John Wiley and Sons, Inc.*

Developmental Significance of Spores

In general, any spore contains the essential information necessary for its development into an adult organism. As a spore is typically a single cell, the information must be packaged within this cell, either in the nucleus, the cytoplasm, or both. A spore contains the same information in its nucleus as the nuclei of the plant that produced it. The problem is why it develops into an organism like the parent if it is a mitospore and unlike the parent if it is a meiospore. A meiospore produced on the sporophyte plant develops into the gametophyte type of plant, never into another sporophyte. Meiosis alone is not responsible for this change in the pattern of development, as we shall see in Chapter 14, in which some additional aspects of the alternation of generations will be considered.

Parthenogenesis

Development of a new individual from an unfertilized egg or gamete occurs naturally in some organisms and can be experimentally accomplished in many plants and animals. According to the strict definition of sexual reproduction—a process involving the participation of two sexually different individuals and thus genetic recombination—parthenogenesis is a special type of asexual reproduction.

However, in most instances, an egg is a gamete that depends on union with a spermatozoan for its activation and development. Perhaps development of an unfertilized egg should be called nonsexual reproduction, because a sex cell is involved but no fusion of this cell with another is necessary for its development. Parthenogenesis has been induced in the eggs of many plants and animals by various artificial means, including temperature shock and treatment with salt solutions and hormones.

Conclusion

In asexual reproduction, all the information necessary for formation of a new individual is present in a single reproductive unit. No genetic recombination such as occurs in sexual reproduction during fertilization is necessary. In reproduction by fragmentation, fission, zooid formation, and strobilation, the major guidelines directing formation of the new individual are directly inherited in a geometrical form from the parent organism. The reproductive unit is essentially a sample of the structure of the parent, and its development is largely an increase by multiplication of the elements of the inherited part of the parent.

In reproduction by buds, gemmules, and gemmae, part of the information for development is in the nuclei of the cells and the remainder is supracellular, present in the cellular architecture of the unit. Some supracellular guidelines are preformed at the time the reproductive unit separates from the parent; others are progressively formed during its development into the new individual.

In reproduction by spores or unicellular buds, the information for development is packaged primarily in the nuclear genome. As far as is known, the nucleus of a spore contains the same genetic information as the nuclei of the organism that produced it. However, epigenetic guidelines for the selective use of this information must also be present, for a meiospore typically develops into an organism different from the one that produced it.

A comparative study of the diverse patterns of asexual reproduction presents all the problems of the control of development found in other patterns of development. The most extreme preformist and epigenetic patterns, as well as those intermediate, are found in asexually reproducing organisms.

References

Allen, C. E. 1919. The basis of sex inheritance in *Sphaerocarpus*. Proc. Amer. Phil. Soc. **58**:289–316.

Arey, L. B. 1965. Developmental Anatomy. W. B. Saunders, Philadelphia.

Berrill, N. J. 1961. Growth, Development, and Pattern. W. H. Freeman, San Francisco.

Brien, P. 1950. Étude d'*Hydra viridis* (Linn.) Ann. Soc. Roy Zoll. Belg. **81**:33–108.

Brown, W. H. 1935. The Plant Kingdom. Ginn, Boston.

Child, C. M. 1902. Studies on regulation: II. Arch. Entw. Mech. **15**:603–637.

Child, C. M. 1941. Patterns and Problems of Development. University of Chicago Press, Chicago.

Dogiel, V. A. 1965. General Protozoology. Clarendon Press, Oxford.

Goette, A. 1907. Vergleichende Entwicklungsgeschichte den Geschlechtsindividuen den Hydropolypen. Zeit. Wiss. Zool. **87**:1–335.

Guilcher. Y. 1950. Contribution à l'étude des ciliés gemmipares, chonotriches et tentaculiferes. Ann. Sci. Nat. Zool. Ser. 11, **13**:33–133.

Hyman, L. H. 1940. The Invertebrates: Protozoa through Ctenophora. McGraw-Hill Book Company, New York.

Hyman, L. H. 1955. The Invertebrates: Echinodermata. The Coelomate Bilateria. McGraw-Hill Book Company, New York.

Johnson, H. P. 1893. A contribution to the morphology and biology of the *Stentor*. J. Morph. **8**:467–562.

Leiby, R. W., and C. C. Hill. 1924. The polyembryonic development of *Platygaster verualis.* J. Agric. Res. **28:**829–840.

Lutz, H. 1949. Sur la production expérimentale de la polyembryonie et de la monstruosité double chez les oiseaux. Arch. Anat. Micr. Morph. Exp. **38:**79–144.

Newman, H. H. 1940. Multiple Human Births. Doubleday, New York.

Okada, Y. K. 1929. Regeneration and fragmentation in the syllidian polychaetes. Arch. Entw. Mech. **115:**542–600.

Patterson, J. T. 1913. Polyembryonic development in *Tatusia novemcincta.* J. Morph. **24:**559–683.

Picken, L. 1960. The Organization of Cells. Clarendon Press, Oxford.

Rasmont, R. 1962. The physiology of gemmulation in freshwater sponges. In D. Rudnick, ed., Regeneration, pp. 3–25. Ronald Press, New York.

Rice, M. E. 1970. Asexual reproduction in a sipunculan worm. Science **167:**1618–1620.

Silvestri, F. 1936. Insect polyembryony and its general biological aspects. Mus. Comp. Zool. Harvard Bull. **81:**469–498.

Sonneburn, T. M. 1930. Genetic studies on *Stenostomon incaudatum:* I. The nature and origin of differences among individuals formed during vegetative reproduction. J. Exp. Zool. **57:**51–108.

Spratt, N. T., Jr., and H. Haas. 1960. Integrative mechanisms in development of the early chick blastoderm: I. Regulative potentiality of separated parts. J. Exp. Zool. **145:**97–137.

Streeter, G. L. 1919. Formulation of single ovum twins. J. Hop. Hosp. Bull. **30:**235–238.

Tartar, V. 1961. The Biology of *Stentor.* Pergamon Press, New York.

Tartar, V. 1968. Microsurgical experiments on cytokinesis in *Stentor coeruleus.* J. Exp. Zool. **167:**21–36.

Waddington, C. H. 1962. New Patterns in Genetics and Development. Columbia University Press, New York.

Waddington, C. H. 1963. Ultrastructure aspects of cellular differentiation. Symp. Soc. Exp. Biol. **17:**85–97.

Wardlaw, C. W. 1955. Embryogenesis in Plants. Methuen, London.

Waris, H., and P. Kallio. 1964. Morphogenesis in *Micrasterias.* Adv. Morph. **4:**45–80.

Waris, H., and I. Rouhiainen. 1970. Permanent and temporary morphological changes in *Micrasterias* induced by gamma rays. Ann. Acad. Sci. Fenn. A. IV. Biologica **167:**1–13.

10 *Patterns of Sexual Reproduction in Plants: General Survey*

Sexual reproduction involves the union of two different genomes, thus a recombination of genetic material. In self-fertilizing plants, there is obviously no biparental inheritance, but there is still the possibility of genetic recombination through meiosis in sporogenesis. Sexual reproduction insures far greater genetic diversity than is possible in organisms that reproduce asexually.

Recall (Chapter 9) that meiosis preceding spore formation (which occurs in fungi, mosses, ferns, and higher plants) insures genetic diversity in the haploid meiospores. The spores develop into genetically different haploid plants and, when cross fertilization occurs between the gametes of these plants, they diversify further. Thus meiosis, which segregates genetic capacities, and fertilization, which recombines these capacities, together are mechanisms for variation among organisms.

Genetic recombination is accomplished in different ways. It can involve the fusion of gametes (specialized sex cells); the fusion of parent cells, which thus function as gametes; the fusion of parts of two parent organisms; the fusion of nuclei of two organisms; an exchange and fusion of nuclei; or an exchange and recombination of DNA, pure genetic material. In this chapter we shall discuss some patterns of recombination, but first we shall discuss plant life cycles in general. Following this, we shall consider some aspects of sexual reproduction, including formation of reproductive organs, egg formation, fertilization, and early development of plant embryos.

Plant Life Cycles

Although several types of life cycles in plants involve sexual reproduction, the gametes are usually products of mitosis in cells of the haploid generation, with apparently two exceptions—Fucales, a group of brown algae, and the diatoms, alga-like unicellular or colonial organisms in which the sex cells are products of mitosis (Allsopp, 1966; Scagel et al., 1966). The life cycle characteristic of most plants is diagramed in Figure 10-1. Note that the haploid, gamete producing generation alternates with the diploid, spore producing generation.

Depending on the kind of plant, gametes are either morphologically similar or dissimilar and are formed on the same or separate gametophytes. The relative durations and sizes of the gametophyte and sporophyte generations vary greatly among different plants. In fungi, the gametophyte constitutes the greater portion of the cycle, the fertilized egg or zygo-

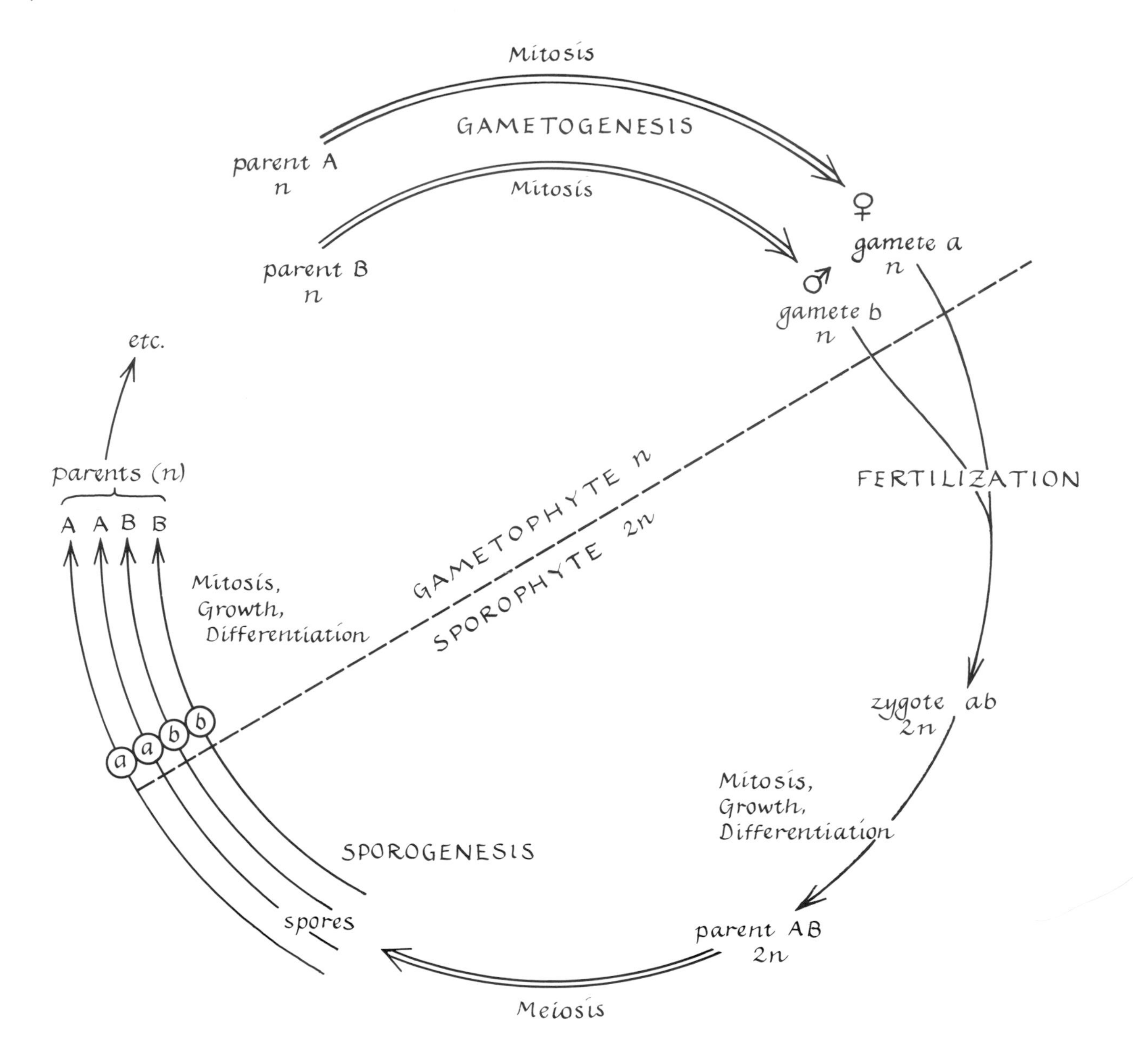

Figure 10-1. *The life cycle characteristic of most plants.*

spore representing the only diploid, sporophyte phase. A similar life cycle occurs in flagellates and other green algae. In some of these, such as *Chlamydomonas* (Figure 10-2), meiosis occurs immediately after fertilization, so that the haploid generation constitutes a much larger arc of the cycle. In other plants such as ferns, the commonly visible generation is the leafy sporophyte; the gametophyte is a much smaller green body called a prothallium that lies flat on the surface of the soil (see Figure 10-21). The life cycle of a fern is shown in Figure 10-3. Note again the distinct alternation of meiospore and gamete producing generations. In liverworts and mosses, the life cycle is similar to that in ferns except

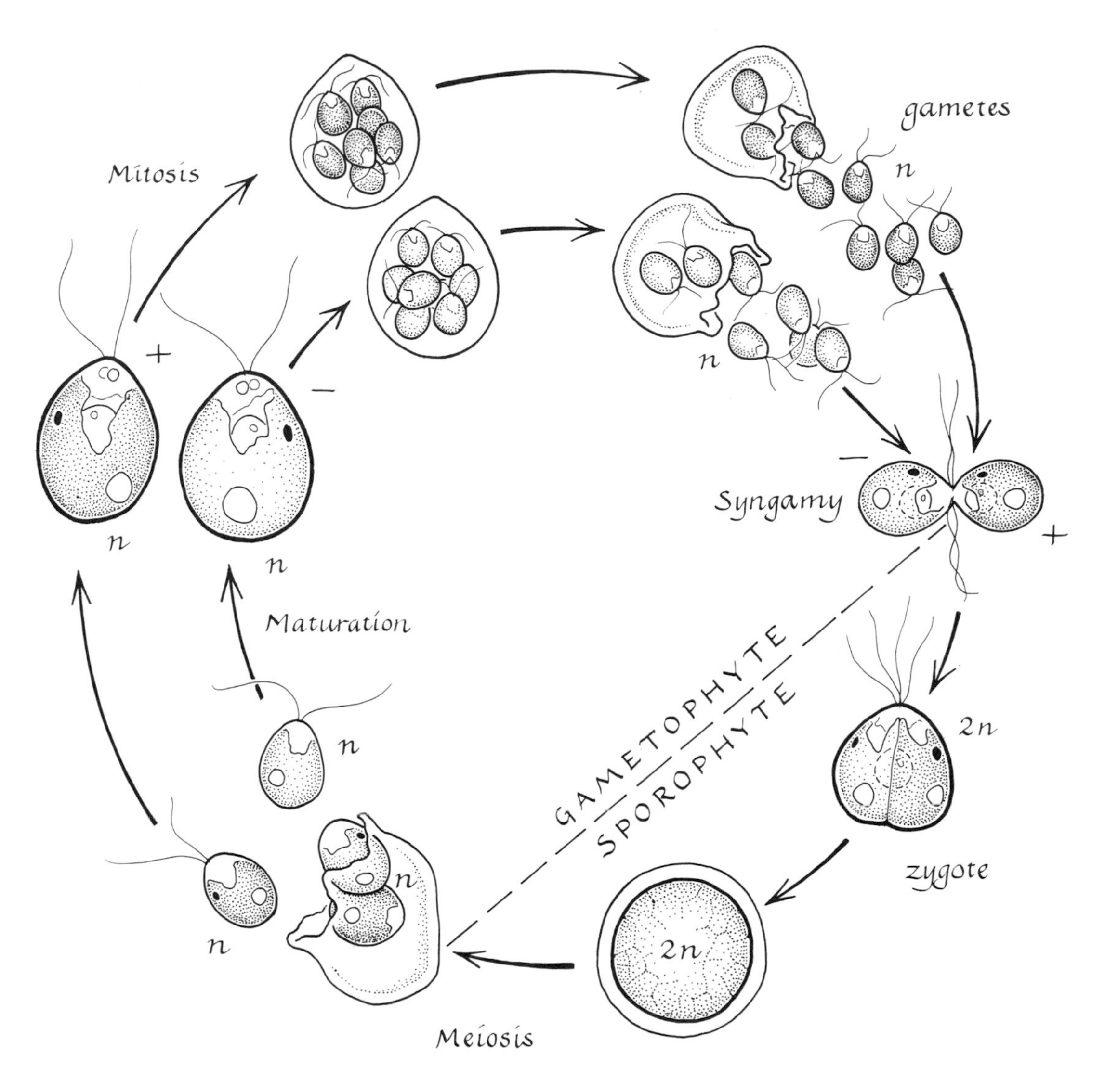

Figure 10-2. *Life cycle of the alga* Chlamydomonas.

that the gametophyte is the larger and more obvious phase of the cycle; the sporophyte is less obvious but still visible to the naked eye.

In higher plants, such as the cone and seed producing gymnosperms and flower and seed producing angiosperms, the gametophyte is usually reduced to a microscopic structure that is encased in the tissues of the dominant sporophyte, the plant we commonly recognize. As noted above, Fucales and the diatoms are exceptions to the generalized life cycle shown in Figure 10-1. In these plants the diploid generation is more obvious, and the gametes are direct products of meiosis. The gametes are the only haploid cells in the cycle; in this respect, the cycle resembles the sexual reproductive pattern in animals.

Types of Sexual Reproduction

Three different types of gametes, representing three types of sexual reproduction, are generally recognized in plants. These are diagramed in Figure 10-4. Differences in the motility of the gametes are also indicated.

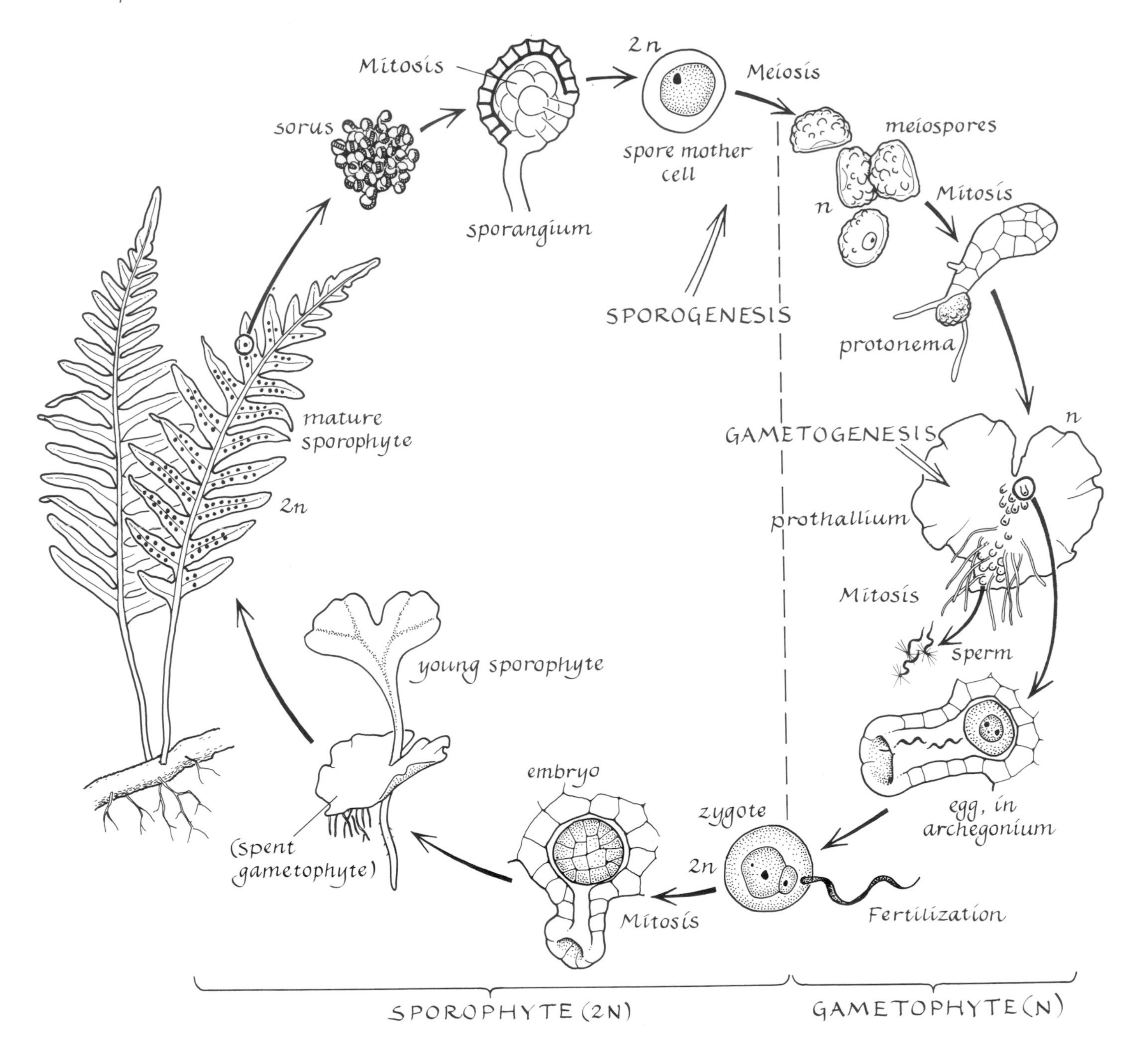

Figure 10-3. *Life cycle of a fern.*

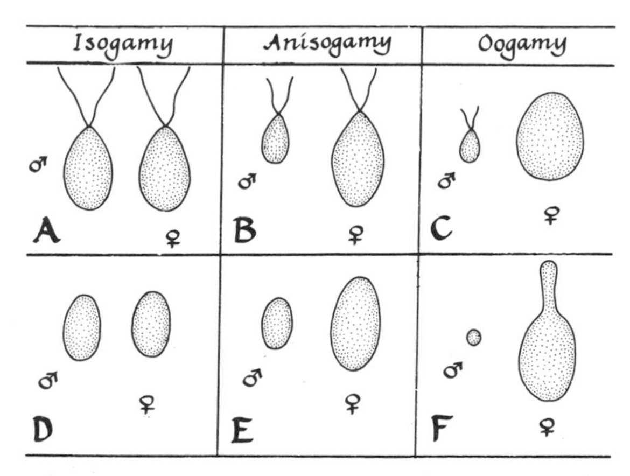

Figure 10-4. *Types of sexual reproduction. In A-C, at least one gamete is motile; in A and B, both gametes are motile; in C, the male gamete is motile; in D-F, neither gamete is motile. From Scagel et al. Copyright © 1966 by Wadsworth Publishing Company, Inc.*

Isogamy

As the term isogamy implies, this sexual process involves the fusion of like gametes. However, fusion usually occurs only between isogametes that are physiologically different. The two different gametes, usually designated as + or − types (it is not possible to assign maleness or femaleness to either of them), are ordinarily produced by two separate adult individuals. The difference between the types is beginning to be understood, and it is generally believed that mating type substances called gamones are involved. For example, in yeast, opposite mating types have complementary macromolecules on their surfaces, which act in the same way as antigens and

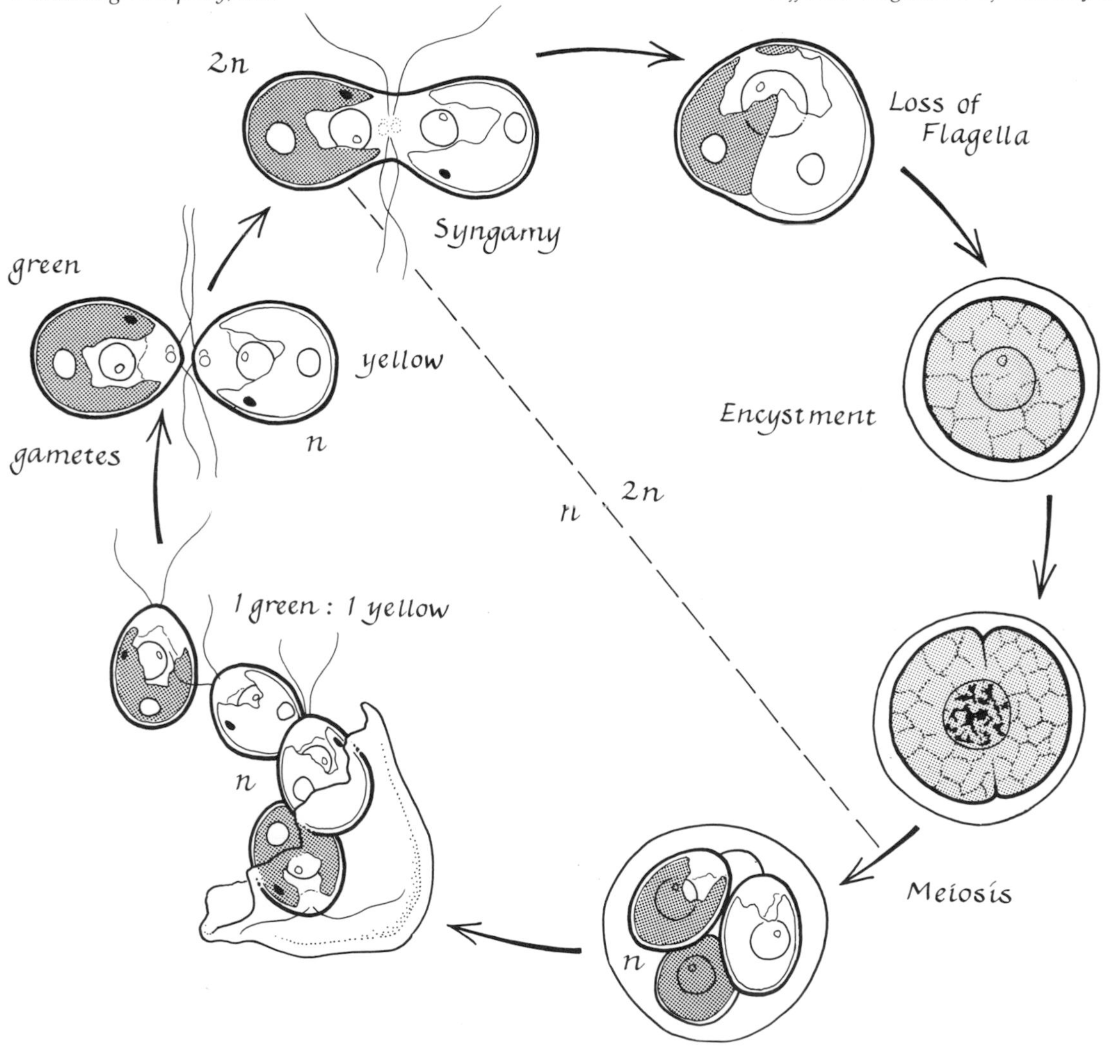

Figure 10-5. *Syngamy and segregation of genetically different isogametes of* Chlamydomonas.

antibodies. The molecules appear to be glycoproteins of relatively low molecular weight (Crandall and Brock, 1968). Probably a similar mechanism of cell surface recognition, leading to adhesion and syngamy (fusion of gametes), occurs throughout both plant and animal kingdoms.

Wiese (1965) has presented evidence that the gamones in *Chlamydomonas* are the surface components responsible for initial flagellar contact between opposite mating types. The process of fertilization in animal eggs also appears to involve an antigen-antibody type of interaction (Tyler and Tyler, 1966). The fusion of genetically different isogametes in *Chlamydomonas reinhardii* is illustrated in Figure 10-5. The first contact of the two gametes is at the tips of their flagella. Fusion of the pair in the papilla region, followed by formation of a protoplasmic bridge, constitutes the pre-zygotic stage. Gamete fusion is complete in 18 to 36 hours, and caryogamy (nuclear fusion) follows (Wiese, 1965). According to Randall et al. (1967), following fusion there is a regression of

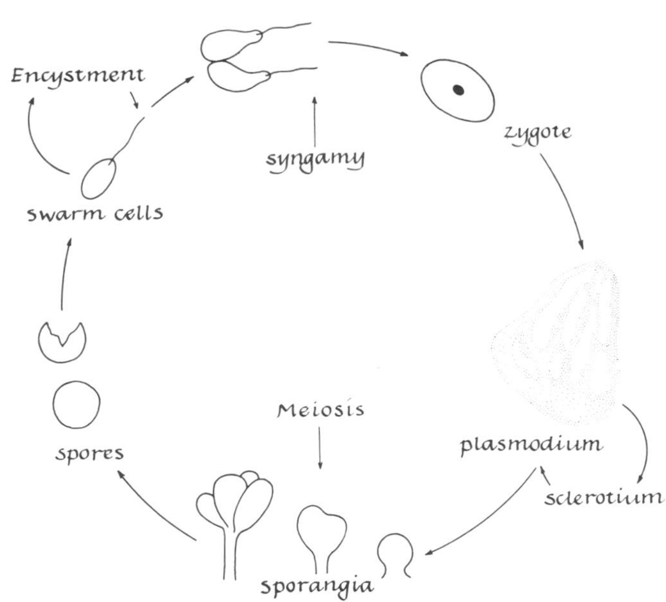

Figure 10-6. *Generalized life cycle of a true slime mold. From Scagel et al. Copyright © 1966, Wadsworth Publishing Company, Inc.*

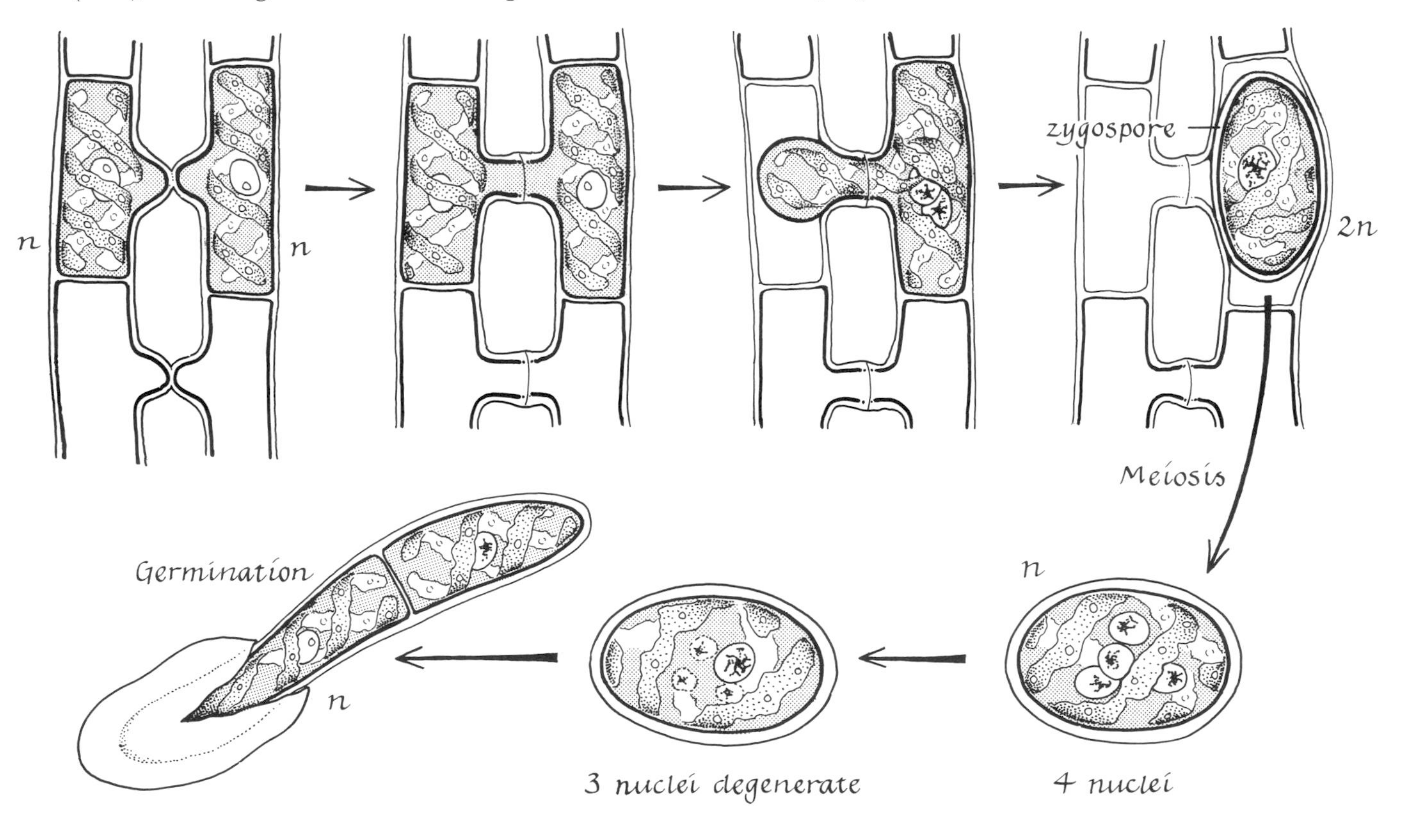

Figure 10-7. *Morphological isogamy and zygospore germination in* Spyrogyra.

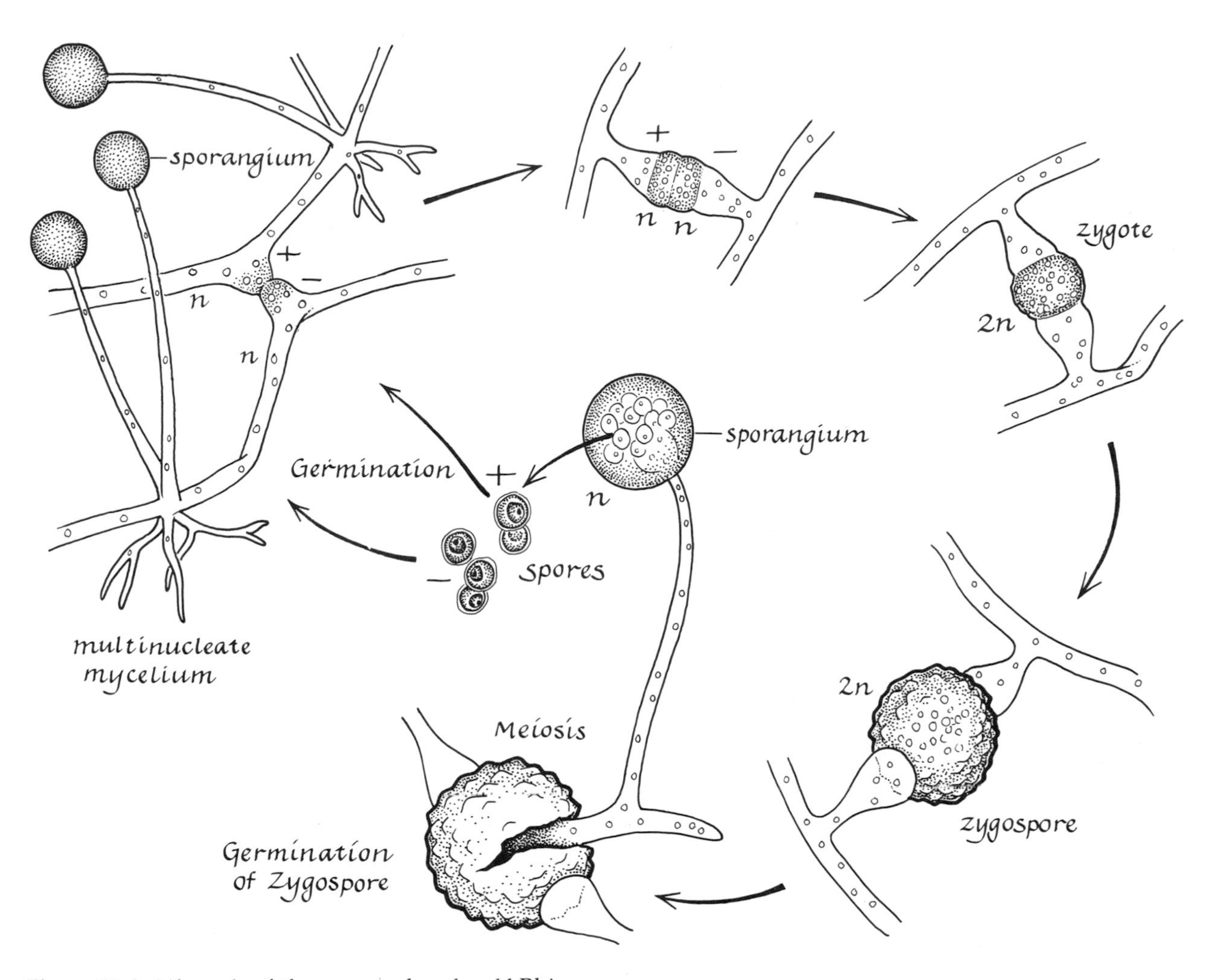

Figure 10-8. *Life cycle of the common bread mold* Rhizopus.

the flagella and their basal bodies (Chapter 25; Figure 10-5). A thick wall around the zygote is then formed, creating a zygospore. After a resting period of a few days, a new basal body is formed. Then meiosis occurs, producing four haploid cells. When fusion occurs between a green and a yellow colored mating type, the offspring of the first generation are in the ratio of one green to one yellow cell. In the second generation, the ratio is still 1 to 1. Were the gametes products of meiosis in a diploid mother cell, a 3 to 1 ratio of green to yellow would occur in the second generation (if green color were dominant).

True slime molds (Myxomycetes) exhibit an isogamous type of reproduction (Bold, 1964; Scagel et al., 1966). The generalized life cycle is shown in Figure 10-6. A flagellated swarm cell emerges from the spore. Under conditions of drought the swarm cell encysts. Sometimes an ameba (myxameba) emerges from the spore, but this ameba later forms a flagellum and becomes a swarm cell. Two swarm cells fuse to produce a zygote that through nuclear division forms the multinucleated plasmodium. Under conditions of relative dryness the plasmodium transforms into a cluster of fruiting bodies (sporangia) in which the spores are products of meiosis. Sexuality, however, has not been demonstrated in all species of true slime molds. In *Didymium nigripes*, the ploidy level of the ameboid and plasmodial stages is the same, and time-lapse movies show the origin of plasmodia from single amebae (Kerr, 1968).

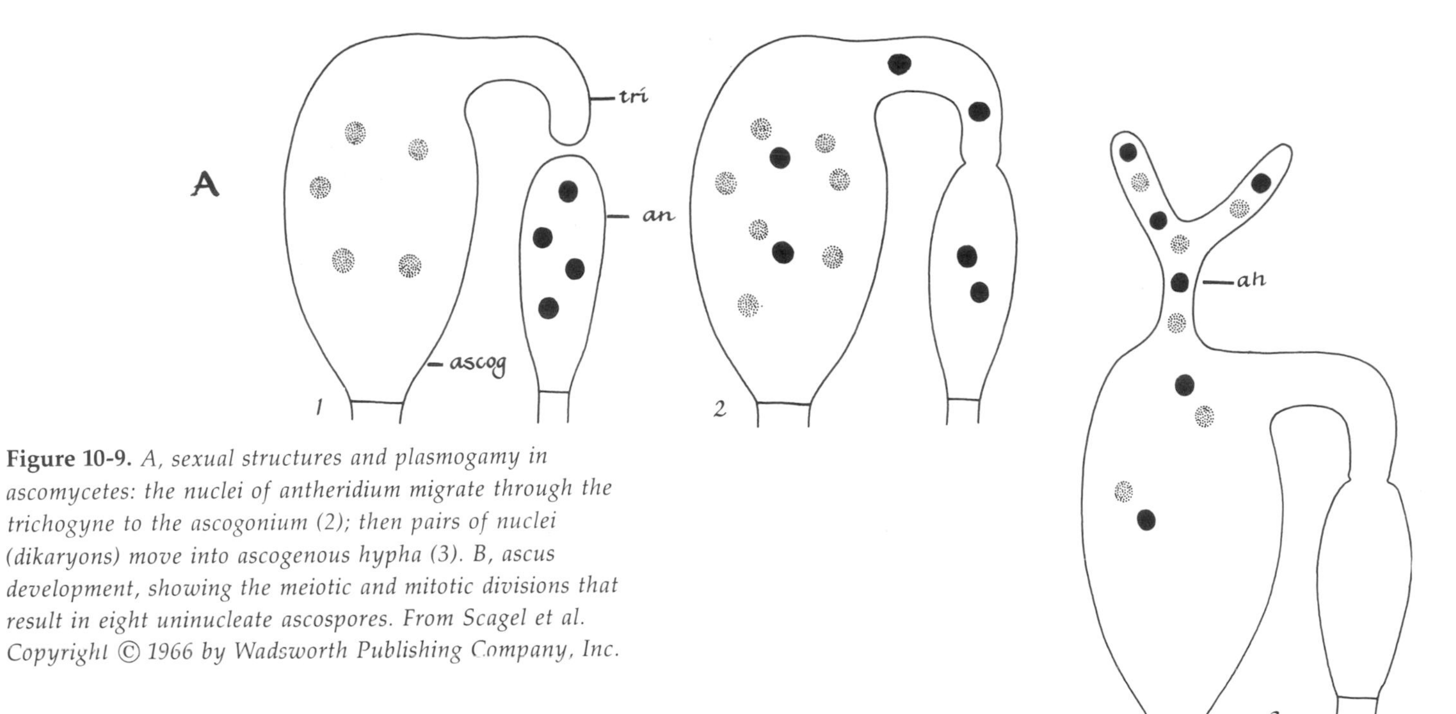

Figure 10-9. *A, sexual structures and plasmogamy in ascomycetes: the nuclei of antheridium migrate through the trichogyne to the ascogonium (2); then pairs of nuclei (dikaryons) move into ascogenous hypha (3). B, ascus development, showing the meiotic and mitotic divisions that result in eight uninucleate ascospores. From Scagel et al. Copyright © 1966 by Wadsworth Publishing Company, Inc.*

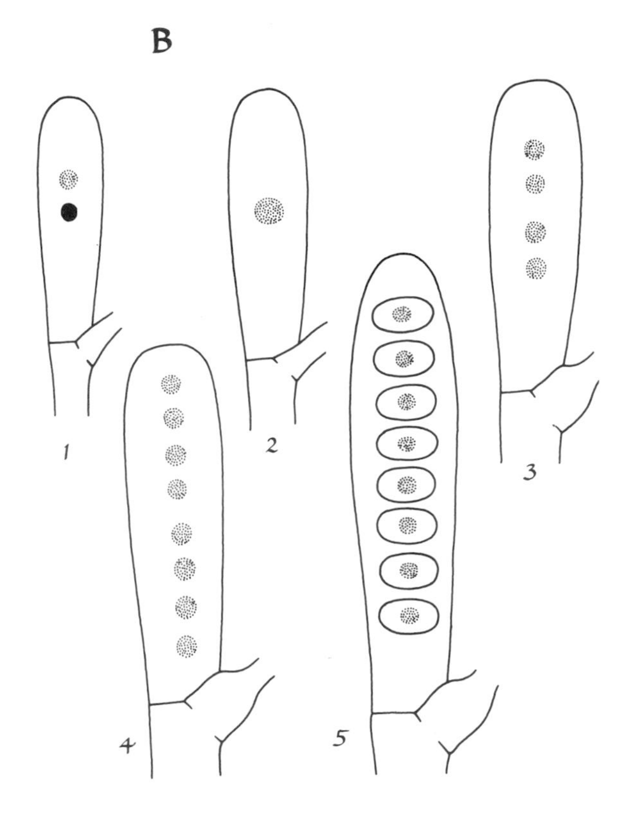

An isogamous type of reproduction involving fusion of amebae has been reported in the cellular slime molds. Both haploid and diploid strains of *Dictyostelium discoideum* occur, and genetic recombination has been observed (Scagel et al., 1966).

Another well-known example of morphological isogamy is exhibited by the filamentous green alga, *Spyrogyra*, so named because of its spirally twisted, ribbon-shaped chloroplast (Figure 10-7). Individual cells of the filament function as gametes, but lack flagella and exhibit ameboid movement. According to Sinnott (1946, 1963), all the cells of one filament usually behave alike; that is, all either receive the protoplasm or all donate it. This is an illustration of physiological heterogamy in morphological isogametes. In this type of isogamy, the parent cells are the gametes rather than the source of gametes as in *Chlamydomonas* (Figure 10-2).

Many fungi, particularly the Ascomycetes and Basidiomycetes, exhibit a type of morphologically isogamous but physiologically heterogamous reproduction in which whole filaments constituting separate mating types of mycelia function like gametes. Nevertheless, in primitive Phycomycetes (water molds) sexual reproduction may occur by fusion of motile isogametes. One of the most familiar examples of the fusion of nonmotile isogametes is found

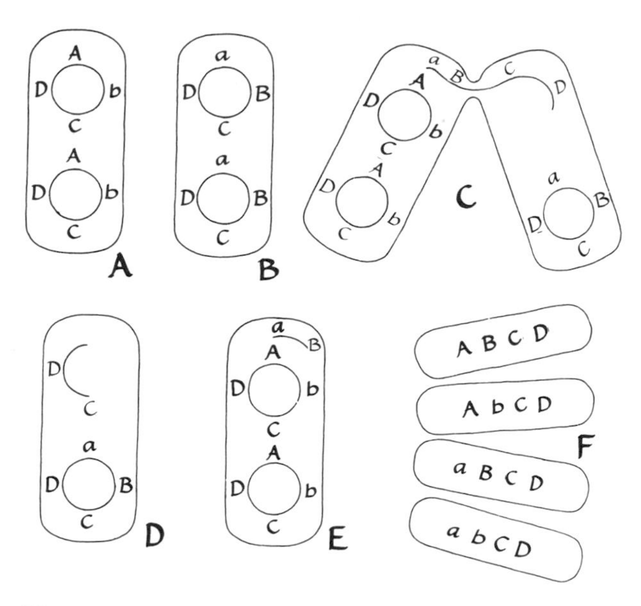

Figure 10-10. *Bacterial conjugation. C, transfer of genetic material from donor to recipient; D-E, cells after conjugation; F, possible descendants of recipient cell. From Scagel et al. Copyright © 1966 by Wadsworth Publishing Company, Inc.*

in the common bread mold *Rhizopus nigricans.* The vegetative body is a coenocytic mycelium composed of branching hyphae, diagramed in Figure 10-8. As short side branches from two separate mycelia approach one another, they begin to fuse. The tips enlarge and each is separated by a cross wall. The walls separating the side branches are absorbed and a single cell, a multinucleated zygote, is formed. The zygote enlarges, synthesizes a thick pigmented cell wall, and becomes a resting zygospore. The zygote and zygospore are the only diploid phases of the cycle. Meiosis, occurring during the germination of the zygospore, forms two types of spores, one producing a +, the other a − type of mycelium.

In some Ascomycetes, hyphal fusion occurs between specialized male and female structures called antheridia and ascogonia. Both of these are usually present on a single mycelium. After fusion of the two structures (Figure 10-9A), an ascogenous hypha is formed. The *ascus,* which contains two haploid nuclei, one from the ascogonium and one from the antheridium, develops from this hypha. The two nuclei in the ascus fuse to produce a diploid zygote nucleus. Then meiosis occurs forming four haploid nuclei, each of which usually divides mitotically to produce eight nuclei. Cell walls surround these nuclei to form the ascospores (Figure 10-9B).

In the pattern of reproduction we have just described (1) specialized unicellular sex organs are involved, (2) cytoplasmic fusion occurs, (3) nuclei are transferred from the male to the female organ, and (4) male and female nuclei associate in pairs but do not fuse until ascus formation. The nuclei and some of the surrounding cytoplasm are the actual isogametes.

In conjugation of bacteria, a process that might be considered isogamous, or agamous, has been demonstrated many times (Hayes, 1952; Jacob and Wohlman, 1961) (Figure 10-10). This process involves an exchange and recombination of DNA. A similar phenomenon occurs in the conjugation of some protozoa in which the nuclei function as gametes (see Chapter 12).

Studies have indicated the possibility of a sexual process in the blue-green algae (Lazaroff and Vishniac, 1962). In *Nostoc,* processes superficially resembling the production of meiotic spores followed by fusion of short filaments suggest a type of isogamous reproduction. Some evidence for sexual reproduction involving genetic recombination in a unicellular blue-green algae has been reported (Kumar, 1962). However, sexual reproduction still remains to be clearly demonstrated in these algae.

Heterogamy (Anisogamy)

The distinction between isogametes and heterogametes is slight. The term heterogamy (or anisogamy) describes the fusion of gametes differing only in size. The term oogamy refers to morphologically different gametes, such as eggs and spermatozoans. Heterogamy is not as common in plants as isogamy or oogamy. *Chlamydomonas* usually forms isogametes, but at least one species forms anisogametes. Heterogamy occurs in the green algae, including *Bryopsis,* some species of *Ulva,* in the marine form, *Codium* (Figure 10-11), possibly in *Pandorina,* in some brown algae, and in the water mold *Allomyces.* In *Allomyces,* the larger "female" gametes formed in a colorless gametangium at the end of a mycelium filament (Figure 10-15) produce a hormone that attracts the smaller "male" gametes, which are formed in an adjacent orange-colored gametangium

(Figure 10-15). As in *Codium,* both gametes are motile and their fusion produces a flagellated zygote, which later germinates to produce a diploid spore-bearing generation (Scagel et al., 1966).

Oogamy

In this type of sexual reproduction the gametes are usually strikingly different in structure, motility, and size. The male gamete is often a flagellated highly motile spermatozoan with a small amount of cytoplasm relative to that of the nucleus. Spermatozoa of several plant groups are illustrated in Figure 10-12. The morphology of the spermatozoa is diverse. The *Cycad* spermatozoan is peculiar because it develops inside the germinated pollen grain, similar to the pollen grain of the flowering plants (Chapter 11).

On the other hand, the female gamete (egg or oosphere) is a nonflagellated, usually nonmotile cell with a large volume of cytoplasm relative to the volume of the nucleus. Although the eggs of different plant groups do not exhibit the morphological diversity of the spermatozoa, they differ in nuclear information and probably in cytoplasmic structure.

Oogamy occurs in many plant groups, from the algae and fungi to the flowering plants. In some of these, the male gametes are motile; in others nonmotile cells are conveyed to the egg by a special tubular structure. One of the simpler patterns of oogamous reproduction is illustrated by *Volvox* (Figure 10-13).

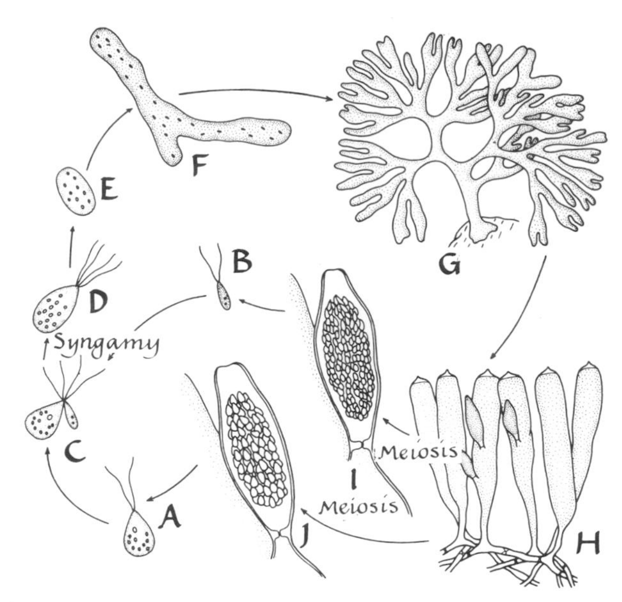

Figure 10-11. *Heterogamy in* Codium. *A, female gamete; B, male gamete; C, fusion of anisogametes (syngamy); D, planozygote; E, zygote; F, filamentous juvenile stage; G, mature diploid thallus; H, vesicles with gametangia; I, male gametangium (meiosis); J, female gametangium (meiosis). Note the formation of gametes by meiosis in the male and female gametangia borne on the diploid thallus* G. *From Scagel et al. Copyright © 1966 by Wadsworth Publishing Company, Inc.*

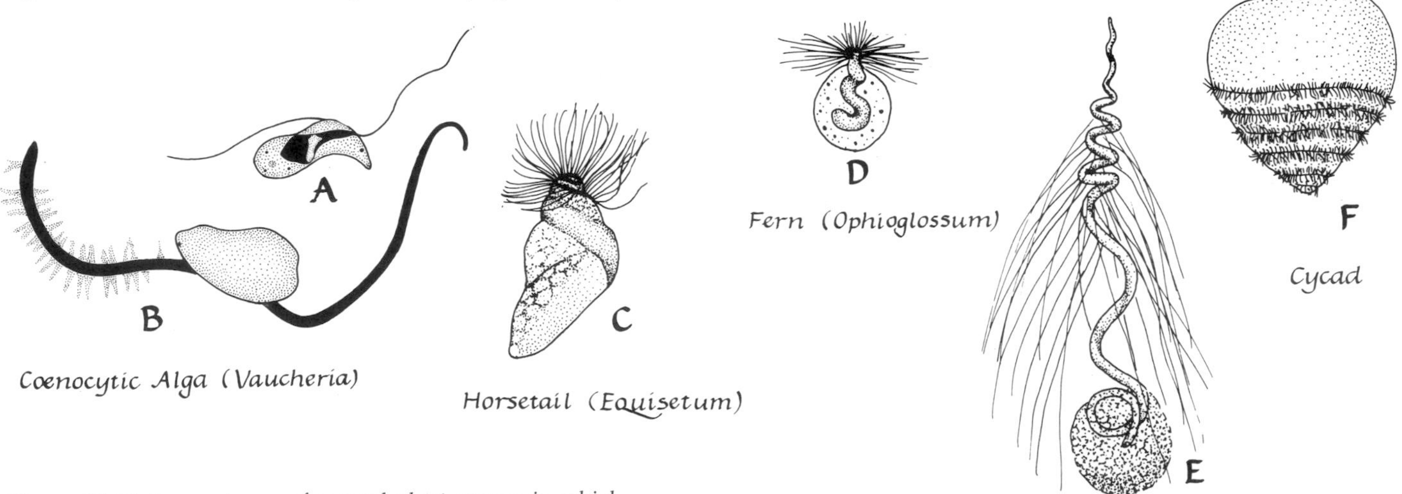

Figure 10-12. *Spermatozoa of several plant groups in which oogamy occurs. From Scagel et al. Copyright © 1966 by Wadsworth Publishing Company, Inc.*

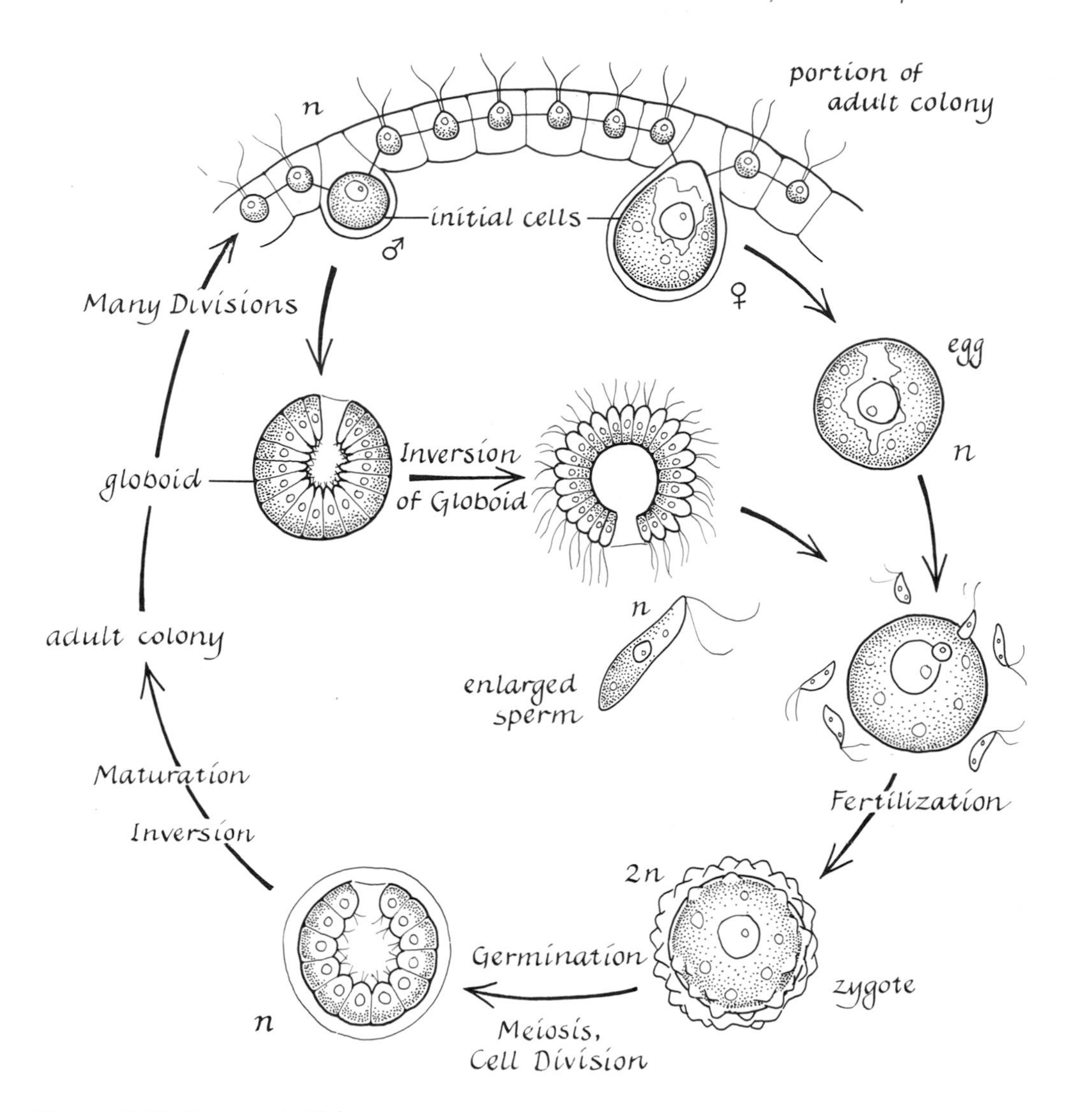

Figure 10-13. *Oogamy in* Volvox.

Some cells in the posterior part of the colony enlarge and develop either as male or female initial cells. In colonies forming both types of gametes, the male initial cells develop long before the female (Brown, 1935; Berrill, 1961). Why some cells differentiate as male, others as female initial cells is as yet an unsolved problem. The female initial enlarges, loses its flagella, synthesizes hematochrome (a red pigment), and develops into the large egg cell. The male initial enlarges, loses its flagella, sinks inward and then rotates 90° to 120°, until its original outer pole points sideways or inwards. The male cell then divides many times to form a globoid in which about 512 spermatozoa are produced. The globoid then undergoes inversion, similar to that occurring in the gonidium (Chapter 9), and the spermatozoa are released.

Fertilization results in a diploid zygote, which becomes surrounded by a thick cell wall. The zygote apparently undergoes meiosis preceding a number of mitotic divisions that form the new haploid colony. In this process, a sphere of cells with flagella pointing inward is first produced, but an inversion process, similar to the other two we have already noted, follows. Unless three of the cells produced by meiosis degenerate (a common occurrence in female type

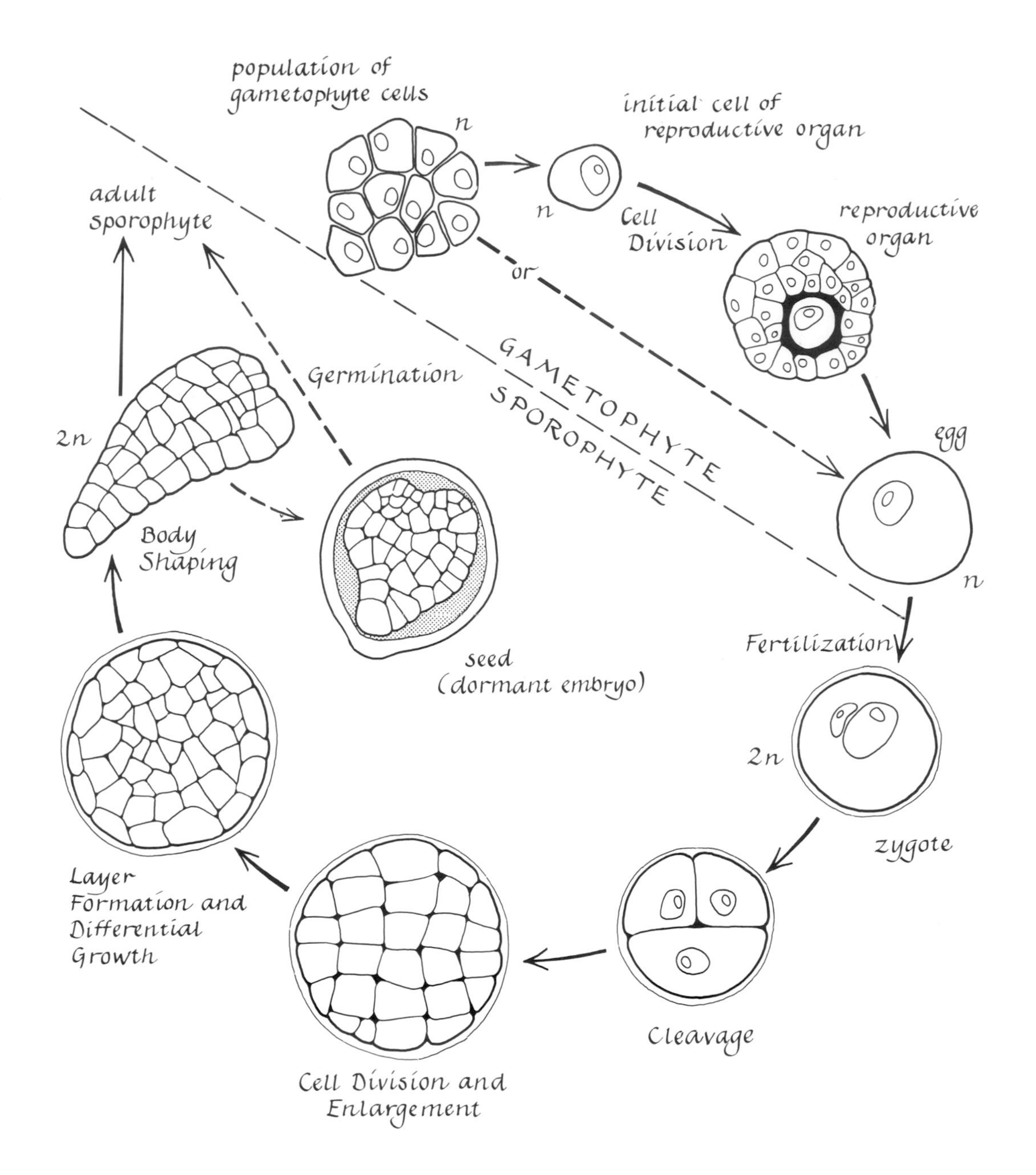

Figure 10-14. *The general chronology of formation and development of a plant egg.*

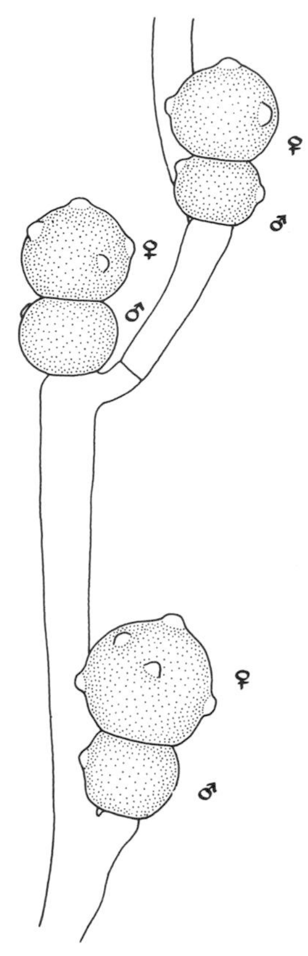

Figure 10-15. *Unicellular, multinucleate sex organs of* Allomyces arbuscula. *From Scagel et al. Copyright © 1966 by Wadsworth Publishing Company, Inc.*

products of meiosis), the new colony must be heterozygous, unlike a colony formed asexually from a gonidium. The persistence of all four products of meiosis could be a mechanism for differentiation of the male and female initial cells.

General Chronology of the Formation and Development of the Plant Egg

Although different plant groups have different patterns of egg formation and plant development, a general sequence of events is common to many kinds of plants (Figure 10-14). In mosses and ferns, one cell of the gametophyte is set aside as the progenitor or initial cell that by division produces a multicellular reproductive organ. Within this reproductive organ, one cell enlarges and becomes an egg. In other plants, (such as *Volvox* and flowering plants) one cell of the gametophyte directly transforms into an egg. In still other plants, the female reproductive organ is a single multinucleated cell inside which several eggs are formed.

Fertilization results in cleavage, essentially a subdivision of the contents of the zygote. This is rapidly followed by further cell division, which forms an increasingly larger cellular aggregate. Layer formation and cellular differentiation occur within the aggregate, and it acquires a characteristic shape as the result of differential growth. At this stage, the pattern diverges in two directions. In some plants, the embryo develops directly into the adult. In others, the embryo becomes encased within a seed and enters a period of dormancy. Under favorable conditions, the seed germinates and the embryo develops into an adult.

Formation of Special Sex Organs and Gametogenesis

In many plants, the gametes develop within unicellular or multicellular sex organs. These exhibit varying degrees of complexity in fungi, green, red, and brown algae, and higher plants. We will discuss only a few examples.

Unicellular Organs

Unicellular sex organs are present in *Volvox,* where the sex organ is actually a gelatinous cell product encasing the male or female initial cell (Figure 10-12). In the water molds *Allomyces* (Figure 10-15) and *Saprolegnia* (Figure 10-16), the sex organs are multinucleate cells that form at the ends of hyphae or as branches of hyphae.

In *Allomyces,* the colorless female gametangium is apparently always distal to the smaller orange-colored male gametangium (Figure 10-15). In *Saprolegnia* (Figure 10-16), the female sex organ (oogonium) develops on a hyphal branch and produces several

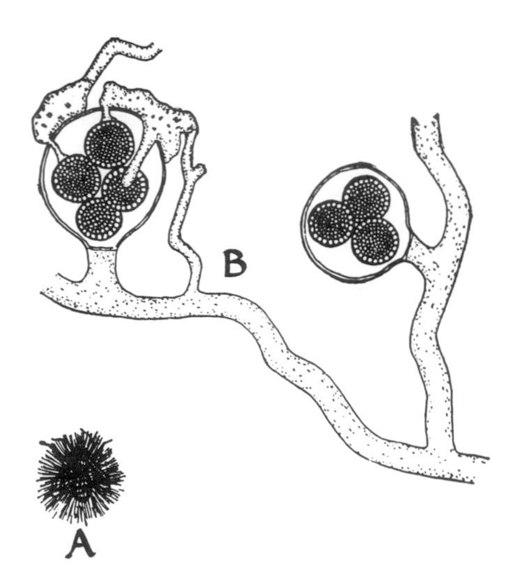

Figure 10-16. *Unicellular, multinucleate sex organs of* Saprolegnia. *A, mycelium; B, fertilization of eggs in an oogonium penetrated by antheridial tubes. From* Botany: Principles and Problems *by Sinnott. Copyright 1946 by McGraw-Hill Book Company. Used with permission of McGraw-Hill Book Company.*

Figure 10-17. *Sex organs and fertilization in* Vaucheria. *From Brown:* The Plant Kingdom, *1935. Used by permission of Mrs. Brown.*

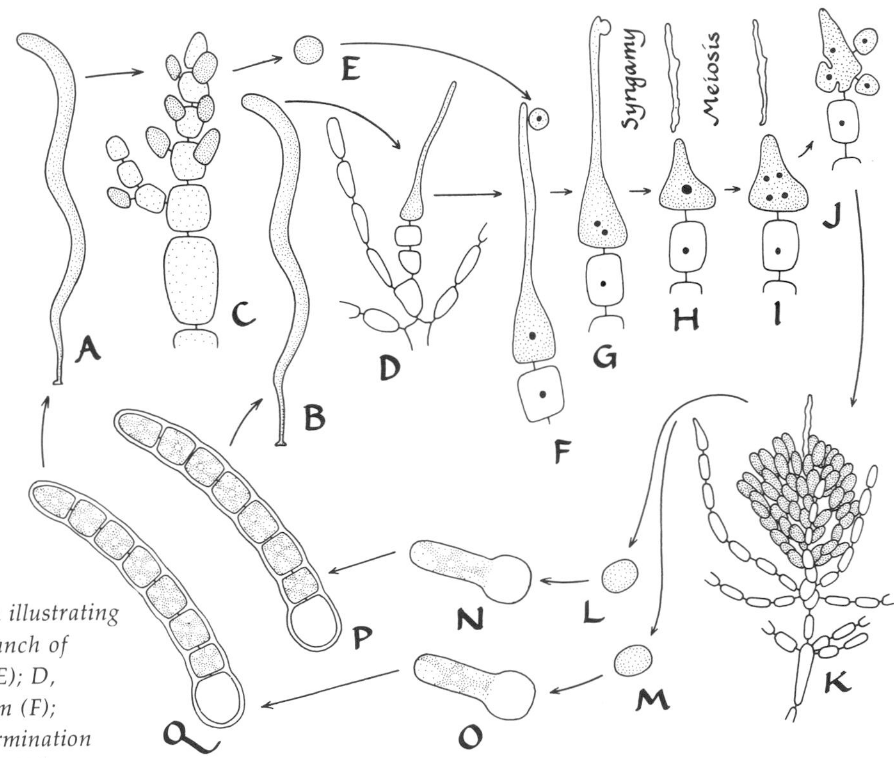

Figure 10-18. *Life cycle of the red alga* Nemalion *illustrating the unique sex organs. A, B, gametophytes; C, branch of male gametophyte and one released spermatium (E); D, branch of female gametophyte with a carpogonium (F); G-J, fertilization; K, mature gonimoblast; L-P, germination of carpospores. From Scagel et al. Copyright © 1966 by Wadsworth Publishing Company, Inc.*

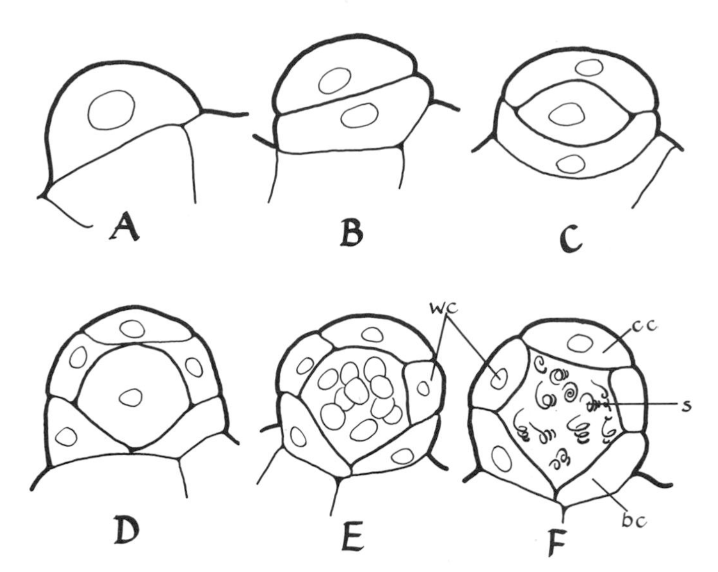

Figure 10-19. *Antheridial development in ferns.* wc, *wall cells;* cc, *cap cell;* bc, *basal cell;* s, *spermatozoa. After Davie from Scagel et al. Copyright © 1966 by Wadsworth Publishing Company, Inc.*

eggs, which are fertilized by the contents of an adjacent hyphal branch called an antheridium. Male nuclei in the anteridium function as nonmotile gametes and are transferred to the eggs by fertilization tubes. The fertilized egg becomes an oospore, which germinates and develops into a group of hyphae constituting the mycelium.

Sex organs of the coenocytic green alga *Vaucheria* are unicellular, but multinucleated oogonia and antheridia are formed on the filamentous gametophyte (Figure 10-17). The sex organs are similar in basic design to those of *Saprolegnia.* Motile sperm are released from the antheridial tube and enter the oogonium through an opening in its wall.

Unique reproductive organs are present in the red algae. In *Nemalion* (Figure 10-18), the sex organs are formed on separate male and female gametophytes. The nonmotile male gametes called spermatia are superficially produced by mitosis on special branches called spermatangia, which correspond to the antheridia of the water molds. The female gamete is essentially the lower portion of a modified oogonium called a carpogonium, which bears an elongated receptive part, the trichogyne.

Note in Figure 10-18 that the male nucleus enters the trichogyne, passes down, and unites with the female nucleus to form the zygote. As meiosis immediately follows fertilization, the zygote is the only diploid phase in the life cycle. The four nuclei resulting from the meiotic divisions are cut off in cells, each of which divides to form a chain of haploid cells. Some chains are male, others female, segregation for sex having occurred during meiosis. Recall (Chapter 9) that segregation for sex also occurs during formation of the meiospores in the liverwort *Sphaerocarpus.* Each chain of cells produces a single spore that develops into either a male or a female gametophyte.

Multicellular Organs

Multicellular sex organs are found in several plant groups above the fungi and algae, including liverworts, mosses, ferns, and some of the "living fossil" plants such as Equisetum and Ginkgo. Motile, multiflagellated spermatozoa are formed in an antheridium (Figure 10-19) when this organ is present, and a single nonmotile egg develops in an archegonium (Figure 10-20). Stages in development of these sex organs on the prothallium (lower surface of the gametophyte) of a fern are illustrated in Figures 10-19 and 10-20. A similar pattern of development occurs in other plants possessing these sex organs.

As shown in Figure 10-19A, the antheridium arises from a single cell, which then divides to form a group of cells (Figure 10-19B-D). In some ferns, the wall of the antheridium consists of only three cells: a cap cell, a circular wall cell, and a basal cell. The central cell of the developing antheridium divides repeatedly, producing the highly specialized spermatozoa (Figures 10-19E and F and Figure 10-21C). The mechanisms controlling differentiation of the products of division of the central cell are not definitely known, but the composition of the microenvironment of the central cell and its division products may be involved. A comparable kind of differentiation occurs in the developing blood islands of vertebrate embryos (Figure 22-8).

The formation of an archegonium is shown in Figure 10-20A-F. The archegonium begins its development by divisions of a single elongated cell of the prothallium. One of the central cells of this group ceases division, enlarges, and develops into the egg. Other cells become the neck canal cells and, by further divisions, still others produce the cells constituting

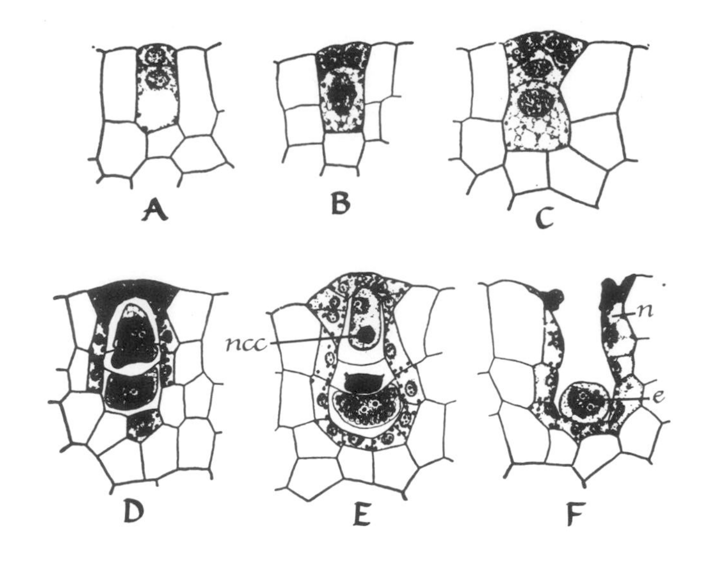

Figure 10-20. *Archegonial development in ferns.* ncc, *neck canal cell;* n, *neck;* e, *egg. From* Botany: Principles and Problems *by Sinnott. Copyright 1963 by McGraw-Hill Book Company. Used with permission of McGraw-Hill Book Company.*

the wall of the archegonium (Figure 10-20D-F). At maturity, the neck of the archegonium opens and the neck canal cells die and disintegrate, producing the cavity of the flask-shaped archegonium. Again, the mechanisms controlling differentiation of the descendants of the initial archegonial cell, only one of which becomes the specialized egg cell, are unknown. The position of the cells may be involved, but this remains to be demonstrated.

In the coniferophytes (cone-bearing plants), archegonia are present, but the nonmotile sperm cells are products of division within a greatly reduced gametophyte which is a germinated pollen grain. Recall the peculiar phenomenon in the cycad, in which a motile spermatozoan develops in a germinating

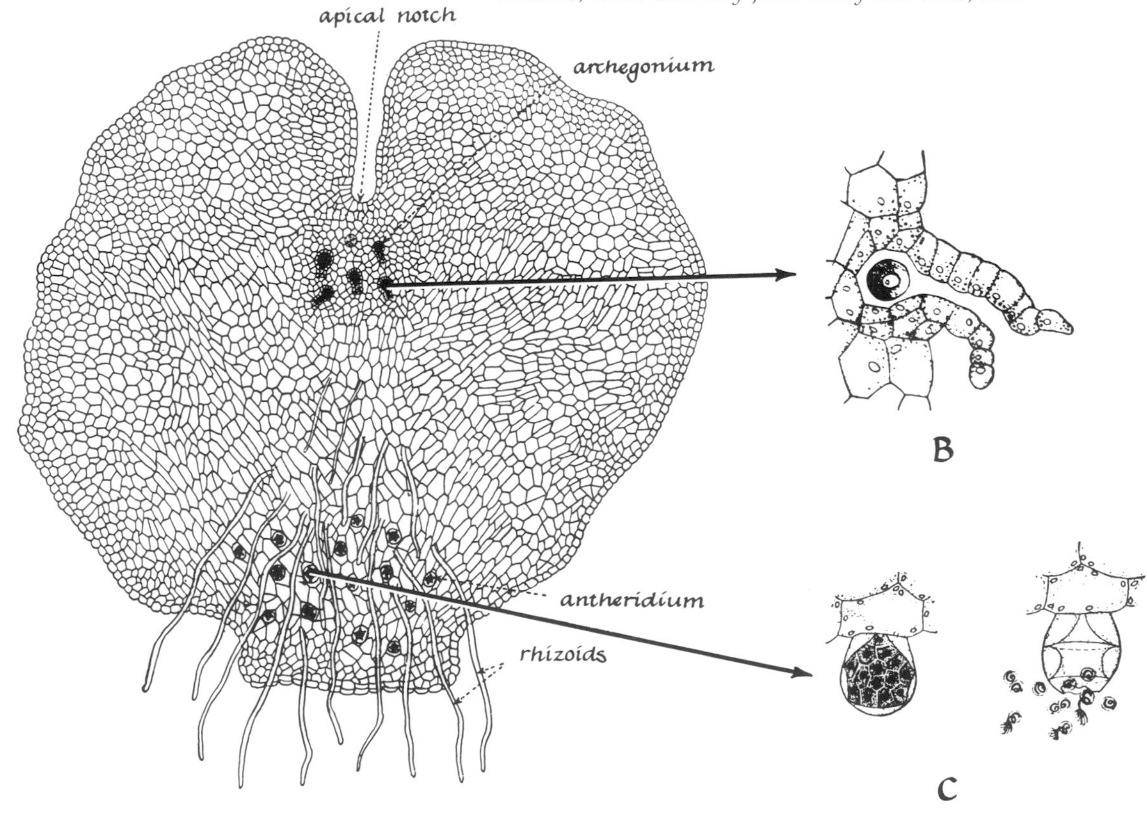

Figure 10-21. *A, Mature gametophyte (prothallium) of a fern as seen from the ventral side;* B, *mature archegonium;* C, *mature antheridia, one showing escape of spermatozoa after rupture of the antheridial wall. From Holman and Robbins, 1940. Courtesy John Wiley and Sons, Inc.*

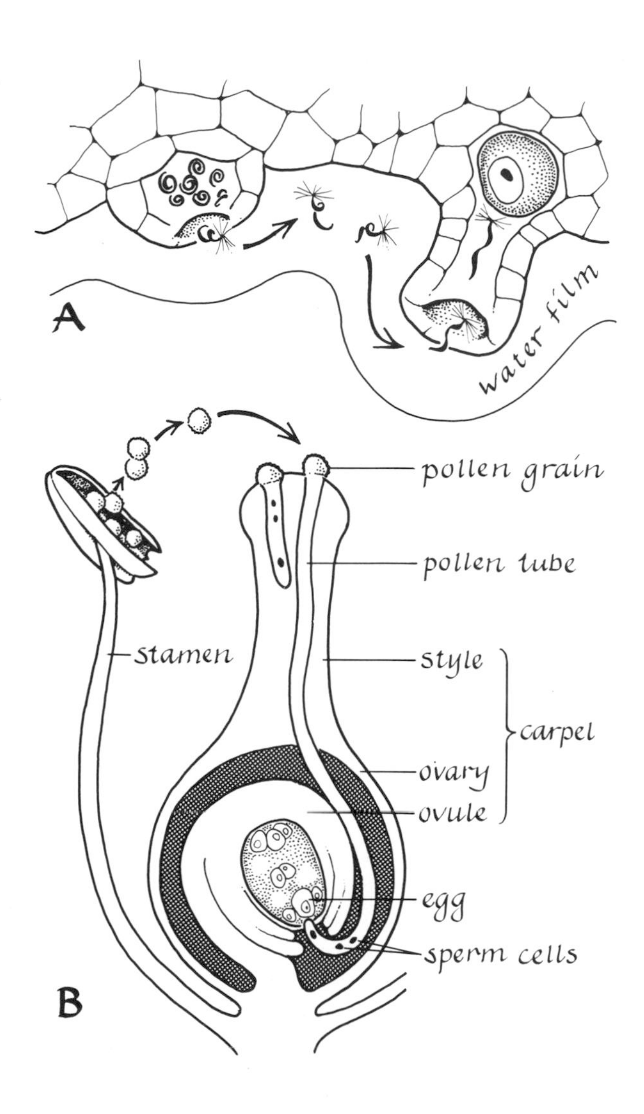

Figure 10-22. *A, movement of spermatozoa from the antheridium to the archegonium; B, method of transfer of the sperm cell to the egg in angiosperms.*

pollen grain. In the angiosperms, there are no multicellular sex organs. The gametes develop directly from cells of the relatively small gametophytes in tissues of the flower. This pattern of reproduction is described in the next chapter.

Differentiation of Male and Female Sex Organs on the Same Gametophyte

The formation of both male and female sex organs on a single gametophyte is also a process not yet understood by biologists. As described in *Allomyces* (Figure 10-15), the distal female and proximal male gametangia are formed in juxtaposition on the same hyphal branch. The gametangia differ not only in the pigments they contain but also in their DNA to RNA ratios, the male gametangium having about twice as much DNA as the female. Therefore, more male than female gametes are produced. The differentiation of the two kinds of gametangia could, in theory, result from a segregation of cytoplasmic constituents (see Chapter 16), or hormone-like substances produced in higher concentration by the tip of the hyphal branch may favor the female differentiation. Alternatively, hormones produced earlier or in higher concentration at the tip might induce differentiation of the male gametangium proximal to the female gametangium (Figure 10-15). It has been demonstrated that sex hormones play a role in sex differentiation in a related water mold *Achlya* (see Chapter 22). The same problem regarding sex differentiation exists in *Saprolegnia* (Figure 10-16) and *Vaucheria* (Figure 10-17).

In many ferns, both antheridia and archegonia develop on a single prothallium only a few millimeters in diameter (Figure 10-21). The archegonia typically develop near the apical notch and the antheridia near the opposite pole (Holman and Robbins, 1940; Sinnott, 1946). The positions of the sex organs are controlled by the polar organization of the prothallium, perhaps by a gradient in the concentration of one or more hormones. The higher concentration of hormones in the apical notch (meristem) region (where the hormones are produced) may stimulate or unmask the portion of the genome responsible for archegonial differentiation. The lower concentration at the opposite pole may permit expression of genes controlling antheridial differentiation. The discovery that a hormone, gibberellin (Chapter 19), when artificially applied to a young prothallium, induces archegonial formation (Voeller, 1964) suggests that hormones may be involved in sex organ differentiation. Antheridium formation can also be induced in cultured prothallia by an extract from the prothallia of related ferns that do develop antheridia in culture (Näf, 1956). Differentiation of male and female sex organs may be controlled by a graded pattern of microenvironmental differences in the gametophyte. The kind of gamete formed within a sex organ would be a further response of the cells to the local microenvironment.

Egg Formation and Its Significance for Later Development

Eggs are specialized reproductive units typically requiring union with a spermatozoan for their development. In Chapter 13 we shall discuss the cytoarchitecture of eggs in more detail. In this chapter it is only necessary to note that besides containing information within its nucleus for its development, the plant egg contains guidelines or directions for the programmatic utilization of this information. These guidelines, built up during formation of the egg, are apparently the components of the cytoplasm. Among the many patterns of cytoplasm components in the eggs of different organisms, the most universal is a polarized pattern that forecasts the polarized structure of the new plant. The polarity of the egg appears to be determined by the polarity of the reproductive organ with which it usually coincides (Chapter 13, Figure 13-12). Thus, this important guideline is directly inherited from the parent.

Fertilization

The union of egg and spermatozoan establishes the biparental inheritance of the new individual and initiates the development of the egg. Fertilization is accomplished in different ways. In liverworts, mosses, ferns, and some other plants, motile spermatozoa (Figure 10-12) are released from the antheridium, swim through a film of water on the gametophyte, and enter the neck of the archegonium (Figure 10-22A). In other plants such as *Vaucheria*, the sperm are deposited by the antheridial tube in the immediate vicinity of the open oogonium (Figure 10-17). In *Saprolegnia* (Figure 10-16), the male nucleus is transferred to the egg by an antheridial filament.

There are no motile sperm in angiosperms. The sperm cell is delivered to the egg by growth of the pollen tube through the tissues of the female organ, the pistil or carpel of the flower (Figure 10-22B and Chapter 11). The pollen tube appears to be guided in its growth through the style of the carpel by a substance produced in the ovules and ovary of the flower and by the high concentration of calcium in the ovary. There is also a pH gradient in the style, with highest acidity in the stigma. This gradient may also direct the growth of the pollen tube (Linskens, 1969).

When motile spermatozoa must swim a distance to reach the archegonium and the egg (Figure 10-22A), the problem of how the spermatozoan "finds" the

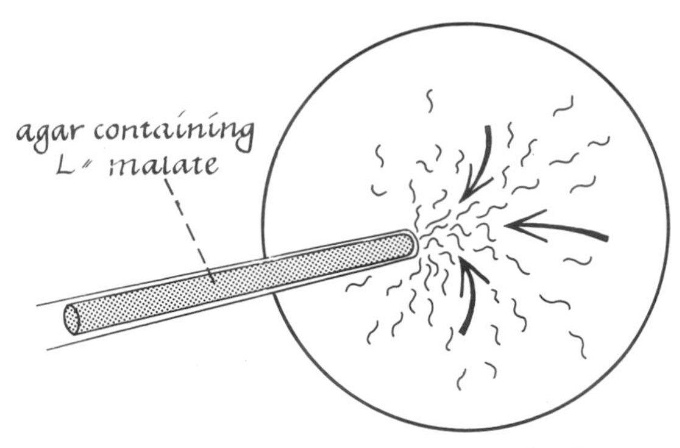

Figure 10-23. *Diagram of experiment demonstrating chemotaxis in fertilization of mosses and ferns.*

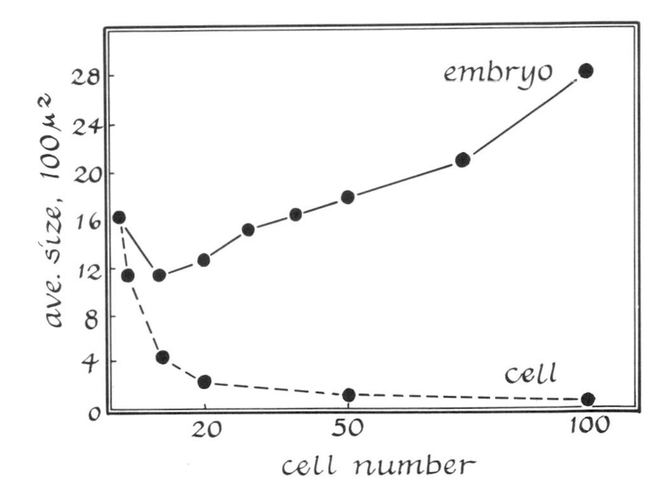

Figure 10-24. *Relation of embryo and cell size to growth of the embryo of a cotton plant. From Jensen, 1964. In* Meristems and Differentiation, *pp. 179–202. Brookhaven Symp. Biol., Upton, N.Y.*

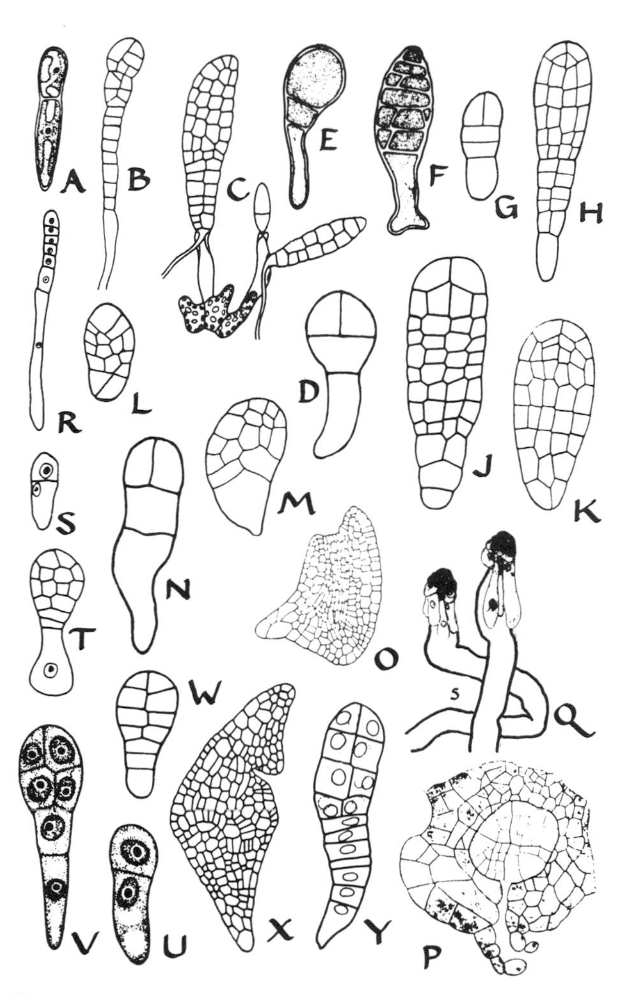

Figure 10-25. *Young embryos of different plant groups. From Wardlaw:* Embryogenesis in Plants, *1955. Courtesy Methuen and Co., Ltd.*

egg arises. Rothschild (1956) established that egg or archegonial secretions in mosses, ferns, horsetails, liverworts, and quillworts have a chemotactic influence on spermatozoa. Chemotaxis was first discovered by Pfeffer (1884) in studies of the fern *Pteridium aquilinum* (see also Shibata, 1911). Studies indicate that L-malic acid, produced by the archegonium, is the effective attracting agent. In artificial diffusion gradients produced by inserting a pipette of 1 percent sodium L-malate into a tapwater suspension of spermatozoa, the spermatozoa move within seconds toward the mouth of the pipette (Figure 10-23). In absence of the malate, the movements of the spermatozoa are random. In further studies, Rothschild discovered that the spermatozoa of some species of plants distinguish between the *cis* and *trans* dicarboxylic acids. Some are attracted by the *cis* but not by the *trans* form, and vice versa. In *Equisetum arvense,* of twenty acids tried, only L-malic and mesotartartic acids were effective.

We have already noted that a hormone produced by the female gametes of *Allomyces* attracts the male gametes. In one species of *Allomyces,* the chemotactic agent appears to be a compound called sirenin after the mythical Greek sirens, a misnomer because the successfully lured male gamete is the only one that lives. Sirenin is an oxygen-containing terpine (Machlis et al., 1966). Chemotactic agents also bring together the gametes of algae in isogamous, anisogamous, and oogamous reproduction. Nevertheless, many nonspecific agents elicit a chemotactic response in algal gametes (Wiese, 1969). As we shall see in Chapter 12, chemotaxis in fertilization of animal eggs is rare.

Cleavage and Early Cell Division

The first events following fertilization of the egg are redistribution of cytoplasmic components, changes in the position and size of vacuoles, and cleavage. Cleavage is usually defined as a subdivision of the fertilized egg or zygote. It results in an increase in cell number but a decrease in the cell size up to a point (Figure 10-24). It thus differs from cell proliferation in tissues, in which the cells do not become significantly smaller. In most plant zygotes, cleavage is of relatively short duration and is followed by an increase in cell number and overall size of the developing embryo. In the cotton egg (Figure 13-5), cleavage does not begin until about four days after fertilization. During this period, cytoplasm, which originally surrounded the large vacuole in the egg, now surrounds the nucleus in a dense mass. Other changes in the cytoplasm include an increase in the number of plastids and an increase in a special tubular endoplasmic reticulum associated with the nucleus (Jensen, 1964). All these changes help establish the polarity of the zygote. Some of the many patterns of cleavage and cell division in various species of plant embryos are illustrated in Figure 10-25.

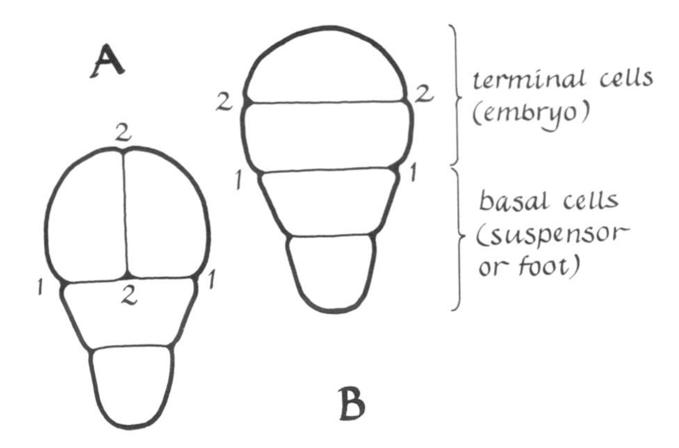

Figure 10-26. *The two basic types of early cleavage in plant eggs.*

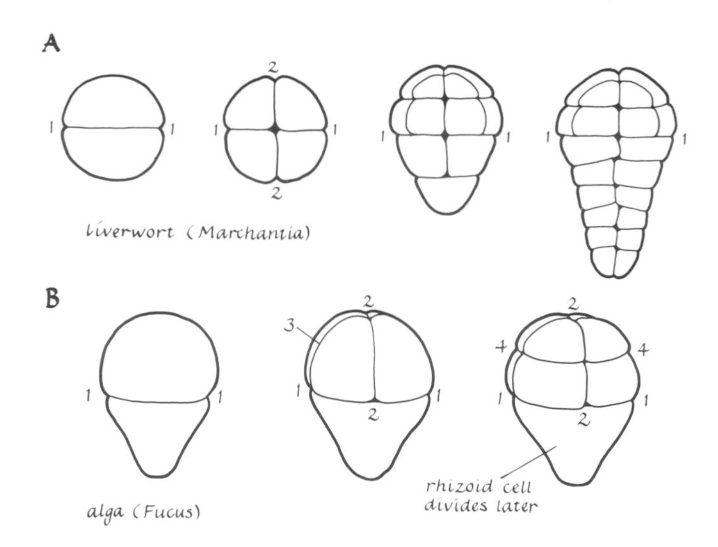

Figure 10-27. *The regular pattern of cleavage in the zygotes of* Marchantia *and* Fucus.

General Features of Early Embryo Development

Although the details of cleavage, cell division, and morphogenesis of plant embryos are diverse, these patterns of early development exhibit some common features (Wardlaw, 1955). For example, there are two basic types of cleavage through the four-cell stage (Figure 10-26).

The two patterns differ mainly in position of the second cleavage plane (denoted by 2-2). The position of the first cleavage plane (denoted by 1-1) is the same in both, at right angles to the polar axis of the egg. One of the two initial cells forms the embryo proper and the other, which is extra-embryonic, functions in nutrition of the embryo (the suspensor or foot cell in Figure 10-26). The diverse fates of the first two cells could be the consequence of their positions (different microenvironments), or more probably segregation of a polarized pattern of cytoplasmic constituents in the egg.

Early cell division in many plant zygotes has a geometric pattern. This is found in some algae, liverworts, and ferns. In Figure 10-27 note that although the pattern of cleavage is fairly regular, cells of different sizes are soon formed (see also the embryos illustrated in Figure 10-25). In ferns, the cleavage pattern is so regular that the fates of cells can be accurately predicted at a very early stage (Wardlaw, 1955). This is shown for several kinds of fern embryos in Figure 10-28.

Although in Figure 10-28 we can see a 1 to 1 relation between the cells and the plant parts they later form, the cells are not necessarily limited to these expressions. Only after experimental manipulation of the cells (changing their positions or isolating them) could we discover whether they have other developmental capacities. Comparing studies of other kinds of plant cells (Chapter 4), we may expect these embryo cells to have capacities other than what they normally express, but this remains to be demonstrated.

Many embryos, such as green algae and angiosperms, exhibit a filamentous or axial development (Figure 10-25). This filamentous arrangement of cells is a basic structural unit in the plant kingdom (Chapter 7) that occurs at early stages in many plant embryos.

A pattern of cell division common to many early embryos is a gradient in cell size (Wardlaw, 1955). This is shown for several embryos from the algae to the angiosperms in Figure 10-29. According to Wardlaw, the apical-basal gradient in decreasing cell population density and increasing cell size is probably a consequence of differential growth, the apical pole of the zygote being the principal locus of synthesis, cell division, and differentiation. This single gradient is clearly supplanted by multiple gradients as the meristematic regions of the plant develop.

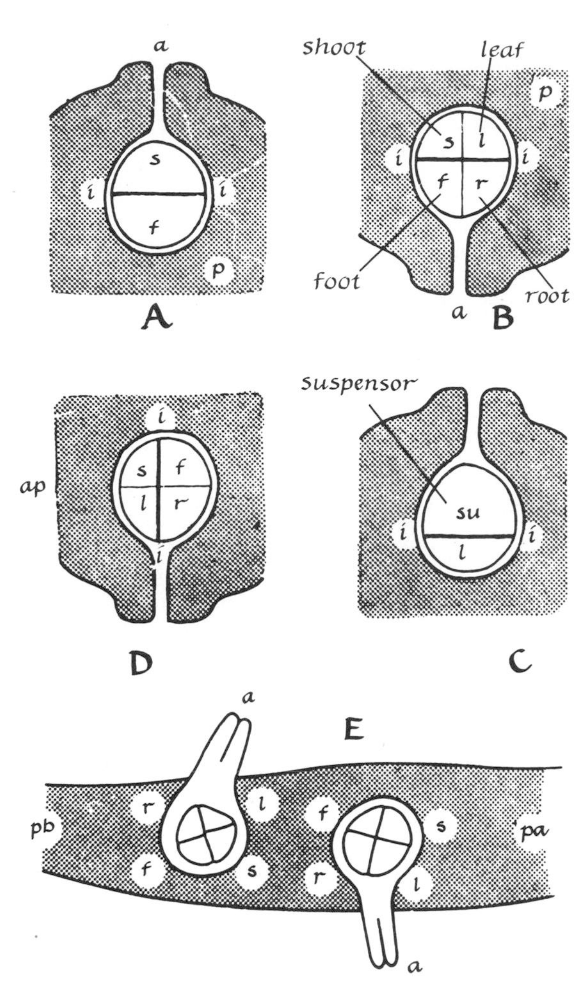

Figure 10-28. *Types of segmentation in the zygotes of ferns: A,* Ophioglossum; *B,* Angiopteris; *C,* Marattiales; *D,* leptosporangiates; *E, positions of embryos induced to develop on both upper and lower surfaces of the prothallium. From Wardlaw:* Embryogenesis in Plants, *1955. Courtesy Methuen and Co., Ltd.*

Seed Formation

In higher plants, the embryo sporophyte develops within a cone or a flower. In both instances, the embryo lies inside a special, protective reproductive structure called a seed. Reproduction by seeds (which are not formed by lower plants) provides an increased chance for the embryo's survival during environmentally unfavorable periods. The formation of seeds will be described in the following chapter.

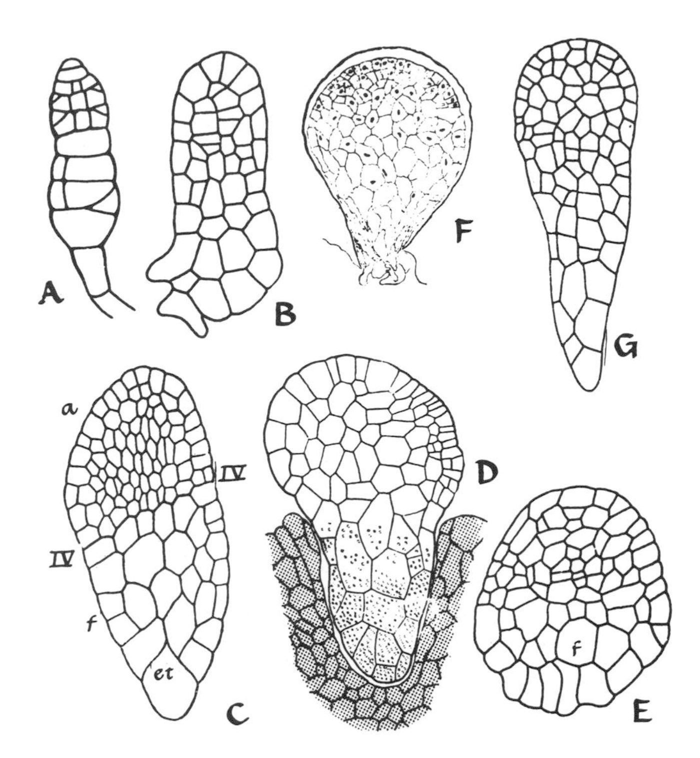

Figure 10-29. *Gradient of cell size in embryonic development of several species of plants. From Wardlaw:* Embryogenesis in Plants, *1955. Courtesy Methuen and Co., Ltd.*

Conclusion

The patterns of sexual reproduction in plants vary greatly. Sexual reproduction insures far more genetic diversity than is possible in asexual reproduction. In plants this diversity is accomplished by genetic segregation in the formation of meiospores and by genetic recombination in the fusion of gametes. The gametes of plants are typically products of mitosis in the gametophyte, and the meiospores are products of meiosis. The gametes of animals are direct products of meiosis. The gametes of many plants are formed in specialized unicellular or multicellular reproductive organs. Of special significance is the formation and composition of the egg, which contains preformed guidelines of development.

The orderly pattern of cell division is a feature of early development in the zygote of many plants. The pattern is so regular in some plants that the fates of cells can be accurately predicted at early stages. Although patterns of early cell division and cleavage vary among the zygotes of different kinds of plants, most can be classed in one of two basic types through the four-cell stage. A gradient in the cell population density and cell size of many plant embryos apparently reflects differential growth and a polarized cytoplasmic organization in the egg.

References

Allsopp, A. 1966. Developmental stages and life histories in the lower green plants. In E. Cutter, ed., Trends in Plant Morphogenesis, pp. 64–87. J. Wiley and Sons, New York.

Berrill, N. J. 1961. Growth, Development, and Pattern. W. H. Freeman, San Francisco.

Bold, H. C. 1964. The Plant Kingdom. Prentice-Hall, Englewood Cliffs, N.J.

Brown, W. H. 1935. The Plant Kingdom. Ginn, Boston.

Crandall, M. A., and T. D. Brock. 1968. Molecular aspects of specific cell contact. Science **161:**473–475.

Hayes, W. 1952. Recombination in *Bact. coli* K 12: Unidirectional transfer of genetic material. Nature **169:**118–119.

Holman, R. M., and W. W. Robbins. 1940. Elements of Botany. Wiley, New York.

Jacob, F., and E. L. Wohlman. 1961. Sexuality and Genetics of Bacteria. Academic Press, New York.

Jensen, W. A. 1964. Cell development during plant embryogenesis. In Meristems and Differentiation, pp. 179–202. Brookhaven Symp. Biol., Upton, N.Y.

Kerr, S. J. 1968. Cytological observations on plasmodial differentiation in the true slime mold, *Didymium nigripes.* (Doctoral thesis, University of Minnesota.)

Kumar, H. D. 1962. Apparent genetic recombination in a blue-green alga. Nature, **196:**1121–1122.

Lazaroff, N., and W. Vishniac. 1962. The participation of filament anastomosis in the developmental cycle of *Nostoc muscorum,* a blue-green alga. J. Gen. Microbiol. **28:**203–210.

Linskens, H. F. 1969. Fertilization mechanisms in higher plants. In C. B. Metz and A. Monroy, eds., Fertilization, Vol. 2, pp. 189–253. Academic Press, New York.

Machlis, L., W. H. Nutting, M. W. Williams, and H. Rapoport. 1966. Production, isolation, and characterization of sirenin. Biochemistry **5:**2147–2152.

Näf, U. 1956. The demonstration of a factor concerned with the initiation of antheridia in polypodiaceous ferns. Growth **20:**91–105.

Pfeffer, W. 1884. Locomotorische Richtungsbewegungen durch Chemische Reize. Unt. Bot. Inst. Tübingen **1:**363–481.

Randall, J., T. Cavalier-Smith, A. McVittie, J. R. Warr, and J. M. Hopkins. 1967. Developmental and control processes in the basal bodies and flagella of *Chlamydomonas reinhardii.* In M. Locke, ed., Control Mechanisms in Developmental Processes, pp. 43–83. Academic Press, New York.

Rothschild, L. 1956. Fertilization. Methuen, London.

Scagel, R. F., R. J. Bandoni, G. E. Rouse, W. B. Schofield, J. R. Stein, and T. M. C. Taylor. 1966. An Evolutionary Survey of the Plant Kingdom. Wadsworth Publishing Company, Inc., Belmont, Calif.

Shibata, K. 1911. Untersuchungen über die Chemotaxis der Pteridophytenspermatozoiden. Jhr. Wiss. Bot. **49:**1–60.

Sinnott, E. W. 1946. Botany: Principles and Problems, 1st ed. McGraw-Hill Book Company, New York.

Sinnott, E. W. 1963. Botany: Principles and Problems, 6th ed. McGraw-Hill Book Company, New York.

Tyler, A., and B. S. Tyler. 1966. Physiology of fertilization and early development. In R. A. Boolootian, ed., Physiology of Echinodermata, pp. 683–741. Wiley, N.Y.

Voeller, B. B. 1964. Gibberellins: their effect on antheridium formation in fern gametophytes. Science **143:**373–375.

Wardlaw, C. W. 1955. Embryogenesis in Plants. Methuen, London.

Wiese, L. 1965. On sexual agglutination and mating type substances (gamones) in isogamous heterothallic Chlamydomonads: I. Evidence of the identity of the gamones with surface components responsible for the sexual flagellar contact. J. Physiol. **1**:45–54.

Wiese, L. 1969. Algae. In C. B. Metz and A. Monroy, eds., Fertilization, Vol. 2, pp. 135–188. Academic Press, New York.

11 *Embryology of Flowering Plants*

Only angiosperms have a complex reproductive apparatus, the flower, which consists of specialized structures in which the sex cells are formed and of auxiliary structures such as petals, which contribute indirectly to the reproductive process (Figure 11-2). As in lower plants, generations alternate, but the gametophyte is much smaller and is incorporated in the sex organs of the flower. The obvious generation in flowering plants is the sporophyte. Another important reproductive structure (which flowering plants share with gymnosperms) is the seed, an embryo sporophyte plant whose development is temporarily arrested, imbedded in nutritive sporophyte tissue, and encased in a protective coat. This adaptation gives the group an excellent chance for survival and dispersal.

Formation of the Flower

When particular changes occur in photoperiod, temperature, moisture, and unknown internal factors (Chapter 22), the shoot apex ceases to form leaves and stem and develops a flower or inflorescence (group of flowers). This striking transformation involves many morphogenetic processes not yet completely understood. Among the first changes in the activity of the shoot apex are an increase in its mitotic activity (Gifford, 1964) and a change in its shape (Hillman, 1962; Wardlaw, 1965). Figure 11-1 illustrates the change in shape, which involves cell division on the flanks of the apex.

The environmental conditions directing the transformation do not act on the apex itself but on the leaves below it. According to the current hypothesis, a hormone-like substance (''florigen'') is synthesized in the leaves and transported to the shoot apex (Heslop-Harrison, 1964; Wardlaw, 1965). The ''florigen'' apparently initiates a program of gene action in the apex leading to flower formation. The organs or parts of a complete flower are illustrated in Figure 11-2.

A complete flower consists of four circles of parts from outside to inside: calyx, corolla, stamens, and carpels. Depending on the species, these parts have a variable number of subunits: sepals in the calyx, petals in the corolla, and variable numbers of stamens and carpels. Although the floral organs have a different structure from the leaves, botanists generally believe that they are homologous (have a similar origin) (Tepfer, 1953; Wardlaw, 1965). This is shown in Figure 11-3.

Despite the homology, the organs of the flower are highly specialized for reproductive processes, and the leaves are specialized for nutrition of the plant.

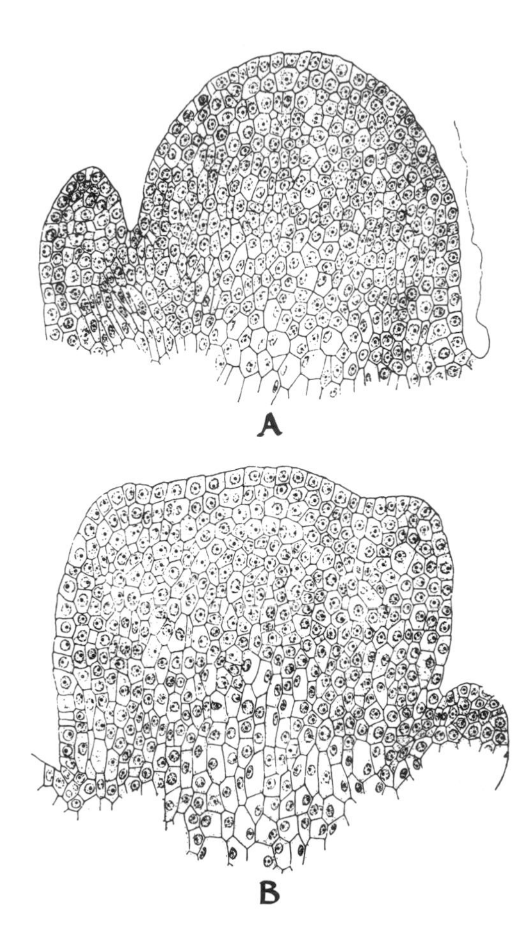

Figure 11-1. *A, vegetative shoot apex of the briar rose* Rubus rosaefolius. *B, the same in transition to a floral apex (× 410). After Engard, from Wardlaw, 1965. Courtesy Longman, London, and author.*

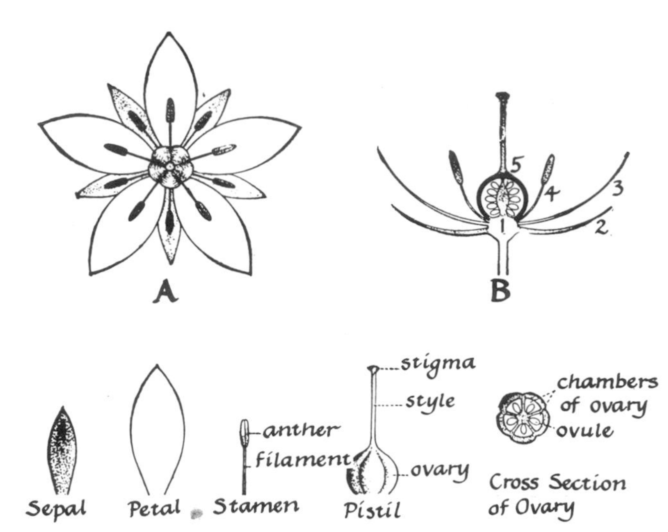

Figure 11-2. *Diagrams of the flower structure of a dicotyledonous seed plant. A, face view of the flower showing the calyx of five sepals, the corolla of five petals, ten stamens, and one carpel. B, longitudinal section showing the relations between the parts: 1, receptacle; 2, calyx; 3, corolla; 4, stamen; 5, carpel. Lower: diagrams of dissected floral organs. From* Botany: Principles and Problems *by Sinnott. Copyright 1946 by McGraw-Hill Book Company. Used with permission of McGraw-Hill Book Company.*

How the flower-inducing stimulus controls the differentiation of the floral organs is an unsolved but challenging problem. A gene relay control model to explain the stages in formation of the flower has been proposed by Heslop-Harrison (1964). The parts of typical flowers develop in a centripetal and basal to apical sequence: bracts (if present), sepals, petals, stamens, and carpels. The model in Figure 11-4 is designed to illustrate how a gene relay system might control this developmental sequence. The model is based on the hypothesis that the "florigen" begins the activity of a gene complex that forms products. These gene products in turn induce the activity of a second gene complex, and so forth.

For example, activity of the first gene complex may initiate development of the sepals, and activity of the second gene complex may initiate development of the petals. In an alternative hypothesis based on a presumed gradient in "florigen" and auxin concentration in the apex, the genome of the cells would respond differentially with the position of the cells. "Florigen" concentration would be higher at the base than at the pinnacle (or center) of the apex, and the auxin gradient would be opposite this. This hypothetical environmental gradient is diagramed in Figure 11-5.

Before we can understand the mechanisms controlling flower formation, more experimental work on the apex must be done. One experiment suggests that a delicate balance exists between the developing floral organs. At an early stage in the transformation from leaf production to flower production, bisection of the apex or removal of the bracts caused a return to leaf-forming activity in the apex; bisection at a

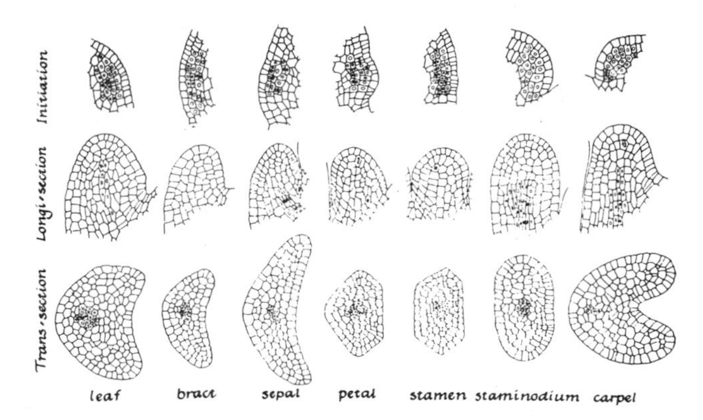

Figure 11-3. *Camera lucida drawings of leaf and floral organs of columbine* (Aquilegia formosa) *showing early stages in their formation at the apical meristem and the homology of the organs. After Tepfer, from Wardlaw, 1965. Courtesy Longman, London, and author.*

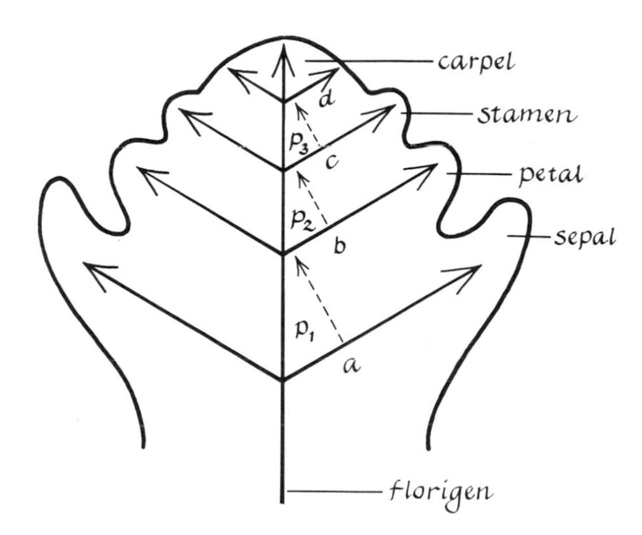

Figure 11-4. *Diagram of the gene relay system that may underlie differentiation of the floral organs. a-d are gene complexes; p1, etc., are gene products.*

later stage resulted in small, separate flowers. In all cases, some prospective floral organs reverted to leaves (Wardlaw, 1965).

Many biochemical changes are associated with the flower-forming process: among them increased RNA synthesis in the tunica cells and increased protein

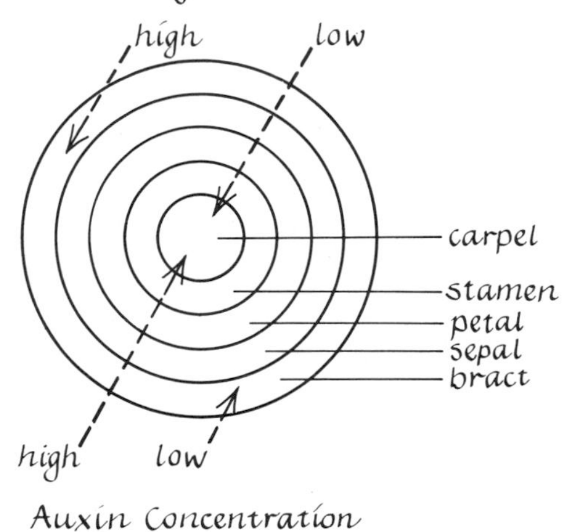

Figure 11-5. *Hypothetical gradients in auxin and florigen concentrations that might underlie differentiation of the floral organs.*

synthesis in tunica and adjacent corpus cells (Gifford, 1964). Inhibiting DNA and RNA synthesis in the cocklebur *(Xanthium pennsylvanicum)* stops the flower-forming activities of the apex (Bonner and Zeevaart, 1962).

An especially interesting aspect of flower development is the differentiation of the sex organs. In some species, male and female flowers are separate but on the same plant; in others, male and female flowers are formed on different plants; but most commonly, flowers are hermaphroditic. Chromosomal sex determining mechanisms underlie sex development in plants such as *Melandrium* (campion) and *Cannabis* (hemp), in which male and female plants are separate (see reviews of Westergaard, 1958; Heslop-Harrison, 1964). In 1907, Correns concluded from the results of crosses in *Melandrium* that the female plant and its egg tend toward femaleness and that the male plant is heterozygous for sex, similar to the sex-determining mechanism in man and some other animals. Sex-linked traits are found in *Melandrium.*

Although chromosomal sex determining mechanisms underlie the development of male and female flowers in some plants, sex expression is susceptible to modification by environmental influences (for example, by photoperiod in the hemp plant) (Chapter 22). Hormones are also known to be involved in

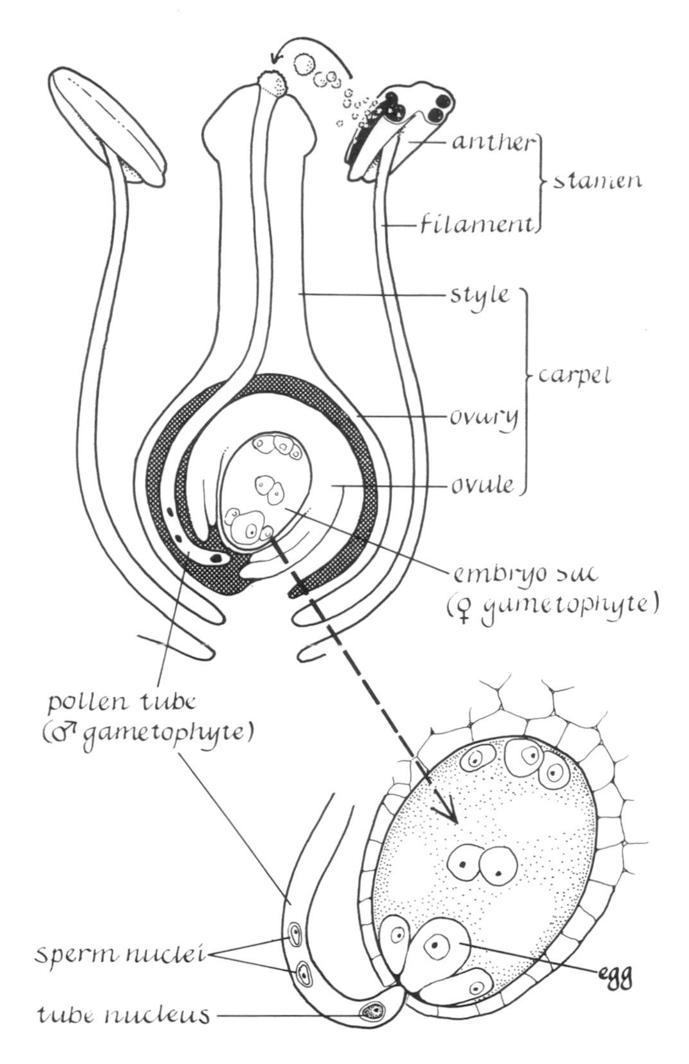

Figure 11-6. *Diagram of the reproductive organs of a flower illustrating their role in fertilization.*

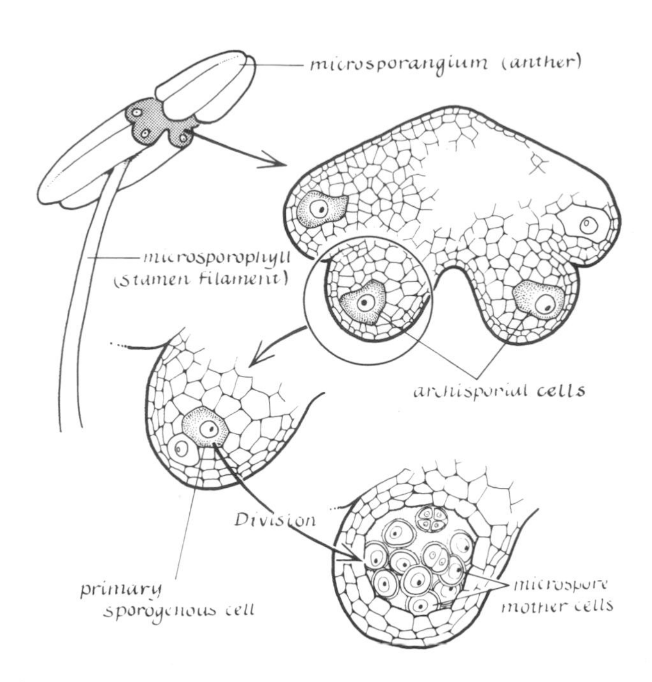

Figure 11-7. *Development of microspore mother cells within the microsporangium.*

sex differentiation in the hemp plant. Complete sex reversal has been achieved by treating genetically male hemp plants with auxins, which shows that hormones can override chromosomal determining mechanisms in plants. Because it has been reported that sex expression in *Melandrium* can be influenced by various animal sex hormones, one might conjecture that auxins function like female hormones in reversing the phenotypic expression of sex in genotypically male plants. However, no male hormone counterpart has been demonstrated (Sinnott, 1960). As we shall see in Chapter 19, sex reversal has been experimentally accomplished in animal embryos.

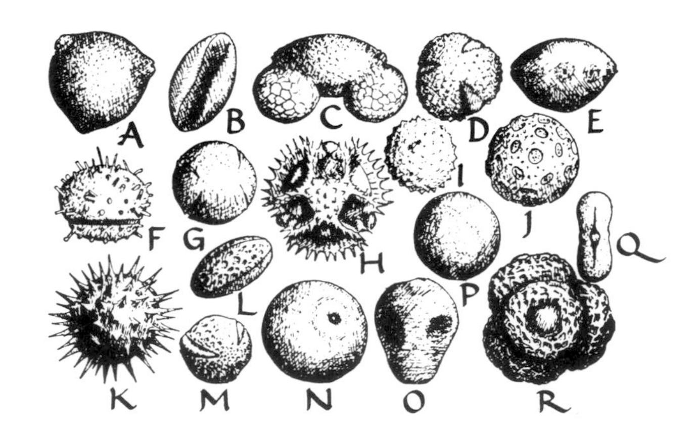

Figure 11-8. *Mature pollen grains (microspores) of various plants. A, birch; B,* Ginkgo*; C, pine; D, mugwort; E, walnut; F, water lily; G, dock; H, dandelion; I, ragweed; J, Russian thistle; K, sunflower; L and M, willow; N and O, grass; P, poplar; Q, wild carrot; R, primitive dicotyledon with four grains united. From* Botany: Principles and Problems *by Sinnott. Copyright 1946 by McGraw-Hill Book Company. Used with permission of McGraw-Hill Book Company.*

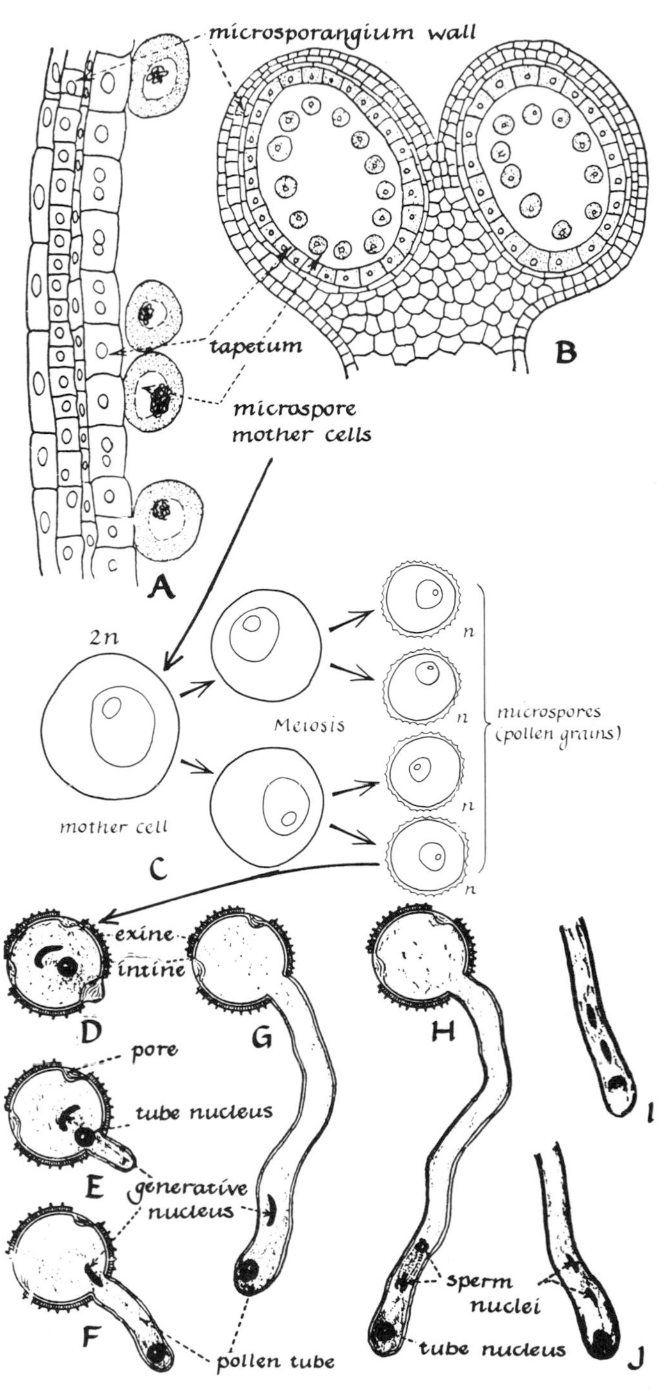

Figure 11-9. *A, longitudinal section of the wall of a wheat anther; B, transverse section of an anther lobe of wheat; C, meiosis in the pollen mother cell forming a tetrad of haploid microspores; D-J, germination of a pollen grain and development of the pollen tube. A, B, and D-J, from Holman and Robbins, 1940. Courtesy John Wiley and Sons, Inc.*

Development of the Reproductive Organs and Gametes

The reproductive organs of a generalized flower are illustrated in Figure 11-6. The male gamete-producing organ of the flower is the stamen, consisting of a filament (microsporophyll) with an anther (microsporangium) in which pollen grains (microspores) are formed. The female gamete-producing organ of the flower (megasporophyll, commonly known as pistil or carpel) consists of a stigma supported by a relatively long style that is attached to an enlarged basal portion, the ovary. In the ovary, depending on the species, are one or more ovules (megasporangia). Megaspores are formed in the megasporangium. In angiosperms, the gametophyte is greatly reduced in size and occupies only a small arc of the cycle. The male gametophyte is the germinated pollen grain. The female gametophyte is a small multinucleated sac (embryo sac) within the ovule.

Development of the Microsporangium and Male Gametophyte

Within the microsporangium are relatively large archesporial cells (Figure 11-7). These divide and the inner daughter cells become primary sporogenous cells, which divide to form microspore mother cells. The microspore mother cells are diploid, like the remaining cells of the plant. Each mother cell meiotically divides to produce four haploid microspores, which become surrounded by a highly resistant coat of sporopollenin—a complex of cellulose, xylan, lipid, and lignin-like substances (Clowes and Juniper, 1968). The coat of the pollen grain is intricately and species-specifically sculptured (Figure 11-8).

The relationship of the microspore mother cell to the microsporangium wall, the meiotic divisions of a microspore mother cell to form a tetrad of microspores, and the germination of the pollen grain to form the male gametophyte are shown in Figure 11-9.

Although little is known about the mechanism that induces meiosis in the mother cells, it has been suggested that the tapetal cells (Figure 11-9A), which show a significant increase in RNA during the induction process, might be involved. Anthers that

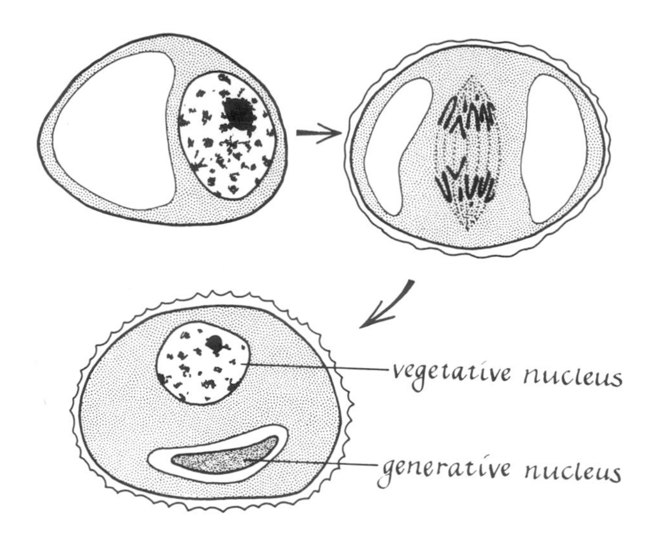

Figure 11-10. *Mitotic division in a pollen grain, forming a vegetative and generative cell.*

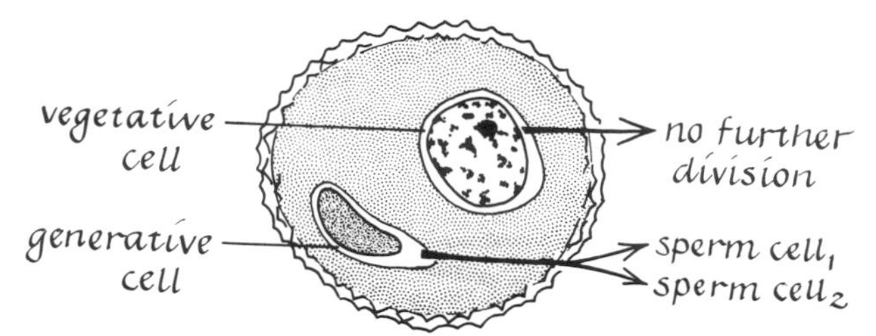

Figure 11-11. *The generative cell divides to form two sperm cells, and the vegetive cell, which does not divide, becomes the tube nucleus cell.*

had been excised and explanted in vitro before meiosis failed to undergo meiotic division, indicating that substances produced by the whole plant are necessary to induce meiosis (Torrey, 1967). As yet it is not known what these are.

A mitotic division occurs inside the pollen grain during its maturation (Figures 11-9 and 11-10). One of the daughter nuclei is called the vegetative or pollen tube nucleus; the other is the generative nucleus. Both nuclei become surrounded by a thin cell wall; therefore, we shall refer to them as the vegetative cell and the generative cell. At a later stage, usually during germination of the pollen grain, the generative cell divides to produce two sperm cells. This is illustrated in Figure 11-11.

The reason only one daughter nucleus divides may be that the nuclei have different cytoplasmic

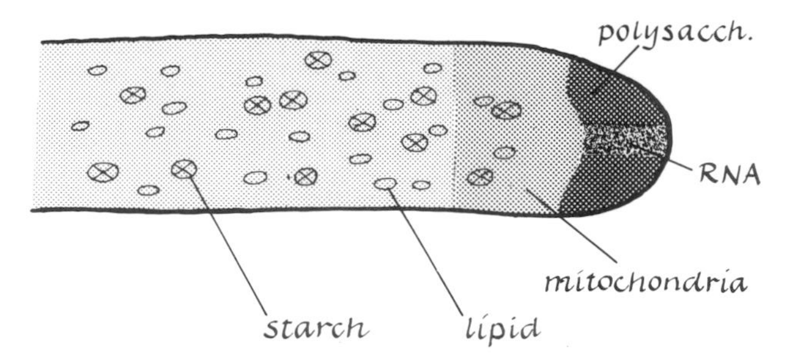

Figure 11-12. *Cytoarchitecture (distribution of organelles and chemical constituents) in the tip of a pollen tube of the lily.*

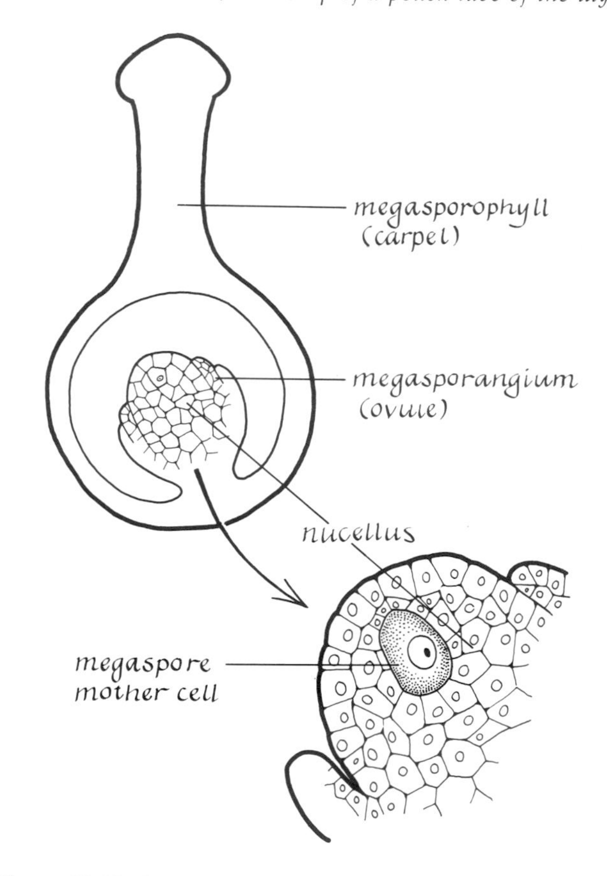

Figure 11-13. *Origin of the megaspore mother cell.*

surroundings. For example, the vegetative nucleus in the pollen grain of *Trandescantia* (spiderwort) is surrounded by cytoplasm relatively rich in RNA, but the generative nucleus is surrounded by cytoplasm containing little RNA (La Cour, 1949). Furthermore, the generative nucleus contains twice as much DNA as the vegetative nucleus (Maheshwari, 1966). Perhaps the cytoplasmic environment of the microspore nuclei determines their behavior. Other influences

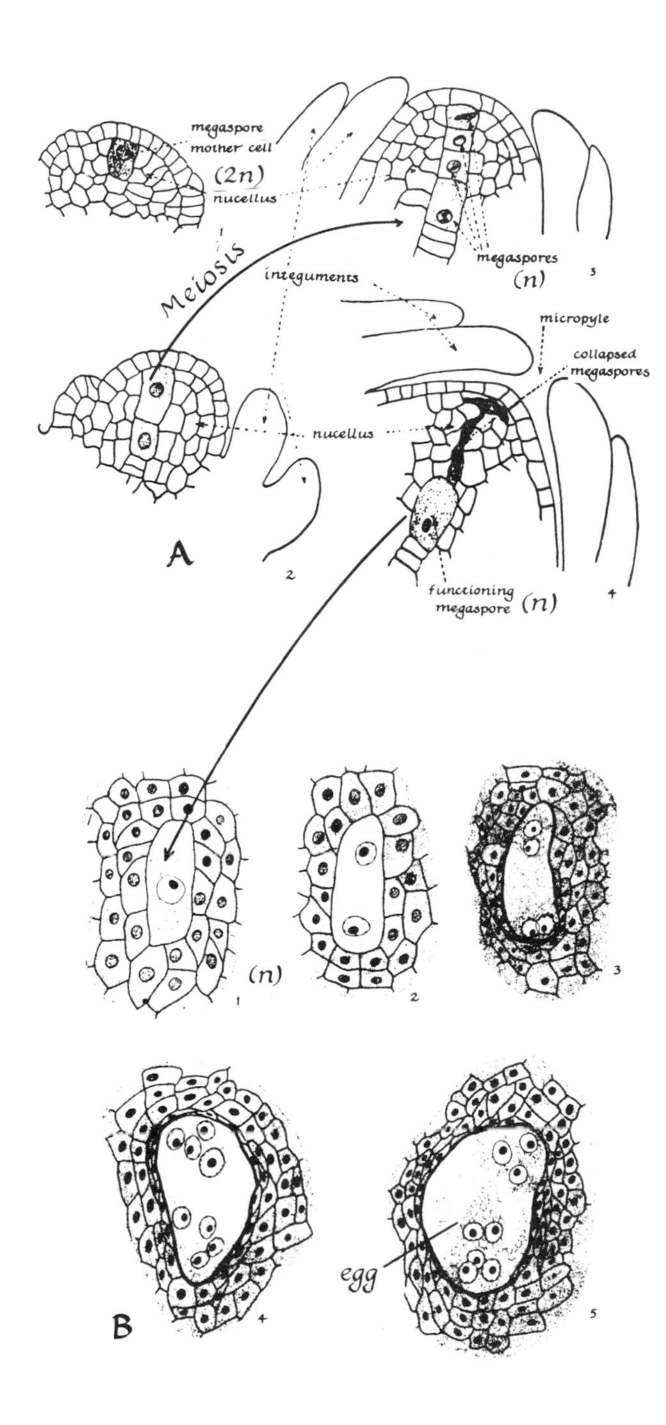

Figure 11-14. *A, median longitudinal sections of the ovule of wheat, showing development of the megaspores and degeneration of all but one of these. B, stages in development of the embryo sac (female gametophyte) of corn. From Holman and Robbins, 1940. Courtesy John Wiley and Sons, Inc.*

of cytoplasm on nuclear activity will be discussed in Chapter 20.

When the pollen grains are mature, they are released by the anther, and by various means (wind, insects, or proximity of the anther and carpel) are deposited on the surface of the stigma at the apex of the carpel. At this time, the germination of the pollen grains begins. This involves both division of the generative cells (Figure 11-9) and formation of a long pollen tube, which grows from the pollen grain through the style to the opening of the ovule (Figure 11-6). Many pollen tubes may grow through the style, but only one fertilizes each embryo sac. Elongation of the tube occurs at its advancing tip, where the tube cell is located. Presumably, the synthetic activities necessary for growth are controlled by this cell. There is an orderly distribution of cytoplasmic particles and chemical constituents such as RNA, starch, lipid, and polysaccharides in the tip of the pollen tube (Rosen et al., 1964). This is illustrated in Figure 11-12. The relatively high concentrations of RNA, polysaccharides, and mitochondria in the tip of the pollen tube are consistent with the view that the tip is the primary locus of tube elongation.

The germinated pollen tube is the male gametophyte. The two sperm cells correspond to the spermatozoa of mosses, ferns, and animals. The stages in the formation of sperm cells are summarized as follows:

microsporophyll ⟶ microsporangium ⟶ archesporial cells ⟶ primary sporogenous cells ⟶ microspore mother cells ⟶ *meiosis* ⟶ four microspores ⟶ germination ⟶ male gametophyte ⟶ two sperm and one vegetative

From the microsporophyll through the microspore mother cell stages, all the nuclei are diploid. The microspores, the male gametophyte, and sperm and vegetative cells are haploid.

In some flowers, the maturation of the pollen grains accompanies maturation of the egg within the embryo sac, and self-fertilization can occur. In other flowers, the two events do not occur simultaneously, preventing self-pollination and resulting in greater genetic diversity.

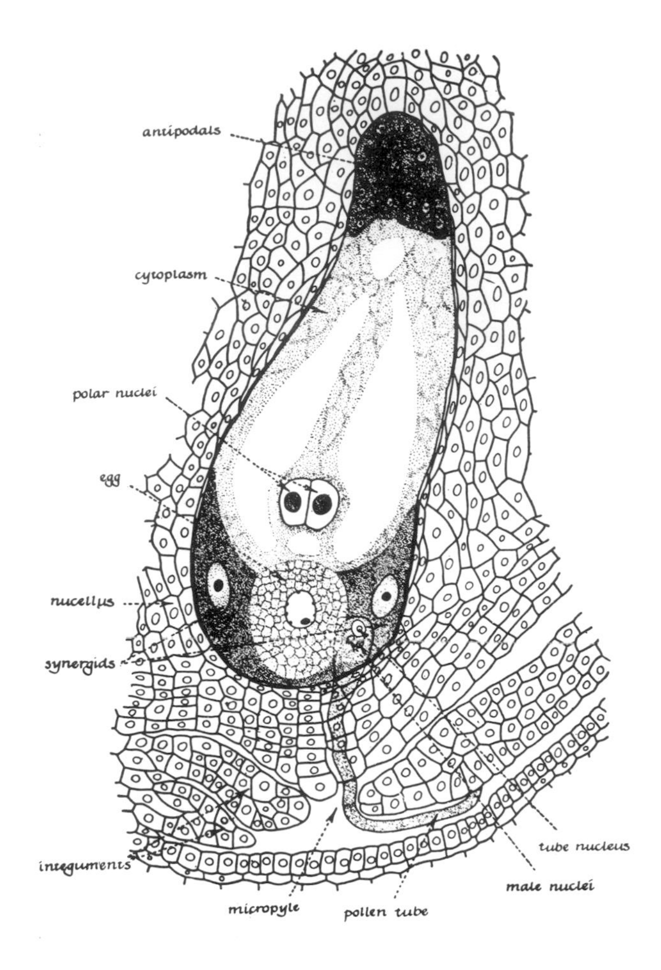

Figure 11-15. *Longitudinal section of mature embryo sac of corn just before fertilization. Note the pollen tube entering the embryo sac. From Holman and Robbins, 1940. Courtesy John Wiley and Sons, Inc.*

Development of the Megasporangium and Female Gametophyte

A group of relatively unspecialized cells called the nucellus is located inside the megasporangium (ovule). One of the nucellus cells enlarges and becomes the megaspore mother cell (Figure 11-13), which undergoes meiotic divisions to produce four haploid megaspores. Unlike the microspores, only one megaspore continues to develop in most angiosperms. The other three degenerate. A similar loss of three meiotic products occurs in the zygospore of *Spirogyra* and also in female animals.

The viable megaspore enlarges and its nucleus divides a variable number of times, depending on the species of plant. Enlargement of the megaspore continues to form the embryo sac. These transformations are shown in Figure 11-14. The megaspore nucleus divides three times to produce eight nuclei. Three of these are located at one pole of the embryo sac and, with the surrounding cytoplasm, are called antipodal cells. Two other nuclei, which lie in the middle bottom half of the embryo sac surrounded by cytoplasm, are called polar nuclei. Two nuclei, each surrounded by cytoplasm, lie on either side of the bottom of the embryo sac. These are called synergids (nurse cells). Of the eight nuclei, the one lying between the nurse cells just above the opening of the ovule (the micropyle) becomes surrounded by a relatively large mass of cytoplasm. This is the *egg cell.*

The mature embryo sac with its egg and other cells is the *female gametophyte* in flowering plants. The female gametophyte of a corn plant is illustrated in Figure 11-15.

The developmental stages leading to formation of the egg can be summarized as follows.

megasporophyll ⟶ megasporangium (ovule) ⟶ one megaspore mother cell ⟶ *meiosis* ⟶ four megaspores (three degenerate) ⟶ enlargement of remaining megaspore ⟶ three mitotic divisions ⟶ female gametophyte (egg plus seven nuclei).

As in the development of the microsporangium, from the megasporophyll through the megaspore

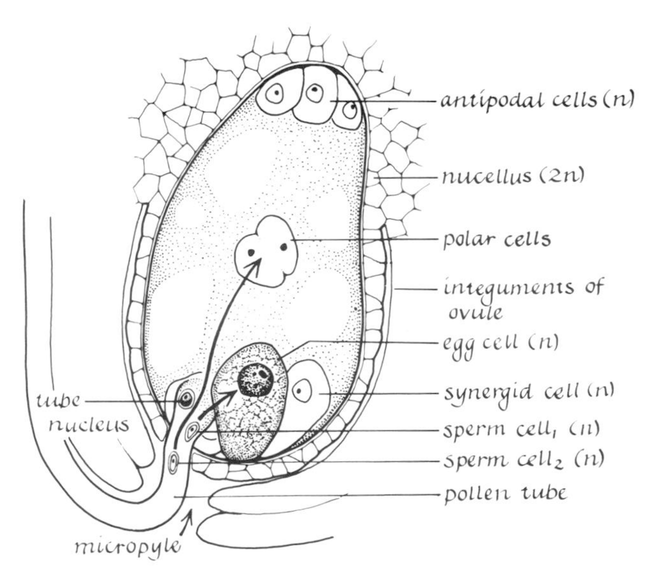

Figure 11-16. *Double fertilization in an angiosperm.*

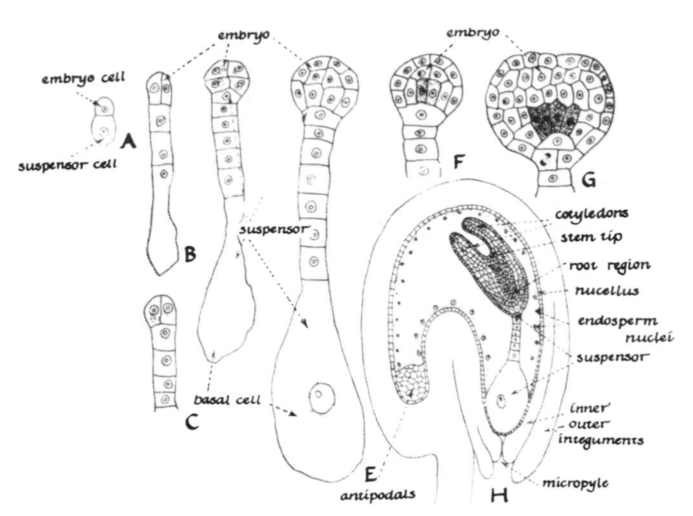

Figure 11-17. *Stages in development of the embryo of shepherd's purse* (Capsella bursa-pastoris). *From Holman and Robbins, 1940. Courtesy John Wiley and Sons, Inc.*

mother cell, all the cells are diploid. The megaspores, the female gametophyte (embryo sac), and its nuclei are haploid.

It is not known why only one of the four megaspores develops into the embryo sac and why only one of the eight nuclei of the embryo sac differentiates into the egg nucleus. Perhaps the positions of the megaspores and the nuclei in the embryo sac determine their fates. Clearly, the megaspore that forms the embryo sac and the nucleus that becomes the egg nucleus are not in the same microenvironment as the other megaspores and nuclei of the embryo sac. As yet no technique has been developed to change the position of the nuclei and test this hypothesis.

Fertilization

Fertilization in flowering plants is a double process (Figure 11-16). The pollen tube grows through the micropyle of the ovule, through the nucellus surrounding the embryo sac, and into the embryo sac. The pollen tube is probably guided by a chemotactic stimulus as noted in Chapter 10. The pollen tube enters the embryo sac by growing into one of the synergid cells, and the two sperm and the vegetative or tube nuclei are released into the cytoplasm of the synergid. By a mechanism still unknown, one of the sperm nuclei enters the egg to form the diploid zygote, while the other enters the central or polar cell containing the polar nuclei to form the triploid endosperm nucleus. This nucleus divides rapidly to form a multinucleate fluid, the liquid endosperm, which bathes the zygote. Later cell walls are laid down in the endosperm. The triploid endosperm tissue is a source of nutrients for the developing zygote. Division of the endosperm nucleus begins shortly after its formation, but division of the zygote is not initiated until hours, days, or even weeks later.

Because the two sperm nuclei are mitotic products of the generative nucleus (Figure 11-11) and all the embryo sac nuclei are mitotic products of the megaspore nucleus, the triploid endosperm nucleus and the diploid zygote nucleus must contain the same kind of genetic information. The endosperm nucleus does contain an extra set of information; but, as we shall see later, redundancy of information alone apparently does not account for differences in cell behavior. Probably differential use of genetic information is a consequence of differences in the microenvironment of the endosperm and zygote nuclei. Unfortunately, this hypothesis has not yet been tested.

Development of the Egg into an Embryo

In the preceding chapter we outlined some general aspects of early embryonic development. Now we can examine in more detail one well-known and carefully studied example of embryogenesis in a commonly occurring plant, *Capsella bursa-pastoris* (Shepherd's purse) (Souèges, 1919; Rijven, 1952). The first nuclear division is followed by formation of a transverse cell wall at right angles to the axis of the egg and embryo sac (Figure 11-17A).

The first division forms a large basal or suspensor cell next to the micropyle, and a small terminal (embryo) cell. The suspensor cell divides further to produce a chain of extra-embryonic cells that feed the early embryo. The embryo cell first divides longitudinally and each of the two cells divides transversely to produce a four-cell stage (Figure 11-17C and D). Further divisions produce a multicellular embryo that develops within the embryo sac and is exposed to influences of the synergids and dividing endosperm cells.

As development progresses, the outer layer of cells becomes morphologically different from the inner cells. The outer layer forms the epidermis of the embryo (Figure 11-17F and G). Near the base of the embryo, where it joins the suspensor cell chain, some cells begin to develop as vascular tissue cells. As further cell division proceeds, the embryo (originally a sphere of cells) begins to change its shape. It flattens, and two projections, the primordia of the cotyledons (first leaves), become evident. Meanwhile a shoot apical meristem and a root apical meristem develop. Differentiating vascular tissue of the future stem connects the two apices (Figure 11-17H). Thus, as the number of cells increases, other kinds of cells appear (Jensen, 1964).

The pattern of cellular differentiation in the early embryo of *Phlox Drummondi* is illustrated in Figure 11-18. The central, fan-shaped core of darkly staining and relatively unvacuolated procambial cells is surrounded by more lightly staining vacuolated cells (Miller and Wetmore, 1946). Two apical growing centers arise in the procambial region and produce the cotyledons (Figure 11-18C). Finally, in both *Phlox* and *Capsella* (and in all other dicotyledons), an embryo plant consisting of the cotyledons (two primary leaves gorged with nutrients) and a very short stem with shoot and root apical meristems is formed within the ovule. The ovule plus the embryo plant within it will develop into a seed. In *Capsella*, development from zygote to mature embryo requires about ten days. At this stage further development of the embryo ceases until the seed germinates, probably because an impervious seed coat, which retards the absorption of water and oxygen, has formed. Factors involved in seed germination and the resumption of embryonic development will be discussed later in this chapter.

During early embryogenesis, important changes in metabolism occur. In the cotton egg, for example, oxygen consumption measured by the Cartesian diver method increases per embryo but the amount per cell decreases, probably because cell size decreased during cleavage (Chapter 10). This is shown in Figure 11-19 (Jensen, 1964). Oxygen consumption per cell begins to increase in embryos containing about 13,000 cells. These changes in metabolism apparently are accompanied by parallel changes in protein synthesis and ultrastructure of the cytoplasm of the cells, including an increase in Golgi bodies.

In the normal development of flowering plants, the embryo sporophyte develops from the fertilized egg within the embryo sac. However, apomixis, aberrant development involving no fertilization

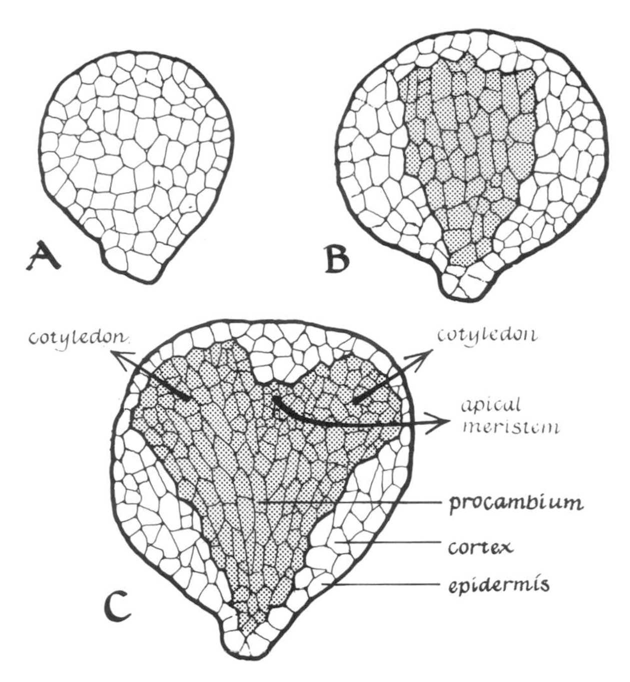

Figure 11-18. *Pattern of cellular differentiation in the early embryo of* Phlox.

process, occurs in many plants. These types of development have been classified by Maheshwari (1963) as follows:

1. Haploid parthenogenesis: The egg develops without fertilization. The result is a haploid embryo sporophyte.
2. Haploid apospory: The embryo develops from a nurse cell or other cell of the embryo sac exclusive of the egg. The result is a haploid embryo.
3. Apospory: The embryo sac develops from the megaspore mother cell without meiosis. The result is a diploid embryo if parthenogenesis occurs.
4. Adventitive embryology: The embryo develops from a nucellus cell. The result is a diploid embryo.
5. Artificial adventitive embryology: The embryo develops from callus tissue formed by cells growing in tissue culture in vitro. The result is a diploid embryo.

In types 1 through 4, the inheritance of the embryo is exclusively maternal. In type 5, the embryo inherits from two parents. Particularly remarkable are instances in which an embryo with a normal cell division pattern develops from a nurse cell, a nucellus cell, or a callus cell, rather than from an egg. These cases illustrate the epigenetic nature of plant embryology and raise the problem of the origin of

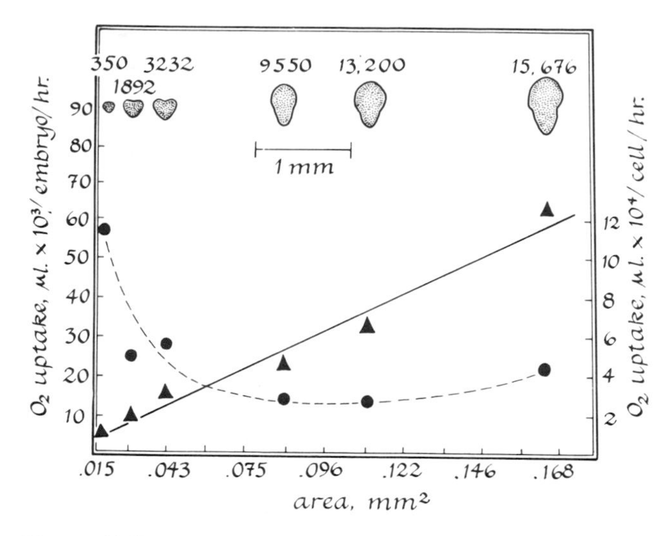

Figure 11-19. *Oxygen uptake per embryo and per cell at various stages in development of the cotton embryo. From Jensen, 1964. In* Meristems and Differentiation, *pp. 179–202. Brookhaven Symp. Biol. No. 16, Upton, N.Y.*

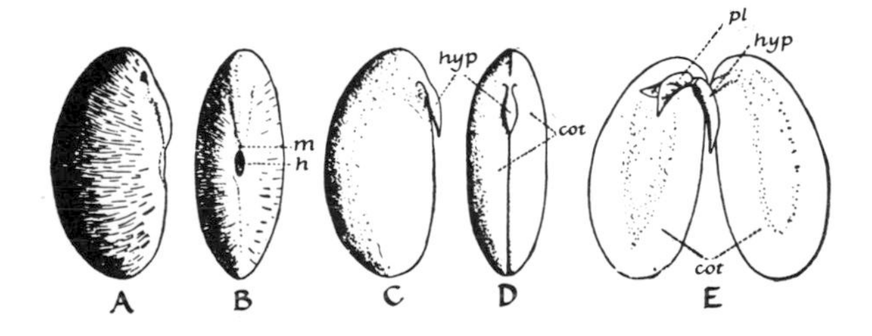

Figure 11-20. *The structure of a bean seed.* m, *micropyle;* h, *hilum;* cot, *cotyledon;* hyp, *hypocotyl;* pl, *plumule. From* Botany: Principles and Problems *by Sinnott. Copyright 1946 by McGraw-Hill Book Company. Used with permission of McGraw-Hill Book Company.*

polarity and symmetry. They provide excellent examples of progressively formed guidelines. Some suggestions about how such guidelines arise during development will be offered in Chapter 22 where we shall discuss the genesis of the internal microenvironment.

Development of the Seed

During development within the ovule, the embryo increases markedly in size. Its integuments (Figure 11-17H) increase in thickness, close over the micropyle, and become hard, woody seed coats. In dicotyledonous plants such as the bean, the cotyledons are by far the largest part of the seed (Figure 11-20). The remainder of the embryo consists of the hypocotyl, the primitive stem and root, and the plumule (bud) inserted between the cotyledons.

In seeds like the bean, the embryo has digested all of the endosperm during seed formation, so all the stored nutrient is in the cotyledons. In other plants (for example, the morning glory), the endosperm still constitutes most of the reserve food. In monocotyledonous plants such as corn and oats, endosperm is present in the seed and the embryo is comparatively small. A seed is thus an embryo plant surrounded by or containing reserve food and covered by protective seed coats. In the preceding chapter, we noted the importance of the seed as an adaptation to protect the plant during periods environmentally unfavorable for growth.

Fruit Development

The ovary at the base of the carpel enlarges to produce the fruit. This enlargement begins shortly after pollination and fertilization of the egg. The number and size of cells in the wall of the ovary increases. Except in rare cases of parthenocarpy (development of fruit from an unfertilized ovary), no fruit will develop unless fertilization has occurred. Thus fertilization is the stimulus for both embryonic development and fruit development. Various types of fruit are shown in Figure 11-21.

Enlargement of the ovary may be tremendous, as in squashes and melons, and its growth extremely differential so that the fruit has a distinct shape characteristic of the species (Figure 11-22). In many fruits there are two periods of growth, cell division and cell enlargement (Sinnott, 1944, 1960, 1963). In more or less spherical gourds (Figure 11-22D and E), cell divisions are equally abundant in all directions, but in elongate gourds (Figure 11-22A and B) the axes of the mitotic spindles are lined up in the direction of elongation. The guidelines controlling the direction of the divisions appear to be supracellular, perhaps involving gradients in hormone-like substances.

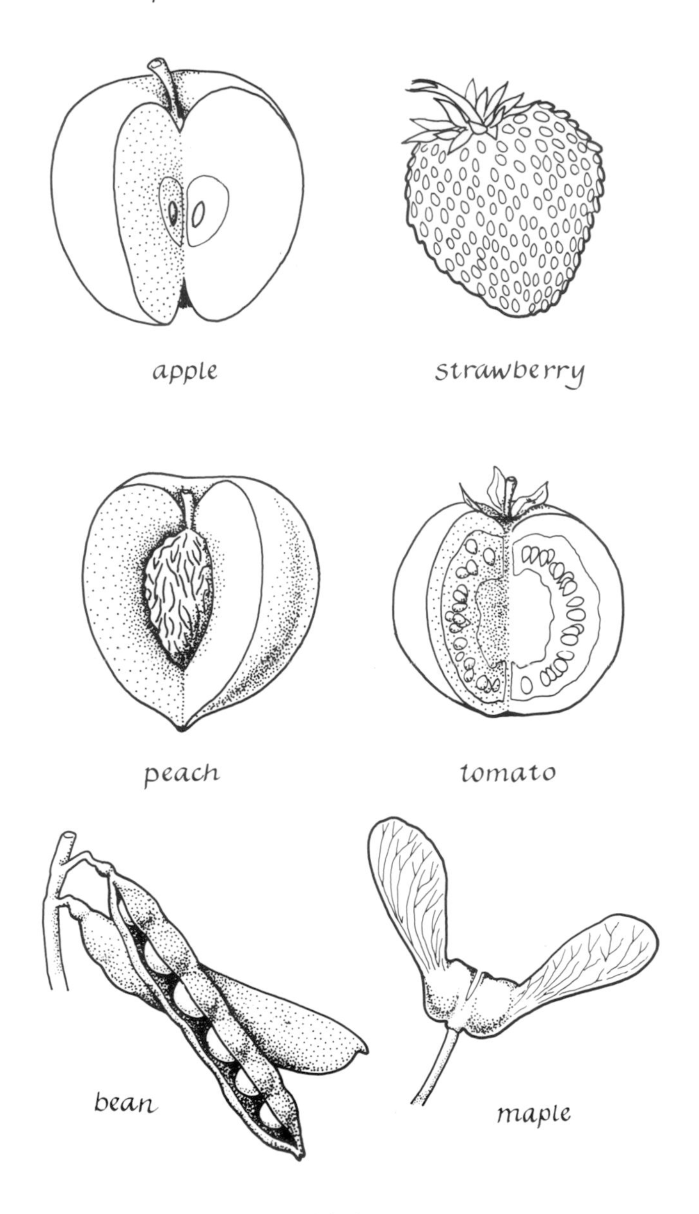

Figure 11-21. *Various types of fruit.*

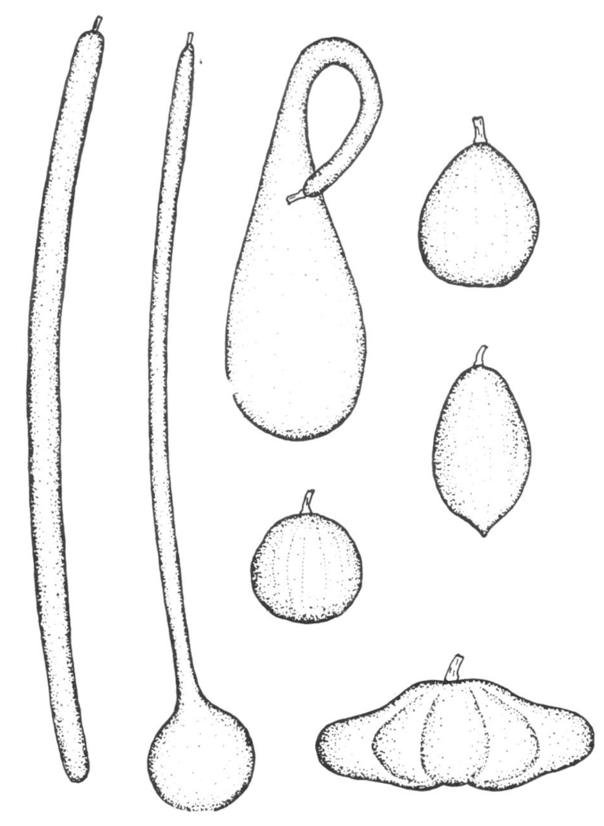

Figure 11-22. *Gourds of various types. From John Tyler Bonner:* Morphogenesis: An Essay on Development *(copyright 1952 by Princeton University Press). Reproduced by permission of Princeton University Press.*

Seed Germination

Seeds normally remain dormant for variable periods of time until favorable conditions such as moisture, temperature, and light stimulate germination. Water is rapidly absorbed by many kinds of seeds, enlargement of the embryo begins, the seed coats rupture, and the young plant sends its root into the soil and its shoot into the air. These events involve both cell elongation and cell division. Germination of the bean seed and formation of the seedling are illustrated in Figure 11-23. Many environmental factors—light, high or low temperature, moisture, oxygen supply—influence seed germination. The biochemical changes accompanying seed germination include changes in enzyme activity (Marré, 1967) and protein synthesis (see Chapter 17).

Plant Embryo Tissue Culture

Many kinds of plant embryos have been removed from the ovule or embryo sac at various stages and cultivated in nutrient media in vitro (Maheshwari, 1960, 1963, 1966; Torrey, 1967). In general, the younger the embryo the more complex are the nutrient requirements for further development in vitro. At present very young embryos have been successfully grown only in a culture medium of sucrose and coconut milk. However, the 32-cell stage of *Capsella* embryos has been successfully cultured in a medium

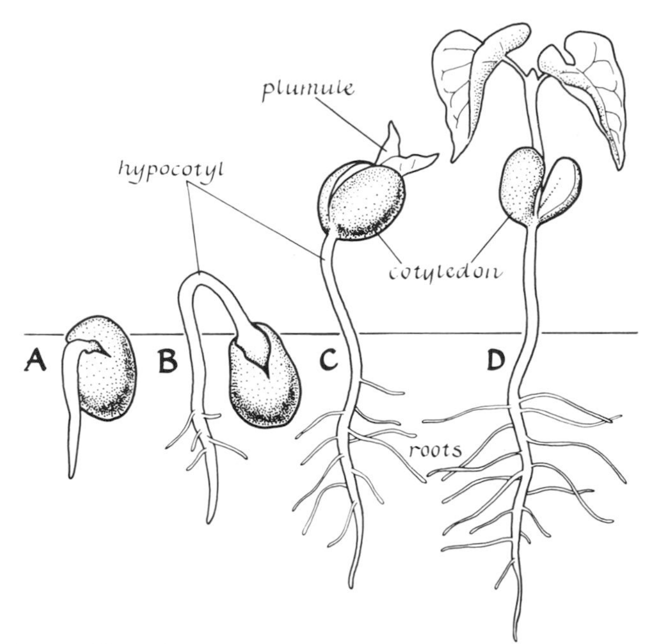

Figure 11-23. *Stages in germination of a bean seed.*

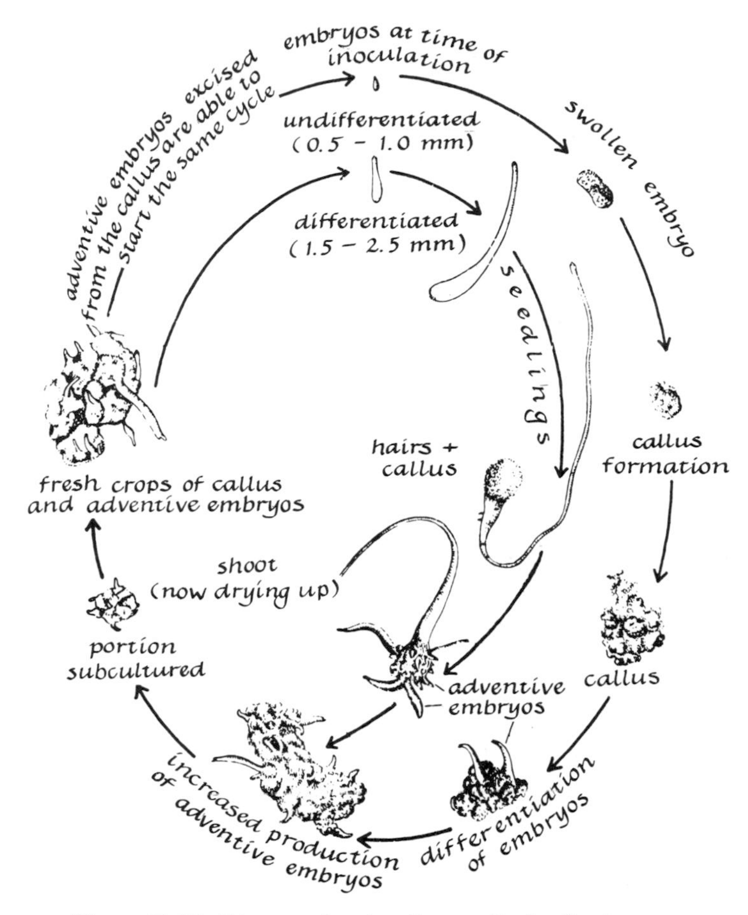

Figure 11-24. *Diagram showing the result of culturing young and older embryos of* Cuscuta. *From Maheshwari and Baldev, 1962. Courtesy Dr. Baldev.*

of inorganic salts, 2 percent sucrose, vitamins, 10^{-7} M indole acetic acid (a synthetic plant hormone), kinetin (a substance that mimics naturally occurring cytokinins, see Chapter 19) and adenine sulfate. In many cases, young plant embryos do not progress normally under culture conditions, but form a callus (Figure 11-24), as for example in *Cuscuta*, the parasitic dodder.

Several embryos develop adventitively from the callus (Maheshwari and Baldev, 1962). In this type of polyembryony, the embryos arise from buds (a similar process in embryo budding occurs in tunicates [Chapter 9]). This epigenetic formation of embryos deserves more careful study, for by such studies we have a good chance of discovering the nature of the influences that convert a mass of unspecialized cells into an organized plant body.

The Plant Egg and Its Significance

Animal eggs are highly specialized reproductive units. As we shall see in Chapter 13, the cytoplasmic structure of animal eggs probably is a primary guideline for differential nuclear activity. Plant eggs also have a cytoplasmic structure, but it appears to be less distinct and complex. With the exception of the eggs of some algae, the plant egg forms and develops in the environment provided by the gametophyte parent. When this happens, a preformed set of guidelines is not as important as when eggs develop independently of the mother organism, because the guidelines can be provided constantly throughout development. It is not known if eggs that develop under the influence of the gametophyte would continue to do so if removed and cultured in vitro. In Chapter 14 an experiment related to this problem is described.

Most plant eggs have a preformed polarity in respect to the distribution of metabolites, and the apical pole of the egg becomes the principal locus of synthesis, growth, and differentiation (Wardlaw, 1955).

The polar distribution of metabolites and other cytoplasmic constituents apparently underlies the unequal divisions of the zygote, beginning with the first division to form a terminal (embryo) and a basal (suspensor or foot) cell. The daughter nuclei are surrounded by cytoplasmically different environments that are probably instrumental in eliciting a differential expression of the genome in these nuclei.

Conclusion

Among the patterns of sexual reproduction in plants, the formation of a new individual from an egg presents special problems, primarily because the plant egg (with some exceptions) develops inside the reproductive organ or body of the maternal plant. Like most animal embryos, the plant embryo is constantly subjected to external influences. The role of these influences in embryology can best be determined by growing the embryo in culture. Older embryos continue to develop normally in culture, suggesting that the embryo eventually becomes independent of maternal influences.

Although there are many patterns of plant embryo development (Chapter 10) all have common features. Plant embryonic development is morphologically simple in comparison with animal embryology, but this apparent simplicity can be misleading. Biochemically, physiologically, and genetically, plant embryology is complex, involving cell differentiation, differential growth, and cellular arrangements. Much of the complexity in plants is produced not by cellular differences but by the arrangement of the cells.

Some features common to all kinds of plant embryonic development have been summarized by Wardlaw (1955):

1. There is a polarized distribution of metabolites in the fertilized egg. When the egg develops inside a reproductive organ, this polar organization is probably induced by the surrounding tissues.

2. The apical pole of the zygote becomes the apical pole of the embryo, the primary locus of growth and differentiation. The basal, nutritive pole develops into the extra-embryonic suspensor system in many embryos.

3. The first division of the zygote is usually at right angles to the polar axis.

4. The positions of the cell division planes vary among different kinds of embryos but generally conform to the principle of division by minimal area cell walls.

5. Cell division progressively becomes differential, resulting in an apical-basal gradient in cell size, with smaller cells at the apex and larger ones in the basal region. Except in algae that have intercallary meristems, the apical cells form the apical meristems of the adult. In many plants, a second center of rapid cell division in the basal region forms the rhizoid or root meristem system. Differential growth is especially important in plant embryology, because it is the primary mechanism underlying changes in the shape of the embryo.

In general, as plant embryos of different species develop to later stages, the originally similar patterns of cell division and growth become very different. Von Baer's Law—that the more general characters of a related group of organisms appear before the species-specific characters—apparently operates in both plant and animal embryology (see Chapter 12). Perhaps the genes common to related species are active earlier in development than the genes controlling formation of species-specific characteristics. To test this hypothesis we need further knowledge of the mechanisms controlling the order in which the genetic information is used as well as of the relationship between the production of gene products and specific developmental events.

As we have noted, all patterns of development are a visible expression of the sequential use of inherited information. This is so obvious that we tend to take it for granted, but the molecular basis of this program remains a central and completely unsolved problem of developmental biology.

References

Bonner, J., and J. A. D. Zeevaart. 1962. Ribonucleic acid synthesis in the bud, an essential component of floral induction in *Xanthium*. Plant Physiol. **37**:43–49.

Clowes, F. A. L., and B. E. Juniper. 1968. Plant Cells. Blackwell, Oxford.

Correns, C. 1907. Die Bestimmung und Vererbung des Geschlechtes nach neuen Versuchen mit höheren Pflanzen. Berlin.

Gifford, E. M. 1964. Developmental studies of vegetative and floral meristems. In Meristems and Differentiation, pp. 126–137. Brookhaven Symp. Biol. Upton, N.Y.

Heslop-Harrison, J. 1964. Sex expression in flowering plants. In Meristems and Differentiation, pp. 109–125. Brookhaven Symp. Biol. Upton, N.Y.

Hillman, W. S. 1962. The Physiology of Flowering. Holt, Rinehart and Winston, New York.

Holman, R. M., and W. W. Robbins. 1940. Elements of Botany. Wiley, New York.

Jensen, W. A. 1964. Cell development during plant embryogenesis. In Meristems and Differentiation, pp. 179–202. Brookhaven Symp. Biol. Upton, N.Y.

La Cour, L. F. 1949. Nuclear differentiation in the pollen grain. Heredity **3:**319–337.

Maheshwari, P., ed. 1960. Plant Embryology—a Symposium. Council of Sci. Ind. Res. New Delhi, M. H. Patwardham, Poona, India.

Maheshwari, P. 1963. Recent Advances in the Embryology of Angiosperms. University of Delhi, Catholic Press, Ranchi, India.

Maheshwari, P. 1966. The embryology of angiosperms—a retrospect and prospect. In E. Cutter, ed., Trends in Plant Morphogenesis, pp. 97–112. Wiley, New York.

Maheshwari, P., and B. Baldev. 1962. In vitro induction of adventitive buds from embryos of *Cuscuta reflexa* (Roxb.). In Plant Embryology: A Symposium, pp. 129–138. Sangum Press, Poona, India.

Marré, E. 1967. Ribosome and enzyme changes during maturation and germination of the castor bean seed. In A. Moscona and A. Monroy, eds., Current Topics in Developmental Biology, pp. 75–105. Academic Press, New York.

Miller, H. A., and R. H. Wetmore. 1946. Studies in the developmental anatomy of *Phlox drummondii* Hook: I. The embryo. Am. J. Bot. **32:**588–599.

Rijven, A. H. G. C. 1952. In vitro studies on the embryo of *Capsella bursa-pastoris.* Acta Bot. Neerl. **1:**157–200.

Rosen, W. G., S. R. Gawlik, W. V. Dashek, and K. A. Siegesmund. 1964. Fine structure and cytochemistry of *Lilium* pollen tubes. Am. J. Bot. **51:**61–71.

Sinnott, E. W. 1944. Cell polarity and the development of form in cucurbit fruits. Am. J. Bot. **31:**388–391.

Sinnott, E. W. 1960. Plant Morphogenesis. McGraw-Hill Book Company, New York.

Sinnott, E. W. 1963. The Problem of Organic Form. Yale University Press, New Haven, Conn.

Souèges, R. 1919. Les premières divisions de l'oeuf et les différencions du suspenseur chez le *Capsella bursa-pastoris* Moench. Ann. Sci. Nat. Bot. Ser. 10, **1:**1–28.

Tepfer, S. S. 1953. Floral anatomy and ontogeny in *Aquilegia formosa* var. *truncata* and *Ranunculus repens.* Univ. Calif. Publ. Bot. **25:**513–647.

Torrey, J. G. 1967. Development in Flowering Plants. Macmillan, New York.

Wardlaw, C. W. 1955. Embryogenesis in Plants, Methuen, London.

Wardlaw, C. W. 1965. Organization and Evolution in Plants. Longman, London.

Westergaard, M. 1958. The mechanism of sex determination in dioecious flowering plants. Adv. Genetics **9:**217–281.

12 *Patterns of Sexual Reproduction in Animals: A Comparative Survey of Animal Embryology*

The Animal Life Cycle

In this chapter we shall survey in a comparative way patterns of sexual reproduction in animals. In animals the gametes are direct products of meiosis and are the only haploid cells. Genetic variation is thus accomplished during gametogenesis, and so differs from what occurs in most plants. The life cycle of an animal is diagramed in Figure 12-1.

The life cycle of the frog is illustrated in Figure 12-2.

Note in Figures 12-1 and 12-2 that meiosis, which occurs in the reproductive organs of the parents when they reach sexual maturity, results in the production of four genetically different cells. In the male, all four products of meiosis in the spermatogonium become functional gametes, spermatozoans. However, in the female (Figure 12-2), only one product of meiosis in the oogonium becomes a functional gamete, an egg; the other three become nonfunctioning *polar bodies.* Like plant life cycles, animal life cycles exhibit the programmatic nature of the developmental processes. As in plants, the sequence of stages in the life cycle of animals is influenced by environmental conditions (Chapter 22).

Three types of reproduction are utilized by animals: *isogamy, heterogamy (anisogamy),* and *oogamy.* As noted in the preceding chapter, isogamy is the fusion of morphologically like gametes which are physiologically and genetically different. Isogamy occurs in protozoa such as rhizopods (for example, *Arcella),* and in some protomonads and sporozoa. The genetic recombination by *conjugation* that occurs in protozoans is a modified type of isogamy. In true isogamy, permanent fusion of gametes occurs; in conjugation, the fusion of the animals is only temporary and an exchange of nuclear material occurs. The best known example of conjugation involves paramecia (Sonneborn, 1954; Beale, 1954). The *micronuclei* of the paramecia function as isogametes. Major events in the conjugation process are outlined and illustrated in Figure 12-3.

Note that three of the haploid micronuclei resulting from the two meiotic divisions in each partner degenerate. Recall that three of the products of meiosis in the females of both plants and multicellular animals also degenerate. Also note (Stage 4, Figure 12-3) that only one of the two micronuclei in each paramecium passes through the conjugation bridge, enters the opposite partner, and fuses with the nonmotile micronucleus to form a diploid zygote nucleus. The *macronucleus,* which contains many copies of the micronuclear DNA, functions in the synthetic, nonsexual activities of the organism. Recall that at least one node of the macronucleus (but not the micronucleus) is necessary for regeneration of a fragment of *Stentor.* In contrast, the DNA of the

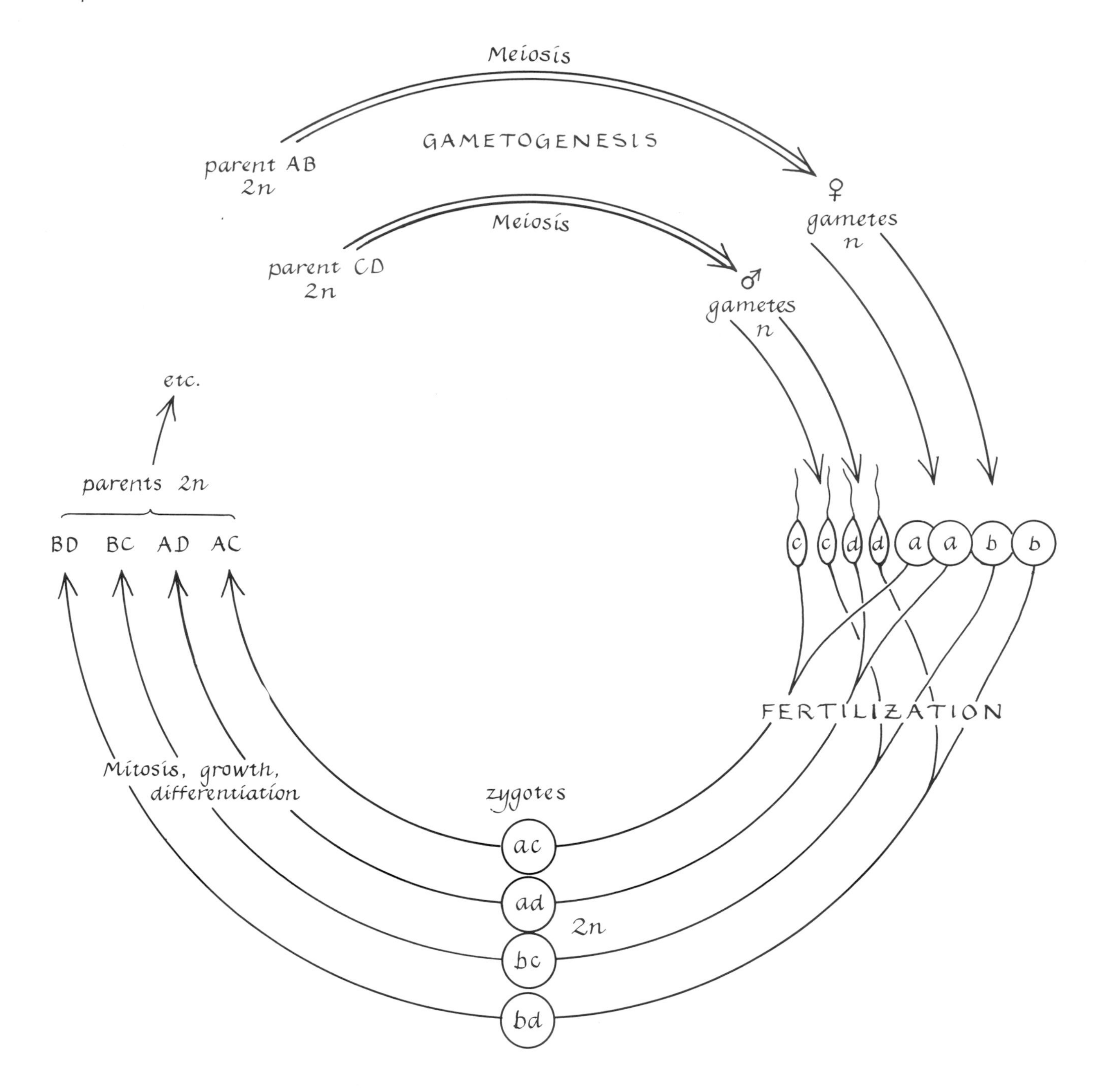

Figure 12-1. *Diagram of the life cycle of an animal.*

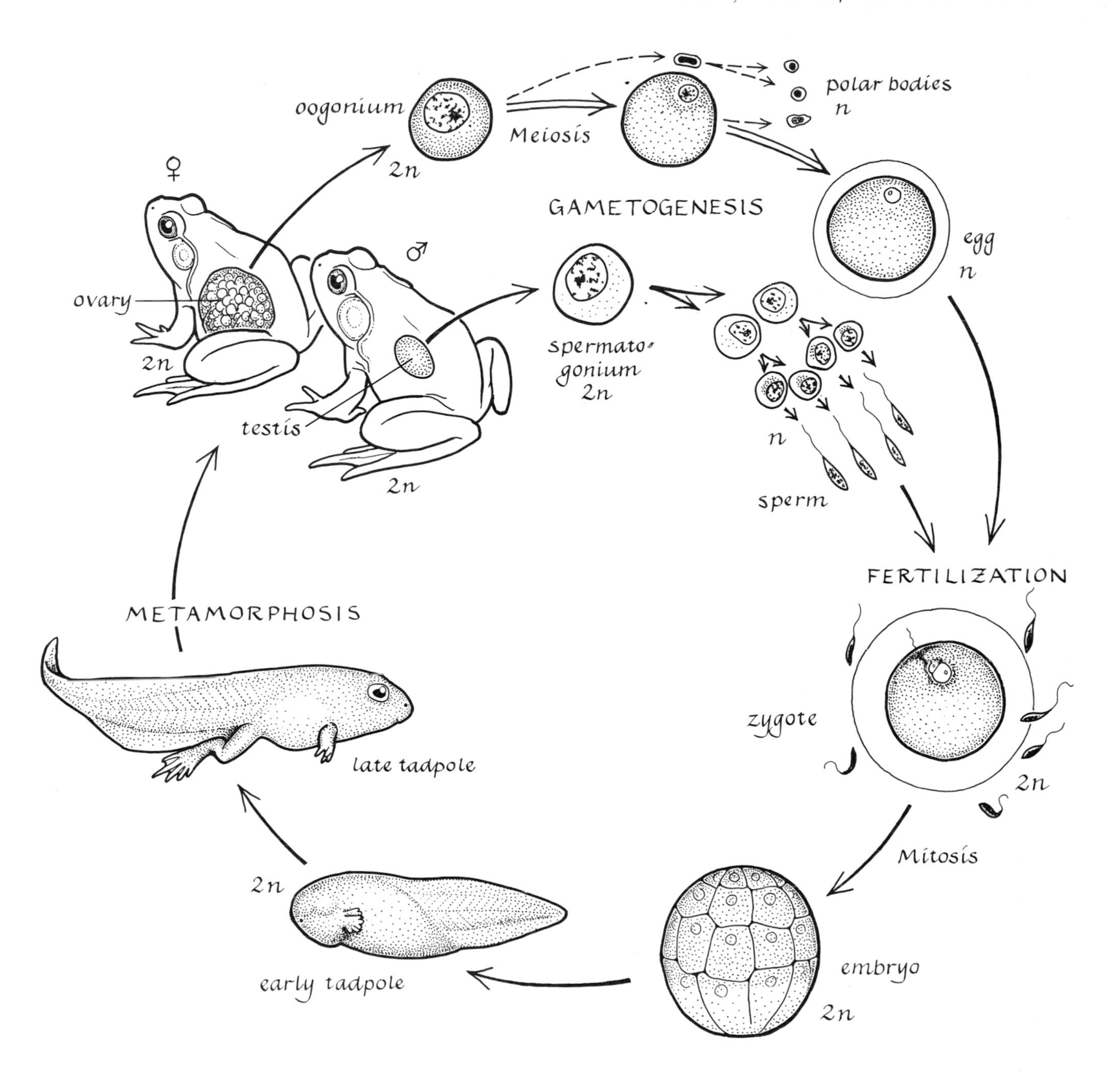

Figure 12-2. *The life cycle of a frog.*

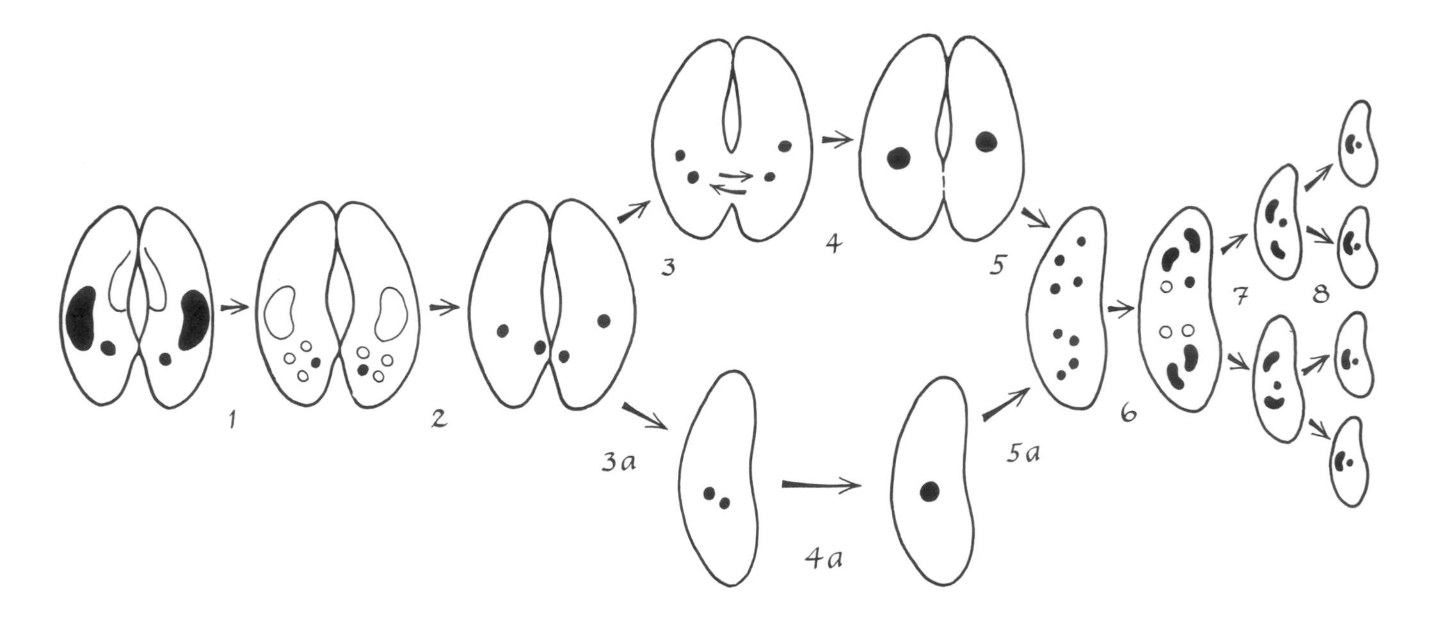

Figure 12-3. *Conjugation and sexuality in* Paramecium.

micronuclei is used primarily for hereditary transmission and genetic recombination. The three mitotic divisions of the zygote nucleus (Stage 5, Figure 12-3) result in formation of eight nuclei, three of which degenerate (Stage 6, Figure 12-3). Of the remaining five nuclei, four become macronuclei and one becomes a micronucleus. This differential behavior of the genetically equivalent nuclei is apparently due to the cytoplasmically different environments surrounding them (Sonneborn, 1954). Finally (Stages 7 and 8, Figure 12-3), cell and micronuclear divisions and parceling out of the macronuclei result in four exconjugant paramecia with normal nuclear complements.

Sexual reproduction by fusion of heterogametes occurs in some colonial ciliate protozoa such as *Zoothamnium alternuns* and *Vorticella* (Summers, 1938a and 1938b). In this case a motile microgamete (♂) fuses with a larger nonmotile and terminal macrozooid which functions as a macrogamete (♀). Major events in sexual reproduction of *Zoothamnium* are diagramed in Figure 12-4.

As indicated in Figure 12-4, *Zoothamnium* also reproduces asexually by the development of microzooids into ciliospores. The ciliospore attaches to the substratum and divides into a terminal macrozooid; further divisions result in a mature branched colony.

Among animals, isogamy and heterogamy are found only in protozoa. In all metazoa including sponges and in some protozoa (for instance, the malarial parasite) development from an egg (oogamy) occurs. Like the plant egg, the animal egg is a single cell, but it has far more developmental potential than do the tissue cells of the mother animal. Perhaps an egg may best be considered a unicellular organism or the unicellular stage of a multicellular organism since, unlike a tissue cell, the egg has the capacity to develop into a whole organism and nothing less. Understanding the nature of the information packed into an egg constitutes a challenge for the student of development. For example, how much of an egg's development is controlled by preformed guidelines, how much by progressively formed guidelines? There is no generally applicable answer since some eggs, especially those of coelenterates, seem to begin development with few clearly preformed guidelines whereas others, such as the eggs of some ascidians, have elaborately built-in preformed guidelines. This problem deserves our detailed examination. In Chapter 13 we shall consider the role played by the cytoplasmic composition of eggs.

Eggs are almost always much larger than tissue

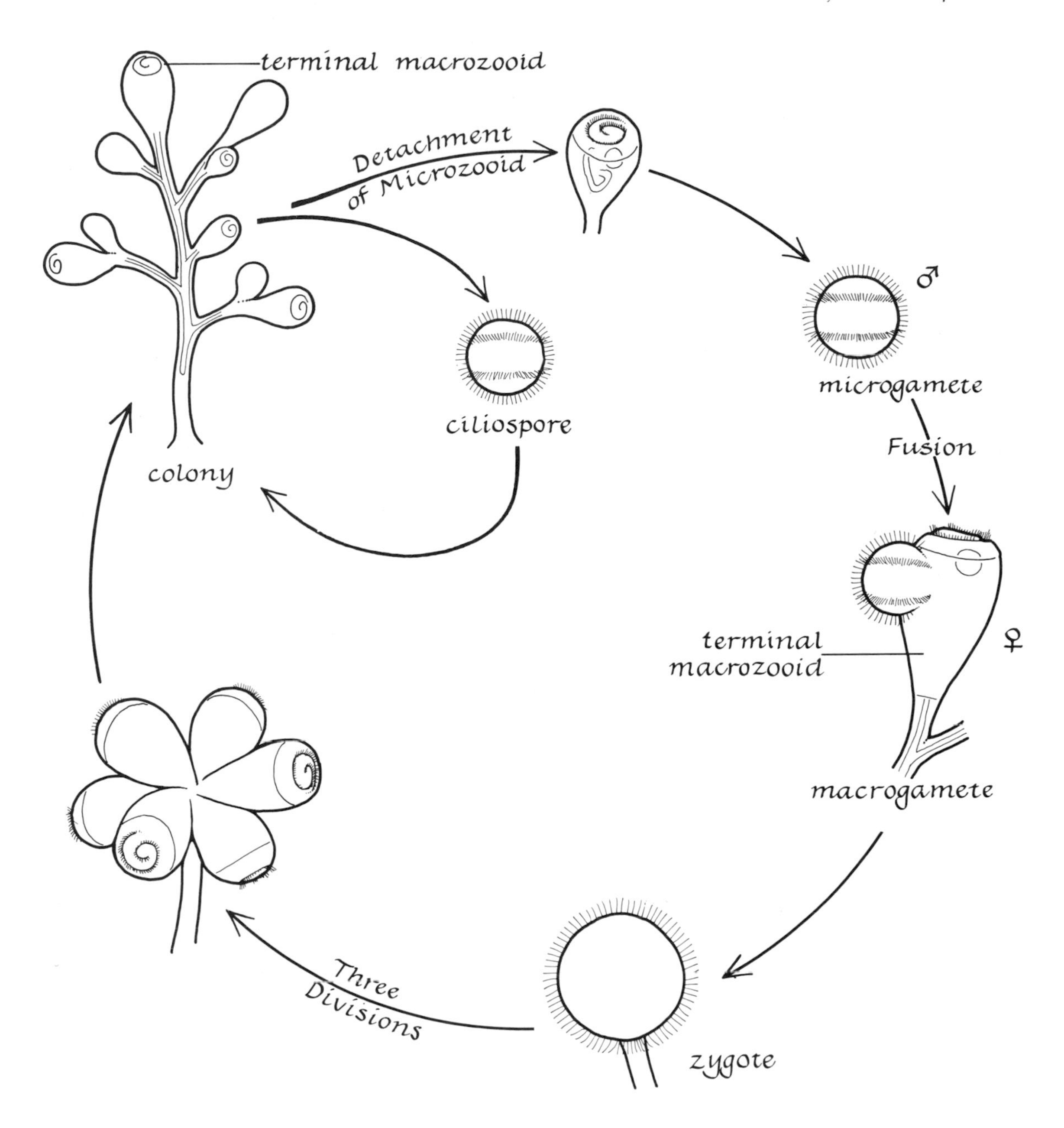

Figure 12-4. *Sexual and asexual reproduction in* Zootham-nium alternans.

cells. This implies a capacity for division into many cells, a capacity which is evident in the rapid rate of cleavage of the egg into many cells. As we shall see in Chapter 13, eggs also contain much more genetic information than most tissue cells (exceptions being gland cells with polytene chromosomes, such as the giant chromosomes of the salivary glands of *Diptera*). Like spores, but unlike reproductive units such as buds and zooids, eggs develop primarily as causally closed developmental systems; their growth is to a large extent independent of the mother organism even in those mammals in which the egg de-

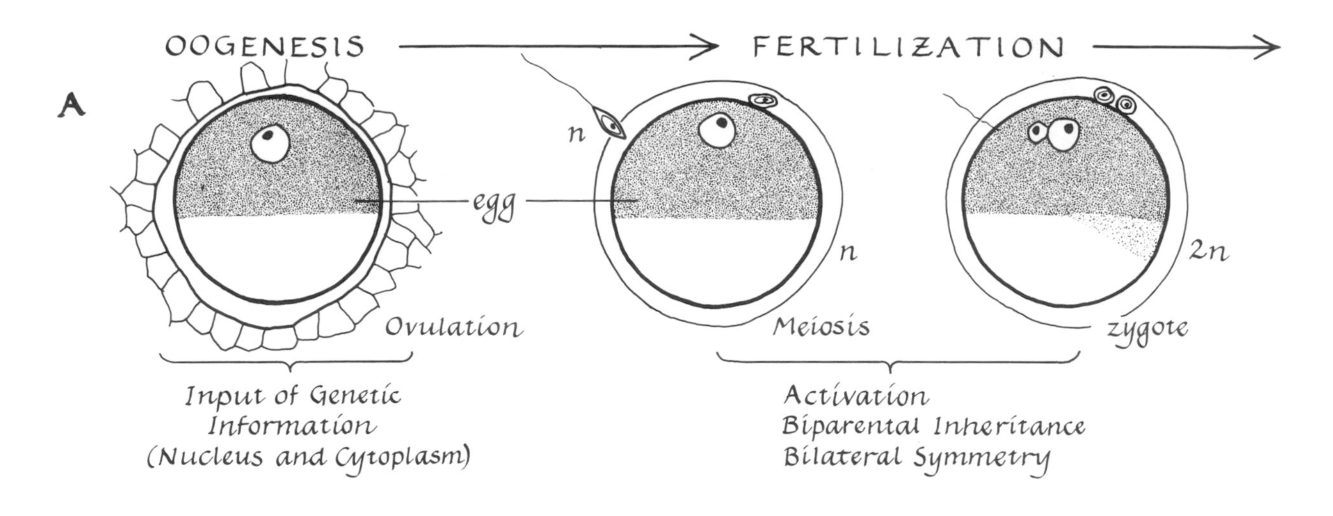

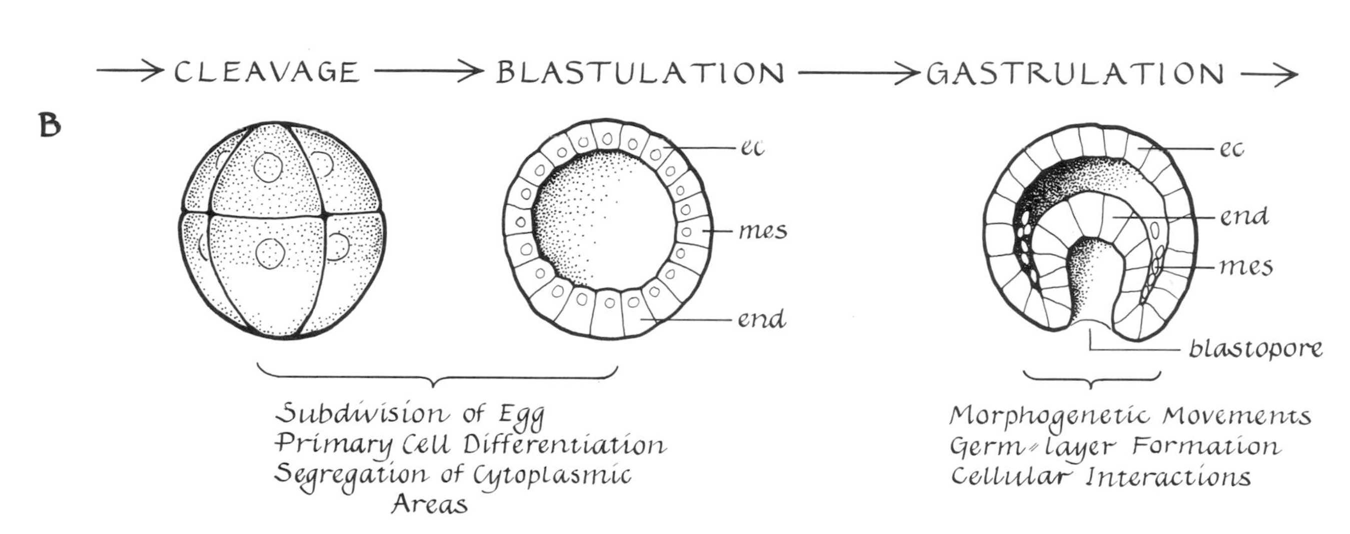

Figure 12-5. *Chronology of animal egg development.*

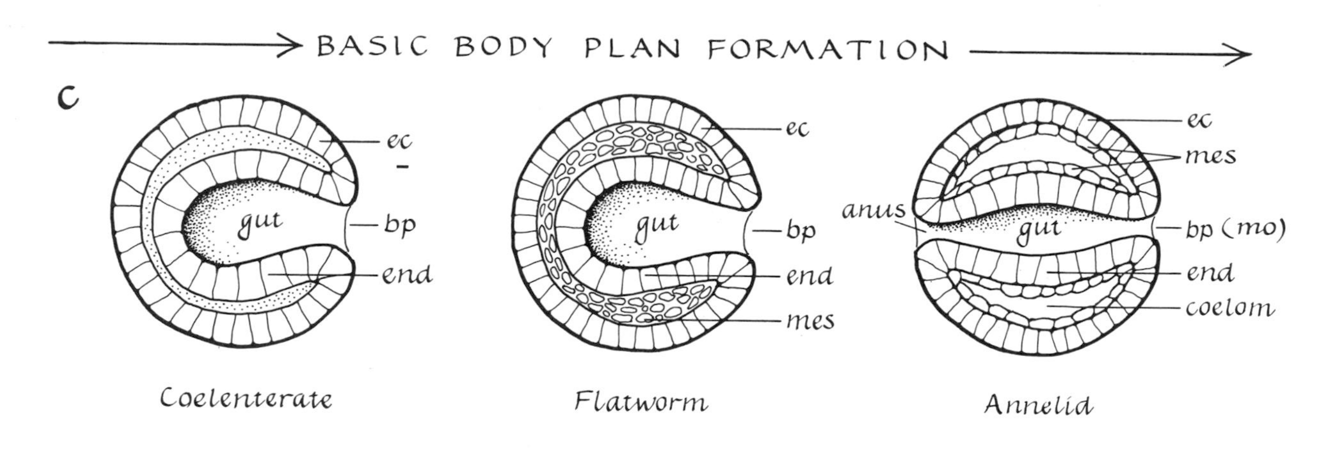
BASIC BODY PLAN FORMATION
C
ec
gut
bp
end
Coelenterate
ec
gut
bp
end
mes
Flatworm
ec
mes
anus
gut
bp (mo)
end
coelom
Annelid

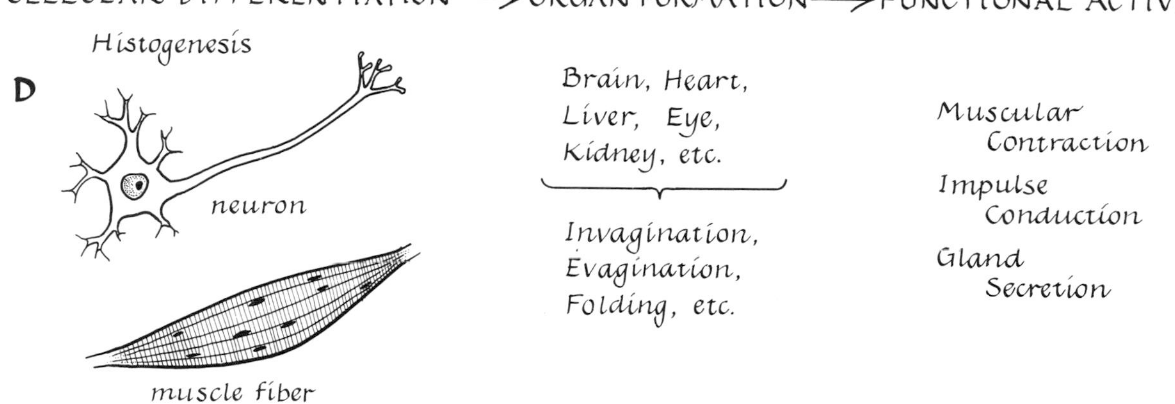
CELLULAR DIFFERENTIATION → ORGAN FORMATION → FUNCTIONAL ACTIVITY →
Histogenesis
D
neuron
muscle fiber
Brain, Heart,
Liver, Eye,
Kidney, etc.
Invagination,
Evagination,
Folding, etc.
Muscular
Contraction
Impulse
Conduction
Gland
Secretion

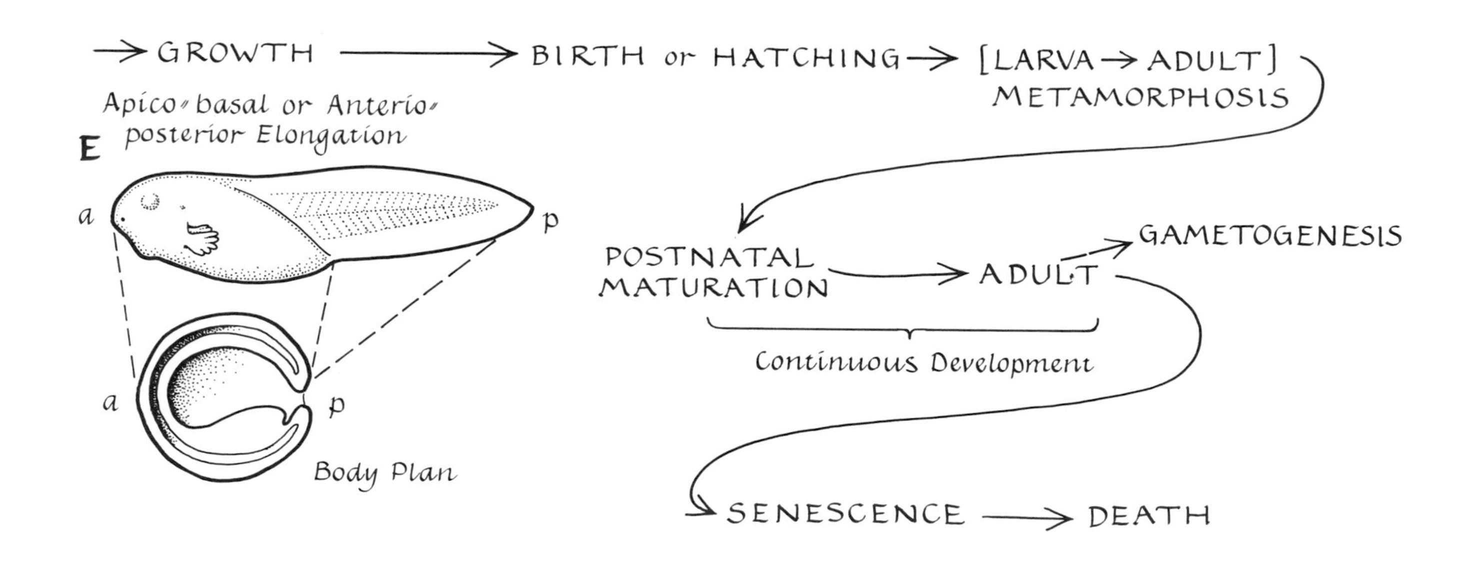
→ GROWTH → BIRTH or HATCHING → [LARVA → ADULT]
METAMORPHOSIS
Apico-basal or Anterio-
posterior Elongation
E
a
p
a
p
Body Plan
POSTNATAL
MATURATION
ADULT
GAMETOGENESIS
Continuous Development
SENESCENCE → DEATH

velops inside the mother's body. Egg formation is, however, influenced to some degree by the mother. The development of animal eggs has been studied so intensively for more than one hundred years that embryology has become a separate science. But we should remember that embryonic development is only one aspect of the larger field of developmental biology.

General Chronology of Egg Development

The basic sequence of events in the development of an animal egg is diagramed in Figure 12-5. Only events which are probably common to the development of all kinds of animal eggs are outlined in the figure. The figure also indicates the developmental significance of most of the events.

Figure 12-5 shows that embryonic development is a complicated process involving cell division, growth, cellular movements, formation of tissue layers, cellular interactions, cellular differentiation, folding of cellular layers, and many other formative and modeling processes. In short, embryonic development is a rather indirect method of producing a new individual. It is primarily an epigenetic process in which inherited information is used step by step. In normal development, the onset of each stage depends on the completion of the preceding stage. No stage is skipped. The entire process is unidirectional. As already noted, the program for use of the information as well as the information itself is inherited.

In most animals, development proceeds uninterruptedly through the stages indicated in Figure 12-5. However, some animals, such as sea urchins, many insects, and frogs, give birth to (or hatch) an incompletely formed but free-living larva. After a variable period of time which depends on the animal, the larva resumes development and metamorphoses into the adult form. We shall discuss this process in greater detail later in this chapter.

Another kind of interruption occurs at the blastula or blastocyst stage in some mammals. This temporary arrest of development, called *diapause*, results in delayed implantation of the egg in the wall of the uterus. The period of diapause ranges from two to three months in fruit bats and armadillos to up to six months in bears and eleven months in otters, fishers, and weasels (Daniel, 1970). The arrested development appears to be due to a deficiency of uterine proteins necessary for nutrition of the embryos.

Specific Events—Variations on the Basic Theme

Gametogenesis

Figure 12-6 illustrates the process of egg and sperm formation in the human.

The diagrams in Figure 12-6 apply also to nonhuman mammals; that is, they apply to all vertebrates in which the male is the heterozygous sex. In other vertebrates, such as amphibians, the female is the heterozygous sex. The gametes of all metazoa are products of meiosis, but in most vertebrates the meiotic process in an egg can only be completed when the egg is either fertilized or artificially activated. In terms of its influence on later development *oogenesis* (egg formation) is far more important than *spermatogenesis* (sperm formation). This will be obvious after we have examined the significance of the cytoarchitecture of the egg as a primary guideline for development in Chapter 13. The sperm primarily contributes information packaged in its nucleus, whereas the egg contributes not only nuclear information but the cytoplasm which constitutes the environment controlling the interaction of the egg and sperm genes.

In most animals, egg formation occurs in a special internal environment, a follicle of the ovary. The cytoplasmic structure of the egg is thus built up under control of surrounding cells of the mother organism. Specific examples of this influence will be cited in Chapter 13, but for the present we must emphasize that the key to a complete understanding of embryonic development will probably be found in the study of oogenesis. Stages in the formation of a mammal egg are illustrated in Figure 12-7 (Arey, 1965).

Growth of the frog *(Rana temporaria)* oocyte during the first three years of the female's life is illustrated in Figure 12-7B. Note that the growth is relatively slow during the first two years but increases markedly during the third year. Note also that there is a

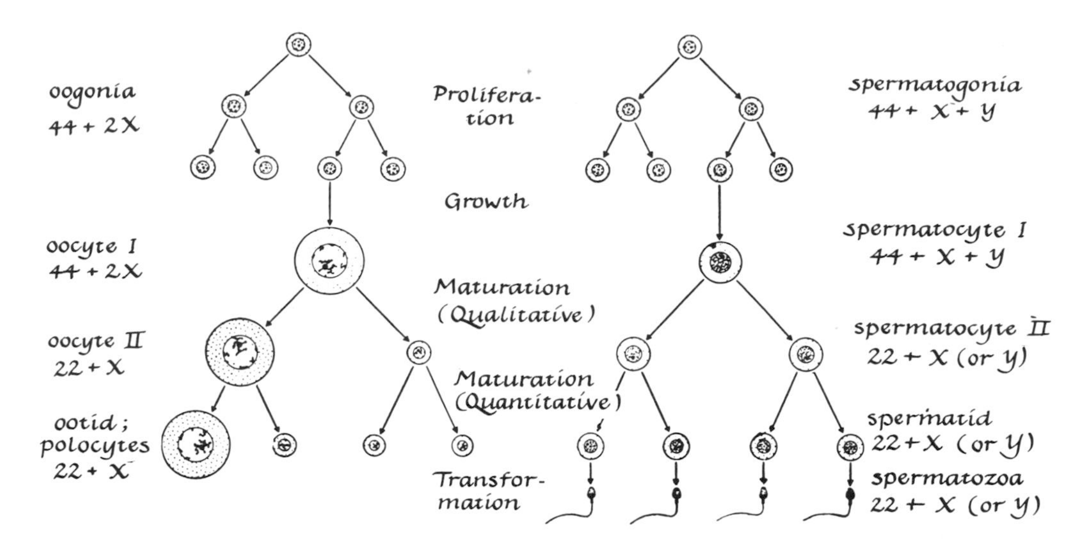

Figure 12-6. *Diagrams comparing oogenesis and spermatogenesis. The assortment of human chromosomes at each stage is also indicated. From Arey:* Developmental Anatomy, *7th edition, W. B. Saunders Company, Philadelphia, 1965.*

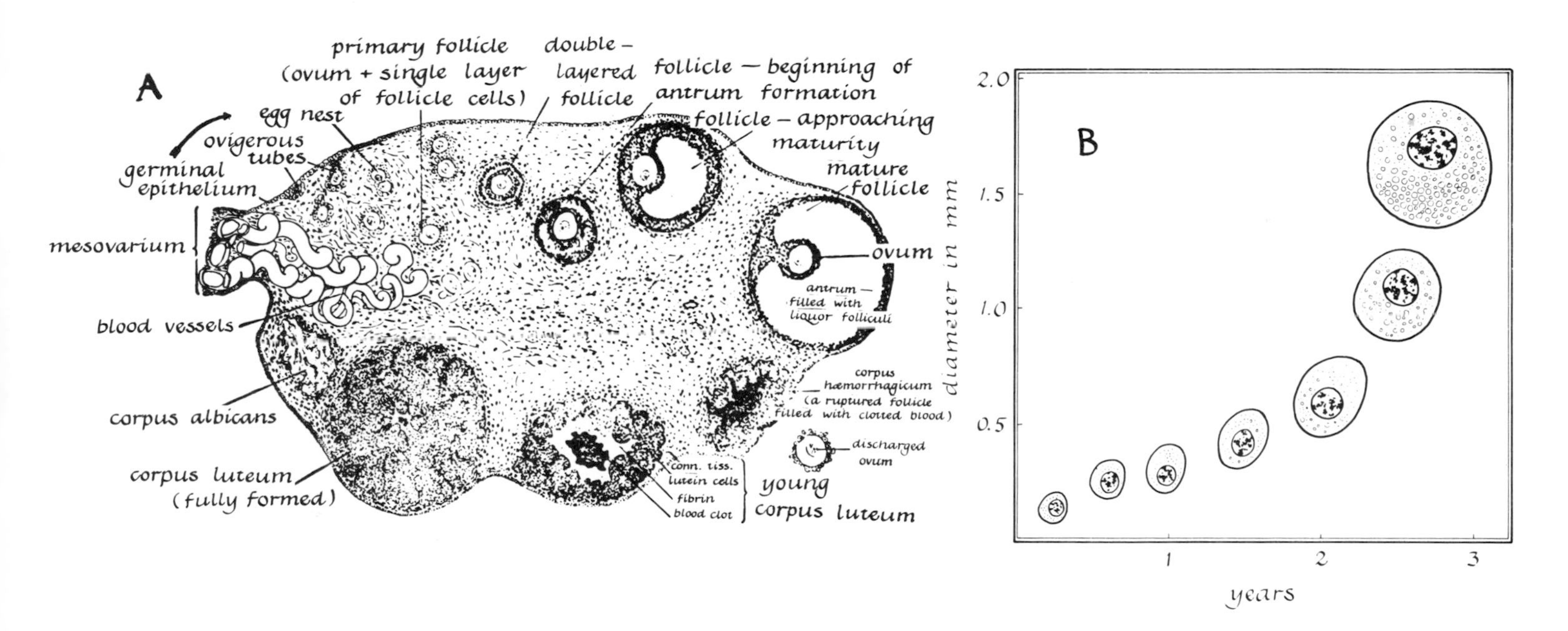

Figure 12-7. *A, life cycle of an egg and its follicle shown in a diagram of the mammalian ovary. Beginning at the arrow, the stages follow clockwise around the ovary. From Arey:* Developmental Anatomy, *7th edition, W. B. Saunders Company, Philadelphia, 1965. B, growth of the frog oocyte.*

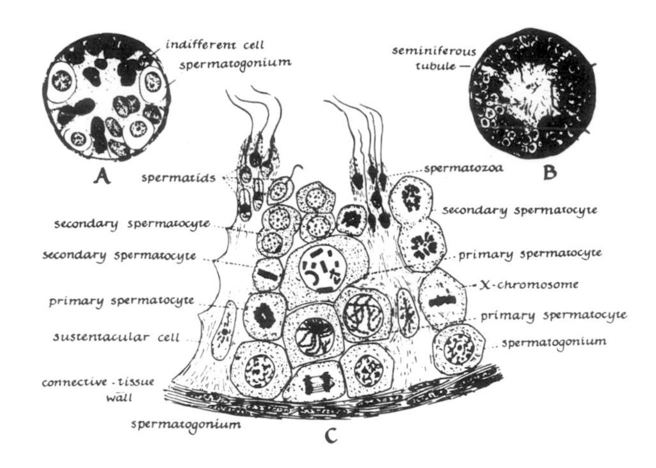

Figure 12-8. *Spermatogenesis in the human. A, section through tubule of a new born (× 400); B, section through tubule of an adult (× 115); C, stages in spermatogenesis in the area outlined in B. From Arey:* Developmental Anatomy, *7th edition, W. B. Saunders Company, Philadelphia, 1965.*

significant increase in the volume of cytoplasm relative to that of the nucleus, and that this is accompanied by the deposition of nutrients in the form of yolk granules, particularly in the vegetal (lower) half.

Major events in spermatogenesis in the human are illustrated in Figure 12-8 (Arey, 1965). Note in Figure 12-8 that the transformation of a spermatogonial cell into a mature spermatozoan involves an intracellular process of cell differentiation. The result is a highly specialized cell capable only of fertilizing an egg. Thus both eggs and sperm are highly differentiated unicellular organisms. Some types of eggs and the structure of the human spermatozoan are exhibited in Figure 12-9.

The mature human egg is approximately 0.14 mm in diameter, or about the size of a period on a printed page. By contrast, a chicken egg (actually only the yolk) is one of the largest single cells known. The human spermatozoan is about 0.06 mm long. In the mature human female, *ovulation* (release of an egg from the ovary) occurs on the average every 28 days, approximately fourteen days after the beginning of the menstrual cycle (Chapter 4). In the human male, about 250 million spermatozoa are present in a single ejaculation. In general, many more spermatozoa than eggs are formed by animals during sexual maturity.

For many years there has been controversy as to the origin of the gametes during embryonic development in animals. Weismann (1892) proposed that the sex cells develop from a special line of cells set aside early in development. This is known as the Theory of the Continuity of the Germ Plasm. There is now much evidence that the gametes develop from *primordial germ cells;* these constitute a cell lineage separate from that of the remaining cells of the embryo, which are called the somatic cells. This is true of many invertebrates—for example, the round worm *Ascaris* and the gall midge *Miastor,* in which the primordial germ cells are cleaved out of a special cytoplasmic region of the egg (Chapter 20). In vertebrates, such as amphibians, birds, and mammals, eggs and spermatozoa probably originate in primordial germ cells that arise outside of and later migrate into the developing gonads. Evidence for this includes the production of 100 percent sterile larvae of *Rana pipiens* by ultraviolet irradiation of the vegetal pole of the egg at the two-cell stage (Smith, 1966); the production of sterile and partly sterile *Xenopus* by pricking the vegetal pole of the egg and thus releasing some or all of the germ plasm located there (Buehr and Blackler, 1970); and tracing the migration of the relatively large and cytologically distinguishable primordial germ cells of birds and mammals from their distant origin outside the gonads into the gonads at a later stage (Clawson and Domm, 1969; Peters, 1970). Further evidence for the origin of the definitive gametes from primordial germ cells is provided by an experiment in which the primordial germ cells of one subspecies of *Xenopus* were removed from the vegetal pole of the egg and transplanted into the vegetal pole region of another subspecies (Blackler, 1962). In some of the toads receiving the graft, all the eggs formed were of the size and color characteristic of the donor subspecies. Thus, the theory of continuity of the germ plasm apparently applies to many vertebrates as well as to many invertebrates.

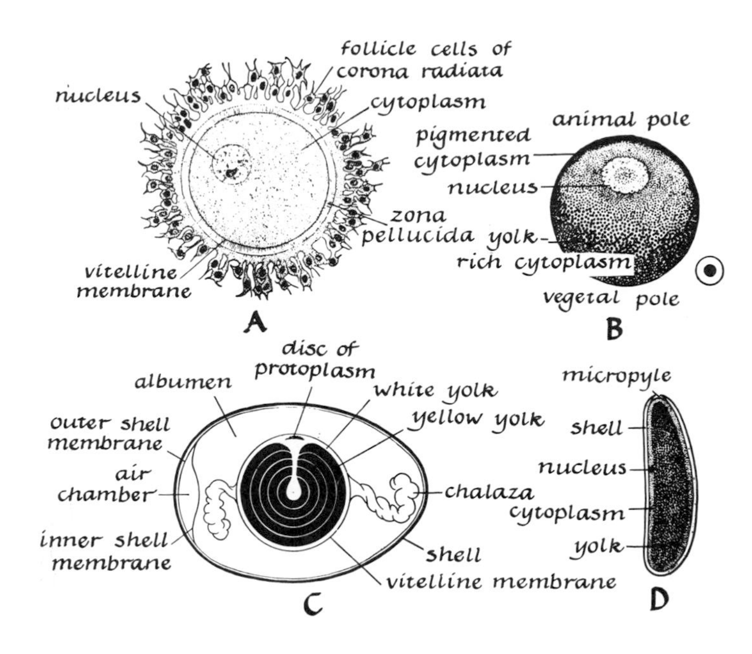

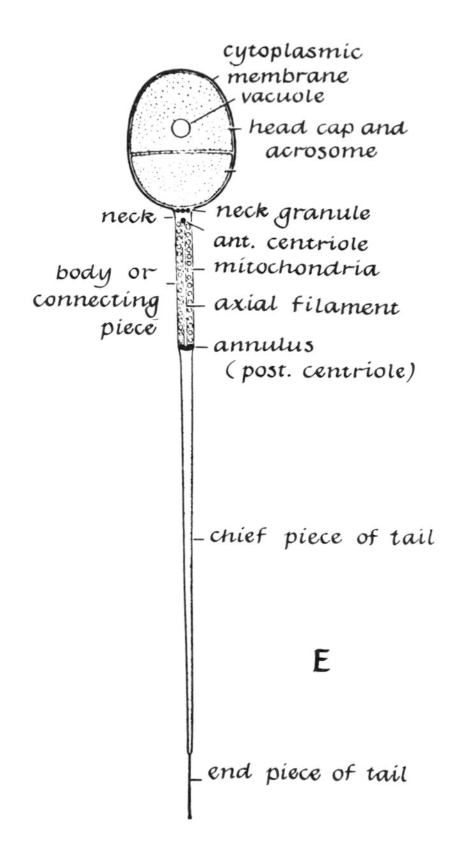

Figure 12-9. *Types of eggs produced by different kinds of animals. A, isolethical human egg (× 200); B, moderately telolecithal egg of the frog (× 15); C, extremely telolecithal egg of the hen; D, centrolecithal egg of the fly; E, structure of the human spermatozoan (× 1700). From Arey:* Developmental Anatomy, *7th edition, W. B. Saunders Company, Philadelphia, 1965.*

Fertilization

Fertilization includes all steps from the approach of the spermatozoan to the fusion of the *pronuclei* within the egg (Tyler and Tyler, 1966a). Fertilization of animal eggs has been studied intensively for many years. For a more detailed exposition of the process, the student should consult one or more of the following references: Rothschild, 1956; Austin, 1965 and 1968; Monroy, 1965; Tyler and Tyler, 1966b; Brachet, 1967; Metz and Monroy, 1967 and 1968. As indicated in Figure 12-5, fertilization is important because it activates the development of the egg and establishes biparental inheritance. In most animals, fertilization is also necessary for the final stages of nuclear maturation of the egg. Animals differ with respect to the process of meiosis and the time at which sperm penetration occurs. This is illustrated in Figure 12-10.

As shown in Figure 12-10, penetration by the sperm induces the meiotic divisions in some invertebrates and protochordates. Sperm penetration completes meiosis in vertebrates. In the sea urchin, the sperm enters after meiosis is completed. Strictly speaking, only in the latter case does the sperm enter a completely mature egg in which both first and second polar bodies have been formed.

In general, a spermatozoan will only penetrate an egg of the same or a closely related species, probably because complementary chemical substances—*fertilizin* on the egg surface and *antifertilizen* on the surface of the sperm—are necessary for the complex interactions involved (Hartman et al., 1947; Hamilton, 1952). However the macromolecular events are not

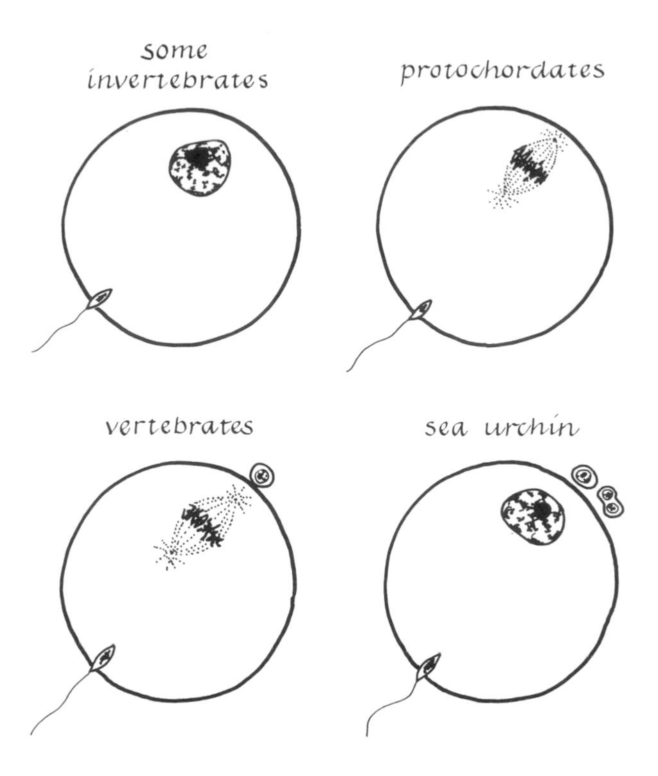

Figure 12-10. *Diagrams of differences in animals with respect to meiosis and sperm penetration.*

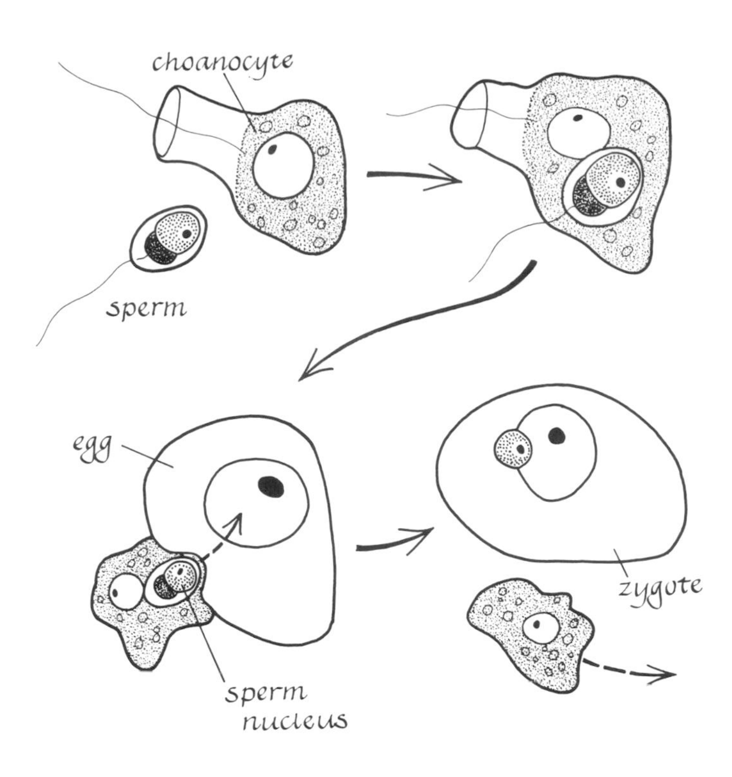

Figure 12-11. *Peculiar method of fertilization found in some sponges.*

well known. Strangely, the sperm of the mussel *Mytilus galloprovincialis* may enter the egg of the sea urchin. The sea urchin egg is activated and develops to the *pluteus* larval stage (Figure 12-42); but following pronuclear fusion, the *Mytilus* chromosomes are extruded, remain compact and do not participate in further development (Kupelwieser, 1909). In this case, the sperm only activates the egg. There are numerous examples of activation of eggs by various artificial, experimental treatments, such as pricking the egg of a frog with a needle which has been dipped in frog blood, or exposing the sea urchin egg to hypo- and hypertonic sea water (see Raven, 1966). Artificial or natural activation of eggs in the absence of a sperm is called *parthenogenesis*.

Rarely, a sperm will penetrate a tissue cell, for instance, the *choanocyte* in sponges (Gatenby, 1919 and 1927; Duboscq and Tuzet, 1937; reviews of Hyman, 1940 and Tyler, 1955). This exceptional phenomenon in the sponge *Grantia compressa*, first described by Gatenby and later confirmed by Duboscq and Tuzet, is illustrated in Figure 12-11.

As shown in Figure 12-11, the sperm penetrates a choanocyte of the sponge *Grantia*. The choanocyte moves to the egg, and the sperm nucleus enters the egg cytoplasm. The choanocyte then moves away. In another instance the sperm enters an amebocyte in the sponge *Cliona*. Sperm-blastomere fusion can occur in the hemichordate *Saccoglossus kowalevskii* (Colwin and Colwin, 1967).

Major events in fertilization are outlined in Figure 12-12. The diagrams are based primarily upon studies of the process in the sea urchin egg but apply in a general way to the process in many other animals.

In contrast to fertilization in many plants, in which either the female gamete or the female reproductive organ releases a sperm-attracting agent (Chapter 10), the operation of a chemotactic mechanism for bringing the animal egg and sperm together is extremely rare. At the present time, the only animals known to possess such a mechanism are two species of the colonial coelenterate *Campanularia* (Miller, 1966; see Figure 12-13). The *funnel* of the *gonangium* (Figure 12-13B), not the egg, is the source of an agent which attracts the spermatozoa. This is apparently not a trapping action since the spermatozoa swimming by the funnel suddenly change their direction, turn sharply and enter the funnel (Figure 12-13C). The agent, which has been partially isolated, is species specific.

An apparent example of the production of a chemotactic agent by a mammal egg has been described by Dickman (1963). When rat and rabbit eggs were transferred into oviducts of newly mated rabbits, many more rabbit spermatozoa were attached to the membrane surrounding the rabbit egg than to that of the rat egg (Dickman, 1963). However, it is well known that the membranes or jelly coats surrounding several kinds of eggs do trap spermatozoa; therefore, the possibility that the trapping activity of the rabbit egg was greater than that of the rat egg cannot be ruled out. Most attempts to demonstrate chemotaxis in animal fertilization have produced negative results (Monroy, 1965), and we are left with the general conclusion that contact between the animal egg and the sperm is a matter of chance collision.

As indicated in Figure 12-12B, fertilizin released by the sea urchin egg causes an increase in swimming activity of the neighboring spermatozoa and then the sudden formation of an acrosomal filament in the first spermatozoan to contact the jelly coat (Figure 12-12C). The acrosomal filament contacts the surface membrane of the egg and induces a series of changes in the ultrastructure of the egg cortex which spread as a wave over the surface of the egg from the point of contact in about twelve seconds (Figure 12-12D). Presumably it is these cortical changes, accompanied later by an elevation of the *vitelline* membrane, which normally prevent the entry of more than one spermatozoan in sea urchin eggs (Rothschild, 1956). *Polyspermy,* the entry of several sperm, does occur normally in some vertebrates, such as chickens, some urodele amphibians, and reptiles, and in some invertebrates, such as mollusks and many insects. In normal polyspermy only one sperm nucleus unites with the egg nucleus; the others degenerate near the egg surface.

How the spermatozoan actually enters the egg cytoplasm is not well known and probably differs from species to species (Colwin and Colwin, 1957, 1964). Electron micrograph studies of the protochordate *Saccoglossus* indicate that fusion of the sperm and egg membranes occurs, followed by entry of the sperm nucleus, as illustrated in Figure 12-12E (Colwin and Colwin, 1964, 1967). In the sea cucumber *Holothuria atra,* contact of the acrosomal filament with the egg surface is followed by formation of a *fertilization cone* by the egg and the engulfment of the entire spermatozoan by this cytoplasmic projection (Colwin and Colwin, 1957). This is illustrated in Figure 12-14.

In the sea urchin, after the spermatozoan or its nucleus has entered the egg, the egg undergoes further changes (Figure 12-12F). Briefly, these include (1) a breakdown of granules located in the egg cortex, (2) extrusion of the substance of the granules into the small space between the surface of the egg and the *vitelline membrane,* and (3) an elevation of the vitelline membrane, which is a noncellular membrane surrounding the unfertilized egg. The vitelline membrane, plus some of the substance of the cortical granules, becomes the *fertilization membrane,* and the space between this and the egg surface is called the *perivitelline space.* These events have been studied by Endo (1961) and Tyler and Tyler (1966B). A similar type of response of the egg is found in all animals which possess cortical granules—for example, bony fishes, frogs, and some mammals.

Fertilization of the sea urchin egg is accompanied by profound changes in its metabolism. These changes are reviewed by Monroy, 1965; Monroy and Maggio, 1966; Tyler and Tyler, 1966B; and Brachet, 1967. Briefly, the changes include a dramatic increase in oxygen consumption; resumption of DNA synthesis and protein synthesis; and activation of proteolytic enzymes. In summary, a mature unfertilized or nonactivated egg is a blocked system. Fertilization is basically a process of cell stimulation, and the discovery of the molecular basis of this stimulation is a fundamentally important aspect of development (Monroy, 1965).

Types of Eggs and Cleavage

On the basis of the amount and distribution of yolk, animal eggs can be classified into four general types. Each type usually exhibits a correspondingly

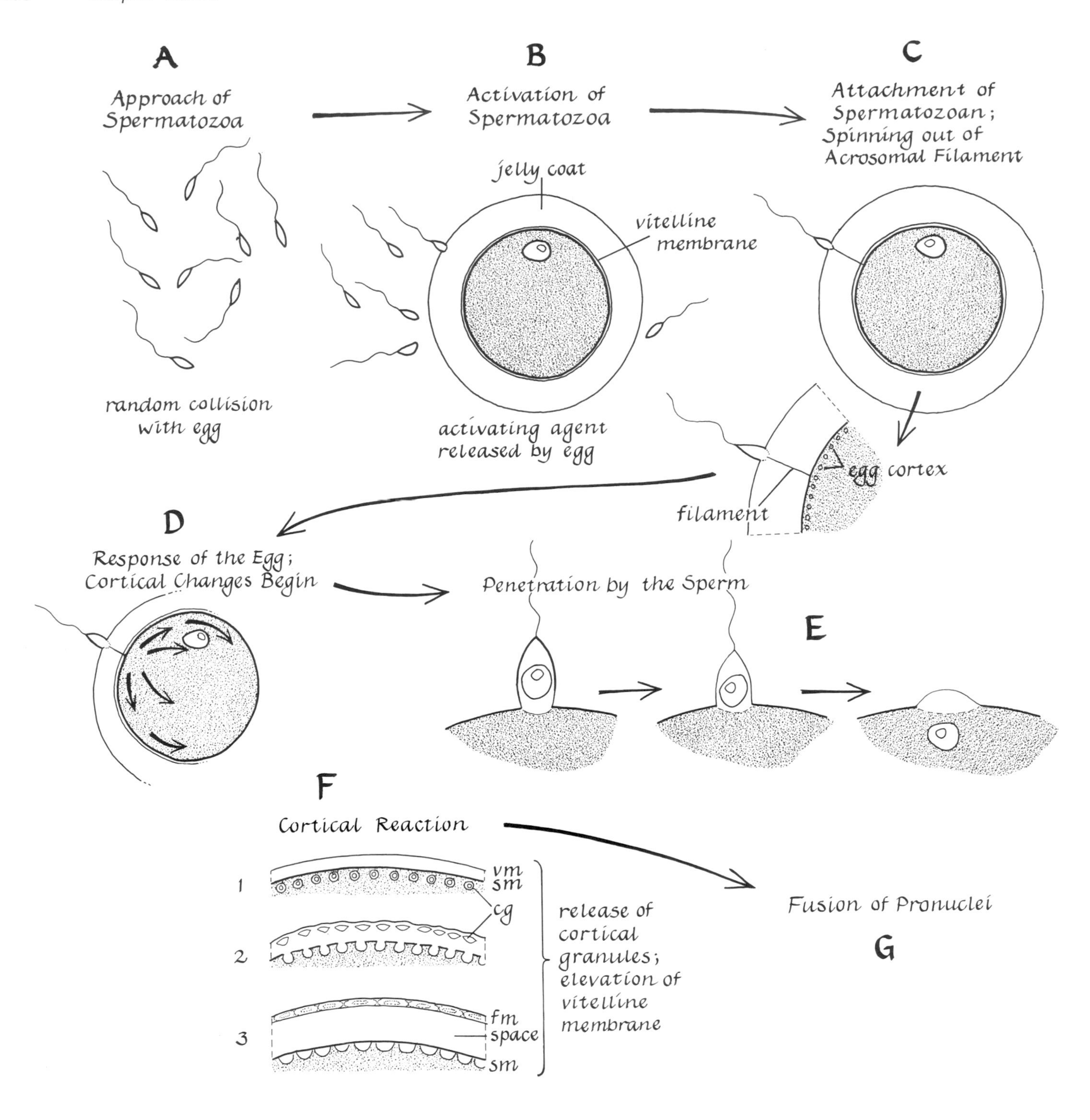

Figure 12-12. *Diagram of the major events in fertilization of an animal egg.*

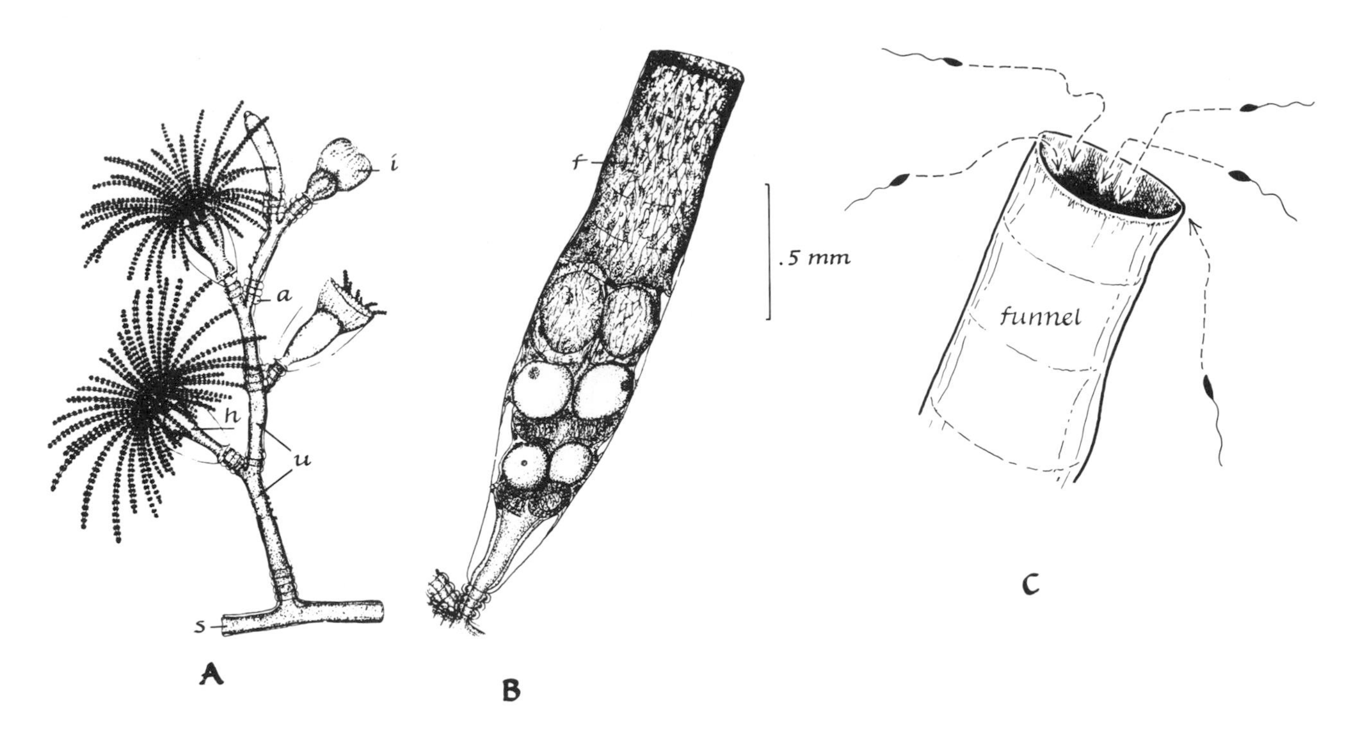

Figure 12-13. *Chemotaxis in* Campanularia flexuosa. *A, part of colony with stolon,* s; *upright,* u; *mature,* h, *and immature,* i, *hydranths. B, female gonangium with funnel,* f, *and eggs below. C, diagram illustrating paths (dotted arrows) of spermatozoa in neighborhood of the funnel opening. A and B from Miller,* J. Exp. Zool., *Volume 162, pp. 23–44, 1966. Courtesy Wistar Press.*

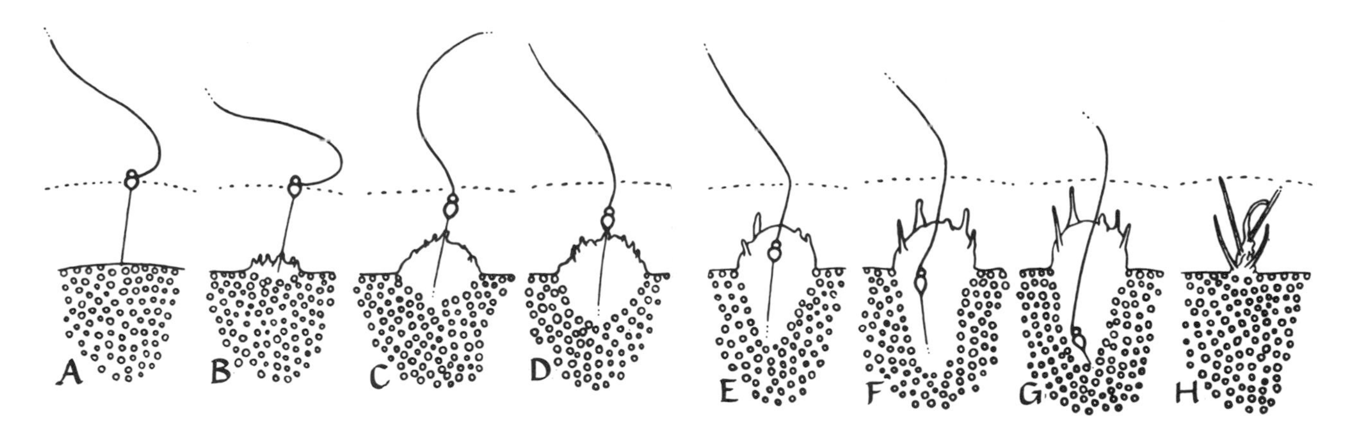

Figure 12-14. *Sperm entry into the egg of a sea cucumber. Note acrosomal filament of sperm and formation of fertilization cone by the egg. From Colwin and Colwin, 1957, p. 152:* Beginnings of Embryonic Development, *Eds., Tyler and VonBorstel. AAAS Publ. No. 18, copyright 1957 by the American Association for the Advancement of Science.*

characteristic cleavage pattern. This is illustrated in Figure 12-15.

The importance of cleavage is that it subdivides the pattern of egg cytoarchitecture (Chapter 13). Consequently, the cleavage cells (called *blastomeres*) contain different samples of the cytoplasm of the egg. The significance of cleavage in animal eggs is that it results in a type of primary cellular differentiation (see Figure 12-5). Cleavage is different from the cell multiplication that occurs in developing buds, gemmules, zooids, gemmae, and, to a large extent, in plant eggs. Both kinds of cell division involve an increase in cell number, but in cleavage the cells become progressively smaller until a minimum cell volume characteristic of the cell type and species is attained. For example, in the mature sea urchin egg before cleavage, the volume of the cytoplasm is 550 times that of the nucleus, whereas at the end of the cleavage process (late *blastula* stage) the cytoplasm volume is only six times that of the nucleus in the individual cells (Brachet, 1950). During cleavage, DNA and (less extensively) cytoplasmic proteins are synthesized.

The mechanism of cleavage has been studied by several investigators. In some eggs, such as the newt egg (Selmanan and Waddington, 1955), the egg cortex is not pulled down into the cleavage furrow. Instead, two new cell membranes are formed between the blastomeres (see also the review of Dan and Kujiyama, 1963). Studies of Goodenough et al. (1968) indicate that the position of the first cleavage furrow is preformed. Electron microscopic preparations reveal numerous parallel filaments less than 100 Å in diameter intimately associated with the inner face of the plasma membrane in the region of the early cleavage furrow of the sea urchin egg. These filaments, which are probably made up of microtubules (Chapter 25), form a belt around the equator of the fertilized egg. A similar mechanism of cleavage by contraction of a preexisting band of 70 Å fibrils has been reported for the meroblastic squid egg *Loligo pealii* (Arnold, 1969). In view of these studies, it is probable that the pattern of cleavage, as well as that of egg cytoplasmic structure, is predetermined.

Cleavage Patterns

There are four distinct patterns of cleavage in animal eggs: biradial, radial, bilateral, and spiral. The first three types are illustrated in Figure 12-16.

A modified type of radial cleavage is found in the many fish, reptile, and bird eggs which are extremely *telolecithal* (Figure 12-15). Telolecithal eggs are those in which the yolk is concentrated at the vegetal pole. The incomplete cleavage of the egg results in formation of a cap of cells called a *blastoderm* on top of the large yolk mass (Figures 12-15 and 12-17).

Note that in parts A, B, and C of Figure 12-17 the cells are not yet complete; that is, they are not separate from one another but are connected by uncleaved surrounding protoplasm of the blastodisc, the *periblast*. Furthermore, as is shown in Figure 12-18, the furrows do not cut through the entire depth of the *blastodisc* (a relatively yolk-free area at the animal pole of the egg). Eventually a group of central cells is formed (Figure 12-17D and E) which are separated from one another by furrows but are still connected by the deeper protoplasm (Figure 12-18).

Further cleavage is irregular, but soon a mass of completely separate central cells is formed. This occurs when central cells in the upper surface are separated from the lower uncleaved cytoplasm by cleavage furrows parallel with the surface of the blastodisc. However, the central cells are still bounded by marginal cells which are connected via the periblast (Figure 12-17E). The population of central cells is increased by addition of cells cut off from the inner ends of the marginal cells, as well as by division of the central cells themselves. As cleavage continues, the blastoderm, along with the marginal cells, increases in diameter as it spreads over the yolk mass. The marginal cells soon cease production of central cells; their nuclei, which continue to divide, move out into the periblast which thus becomes a *syncytium*.

The nuclei in the periblast divide rapidly, and some of them migrate into the uncleaved protoplasmic region below the blastodisc (Figure 12-18). The exact role played by the periblast in expansion of the blastoderm over the yolk is not clear. It has been suggested that cells which are cut off from the periblast are added to the margin of the blastoderm (Hamilton, 1952). However, this particular aspect of development of the egg requires more study.

In birds and reptiles, cleavage leading to formation of the blastoderm occurs while the egg passes down the oviduct of the mother. Also, *albumen*, shell membranes, and a shell are deposited around the egg during this period.

Spiral cleavage, which occurs in mulluscs, anne-

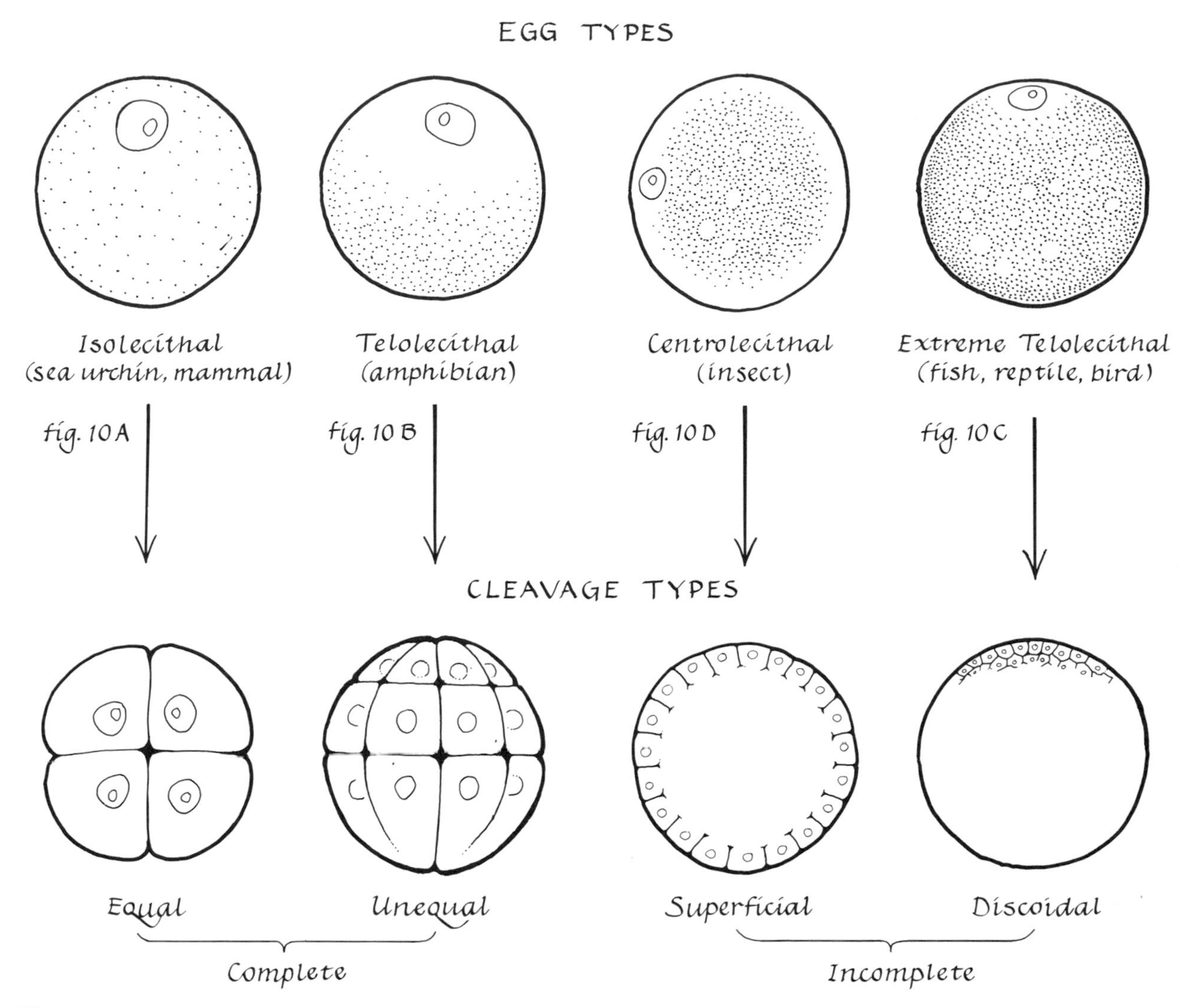

Figure 12-15. *Diagrams of egg types and corresponding cleavage types. Stippling indicates distribution and amount of yolk.*

lids, and nemerteans (Conklin, 1917; Costello, 1948), is a particularly complex type of cleavage (Figure 12-19).

As shown in the figure, the cleavage spindles in each of the four *macromeres* of the four-cell stage oscillate alternately to the right and left and right in cutting off the successive quartets of *micromeres* (1d, 2d, 3d, 4d; 1a, 2a, 3a, and so on) at the animal pole (Figure 12-19E). The spindles in the quartets of micromeres also oscillate to right and left in forming additional micromeres (Figure 12-19D). As cleavage proceeds, more rapid division at the animal pole *(An)* results in formation of a cap of many micromeres on top of a few slowly cleaving macromeres (Figure 12-19F). As we shall see in Chapter 13, the particular pattern of spiral cleavage in snail eggs is correlated with the right- or left-handed coiling of the shell.

The geometrical regularity of the cleavage pattern in many eggs is impressive, especially in early stages. This feature in tunicates, echinoderms, annelids, and molluscs has made it possible to study cell lineage in detail—that is, to trace the fate of early blastomeres and their division products in forming later embry-

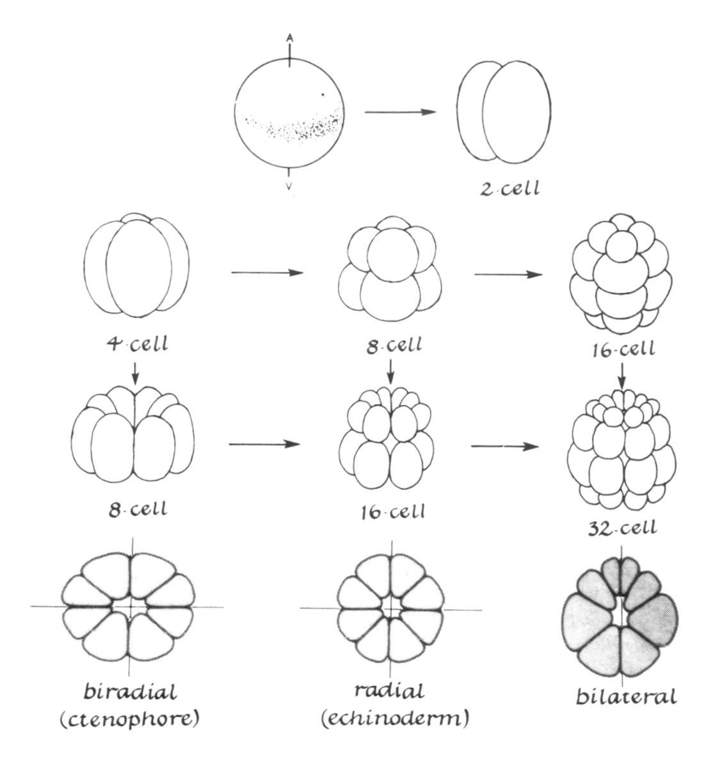

Figure 12-16. *Cleavage patterns. First three rows, side view; bottom row, sections viewed from animal pole. From* The Science of Zoology *by Weisz. Copyright 1966 by McGraw-Hill Book Company. Used with permission of McGraw-Hill Book Company.*

onic or larval organs or parts (Conklin, 1905, 1917; Wilson, 1925; review of Costello, 1948). The regularity of cleavage in many kinds of animal eggs raises the problem of understanding the mechanism of its control. Recall that we noted the same problem in regard to the control of cell division patterns in colonial algae. Whatever the mechanism, it is clear that it is part of the information inherited by the egg or other reproductive unit.

Influences of egg size, distribution and amount of

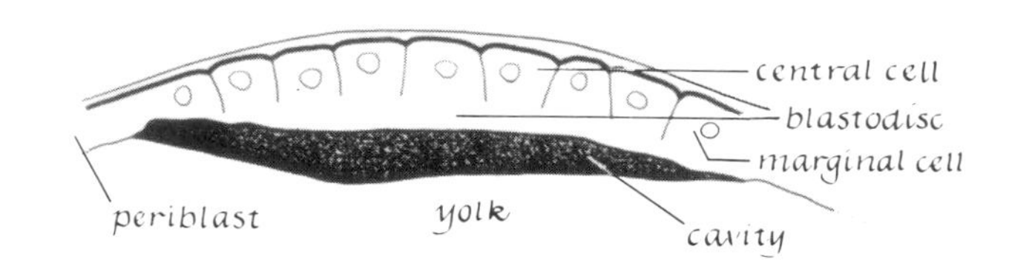

Figure 12-18. *Diagram of a transverse section through the blastodisc of a telolecithal egg showing incomplete cleavage.*

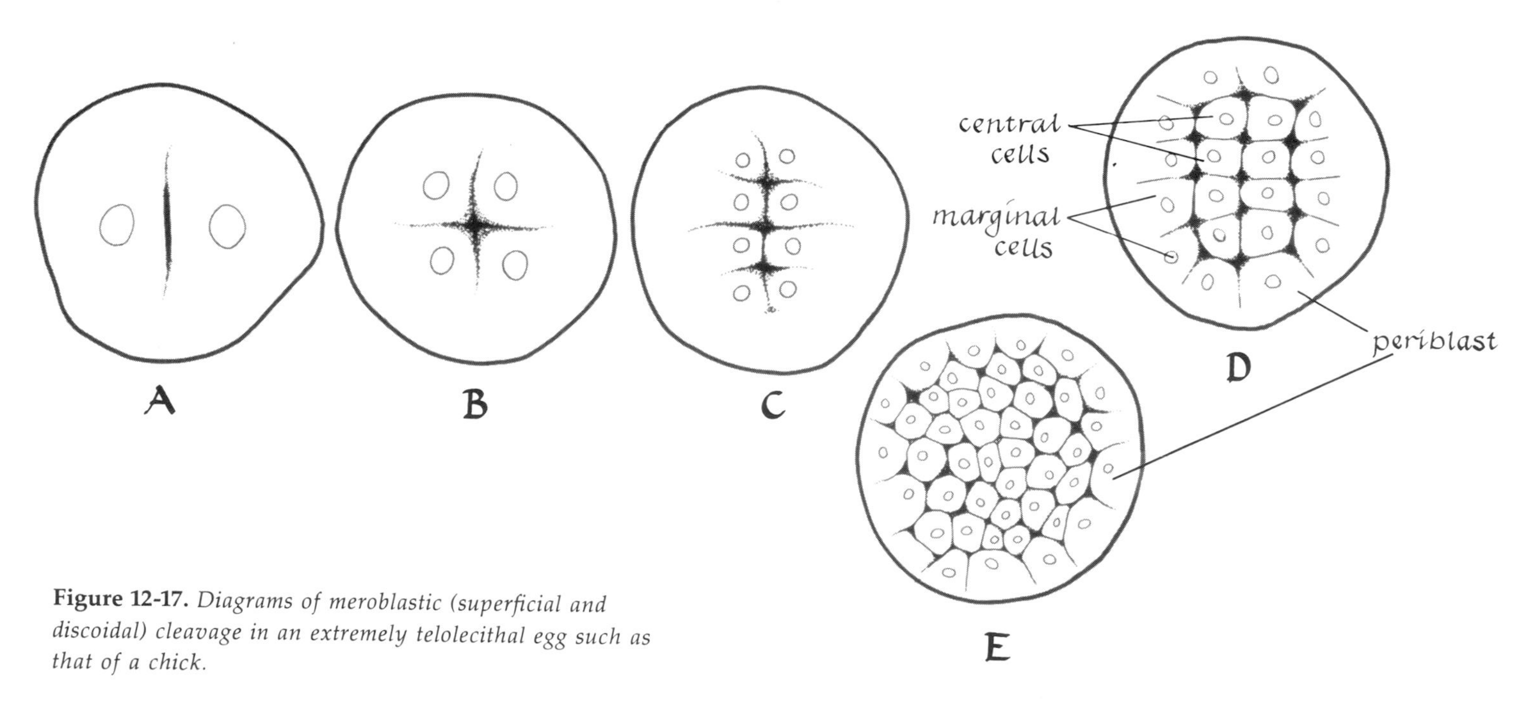

Figure 12-17. *Diagrams of meroblastic (superficial and discoidal) cleavage in an extremely telolecithal egg such as that of a chick.*

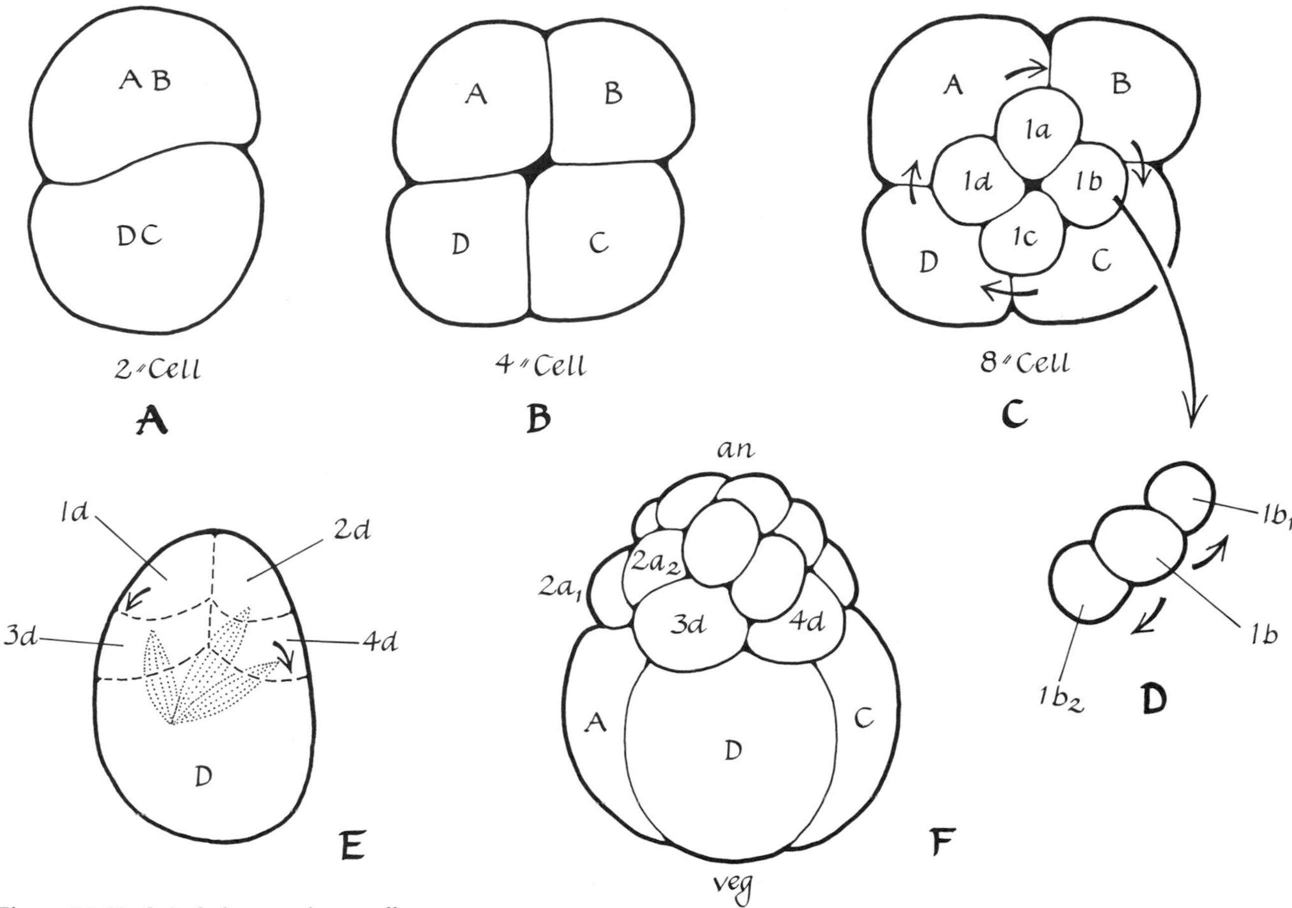

Figure 12-19. *Spiral cleavage in a mollusc egg.*

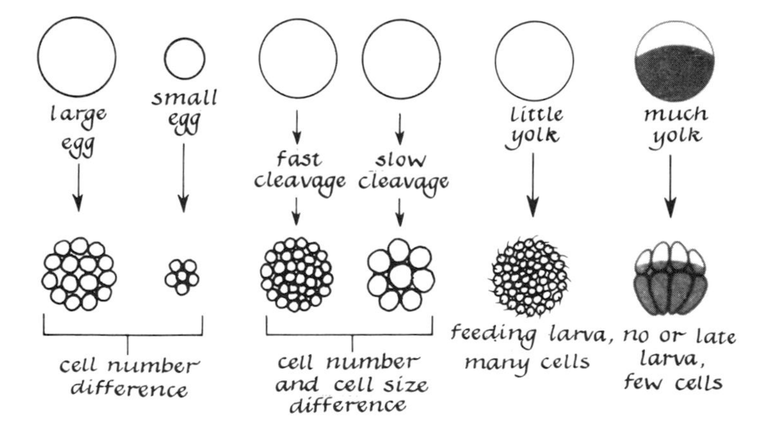

Figure 12-20. *The effects on development of initial egg sizes, cleavage rates, and amount and distribution of yolk. From* The Science of Zoology *by Weisz. Copyright 1966 by McGraw-Hill Book Company. Used with permission of McGraw-Hill Book Company.*

yolk, and cleavage rates upon subsequent development have been pointed out by Weisz (1960) and are summarized in Figure 12-20.

The rate of cleavage also significantly influences whether the later stages of development are primarily capable or incapable of repairing experimentally imposed defects. Generally, the more rapid the rate, the more capable is the embryo of repairing defects. The size and yolk content of an egg seem to be correlated with whether or not a larval stage occurs.

Blastula Patterns

Cleavage results in a multicellular stage called a *blastula*, which typically is a hollow sphere bounded by a single layer of cells. There are many variations of this basic theme throughout the animal kingdom. Some of the well-known blastula patterns, which

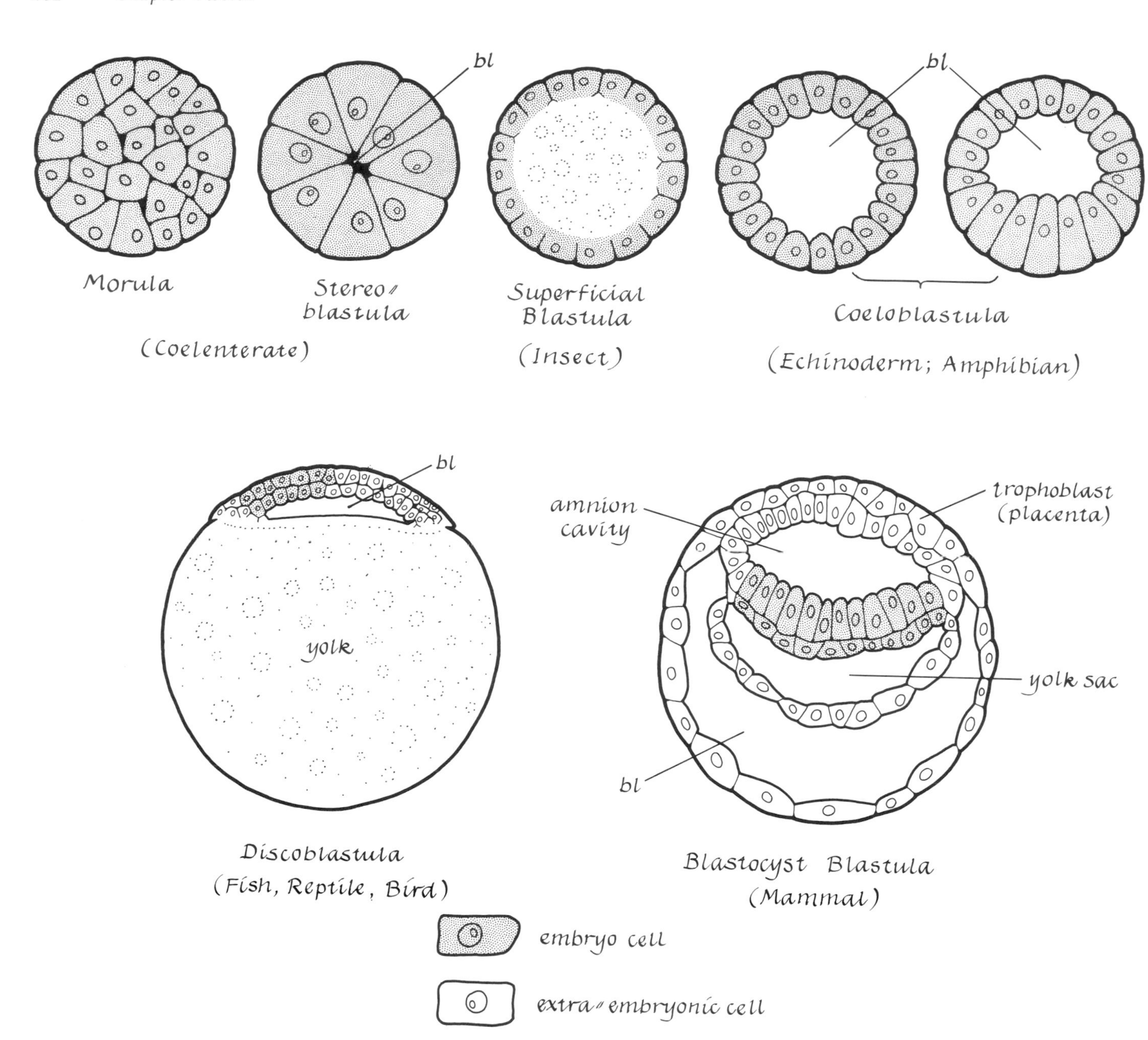

Figure 12-21. *Blastula patterns.*

seem to be largely a consequence of the type of egg and pattern of cleavage, are illustrated in Figure 12-21.

Gastrula Patterns

The word gastrulation means formation of a little stomach, that is, a gut, the primordium of a digestive tube. In modern embryology, it has come to mean the formation of the *germ layers,* the primordial tissues of the embryo. As one might expect from what we have noted about the influence of egg type on cleavage pattern and the influence of cleavage and yolk content upon blastula pattern, patterns of gastrulation vary among different kinds of animals. Some general patterns of gastrulation are shown in Figure 12-22.

As indicated in Figure 12-22, a multilayered archi-

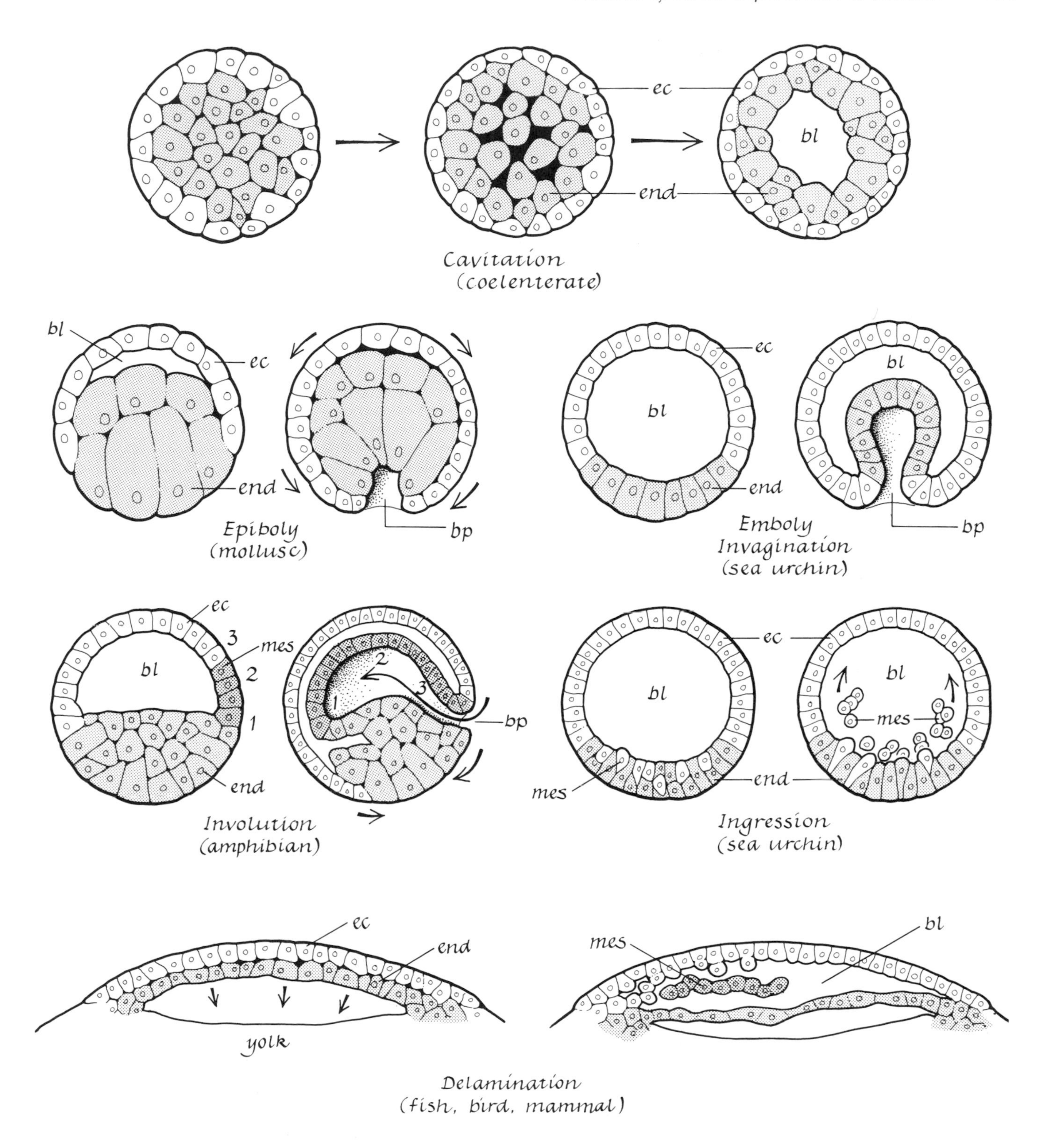

Figure 12-22. *Gastrula patterns and the methods of gastrulation.*

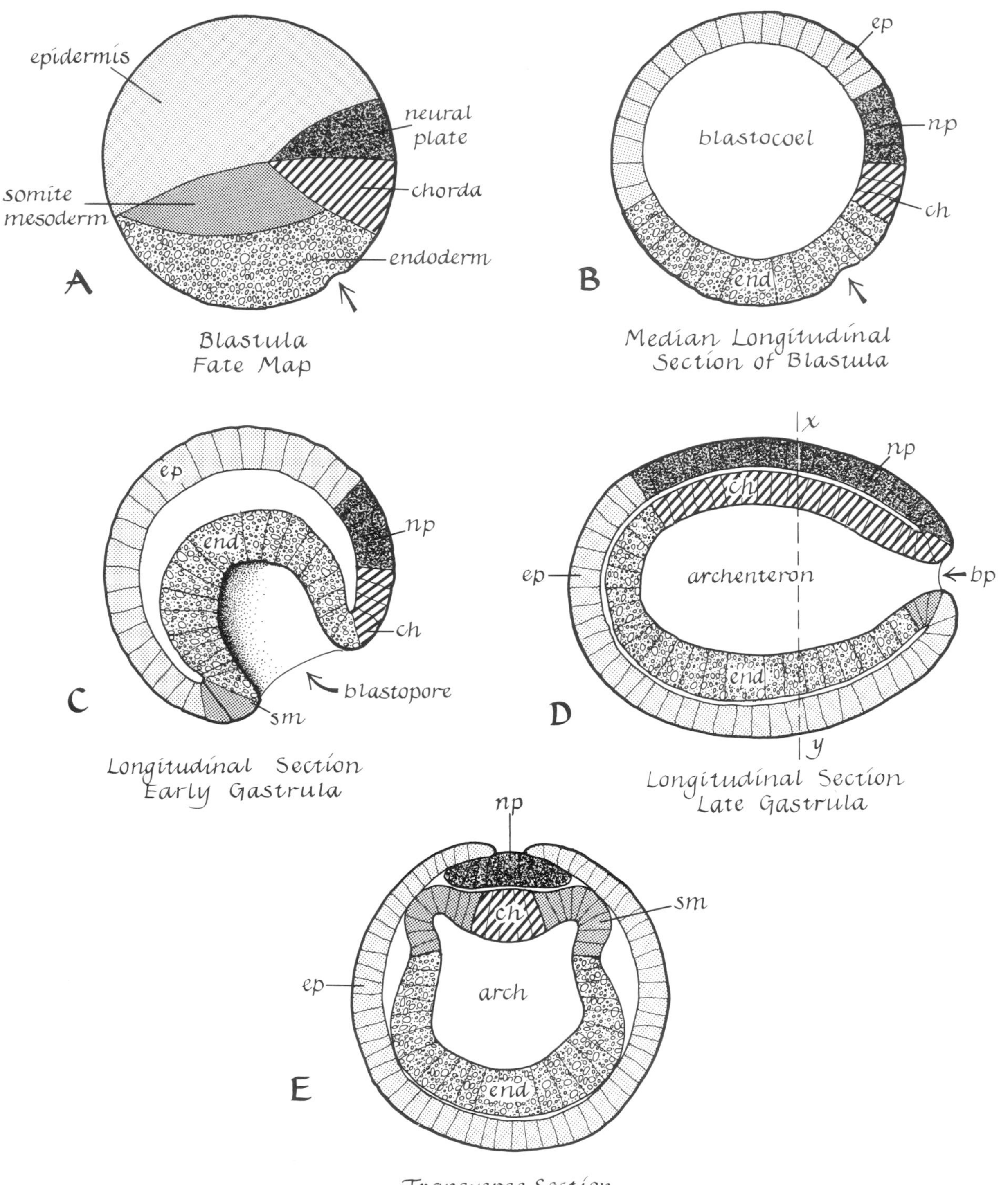

Figure 12-23. *Diagrams of germ layer formation in protochordates. See text for explanation of symbols used in this figure and also in figures 12-24, 12-25, 12-26, 12-27, and 12-37.*

tecture can be attained in several ways. *Cavitation* is the hollowing out of an originally solid sphere of cells. *Epiboly* is the spreading of smaller cells at one pole of the blastula (the animal pole, *An*) over larger cells at the opposite pole (the vegetal pole, *Veg*). Epiboly occurs in mollusc, annelid, and amphibian eggs. *Emboly (invagination)* is a pushing of the vegetal pole region into the cavity of the blastula, the *blastocoel* (bl, Figure 12-22). *Involution* is the rolling of surface areas at the animal pole into the blastocoel. *Ingression* is the movement of individual cells from the vegetal pole into the blastocoel. *Delamination* is the separation of cellular layers which have been produced by vertically oriented mitotic spindles. Both cell division and cell or tissue movements are involved during gastrulation, but some questions still remain regarding the relative roles played by these basic component processes.

The significance of the layered type of architecture in contrast to the *morula* type (Figure 12-21) resides in the fact that not only is a degree of cellular isolation conducive to cell specialization attained, but also that different kinds of cells are brought into closer relationship by morphogenetic movements which favor interaction and further cell specialization.

Methods of Germ Layer Formation in Vertebrates

Although the germ layers of embryos are formed in different ways (Figure 12-22), most of the ways in which vertebrates form germ layers can be classified into the basic types diagramed in Figures 12-23 through 12-27 and Figure 12-37. In all these figures the following symbols are used to denote the germ layers and/or their major subdivisions at the end of gastrulation:

ep = epidermis (skin)
np = neural plate (primordium of the nervous system)
ch = chorda (the notochord characteristic of chordates)
m = mesoderm
sm = somite mesoderm
lm = lateral mesoderm
end = endoderm

In all of these figures, comparable stages are denoted by the letters A through E. When a *blastopore* (an opening connecting the *blastocoel* with the outside of the blastula) is known to be present, its position is indicated by an arrow. In order to provide an overview of the comparative processes, only the most general features of germ layer formation are illustrated. The diagrams are largely based on the results of vitally staining or systematically marking cells on the surface of the blastula stage with various nontoxic agents. By this means a map of the normal fate of the surface cells can be constructed. More detailed descriptions can be found in a number of excellent texts on animal embryology: *Lillie's Development of the Chick*, revised by Hamilton, 1952; *Developmental Anatomy*, Arey, 1965; *Morphogenesis of the Vertebrates*, Torrey, 1967; *Comparative Embryology of the Vertebrates*, Nelsen, 1953; *Foundations of Embryology*, Patten, 1964.

Protochordate Type (Amphioxus)

Amphioxus produces an isolecithal egg which undergoes complete cleavage resulting in a *coeloblastula* (a hollow-ball type of blastula) (Figures 12-21 and 12-23).

Gastrulation is accomplished by invagination of the endoderm area at the vegetal pole and involution of the chorda over the dorsal lip of the blastopore (Figure 12-23C and D). Mesoderm adjacent to the chorda is involuted around the lateral lips of the blastopore. The *somites* (blocks of mesoderm) arise from mesoderm pouches (sm) lateral to the chorda (Figure 12-26E). The archenteron later becomes a closed endodermal tube by growth and fusion of its dorso-lateral edges.

Primitive Fish—Amphibian Type: Lungfish, Sturgeon, Cyclostome, Frog, Salamander

These animals produce moderately telolecithal eggs which undergo complete but unequal cleavage resulting in a coeloblastula with larger cells at the vegetal pole (Figure 12-21). The blastocoel is thus displaced toward the animal pole (Figure 12-24B). As in *Amphioxus*, gastrulation is accomplished by invagination of a small portion of endodermal area just ventral to the chorda area, followed by involution of the chorda, somite, and lateral mesoderm around the dorsal and lateral lips of the blastopore. The result is a cup of mesoderm composed of chorda,

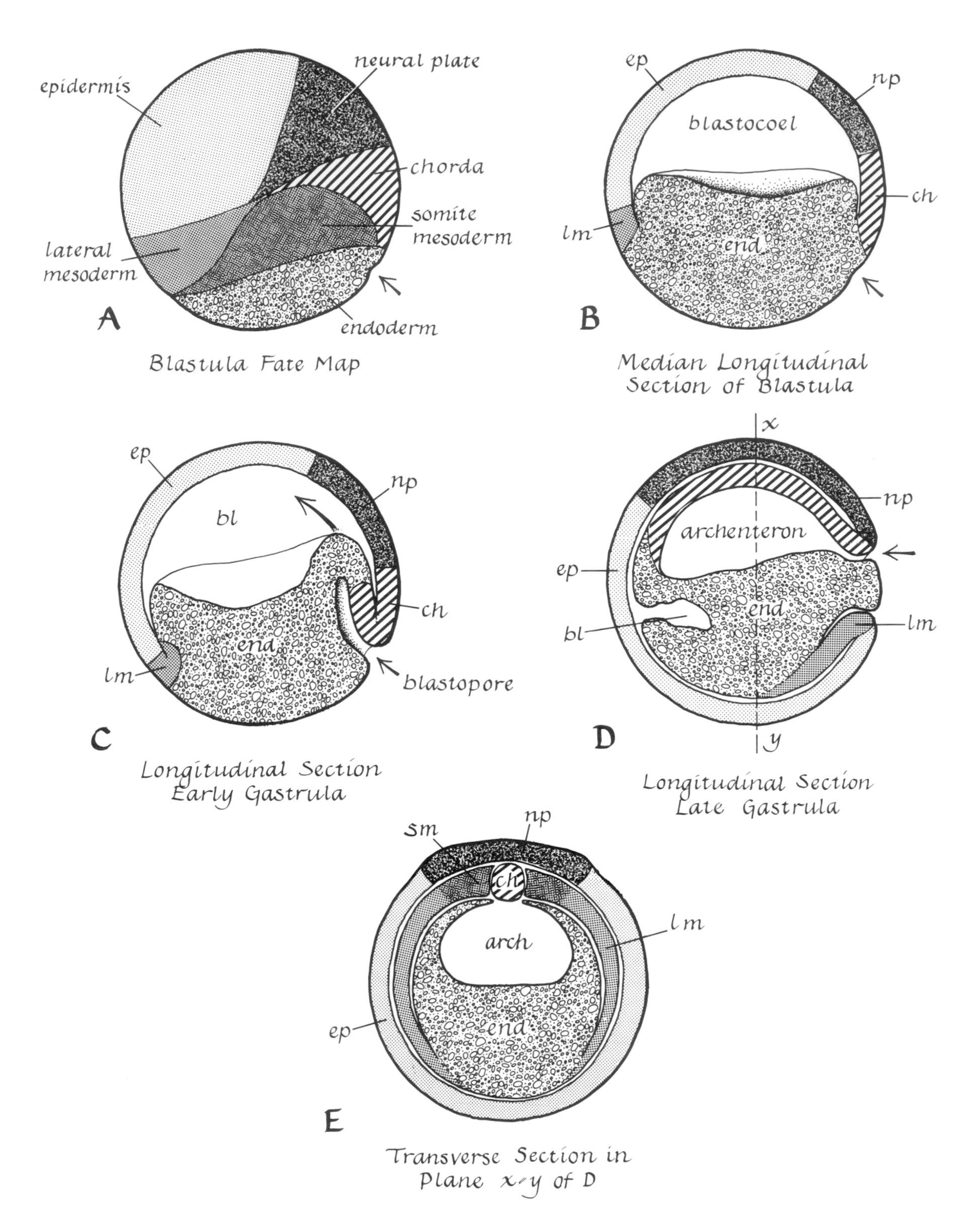

Figure 12-24. *Diagrams of germ layer formation in primitive fishes and most amphibians.*

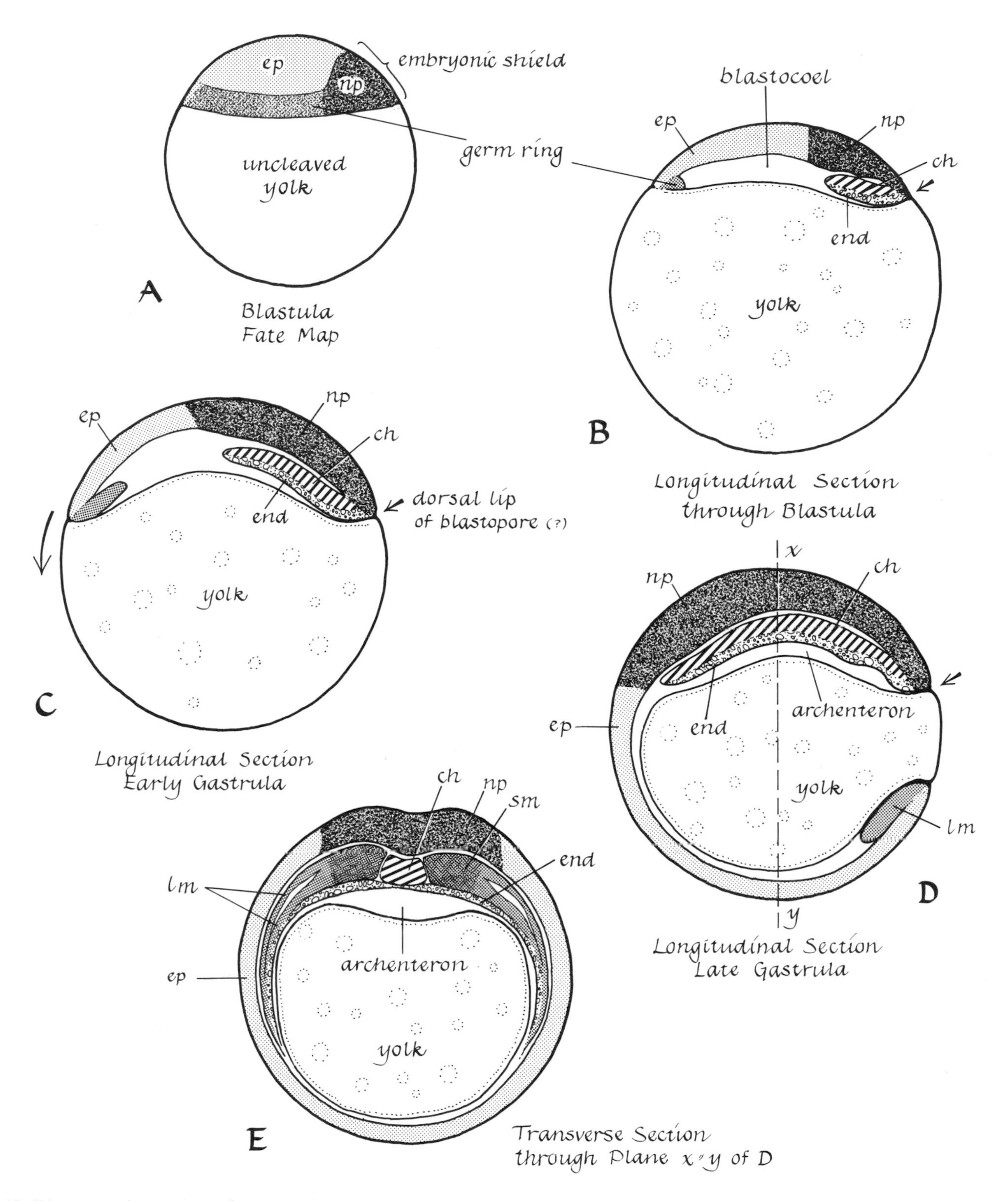

Figure 12-25. *Diagrams of germ layer formation in elasmobranches and advanced fishes.*

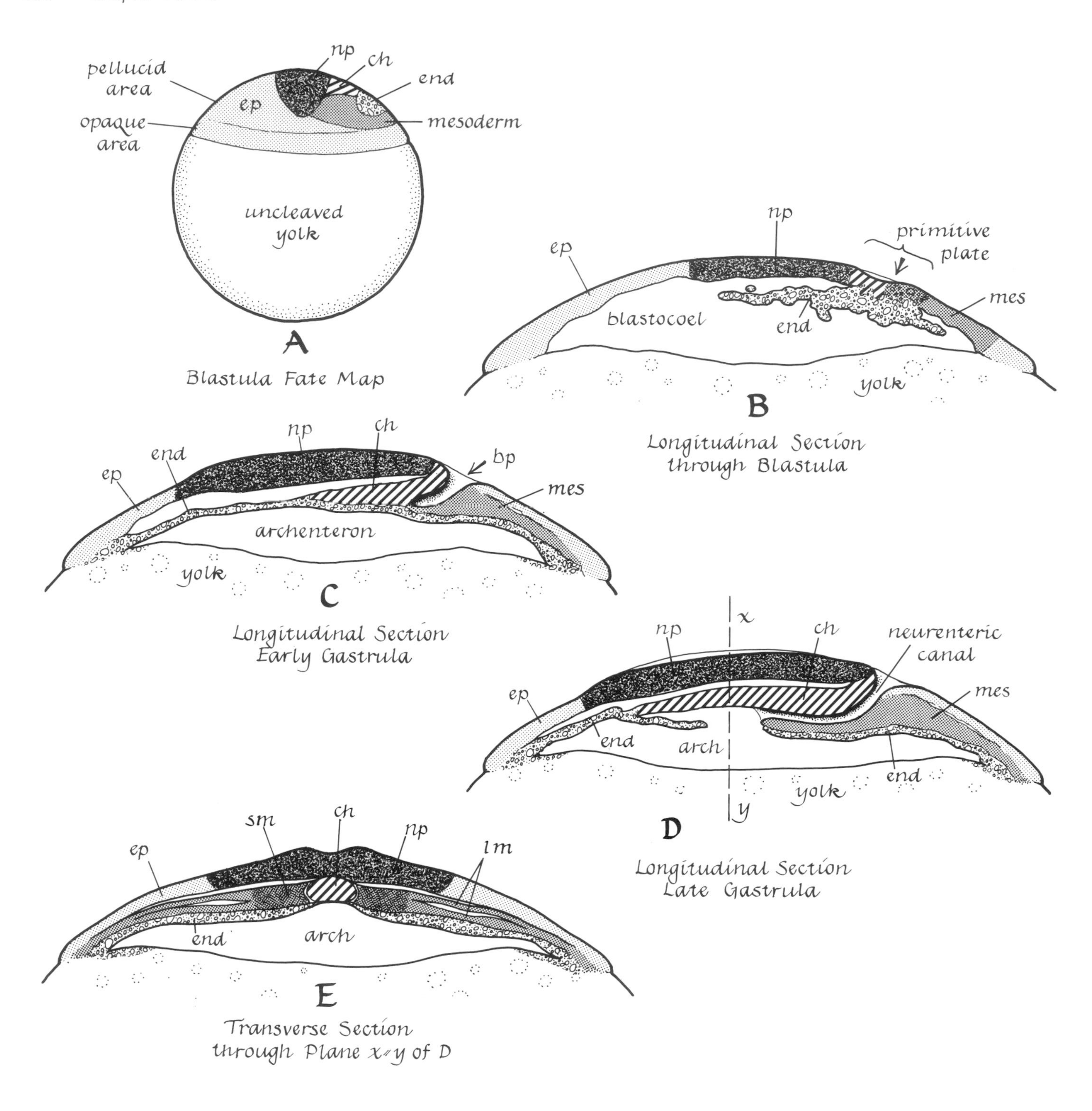

Figure 12-26. *Diagrams of germ layer formation in reptiles and some primitive birds.*

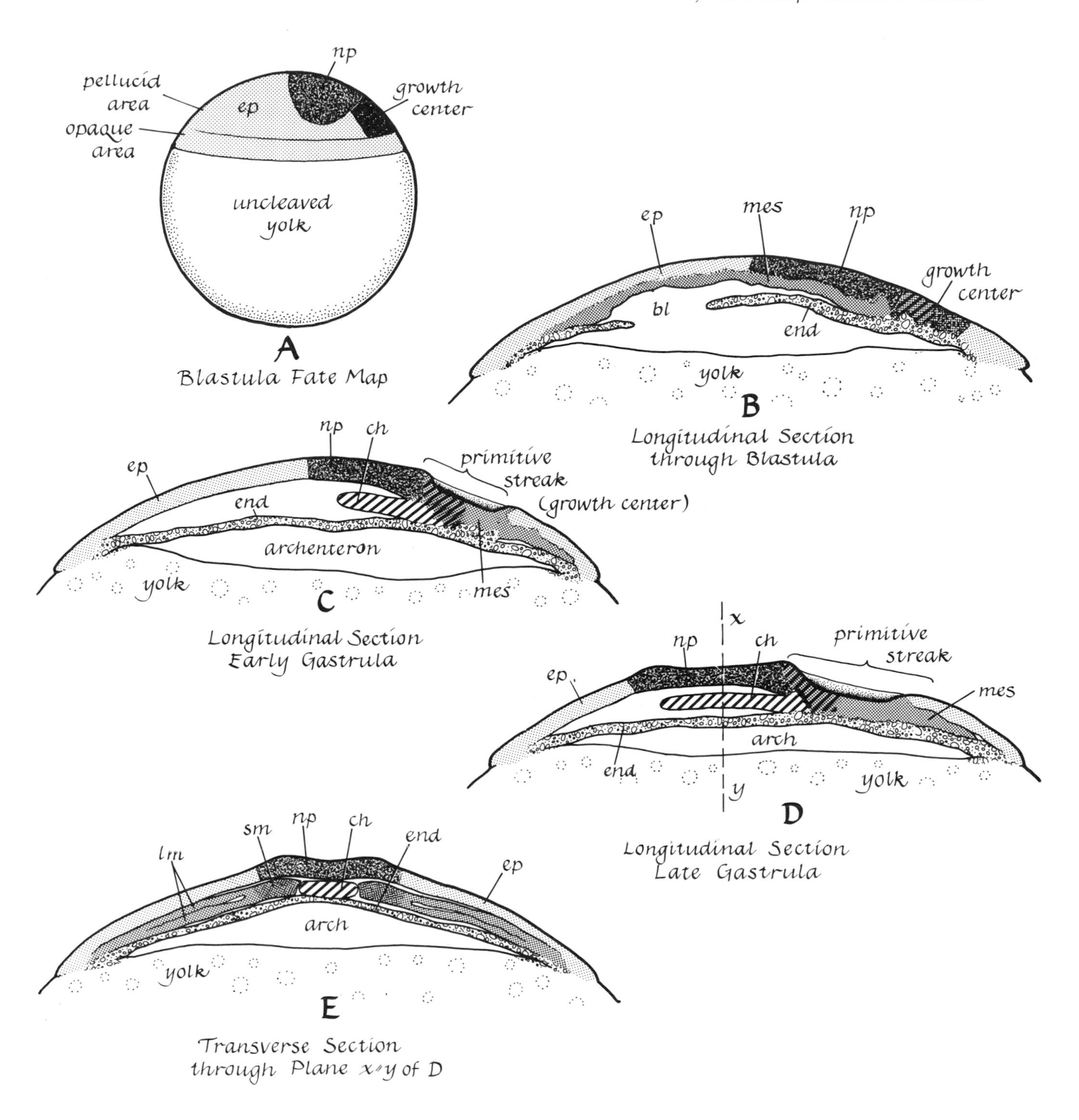

Figure 12-27. *Diagrams of germ layer formation in advanced birds, such as chick and duck.*

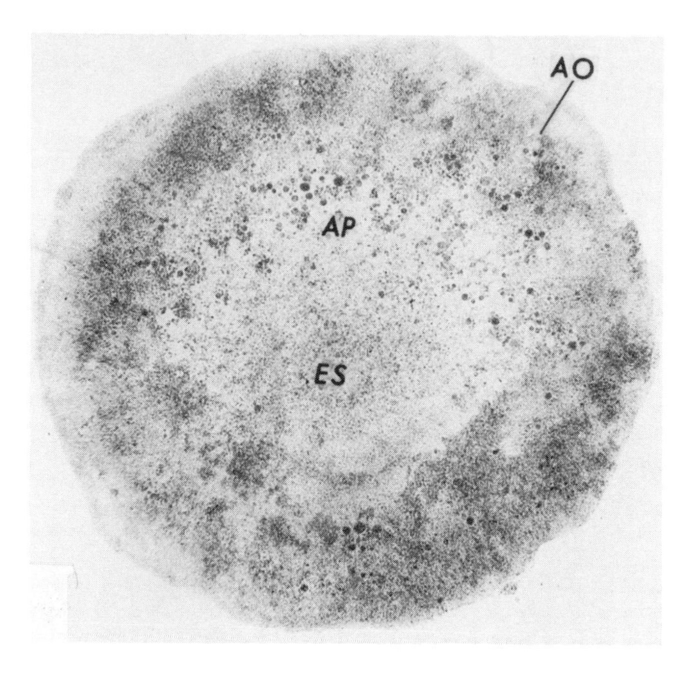

Figure 12-28. *Photograph of a fixed and stained unincubated chick blastoderm.* AO, *opaque area;* AP, *pellucid area;* ES, *embryonic shield. From Spratt,* J. Exp. Zool. *Volume 120, p. 110, 1952. Courtesy Wistar Press.*

somite, and lateral mesoderm inverted over the endoderm, which in this case is an upward-facing cup mostly filled with yolk-rich cells (Figure 12-24E). Later the rim of the cup grows below the chorda to form the roof of the archenteron. The somites arise as blocks of mesoderm lateral to the chorda. Still more laterally, the lateral mesoderm grows down between the endodermal cup and the outer epidermis. Eventually a three-layered embryo is formed; it consists of (1) an endodermal tube (the archenteron) enclosed within (2) a mesodermal tube composed dorsally of chorda and somite cells and laterally and ventrally of lateral mesoderm cells; and (3) an ectodermal tube, composed dorsally of the neural plate and laterally and ventrally of epidermis cells, which encloses both endoderm and mesoderm.

Elasmobranch — Advanced Fish Type: Shark, Teleosts

In the extremely telolecithal eggs formed by elasmobranchs and bony fishes, cleavage is meroblastic; this results in a modification of the methods of gastrulation found in the preceding types and the formation of a blastula which is not a hollow sphere but a blastoderm on top of the yolk. The blastoderm consists of a marginal region several cell diameters thick called the *germ ring* and a thick posterior region, the *embryonic shield* (Figure 12-25). The blastula consists of a coherent upper layer composed of prospective epidermis and neural plate cells, an incomplete mesodermal layer comprised of a small group of notochord and lateral mesodermal cells, and an incomplete endodermal layer. Germ layers are thus formed by delamination (Figure 12-22), but whether or not there is any blastopore through which surface cells are invaginated remains to be definitely demonstrated. The blastoderm eventually expands over and around the uncleaved yolk mass.

Reptile — Primitive Bird Type: Turtle, Emu, Puffin, Goose

These animals, like the preceding type, produce extremely telolecithal eggs which undergo meroblastic cleavage resulting in formation of a blastoderm on top of the yolk (Figure 12-26). Note in the figure that the fate map at the blastula stage is similar to that in protochordates, primitive fishes, and amphibians; that is, in the two maps the prospective areas in the surface of the blastula are distributed similarly (Figure 12-26A and B). Prospective chorda and some endoderm cells are apparently invaginated through a blastopore (Figure 12-24C). Later, a *neurenteric* canal develops and connects the space above the blastoderm (later enclosed by the folding of the neural plate to form a neural tube) with the archenteron (Figure 12-26D). The neurenteric canal later closes. The canal's significance remains obscure.

Advanced Bird Type (Duck, Chick)

The extremely telolecithal eggs of the chick and duck undergo meroblastic cleavage resulting in formation of a blastoderm four or five cell diameters thick (Figure 12-27B). The upper layer of the blastoderm contains epidermis cells and neural plate cells; behind these is a region which appears to function as a growth center (Figure 12-27B). Vertically oriented mitotic spindles found in this area indicate that cells are probably added to the incompletely formed middle and lower layers. However, completion of the middle and lower layers also involves cell movements which will be described soon.

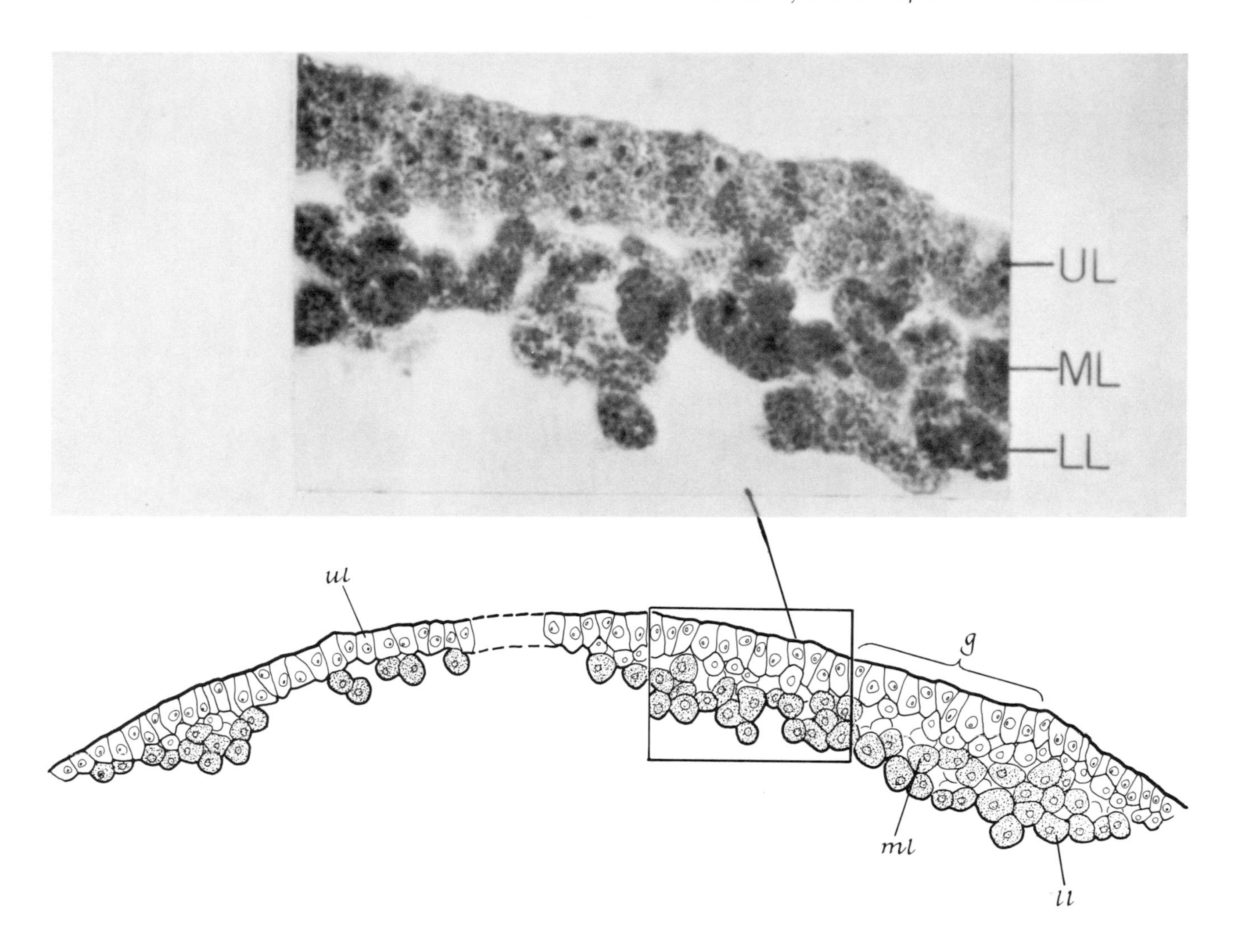

Figure 12-29. *Fraction of a median longitudinal section of an unincubated chick blastoderm, the general morphology of which is illustrated by the diagram below the photograph.* ml, *middle layer;* ul, *upper layer;* ll, *lower layer;* g, *growth center;* es, *embryonic shield. From Spratt and Haas,* J. Exp. Zool., *Volume 158, p. 15, 1965. Courtesy Wistar Press.*

At the time of egg laying, the blastoderm has reached a late blastula or early gastrula stage containing between 20,000 and 60,000 cells. A photograph of a stained, unincubated chick blastoderm is shown in Figure 12-28.

A photograph of a portion of a longitudinal section through a similar unincubated chick blastoderm, along with a diagram showing its general morphology, is shown in Figure 12-29.

During the first eight to ten hours of incubation, the sheet of middle and lower cells spreads in an anterior direction until it completely covers the underside of the pellucid area. Marking the spreading sheet with colored particles or vital dyes (Spratt and Haas, 1965) reveals that the pattern of movement of cells in the sheet is similar to that of water in a fountain. As a result of the movement of the double-layered sheet, the middle (mesodermal) and lower (endodermal) germ layers are formed as coherent sheets below the upper (ectodermal) layer. Some cells are also added to the middle and lower germ layers by proliferation and centripetal movement of cells in the remainder of the ring. The pattern of movements by which the middle and lower germ

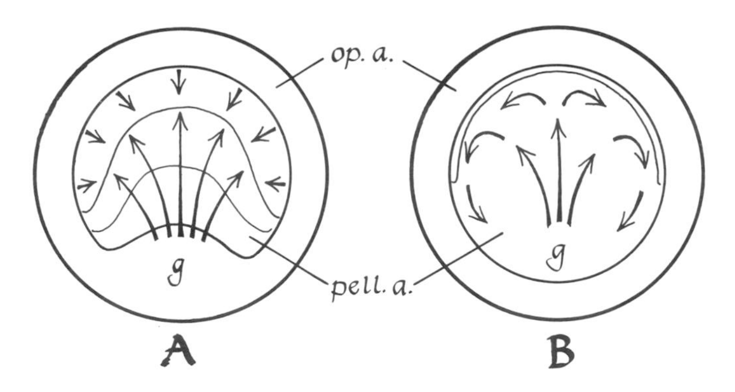

Figure 12-30. *Diagrams of the movements by which the middle and lower layers are completed during the first 8-10 hours of incubation of the chick blastoderm.*

layers are completed is shown in Figure 12-30.

As indicated in the figure, where the centrifugally expanding sheet of cells meets the centripetally moving cells from the ring, movement in the sheet becomes circumferential. This region is essentially the boundary between the pellucid and opaque areas (see also Figure 12-28).

Bird and mammal embryos contain a structure called the *primitive streak* (Figure 12-31). The streak first appears after about eight hours of incubation as a triangular-shaped group of middle-layer cells in the region from which the middle and lower-layer cells have been spreading out (*g* in Figures 12-29 and 12-30). The streak apparently forms because cells are being added to the middle layer faster than the middle-layer cells are spreading out. The result is a densely packed group of cells (Figure 12-31).

During the next 10 to 15 hours of incubation the streak elongates by addition of cells to its anterior and posterior ends and by addition all along its length. After about twenty hours of incubation the streak is completely formed; it is a sharply delineated rod-shaped structure lying in the midline of the pellucid area, which is now pear shaped (Figure 12-32).

During elongation of the streak, mesoderm cells spread out laterally and anteriorly from its sides. The streak also appears to be a source of cells which are added to the ectoderm by cell proliferation in the ectodermal layer of the streak, and of cells which are added to the endoderm. Mitotic index studies indicate that continuous and relatively greater cell proliferation occurs in the primitive streak than in its

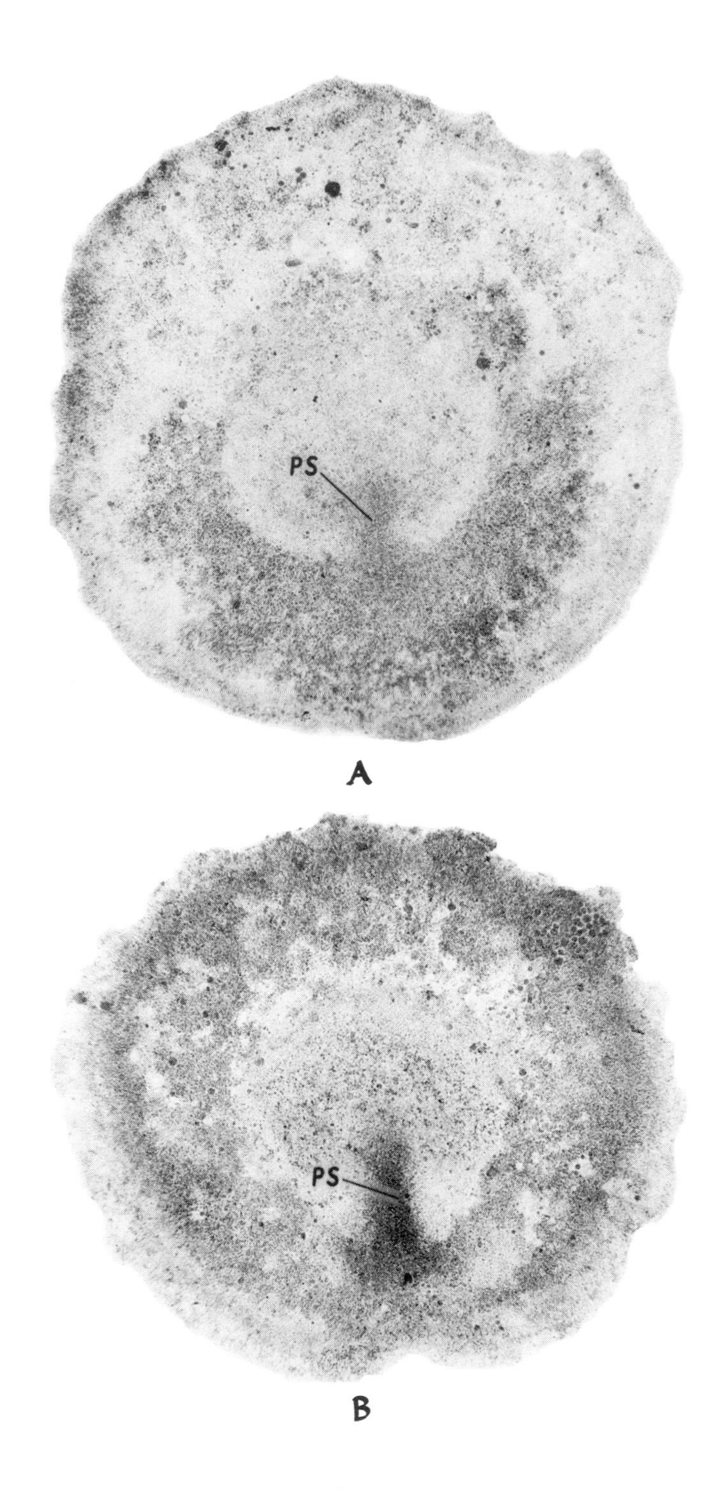

Figure 12-31. *Photographs of fixed and stained early primitive streak chick blastoderms. A, slightly younger than B.* PS, *primitive streak. From Spratt,* J. Exp. Zool., *Volume 120, p. 110, 1952. Courtesy Wistar Press.*

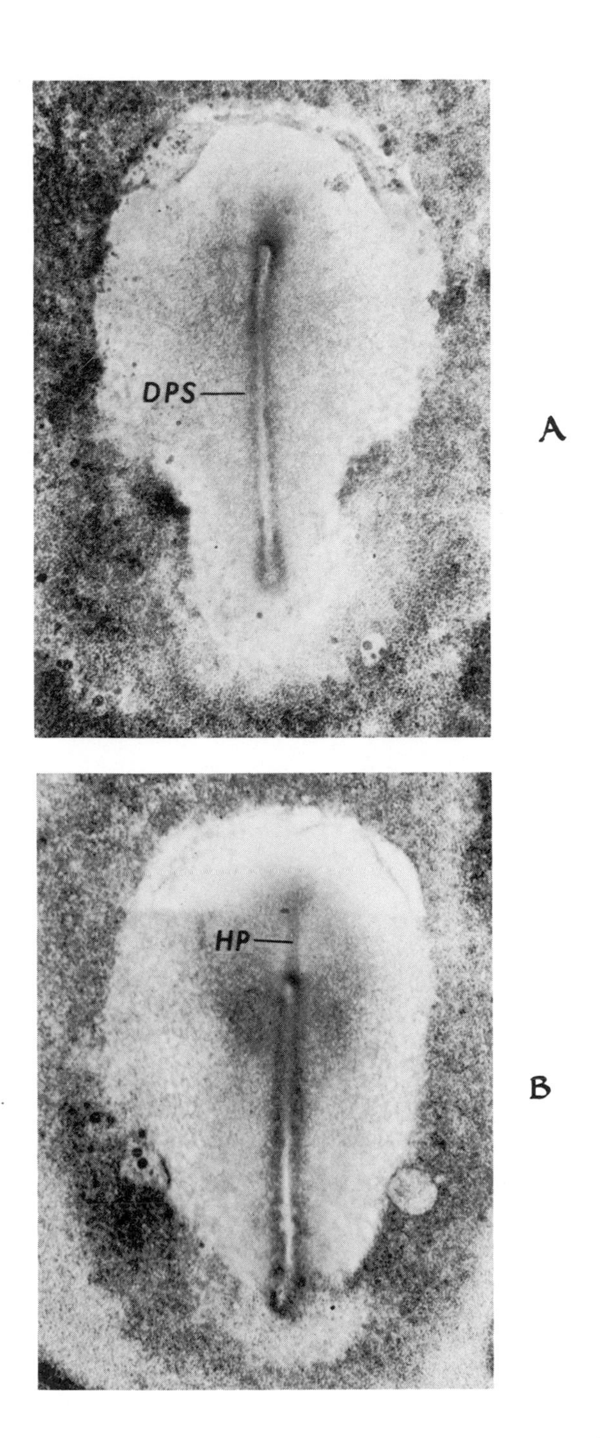

Figure 12-32. *Photographs of fixed and stained pellucid area portions of Hamburger-Hamilton stage 4 and 5 chick blastoderms. A, full length primitive streak; B, regressing primitive streak with notochord extending from its anterior (node) end (see also Figure 12-34). From Spratt,* J. Exp. Zool., *Volume 120, pp. 122 and 123, 1952. Courtesy Wistar Press.*

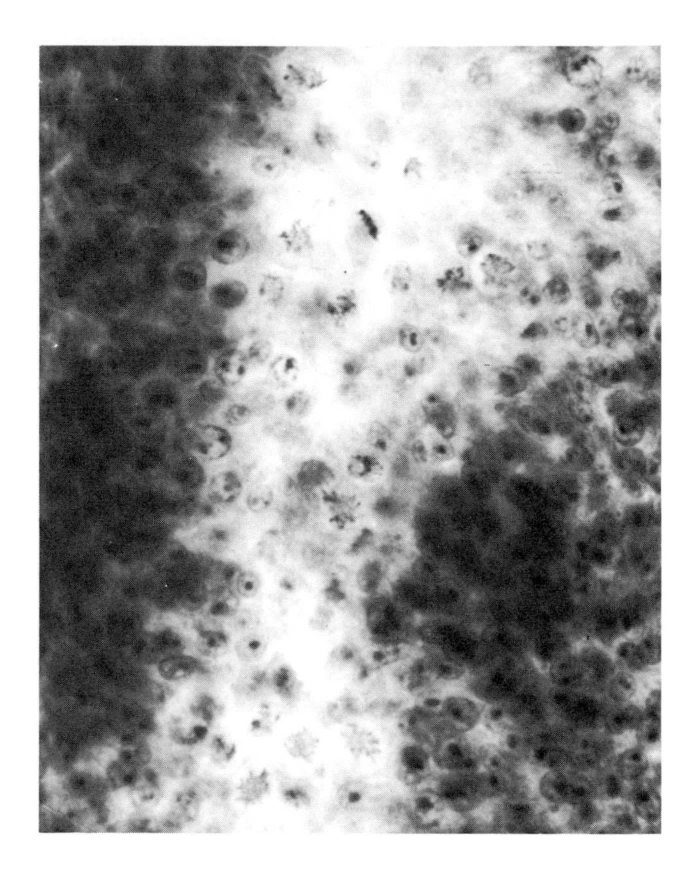

Figure 12-33. *Photograph of the anterior end of the primitive streak of the blastoderm illustrated in Figure 12-32B. Note the vertical position of most of the metaphase spindles.*

surrounding areas (Emanuelson, 1961; Spratt and Haas, 1965; Spratt, 1966). The streak thus functions as a source of cells which are added to all three germ layers. Figure 12-33 is a photograph of the anterior end of the streak shown in Figure 12-32B. It shows that most of the metaphase spindles are vertically oriented (Spratt, 1971).

After about twenty hours of incubation, the streak begins to move in a posterior direction as the embryo body axis progressively forms and elongates anterior to the streak. This movement, called regression, is a complicated process, involving an actual displacement in space and shortening of the streak by dissolution at its posterior end. At first glance the embryo body appears to form in front of rather than out of the regressing streak. However, experiments in which all layers of the streak were marked with vital

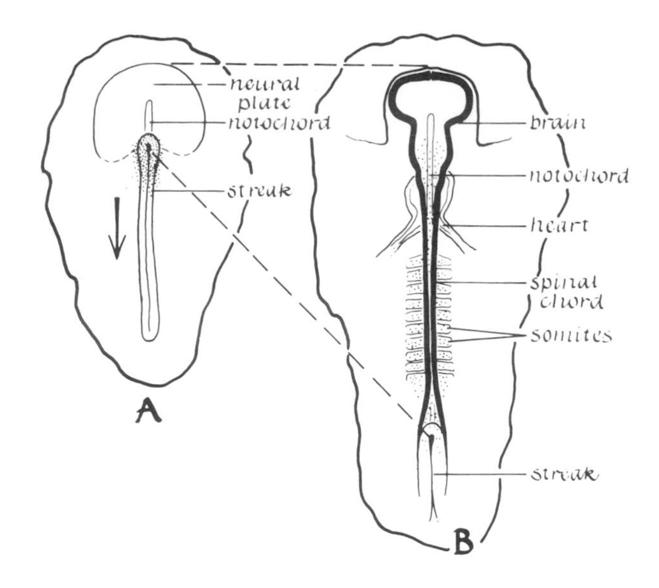

Figure 12-34. *Diagram of the regression of the primitive streak and formation and elongation of the embryo body axis. A, beginning of regression; B, continued regression of the streak.*

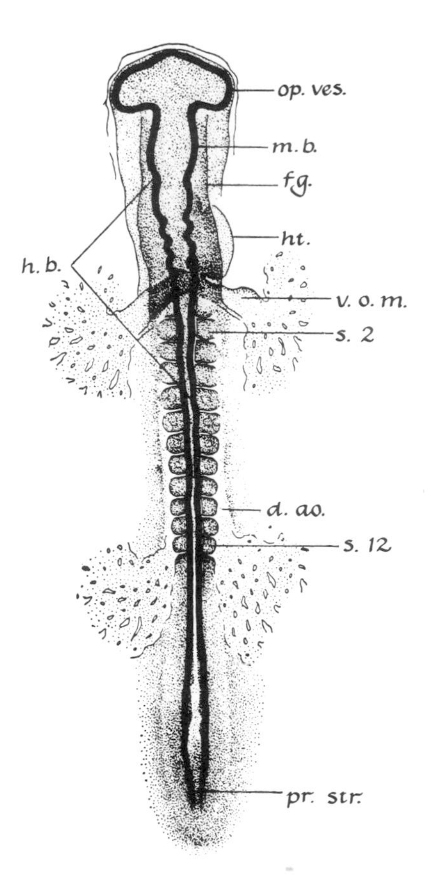

Figure 12-35. *Dorsal view of a 12-somite embryo drawn as a transparent object with transmitted light.* op. ves., *optic vesicle;* m.b., *midbrain;* f.g., *foregut;* ht., *heart;* h.b., *hindbrain;* v.o.m., *vitelling vein;* s.2, *second somite;* d.ao., *dorsal aorta;* s.12, *twelfth somite;* pr.str., *primitive streak. From Lillie's* Development of the Chick, *3rd edition, Rev. by Hamilton, copyright 1952 by Holt, Rinehart and Winston, Inc.*

dyes—formazan, carbon, or carmine particles—indicate that axial organs (spinal cord, notochord, somites) are formed out of cells "left behind" as the streak regresses (Spratt and Haas, 1965). This is diagramed in Figure 12-34.

Note in the figure that stained or marked cells in the anterior end of the streak (stippled region in Figure 12-35A) are incorporated into the notochord, spinal cord, and somites as the streak regresses. After forty to fifty hours of incubation the anterior end of the streak becomes incorporated into the *tail bud* (Figures 12-35 and 12-36A), which continues to regress and contribute cells to the axial organs. Regression of the streak and its contribution to the embryo body may be likened to the drawing of a line with a piece of chalk. The piece of chalk would correspond to the streak; the line would correspond to the axis of the embryo.

Later stages in development of the chick are illustrated in Figure 12-36B (Hamburger and Hamilton, 1951).

Mammal Type: Human

Most mammal eggs, including those of the human, are isolecithal and undergo complete cleavage resulting in formation of a blastocyst type of blastula (Figures 12-21 and 12-37). The *blastocyst*, a special adaptation for intrauterine development, consists of (1) an outer vesicle of extraembryonic cells, the *trophoblast*, enclosing (2) a fluid-filled vesicle, the *amnion*, below which is (3) a fluid-filled *yolk sac*. The blastoderm (embryonic disc) lies between the amnion and the yolk sac. The upper layer of the blastoderm is continuous with the amnion, and the

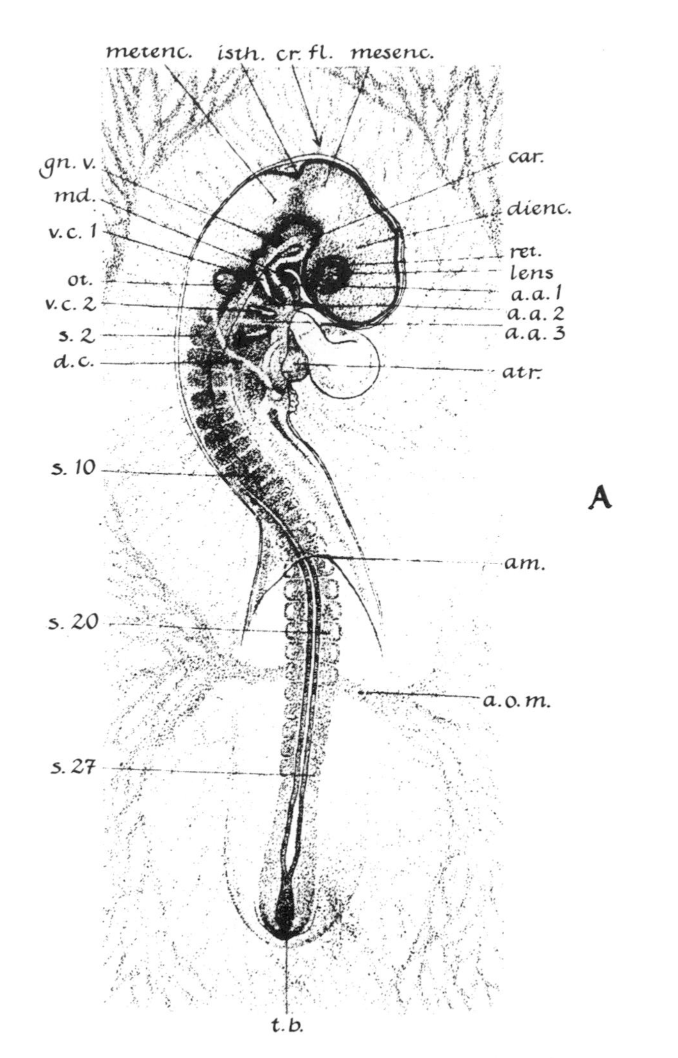

lower layer is continuous with the yolk sac (Figure 12-37A and B). The embryo receives nutrients and oxygen and eliminates its gaseous and liquid waste products, through the *placenta,* which develops from a portion of the trophoblast after the blastocyst has implanted in the mother's uterine wall. Note in Figure 12-37 that a primitive streak and neurenteric canal are present in the blastoderm. Since no one has yet succeeded in cell-marking human or other mammal embryos, it is not known whether germ layer formation is by a delamination or by an invagination process. Later development of the mammal embryo is similar to that of the chick in respect to regression of the primitive streak and elongation of the embryo body (Figure 12-34).

Figure 12-36. *A, a 27-somite embryo viewed as a transparent object from above.* a.a.1,2,3, *aortic arches;* am., *posterior edge of head fold of the amnion;* car., *carotid loop;* gn.v., *trigeminal ganglion;* ret., *retina;* v.c.1,2, *visceral clefts. Other abbreviations as in Figure 12-5. From Lillie's* Development of the Chick, *3rd edition, Rev. by Hamilton, copyright 1952 by Holt, Rinehart and Winston, Inc. B, Later stages in development of the chick. Numbers indicate Hamburger-Hamilton stages: 26 (ca. 5 days); 35 (ca. 9 days); 43 (ca. 17 days). From Hamburger and Hamilton,* J. Morphol., *Volume 88, pp. 49–92, 1951. Courtesy Wistar Press.*

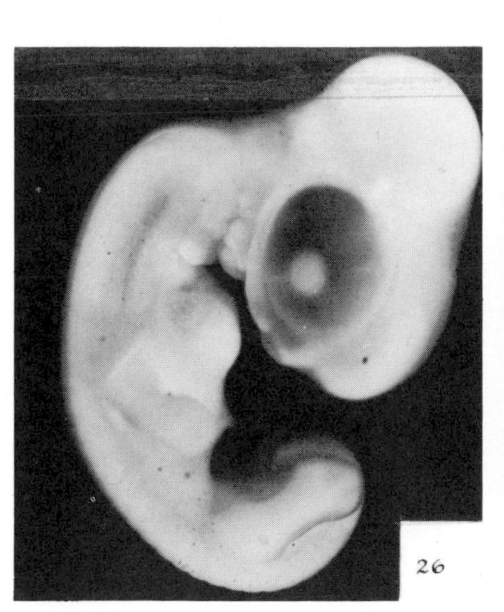

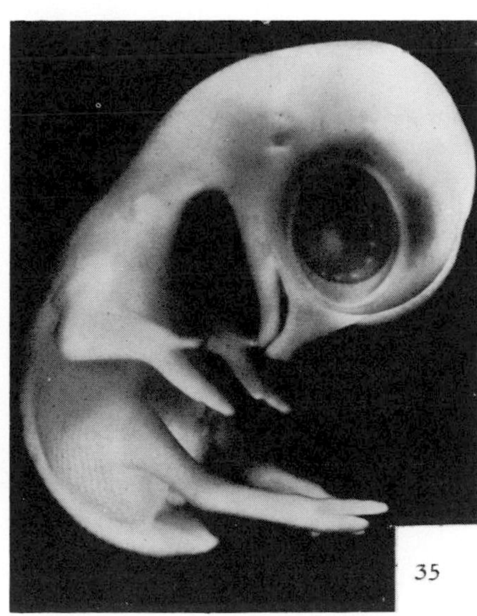

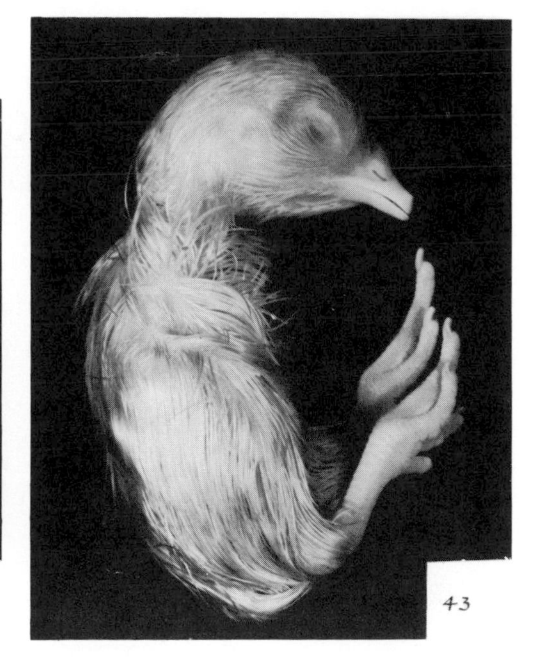

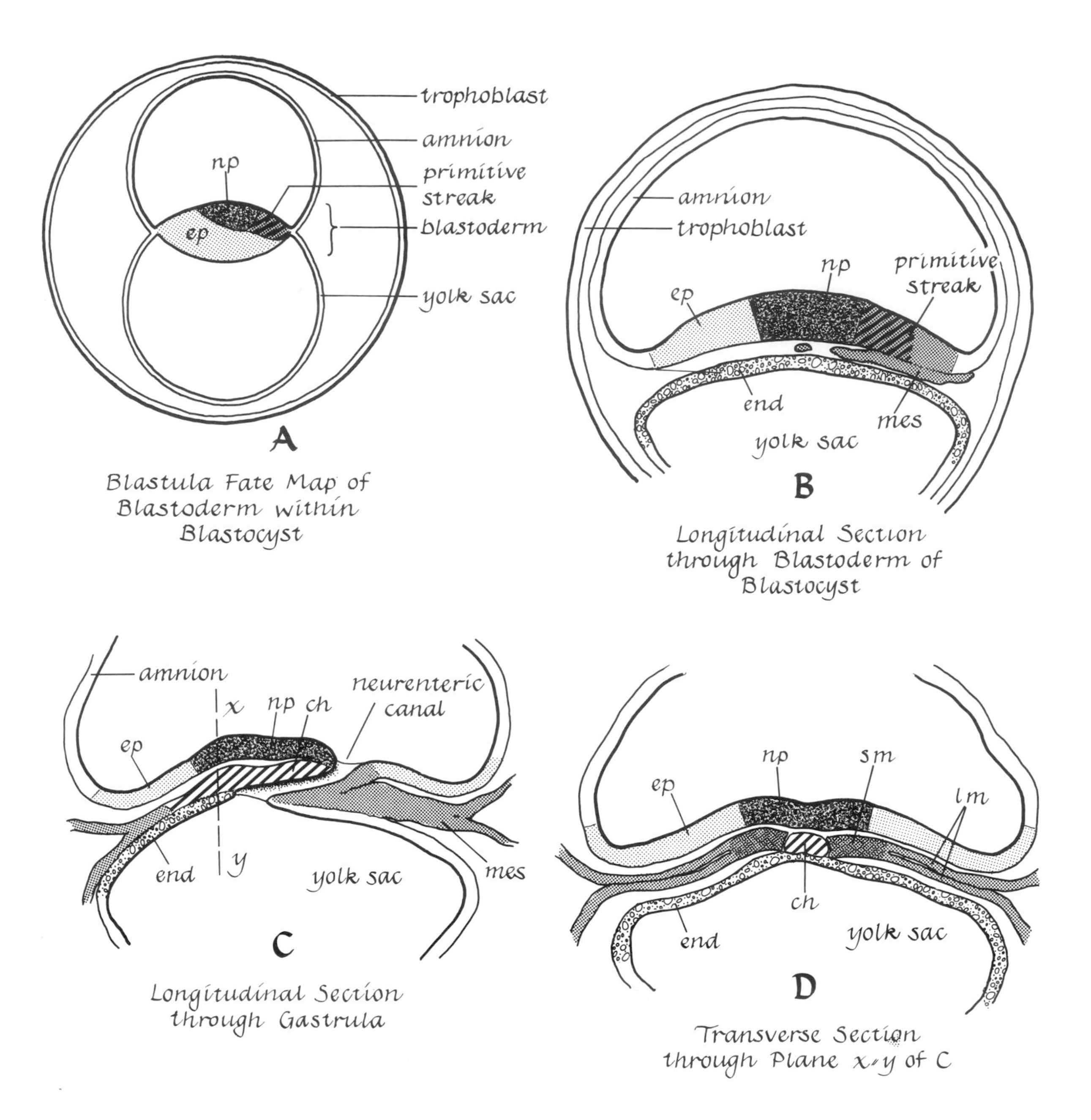

Figure 12-37. *Diagrams of germ layer formation in mammals (human).*

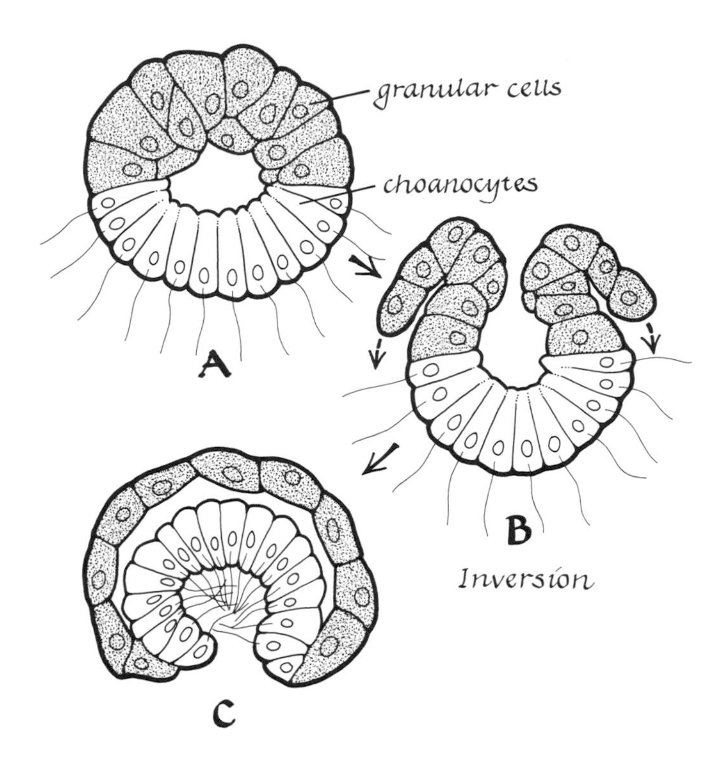

Figure 12-38. *Inversion and formation of the basic body plan in a sponge.*

Much remains to be discovered about the methods of germ layer formation in vertebrates. The descriptions we have just presented are only tentative.

Basic Body Plan Formation

Another primary result of gastrulation is establishment of the basic body plan upon which further development is based. In all metazoa, the plan consists of the arrangement of germ layers in a design out of which, by processes of differential growth, folding, invaginations, and evaginations, the detailed anatomy of the adult (or the larva in some animals) is derived. A few types of body plans are discussed in what follows.

Porifera

In some species of the class *Demospongia,* the body plan is derived by a process of inversion of the germ layers; flagellated cells which originally were on the surface become covered by the outward migration of granular cells which originally were below the surface. This process is similar to the inversion which occurs in a hydra which has been turned inside out, but is not like the inversion of daughter colonies or of the sperm globoid of *Volvox* (MacBride, 1914; Hyman, 1940). In sponges of the class *Calcaria,* inversion like that in *Volvox* has been described (Dubosque and Tuzet, 1935). The latter process, as well as the basic body plan of a sponge, is illustrated in Figure 12-38.

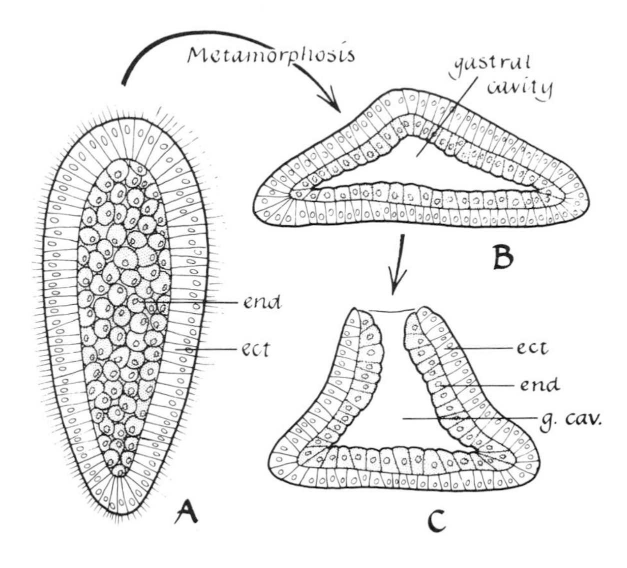

Figure 12-39. *Formation of the basic body plan in* Tubularia.

Coelenterata (Cnidaria)

In some hydrozoa, cleavage results in formation of a *planula larva,* consisting of a more or less solid mass of endoderm cells covered by ciliated ectoderm cells (Figure 12-39A). After a period of movement, the larva attaches to a substratum (Figure 12-39B). By a cavitation process of gastrulation, the larva metamorphoses into the basic body plan (Figure 12-39B and C), which consists of an outer layer of ectoderm and an inner layer of endoderm surrounding a gastral cavity. In other hydrozoa, for example, *Tubularia,* the body plan is a direct product of gastrulation in which there is no larval stage (Hyman, 1940).

Platyhelminthes (Flatworms)

In most flatworms, gastrulation results in a three-layered larval body plan. The blastopore through

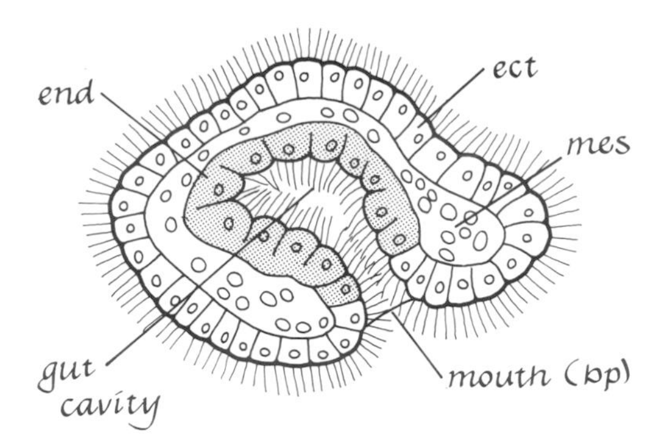

Figure 12-40. *Larval body plan of a flatworm.*

which surface cells were invaginated becomes the mouth of the larva and, after the larva metamorphoses, of the mature worm (MacBride, 1914). In parasitic flatworms (for example, the tapeworm), the body plan is modified so that there is no gut cavity. The larval body plan of flatworms is illustrated in Figure 12-40. Note in the figure that a third germ layer, the mesoderm, is present between the ectoderm and endoderm.

Annelida

In some segmented worms, the body plan is a direct result of gastrulation; in others, a larval body plan is formed first. The larva, called a *trochophore,* is characteristic of both annelids and molluscs. The structure of a trochophore, as well as the basic body plan of annelids, is illustrated in Figure 12-41.

Note in Figure 12-41 that the blastopore of the gastrula becomes the anus of both the larva and the adult worm, and that the mouth is a secondary and new opening to the digestive tract (gut). Also note that the mesoderm has split into outer and inner layers, enclosing a space which becomes the body cavity *(coelom).* During metamorphosis, most of the trochophore becomes the head of the adult. The segmented body is formed by the proliferation of stem cells in the anal region of the larva (Figure 12-54).

Echinodermata (Sea Urchins)

In most echinoderms (for example, *Paracentrotus lividus*), gastrulation results in formation of the basic body plan of the larva. The completed gastrula itself

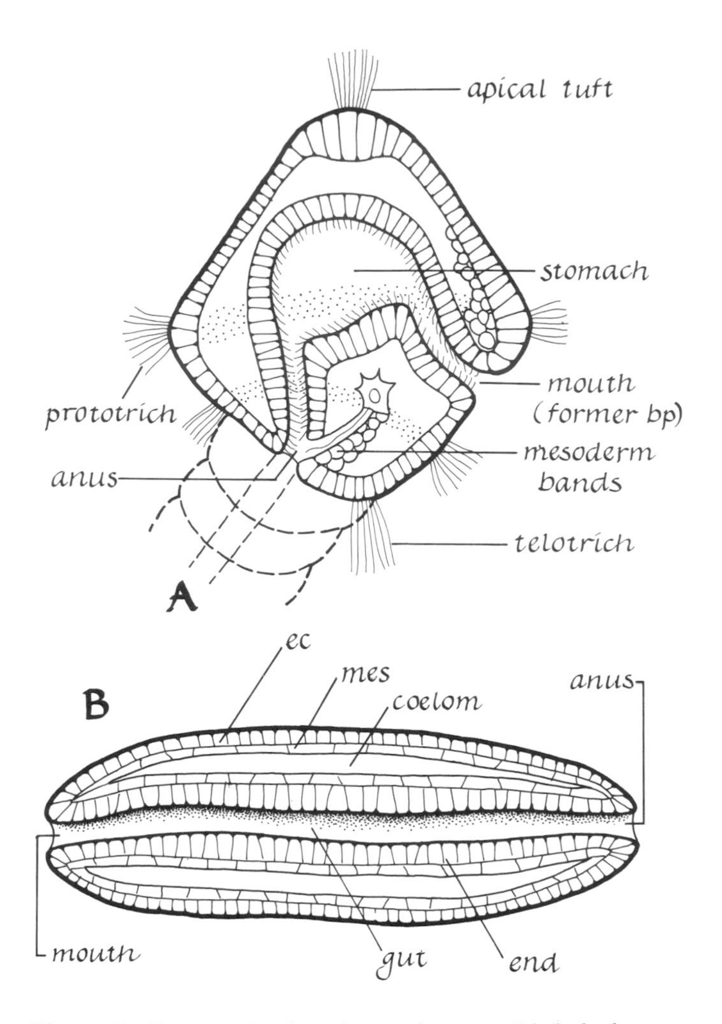

Figure 12-41. *A, trochophore larva of an annelid; B, body plan of an annelid.*

constitutes the body plan (Figure 12-42M and N), which transforms into the pluteus larva by a modeling process (Figure 12-42O and P). Also shown in Figure 12-42 are cleavage, blastula, and gastrula stages of the sea urchin. The basic body plan of the adult sea urchin is the *echinus rudiment,* an approximately radially symmetrical bud-like structure which develops on the left side of the bilaterally symmetrical larva (Figures 12-51 and 12-52). The development of the echinus rudiment will be described later in this chapter.

Insecta

As illustrated in Figures 12-9D and 12-15, the insect egg is centrolecithal. The zygote nucleus divides

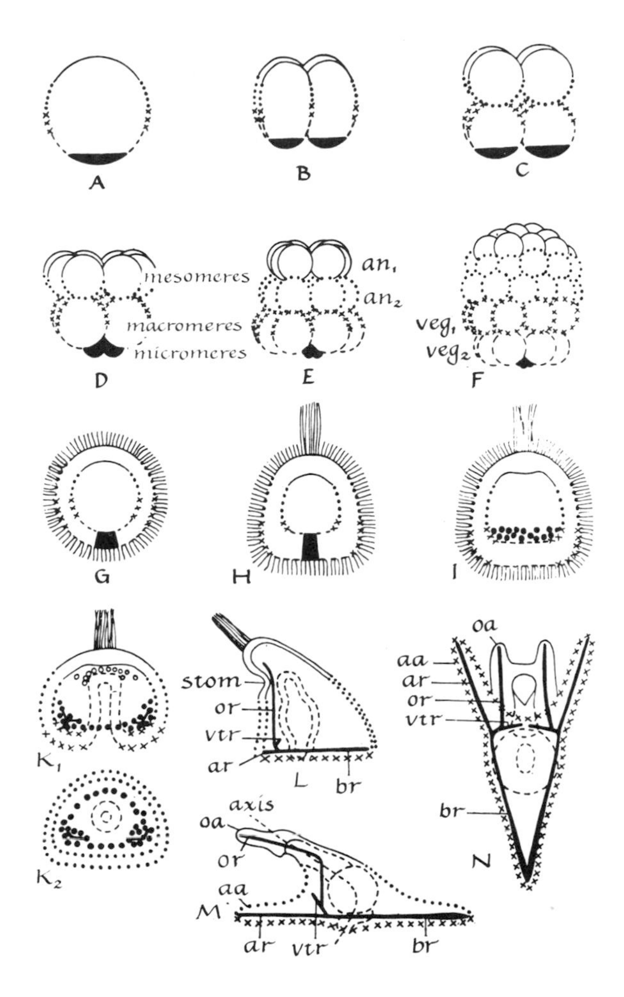

Figure 12-42. *Early development of the sea urchin. A-F, cleavage stages; and H, blastula states; I and K_1,K_2, gastrula stages; L, prism larva; M and N, side and ventral views of the pluteus larva. From Horstadius, 1939. Courtesy Cambridge University Press.*

many times, producing many nuclei with no separating cell membranes. Most of the nuclei migrate into the relatively yolk-free layer of cytoplasm at the surface of the egg where they become separated from one another by cell membranes. The result is a blastoderm formed by superficial cleavage (Figure 12-15). The blastoderm thickens along one side of the egg, forming an elongate area called the germ band (Figure 12-43A), which invaginates and involutes into the interior of the yolk mass (Figure 12-43B and C). Later it divides into segments which correspond to those of the larval or the adult body (Figure 12-43D).

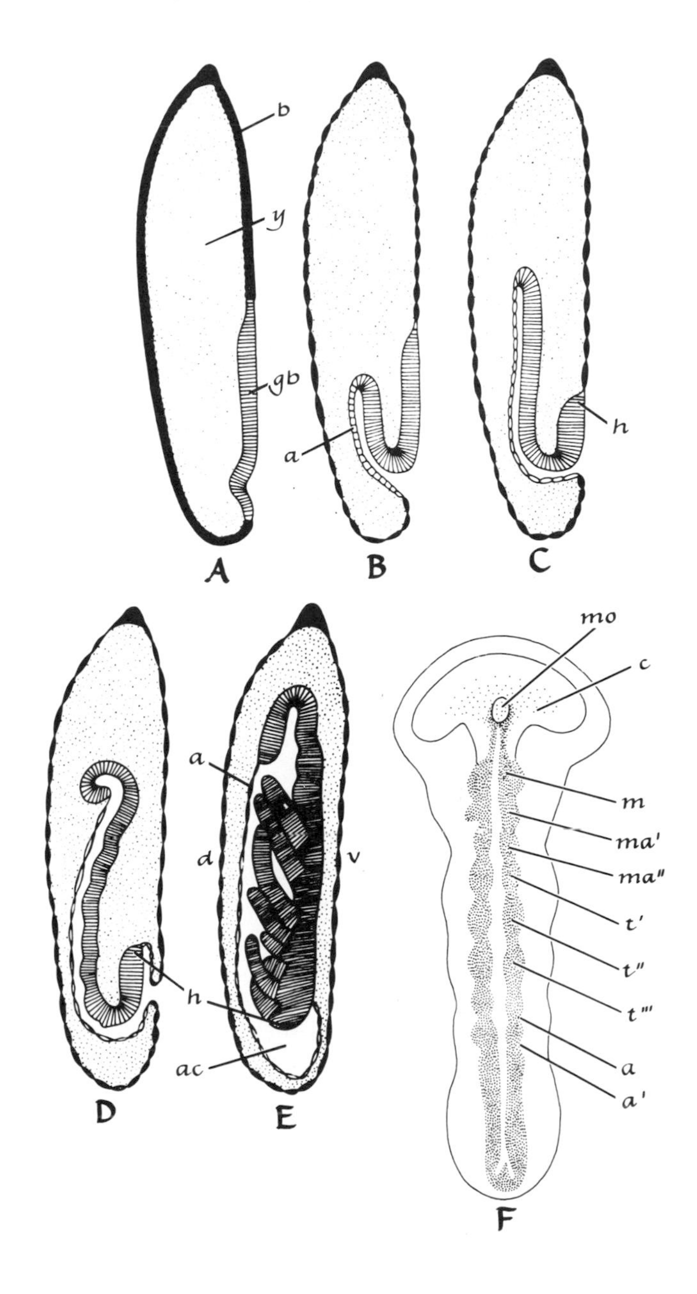

Figure 12-43. *A-E, early development of the embryo of the dragonfly.* a, *amnion;* ac, *amniotic cavity;* b, *blastoderm;* d, *dorsal;* h, *head;* gb, *germ band;* v, *ventral;* y, *yolk.* f, *segmentation of the germ band in a beetle embryo* Lina. a', *first and second abdominal segments;* c, *cerebral lobe;* t', t'' *and* t''', *thoracic segments;* m, *mandibular segment;* ma', ma'', *maxillary segments;* mo, *mouth. Reprinted with permission of The Macmillan Company, from* Embryology of Invertebrates *by Korschelt and Heider. Copyright © by The Macmillan Company, 1899.*

The segmented germ band constitutes the basic body plan. Appendages develop later from certain segments (Figure 12-43E and F).

Vertebrata

The basic body plans of all vertebrates are similar, even though they are derived by different methods of gastrulation, as we have already seen. The typical plan, based on that of the frog embryo, is shown in Figure 12-44.

Characteristic of the vertebrate plan is the dorsal position of the *neural plate,* which later folds up to form the hollow *neural tube,* the primordium of the brain and spinal cord. Also characteristic is the solid axial rod of mesoderm—the chorda or notochord; inner portions of the somites later form the vertebral skeleton around this rod.

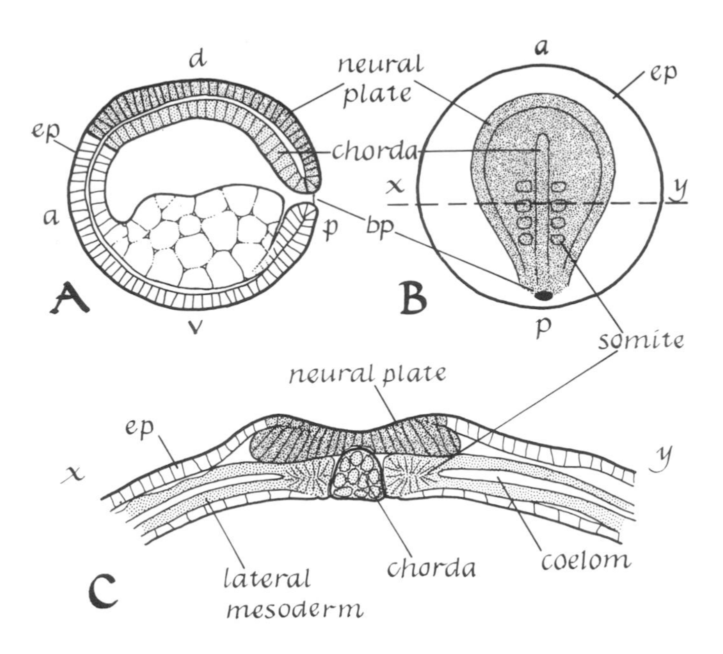

Figure 12-44. *The basic body plan of a vertebrate.*

Cellular Differentiation

The germ layers of animals are provisionally differentiated tissues. In normal development, these layers have definite morphological and functional fates. Here is a brief account of which germ layers give rise to which specialized cells:

Ectoderm:
- Skin and its derivatives: hair, feathers, scales, sweat and sebaceous glands.
- Nervous system: brain; spinal cord; peripheral nerves; ganglia; sensory receptors—such as rods and cones of the retina of the eye, various pain, temperature, and tactile receptors in the skin, auditory receptors in the ear, taste receptors in the tongue.
- Neural crest: pigment cells of the hair, feathers, and skin; ganglia; cartilages in some animals; adrenal gland medulla.

Mesoderm:
- Skeleton
- Muscles
- Circulatory system
- Connective Tissues

Endoderm:
- Digestive tract and its derivatives: thyroid, liver, pancreas, lungs, and so on.
- Primordial germ cells (in some animals).

In terms of the functional role they play in the adult, the primary germ layers have the following fates:

Ectoderm: protection, sensation, and coordination.
Mesoderm: support, motion, conduction, coordination, and connection.
Endoderm: nutrition, coordination, and reproduction.

The development of the germ layers illustrates the principle of progressive determination (differentiation). This is illustrated in Figure 12-45 for some of the cell types derived from the ectoderm.

Although the germ layers have definite fates in normal development, these are not immediately fixed: it is experimentally possible in some kinds of embryos to change a cell's fate by moving it from one layer into another, provided this is done at an early stage. However, it is possible to change the fates of cells within a germ layer at a later stage. Note, for example, in Figure 12-45, that the fates of many types of ectoderm cells are interchangeable. As shown in the figure, it is possible to change the fate of prospective nerve or skin cells, prospective head or trunk skin, brain or spinal cord cells, and so on. Eventually the fates of the cells become fixed, and it is almost

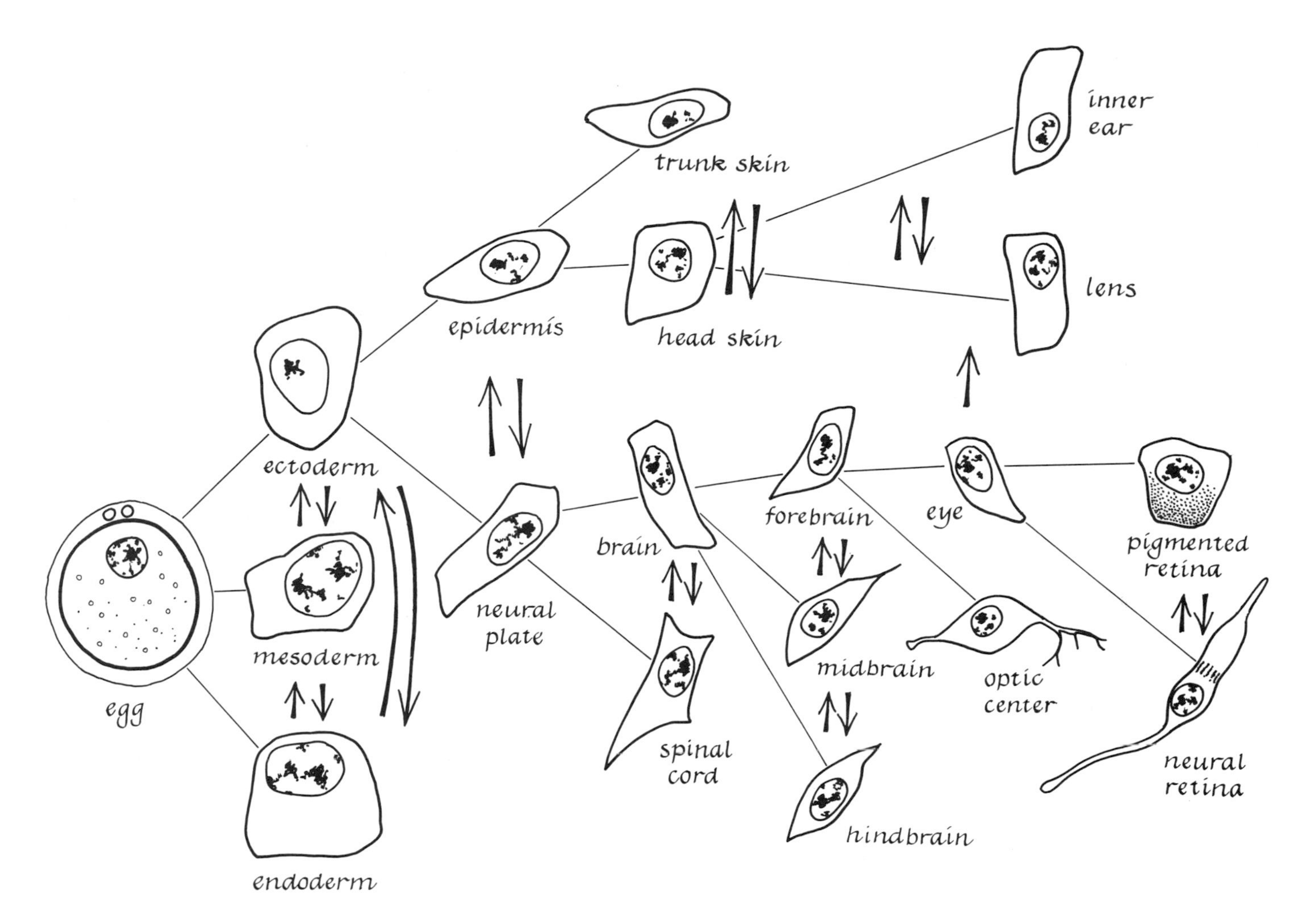

Figure 12-45. *Diagram of the progressive determination (differentiation) of some cell types derived from the ectodermal germ layer. The symbol ⇅ indicates that the cells in question have interchangeable fates, if they are experimentally manipulated in the right way at the right time.*

impossible to change their fates experimentally. For example, no one has yet been able to cause hind-brain, forebrain, or midbrain cells to become skin cells, or even neural plate cells to become mesoderm cells.

Unlike what occurs in plants, cellular differentiation in animals ultimately and usually results in irreversible specialization of the cells. In terms of genetic potentiality, one might expect differentiated animal cells to be capable of dedifferentiation and redifferentiation into new cell types, but this has yet to be clearly demonstrated except in some instances (see Chapter 5).

Organogenesis

Formation of the major body regions (head, trunk, tail) and organs of the embryo is accomplished by one or more basic morphogenetic processes, some of which are outlined below (see Arey, 1965; Balinsky, 1970).

Process	Organ
(a) Invagination	nose, ear, lens of the eye
(b) Evagination	eye, liver, pancreas, thyroid
(c) Folding	neural tube (in some animals), gut
(d) Cell migration and/or aggregation	ganglia; neural crest derivatives, including visceral cartilages, adrenal medulla, and pigment cells; blood vessels; primordial germ cells
(e) Cavitation	neural tube (in some animals), ducts
(f) Localized growth	brain, limb buds
(g) Cell death	appendages: wing, hand

Normal stages in development of the fertilized frog egg into a tadpole illustrate some externally visible modeling processes (Figure 12-46).

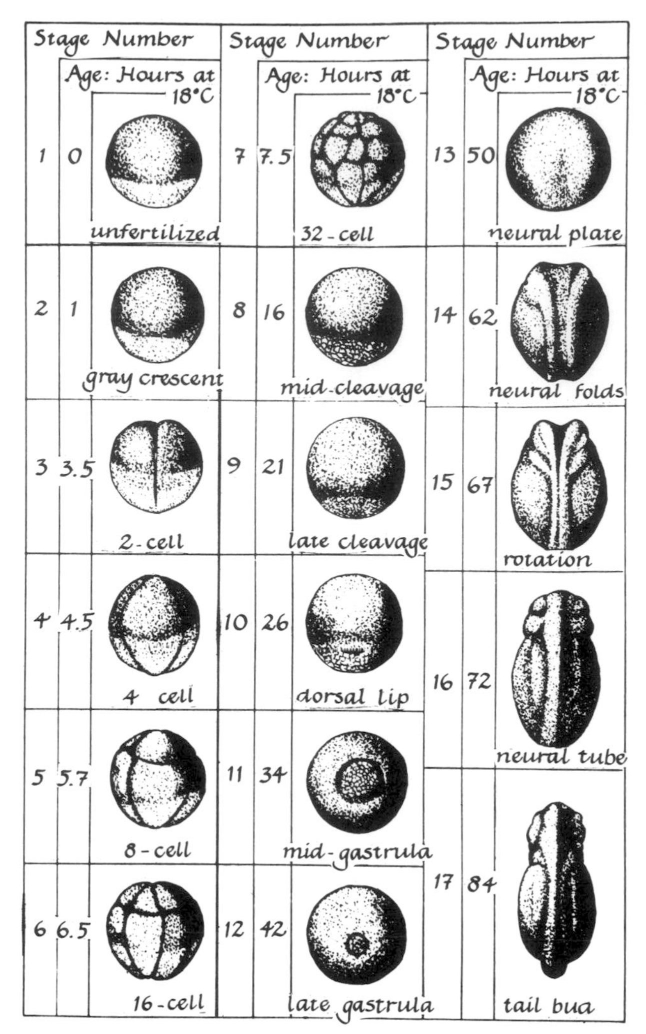

Figure 12-46. *Normal stages in development of the egg of the frog* Rana pipiens *into the tadpole. Shumway,* Anat. Record, *Volume 78, p. 143, 1940. Courtesy Wistar Press.*

Growth

Growth is increase in volume and mass. The period of embryonic development following cellular differentiation and organ formation is mainly characterized by growth of the organs and tissues of the body. In man, for example, the basic body plan is established after eighteen days of gestation. Almost all the major organs are formed by the end of six weeks of gestation. From this time on, growth is the most obvious aspect of development although further details of organogenesis continue for some time after birth (Arey, 1965). Figure 12-47 illustrates the development of a human embryo from age fourteen days to age fifteen weeks.

Differential growth is particularly important because it plays a profound role in embryonic development of animals as well as of plants, especially in shaping the organs, body parts, and proportions. This is illustrated in Figure 12-48.

Birth, Hatching

Birth or hatching is important because it marks the beginning of a relatively independent existence for the new individual, even if it has not yet attained the adult form. For example, in many animals, birth or hatching results in the release of a larval form whose transformation into an adult may be months away. In the birth and hatching processes of some animals (such as mammals and birds) dramatically sudden changes occur in functional activities of the circulatory, respiratory, and excretory systems. For a detailed description of these, consult the books of Arey (1965) and Hamilton (1952).

Metamorphosis

After birth or hatching, some animals continue to develop directly into the adult form. In many other

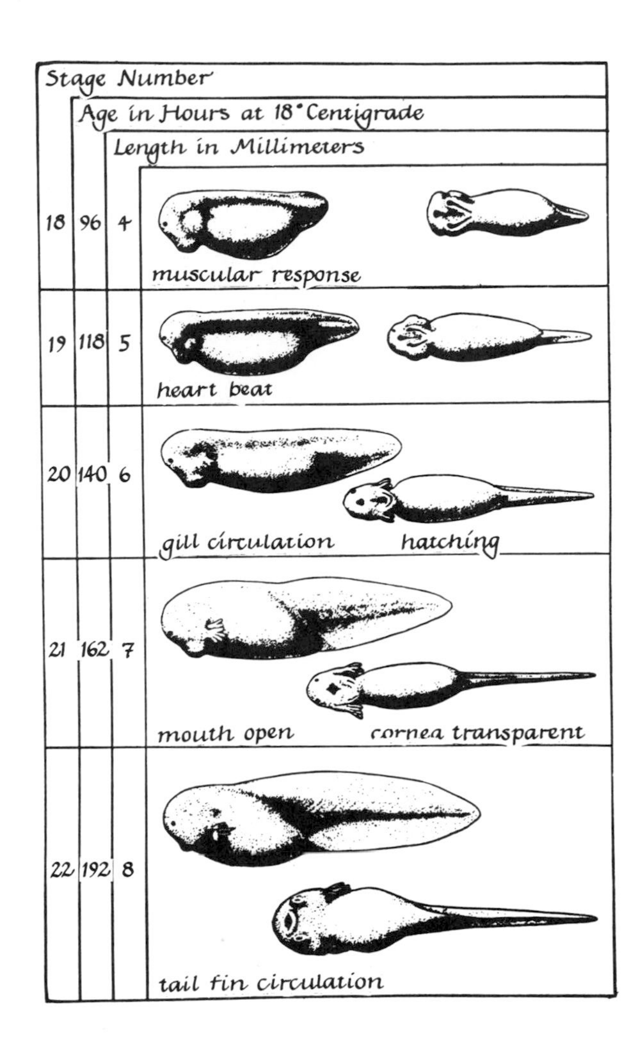

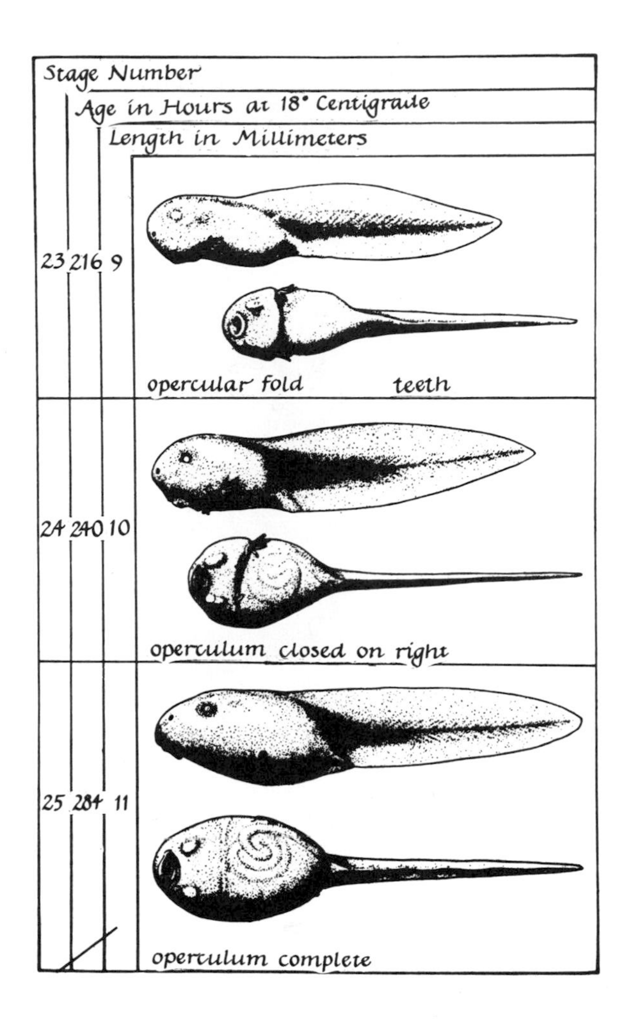

animals birth or hatching results in the release of a *larva,* a free-living and free-feeding animal which transforms (metamorphoses) into an adult after a variable period of existence. Familiar examples of this indirect kind of development include the tadpole, which transforms into the adult frog or toad (Figure 12-49); the swimming tunicate (sea squirt) tadpole (Figure 13-1), which transforms into the sessile adult (Figure 5-24); the caterpillar, which transforms into the butterfly or moth (Figure 12-50B); the pluteus larva of the sea urchin (Figure 12-52), which transforms into the adult that looks like a pincushion. Other examples are illustrated in Figure 12-54.

Sometimes (as in grasshoppers and caterpillars) the transformation of a larva into an adult is a gradual process involving a series of stages called *instars* in which the larva grows, sheds (molts) its skin (cuticle), and grows again (see Figure 12-50). Note in Figure 12-50 that the structure of the *nymph* (juvenile stage of the grasshopper) is very similar to that of the adult, whereas the structure of the caterpillar is very different from that of the moth. In the latter instance, the caterpillar stops feeding and transforms into a *pupa.* The adult moth is formed inside the pupa by a complex series of changes involving death of most of the larval tissues and development of the adult structures such as wings and legs from groups of organ-specific cells called *imaginal discs.*

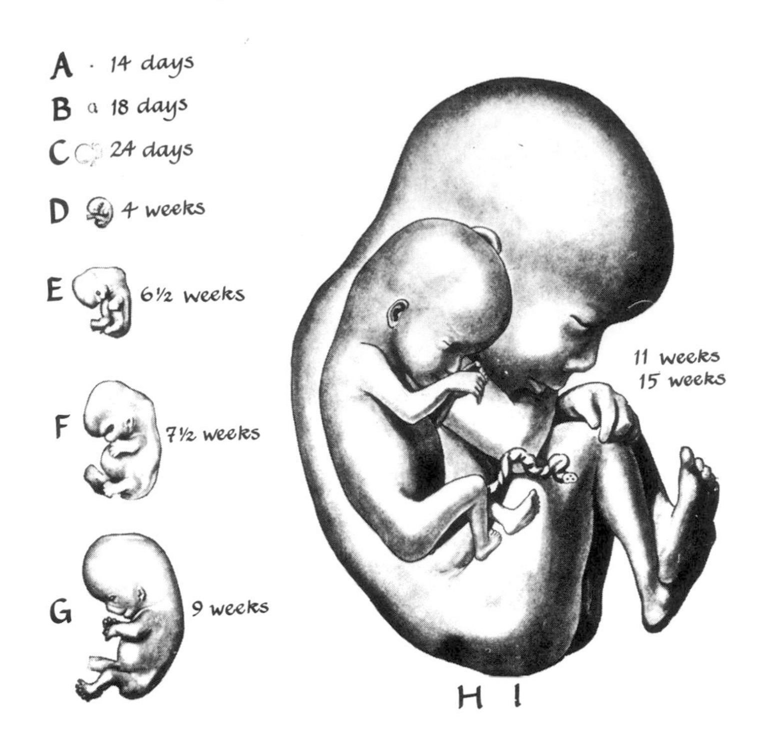

Figure 12-47. *A graded series of human embryos at natural size. From Arey:* Developmental Anatomy, *7th edition, W. B. Saunders Company, Philadelphia, 1965.*

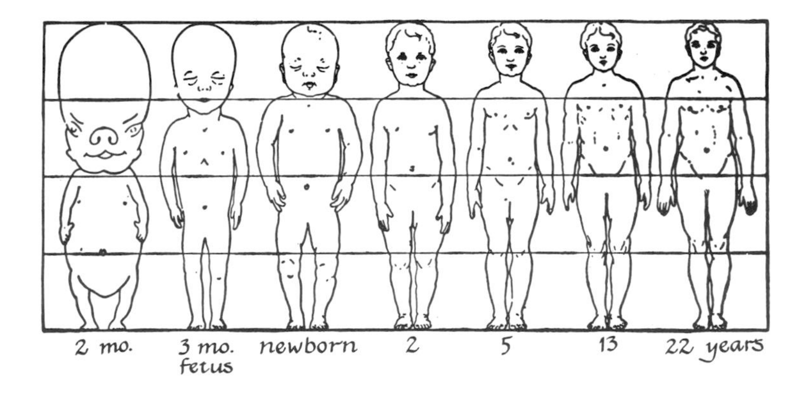

Figure 12-48. *Diagram of changes in the proportions of the human body during prenatal and postnatal growth. Stages are shown to the same total height. From Arey:* Developmental Anatomy, *7th edition, W. B. Saunders Company, Philadelphia, 1965.*

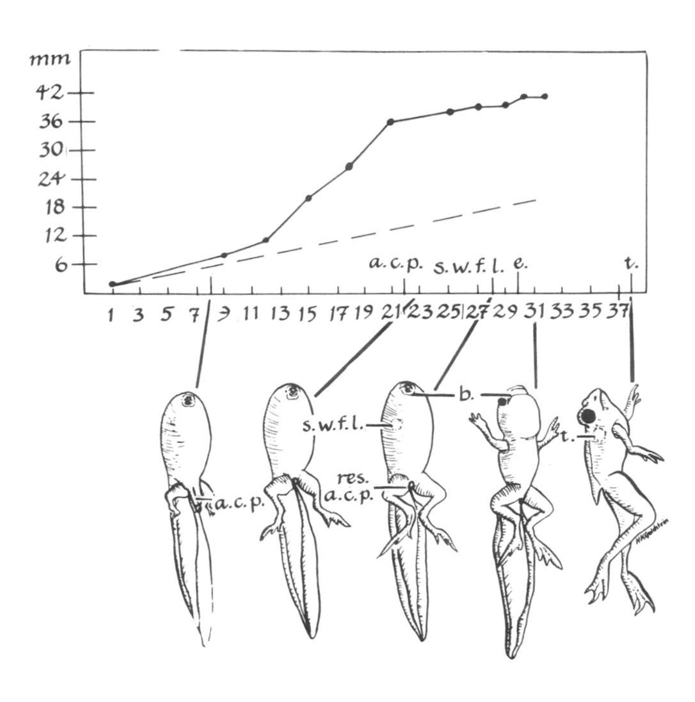

Figure 12-49. *Pattern of metamorphosis in the frog* Rana pipiens. *Graph shows growth of the hind legs (solid line) compared to that of the body (dashed line). The scale on the ordinate is for hindleg length; the abscissa scale is in days.* a. c.p., *anal canal piece;* res. a. c.p., *resorption of anal canal piece;* s.w. f. l., *skin window for the forelegs;* b, *horny teeth and beaks;* t, *tympanum. After Etkin, Taylor, and Kollros in Willier, Weiss, and Hamburger:* Analysis of Development, *W. B. Saunders Company, Philadelphia, 1955.*

In the sea urchin and other echinoderms which have larval stages, the adult develops from a special bud-like structure (in almost all instances, the echinus rudiment) on the left side of the larva. The process is complicated, but its general features are indicated in Figures 12-51 and 12-52.

Note in Figures 12-51A and 12-52 that the pluteus larva is bilaterally symmetrical but asymmetrical in respect to the echinus rudiment. This asymmetry undoubtedly has its origin in the cytoplasmic structure of the mature egg (Chapter 13), although it is not visibly evident in the radially symmetrical cleavage of the egg (Figure 12-16).

In general, the nature of the metamorphic changes in an animal depends largely on the time during the developmental period at which metamorphosis occurs (Weisz, 1966). This is illustrated in Figure 12-53.

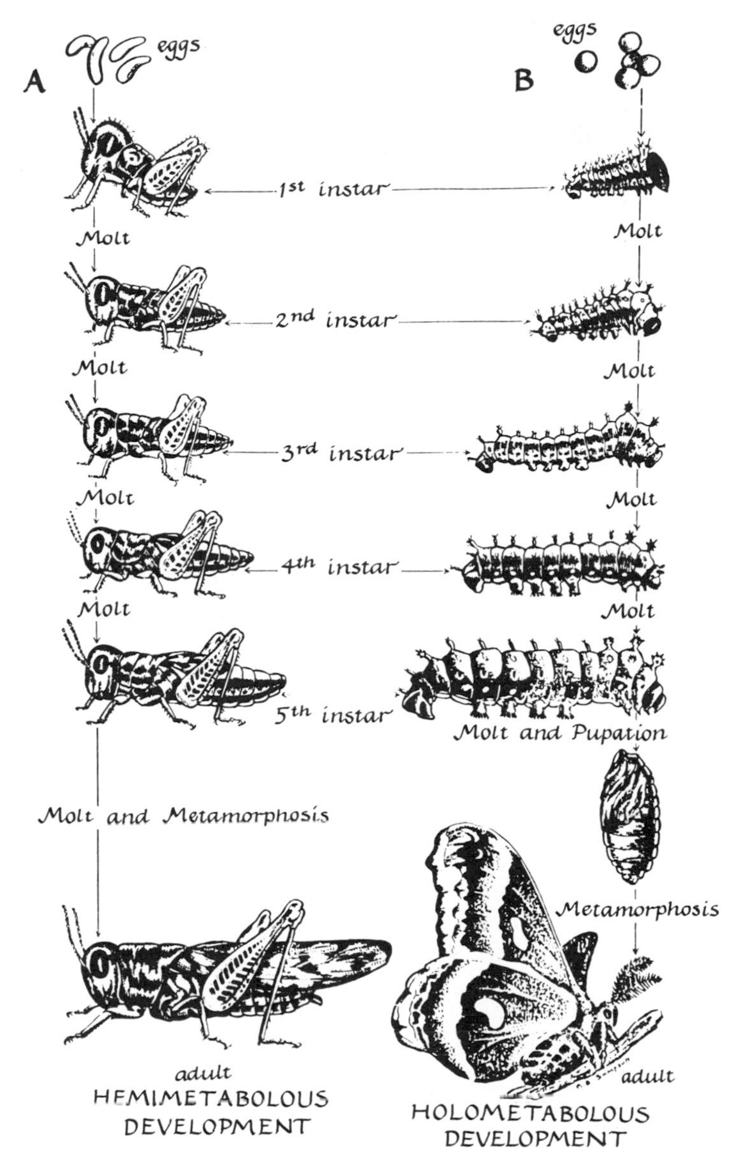

Figure 12-50. *A, Metamorphosis of a hemimetabolous insect, the grasshopper* Melanoplus differentialis, *and B, a holometabolous insect, the giant silkworm moth* Platysamia cecropia. *After Turner in Willier, Weiss, and Hamburger:* Analysis of Development, *W. B. Saunders Company, Philadelphia, 1955.*

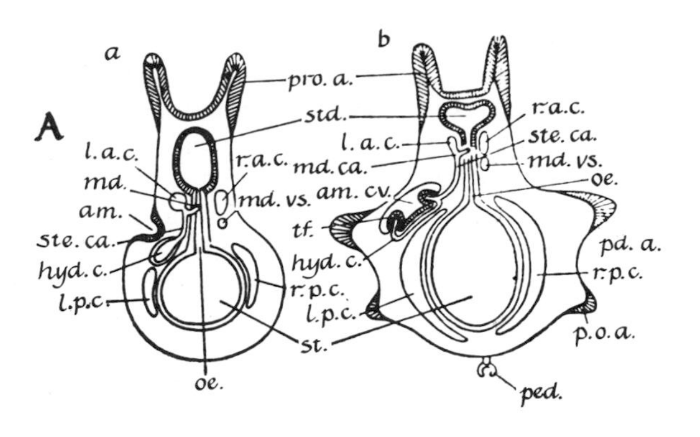

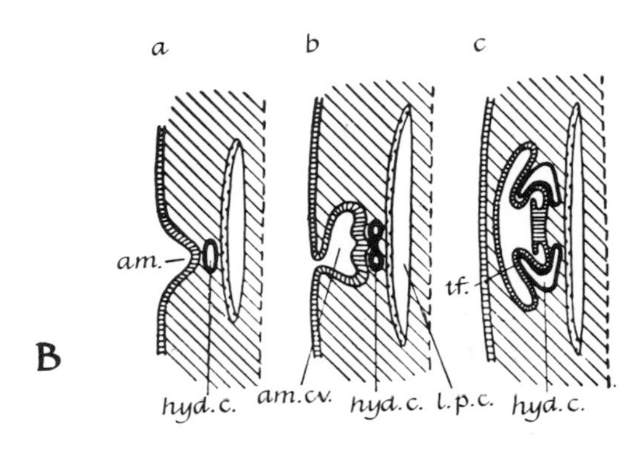

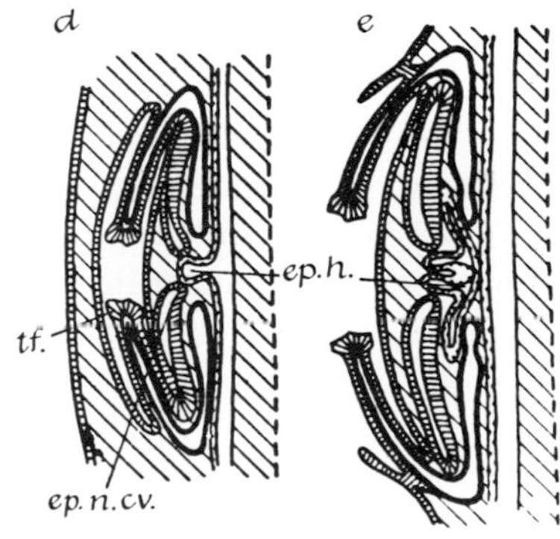

Figure 12-51. *A, diagrams showing early formation of the echinus rudiment,* am *in* a *and* am.cv. *in* b; *B, diagrams illustrating development of the echinus rudiment.* am, *amniotic invagination;* am.cv., *amniotic cavity;* hyd.c., *hydrocoel;* l.p.c., *left posterior coelum;* t.f., *tube foot rudiment;* ep.n.cv., *epineural space;* ep.h., *perihaemal system. From Kumé and Dan, 1968.*

As shown in the figure, the larval phase may be long, as in some species of crustacea and echinoderms, or completely suppressed, as in other echinoderms. Whether or not there is a larval phase seems to depend upon either early hatching or late metamorphosis.

Typically, animals which extrude small and relatively yolk-free eggs into water have long larval stages. Animals which lay large yolky eggs in water tend to have a very short or no larval stage (Weisz, 1966). Larval tissues are characteristically limited in growth potential and are generally incapable of repair or regeneration compared with adult-forming tissues. The sand dollar larva, for instance, cannot regenerate missing parts, though the echinus rudiment can.

Metamorphosis involves any number of degrees of reorganization of the larval body, a few of which are shown in Figures 12-49, 12-50, and 12-52. Weiss (1966) has classified these into four types, which are shown in Figure 12-54.

Note in the figure that at one extreme (Figure 12-54A), essentially all the larval tissues are used to form the adult: there are no separate larval and adult tissues in the larva. At the other extreme, the adult itself is a larva (Figure 12-54D) except for the later development of reproductive organs. In between (Figure 12-54B and C), the larvae contain differing proportions of larval and adult tissue. (For instance, in the caterpillar, the imaginal discs constitute adult tissues.)

Metamorphic transformations are accompanied by biochemical and metabolic changes (Etkin, 1955; Weber, 1967; Etkin and Gilbert, 1968), and involve the basic processes of growth, cell differentiation, cell death, and so on. In many animals, such as amphibians and insects, these processes are controlled by hormones. This aspect of metamorphosis will be discussed in Chapter 19. Metamorphic patterns present the student of development with the universal problem of how differential use of the genome is controlled. How can the same genome produce at one time a caterpillar and at another a butterfly, at one time a tadpole and at another a frog? We shall consider some current ideas about the mechanisms controlling gene action in Chapters 17, 20, and 22.

Embryonic Development and Evolution

The sequence of stages in animal embryonic development described in this chapter corresponds roughly to the sequence of changes which is presumed to have occurred in the evolution of animals.

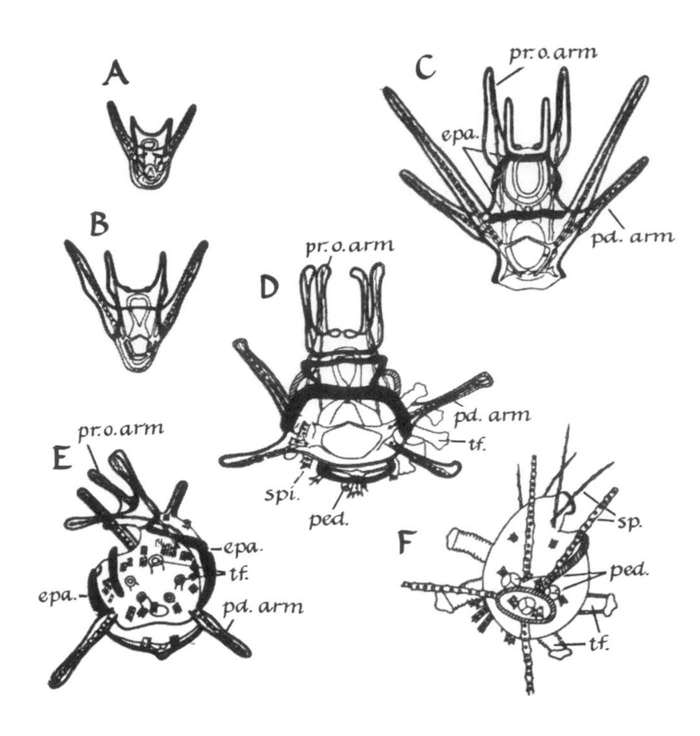

Figure 12-52. *Late larval stages in the sea urchin* Anthocidaris crassispina. *a-c, development of the pluteus larva; d-f, stages in metamorphosis. Note the projection of tube feet,* tf, *from the echinus rudiment in* D *and the young sea urchin,* F, *which begins its independent existence after discarding the larval structures. From Kumé and Dan, 1968.*

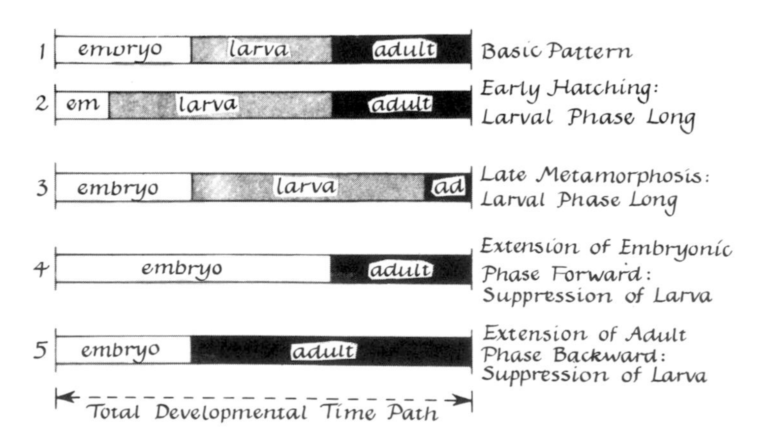

Figure 12-53. *Types of metamorphic patterns. From* The Science of Zoology *by Weisz. Copyright 1966 by McGraw-Hill Book Company. Used with permission of McGraw-Hill Book Company.*

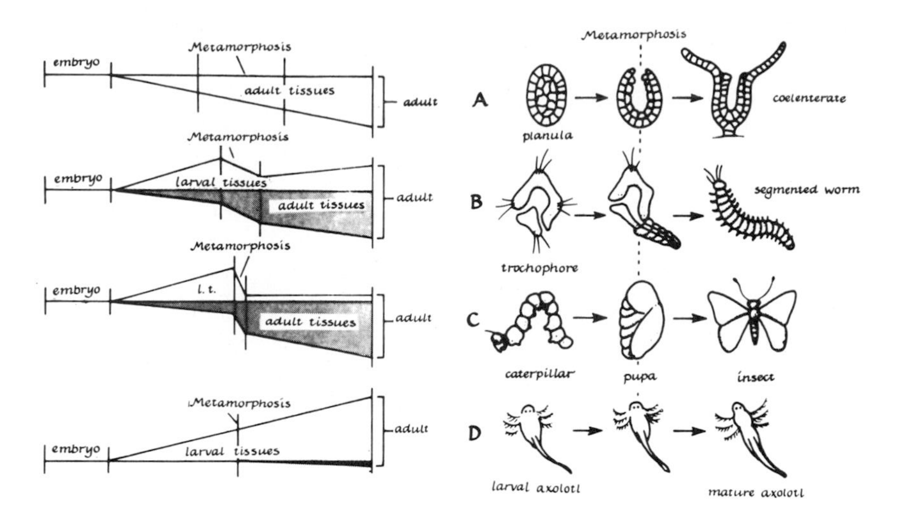

Figure 12-54. *Types of larval patterns, showing various degrees of reorganization of the larval body during metamorphosis. From* The Science of Zoology *by Weisz. Copyright 1966 by McGraw-Hill Book Company. Used with permission of McGraw-Hill Book Company.*

Features common to large groups of animals, such as the presence of gill furrows in the pharyngeal region of both fish and human embryos, appear early in development. It is generally believed that such features indicate the common ancestry of animals possessing them. On the other hand, features which distinguish animals from one another appear later in development, apparently indicating that they were acquired later in the course of their evolution. The similarity at early stages and dissimilarity at later stages in seven vertebrate embryos are illustrated in Figure 12-55. This general correspondence of embryonic and evolutionary changes within groups of animals probably indicates that they share a significantly large portion of genetic information. The correspondence is not unexpected, for the program of development itself is a product of evolution (Hyman, 1940). The similarity between embryonic stages of an organism and the presumed stages in its evolution is called the Theory of Recapitulation, which states that ontogeny recapitulates phylogeny. However, this concept is only of limited application, since many of the stages and structures present in embryos are not known to have been present in their evolution.

Conclusion

Although there are many diverse patterns of sexual reproduction, there is a chronological sequence of events which is essentially common to all. Embryonic development is a special kind of reproduction. Unlike the development of an adult from a bud, zooid, or spore, the process is relatively complex and indirect. Although the egg has detailed guidelines in its cytoplasm, its development is primarily epigenetic. This is clearly evident in the progressive increase in structural and functional complexity of a developing embryo. Perhaps the egg is best considered as the unicellular stage of a multicellular organism.

The sequence of stages through which the embryo progresses are probably programmed in the mature egg, at least in terms of genetic capacities, but it is equally clear that the acting out of the program is an epigenetic process. In a general way, this sequence is similar to the presumed sequence of evolutionary changes by which each species of animal evolved. The program for development of an egg is itself a product of evolution, and the existence of ancestral resemblances during embryonic development is a general biological principle.

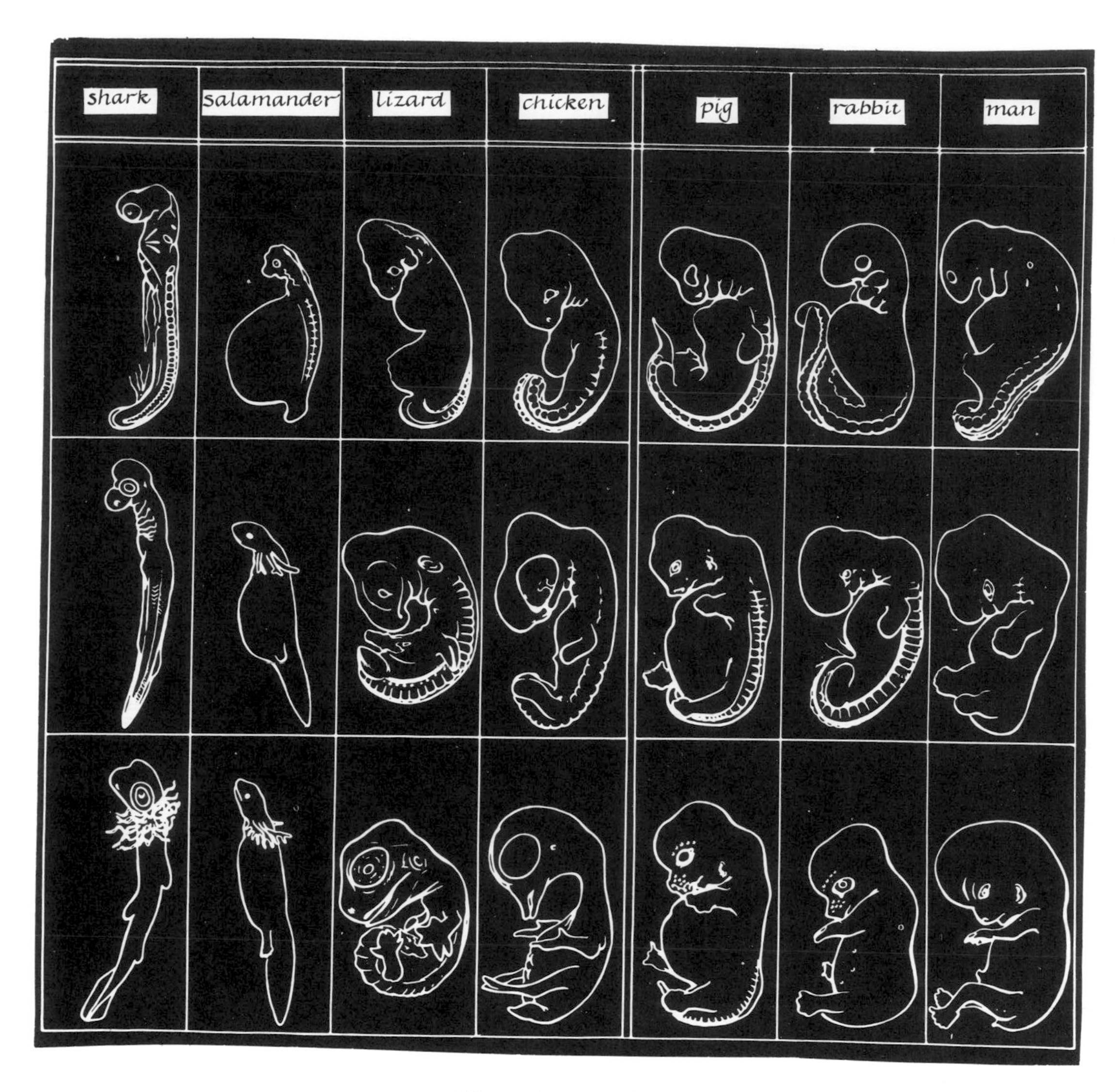

Figure 12-55. *The similarity at early stages and dissimilarity at later stages in vertebrate embryos. From Torrey, 1962. Courtesy John Wiley and Sons, Inc.*

References

Arey, L. B. 1965. Developmental Anatomy. 7th ed. W. B. Saunders, Philadelphia.

Arnold, J. M. 1969. Cleavage furrow formation in a telolecithal egg *(Loligo pealii):* I. Filaments in early furrow formation. J. Cell. Biol. **41**:894–904.

Austin, C. R. 1965. Fertilization. Prentice-Hall, Englewood Cliffs, N.J.

Austin, C. R. 1968. Ultrastructure of Fertilization. Holt, Rinehart and Winston, New York.

Balinsky, B. I. 1970. An Introduction to Embryology. 3rd ed. W. B. Saunders, Philadelphia.

Beale, G. H. 1954. The Genetics of *Paramecium aurelia.* Cambridge University Press, London.

Blackler, A. W. 1962. Transfer of primordial germ-cells between two subspecies of *Xenopus*. J. Embryol. Exp. Morph. **10**:641–651.

Brachet, J. 1950. Chemical Embryology. Interscience, New York.

Brachet, J. 1967. Biochemical changes during fertilization and early embryonic development. In A. V. S. DeReuck and J. Knight, eds., Cell Differentiation, pp. 39–64. Ciba Found. Symp., Little, Brown Boston.

Buehr, M. L., and A. W. Blackler. 1970. Sterility and partial sterility in the South African clawed toad following pricking of the egg. J. Embryol. Exp. Morph. **23**:375–384.

Clawson, R. C., and L. V. Domm. 1969. Origin and early migration of primordial germ cells in the chick embryo: a study of the stages definitive primitive streak through eight somites. Am. J. Anat. **125**:87–111.

Colwin, A. L., and L. H. Colwin. 1957. Morphology of fertilization: acrosome filament formation and sperm entry. In A. Tyler, R. C. von Borstel, and C. B. Metz, eds., The Beginnings of Embryonic Development, pp. 135–168. Am. Assoc. Adv. Sci., Washington, D.C.

Colwin, A. L., and L. H. Colwin. 1964. Role of the gamete membranes in fertilization. In M. Locke, ed., Cellular Membranes in Development, pp. 233–279. Academic Press, New York.

Colwin, A. L., and L. H. Colwin. 1967. Behavior of the spermatozoan during sperm-blastomere fusion and its significance for fertilization *(Saccoglossus Kowalevskii: Hemichordata)*. Zeit. f. Zellforsch. **78**:208–220.

Conklin, E. G. 1905. The organization and cell-lineage of the ascidian egg. J. Acad. Nat. Sci., Philadelphia, **13**:1–119.

Conklin, E. G. 1917. Effects of centrifugal force on the structure and development of the eggs of *Crepidula*. J. Exp. Zool. **22**:311–419.

Costello, D. P. 1948. Ooplasmic segregation in relation to differentiation. Ann. N.Y. Acad. Sci. **49**:663–683.

Dan, K., and M. K. Kujiyama. 1963. A study of the mechanism of cleavage in the amphibian egg. J. Exp. Biol. **40**:7–14.

Daniel, J. C., Jr. 1970. Dormant embryos of mammals. Biol. Sci. **20**:411–415.

Dickman, Z. 1963. Chemotaxis of rabbit spermatozoa. J. Exp. Biol. **40**:1–6.

Duboscq, O., and O. Tuzet. 1937. L'ovogénèse, la fécondation, et les premiers stades du développement des éponges calcaires. Arch. Zool. Exp. Gén. **79**:157–316.

Emanuelson, H. 1961. Mitotic activity in chick embryos at the primitive streak stage. Acta. Physiol. Scand. **52**:211–233.

Endo, Y. 1961. Changes in the cortical layer of sea urchin eggs at fertilization as studied with the electron microscope: I. *Clypeaster japonicus*. Exp. Cell Res. **25**:383–397.

Etkin, W. 1955. Metamorphosis. In B. H. Willier, P. Weiss, and V. Hamburger, eds., Analysis of Development, pp. 631–663. W. B. Saunders, Philadelphia.

Etkin, W., and L. I. Gilbert, eds. 1968. Metamorphosis: a problem in developmental biology. Appleton-Century-Crofts, New York.

Gatenby, J. B. 1919. Germ cells, fertilization and early development of *Grantia*. J. Linn. Soc. Lond. Zool. **34**:261–297.

Gatenby, J. B. 1927. Further notes on the gametogenesis and fertilization of sponges. Quart. J. Micr. Sci. **71**:174–188.

Goodenough, D. A., S. Ito, and J. P. Revel. 1968. Electron microscopy of early cleavage stages in *Arbacia punctulata*. Biol. Bull. **135**:420 (abstr.).

Hamburger, V., and H. L. Hamilton. 1951. A series of normal stages in the development of the chick embryo. J. Morph. **88**:49–92.

Hartman, M., F. C. Medem, R. Kuhn, and H. J. Bïelig. 1947. Untersuchungen über die Befruchtungsstoffe der Regenbogenforelle. Zeit. f. Naturforsch. **26**:330–349.

Horstadius, S. 1939. The mechanics of sea urchin development studied by operative methods. Biol. Rev. Camb. **14:**132–179.

Hyman, L. H. 1940. The Invertebrates: Protozoa through Ctenophora. McGraw-Hill Book Company, New York.

Korschelt, E., and K. Heider. 1899. Text-book of the Embryology of Invertebrates. Vol. 3, Macmillan, New York.

Kumé, M., and K. Dan, eds. 1968. Invertebrate Embryology. Nolit Publishing House, Belgrade.

Kupelwieser, H. 1909. Entwicklungserregung bei Seeigeleien durch Molluskensperma. Arch. Entw. Mech. **27:**434–462.

Lillie, F. R. 1952. Development of the Chick. 3rd ed. rev. H. L. Hamilton. Holt, Rinehart and Winston, New York.

MacBride, E. W. 1914. Embryology. In W. Heape, ed., Invertebrata. Vol. 1. Macmillan, London.

Metz, C. B., and A. Monroy, eds. 1967, 1968. Fertilization: Comparative Morphology, Biochemistry and Immunology. Vols. 1 and 2. Academic Press, New York.

Miller, R. L. 1966. Chemotaxis during fertilization in the hydroid, *Campanularia.* J. Exp. Zool. **162:**23–44.

Monroy, A. 1965. Biochemical aspects of fertilization. In R. Weber, ed., The Biochemistry of Animal Development, pp. 73–135. Academic Press, New York.

Monroy, A, and R. Maggio. 1966. Amino acid metabolism in the developing embryo. In R. A. Boolootian, ed., Physiology of Echinodermata, pp. 743–756. John Wiley, New York.

Peters, H. 1970. Migration of gonocytes into the mammalian gonad and their differentiation. Phil. Trans. Roy. Soc. London. Ser. B. **259:**91–101.

Raven, C. P. 1966. Oogenesis: The Storage of Developmental Information. Pergamon Press, New York.

Rothschild, L. 1956. Fertilization. Methuen, London.

Selmann, G. G., and C. H. Waddington. 1955. The mechanism of cell division in the cleavage of the newt's egg. J. Exp. Biol. **32:**700–733.

Shumway, W. 1940. Stages in the normal development of *Rana pipiens.* Anat. Rec. **78:**139–147.

Smith, L. D. 1966. The role of a "germinal plasm" in the formation of primordial germ cells in *Rana pipiens.* Devel. Biol. **14:**330–347.

Sonneborn, T. M. 1954. Patterns of nucleocytoplasmic integration in Paramecium. In Proc. 9th Int. Congr. Genetics, Caryologia, Suppl. Vol. 6, pp. 307–325.

Spratt, N. T., Jr. 1966. Some problems and principles of development. Am. Zool. **6:**9–19.

Spratt, N. T., Jr., and H. Haas. 1965. Germ layer formation and the role of the primitive streak in the chick: I. Basic architecture and morphogenetic tissue movements. J. Exp. Zool. **158:**9–38.

Spratt, N. T., Jr. 1971. The primitive streak: is it a blastopore or a blastema? In Topics in the Study of Life, pp. 98–105. Harper and Row, New York.

Summers, F. M. 1938a. Some aspects of normal development in the colonial ciliate, *Zoothamnium alternans.* Biol. Bull. **74:**117–129.

Summers, F. M. 1938b. Form regulation in *Zoothamnium alternans.* Biol. Bull. **74:**130–154.

Tyler, A. 1955. Gametogenesis, Fertilization and Parthenogenesis. In B. H. Willier, P. Weiss, and V. Hamburger, eds., Analysis of Development, pp. 170–212. W. B. Saunders, Philadelphia.

Tyler, A., and B. S. Tyler. 1966a. The gametes; some procedures and properties. In R. A. Boolootian, ed., Physiology of Echinodermata, pp. 639–682. Interscience, John Wiley, New York.

Tyler, A., and B. S. Tyler. 1966b. Physiology of fertilization and early development. In R. A. Boolootian, ed., Physiology of Echinodermata, pp. 683–741. Interscience, John Wiley, New York.

Weber, R. 1967a. Biochemistry of amphibian metamorphosis. In R. Weber, ed., Biochemistry of Animal Development, pp. 227-301. Academic Press, New York.

Weber, R. 1967b. Biochemistry of amphibian metamorphosis. In M. Florkin and E. Stotz, eds., Comprehensive Biochemistry, pp. 145–198. Elsevier, New York.

Weismann, A. 1892. Daskeimplasma. Eine Theorie der Vererbung. Jena.

Weisz, P. B. 1966. The Science of Zoology. McGraw-Hill Book Company, New York.

Wilson, E. B. 1925. The Cell in Development and Heredity. 3rd ed. Macmillan, New York.

13 *Cytoarchitecture of the Egg and Its Role in Development: The Significance of Oogenesis*

It is customary to begin the study of development of a new individual with the mature reproductive unit, but that unit (egg, spore, bud, or whatever) is itself the outcome of considerable development during its formation by the mother organism. Eggs and spores are particularly interesting units in that they are designed solely to form a whole new individual. Here we shall be primarily concerned with eggs, since their formation and structure has been studied more intensively than that of other reproductive units.

As noted in Chapter 12, an egg is probably best considered as the unicellular stage of a multicellular organism. The cytoarchitecture of the mature egg, particularly its polarity and symmetry, probably constitutes the primary framework within which inherited potentialities are selectively realized. It appears that much of the program for development of an egg resides in the *cytoplasm,* not exclusively in its nucleus. Since the nucleus of a differentiated tissue cell can be experimentally substituted by nuclear transplantation for the egg nucleus and will support normal development, there seems to be nothing intrinsically special about the egg nucleus. This does not mean that the nucleus of the egg plays no role in egg formation. Indeed, the egg nucleus and nuclei of cells surrounding the developing egg probably play important roles in the genesis of the cytoplasmic structure, but once this has been formed, almost any nucleus from the same species can take over. This will be discussed in Chapter 20.

In a strict sense, probably no egg is cytoplasmically homogeneous, and the composition of the nucleus differs from that of the cytoplasm. In most eggs that have been examined, there is a geometrically patterned distribution of substances in the cytoplasm, primarily in the surface cytoplasm or cortex. In some eggs this cortical pattern is more obvious (though not necessarily more important) than in others (Picken, 1960).

Correspondence of Egg and Adult Body Organization

It has long been recognized by embryologists (Conklin, 1905; Wilson, 1925) that mature eggs are morphological and physiological systems, possessing certain differentiations which are correlated with corresponding differentiations of the adult or larva. These differentiations exist in eggs as cytoplasmically different regions, often as visible differences in the distribution of various inclusions such as granules, vacuoles, and yolk spheres. The correspondence between cytoplasmic differentiation and cellular differentiation as pointed out by Conklin is illustrated for two topographically different patterns of cytoarchitecture in Figures 13-1 and 13-2.

Wilson (1925), Conklin (1905), Morgan (1927), and, more recently, Raven (1961, 1964) and others, have apparently established that an egg's early development is largely under control of a preformed distribution of cytoplasmic components. At least it seems clear that cytodifferentiation of the egg involves the control and direction of preexisting mechanisms producing specific populations of molecules rather than the creation of the mechanisms themselves (Williams, 1965).

General Patterns of Egg Cytoarchitecture

Three general types of cytoplasmic patterns have been recognized (Harrison, 1945). These are illustrated in Figure 13-3.

The patterns illustrated in Figure 13-3 have, in some cases, been deduced from studies of the development of separated parts of eggs or of separated early blastomeres (Morgan, 1927; Reverberi and Ortolani, 1962; see also Fankhauser, 1955, and Weiss, 1939, for reviews). The studies of Horstadius (1939) on the development of separated parts of uncleaved and early cleaving sea urchin eggs illustrate how a partially invisible pattern of cytoarchitecture can be deduced (Figure 13-4). In some eggs the cytoplasmic pattern is directly visible in the living state (for instance, in the *Ascidian* egg) or can be visualized by histochemical staining methods (see Needham, 1950, and Reverberi, 1957, for reviews).

Separation of the egg or the blastula into two halves along the animal-vegetal axis, or of the two blastomeres at the two-cell stage (Figure 13-4A), results in the development of each half into a normal but half-sized pluteus larva. However, separation of the egg or of the eight-cell stage into halves by a cut or constriction transverse to the polar axis results in abnormal development of both parts (Figure 13-4B). The experiments indicate a difference in cytoplasmic composition along the polar axis. Other experiments (Figure 13-4C) have shown this difference between the animal and vegetal poles of the egg to be gradient in nature.

There are many modifications of the basic patterns throughout the animal and plant kingdoms.

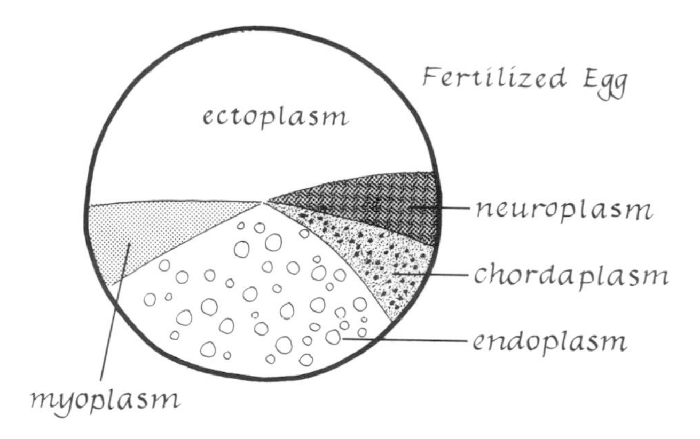

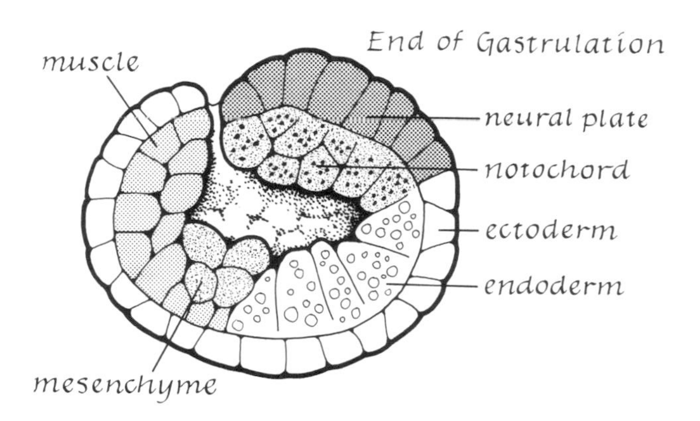

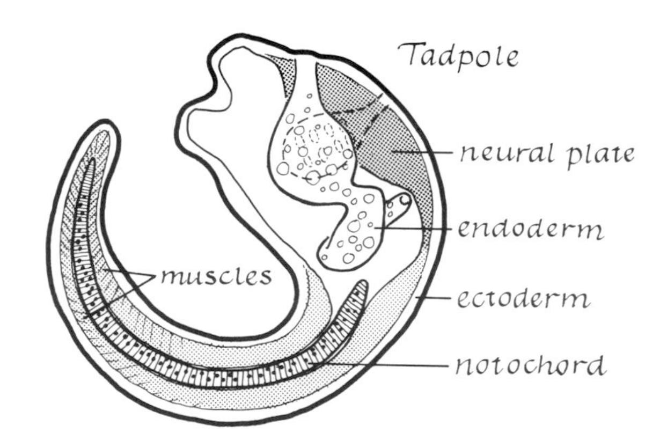

Figure 13-1. *Diagrams of the correspondence between localized cytoplasmic differentiation in the egg and cellular differentiation in the larva of an ascidian.*

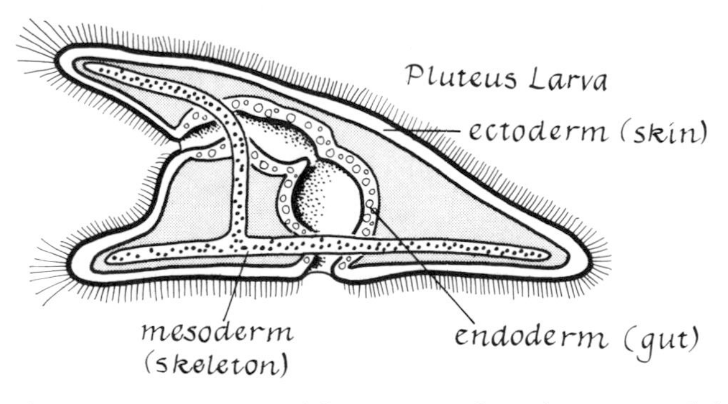

Figure 13-2. *Diagrams of the correspondence between graded cytoplasmic differentiation in the egg and cellular differentiation in the pluteus larva of an echinoderm.*

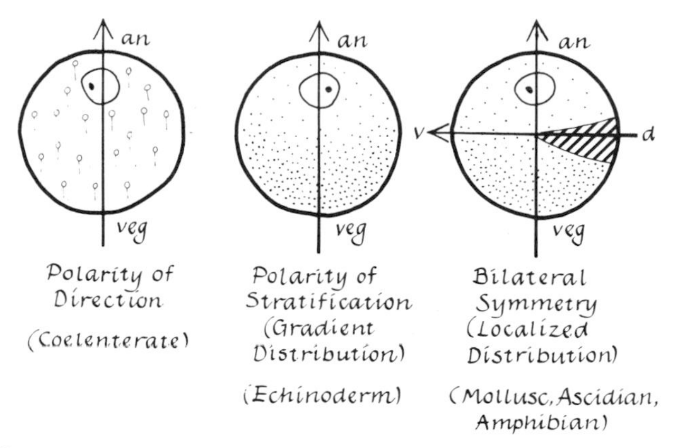

Figure 13-3. *Three general types of cytoplasmic patterns in uncleaved eggs.*

Two such modifications are illustrated below—one for a bracken fern *Pteridium aquilinum* (Bell, 1963), in Figure 13-5A, and one for the fertilized egg of a cotton plant (Jensen, 1964), in Figure 13-5B and C. Note the distinct polarity of both eggs. This is typical of plant eggs. Note the concentration of cytoplasmic DNA around the nucleus of the fern egg. The cotton egg has a unique type of endoplasmic reticulum, consisting of internal tubes surrounding the nucleus. Following fertilization, the quantity of this type of endoplasmic reticulum increases (Jensen, 1964).

Patterns of Distribution of Cytoplasmic Components

The nature of the developmentally significant components of the egg cytoplasm is a problem that remains unsolved. Visible patterns of distribution of yolk, pigment, oil droplets, ribosomes, and the like are present in many kinds of eggs (Brachet, 1960). It is not known whether or not these distributions are significant for normal development. In some eggs the visible inclusions can be displaced by centrifugation without serious influence upon later development. In other eggs, centrifugation or isolation of blastomeres containing different concentrations of components results in abnormal development. We shall consider the meaning of these results later in this chapter.

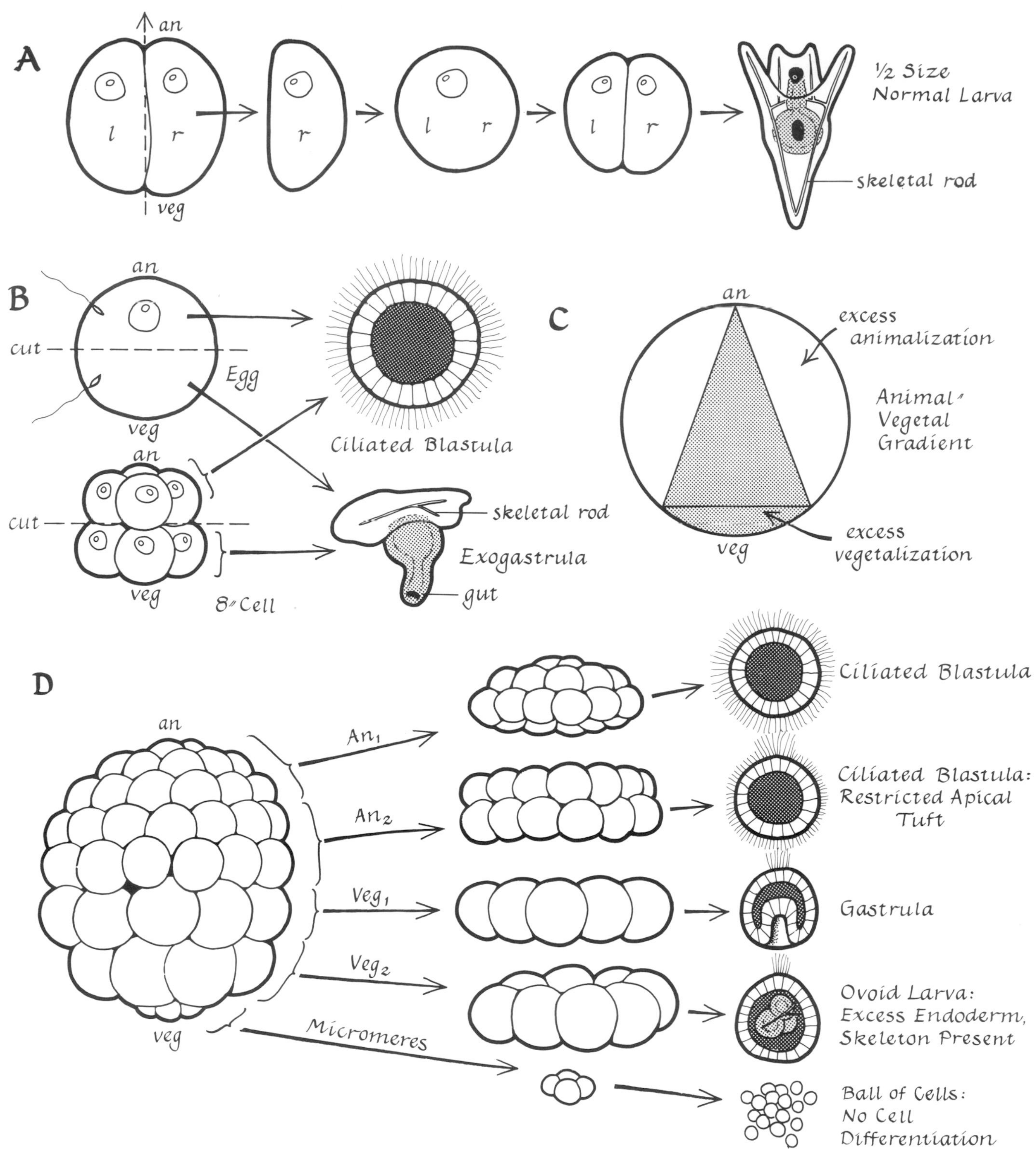

Figure 13-4. *Experiments illustrating how the general cytoarchitecture of the sea urchin egg can be deduced. A, separation of blastomeres at the two cell stage; B, division of the unfertilized egg in the equatorial plane transverse to the animal-vegetal axis, and separation of the animal and vegetal quartets of blastomeres at the eight cell stage. C, diagram of the quantitative differences along the animal-vegetal axis. D, development of isolated tiers of cells of the 64-cell stage. Based on studies of Horstadius, 1939.*

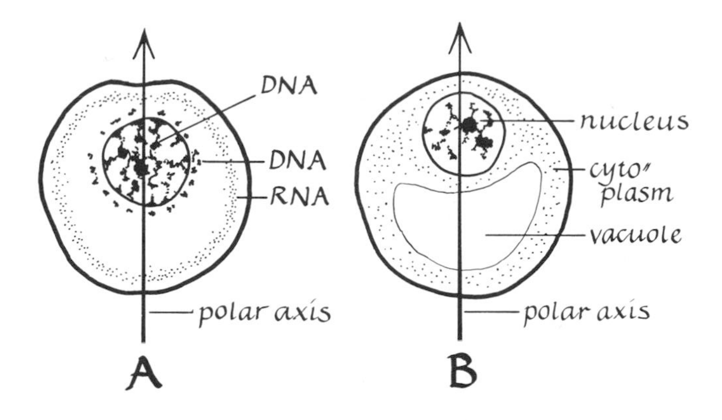

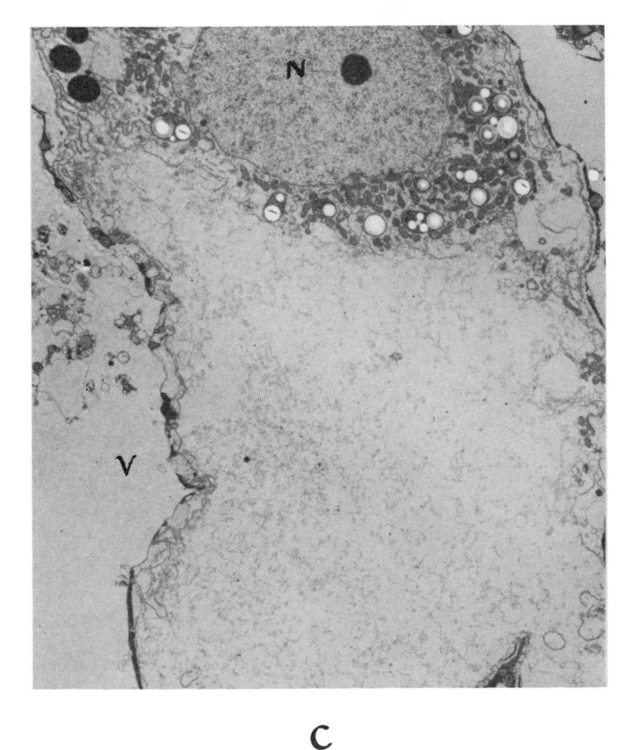

Figure 13-5. *A, pattern in the egg of the fern* Pteridium; *B, pattern in the egg of the cotton plant* Gossypium; *C, electron micrograph of the young zygote of the cotton plant showing the nucleus,* N, *surrounded by dense cytoplasm containing starch and the vacuole,* V *(× 4800). C, photograph courtesy Dr. Wm. Jensen.*

Among the components of egg cytoplasm are the *mitochondria;* these contain enzymes involved in oxidative phosphorylation, which results in the production of energy-rich ATP. All mitochondria contain a complete protein-synthesizing system (Chapters 23 and 25). Mitochondria of the developing oocyte of *Xenopus* contain not only circular molecules of DNA but also a polymerase necessary for the synthesis of RNA copies (Dawid, 1968). There is also some evidence that, within an organism, different kinds of cells contain different kinds of mitochondria (Porter, 1963) or different numbers of mitochondria (for instance, in the variegated leaves of plants, Du Buy et al., 1950).

Studies of Dawid (1968) indicate that there are no general homologies in base sequences between mitochondrial and nuclear DNA; that is, different genes are present in the mitochondria and the nucleus. In view of this evidence, mitochondria may constitute important units in the pattern of cytoplasmic differentiation (Weber, 1962). In general, when cells cleaved out of mitochondrial-rich parts of the cytoarchitecture are tested by isolation from the egg, they have greater and/or different capacities for further development than those arising from mitochondrial-poor areas. Figure 13-6 illustrates a few patterns of mitochondrial distribution during early cleavage in several types of eggs (see review by Reverberi, 1961). The patterns shown in the figures have been observed mostly by using histochemical or other staining methods; this makes critical evaluation of the observations difficult.

In coelenterate eggs (Figure 13-6A) the mitochondria are approximately evenly distributed, each blastomere containing about the same number through the sixteen-cell stage. In some coelenterates each blastomere of the sixteen-cell stage can continue to divide when isolated, giving rise to a whole planula larva. During cleavage of the ctenophore egg, *Beroe ovata*, there is an orderly segregation of most of the cortically located mitochondria of the egg into the micromeres of the sixteen-cell stage. These mitochondrial distributions were detected by Janus green staining and the Nadi reaction (Reverberi, 1957a). Through further division, the micromeres produce the comb or swimming plates of the larva. As shown in Figure 13-6B, when the sixteen-cell stage is divided into two groups of cells, two larvae form; the number of comb plate rows in each then corresponds to the number of micromeres present in the groups of cells (Fischel, 1898). The total number of comb plate rows in the two larvae is eight—the number in a normal larva. In the annelid *Tubifex rivulorum* mitochondria, as indicated by the Nadi staining reaction for cytochrome oxidase, are concentrated in animal and vegetal polar caps of the egg (Figure 13-6C). During cleavage most of the mitochondria

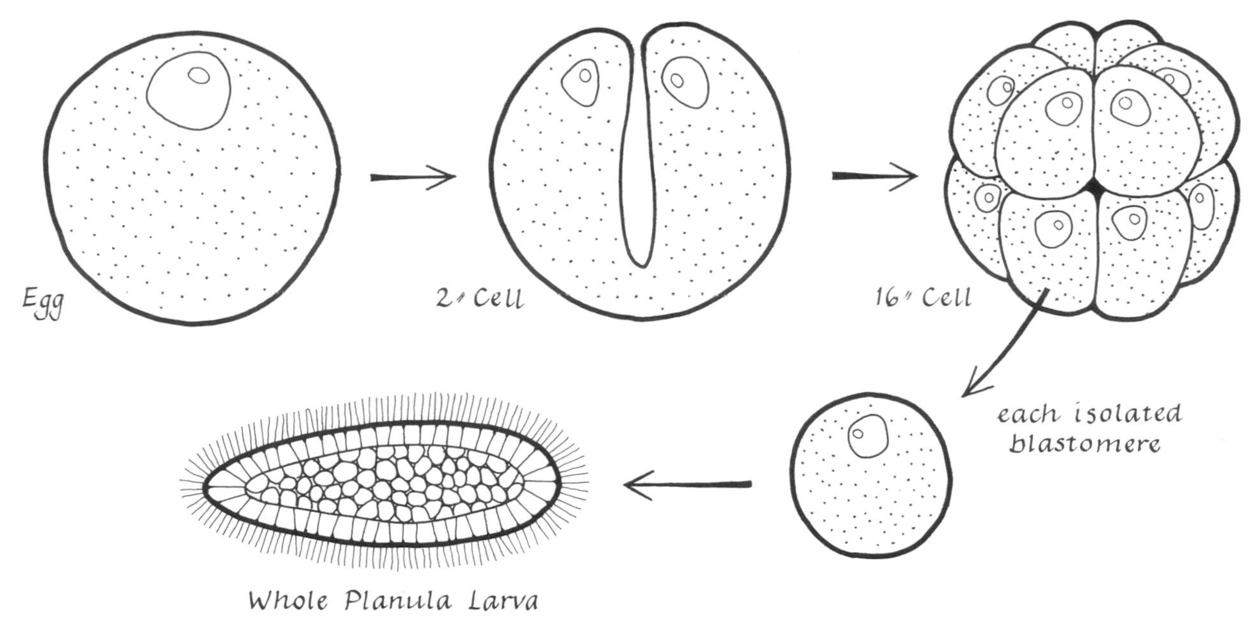

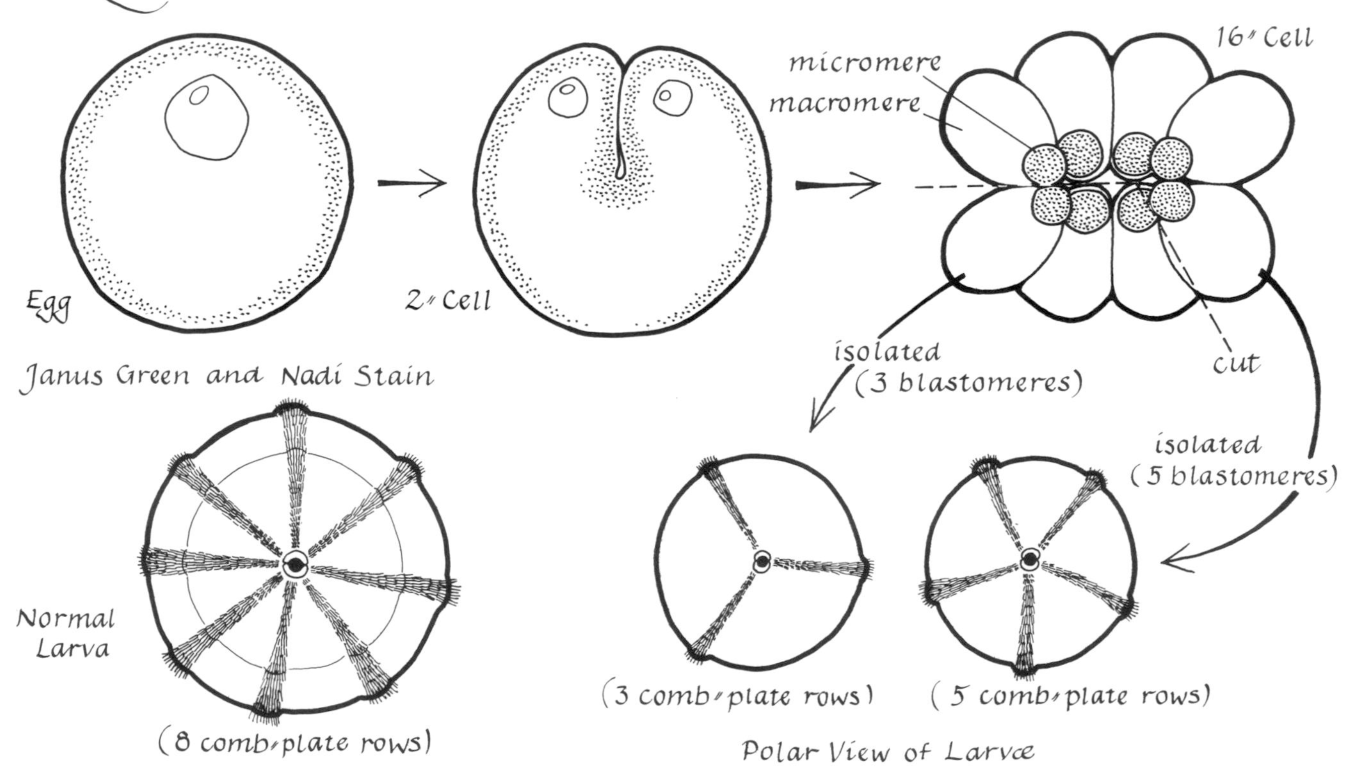

Figure 13-6. *Patterns of mitochondrial distribution in eggs and early cleavage stages of several invertebrates and a protochordate.*

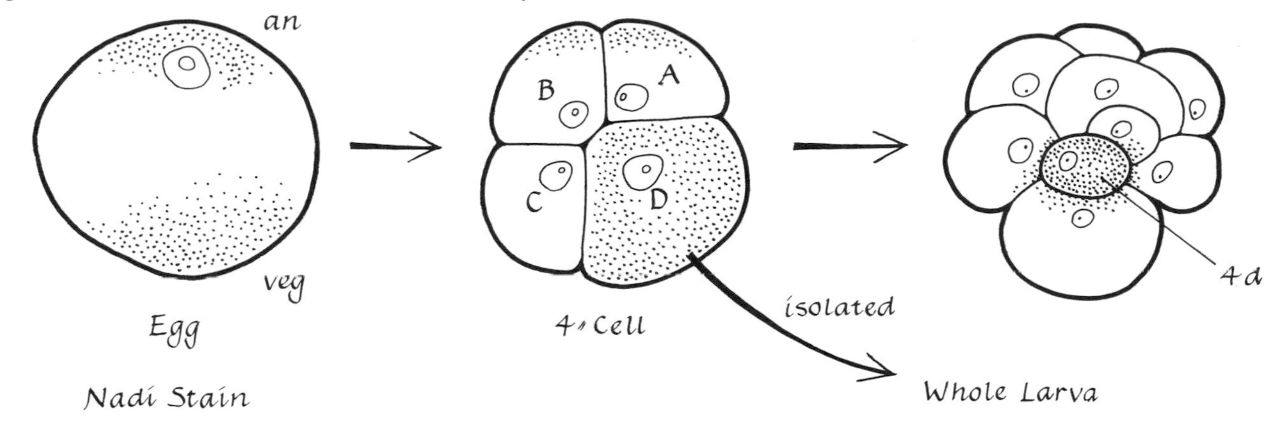
C. POLAR DISTRIBUTION (ANNELID)
an
B
A
C
D
veg
Egg
4-Cell
isolated
4d
Whole Larva
Nadi Stain

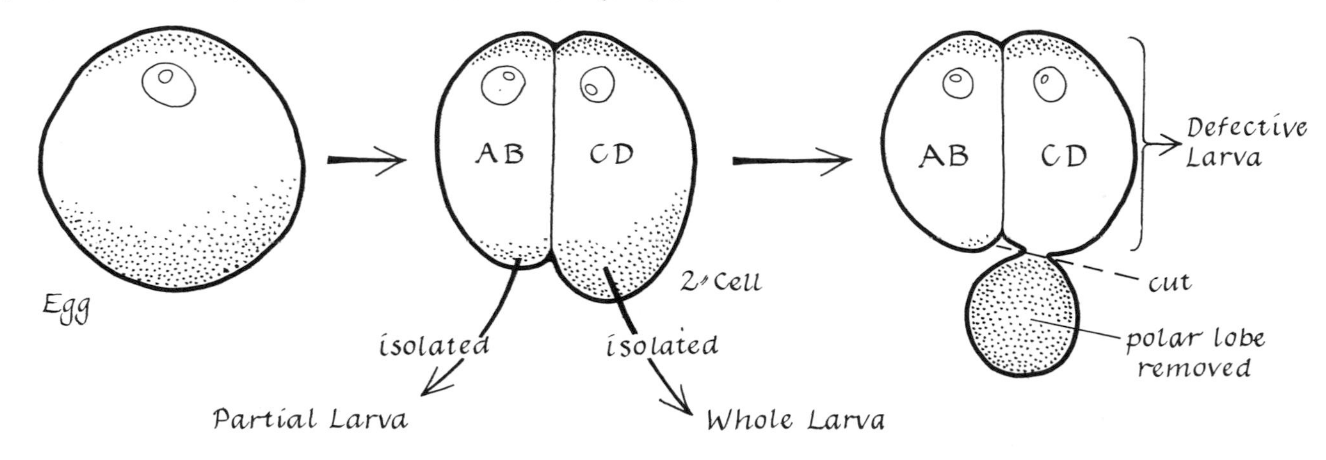
D. POLAR LOBE DISTRIBUTION (MOLLUSC)
AB
CD
AB
CD
Defective Larva
Egg
2-Cell
cut
isolated
isolated
polar lobe removed
Partial Larva
Whole Larva

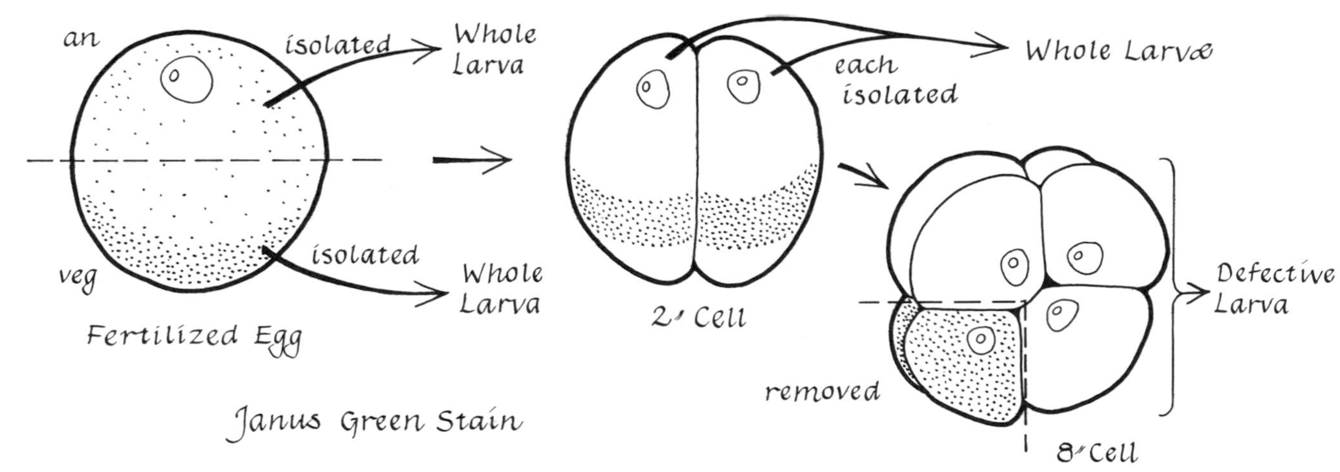
E. CRESCENT DISTRIBUTION (ASCIDIAN)
an
isolated
Whole Larva
Whole Larvæ
each isolated
veg
isolated
Whole Larva
2-Cell
Defective Larva
Fertilized Egg
removed
Janus Green Stain
8-Cell

are segregated into the D cell at the four-cell stage, and into the 4d cell derived from this at a later stage (Lehmann, 1948, 1956; Reverberi, 1961). Electron microscopy reveals that mitochondria are concentrated in the 4d cell (Lehmann, 1956; Weber, 1962). Note that an isolated D cell can develop into a whole worm. In molluscs like *Dentalium entalis* and *Ilyanassa obsolota* the polar lobe, a cytoplasmic projection from the CD cell at the two-cell stage, has a relatively large mitochondrial population (Figure 13-6D). Removal of the polar lobe results in a defective larva (Wilson, 1925; Clement, 1962; Raven, 1964). One of the best examples of differential segregation of mitochondria is found in the egg of the ascidian *Phalusia*. Following fertilization, the mitochondria, revealed by Janus green staining, form a crescent at the vegetal pole. The first three cleavages of the egg segregate the mitochondria into the two posterior vegetal blastomeres (Figure 13-6E). In ascidian tadpoles originating from Janus green stained eggs, the muscle cells are colored green. Removal of the posterior vegetal blastomeres (Chapter 15, Figure 3) results in a larva which lacks muscles (Reverberi, 1957b; 1961). The histochemically detected mitochondrial localizations in *Ciona*, another tunicate, have been confirmed by a quantitative method for detection of cytochrome oxidase (Berg, 1956). Further evidence of the possible importance of mitochondria as components of the cytoplasm of eggs is revealed by centrifuging the unfertilized egg of *Ciona intestinalis*. After the centrifuged egg is fertilized, only the half containing most of the mitochondria develops into a complete larva (Reverberi and Spina, 1959). However, in another ascidian, *Ascidia malaca*, all types of egg halves develop into complete tadpoles after fertilization (Reverberi and Ortolani, 1962). Despite the results shown in Figure 13-6, it cannot be definitely concluded that the distribution of mitochondria is a developmentally significant aspect of the cytoarchitecture. Ribosomal localizations or gradients in these and other kinds of eggs (Figure 17-5) are also of possible significance (Brachet, 1967).

Origin of Cytoarchitecture: Oogenesis

Studies of the formation of an egg are just as important to an understanding of development as studies of the development of the mature egg into an adult. Most early studies of oogenesis have been reviewed by Raven (1961). With some exceptions (for example, eggs of ferns or mosses removed from the archegonium), all the information necessary for development is built into the egg during oogenesis. The environment external to the mature egg is largely permissive, in that it does not control the program of stages. The nature of the information and how it is put into the egg is thus an extremely important problem in embryonic development. A few aspects of the process of egg formation are examined below.

Origin of the Egg

In plants, as we saw in Chapter 10, eggs are formed in the gametophyte within some reproductive organ —for instance, an *oogonium* in certain fungi, an *archegonium* in liverworts, mosses, and ferns, and an embryo sac in *angiosperms*. In some brown algae, *Fucales*, the egg is a direct product of meiosis as in animals (Chapter 12). The influences that cause one gametophyte cell to be set apart to become an egg are unknown. (We noted this in Chapter 10.) In many invertebrates and vertebrates, as we saw in Chapter 12, but not in plants, the eggs develop from special progenitor cells, the *primordial germ cells*, set apart relatively early in development of the egg. The primordial germ cells of frogs and *Xenopus* are cleaved out of a special cytoplasmic region rich in DNA at the vegetal pole of the egg and migrate later into the developing ovary (Bounoure, 1934; Nieukoop, 1946; Blackler, 1962). In vertebrates in general there is good evidence for the extragonadal origin of primordial germ cells. In some invertebrates (*Ascaris* and many insects) there is also good evidence for continuity of germ plasm from generation to generation. There is evidence that testicular and nongonadal tumors in mice (examples of *teratomas*), in which embryo-like bodies resembling five- to eight-day mouse embryos may form, have arisen from primordial germ cells (Stevens, 1967). There are apparent but doubtful exceptions to the origin of animal eggs from primordial germ cells. Eggs have been said to arise from choanocytes in sponges (Raven, 1961). However, in most animals, eggs develop within the ovary from primordial germ cells of extragonadal origin (Chapter 12).

As already noted in Figures 12-2 and 12-6, meiosis in an oocyte derived from an oogonium typically

Figure 13-7. *Diagram based on electron micrographs of the guinea pig oocyte showing the oocyte-follicle cell relationship. Note the extensions of the follicle cells,* PFE, *within the* zona pellucida, ZP. *Where extensions of follicle cells touch the surface of the oocyte, desmosomes are present. Microvilli,* MV, *are present on the surface of the oocyte.* M, *mitochondria;* G.B., *Golgi bodies;* ER, *endoplasmic reticulum;* MVB, *multivescular bodies. From Beams, in* Cellular Membranes in Development. *Copyright 1964 by Academic Press.*

results in formation of four cells, one of which becomes the egg, the others, polar bodies. In craneflies, each oogonium undergoes four mitotic divisions to produce sixteen cells, one of which becomes an oocyte, the others, nurse cells. In this process, Feulgen-positive bodies are preferentially distributed to the oocyte—an example of differential mitosis (Bayreuther, 1956). A similar process in *Dytiscus* (water beetle) results in an oocyte with extra DNA. The oocytes in craneflies and water beetles then undergo typical meiosis to form four cells, only one of which becomes an egg.

Contribution to the Egg by Follicle or by Nurse Cells

All eggs are formed in relation to the body of the mother organism, even in algae and fungi where the egg initial may not be surrounded by a special reproductive structure. As already noted (Chapters 10 and 11), the egg of more complex plants forms in a multicellular reproductive structure which may be an archegonium or an embryo sac. In many animals, the egg forms in a multicellular vesicle called a *follicle*. In cases where there is no multicellular reproductive organ, one or several nurse cells may come in contact with a developing egg and may inject nutrient inclusions into the forming oocytes. Autoradiographic studies suggest that RNA of nurse cells is fed into the oocyte of the housefly (Bier, 1963). Mitochondria and DNA are transferred to some kinds of eggs by surrounding nurse cells. Direct transfer of substances through cytoplasmic bridges from follicle cells has been described in cephalopods, insects, fishes, lizards, chicks, and rabbits (see review of Raven, 1961). An extreme type of contribution of follicle cells to the oocyte occurs in the snail *Helix* and some coelenterates—the phagocytosis by the oocyte of whole follicle cells. In *Drosophila,* there are direct connections between nurse cells and oocyte by means of special apertures, "fusomes," which may be open or closed. The apertures are large enough to allow mitochondria to enter the egg cytoplasm (see review of Williams, 1965). Williams cites a claim that labeled DNA of follicle cells may enter the egg cytoplasm, but this requires confirmation. There is evidence for the transfer of antigens from the blood to oocytes of insects (Telfer, 1961), birds (Nace, 1953; Schechtman, 1956), and amphibians and mammals (Williams, 1965). One example of the close relationship between follicle cells and the oocyte of the guinea pig (Beams, 1964) is illustrated in Figure 13-7.

That follicle or nurse cells contribute significantly beyond the supply of nutrients and possibly extra DNA and RNA remains to be demonstrated. Also, the relative roles played by the oocyte nucleus and the follicle or nurse cells are not clearly understood. There is evidence that the oocyte nucleus transfers material to the cytoplasm in the snail, *Limnaea,* and in amphibians, sea urchins, and insects (Raven, 1961). A similar transfer has been described for the egg of a fern (Bell, 1963). A possible indication of transfer from nucleus to cytoplasm is the common appearance of nuclear extrusions during oogenesis (Raven, 1961; Bell, 1963; Grant, 1965). The complex hood-shaped evaginations of the nuclear membrane in the egg of the fern *Pteridium aquilinum* are illustrated in Figures 13-8 and 13-9. There is evidence that these evaginations become detached and that some might develop into mitochondria, others into proplastids, but this remains to be confirmed. Thus, the nucleus in this egg appears to play an important role in development of the cytoarchitecture of the egg. The membrane surrounding the nucleus of the cotton egg also has numerous projections into the cytoplasm (Jensen, 1964).

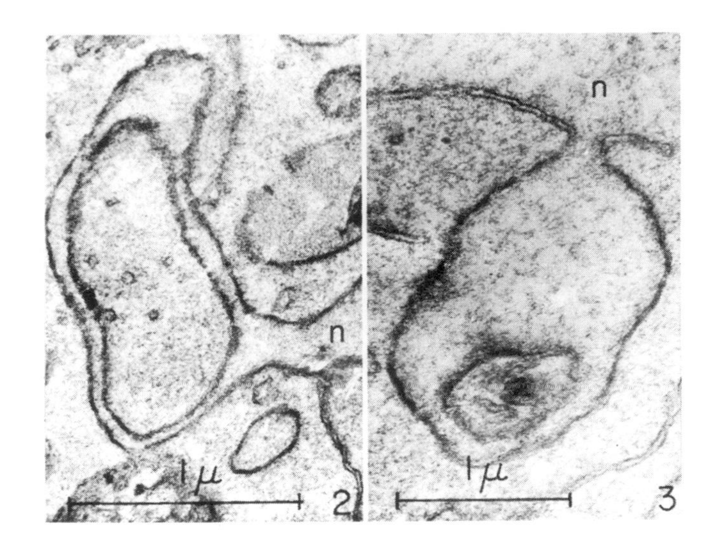

Figure 13-8. *Electron micrograph of complex evaginations of the nucleus of the maturing egg of the fern* Pteridium. n, *nucleus. Evaginations are probably sections in different planes of a hooded protrusion of the nucleus similar to that illustrated in Figure 13-9. From Bell, in* Cellular Membranes in Development. *Copyright 1964 by Academic Press.*

Origin of Polarity and Symmetry of the Oocyte

All animal and most plant eggs are polarized, and this polarity constitutes a basic guideline for later development. Polar organization may be exhibited in the shape of the egg, as in many insects; in the distribution of yolk and pigment, as in amphibian eggs; or in the eccentric position of the nucleus (Raven, 1961). Some oocytes (such as those of rats) may have a bilaterally symmetrical pattern involving the distribution of basophilic (RNA) granules (Dalcq and Seaton-Jones, 1949). This is shown in Figure 13-10. Unlike other blastomeres, single isolated blastomeres cleaved out of the dorsal basophilic region of the rabbit egg can develop into embryos, in some cases into complete rabbits, when implanted in a foster mother (Chapter 15).

In some eggs fertilization or activation initiates *ooplasmic segregation,* which involves a relatively simple (as in frogs) or complex (as in tunicates) shifting about of cytoplasmic components into a symmetrical pattern that did not exist before. In some species of salamanders and frogs, the entry of the sperm initiates streaming movements that cause the *grey crescent* to appear opposite the point of sperm entry. The grey crescent forms on the future dorsal side of the egg and visibly denotes its bilateral symmetry.

Because polarity constitutes a basic guideline for later development, its source in those mature eggs which exhibit it before fertilization is an important problem. In *Volvox globitor* the point at which the egg is attached to the parent determines its polarity. In mosses and ferns the plane of polarity of the egg coincides with the long axis of the archegonium. This occurs even in fern prothallia in which archegonia develop on the upper surface following illumination of the lower surface (Figure 10-28). The relationship of the polar axis of the fern or moss egg to the axis of the archegonium is illustrated in Figure 13-11.

The first cleavage plane is not random. As indicated in the figure, the plane of the egg's first division either coincides with the long axis of the archegonium or, more frequently, is at right angles

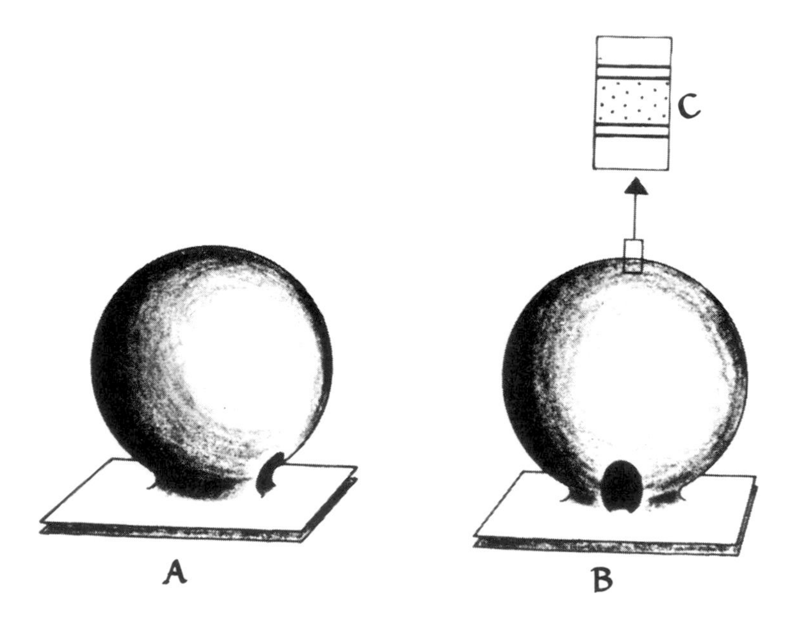

Figure 13-9. *Reconstruction of a hooded protrusion of the nucleus of* Pteridium. A *and* B, *side and face views;* C, *section of boundary of hood showing how it is formed by four unit membranes. The pore is usually adjacent to the nucleus. From Bell, in* Cellular Membranes in Development. *Copyright 1964 by Academic Press.*

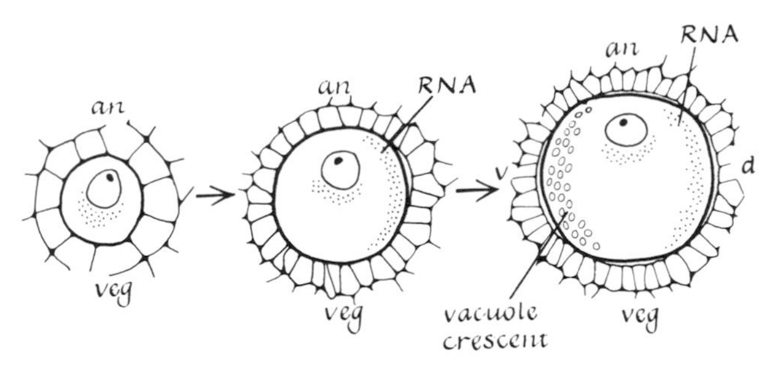

Figure 13-10. *Development of bilateral symmetry in the egg of the rat.*

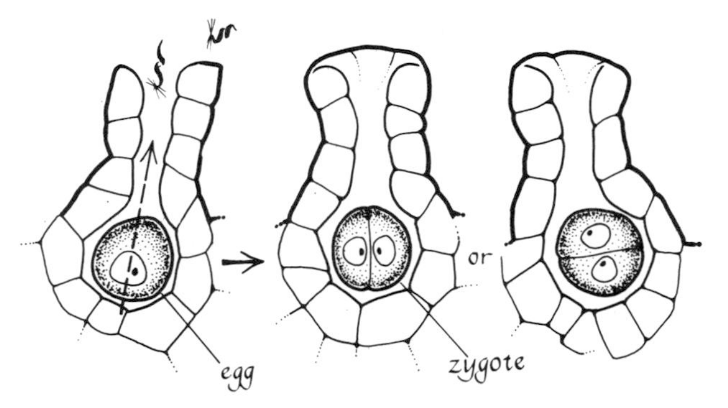

Figure 13-11. *Relationship of the axis of the fern or moss egg to the axis of the archegonium.*

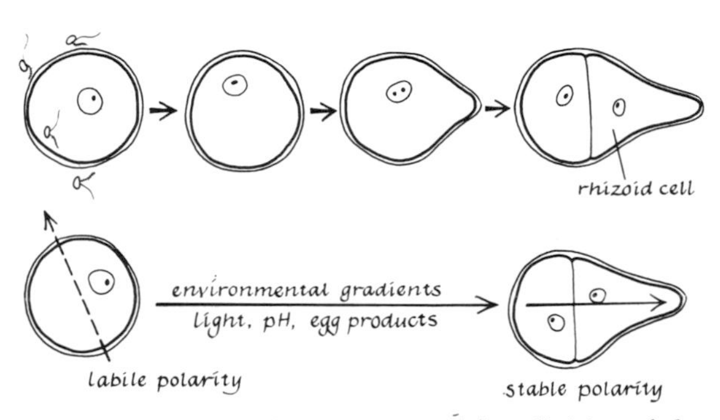

Figure 13-12. *Upper: fertilization and first division of the* Fucus *egg. Lower: possible role of environmental agents in stabilizing the polarity of the egg.*

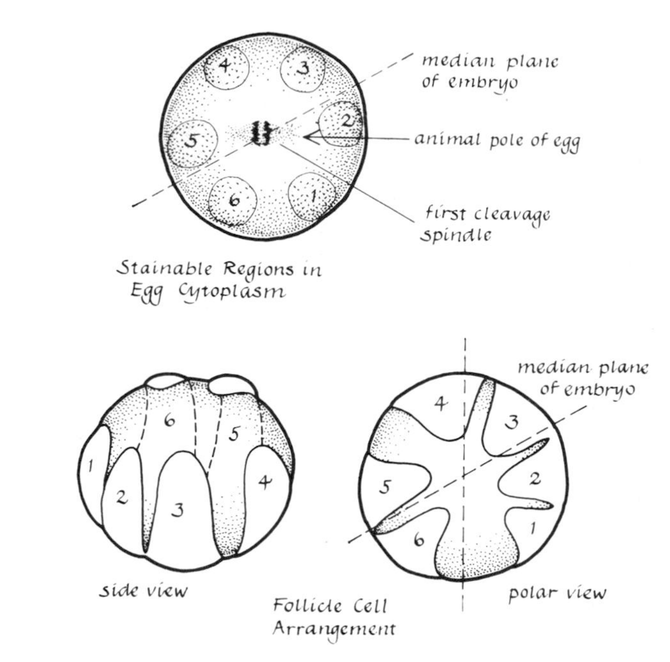

Figure 13-13. *The correspondence of the asymmetrical pattern of stainable spots on the surface of the* Limnaea *egg, upper to the asymmetrical arrangement of follicle cells, lower diagram.*

to it (Wardlaw, 1955). The archegonium, as part of the parent organism, thus provides direct continuity between parent and offspring. Recall (Chapter 11) that in seed plants the polarity of the egg coincides with the long axis of the embryo sac.

Even in the egg of *Fucus furcatus* (Figure 13-12) where external environmental influences (CO_2, pH gradients, polarized light, and so on; see Whitaker, 1940; Jaffe, 1968) may determine the polarity of the mature egg, it is still possible that a preformed polarity is present and becomes directionally stabilized by the external agents. This is indicated by the fact that no external influences are needed for normal development of *Fucus* eggs and that polarity induced by light can be reversed by counterillumination if applied before the polarity becomes stabilized (Jaffe, 1968). In an alga, *Laminaria,* related to *Fucus,* the egg develops into an embryo whose axis coincides with that of the oogonium (Wardlaw, 1955).

In general, polar organization in animals is established during oogenesis; this is true, for example, in sea urchins (Horstadius, 1939), amphibians (Dalcq, 1957), and birds (Bartelmez, 1918). In the case of the frog egg, the sperm entrance point probably stabilizes the position of a preexisting bilaterally symmetrical organization. A grey crescent-like area develops even in unfertilized, nonactivated, and aging frog or toad eggs. We shall have more to say about this later in this chapter.

The Egg of Limnaea

How the egg of *Limnaea stagnalis,* a freshwater snail, acquires a spiral pattern of organization has been described by Raven (1948, 1964, 1967). The pattern appears to be imprinted upon the egg by the surrounding follicle or nurse cells. Six cytoplasmic regions in the mature egg stain differently from the rest of the cytoplasm. The stainable spots are asymmetrically arranged, but the pattern of asymmetry apparently is not reversed in right- and left-handed shell types (Ubbles et al., 1969). The significant point is that the asymmetrical pattern of spots on the egg corresponds to the asymmetrical arrangement of the follicle cells. This is illustrated in Figure 13-13.

There is not only a definite relationship between the follicle cell arrangement and the cytoplasmic pattern in the egg, but also a definite relationship between the cytoplasmic pattern and the polarity and dorso-ventrality of the later embryo (Raven, 1967). In this case the genes of the mother organism, rather than those of the male or the combination of genes in the zygote, control the pattern of cytoplasmic asymmetry and subsequent development. This is illustrated in crosses between snails with right- and left-handed forms of the spiral shell, since all the F_1 offspring exhibit the asymmetry of the mother—even when the father is a right-handed, dominant snail (Boycott, et al., 1931). This is an example of maternal inheritance (Chapter 18).

Although there are no such clearly demonstrated examples of maternal influences on the establishment of the cytoarchitecture of the egg in other organisms, a reasonable conclusion is that such an influence is a general principle of embryology. The follicle or nurse cells and their products constitute the microenvironment, the patterned composition of which is progressively imprinted upon the developing egg (Raven, 1964). This is illustrated in Figure 13-14.

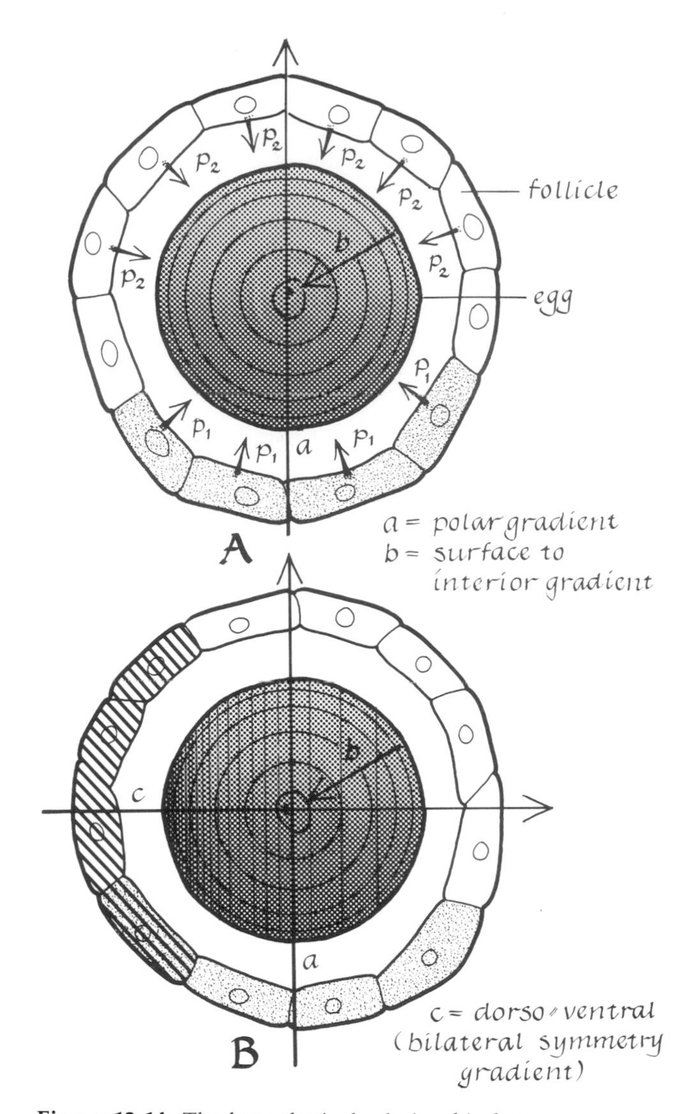

Figure 13-14. *The hypothetical relationship between an egg and the reproductive organ in which it is formed. A, origin of surface-interior and apico-basal (polar) differences in the egg; B, origin of bilateral symmetry (dorso-ventral axis) in the egg.*

The diagrams of Figure 13-14 illustrate a hypothetical relationship between an egg and the reproductive organ in which it is formed. In Figure 13-14A, *a* indicates a gradient in the concentration of products such as nutrients, hormones, or other substances released by the basal cells of a reproductive organ. The side of the egg next to the basal cells which release more of the same, or different kinds of products, would be exposed to a higher concentration of these products or to products different from those released at the opposite side of the egg. This could establish an apico-basal or antero-posterior gradient and thus a polar difference in the egg, since the amount of the products inside the egg would be greater on one side than on the other. In addition, in contrast to the interior of the developing egg, the entire surface is exposed to a higher concentration of products. This could result in a graded surface-interior difference in the egg. The intersection of the apico-basal or antero-posterior and surface-interior differentials would result in a radially symmetrical organization in the egg, a reflection of the radial symmetry of the reproductive organ. Bilateral symmetry of the egg could arise as a result of a second, dorso-ventral gradient (Figure 13-14B, *c*) in concentration of products released by a bilaterally symmetrical reproductive organ. Asymmetry in the right and left halves of the parent organism is also imprinted upon the egg, but in this case the asymmetry of the egg is not as stable as is either polarity or bilateral symmetry. We shall return to this problem in Chapter 26.

In summary, it appears that although the pattern of cytoarchitecture in the egg is epigenetically built up, it is under control of a preexisting or preformed pattern in the structure or activities of the reproductive organ. Consequently, the mother organism constructs a sample of her own organization and symmetry in the egg, thus preserving continuity between parent and offspring. The correspondence in symmetry of a frog egg and of a mature frog is not obvious; however, it is clearly established that the symmetry of the offspring is derived from that of the egg. Correspondingly, the symmetry of the egg is derived from that of the parent. In this respect, reproduction by means of an egg is not different in principle from reproduction by fission, zooid-formation, or bud-formation—in which cases the continuity between parent and offspring is more obvious.

Nucleoprotein Synthesis in Oocytes

As informational molecules bearing inherited instructions, nucleoproteins might play a significant role in the genesis of cytoarchitectural patterns. Exactly how nucleoproteins could constitute components of the cytoarchitecture is not known, but there is increasing evidence (Raven, 1961; Grant, 1965; Gross et al., 1965; Williams, 1965; Brachet, 1967; and others) that mature eggs in both plants (Bell, 1963) and animals (Dawid, 1966) generally contain much more DNA and RNA than most tissue cells. The scale of nucleoprotein synthesis appears to be unique to developing oocytes. DNA as well as RNA is present in both the nucleus and the cytoplasm of eggs. According to Brachet (1967), the sea urchin egg contains enough DNA to provide for the formation of 400 blastomeres; an amphibian egg contains enough DNA to form 5,000 to 10,000 cells. Much of the DNA of the sea urchin egg is in the cytoplasm, as demonstrated by the fact that the anucleated half of a sea urchin egg contains as much DNA as the nucleated half (Chapter 17). Histochemical studies indicate that a large amount of the DNA of an amphibian egg is in the cytoplasm (Grant, 1965).

DNA in the cytoplasm of sea urchin oocytes seems to be associated with yolk platelets (Brachet, 1967) and mitochondria (Piko et al., 1967). However, in a study of the *Xenopus* egg, Dawid (1966) and Brown and Dawid (1968) found all the DNA in the mitochondria. Possibly there are two kinds of cytoplasmic DNA: DNA with high molecular weight in the mitochondria, and DNA having low molecular weight in yolk granules. Brachet makes the interesting suggestion that DNA molecules primarily located in the cortex of the egg may function as self-replicating bodies, but there is no proof of this as yet.

During oogenesis there is also synthesis of RNA (Brachet, 1967; Piatigorsky et al., 1967; Davidson, 1968)—particularly of ribosomal RNA, which constitutes about 95 percent of the total RNA in amphibian eggs (Brown and Dawid, 1968). Studies of four species of amphibians, an echiuroid worm, and the surf clam have revealed that the oocytes of these animals contain many extra copies of the genes for the 28s and 18s subunits of ribosomal RNA. The synthesis of ribosomal RNA is carried out on the large number (600 to 1,000) of nucleoli within the germinal vesicle. The large quantities of ribosomes synthesized by the oocyte are stored for future use during development of the egg (Brown and Dawid, 1968). This amplification of genes in the nucleoli specific for ribosomal synthesis is characteristic of oocytes.

The other two kinds of RNA, transfer and messenger, are also synthesized during oogenesis. In the sea urchin about 2 percent of this RNA is template (that is, messenger RNA) according to Slater and Spiegelman (1966). Although there is much evidence (see Chapter 17) for the synthesis of cytoplasmic nucleoproteins during oogenesis, we have no evidence that these molecules constitute a structural component of the egg cytoarchitecture. Perhaps their primary role is to control the synthesis of cytoplasmic components (enzymes and other proteins) rather than directly to associate the proteins into a geometrical pattern of egg architecture. In Chapter 17 we shall consider in more detail the program of nucleoprotein synthesis underlying and controlling the early stages of development. In this section we have only pointed out that nucleoprotein synthesis in oogenesis constitutes a functional aspect of egg cytoarchitecture, in that much of the information for later control of development is preformed during oogenesis.

Experiments Illustrating the Importance of the Egg Cortex

The most complex and stable part of the cytoarchitecture of an egg resides in the *cortex*, the relatively more gelated, subsurface part of the egg. This is why centrifugation, which usually redistributes endoplasmic but not cortical inclusions, has little influence upon later development. Several kinds of experiments demonstrate this.

Centrifuge Experiments (Morgan, 1927; Harvey, 1951)

The distribution of visible egg components (such as granules, globules, yolk, and pigment) does not constitute the basic egg organization. A newly created axis of stratification in the centrifuge does not change the original polar axis of development. This is illustrated for the egg of the mollusc, *Cumingia*, and the egg of the sea urchin in Figure 13-15.

In the molluscs *Ilyanassa* and *Dentalium*, removal of the "first" polar lobe at the two-cell stage (Figure 13-16) results in a *trochophore* larva which lacks the apical tuft and the posttrochal region. This was discovered by Crampton (1896) in studies on *Ilyanassa*, the mud snail. Removal of the polar lobe in *Dentalium* results in a larva which develops neither a foot nor a shell (Wilson 1904).

However, if fertilized *Dentalium* eggs, prior to removal of the lobe, are centrifuged at 300 g for ten minutes—thereby producing stratification of the cytoplasm—the above result is not altered. It thus appears that the cortex of the lobe, rather than the lobe's endoplasmic structure (Verdonk, 1968), is responsible for apical tuft and posttrochal development.

Even when the centrifugal force is great enough (10,000 g) to pull the egg of a sea urchin into halves and the halves into quarters, any of these quarters if fertilized may develop into pluteus larvae—although each quarter contains a different fraction of the egg content (Harvey, 1951). See Figure 13-17.

In the experiment illustrated in Figure 13-17, each quarter contains a different part of the egg cortex.

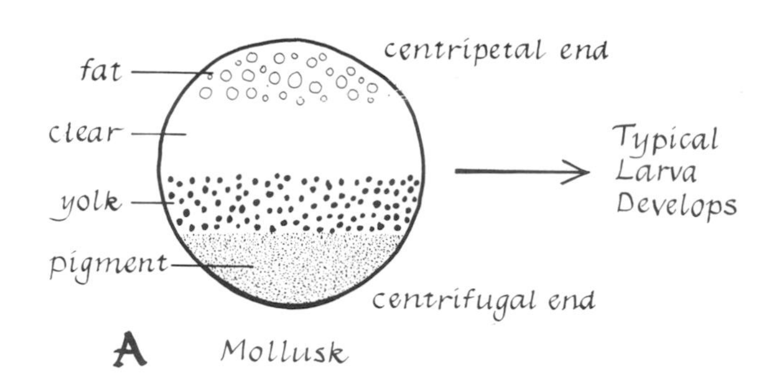

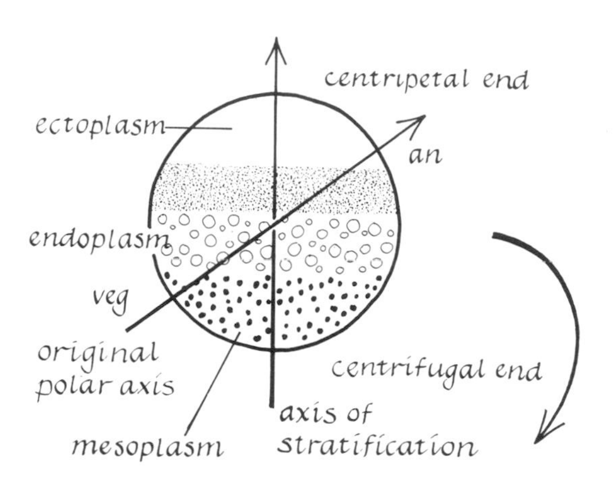

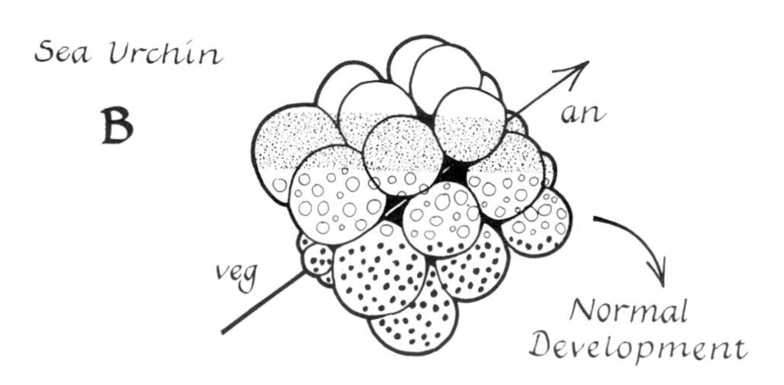

Figure 13-15. *Diagrams illustrating that visible inclusions in the egg cytoplasm do not constitute the basic egg organization in respect to later development.*

Some *regulation* of cortical structure must have occurred. The cortical organization of the sea urchin egg before fertilization is thus not irreversibly differentiated. Such an egg is described as a regulative egg, a system part of which can reconstitute the pattern of the whole (see Chapter 15).

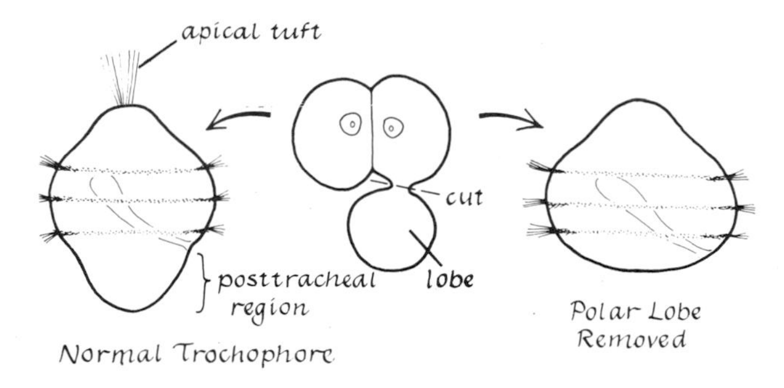

Figure 13-16. *Removal of the first polar lobe of the two-cell stage of the molluscs* Dentalium *and* Ilyanassa *results in defective development.*

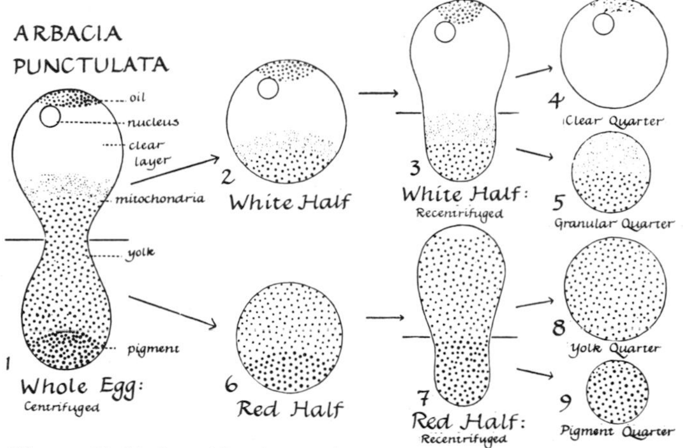

Figure 13-17. *Stratification and separation of the egg of* Arbacia punctulata *into quarters by centrifugal force (3 minutes at 10,000 × g). From E. B. Harvey, 1951,* Annals of the New York Academy of Sciences, *Volume 51, p. 1337, Figs. 1–9. Reproduced by permission.*

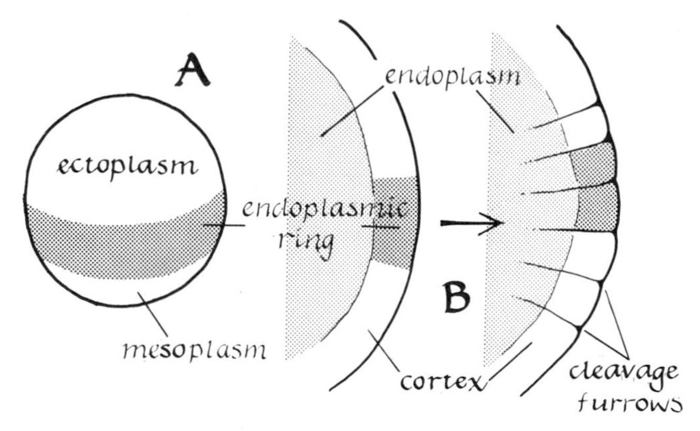

Figure 13-18. *Cortex of the egg of the sea urchin* Paracentrotus *remains on the surface during cleavage.*

One must interpret the results of centrifuging eggs carefully. Relatively low g forces may stratify the inclusions in the endoplasm of the egg without disturbing the cortical structure. It would be erroneous to conclude that no stable cytoarchitecture is present. Thus in the ascidian egg of the species *Styela partita,* strong but not weak centrifugation displaces entire areas of the cytoplasm, including cortical areas differing in staining properties. The result is the development of monsters whose organs are entirely out of proper relation to one another (Conklin, 1929).

Removal of Endoplasm (Horstadius et al., 1950)

The volume of the egg of the sea urchin *Echinus esculentus* was reduced by 50 percent by withdrawing fluid endoplasm with a micropipette. Twenty of these eggs developed into complete but half-sized pluteus larvae. In contrast, when a sea urchin egg was cut into animal and vegetal halves, as we noted above, development was abnormal.

Stability of the Position of the Cortex during Cleavage (Runnstrom, 1928; Motomura, 1935)

The egg of the sea urchin *Paracentrotus lividus* has an orange-red girdle of cortical cytoplasm just below its equator. This cytoplasm is incorporated into gut endodermal cells during cleavage. However, during cleavage the furrows initiated at the surface of the egg do not pull the cortical cytoplasm into the interior (Figure 13-18).

As Figure 13-18 shows, the cortex of the egg remains on the surface; thus, cleavage does not disrupt the cortical pattern.

Transplantation of the Cortex of the Xenopus Egg (Curtis, 1960, 1962)

Grafts of grey crescent cortex from eight-cell embryos, when placed in the ventral margin of uncleaved fertilized eggs, induced secondary embryonic axes. Also, excision of the grey crescent cortex of uncleaved eggs prevented embryo formation. These results are illustrated in Figure 13-19C and D.

Note in Figure 13-19A that removing the grey crescent cortex from an eight-cell stage does not prevent embryo formation since by this time the crescent has already exerted its influence on the remainder of the embryo. Also, grafting the grey crescent cytoplasm of an uncleaved egg into an eight-cell stage (Figure 13-19B) does not produce a secondary embryonic axis; this is presumably because the host grey crescent has already exerted its organizing influence over the embryo, making the nuclei of the embryo incompetent to respond to influences of the transplanted grey crescent cytoplasm in which they reside. Although these results appear to demonstrate clearly that the cytoarchitecture of this egg carries information essential to later development (Pasteels, 1964), they have not been confirmed; until they are, their validity remains in doubt.

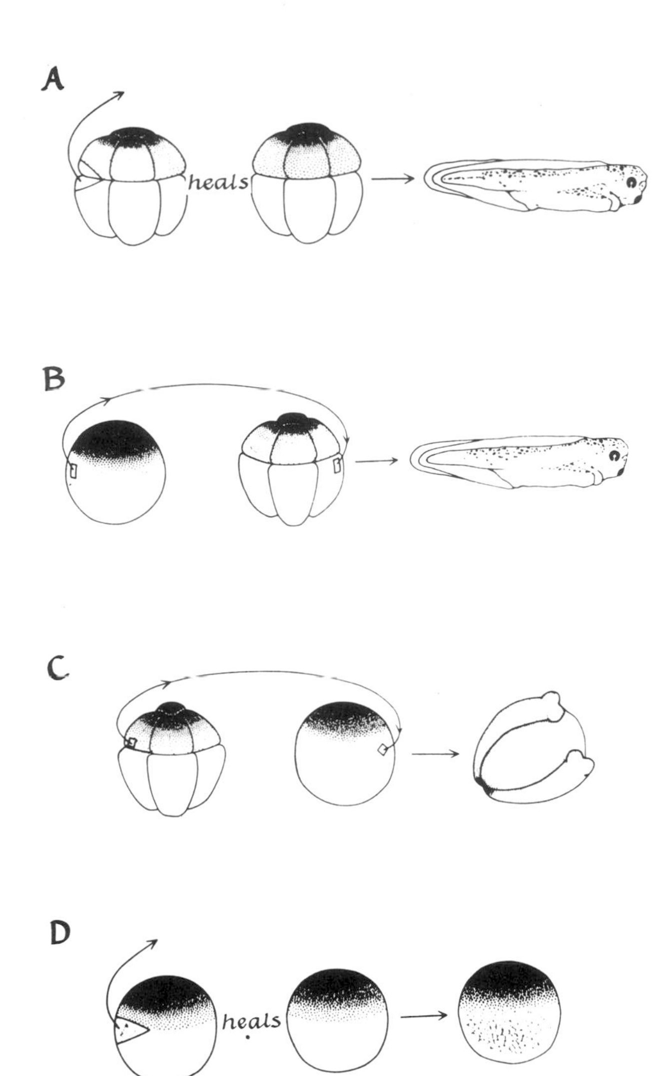

Figure 13-19. *Transplantation and excision of the grey crescent cortical cytoplasm of the egg of* Xenopus. *From Curtis, 1962. Courtesy Clarendon Press, Oxford.*

Cortical Pattern in Cephalopods

Arnold (1963, 1968) has described a mosaic-type prepattern in the cortex of squid and octopus eggs. Defects in the cortex produced by ionizing irradiation are not repaired. Furthermore, there is no disruption or displacement of the cortex by ooplasmic streaming movements within the egg. Recall (Chapter 12) that there is evidence for preformed cleavage furrows in the cortex of the squid egg (Arnold, 1968B).

Relationship of Cytoarchitectural and Cleavage Patterns

In many animal and plant eggs the cytoarchitectural pattern appears to control or direct the pattern of early cleavage. This is suggested by the progressive and orderly separation of the diverse cytoplasmic regions by the cleavage planes. Cleavage of the egg of an ascidian is one example (Conklin, 1905 and Weiss, 1939), and is shown in Figure 13-20.

We have already cited an example of the correspondence of the egg pattern and spiral cleavage in snail eggs. In some eggs, particularly in vertebrates, the cleavage pattern may be irregular and apparently without orderly relation to the symmetry or polarity of the egg. Nevertheless, it seems clear (indeed, the orderly nature of development demands it) that cleavage does not disrupt or destroy the pattern of egg structure. For example, cleavage transforms the *intracellular* pattern of cytodifferentiation into an *intercellular* pattern of diverse cell types in a cell population. This does not mean that the cytoplasmic components are rigidly fixed in position. There may be extensive redistribution of components just prior to cleavage and within the blastomeres during cleavage. However, this redistribution is always orderly.

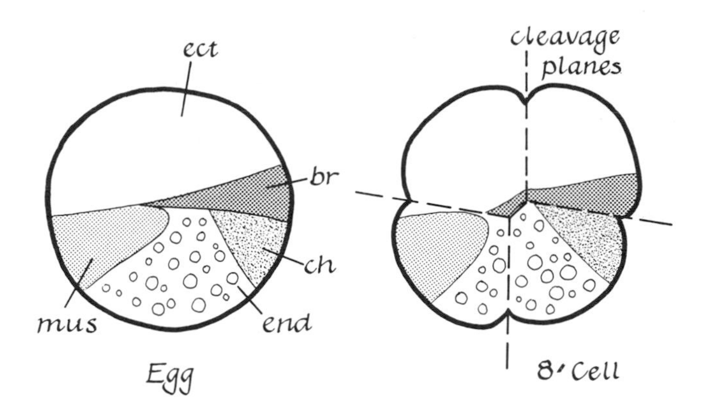

Figure 13-20. *Relationship of cytoarchitectural and cleavage patterns in an ascidian egg.*

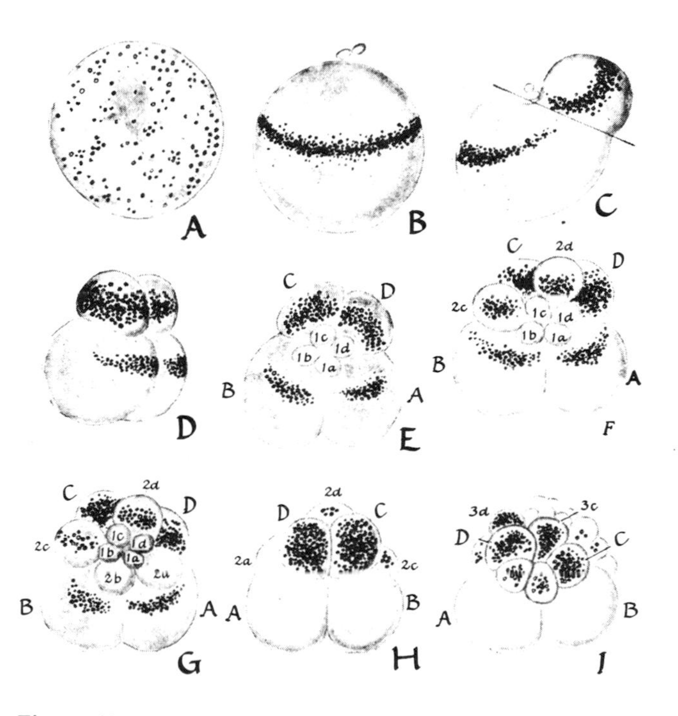

Figure 13-21. *Orderly distribution of indophenol blue staining particles during cleavage of the egg of* Aplysia. *From Needham, 1950. Courtesy Cambridge University Press.*

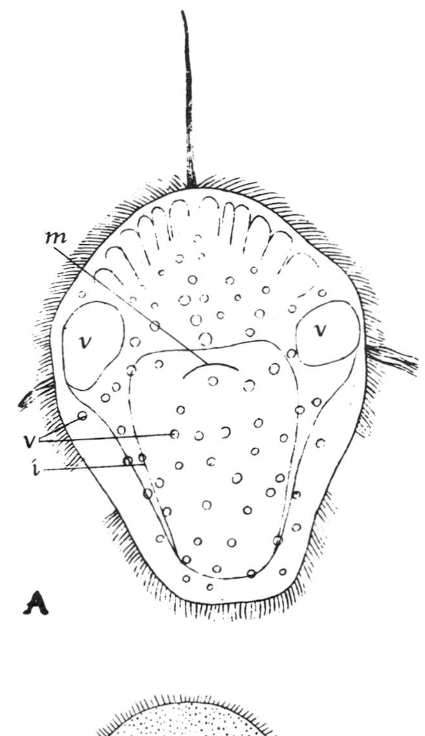

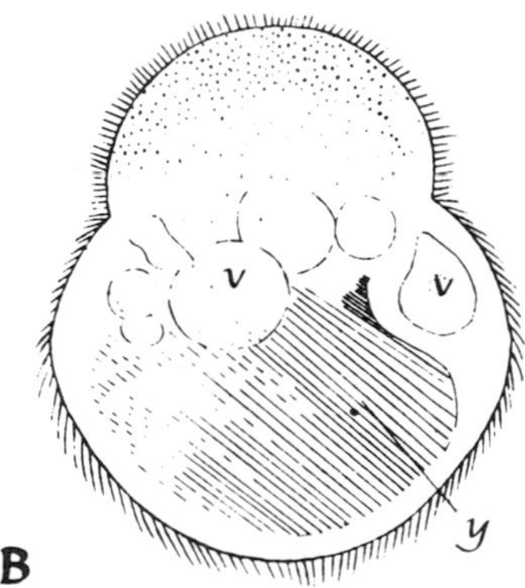

Figure 13-22. *Differentiation without cleavage in the egg of* Chaetopterus pergamentaceus. *A, normal trochophore larva; B, larva developing from activated but undivided egg. y, yolk;* v, *vacuoles. From Lillie, 1902. Courtesy Springer-Verlag.*

An example is the distribution of ascorbic acid (Vitamin C) during cleavage of the egg of the mollusc *Aplysia* (Needham, 1950), shown in Figure 13-21. Presumably these small water-soluble molecules are attached to or are present in larger, less randomly diffusing particles. However, it remains uncertain that ascorbic acid was actually being detected by the methods used. Nevertheless, the experiments reveal an orderly distribution of a stainable component of the cytoplasm during cleavage.

As noted above, most plant eggs exhibit a polarized pattern of cytoarchitecture (Figure 13-5) before or after fertilization. The first division of the zygote is typically by a wall at right angles to the polar axis (Chapter 10) and results most commonly in a smaller, densely protoplasmic distal cell and a larger, frequently vacuolated basal cell. This division is thus largely determined by the constitution of the zygote (Wardlaw, 1955).

Cytoarchitecture as a Primary Guideline

An instructive experiment was performed many years ago on eggs of the worm *Chaetopterus pergamentaceus* by Frank Lillie (1902). Unfertilized eggs were activated by treatment with KC1 in sea water. When returned to ordinary sea water, the eggs underwent progressive changes in the undivided cytoplasm which mimicked the development of a normal egg up to the trochophore larval stage. This is illustrated in Figure 13-22.

The phenomenon shown in Figure 13-22, which has been called *cytodifferentiation without cleavage*, demonstrates the presence of a specific plan of development in the cytoplasm of an egg. However, as one might suspect, given both the absence of normal nucleo-cytoplasmic interactions and the partial isolation of cytoplasmic areas by the membranes of cleavage, the development stops at the pseudolarval stage and is never as orderly as in normally cleaving eggs.

There are a number of other examples of egg development in the absence of cleavage. A bilateral organization characterized by the appearance of the grey crescent has been observed in unfertilized frog and other amphibian eggs (Morgan, 1902, 1927; Holtfreter, 1943). More surprisingly, cortical streaming movements in uncleaved frog eggs simulate the epibolic cell movements in a normal blastula. Furthermore, even gastrulation is simulated, and a pseudocellular pattern is engraved on the egg surface. The pattern may be so much like that of a blastula that only by dissection or histological sectioning of the egg can one be sure that it is actually unsegmented. Even large wounds in the egg cortex are closed in uncleaved eggs aged for up to five days. Still more significant is the observation that the events simulating epiboly, blastulation, invagination, and involution (gastrulation) coincide temporally with corresponding events in normal, cleaving eggs (Figure 13-23).

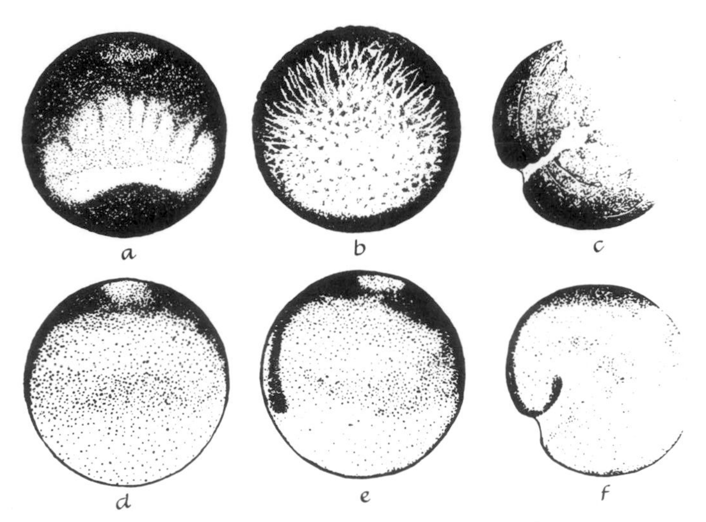

Figure 13-23. *Cytoplasmic movements simulating gastrulation movements in an unfertilized egg of* Rana pipiens *kept for 2-3 days in tap water,* a *and* b, *or in frog Ringer,* c. d-f, *median sections of eggs in which gastrulation movements were simulated. From Holtfreter,* J. Exp. Zool., *Volume 93, p. 258, 1943. Courtesy Wistar Press.*

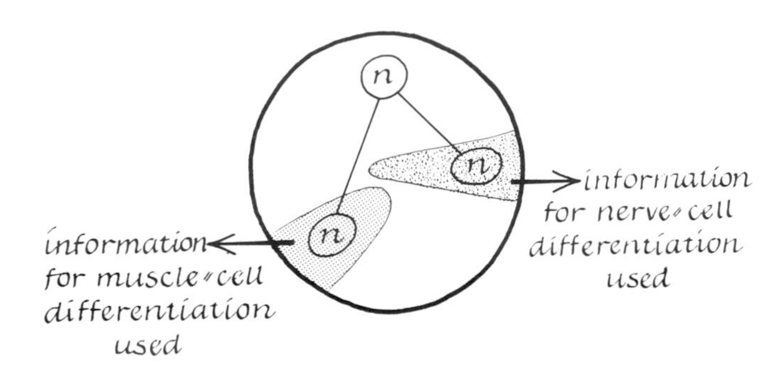

Figure 13-24. *Diagram illustrating the hypothesis that cleavage nuclei respond differentially to different cytoplasmic microenvironments.*

A similar synchrony of shape changes in the isolated, anucleate polar lobe of the eggs of the molluscs *Dentalium* and *Ilyanassa* (Figure 13-16), and cleavage of the remainder of the eggs has frequently been observed (Wilson, 1925). Clement and Tyler (1967) have shown that the detached polar lobe of *Ilyanassa* continues its protein-synthesizing activity, as shown by the incorporation of labeled amino

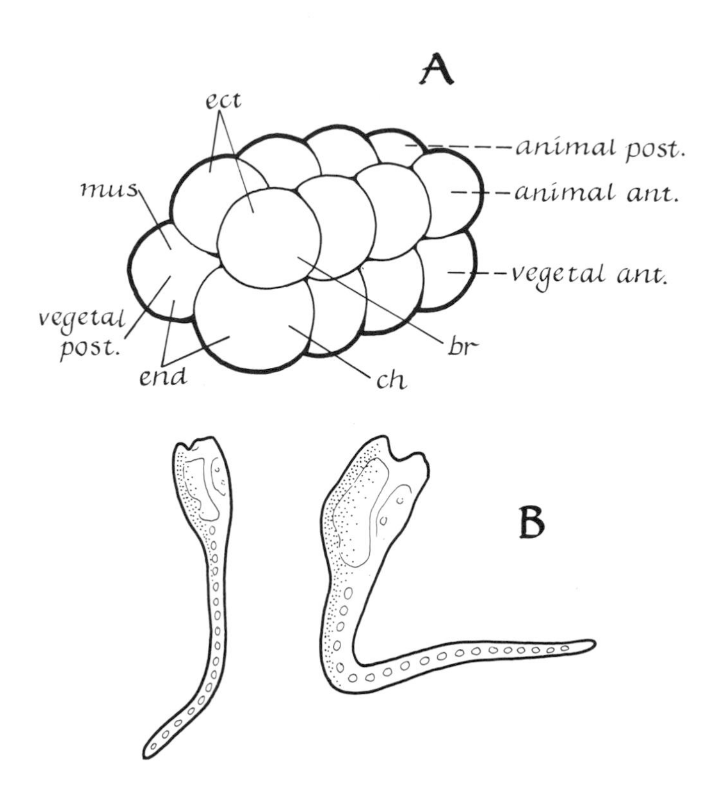

Figure 13-25. *A, diagram of two ascidian eggs fused at the eight-cell stage; B, normal and giant tadpoles.*

acids into proteins. Clearly, a program of development is built into the egg cytoplasm during oogenesis—a program that may be acted out, at least in its general features, quite independently of the cellulation process.

We have suggested that the cytoplasm of an egg might function in selecting inherited information. For example, in the egg of *Drosophila,* a cleavage nucleus may divide so that one daughter nucleus lies in a special cytoplasmic region (the pole plasm), the other outside it. The former nucleus becomes the nucleus of a germ cell, the latter, that of a somatic cell (Poulson, 1947). Thus, the kind of cytoplasm surrounding a nucleus controls its activity. Recall that nuclei entering the grey crescent cytoplasm grafted to an uncleaved *Xenopus* egg probably participate in the development of a secondary embryo body.

As a general hypothesis it is suggested that the spatial pattern of cytoplasmic differentiation within an egg constitutes a pattern of diverse, biochemically different microenvironments to which the cleavage nuclei respond differentially (Figure 13-24). Specific examples of cytoplasmic influence on nuclear activities in eggs and other cells will be discussed in Chapter 20.

Regulation of Cytoarchitectural Patterns

An egg which contains a visible or experimentally demonstrable cytoarchitectural pattern may nonetheless be able to repair defects in its pattern. The repair capacities of mosaic eggs are more limited than those of regulative eggs (Chapter 15). But even in the relatively mosaic egg of an ascidian which has localized, cytoplasmically different regions, fusion of two eggs of *Ascidiella* or *Ciona intestinalis* at the eight-cell stage may lead to the formation of a single, giant tadpole (Reverberi and Gorgone, 1962). However, this type of integration is realized only if the two eight-cell stages are pressed together so that their homologous regions come into contact (Figure 13-25).

As seen in the illustration, ectodermal, chordal, muscular, and endodermal regions constitute a topographical continuum. The necessity of preserving this continuum attests to the importance of the original cytoarchitectural pattern of the egg.

Fusion of two eggs to produce a single individual has also been demonstrated in other eggs, such as those of echinoderms (Bierens de Haan, 1913), urodeles (see Raven, 1954), and mice (Mintz, 1965). Even though echinoderm eggs are regulative, unification to produce a giant larva occurs only when the polar axes of the two cleaving eggs are parallel and coincide in direction.

The egg fusion experiments thus demonstrate that even though a preformed pattern of cytoarchitecture is present it is not rigid and unchangeable. At the same time it must be understood that there are limits to the regulative capacities of eggs. We should avoid concluding that regulation is a demonstration of the absence of cytoarchitecture in an egg.

Another way of demonstrating the degree of repair capacity in the cytoplasmic pattern of an egg before and during early cleavage is to cut the egg in half or separate the blastomeres (Figure 13-4). We shall consider this in Chapter 15, where we shall compare regulative and mosaic developmental patterns.

Conclusion

If we accept the hypothesis that all nuclei of a developing organism contain the same kinds of genes, then selective use of portions of the genome must be accomplished by influences outside the nucleus. In egg development the most directly effective influences emanate from the cytoplasm. Furthermore, the egg cytoplasm serves as a differentiated environment for selective gene activation in the nuclei that reside in the various environments constituting the cytoarchitecture of the egg.

As we noted earlier, differential activities of the genes in different follicle cells probably contribute to the establishment of the geometrical pattern of cytodifferentiation in the oocyte. Figure 13-26 is a hypothetical scheme illustrating how a gene product, *a*, released by follicle cell nuclei might combine with a gene product, *b*, of the oocyte nucleus to constitute a pattern of cytoarchitecture in the oocyte. As indicated in the figure, the combined product, *ab*, might then activate another oocyte nuclear gene, *2*, the product of which could combine with the product *d* resulting from the action of another follicle cell nuclear gene, *5*. This latter combined product might activate oocyte gene *4*, whose product would become an additional component of the cytoarchitecture of the developing egg. Thus, a program of sequential gene activation induced by products of activity of the cells in the reproductive organ could result in the development of a cytoarchitectural pattern.

The cytoarchitecture of an egg, structured by prior gene activity, is the basic guideline for subsequent gene action and thus for early and probably all development of the egg. The orderly nature of development of eggs seems to require such a conclusion.

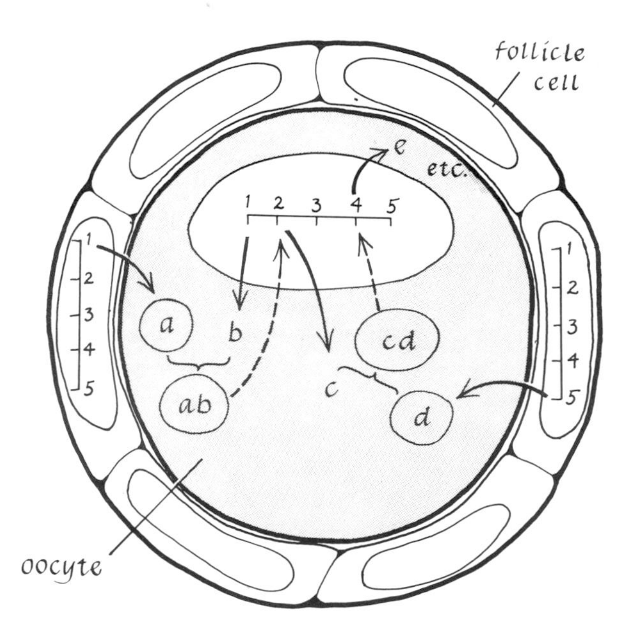

Figure 13-26. *How gene products of follicle cell nuclei and the egg nucleus might combine to form components of an egg's cytoarchitecture.*

References

Arnold, J. M. 1963. An embryological study of the squid, *Loligo pealii.* Doctoral thesis, University of Minnesota, Minneapolis.

Arnold, J. M. 1968a. The role of the egg cortex in cephalopod development. Devel. Biol. **18:**180–197.

Arnold, J. M. 1968b. An analysis of cleavage furrow formation in the egg of *Loligo pealii.* Biol. Bull. **135:**413 (Abstr.).

Bartelmez, G. W. 1912. The bilaterality of the pigeon's egg. J. Morph. **23:**269–305.

Bartelmez, G. W. 1918. The relation of the embryo to the principal axis of symmetry of the bird's egg. Biol. Bull. **35:**319–361.

Bayreuther, K. 1956. Die oögenese der *Tipuliden.* Chromosoma **7:**508–557.

Beams, H. W. 1964. Cellular membranes in oogenesis. In M. Locke, ed., Cellular Membranes in Development, pp. 175–219. Academic Press, New York.

Bell, P. R. 1963. The cytochemical and ultrastructural peculiarities of the fern egg. J. Linn. Soc. Lond. Bot. **58:**353–359.

Bell, P. R. 1964. The membranes of the fern egg. In M. Locke, ed., Cellular Membranes in Development, pp. 221–231. Academic Press, New York.

Berg, W. E. 1956. Cytochrome oxidase in anterior and posterior blastomeres of *Ciona intestinalis.* Biol. Bull. **110:**1–6.

Bier, K. 1963. Synthese, interzellulärer Transport, und Abbau von Ribonuklein saüre im Ovar der Stubenfliege *Musca domestica*. J. Cell Biol. **16:**436–440.

Bierens de Haan, J. A. 1913. Über homogene und heterogene Keimreschmelzungen bei Echiniden. Arch. Entw. Mech. **36:**473–536. See also ibid. **37:**420–432.

Blackler, A. W. 1962. Transfer of primordial germ cells between two subspecies of *Xenopus laevis*. J. Embryol. Exp. Morph. **10:**641–651.

Bounoure, L. 1934. Recherches sur la lignée germinale chez la grenouille rousse aux premier stades du développment. Ann. Sci. Natur. Zool. 10 Ser. **17:**67–248.

Boycott, A. E., C. Diver, and S. L. Garstang. 1931. The inheritance of sinistrality in *Limnaea peregra*. Phil. Trans. Roy. Soc. Lond. Ser. B. **219:**51–130.

Brachet, J. 1960. The Biochemistry of Development. Pergamon Press, New York.

Brachet, J. 1967. Biochemical changes during fertilization and early embryonic development. In A. V. S. DeReuck and J. Knight, eds., Cell Differentiation, pp. 39–64. Ciba Found. Symp., Little, Brown, Boston.

Brown, D. D., and I. B. Dawid. 1968. Specific gene amplification in oocytes. Science **160:**272–280.

Clement, A. C. 1962. *Ilyanassa* following removal of the D macromere at successive cleavage stages. J. Exp. Zool. **149:**193–215.

Clement, A. C., and A. Tyler. 1967. Protein-synthesizing activity of the anucleate polar lobe of the mud snail, *Ilyanassa obsoleta*. Science **158:**1457–1458.

Conklin, E. G. 1905. Mosaic development in ascidian eggs. J. Exp. Zool. **2:**145–223.

Conklin, E. G. 1929. Problems of Development. Am. Nat. **63:**5–36.

Crampton, H. E. 1896. Experimental studies on gastropod development. Arch. Entw. Mech. **3:**1–26.

Curtis, A. S. G. 1960. Cortical grafting in *Xenopus laevis*. J. Embryol. Exp. Morph. **8:**163–173.

Curtis, A. S. G. 1962. Morphogenetic interactions before gastrulation in the amphibian, *Xenopus laevis*—the cortical field. J. Embryol. Exp. Morph. **10:**410–423.

Dalcq, A. M., and A. Seaton-Jones. 1949. La répartition des éléments basophiles dans l'oeuf du rat et du lapin et son intéret pour la morphogénèse. Bull. Acad. Roy. Belg. Cl. Sci. (v) **35:**500.

Davidson, E. H. 1968. Gene Activity in Early Development. Academic Press, New York.

Dawid, I. B. 1966. Evidence for the mitochondrial origin of frog egg cytoplasmic DNA. Proc. Nat. Acad. Sci. U.S. **56:**269–276.

Dawid, I. B. 1968. Studies on frog oocyte mitochondrial DNA. Carnegie Inst. Year Book **66:**20–25.

DuBuy, H. G., M. W. Woods, and M. D. Lackey. 1950. Enzymatic activities of isolated normal and mutant mitochondria and plastids of higher plants. Science **111:**572–574.

Fankhauser, G. 1955. The role of the nucleus and cytoplasm. In B. H. Willier, P. Weiss, and V. Hamburger, eds., Analysis of Development, pp. 126–150. W. B. Saunders, Philadelphia.

Fischel, A. 1898. Experimentelle Untersuchungen am Ctenophorenei: I. Von der Entwicklung isolirter Eitheile. Arch. Entw. Mech. **6:**109–130.

Grant, P. 1965. Informational molecules and embryonic development. In R. Weber, ed., The Biochemistry of Animal Development, Vol. 1, pp. 483–593. Academic Press, New York.

Gross, P. R., L. I. Malkin, and M. Hubbard. 1965. Synthesis of RNA during oogenesis in the sea urchin. J. Mol. Biol. **13:**463–481.

Harrison, R. G. 1945. Relations of symmetry in the developing embryo. Trans. Conn. Acad. Arts Sci. **36:**277–330.

Harvey, E. B. 1951. Cleavage in centrifuged eggs, and in parthenogenetic merogones. Ann. N.Y. Acad. Sci. **51:**1336–1348.

Holtfreter, J. 1943. Properties and functions of the surface coat in amphibians. J. Exp. Zool. **93:**251–323.

Horstadius, S. 1939. The mechanics of sea-urchin development, studied by operative methods. Biol. Rev. Camb. **14:**134–179.

Horstadius, S., I. J. Lorch, and J. F. Danielli. 1950. Differentiation of the sea urchin egg following reduction of the interior cytoplasm in relation to the cortex. Exp. Cell Res. **1:**188–193.

Jaffe, L. F. 1968. Localization in the developing *Fucus* egg and the general role of localizing currents. Adv. Morph. **7:**295–328.

Jensen, W. A. 1964. Cell development during plant embryogenesis. In Meristems and Differentiation, pp. 179–202. Brookhaven Symp. Biol., Upton, New York.

Lehmann, von F. E. 1948. Zur Entwicklungs-physiologie der Polplasmen des Eies von *Tubifex*. Rev. Suisse Zool. **55:**1–43.

Lehmann, von F. E. 1956. Plasmatische Eiorganisation und Entwicklungsleistung beim Keim von *Tubifex* (Spiralia). Naturwiss. **43:**289–296.

Lillie, F. R. 1902. Differentiation without cleavage in the egg of the annelid *Chaetopterus pergamentaceus*. Arch. Entw. Mech. **14:**477–499.

Mintz, B. 1965. Experimental mosaicism in the mouse. In Preimplantation Stages of Pregnancy, p. 194. Ciba Found. Symp., Little, Brown, Boston.

Morgan, T. H. 1902. The dispensability of gravity in the development of the toad's egg. Anat. Anz. **21:**313–316.

Morgan, T. H. 1927. Experimental Embryology. Columbia University Press, New York.

Motomura, I. 1935. Determination of the embryonic axis in the eggs of amphibia and echinoderms. Sci. Rep. Tokoku University **10:**211–245.

Nace, G. W. 1953. Serological studies of the blood of the developing chick embryos. J. Exp. Zool. **122:**423–448.

Needham, J. 1950. Biochemistry and Morphogenesis. Cambridge University Press, London.

Nieukoop, P. D. 1946. Experimental investigations on the origin and determination of the germ cells, and on the development of the lateral plates and germ ridges in urodeles. Arch. Néerl. Zool. **8:**1–205.

Pasteels, J. J. 1964. The morphogenetic role of the cortex of the amphibian egg. Adv. Morph. **3:**363–388.

Piatigorsky, J., H. Ozaki, and A. Tyler. 1967. RNA- and protein-synthesizing capacity of isolated oocytes of the sea urchin *Lytechinus pictus*. Devel. Biol. **15:**1–22.

Picken, L. 1960. The Organization of Cells and Other Organisms. Clarendon Press, Oxford.

Pikó, L., A. Tyler, and J. Vinograd. 1967. Amount, location, priming capacity, circularity and other properties of cytoplasmic DNA in sea urchin eggs. Biol. Bull. **132:**68–90.

Porter, K. R. 1963. Diversity at the subcellular level and its significance. In J. M. Allen, ed., The Nature of Biological Diversity, pp. 121–163. McGraw-Hill Book Company, New York.

Poulson, D. F. 1947. The pole cells of Diptera, their fate and significance. Proc. Nat. Acad. Sci. U.S. **33:**182–184.

Raven, C. P. 1948. The chemical and experimental embryology of *Limnaea*. Biol. Rev. Camb. **23:**333–369.

Raven, C. P. 1954. An outline of Developmental Physiology, McGraw-Hill Book Company, New York.

Raven, C. P. 1961. Oogenesis: The Storage of Developmental Information. Pergamon Press, New York.

Raven, C. P. 1964. Mechanisms of determination in the development of gastropods. Adv. Morph. **3:**1–32.

Raven, C. P. 1967. The distribution of special cytoplasmic differentiations of the egg during early cleavage in *Limnaea stagnalis*. Devel. Biol. **16:**407–437.

Reverberi, G. 1957a. Mitochondrial and enzymatic segregation through the embryonic development of ctenophores. Acta. Embryol. Morph. Exp. **1:**134–142.

Reverberi, G. 1957b. The role of some enzymes in the development of ascidians. In A. Tyler, R. C. von Borstel and C. B. Metz, eds., The Beginnings of Embryonic Development, pp. 319–340. Am. Assoc. Adv. Sci., Washington, D.C.

Reverberi, G. 1961. The embryology of ascidians. Adv. Morph. **1**:55–101.

Reverberi, G., and I. Gorgone. 1962. Gigantic tadpoles from ascidian eggs fused at the 8-cell stage. Acta. Embryol. Morph. Exp. **5**:104–112.

Reverberi, G., and R. La Spina. 1959. Normal larvae obtained from dark fragments of centrifuged *Ciona* eggs. Experentia **15**:122

Reverberi, G., and G. Ortolani. 1962. Twin larvae from halves of the same egg in Ascidians. Devel. Biol. **5**:84–100.

Runnström, J. 1928. Plasmabau und Determination bei den Ei von *Paracentrotus lividus* Lk. Arch. Entw. Mech. **113**:556–581.

Schechtman, A. M. 1956. Uptake and transfer of macromolecules by cells with special reference to growth and development. Intern. Rev. Cytol. **5**:303–322.

Slater, D. W., and S. Spiegelman. 1966. An estimation of genetic messages in the unfertilized echinoid egg. Proc. Nat. Acad. Sci. U.S. **56**:164–170.

Stevens, L. C. 1967. The biology of teratomas. Adv. Morph. **6**:1–31.

Telfer, W. H. 1961. The route of entry and localization of blood proteins in oocytes of saturniid moths. J. Biophys. Biochem. Cytol. **9**:747–759.

Ubbels, G. A., J. J. Bezem, and C. P. Raven. 1969. Analysis of follicle cell patterns in dextral and sinistral *Limnaea peregra*. J. Embryol. Exp. Morph. **21**:445–466.

Verdonk, N. H. 1968. The effect of removing the polar lobe in centrifuged eggs of *Dentalium*. J. Embryol. Exp. Morph. **19**:33–42.

Wardlaw, C. W. 1955. Embryogenesis in Plants. Methuen, London.

Wardlaw, C. W. 1965. Organization and Evolution in Plants. Longman, London.

Weber, R. 1962. Electron microscopy in the study of embryonic differentiation. In R. Harris, ed., The Interpretation of Ultrastructure, pp. 393–409. Academic Press, New York.

Weiss, P. 1939. Principles of Development. Henry Holt, New York.

Whitaker, D. M. 1940. Physical factors of growth. Growth Symp. **4**:75–90.

Williams, J. 1965. Chemical constitution and metabolic activities of animal eggs. R. Weber, ed., The Biochemistry of Animal Development, Vol. 1, pp. 13–71. Academic Press, New York.

Wilson, E. B. 1904. Experimental studies on germinal localization: I. The germ regions in the egg of *Dentalium*. J. Exp. Zool. **1**:1–72.

Wilson, E. B. 1925. The Cell in Development and Heredity. 3rd ed. Macmillan, New York.

14 *Alternation of Asexual and Sexual Patterns*

Many organisms reproduce sexually at some time in their life cycles and asexually at another. Often the two phases of the organism are so different morphologically from one another that they appear to represent different organisms. The basic problem is understanding the mechanism that controls the alternation. It is well established that within any one species the same genetic information is used in different ways at different times. This selective and programmatic use of inherited information is basic to the transformations from one stage into another that all organisms undergo. Thus, the sexually and asexually reproducing forms of an organism may reasonably be considered stages in development, not different in principle from, for example, larval and adult stages or blastula and gastrula stages. Some stages, such as those in egg or bud development, are of relatively short duration, whereas others, such as larval stages in some animals, are long lasting. Similarly, the sexually reproducing stage in angiosperms is relatively short and the asexually reproducing stage is of much longer duration. The extent to which the succession of stages is intrinsically controlled or alterable by external environmental influences is of particular interest to the student of development.

The alternation of a spore- or bud-producing generation with a gamete-producing generation in both plants and animals is remarkably automatic. In plants such as liverworts, mosses, ferns, gymnosperms, and angiosperms, a meiospore (Chapters 10 and 11) develops only into the gametophyte type of plant, and an egg only into a sporophyte. Why a gametophyte produces only gametes and never spores, and why the medusa of a coelenterate will form gametes at one time and buds which develop directly into more medusae at another has never been satisfactorily explained.

No change in the chromosome number is involved in the alternation of sexually and asexually reproducing generations in animals. The alternation in animals, as in the examples given below, is between a gamete- and a bud-producing form of the organism, both of which are diploid. Alternation between a haploid, gamete-producing form and a diploid, meiospore-producing form occurs only in plants. The term *spore,* used in reference to alternation in plants, refers exclusively to meiospores as defined in Chapter 9.

Alternation of Sexual and Asexual Reproduction in Animals

Only a few examples of alternation of generations in animals are discussed here. Many more exist among the invertebrates. It appears that alternation in both animals and plants is a response to changing environmental conditions.

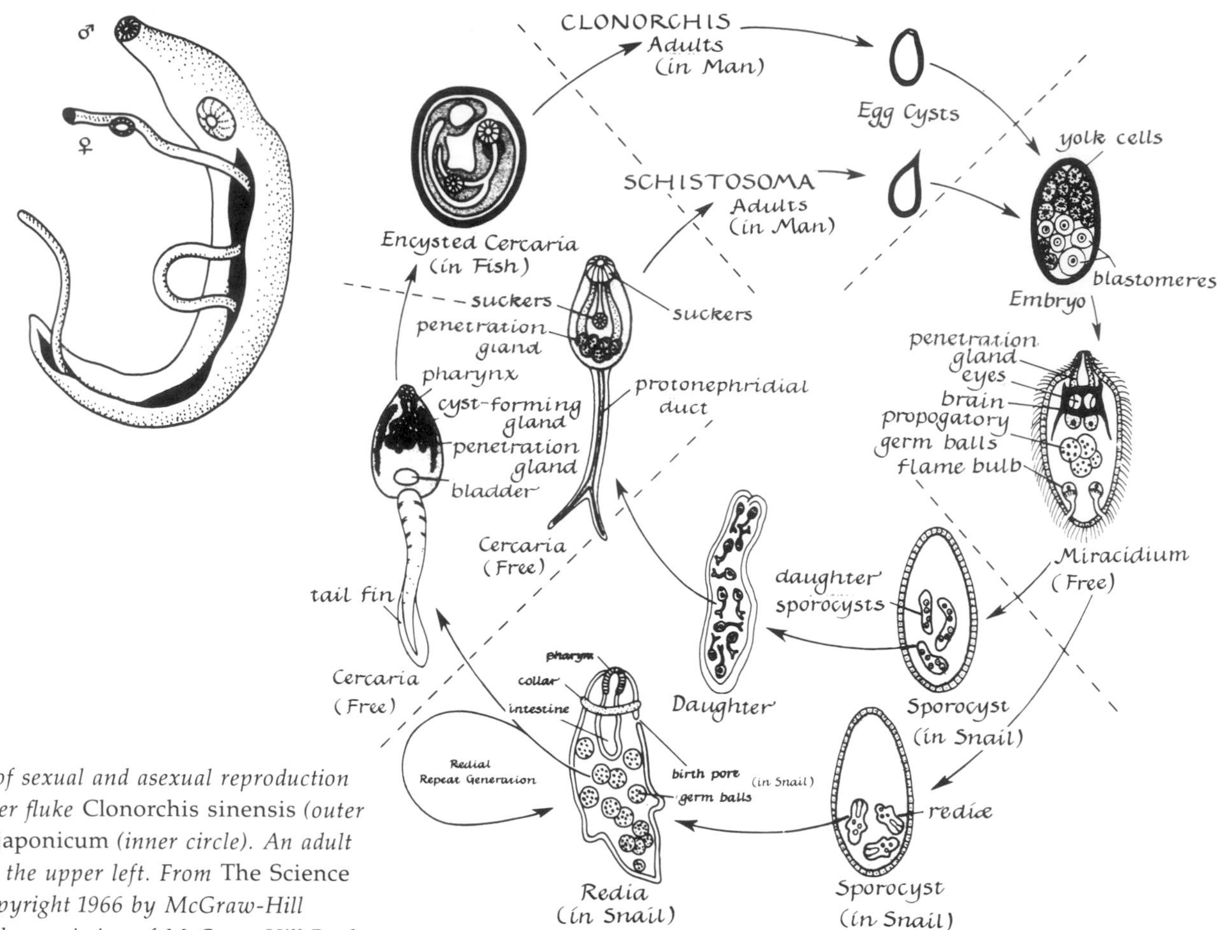

Figure 14-1. *Alternation of sexual and asexual reproduction in the life cycle of the liver fluke* Clonorchis sinensis *(outer circle) and* Schistosoma japonicum *(inner circle). An adult* Schistosoma *is shown in the upper left. From* The Science of Zoology *by Weisz. Copyright 1966 by McGraw-Hill Book Company. Used with permission of McGraw-Hill Book Company.*

Protozoa

An example of the possible environmental control of the alternation of reproductive patterns is found in flagellates (for instance, *Trichonympha* and related genera) which parasitically inhabit the gut of the wood-eating cockroach *Cryptocercus.* The flagellates reproduce sexually only while the host molts. Apparently the molting hormone, *ecdysone,* a steroid similar in basic structure to sex hormones of vertebrates (Chapter 19), might be the agent which induces the change to a sexual reproductive pattern, since injecting the host with ecdysone induces gametogenesis in the flagellates without inducing metamorphosis in the host (Cleveland, 1956; Cleveland et al., 1960).

Helminths

Many parasitic animals (the liver fluke, for instance) exhibit an alternation of sexual and asexual reproductive processes during their life cycle; during each stage, the individuals are strikingly different in morphology and functional activity (Figure 14-1). Apparently the parasite's environment initiates and determines the kind of reproductive process in which it engages.

Coelenterates

Among animals, the coelenterates exhibit the best-known examples of polymorphism and alternation of sexual with asexual reproduction. In Chapter 7

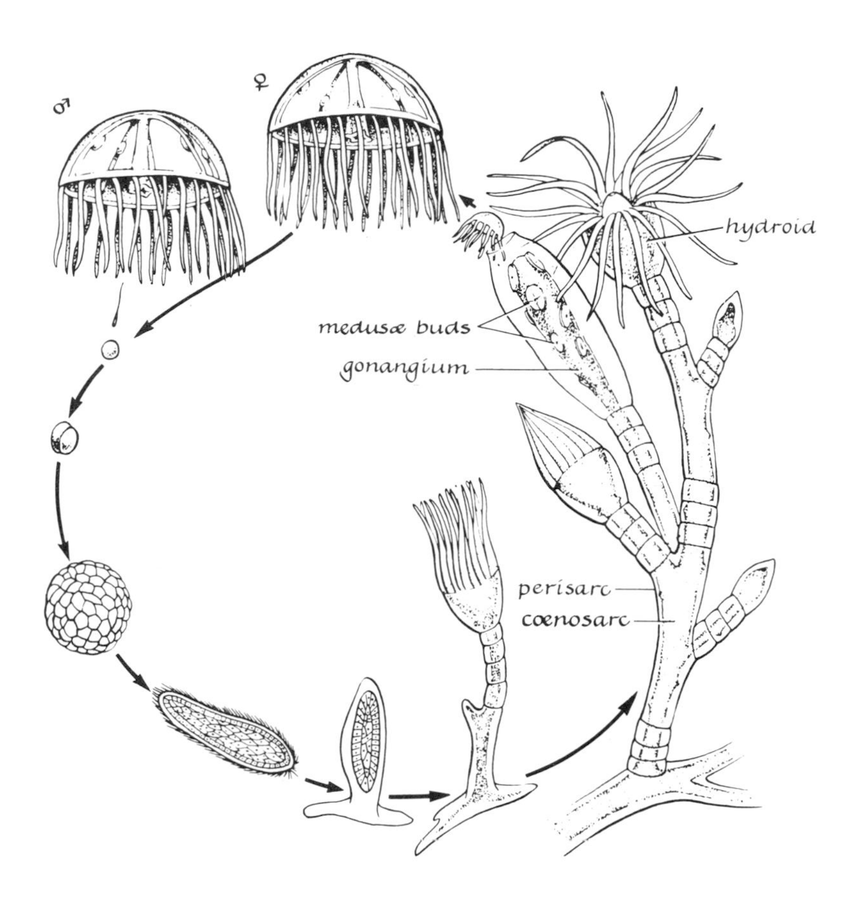

Figure 14-2. *Life cycle of the colonial marine hydrozoan* Obelia. *From Villee:* Biology, *5th Edition, W. B. Saunders Company, Philadelphia, 1967.*

we discussed examples of polymorphism in colonial coelenterates. An alternation of larva and adult, as well as structural differences among adults (polyp and medusa), is characteristic of many coelenterate life cycles. A typical example of the alternation of sexually and asexually reproducing generations in *Obelia* is illustrated in Figure 14-2. As shown in the figure, at one stage the adult is a bud-producing polyp (hydranth) similar to *Hydra.* However, the buds do not develop into more polyps, as in *Hydra,* but into medusae (jelly-fishes). The medusae are either male or female, and produce sperm or eggs respectively. The fertilized eggs develop into planula larvae which later metamorphose into new polyps.

Among the coelenterates there are several modifications of this typical pattern of alternating egg- and bud-producing generations (Hyman, 1940; Berrill, 1961). Some of the distinct life cycle patterns are enumerated below.

Asexual Patterns:

(1) Polyp ⟶ bud ⟶ polyp (Hydra-type)
(2) Polyp ⟶ bud ⟶ medusa (see, for instance, Figure 14-2)
(3) Medusa ⟶ bud (on manubrium, radial canal, tentacle, or bell margin) ⟶ medusa
(4) Polyp fragment ⟶ polyp (always)
(5) Planula ⟶ bud ⟶ medusa
(6) Planula ⟶ bud ⟶ planula

Sexual Patterns:

(1) Polyp ⟶ egg and/or sperm ⟶ polyp (Hydra-type)
(2) Medusa ⟶ egg and/or sperm ⟶ medusa
(3) Medusa ⟶ egg and/or sperm ⟶ planula ⟶ medusa
(4) Medusa ⟶ egg and/or sperm ⟶ planula ⟶ polyp (see, for instance, Figure 14-2)

These diverse patterns illustrate the programmatic nature of development. For example, in *Obelia* a fertilized egg produced by the medusa always develops into a larva, then a polyp, never directly into another medusa. A polyp of *Obelia* always forms a medusa bud, never an egg or sperm. A bud formed on a medusa always develops into a medusa, whereas a bud formed on a polyp develops either into another polyp (as in *Hydra*) or into a medusa (as in *Obelia*), depending upon the species.

There is evidence (Berrill, 1961) that, in some coelenterates, which structure (polyp, stolon or medusa; see Figure 7-30) develops is environmentally controlled. In *Bougainvillia* the type of structure developing from the growth zone at the base of a polyp's tentacles depends upon the temperature of the sea water. A stolon forms at temperatures above 20° C. At slightly below 20° C, the outgrowth is usually a hydranth, and at still lower temperatures a medusa may develop. Berrill (1961) suggests that the relative growth rate of ectoderm and endoderm underlies this poly-

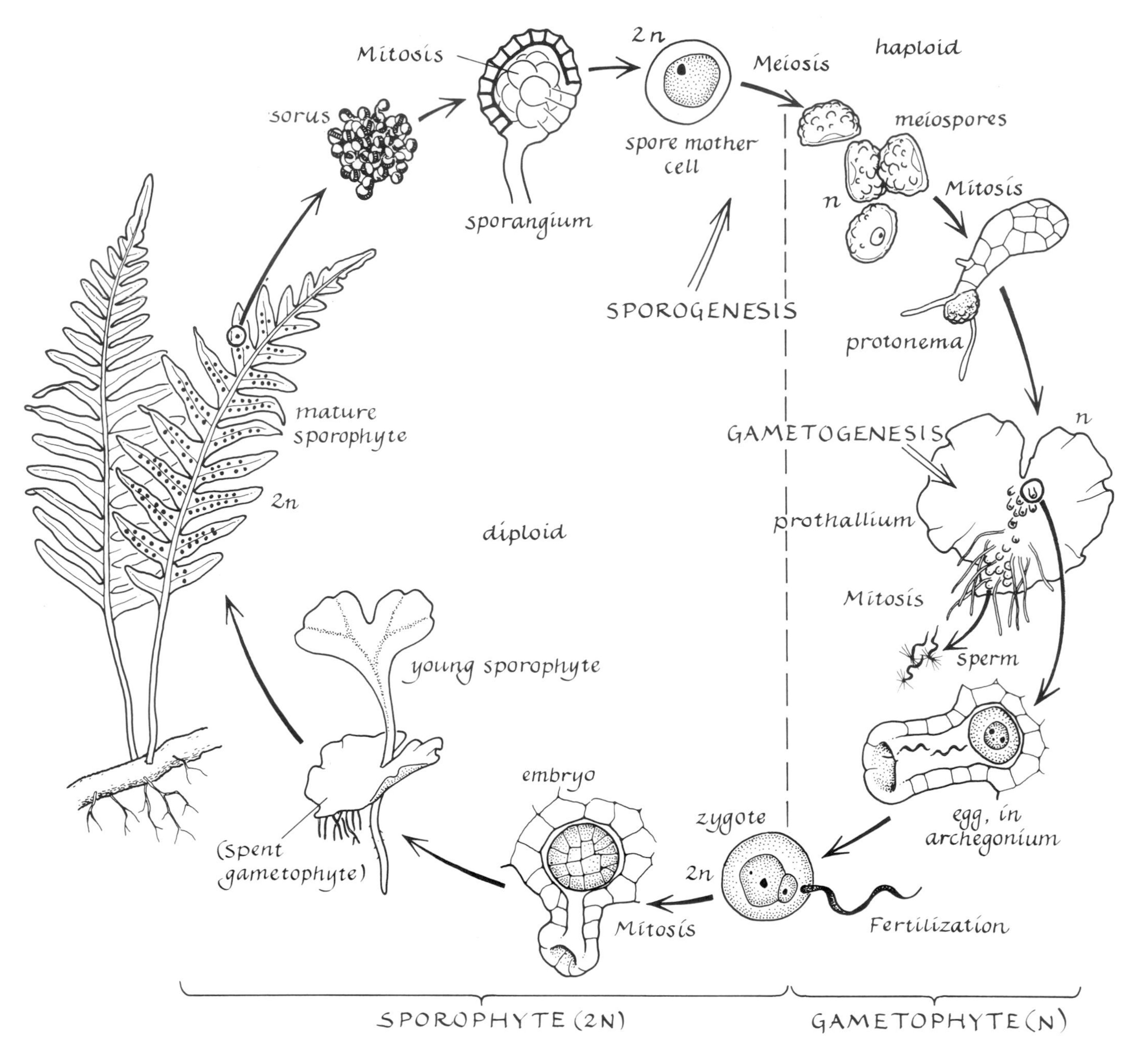

Figure 14-3. *Life cycle of a fern.*

morphism. Whether the reproduction is sexual or asexual in many coelenterates such as *Hydra littoralis* (Loomis, 1961) is also influenced by environmental conditions such as light, temperature, and/or nutrients. Such environmental influences can select the type of development, but once selected, the program proceeds automatically.

Alternation of Sexual and Asexual Reproduction in Plants

A typical example of alternation of generations in higher plants is illustrated by the life cycle of a fern, shown in Figure 14-3. Not all plants conform to this

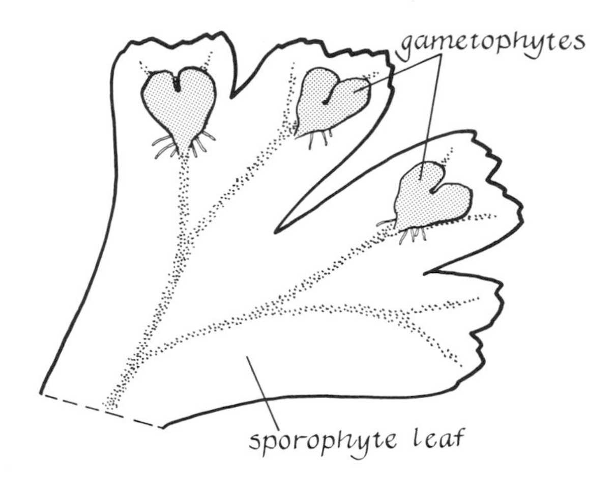

Figure 14-4. *Aposporous development of gametophytes on an isolated sporophyte leaf of the fern* Gymnogramme sulphurea.

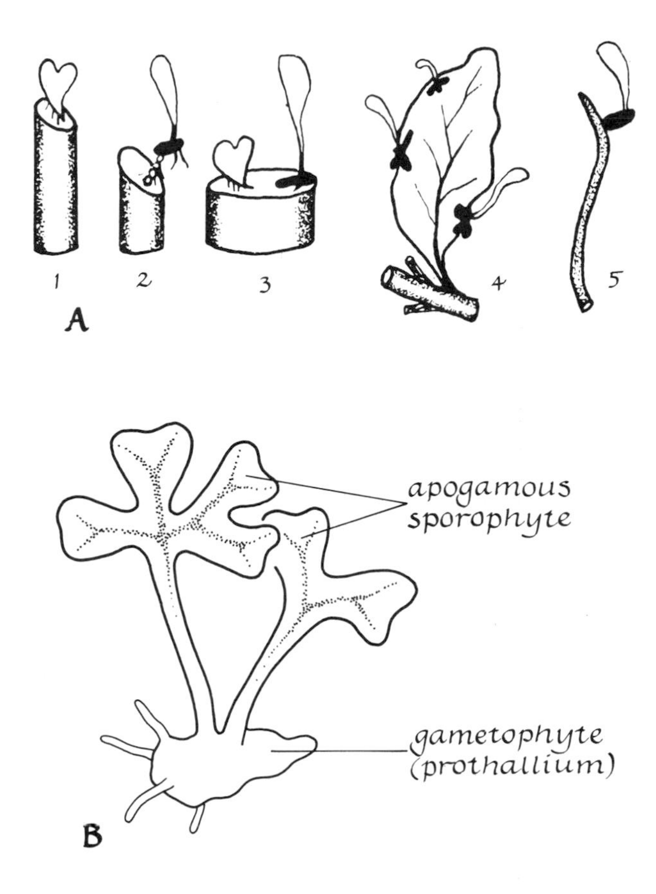

Figure 14-5. A, Phlebodium aureum. *1, aposporous formation of a gametophyte on a petiole segment of the sporophyte; 2, regeneration of a sporophyte from a petiole segment of a sporophyte; 3, gametophyte and sporophyte developing on a section of the sporophyte rhizome; 4, sporophytic type plants forming on the leaf of an apogamously produced sporophyte; 5, sporophyte developing on an outgrowth of an apogamous sporophyte.* B, *apogamous formation of a sporophyte on the cultured gametophyte of the fern* Osmunda cinnamomea. a, *from* J. Linn. Soc. (Bot.), *Volume 58, 1963. Published by permission of the Linnean Society of London.*

model. In some algae and fungi, the zygospore is the only diploid cell (Chapter 10). Recall that a diploid plant has two sets of *homologous* chromosomes, not just twice as many chromosomes as a haploid plant.

As shown in the figure, under natural conditions the sporophyte generation produces only spores, the gametophyte only gametes. Fertilized eggs develop only into sporophytes, spores only into gametophytes (Allsopp, 1966). Thus, one pattern of development leads to another. Under conditions of culture *in vitro,* the normal alternation can be modified so that stages are skipped, but apparently the cycle cannot be made to run backwards. For example, pieces of some fern sporophytes (2n) can regenerate into diploid gametophytes, skipping the spore-forming process (Goebel, 1905; Wetmore et al., 1963). This is called *apospory.* Aposporously formed gametophytes on an isolated leaf of a sporophyte placed in culture (Vladesco, 1935) are illustrated in Figure 14-4. See also Figure 14-5A (1 and 3) (Ward, 1963).

In the ferns *Adiantum* and *Phlebodium,* juvenile leaves of a young sporophyte in culture will form gametophytes, but adult leaves of cultured apical meristems will regenerate only sporophytes (Ward, 1963). Ward interprets this to mean that some factor *X* determining the gametophyte type of development is present in the juvenile leaves that is lost when they mature. Cultured pieces of several species of fern gametophyte may regenerate a haploid sporophyte, skipping the egg- and sperm-forming process (Goebel, 1905; Whittier and Steeves, 1960; Wetmore et al., 1963). This is called *apogamy.*

Note in Figure 14-5A that a piece of the sporophyte rhizome produced in proximity both a gametophyte (the result of apospory) and a sporophyte (the result of apogamy). This finding is not easily interpreted unless one assumes that very slight differences in

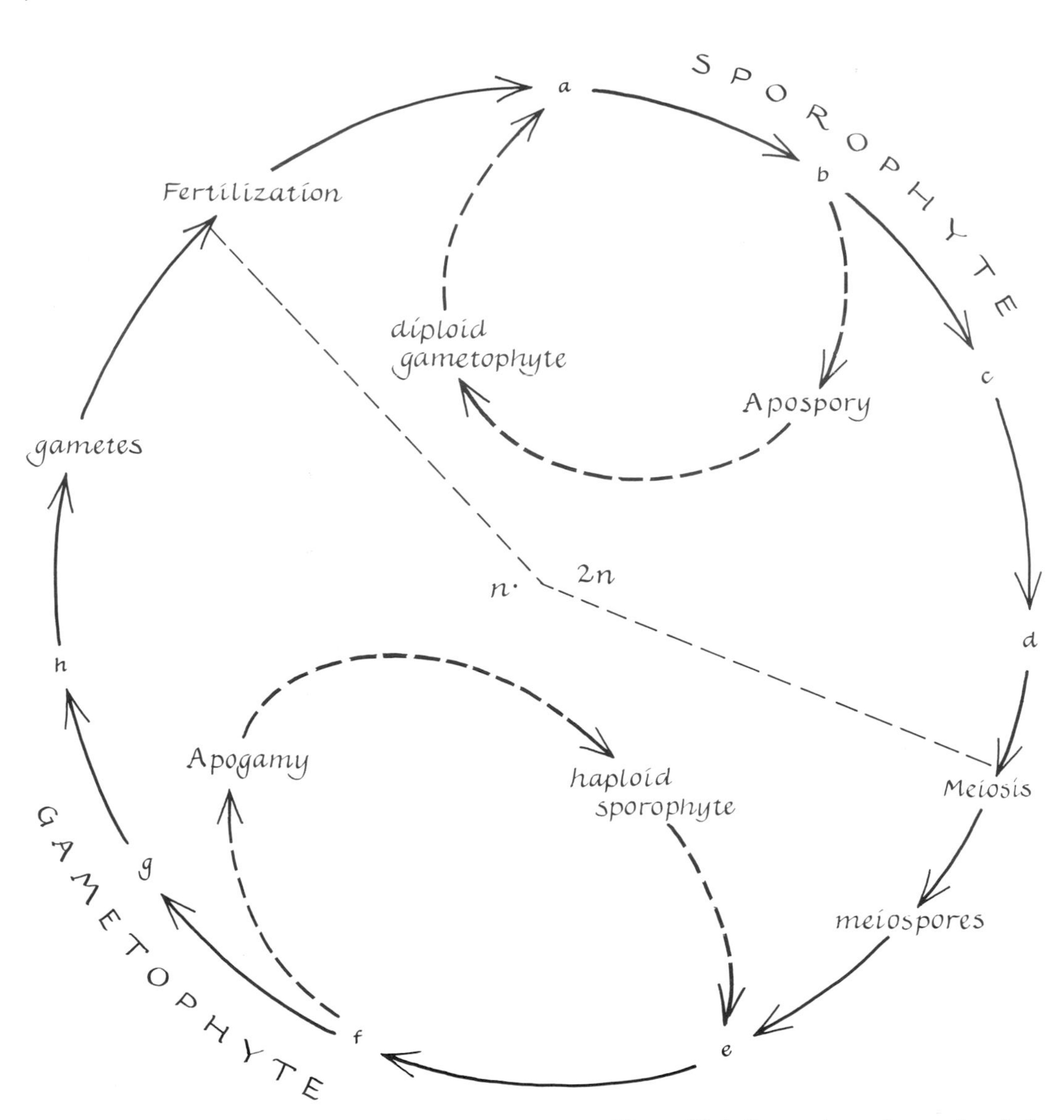

Figure 14-6. *Comparison of normal and abnormal life cycles in mosses and ferns.*

the environment can select programs of development. There is evidence that environmental agents can influence the development of fern fragments growing *in vitro* (Sussex and Clutter, 1960). In the fern *Pteris cretica*, cultured sporophyte cells form a callus. In the presence of sucrose and auxins the callus remains undifferentiated, but when the callus is grown on a medium containing only mineral nutrients a gametophyte develops. Addition of sucrose to the mineral medium results in development of a sporophyte (Bristow, 1962). Sucrose is also an important constituent of the culture medium in the apogamous production of sporophytes on the prothallia of several species of ferns (Whittier and Steeves, 1960). Apogamy is illustrated in Figure 14-5B.

The fact that a sporophyte or a gametophyte can develop from a piece of a gametophyte, whereas normally the sporophyte develops from an egg inside the archegonium (Figure 14-3), proves that gametophyte cells other than the egg can carry the information selected for this particular pattern of development. Even when an egg or older embryo is partially removed from the archegonium and grown in culture, it usually develops into a sporophyte (Ward and Wetmore, 1954). It has been reported that the isolated, cultured fertilized egg of the fern *Todea barbara* can form a gametophyte-like structure resembling a prothallus (DeMaggio and Wetmore, 1961; DeMaggio, 1963). Since no distinct sex organs (archegonia or antheridia) developed, this finding probably does not represent the formation of a gametophyte by a fertilized egg.

Under natural conditions the sporophyte is diploid and the gametophyte haploid; however, diploid and tetraploid gametophytes and tetraploid and octoploid sporophytes of the ferns *Adiantum pedantum, Osmunda cinnomomea,* and *Todea* have been produced experimentally. When an excised leaf from the sporophyte of the fern *Todea barbara* was grown in culture, diploid gametophytes were formed aposporously on the leaf margins. When the diploid eggs and spermatozoa of these gametophytes fused in fertilization, tetraploid sporophytes developed from the zygotes. Repetition of this procedure resulted in octoploid sporophytes. Tetraploid gametophytes have also been produced as a result of apospory in tetraploid sporophytes. Haploid sporophytes are formed when a piece of a gametophyte regenerates a sporophyte (Morel, 1963; Ward, 1963). Diploid gametophytes have also been produced in mosses (Maltzahn and MacNutt, 1960). It is significant that the 1n, 2n, and 4n gametophytes are all phenotypically (that is, morphologically) alike. The 2n, 4n, and 8n sporophytes are also phenotypically alike. Clearly the different morphology of the two generations is not simply a matter of chromosome number or kinds of genes present. It has been suggested that differences in mRNAs might be involved (Morel, 1963; Wetmore et al., 1963). The normal and abnormal life cycles of plants such as mosses and ferns are diagrammatically compared in Figure 14-6. The normal life cycle is indicated with heavy arrows, the abnormal cycles with dashed arrows. The letters *a, b, c* and *e, f, g* indicate stages in development of the sporophyte and gametophyte prior to formation of the reproductive cells.

The normal and experimentally induced abnormal paths of development for sporophyte and gametophyte cells are summarized below.

G = gametophyte S = sporophyte	Somatic cells
E = egg Sp = meiospore	Reproductive cells

S cells may form Sp cells (normal)
S cells may form S cells (growth and regeneration)
S cells may form G cells (apospory)
G cells may form E and/or sperm (normal)
G cells may form G cells (growth and regeneration)
G cells may form S cells (apogamy)
Sp cells only form G cells (normal)
E cells only form S cells* (normal)

*Note possible exception described above.

This summary makes it clear that the reproductive cells, egg and spore, have a more restricted potential than the somatic cells. They are more specialized in that a rigid program appears to be packaged in them. Sporophyte cells can give rise to more sporophyte cells, spores, or even gametophyte cells; spores and eggs have no such alternatives (Allsopp, 1966). In Chapter 13 we examined the role which the cytoarchitecture of the egg plays in development. Probably the spore also possesses a special cytoplasmic structure that determines how genetic potentialities will be expressed.

Conclusion

Recall (Chapter 9) that asexual reproduction is an adequate method of continuing the plant or animal organism through many generations, but it does not provide for as much genetic variability as does sexual reproduction (Chapters 10, 11, and 12). In animals, this variability is provided by the meiotic process during gametogenesis. In plants, the variability is provided by meiosis in sporogenesis.

A particularly important aspect of sexual reproduction is the formation of the specialized sex cells. When the parent organism produces both eggs and sperm, as in many invertebrates, some vertebrates (Chapter 19), and fern prothalli of some species, the progenitor cells must somehow make differential use of the genome common to the eggs and sperm. The two different programs of genome use are presumably selected by influences external to the genes.

In the alternation of structurally different adult forms in both plants and animals (sexual, gamete-producing and asexual, spore- or bud-producing), the same problem of understanding the differential use of the genome arises. When the influences external to the genes are discovered, we may better understand how genetic expression in development is controlled. It seems highly significant that each new generation is morphologically unlike the one that precedes it, and that this program is apparently irreversible.

References

Allsopp, A. 1966. Developmental stages and life histories in the lower green plants. In E. Cutter, A. Allsopp, F. Cusick, and I. M. Sussex, eds., Trends in Plant Morphogenesis, pp. 64–87. Longman, London.

Berrill, N. J. 1961. Growth, Development and Pattern. W. H. Freeman, San Francisco.

Bristow, J. M. 1962. The controlled *in vitro* differentiation of cells derived from a fern, *Pteris cretica* L., into gametophytic or sporophytic tissues. Devel. Biol. **4:**361–375.

Cleveland, L. R. 1956. Brief accounts of the sexual cycles of the flagellates of *Cryptocercus.* J. Protozool. **3:**161–180.

Cleveland, L. R., A. W. Burke, and P. Karlson. 1960. Ecdysone induced modification in the sexual cycles of the protozoa of *Cryptocercus.* J. Protozool. **7:**229–239.

DeMaggio, A. E. 1963. Morphogenetic factors influencing the development of fern embryos. J. Linn. Soc. Lond. Bot. **58:**361–376.

DeMaggio, A. E., and R. H. Wetmore. 1961. Morphogenetic studies on the fern *Todea barbara:* III. Experimental embryology. Am. J. Bot. **48:**551–565.

Goebel, K. 1905. Organography of Plants. Clarendon Press, Oxford.

Hyman, L. H. 1940. The Invertebrates: Protozoa through Ctenophora. McGraw-Hill Book Company, New York.

Loomis, W. F. 1961. Feedback factors affecting sexual differentiation in *Hydra littoralis.* In H. M. Lenhoff and W. F. Loomis, eds., The Biology of Hydra, pp. 337–362. University of Miami Press, Coral Gables, Fla.

Maltzahn, von K. E., and M. M. MacNutt. 1960. Intracellular configuration and dedifferentiation in *Splachnum ampullaceum* L. Nature **187:**344–345.

Morel, G. 1963. Leaf regeneration in *Adiantum pedatum.* J. Linn. Soc. Lond. Bot. **58:**381–383.

Sussex, I. M., and M. E. Clutter. 1960. A study of the effect of externally supplied sucrose in the morphology of excised fern leaves *in vitro.* Phytomorphol. **10:**87–99.

Vladesco, M. 1935. Recherches morphologiques et expérimentales sur l'embryogénie et l'organogénie des fougères leptosporangiées. Rev. Gén. Bot. **47:**684–771.

Ward, M. 1963. Developmental patterns of adventitious sporophytes in *Phlebodium aureum,* J. Sm. J. Linn. Soc. Bot. **58:**377–380.

Ward, M., and R. H. Wetmore. 1954. Experimental control of development in the embryo of the fern, *Phlebodium aureum.* Am. J. Bot. **41:**428–434.

Wetmore, R. H., A. E. DeMaggio, and G. Morel. 1963. A morphogenetic look at the alternation of generations. J. Indian Bot. Soc. **42A:**306–320.

Whittier, D. P., and T. A. Steeves. 1960. The induction of apogamy in the bracken fern. Canad. J. Bot. **38:**925–930.

15 *Mosaic and Regulative Patterns of Development*

A distinctive property of both adult and developing organisms is their capacity to repair injuries. The ability to repair defects or replace lost parts during development has been customarily termed *regulation* by animal embryologists (Weiss, 1939); the word *regeneration* is usually used when referring to the adult organism (see Chapter 5). However, the distinction between the two terms is not based entirely upon whether the repair process occurs in an adult or in a developing organism. Regulation refers primarily to the capacity of a part to reconstitute the whole organism, whereas regeneration is the capacity of the whole organism to replace a lost part. Perhaps a more meaningful distinction between regulation and regeneration would take into account the fact that different kinds of guidelines or control mechanisms operate in each of them. In regeneration, control mechanisms operating in the whole organism determine the nature of the regenerated part; in regulation, control mechanisms present in the part direct its development into a whole organism. Thus in regeneration the instructions are largely preformed, whereas in regulation the guidelines are mostly progressively formed. Nevertheless, both preformed and progressively formed guidelines occur in both kinds of repair processes.

Developing organisms have widely differing regulative capacities. Some have almost no regulative abilities. Such patterns of development are described as *mosaic;* each part of the developing organism is so different and stable that it cannot transform into any other part, nor can it produce by additive growth a different kind of missing part. At the other extreme are developing organisms with almost unlimited capacities to replace missing parts by renewed growth or to reorganize a fragment of the organism into a new whole organism. Such patterns are described as *regulative.*

The repair capacities of a developing organism are revealed by experimentally injuring or removing a part, or even by cutting the organism into fragments at various stages of development. Such experiments, carried out extensively in the first half of this century, have provided much information about developmental processes, especially about the relative roles of preformed and progressively formed guidelines for development. It has become clear that organisms differ in respect to the rigidity of the program of development packaged in their reproductive units, particularly when the units are eggs, and that no pattern of development is exclusively regulative or mosaic.

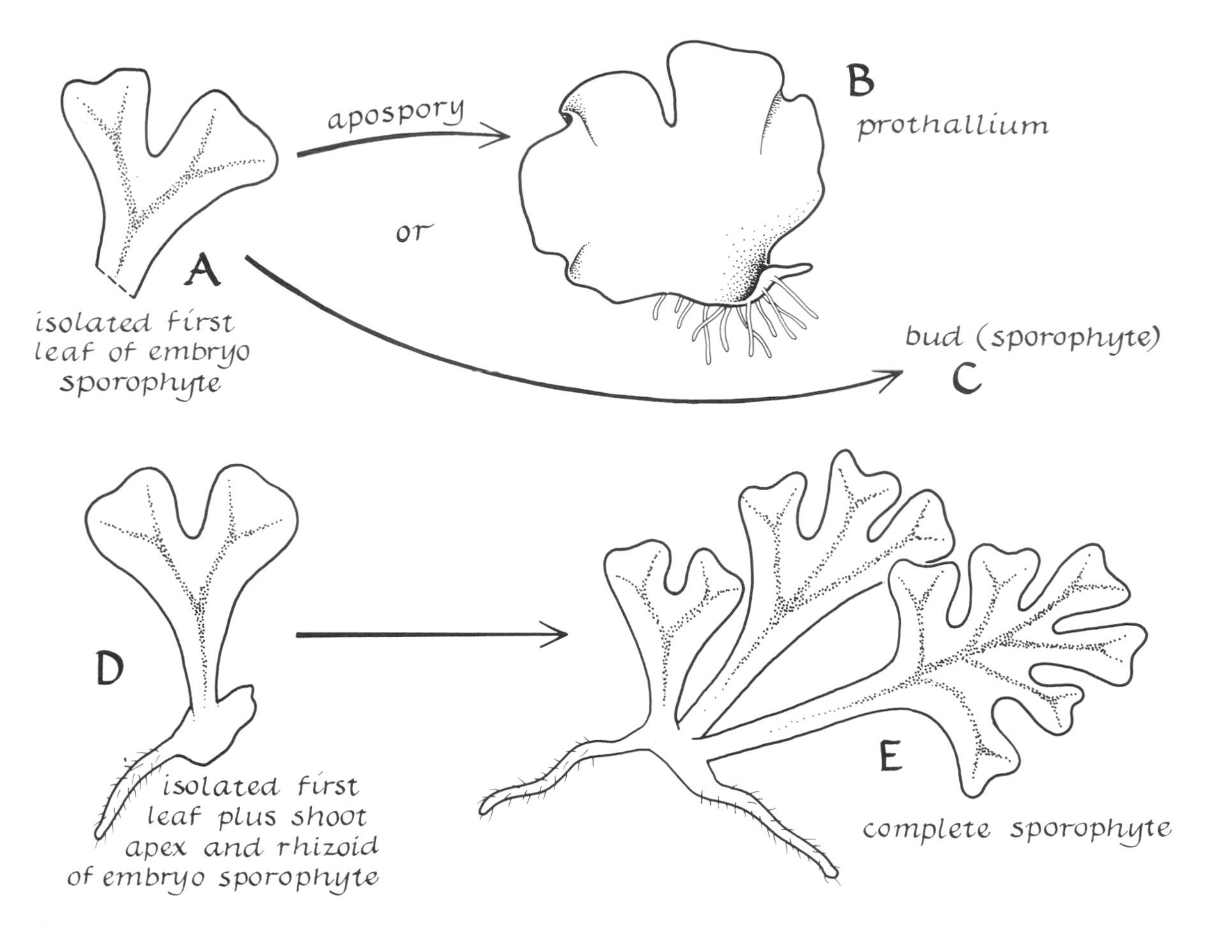

Figure 15-1. *Regulation in the embryo sporophyte of a fern.*

Regulation in Plant Embryos

Experiments on the embryo sporophyte of certain ferns (Goebel, 1908; Vladesco, 1935; Wardlaw, 1955) reveal that its pattern of development is remarkably regulative (Figure 15-1). When the prospective first leaf (Figure 15-1A) was excised and placed on moist soil, it developed either into a prothallium (2n gametophyte, Figure 15-1B) or a bud (2n sporophyte, Figure 15-1C). When the first leaf of the fern *Gymnogramme sulphurea* plus the shoot apex of the young sporophyte (Figure 15-1D) were isolated and cultured, a complete, almost normal plantling sporophyte developed (Figure 15-1E).

Partial regulation also occurs in embryo sporophytes of the sesame plant *(Sesamum indicum)* that have been longitudinally divided in half. However,

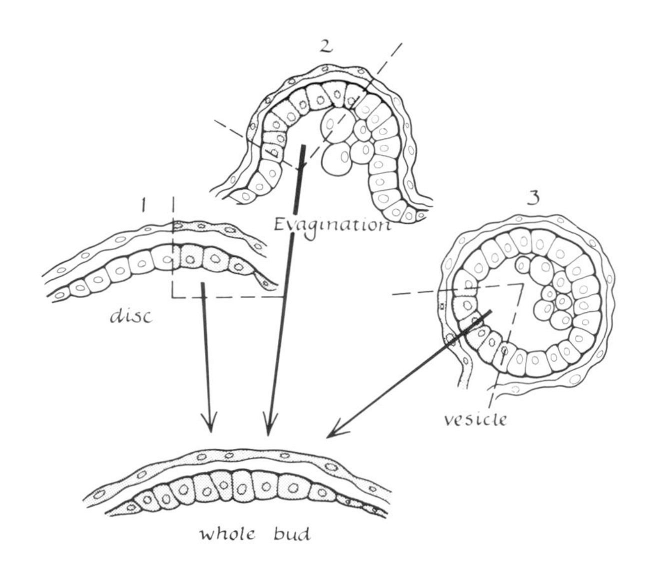

Figure 15-2. *Regulation in the bud disc of a tunicate.*

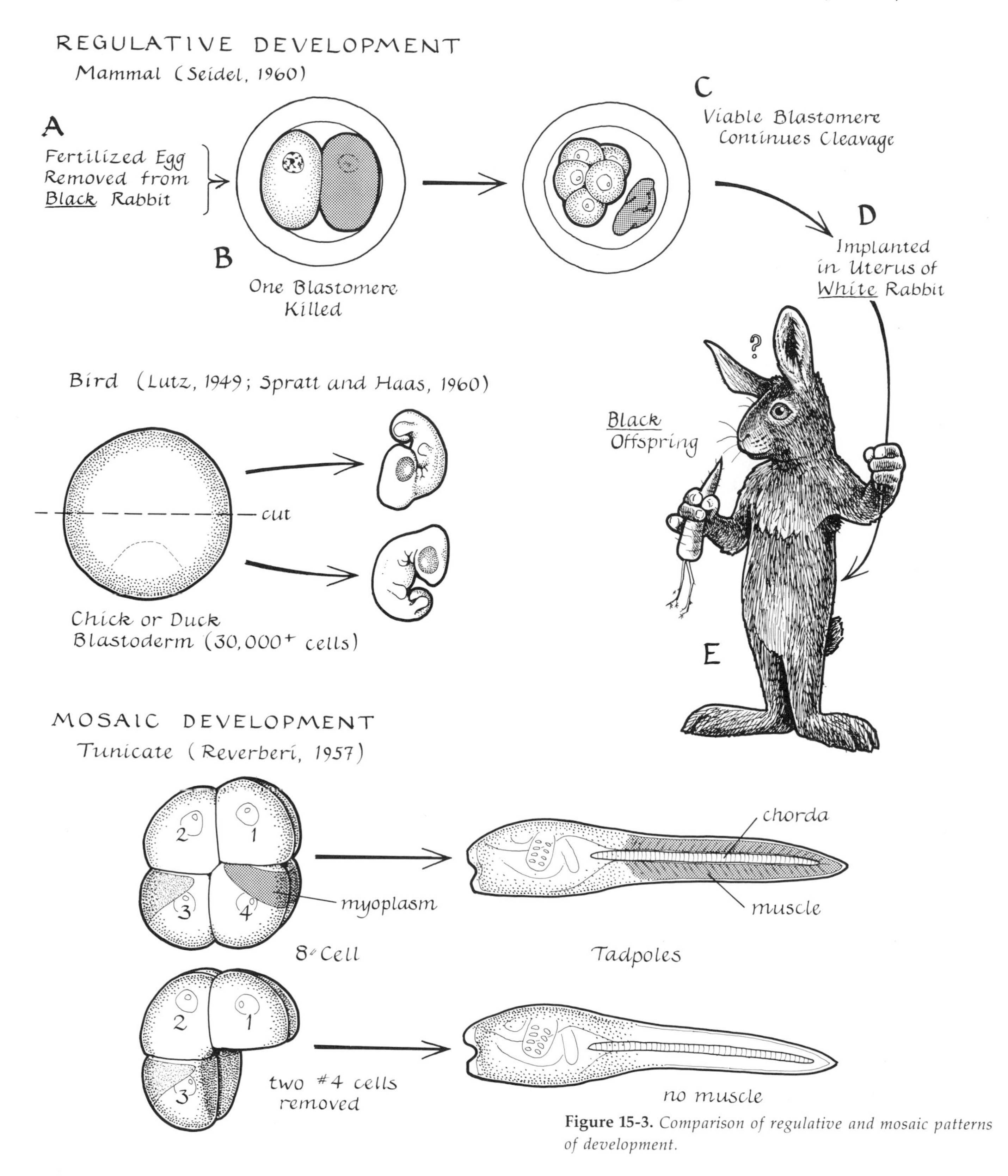

Figure 15-3. *Comparison of regulative and mosaic patterns of development.*

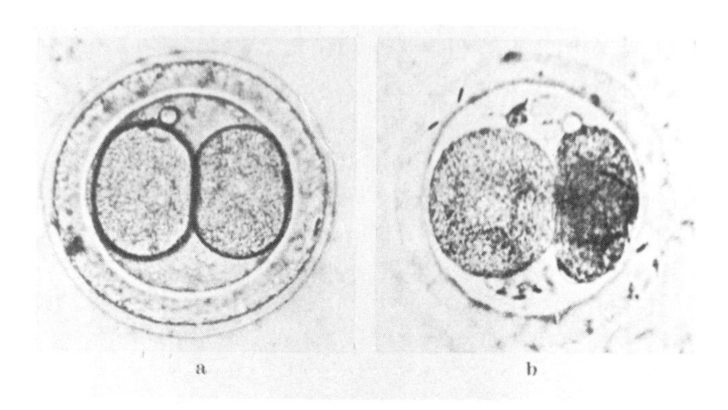

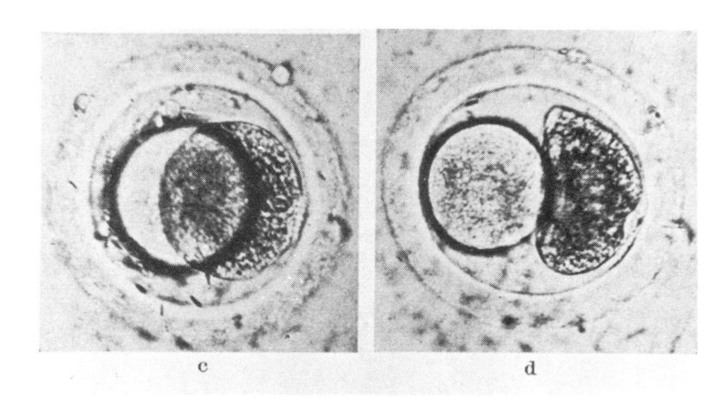

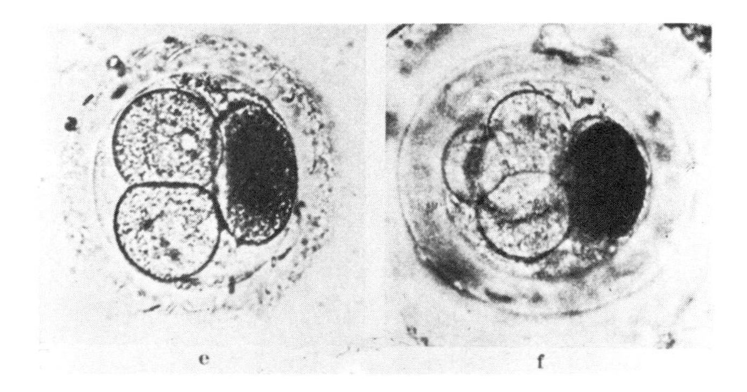

Figure 15-4. *Regulation at the two-cell stage of the rabbit* (Oryctolagus cunniculus) *egg. A, normal two-cell stage; B and D, destruction of one blastomere; E and F, cleavage in living blastomere. From Seidel, 1960. Courtesy Springer-Verlag.*

when unequally divided, the smaller part of the embryo failed to reconstitute an apical growing point, presumably indicating mosaic properties in these embryos (see Wardlaw, 1968).

Figure 15-5. *Chinchilla foster mother and four black Alaska rabbits, each of which was derived from one-half of an implanted black Alaska egg. From Seidel, 1960. Courtesy Springer-Verlag.*

Regulation in Tunicate Buds

An example of regulation in an asexual pattern of development of a tunicate is illustrated in Figure 15-2. Fractions of bud discs or of buds through stage *c* regulate to form a whole bud which then forms the adult. No regulation occurs after stage *c* (Berrill, 1961). Recall that the adult tunicate possesses great regenerative capacity (Chapter 5). In those organisms which possess it, regulative capacity is most pronounced at early stages of development, lost at later stages, and often replaced by regenerative capacity in the adult.

Mosaic and Regulative Patterns in Animal Egg Development

Mosaic and regulative types of development revealed by experiments on animal eggs are illustrated in Figure 15-3. Note that the mammal egg and bird embryos at advanced stages of development (several thousand cells) possess marked regulative capacities. Characteristically, vertebrate embryos, excluding amphibians, are highly regulative. At the other extreme are tunicate, mollusc, and annelid embryos which are primarily mosaic in their development. Also note in the figure that part of a rabbit or chick embryo can form a whole organism, whereas the tunicate embryo at the eight-cell stage cannot replace a missing part.

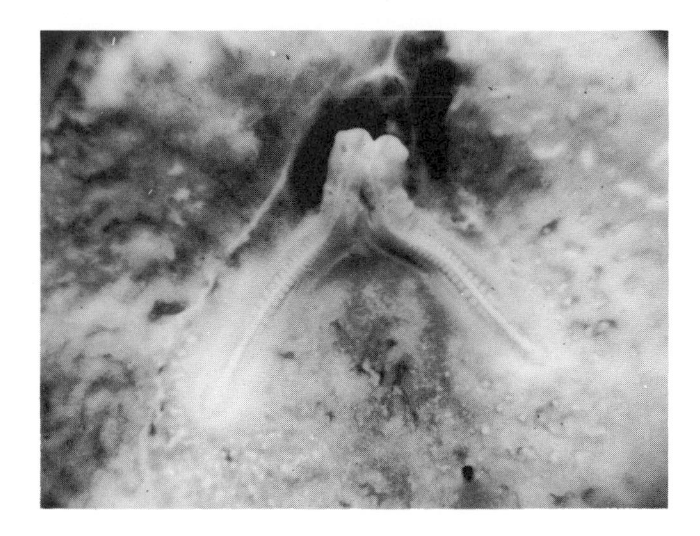

A

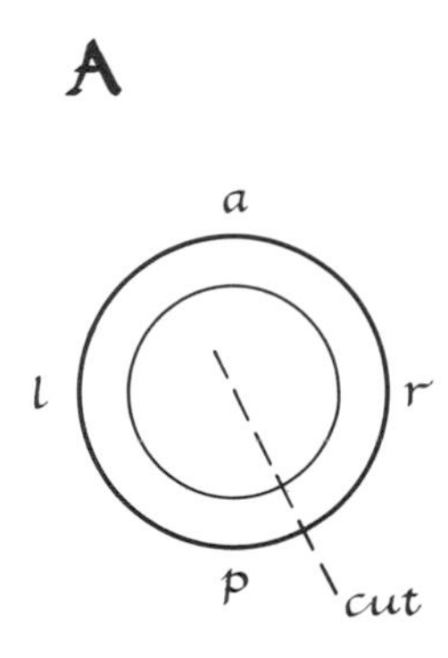

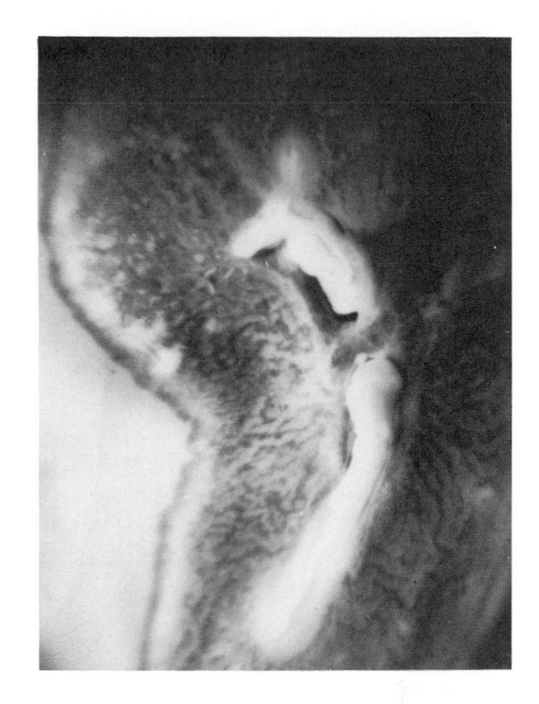

B

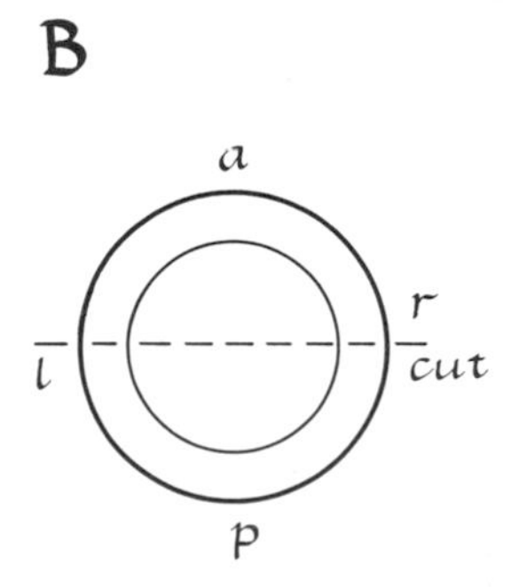

Figure 15-6. *Regulation in the unincubated chick blastoderm. From Spratt and Haas,* J. Exp. Zool., *Volume 164, p. 35, 1967. Courtesy Wistar Press.*

Figure 15-4 is a photograph of two living rabbit eggs in which one blastomere of the two-cell stage was killed by pricking with a needle through the surrounding egg membrane, the *zona*. Note that the killed blastomere becomes dark and fails to undergo further cleavage, but that the remaining blastomere (Figure 15-4E and F) cleaves and forms a four-cell stage.

Figure 15-5 is a photograph of a Chinchilla foster mother in whose uterus four black Alaska rabbit eggs were implanted; in each egg one blastomere at the two-cell stage had been killed. Also shown are the four black Alaska rabbits which developed from the four remaining blastomeres (Seidel, 1960, 1969).

Blastomeres cleaved out of the dorsal basophilic region of the rabbit egg's four-cell stage can develop into five somite embryos when the remaining blastomeres are killed. This capacity of an isolated blastomere appears to be correlated with its having more ribosomes and polysomes than the other blastomeres (Figure 13-10 and Seidel, 1969). The mammal egg thus has mosaic properties in respect to the regulative capacities of its different parts.

The extensive regulative capacity of bird embryos even as late as a 30,000-cell stage is illustrated in Figure 15-6. When an unincubated chick blastoderm was cut into approximately right and left halves through a window in the egg shell (Figure 15-6A), each half formed a whole embryo body (Spratt and Haas, 1967). When cut into anterior and posterior halves (Figure 15-6B), again each half formed a complete embryo body. Earlier experiments of Lutz (1949) and Spratt and Haas (1960) clearly demonstrate the remarkable regulative capacities of bird embryos. Fragments as small as one-eighth of the unincubated blastoderm are able to form whole embryo body systems when explanted on a nutrient medium *in vitro* (Spratt and Haas, 1960). As explained in Chapter 5, interruption of the continuity of the organism is often a necessary stimulus for regeneration. This is also true for regulation in explanted, unincubated blastoderms: the continuity between the group of cells normally forming the embryo body and other cells must be interrupted in order to stimulate regulation in the latter (Figure 15-7; see also Figures 26-5 and 26-6). The experiment illustrated in Figure 15-7 also demonstrates the dominance of the prospective embryo-forming region.

As noted earlier in this chapter, no egg is exclusively regulative or mosaic. The way in which an egg is cut in half often determines whether it will exhibit either regulative or mosaic features. This is true of amphibian and sea urchin eggs, shown in Figure 15-8. Generally, these eggs are capable of regulation only if each isolated part contains samples of all cytoplasmically different regions (Figure 15-8B and C). Recall (Chapter 13) that separated blastomeres of the two-cell stage of a sea urchin may develop into whole pluteus larvae, but that separation of the eight-cell stage into four animal and four vegetal blastomeres results in abnormal development of each half. The realization of regulative capacity also depends upon growth capacity, which in turn is influenced by availability of nutrients. This has been shown to be the case in explanted chick blastoderms (Spratt and Haas, 1967).

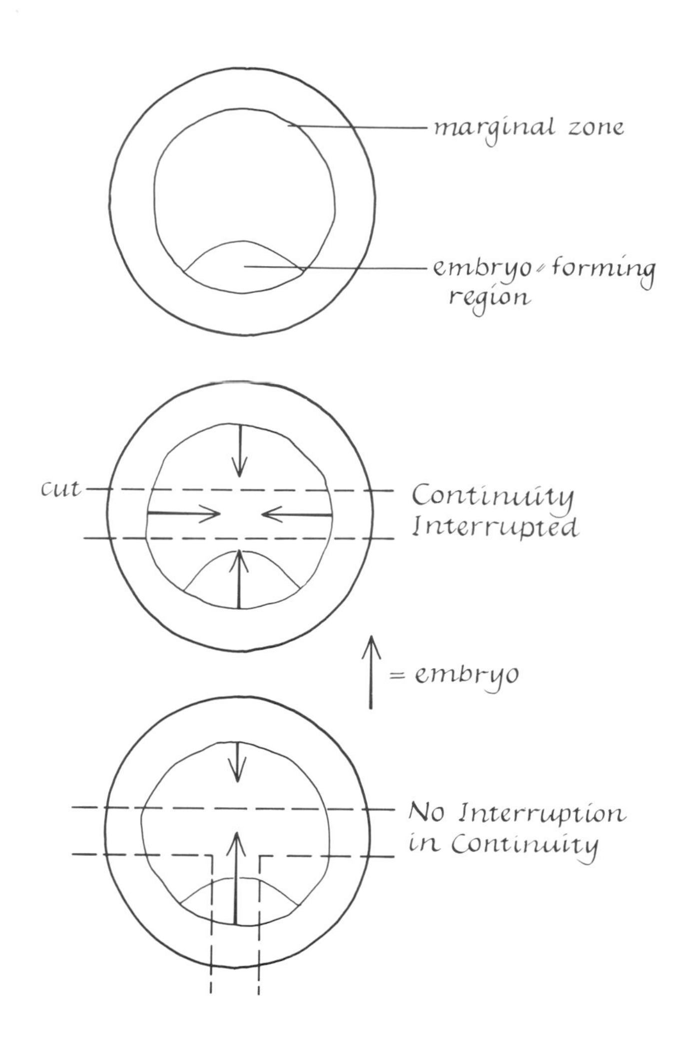

Figure 15-7. *Diagrams of the result of interrupting continuity between the prospective embryo-forming cell group (right) and other cells (center) and of not interrupting the continuity (left) in explanted unincubated chick blastoderms.*

Comparison of Mosaic and Regulative Patterns

The continuum between extremely mosaic and extremely regulative patterns of development is illustrated in Table 15-1. The primary developmental patterns of groups of animals are compared only in a general way. The distinction between these patterns is largely a matter of degree. However, not all organisms are guided in their development to the same degree by preformed or progressively formed (emergent) guidelines. The dashed lines in the table indicate mosaic properties in the animal groups; the solid lines indicate regulative properties.

There are exceptions to the data of Table 15-1. For example, there is some evidence of interaction between blastomeres of the eight-cell stage of the ascidian *Ciona intestinalis* (Pucci-Minafra and Ortolani, 1968); also equatorial meridional or oblique halves

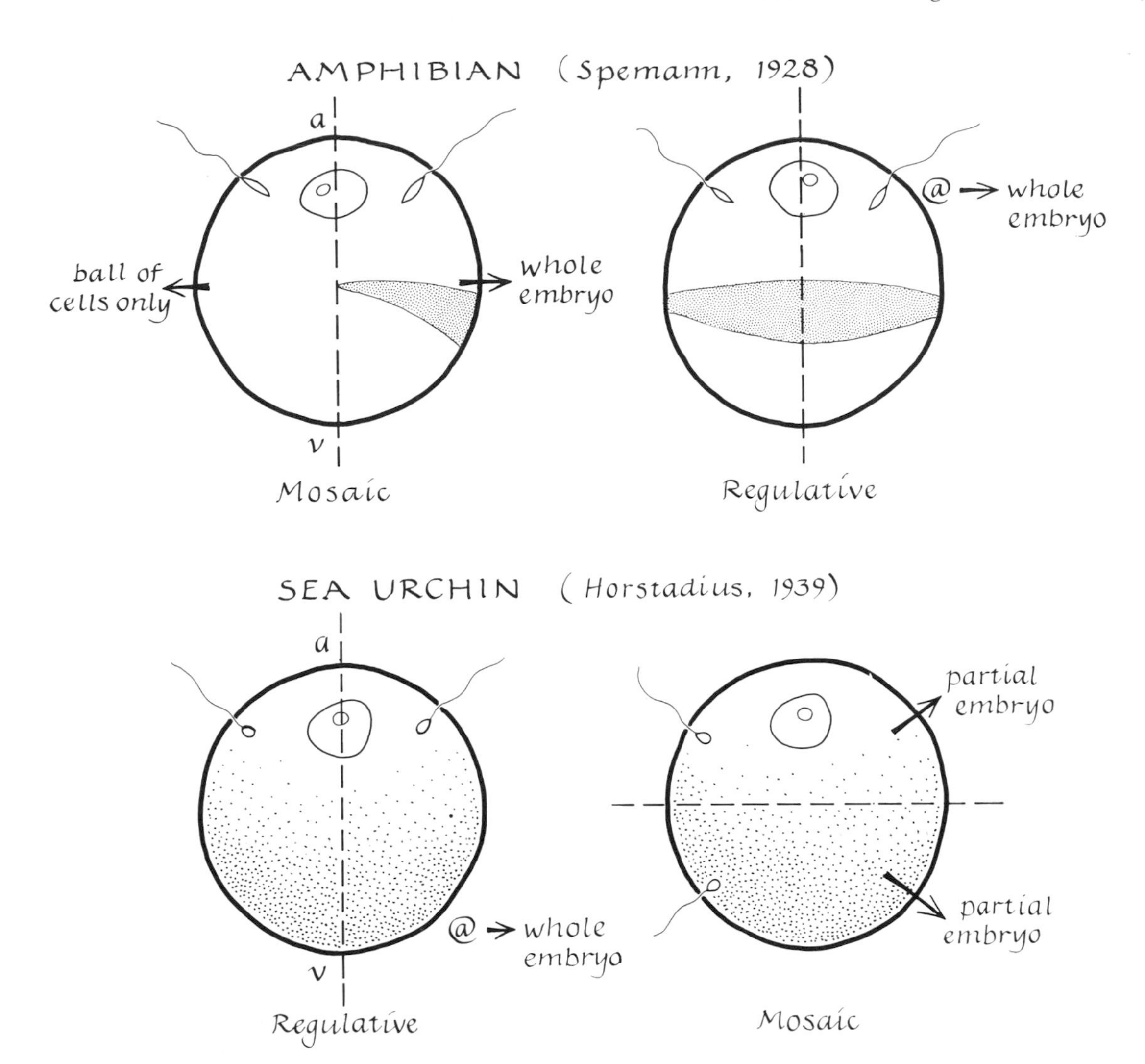

Figure 15-8. *How mosaic or regulative properties of an egg are revealed by the way in which the egg is divided.*

of the unfertilized eggs of *Ascidia malaca* develop into whole larvae after being fertilized (Reverberi and Ortolani, 1962). Recall that two ascidian eggs experimentally fused at the eight-cell stage developed into a single tadpole, a further indication of regulative capacity (Chapter 13).

Paterson (1957) reported that the isolated middle third of a transected early gastrula of the frog *(Rana pipiens)* is capable of extensive regulation. After removal of the animal and vegetal thirds with a fine glass needle, the middle third (which did not include the dorsal blastopore lip) developed into a swimming tadpole in forty of eighty-two cases. However, when the animal third was not removed from the middle third, development did not occur, indicating that a balance between animal and vegetal portions is essential for expression of regulative capacities. Horstadius (1939) found that this balance is also essential for regulation in isolated regions of the sixty-four-cell stage sea urchin embryo. The distinction between extremely regulative and mosaic patterns of development is shown in Figure 15-9.

Table 15–1. Developmental Patterns of Animal Embryos in Terms of Their Mosaic and Regulative Features

Mosaic ——— *Primary Pattern of Early Egg Development* ——— *Regulative*	*Ascidian*	*Mollusc*	*Echinoderm*	*Coelenterate*	*Vertebrate (excluding Amphibian)**
Blastomere Properties	largely cytoplasmically inherited				largely acquired
Cytoplasmic Localization	restricted				diffuse
	discrete				graded
Cleavage planes and Cytodifferentiation	coincide				transcend
	determinate				indeterminate
Rate of Cytodifferentiation	rapid				slow
Rate of Cleavage	slow				rapid
Number of Cells in Blastula	64	64-128	500	1,000 (approximately)	5,000-60,000 (approximately)
Size of Blastula Cells	large				small
Cellular Interactions	rare (localized)				common (widespread)
Time from Fertilization to Gastrulation	2½ hours	4-6 hours	10 hours	20 hours (approximately)	1½-14 days
Cell Division Potential	very limited				very great
Synthetic Capacity	slight				pronounced
	(localized)				(widespread)
Level of Early Differentiation	cellular				supracellular
Cleavage Pattern	complete				Partial (complete in mammal)
Initial Result of Embryonic Development	larva				adult
Mother organism provides instructions	detailed				general (only)

From *Introduction to Cell Differentiation* by Nelson T. Spratt, Jr., copyright © 1964 by Reinhold Publishing Corporation, by permission of Van Nostrand Reinhold Company.
*similar to Ascidian

Conclusion

Figure 15-10 points out the difference between mosaic and regulative patterns of development. The basic nature of development is the progressive utilization of inherited information, that is, progressive reading of the DNA code; therefore, in mosaic development patterns, relatively more information has been used by the time the egg reaches maturity than in regulative patterns. In the vascular cambial cells of plants relatively little information has been used.

Classifying a pattern as regulative or mosaic de-

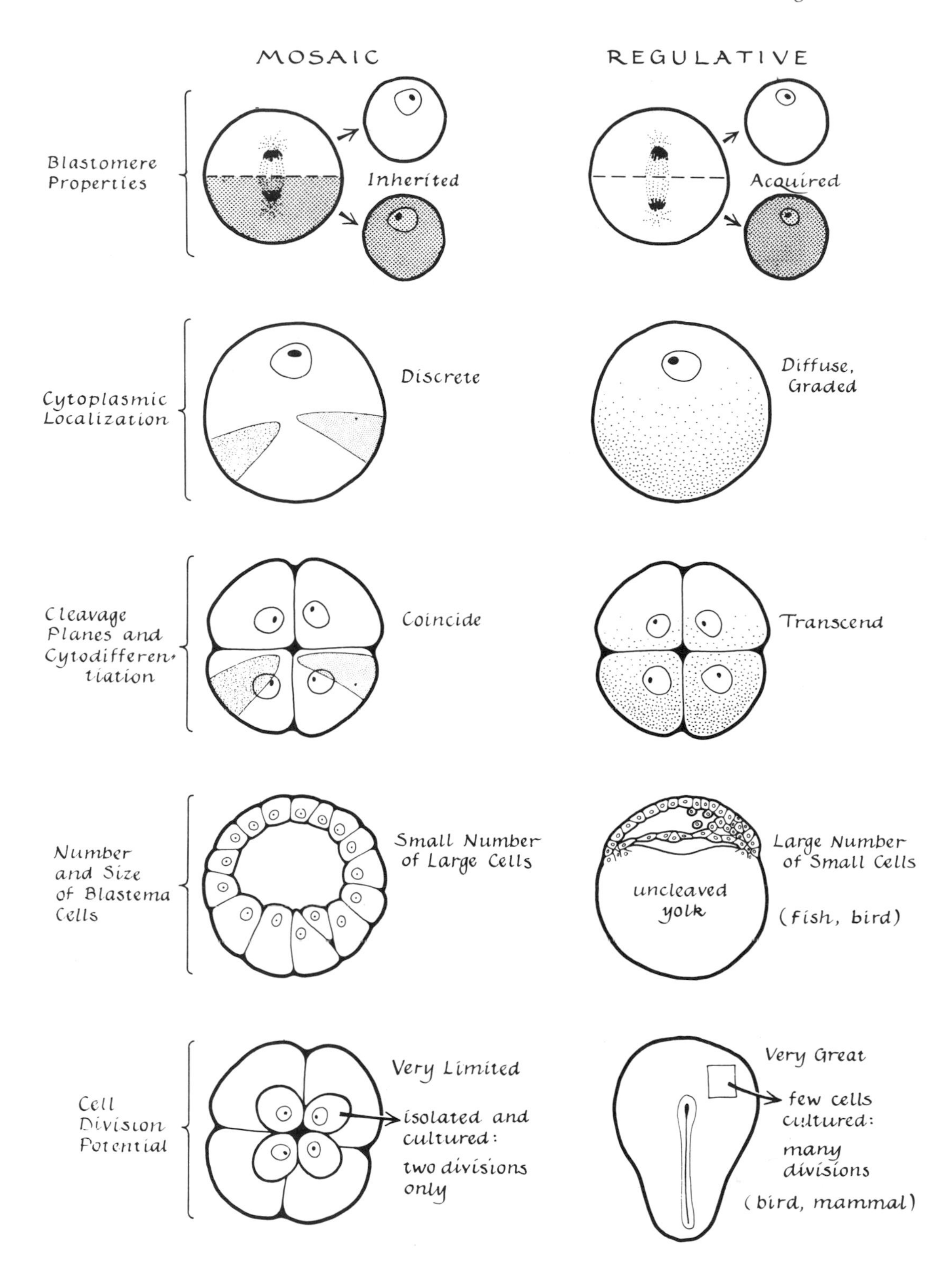

Figure 15-9. *The distinction between extremely regulative and mosaic patterns of development.*

pends upon the stage in the use of the genome at which we begin our observations. Note in the figure that there are stages of development in an egg prior to ovulation.

In patterns classified as mosaic, a portion of the genetic information has been used to build up a complex cytoarchitecture, a cytoplasmic environment which guides the further use of the genome. If not too much information has been used to establish the normal fate of a part, this fate may not be realized if the part is placed in a new environment (Chapters 16, 21, and 22). At later stages when more information has been used (for instance, hypothetical stage 6 in the diagram), the parts may be so specialized that selection of a new gene program may be essentially impossible.

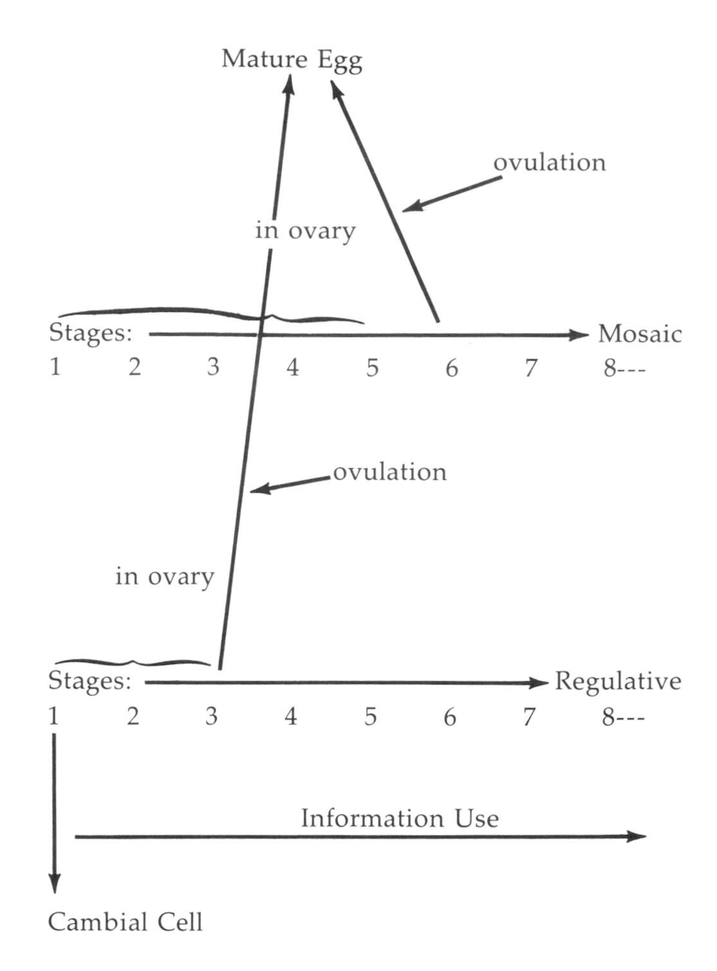

Figure 15-10. *Difference between mosaic and regulative patterns in relation to the use of inherited information.*

References

Berrill, N. J. 1961. Growth, Development and Pattern, W. H. Freeman, San Francisco.

Goebel, K. 1908. Einleitung in die experimentelle Morphologie der Pflanzen. Teubner, Leipzig.

Horstadius, S. 1939. The mechanics of sea urchin development, studied by operative methods. Biol. Rev. Camb. **14:**134–179.

Lutz, H. 1949. Sur le production expérimentale de la polyembrionie et la monstruosité double chez les oiseaux. Arch. Anat. Micro. Morph. Exp. **38:**79–144.

Paterson, M. C. 1957. Animal-vegetal balance in amphibian development. J. Exp. Zool. **134:**183–205.

Pucci-Minafra, I., and G. Ortolani. 1968. Differentiation and tissue interaction during muscle development of ascidian tadpoles. An electron microscope study. Devel. Biol. **17:**692–712.

Reverberi, G. 1957. The role of some enzymes in the development of ascidians. In A. Tyler, R. C. von Borstel, and C. B. Metz, eds., The Beginnings of Embryonic Development, pp. 319–340. Am. Assoc. Adv. Sci., Washington, D.C.

Reverberi, G., and G. Ortolani. 1962. Twin larvae from halves of the same egg in ascidians. Devel. Biol. **5:**84–100.

Seidel, F. 1960. Die Entwicklungsfähigkeiten isolierter Furchungszellen aus dem Ei des Kaninchens, *Oryctolagus cuniculus.* Arch. f. Entw. Mech. **152:**43–130.

Seidel, F. 1969. Entwicklungspotenzen des frühen Saugetierkeims. Arb. Forsch. Landes Nordrhein-Westfalen. **193:**7–91.

Spemann, H. 1928. Die Entwicklung seitlicher und dorsoventraler Keimhaften bei verzoegerter Kernversorgung. Zeit. f. Wiss. Zool. **132:**105–134.

Spratt, N. T., Jr., and H. Haas. 1960. Integrative mechanisms in development of the early chick blastoderm: I. Regulative potentiality of separated parts. J. Exp. Zool. **45**:97–137.

Spratt, N. T., Jr., and H. Haas. 1967. Nutritional requirements for the realization of regulative (repair) capacities of young chick blastoderms. J. Exp. Zool. **164**:31–46.

Vladesco, M. 1935. Recherches morphologiques et expérimentales sur l'embryogénie et l'organogénie des fougères leptosporangiées. Rev. Gén. Bot. **47**:684–771.

Wardlaw, C. W. 1955. Embryogenesis in Plants. Methuen, London.

Wardlaw, C. W. 1968. Morphogenesis in Plants. Methuen, London.

Weiss, P. 1939. Principles of Development. Henry Holt, New York.

Part Three:

Universal Control Mechanisms of Development

16 *Control of the Basic Processes of Development*

Though the mechanisms which underlie and control the basic processes of development (Chapter 2) are only partially known, we can formulate some conclusions about them. In this chapter we shall present in a general way evidence for the control mechanisms of the component processes: growth, differentiation, cellular interaction, movement, and metabolism. A more detailed discussion of some control mechanisms will be presented in the following chapters.

Control of Growth

Growth, particularly increase in cell number accompanied by increase in cell size, plays a dominant role not only in increasing an organism's size or structure but also in changing its shape (Thompson, 1942). In plants, growth, involving increase in cell size, and the direction in which this occurs (Chapter 4) are particularly important in shaping the plant body and its organs (Sinnott, 1960).

Role of Hormones

As we shall see in Chapter 19, hormones help integrate growth processes in both animals and plants. The hormones controlling plant growth are relatively simple compounds, the most important being β-indole acetic acid, an auxin. This auxin primarily stimulates cell elongation. Cytokinins, which are adenine derivatives, stimulate cell division. Plant growth is controlled by a balance in the concentration of different plant hormones rather than by the total concentration (Skoog and Miller, 1957; Zeevaart, 1966).

In animals, hormones such as those released by the pituitary, thyroid, adrenal glands, and gonads can stimulate or suppress cell proliferation in developing organs. The thyroid, adrenal, and sex hormones are relatively simple molecules, whereas the pituitary hormones are complex proteins or peptides. Another protein which functions as a hormone in stimulating growth is the *nerve growth-promoting protein* which has been isolated from certain mouse sarcomas, mouse salivary glands, and snake venom (Levi-Montalcini, 1958, 1964). This protein (having a molecular weight of 20,000 or 44,000) specifically stimulates RNA synthesis and cell division in sympathetic and sensory ganglia of birds and mammals (Figure 16-1). The protein is a normal constituent of sympathetic nerve cells and is present in the body fluids of several vertebrates, including man. An antiserum to the protein destroys the sympathetic nerve cells of newborn animals while having no effect on other cells, thus constituting what can be called chemical surgery. It has been suggested that this hormone depresses genes which regulate growth processes,

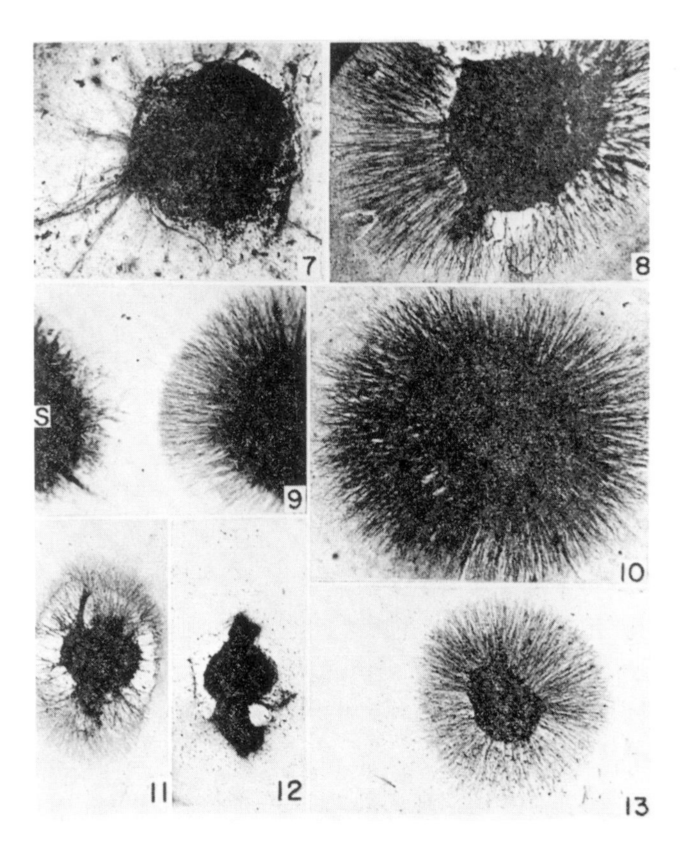

Figure 16-1. *Microphotographs of silver-impregnated sensory ganglia of 7-day chick embryo* in vitro: *Control, 7; effects of extract of mouse tumor, 8; fragment of mouse tumor, 9; snake venom, 10; and mouse salivary gland extract,* 13. *Rat embryo ganglia: Control,* 12; *effect of salivary gland on ganglion of rat embryo,* 11. *From Levi-Montalcini, 1958. Reprinted by permission of the Johns Hopkins Press.*

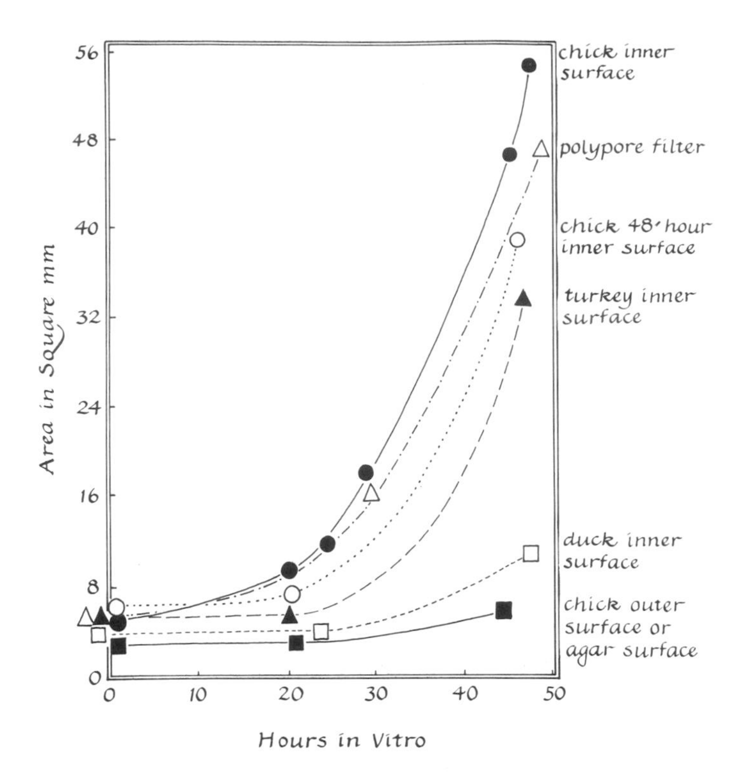

Figure 16-2. *Increase in area of unincubated chick blastoderms explanted on various substrates* in vitro. *"Inner surface" or "outer surface" refers to that of the vitelline membrane.*

resulting in disproportionate cell multiplication of the sympathetic nerve cells. This protein is characteristic of many animal hormones in that it acts on a specific target type of cell. In plants hormones appear to be less specific in respect to target tissues.

Role of Cell Products

During development, organs grow to a size characteristic of the species, then stop growing. It has been suggested that the growth of an organ is limited by a type of negative feedback mechanism. Weiss and Kavanau (1957) postulate certain cell products, "templates," which are produced and remain inside cells, and other cell products, "antitemplates," which are released into the surrounding environment. The "antitemplates" can combine with the "templates" and inactivate them. When the concentration of the "antitemplates" reaches a certain level, all the "templates" are inactivated and growth stops. Tests of this hypothesis by intraperitoneal injection of homogenates or implantation of organs (livers, for example) into rats have been inconclusive, and thus Goss (1964) concludes that specific feedback mechanisms have not been definitely proved to exist. Instead of specific growth-controlling substances, a balance of many metabolites may control growth. An example of such balance controlling plant growth is cited in Chapter 19.

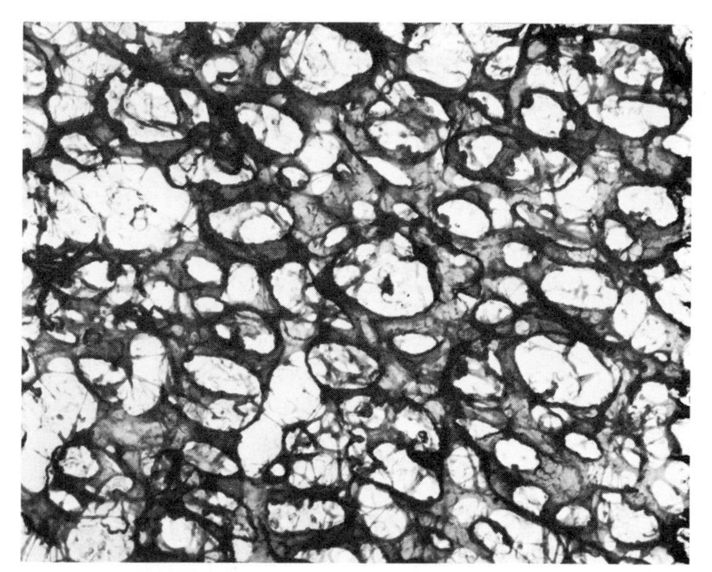

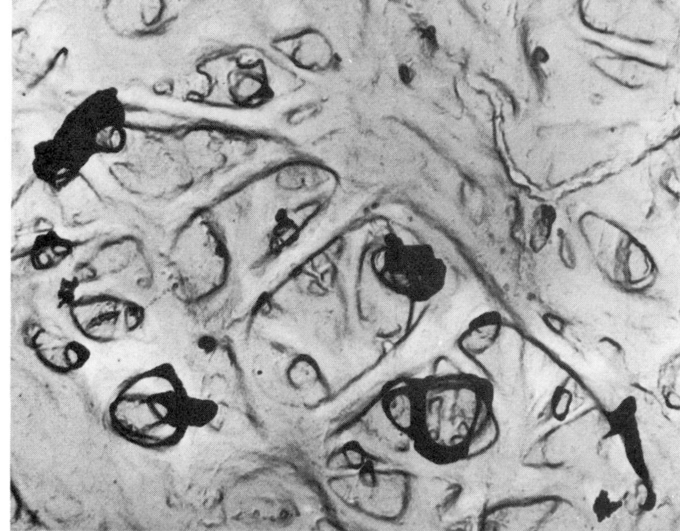

Figure 16-3. *Carbon replicas of polypore filter with 0.45 M pores,* P, *and inner surface of chick vitelline membrane with interstices,* P, *in mesh. Both × 8,000. Photographs by Jensen, 1969.*

Role of Cell Population Density and Physical Properties of the Environment

Increasing density stimulates growth of cultured cells, but ultimately a concentration is reached at which growth slows or ceases. This is not due entirely to depletion of nutrients or to the accumulation of inhibitory cell products, since medium removed from nongrowing cultures supports the growth of actively growing cultures (Stoker, 1967). Other mechanisms involved have yet to be discovered.

The growth of cells is influenced by their immediate environment. For example, under culture conditions, cells of multicellular animals will grow and apparently will divide only when attached to a solid substratum. When the available space (for instance, the area of the culture dish) is filled, growth usually ceases. Addition of more nutrients may affect the density at which inhibition occurs but does not lead to a resumption of active growth.

Growth is influenced by the texture as well as the presence of the substratum. For example, rapid growth and outward movement occurs in chick blastoderms explanted to the inner surface of a piece of chick vitelline membrane (the normal substratum), but much less growth and movement occurs when the explant is placed on the outer surface, or on a duck vitelline membrane or an agar surface (Spratt, 1963). The vitelline membranes of different breeds of birds have different ultrastructural textures (Jensen, 1966, 1969), but a polypore filter with a pore size of about 0.45 μ can substitute for the chick vitelline membrane in supporting an almost equivalent rate of blastoderm expansion. This demonstrates that the texture of the membrane rather than its biochemical composition is the significant growth-controlling factor (Figure 16-2). The similarity between the textural coarseness of the inner surface of the vitelline membrane and that of the polypore filter is illustrated in Figure 16-3.

Further examples of the control of growth by physical factors are the stimulation of cell division below a weight placed on an explanted chick blastoderm (Spratt and Haas, 1960), and stimulation of the growth of explanted long bones of ten-day chick embryos forced to push weights (Haaland, 1968).

The direction of growth (for instance, the directional outgrowth of processes from nerve cell bodies and the growth of twining plant vines) is largely controlled by contact with a physical substratum. In nerve cells the processes not only depend upon the presence of a substratum, but the orientation of the substratum guides the outgrowth. This is called contact guidance (Weiss, 1941) or *thigmotropism* (Chapter 24).

Contact Inhibition

Growth usually ceases in cultures of freshly isolated cells when the growing sheet of cells forms a confluent monolayer. Increasing the supply of nutrient medium allows further growth until (in cultures of human cells) five times the number of cells required to form a monolayer is produced (Stoker, 1967).

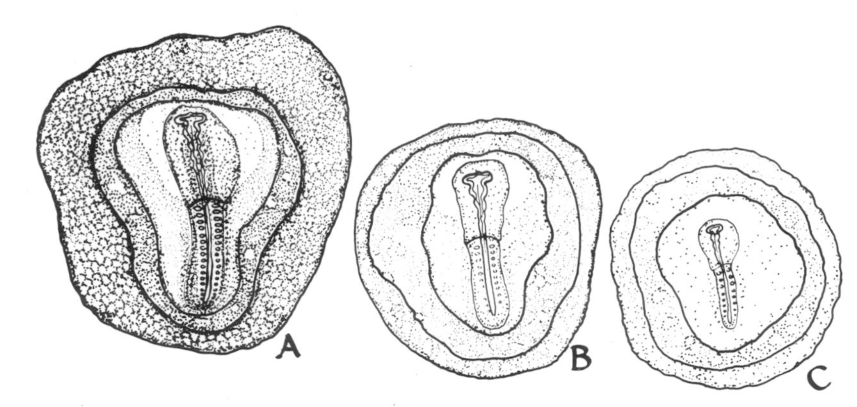

Figure 16-4. *Camera lucida drawings of living, whole unincubated chick blastoderm explants after 40 hours in* vitro. *A, on egg-extract culture medium; B, on White's synthetic medium; C, on glucose medium. The egg-extract medium is the most nutritionally adequate, the glucose medium the least adequate. From Spratt and Haas,* J. Exp. Zool., *Volume 164, p. 35, 1967. Courtesy Wistar Press.*

Role of Nutrients

Growth depends upon the kind as well as the amount of nutrient available. The growth rate of explanted, unincubated chick blastoderms is influenced by the nutritional adequacy of the culture medium (Spratt and Haas, 1967). This is illustrated in Figure 16-4.

Furthermore, different organs or tissues in developing embryos (for instance, chick embryos) have different nutrient requirements for their optimal growth (Spratt, 1950). Limited nutrition may result in differential growth—in growth of one cell type at the expense of another less efficient in using the available nutrient. As noted in Chapter 15, realizing regulative capacities depends upon the availability of nutrients.

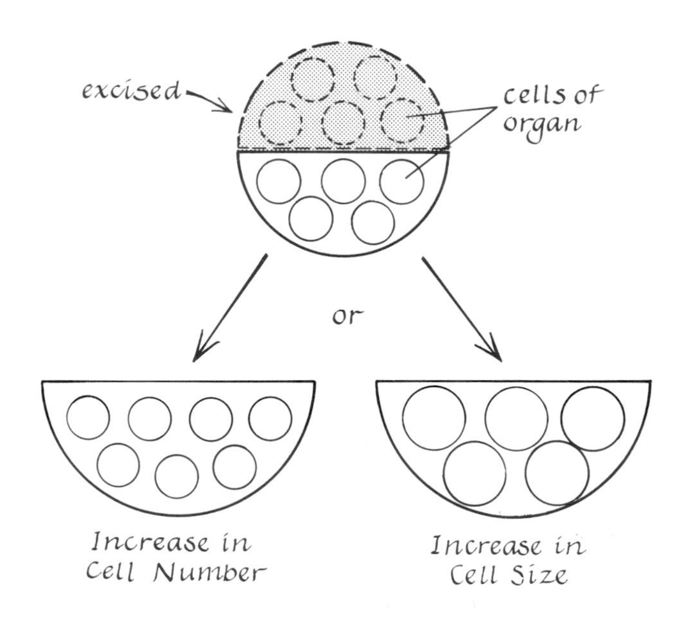

Figure 16-5. *Compensatory growth involving increase in cell size (left) or increase in cell number (right).*

Functional Demand

Functional demands upon an organ or tissue can regulate growth. Removal of one kidney results in enlargement of the remaining kidney. This is called *compensatory growth.* Other examples include the increase in blood cell production under conditions of oxygen deficiency and the enlargement of lymphatic organs when challenged antigenetically. Compensatory growth also occurs in higher plants. Thus, feedbacks by which proper functional levels are maintained regulate organ and tissue growth. Compensatory growth can involve either increase in size of functional units (such as cells) or increase in number of units (Figure 16-5).

Control of Differentiation

As we have defined it, cellular differentiation is cell specialization, apparently resulting from the selected use of only a portion of the cell's genetic repertory. Two basic questions confront us when we try to understand cellular differentiation: (1) What do cells normally do; that is, what mechanisms are used in the course of normal development? (2) What can the cells do; that is, what mechanisms or capacities do the cells have for responding to abnormal

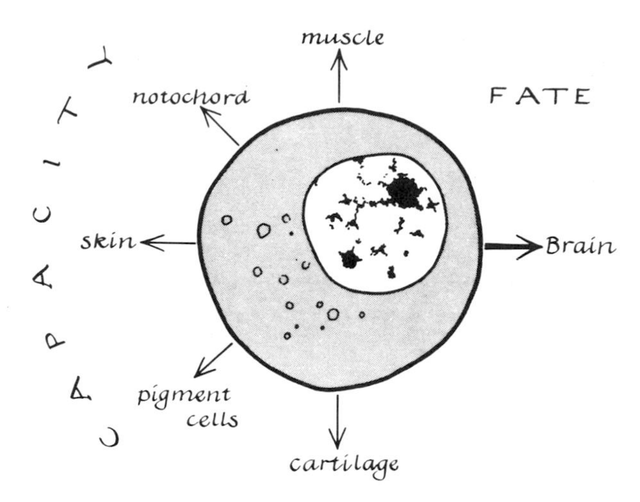

Figure 16-6. *Diagram of the distinction between prospective capacity and normal fate of a cell in a young embryo. From* Introduction to Cell Differentiation *by Nelson T. Spratt, Jr., Copyright © 1964 by Reinhold Publishing Corporation, by permission of Van Nostrand Reinhold Company.*

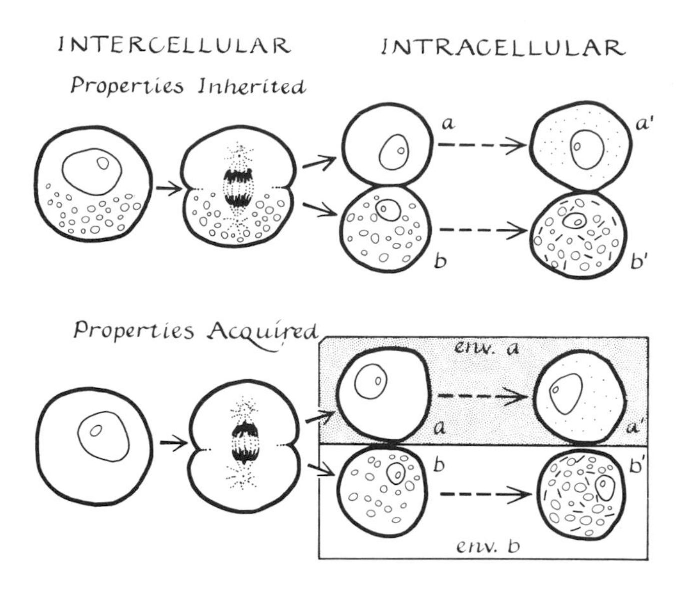

Figure 16-7. *Two methods of intercellular and intracellular differentiation. From* Introduction to Cell Differentiation *by Nelson T. Spratt, Jr., Copyright © 1964 by Reinhold Publishing Corporation, by permission of Van Nostrand Reinhold Company.*

conditions? It has so far been proven much easier to discover what cells can be made to do than to discover why (or how) they behave in a circumscribed way in undisturbed, normal development. The distinction between what cells normally do in nature and what they can be made to do in the laboratory has been commonly described by embryologists as the difference between the prospective *fate* and the prospective *capacity* of the cells. This is illustrated in Figure 16-6.

In normal development the expression of genetic potential is controlled in a preestablished, orderly way and results in realization of the normal fate of the cell (in Figure 16-6, a brain cell). However, artificial influences on the cell, such as centrifugation, transplantation to a strange location, or explantation *in vitro,* can channel its differentiation in another direction, particularly during early stages of development. (Figure 16-6 indicates that a potential brain cell can develop into a muscle, notochord, skin, pigment, or cartilage cell.) Thus abnormal influences upon the genome of the cell result in selection of a different program of gene activity and lead to expression of one of the alternative capacities open to the cells.

General Mechanisms of Cell Differentiation

Diverse cell types are formed in two general but not mutually exclusive ways during development: by inheriting different kinds of cytoplasm during cell division and by responding to different kinds of extracellular microenvironments. In the first case, two cells are different because they have inherited different preformed guidelines; in the second, they become different because they are surrounded by different environments. This is diagramed in Figure 16-7.

There are numerous examples of the role played by the cytoplasm in cell differentiation following differential mitosis. Among these are the degeneration of daughter cells which receive no mitochondria during spermatogenesis in aphids (Goldschmidt, 1955), the formation of slow-growing, small colonies of yeast bud cells which contain abnormal mitochondria (Ephrussi, 1953), and the segregation of certain specific enzyme activities into a cell which becomes a root hair cell in the grass *Phleum.* In the last example, the other daughter cell lacks the necessary enzymes and consequently is hairless. See Torrey (1967) for a review of this and other examples in plants. Some of the best examples of cell differentiation resulting from differential distribution of cytoplasmic substances are found in the cleavage of mosaic eggs (Chapters 13 and 15). In this case the problem is whether inheritance of cytoplasmically different substances influences cell differentiation in post-cleavage stages of animal embryos. Possibly it does in some instances, such as in the development of the bristle organ of the fruit fly. In this case, a giant cell in the epidermis divides, one daughter cell becomes a nerve cell, and the other divides again to produce a bristle cell and a socket cell (Stern, 1954). The possibility that a cytoplasmic segregative mechanism operates in developing fruit flies has been discussed by Hadorn (1966).

Cellular differentiation often results from extracellular environmental influences. The history of experimental embryology provides many demonstrations that the differentiation of cells is controlled in many instances by their position (their local microenvironment) in the embryo. When a group of cells which normally would have participated in formation of the neural plate and later the brain of an amphibian embryo is transplanted at an early gastrula stage into the prospective skin-forming region, the cells differentiate into skin cells. Conversely, prospective skin cells placed in the brain-forming region develop into brain cells (Spemann, 1918). See also Chapter 21.

Cell death, which we have considered a type of cell differentiation, also seems often to be controlled by influences external to the cells. This is probably true of cell death during metamorphosis in insects and amphibians, during morphogenesis of the chick wing (Figure 2-11), and in sensory neurons of spinal ganglia where the number of dying neurons is related to the peripheral load encountered by the outgrowing nerve processes (review of Saunders and Fallon, 1966).

Interaction of Genes and Environment

Many years ago, Roux (1883, 1905) and Weismann (1892, 1904) suggested that cellular differentiation was based upon a progressive segregation or sorting out of nuclear determinants (genes) in the germ cells into the blastomeres during cleavage of the egg. In modern terminology this would be called a segregation of genes (Figure 16-8).

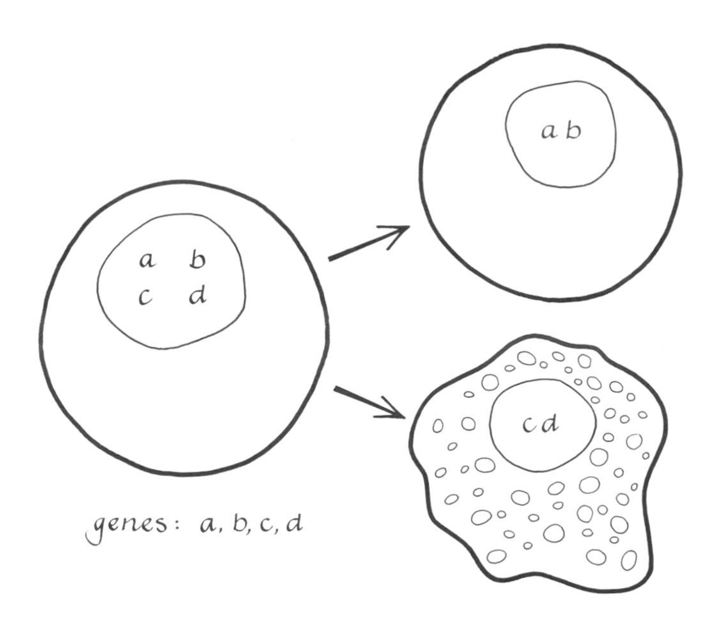

Figure 16-8. *Segregation of determinants (genes) during cell division.*

Somatic mutation, involving a change in cell genotype, could in theory be considered a mechanism of differentiation. As we shall see in Chapters 18 and 26, such mutations occur in the plastids of some plant cells and thus are involved in their differentiation. Somatic mutation has also been invoked as the cause of the change from the normal to the neoplastic state in cells, but serious doubt has been cast upon this claim (Pierce, 1967). The irregularity in occurrence of somatic mutations appears to eliminate them as a mechanism of orderly cellular differentiation.

One manifestation of cellular differentiation in plants is an amplification of the genome; this is exhibited as an increase in ploidy level. The farther a cell in a root is from the meristem, the higher is the ploidy in its nucleus (Clowes, 1961). Different chromosome numbers in different somatic organs of the same individual plant occur in several species of plants, and may involve either an increase or a decrease in the diploid number (Lewis, 1970). An amplification of the genome occurs in the differentiation of the salivary glands in insects, and is expressed as polyteny of the giant chromosomes (Chapter 3).

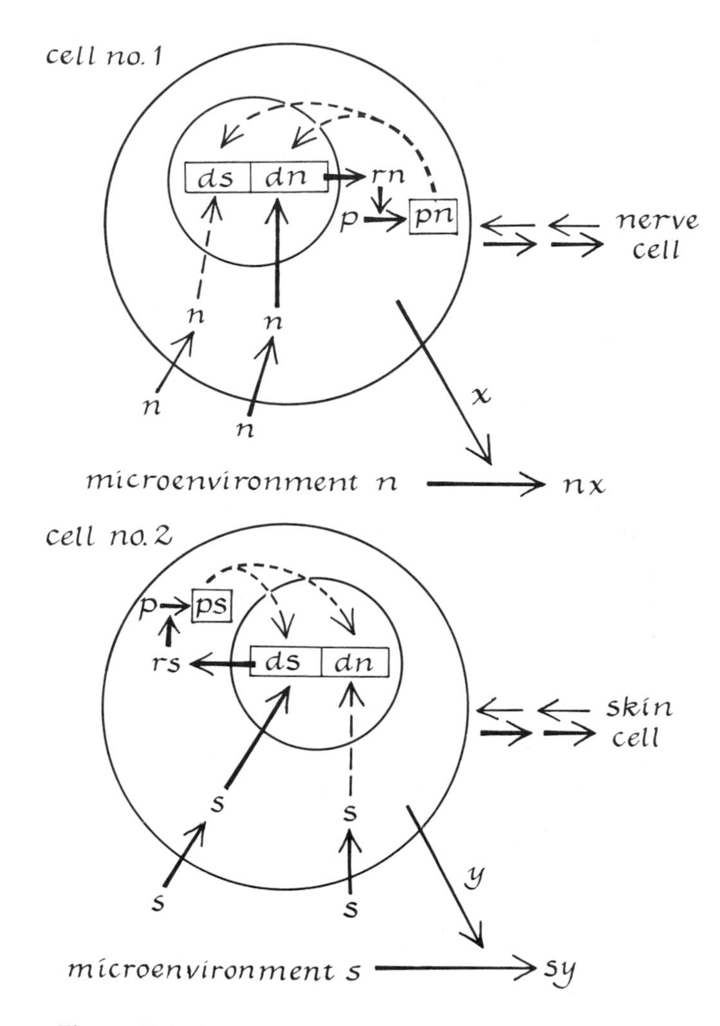

Figure 16-9. *Simple model illustrating how the gene activity pattern of a cell is controlled by the microenvironment of the cell. From* Introduction to Cell Differentiation *by Nelson T. Spratt, Jr., Copyright © 1964 by Reinhold Publishing Corporation, by permission of Van Nostrand Reinhold Company.*

However, these are quantitative, rather than qualitative changes in the genome of the cells and are probably manifestations of cellular differentiation, not causes of it.

Although theoretically attractive, Weismann's hypothesis is contradicted by the fact that when blastomeres are separated at or even after the two-cell stage of many embryos, each cell forms a complete embryo (Chapter 15). Furthermore, during mitosis a dividing cell transmits to each daughter cell a presumably identical set of genes. By the test of developmental capacity, the daughter cells, at least at early stages of development, receive the same genetic information. As described in Chapter 5, different kinds of single tissue cells isolated from an adult plant are capable of proliferating in culture and eventually giving rise to a complete plant. We also noted in Chapter 5 that a small group of pure ectoderm cells of some coelenterates can form a complete animal with all cell types represented. Still more convincing evidence that differentiation is not achieved by modification of the genetic information (that is, by changes in the nucleotide sequences of DNA) is provided by the discovery that nuclei of some differentiated cells can be substituted for the egg nucleus and support the development of a whole plant or animal (Chapter 20). Understanding the role of genes in cellular differentiation thus involves understanding not how genetic information is inherited but how two or more cells with the same inherited information in their nuclei make different use of this information. As noted in earlier chapters, there is much evidence that the microenvironment surrounding a cell (for instance, the position of a cell in the developing organism) determines the pattern of utilization of its inherited information. Consequently, understanding the mechanisms controlling cell differentiation depends not only upon knowing the composition of the spectrum of environments beginning with the extracellular environment and extending inwards to the immediate environment of the genes, but also upon knowing how environmental agents influence the cell. This will be discussed in greater detail in Chapter 22. A simplified model for the control of gene action by the environment of a cell is illustrated in Figure 16-9.

The sequence of events leading to progressive differentiation of cell Number 1 is summarized as follows: specific molecules, n, in the environment enter the cell, pass through the cytoplasm (or possibly interact with cytoplasmic molecules), and enter the nucleus. One or more n molecules stimulate the gene, dn, to direct the synthesis of a correspondingly specific rn molecule (such as a messenger RNA molecule). Other n molecules may suppress the activity of the ds gene, which tends to direct the synthesis of keratin, characteristic of skin cells. The rn molecules, bearing a copy of the dn gene, pass into the cytoplasm and direct synthesis of a specific kind of protein molecule, pn, from precursors, p. The protein, pn, is characteristic of a nerve cell. The pn molecules can promote further activation of the dn gene or aid in repressing the s gene. The newly synthesized pn molecules can also bring about a change in the kinds or quantities of products, y, released by the cell into its environment. In this way the cell would begin to control its own local environment and its own further differentiation. The same kind of mechanism would operate in controlling differential gene activation and/or suppression in cell Number 2, resulting in differentiation of this cell as a skin cell.

The model assumes for simplicity that the two cells diagramed in the figure have inherited cytoplasmic compositions insufficiently diverse to play any directive role in their differentiation. This is clearly not true for all cells (Chapter 13). However, the model can be made to apply to cells with significantly different cytoplasmic compositions by substituting "cytoplasm n" and "cytoplasm s" for microenvironments n and s. Cell Number 1 is normally in or experimentally placed in microenvironment type n, whose composition directs its differentiation as a nerve cell. Cell Number 2 is in a different microenvironment, s, which directs its differentiation as a skin cell. Only two genes and one environmental component are indicated in the model—an extreme simplification of a mechanism probably involving many genes and many environmental components.

The model illustrates how a single kind of specific molecule might direct cell differentiation. Some evidence does exist for the decisive role of specific molecules in differentiation—for example, ecdysone, the molting hormone in insects (Karlson and Sekeris, 1967), vitamin A in suppressing keratin synthesis in skin (Weiss and James, 1955), and phenylalanine in promoting melanin synthesis (Wilde, 1958). However, differences in concentration of a single environmental agent—for instance, the plant hormone indoleacetic acid (IAA), which is one of the auxins, or several animal hormones (Karlson, 1965)—may cause the selection of different paths of cell specialization.

Stability of Cell Types

Closely related to mechanisms that control cellular differentiation are those that maintain the cell in its specialized state, for they might also operate in transforming a differentiated cell of one type into another type. In this case the problem is whether functionally and structurally specialized cells can dedifferentiate and redifferentiate along a new path, or whether a differentiated cell remains stable as to type. Stability of cell type is characteristic not only of normal development, but also of regeneration and of reaggregating, dispersed tissue cells. These latter two subjects have already been discussed in Chapters 5 and 8. Theoretically, stability of the differentiated state of cells could be maintained by mechanisms intrinsic to the cell: a cell might be independent of microenvironmental influences. Stability could also be maintained by the stability of the microenvironment surrounding the cell. Possibly it could depend upon both intrinsic and extrinsic mechanisms (Grobstein, 1959).

Perhaps the best examples of intrinsically stabilized cell types are cells which have attained a terminally differentiated state. Among these are erythrocytes of mammals, lens fiber cells that have lost their nuclei, and the highly specialized egg and sperm. Examples of presumably irreversibly differentiated cells include the blastomeres of mosaically developing organisms, the reaggregating cells of older amphibian, chick, and mammal embryos (Chapter 8), tissue-cultured cells and sieve-tube elements, and companion cells and root-hair cells of plants. Clones of pigment- and cartilage-forming cells have been grown *in vitro* for many cell generations, and thus appear to be irreversibly differentiated. However, conditions which usually confront cultured cells do not constitute a rigorous test of cell stability, since the culture conditions provide a stable environment. Furthermore, in the case of tissue cultures the cells have not been released from the tissue system. A rigorous test should involve the culture of single cells freed from their normal cellular associations or, better, single cells reassembled in new strange associations.

Although cell properties such as shape, motility, and perhaps mitotic activity seem to depend largely upon the continuous presence of extracellular influences, it is not established that these properties are expressions of the differentiated state of the cell. Changes in the shape of mesenchyme cells in response to the texture of the substratum on which they move are better considered *modulations* (temporary differentiations) than cells in diversely differentiated states (Figure 16-10).

Crown gall development is an example of how extrinsic conditions can stabilize cell types in plants. These galls, which form in many kinds of plants, are

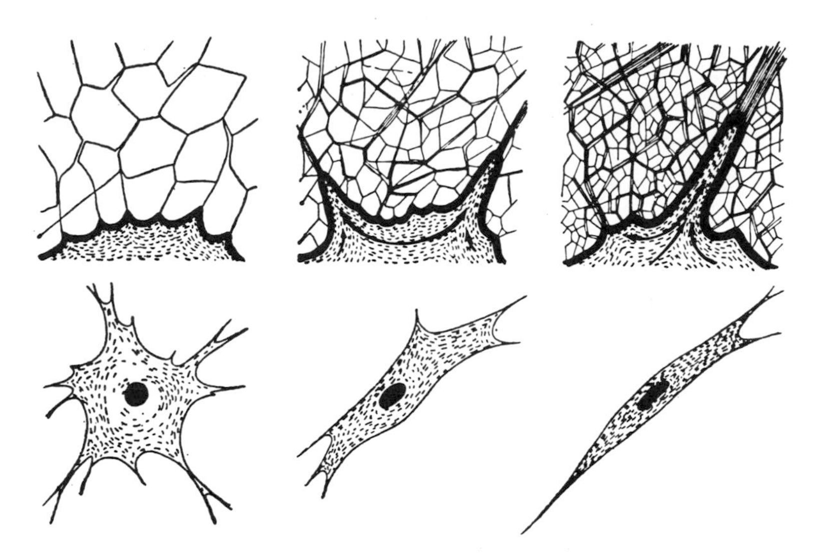

Figure 16-10. *Modulations in the shape of mesenchyme cells induced by differences in coarseness of the texture in a fibrin clot. From Picken, 1960. Courtesy Clarendon Press, Oxford.*

Figure 16-11. *A crown gall on sunflower. Courtesy Department of Plant Pathology, University of Wisconsin.*

amorphous masses of relatively undifferentiated tissue (Figure 16-11). A specific bacterium, *Agrobacterium tumefaciens,* induces the transformation of normal plant cells into crown gall tumor cells; these tumor cells, for the sake of rapid growth, synthesize all hormonal metabolites required by the normal cell. This transformation is heritable in the absence of the inciting bacteria (Braun, 1963, 1968). Treating the infected plant with heat can eliminate the bacteria. The cells grow continuously and cannot form structures such as leaves, roots, or buds.

Incomplete transformation results in formation of a *teratoma,* a chaotic assembly of tissues and organs, in this case including abnormal leaves and buds. These tumors are transplantable. When tumor shoots derived from tumor buds were forced into rapid but organized growth by serial grafting to normal plants, the shoots gradually recovered and produced normal plants. Although the tumor state of the cells is inherited in the absence of the bacteria, the possibility still exists that a bacteriophage introduced with the bacteria might remain and become attached loosely to the chromosomes of the tumor cell. The bacterial virus might thus cause the tumor state. Rapid growth might dilute the viral population and result in normal plants. The stability of the tumor cell would then depend upon the added information in the DNA of the virus.

In animals, there are fewer examples of the stabilization of cell type by exclusively extracellular influences. One example is that of the pigmented retinal and iris cells in adult salamander eyes (Stone, 1950). The persistence of pigment granules in the cytoplasm of these cells depends upon the continued presence of the adjacent unpigmented layer of the nervous retina cells. If the latter cell layer is surgically removed or degenerates as a result of cutting the optic nerve (Chapters 5 and 24), the pigmented cells not only lose their pigment granules but, after mitotic divisions, transform into the various complex nerve cell types of the nervous retina. Another example of the stabilization of cell type by extrinsic influences has been described in a study of *Hydra* by Burnett (1968). Differentiated cells, such as the gland cell and digestive cell, propagate the differentiated state, but if exposed to an ionic environment with the proper Na^{+}-Ca^{++} ratio, the gland cells dedifferentiate to form interstitial cells, and the latter redifferentiate into cnidoblasts, nerve cells, or gametes. Under normal conditions attaining and maintaining the differentiated state depends upon an inducer produced by the neurosecretory cells in the hypostomal region. Burnett postulates that a quantitative apico-basal gradient of the inducer controls all cell differentiation in *Hydra* and that the ratio of the inducer to an inhibitor (produced by dividing cells) at any body level determines the differentiation of cells at that level. Since ions control cell differentiation, it is further postulated that the inducer acts by altering the permeability of the cell membrane to ions.

An example of conservation of genetic potentiality coupled with a limited stability of the differentiated state is found in the differentiation of imaginal discs in *Drosophila* (Chapter 12) when implanted in adult and larval hosts (Gehring, 1967, 1968; Hadorn, 1966). Single cells of a disc were first marked by x-ray-induced crossing-over. The disc was then grafted into an adult female fly. Cell proliferation in the graft led to formation of a genetically marked cell clone (Figure 16-12).

Later, the disc was removed and implanted in a larval host. When the larva metamorphosed, the clonal population of cells (called a blastema) differentiated. When prospective palp cells were marked, the clone differentiated either as wing, head, or antennal cells, but not as tarsus cells. When antennal cells were marked, they differentiated as wing or tarsus but not as palp or head cells. The change in fate of the transplanted cells thus is not random. A specific frequency of each type of *transdetermination*, sometimes a specific transdetermination sequence, occurs for the different steps. The probability of the transdetermination steps is independent of the kind of original imaginal disc from which the *in vitro* culture was started, but appears to depend only upon the determination state of the blastema. These studies show that a specific state of differentiation can be propagated over many cell generations (for instance, in the adult fly). However, the stability of the cells is limited, since clones derived from determined cells give rise to a limited array of other cell types. The problem remaining is why certain transformations occur frequently and others infrequently or not at all. In any case, no change in genetic information seems to be involved, and we are left with the problem of the nature of the environment that controls gene expression in the cells it surrounds.

Studies of differentiated cells cultured *in vitro* provide much evidence that the differentiated state is

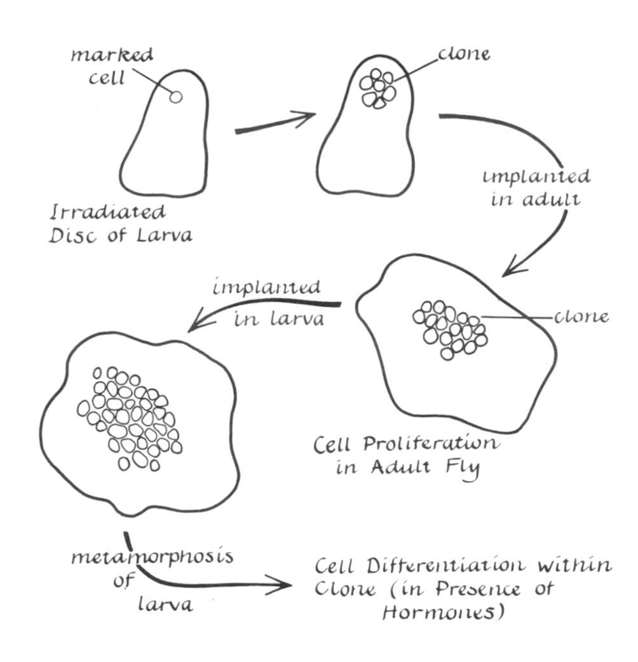

Figure 16-12. *Diagrams illustrating the method of producing a genetically marked clone of imaginal disc cells of* Drosophila.

maintained even if the cells undergo modulations like those described above (see review of Grobstein, 1959). For example, under dilute, clonal culture conditions, embryonic chick cartilage cells retain their stable phenotype for thirty-five or more generations (Coon, 1966), demonstrating that the differentiated state is heritable.

Control of Cellular Reactions

Little is known about cellular interaction or the mechanisms which control it. How much influence one group of cells has upon another probably depends on many conditions, such as the numbers and kinds of interacting cells and the distance between them. The cell surface also must be important in controlling cellular interactions (Stoker, 1967). Products of one group of cells influence neighboring or even distant cells. Not only the concentrations but also the proportion of these products must be involved in controlling the processes. Until more is known about the microenvironments surrounding cells,

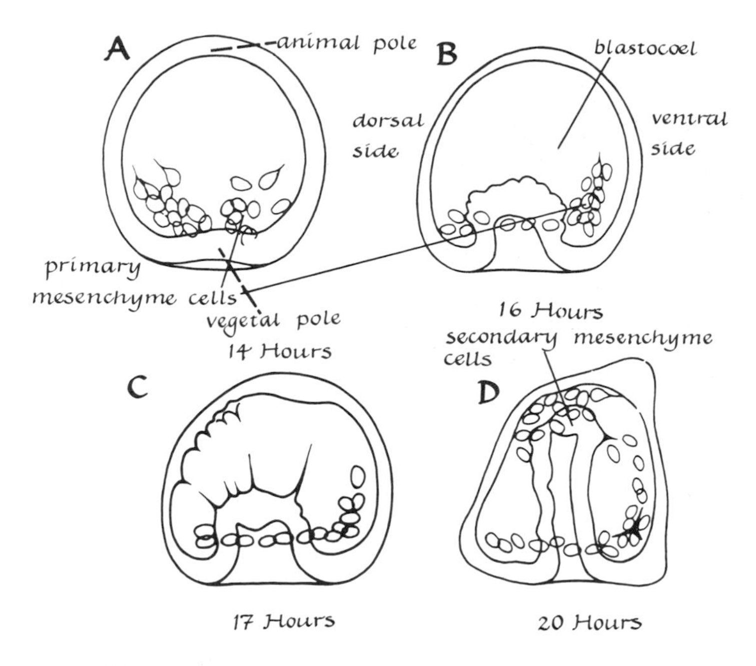

Figure 16-13. *Diagrams illustrating the role of primary and secondary mesenchyme cells during invagination in a sea urchin gastrula. From Gustafson and Wolpert in* International Review of Cytology. *Copyright 1963 by Academic Press.*

little can be concluded about the control of cellular interactions. In Chapter 21, cellular interactions will be described in greater detail.

Control of Cell and Tissue Movements

Very little is known about the mechanisms controlling cell and tissue movements. We can cite only a few processes which, to a greater or lesser extent depending upon the kind of organism, appear to underlie morphogenetic cell movements in animal development. Tissue movements are of minor importance in plant development, and will not be discussed.

Differential Growth

The importance of differential growth in animal development, particularly in embryology, is usually denied (Trinkaus, 1965). However, growth patterns in animal embryos, with some exceptions (Haas and Spratt, 1968), have not been studied carefully. The difficulty is how to distinguish between the role of growth in producing cells for movement (for example, in epiboly), and the role of cell addition in spreading out a sheet of cells. There is evidence for growth centers in early chick embryos (Spratt, 1963), but it has not been proved that the addition of cells to a spreading cell sheet is the underlying cause of the sheet's movement. However, there is no evidence for a chemotactic mechanism in the spreading of germ layers in chick embryos; further, the movement of the endodermal layer below the upper layer ceases when the source of cells in the posterior part of the blastoderm is removed (Chapter 12).

Chemotaxis

In this section we are not concerned with chemotactic mechanisms in fertilization (Chapters 10 and 12), but with chemotaxis during embryonic development. To what extent chemotaxis directs cell and tissue movements in animal embryos remains an unsolved problem (Trinkaus, 1965). In the reaggregation and sorting out of dispersed tissue cells, random rather than directed cell movement seems to be the rule (Steinberg, 1964). In aggregation and possibly in morphogenesis in cellular slime molds, the cells move in response to a gradient in concentration of acrasin (Chapter 6), but no chemotactic substance has been identified and shown to have an influence in controlling morphogenetic movements in animal embryos. However, the primordial germ cells (Chapter 13) of amphibians and mammals may be guided in their movement to the developing gonad by a chemotactic mechanism, but this remains to be demonstrated. In birds the primordial germ cells reach the gonads via the circulatory system. Although they may collect in the gonad as the result of a "trapping" mechanism, *in vitro* association of a sterile gonad region and the germinal crescent results in migration of the primordial germ cells into the gonad region. This suggests the operation of a chemotactic mechanism (Dubois and Croisille, 1970).

Contact Guidance and Texture of the Substratum

There is much evidence (Weiss, 1961; Spratt, 1963; Trinkaus, 1965) that the directional migration of cells and cell sheets is guided by contact of the cells with an oriented substratum. Pigment cells move along blood vessels, schwan cells move along nerve cell processes, heart-forming cells move along an oriented endodermal substratum (DeHaan, 1963). In *Hydra*, epitheliomuscular cells move directionally along the mesolamella (Shostak and Globus, 1966).

Except for cells with flagella or cilia, movement requires contact of the cell with a solid substratum. Furthermore, the texture of the substratum controls both the rate and direction of cell movement. There is some evidence that the rate at which cells move over a substratum depends upon the capacity of the cells to adhere to the substratum, close adherence being correlated with more rapid movement (Carter, 1967). Expansion of the young chick blastoderm involves not only cell proliferation but also cell movement. The rate at which cells near the edge of the blastoderm move is influenced by the texture of the substratum (Figures 16-2 and 16-3).

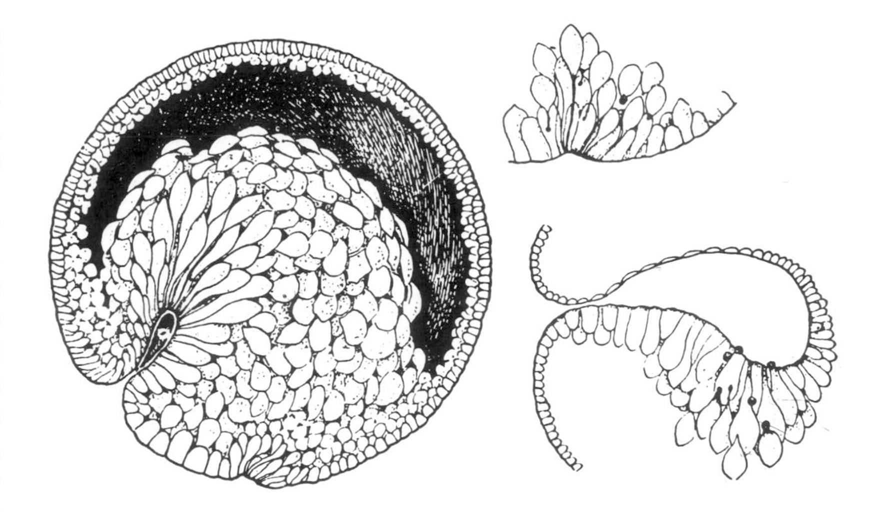

Figure 16-14. *Schematic sections through an advanced gastrula stage of an amphibian. Left, whole embryo; upper right, ingression of cells at the ventral blastopore lip; lower right, endoderm cells lining the archenteron. Note the bottle-shaped cells. From Holtfreter,* J. Exp. Zool., *Volume 94, p. 274, 1943. Courtesy Wistar Press.*

Contact Inhibition

Fibroblasts in tissue culture tend to stop migrating when cells contact one another, the general result being a monolayer on the surface of the culture medium (Abercrombie, 1961). Contact inhibition could thus be a means of controlling movement of cell sheets such as occurs in morphogenetic tissue movements in gastrulation. However, certain mouse tumor cells are not inhibited by contact but move over one another (Abercrombie and Heaysman, 1954, 1957). Tumor cells appear to be subject to contact inhibition when in contact with normal cells but not when in contact with one another (Stoker, 1967).

Pseudopod Contraction

Dan and Okazaki (1956), Gustafson and Kinnander (1956), and Gustafson and Wolpert (1963, 1967) have presented evidence that the contraction of filapodia spun out by secondary mesoderm cells in the tip of the invaginating archenteron of the sea urchin continues the inward movement of the archenteron. This is diagramed in Figure 16-13D.

Specialized Motile Cells

Special "bottle cells" may initiate invagination in amphibian gastrulation (see Holtfreter, 1943; Rhumbler, 1902). These cells, attached by long processes to the surface cells, move inward into the blastocoel; in so doing, they appear to pull in the outer surface. Either by contraction of their processes or by continued inward movement, the "bottle cells" bring about continued invagination and involution of the surface.

Changes in Cell Shape

The ingression of primary mesenchyme cells in sea urchin gastrulation (Figure 16-13A) appears to involve a change in cell shape and a decrease in cell adhesiveness (Gibbins et al., 1969). Shortly after ingression of the primary mesenchyme cells, the invagination of the vegetal plate begins (Figure 16-13B and C). This is also believed to result from a change in the shape of cells in the vegetal plate of the blastula (Gustafson and Wolpert, 1963). A similar mechanism

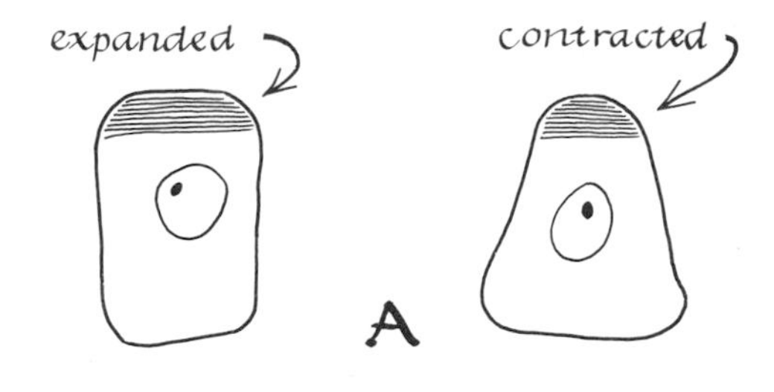

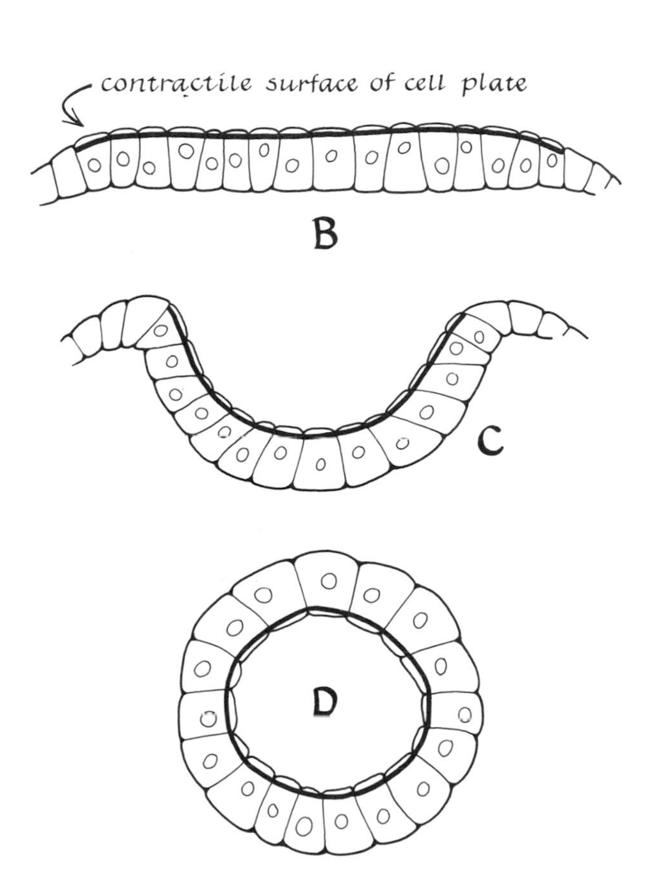

Figure 16-15. *Diagram of the probable role of microfibrils in causing changes in cell shape and consequent rolling up of a flat plate of cells to form a tube.*

underlying neurulation and probably initiating gastrulation in the tree frog *Hyla regilla* has been reported by Baker (1965) and Baker and Schroeder (1967). Electron micrographic studies have revealed intracellular microfibrils in the apical cytoplasm (next to the future lumen of the neural tube) of the neural plate cells. It is postulated that contraction of the microfibrils changes the cell shape from more or less cylindrical to cone-like, and so causes the flat plate to roll up into a tube (Figure 16-15). Changes in cell shape also appear to underlie cell movements which occur in the neural plate of the newt *Taricha torosa* prior to neurulation (Burnside and Jacobson, 1968). In the mouse embryo, changes in cell shape involving contraction of microfilaments just below the outer surface of the cells appear to cause the lens placode to invaginate to form the lens vesicle of the eye (Wren and Wessells, 1969).

Control of Metabolism

Although of obvious importance, a discussion of the control of metabolic processes underlying developmental events is far beyond the scope of this text. We can only emphasize that the interaction of genes and the environment changes during development and that the biochemical processes are controlled to a large degree by the organism's structural composition (see reviews of Needham, 1950; Boell, 1955; Steinbach and Moog, 1955; Halvorson, 1964).

Growth and Cell Differentiation

Generally it has been claimed that cell division and cell specialization are antagonistic—cannot occur at the same time—but recent studies of Davies et al. (1968) provide evidence that chromosomal replication (DNA synthesis) and the synthesis of a specific cell-product, collagen, by diploid rat and fetal human cells (Figure 16-16) can occur simultaneously. Similar examples have been reviewed by Cahn (1968). However, replication of DNA may be a necessary precursor for a new program of DNA transcription (Gehring, 1968; review of Ebert and Kaighn, 1966). Apparently, certain cells under certain conditions may engage in simultaneous replication and transcription of the genome. Studies of gene action in somatic cell hybrids (Chapter 20) and differential gene amplification in oocytes of amphibians (Chapter 17) suggest that complete or partial replication of the genome might be required in passing from one stage of development to the next (Ebert, 1968).

Conclusion

Although much remains unknown about control of the component processes in development, some conclusions can be drawn. Growth is controlled by the chemical and physical properties of the environment surrounding the cells. Cellular differentiation may result from the inheritance of different cytoplasmic guidelines or from the influence of different extracellular microenvironments upon the expression of the genome. Once a cell has become specialized morphologically and functionally, its differentiation is maintained by either intrinsic or extrinsic mechanisms or both. The stability of the differentiated state is impressive but does not appear to involve any change in information content *per se*. Instead, this stability appears to depend primarily upon the stability of the surrounding microenvironment which controls the pattern of information use.

Cellular interaction is a mechanism of cell differentiation. Interactions between cells apparently are controlled by properties of the cell surface, the number of interacting cells, and the concentration of cell products; the last-mentioned factor is of considerable influence in the mediation of the interactions.

Mechanisms controlling or underlying morphogenetic cell and tissue movements are not well known but seem to involve contact of cell with cell, contact of the cell with a substratum, texture of the substratum, changes in cell shape, and possibly differential growth.

In the final analysis, understanding the mechanisms that control the program of development is equivalent to understanding the mechanisms controlling gene action (Allen, 1965; Schultz, 1965; Waddington, 1962; and many others). In studying the control of gene action, the developmental biologist must take into account not only gene action during a particular stage of development but also the gene action which has led to the attainment of the stage in question. Both the internal and external environments of the cell influence the selection of programs of gene action. The composition of a developing organism's internal environment undergoes a sequence of changes which are just as orderly as the changes in the organism's cells. This internal environment is itself a product of orderly gene action.

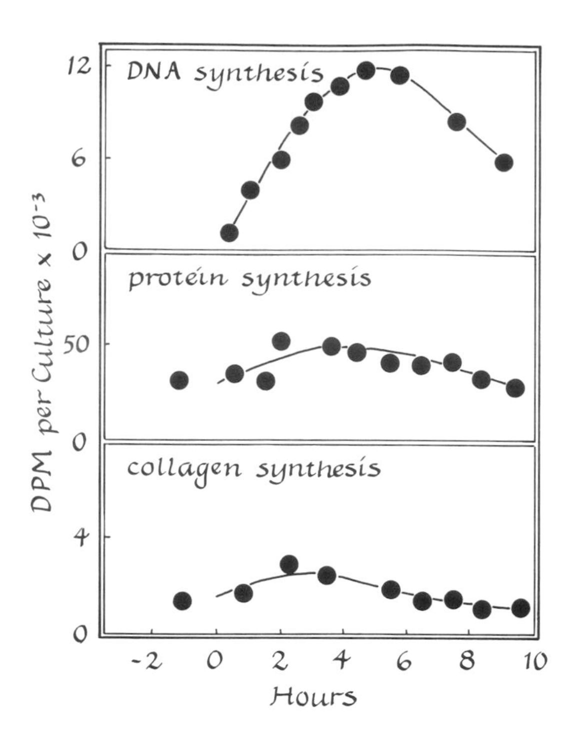

Figure 16-16. *DNA, protein, and collagen synthesis in a 3T6 aneuploid mouse line. Note that collagen synthesis continues through the period of DNA synthesis. From Davies et al.,* Science, *Volume 159, pp. 91–93, copyright 1968 by The American Assn. Adv. Science.*

References

Abercrombie, M. 1961. The basis of the locomotory behavior of fibroblasts. Exp. Cell Res. Suppl. **8**:188–189.

Abercrombie, M., and J. Heaysman. 1954. Observations on the social behavior of cells in tissue culture: II. Monolayering of fibroblasts. Exp. Cell Res. **6**:293–306.

Abercrombie, M., and J. Heaysman. 1957. Ibid.: III. Mutual influence of sarcoma cells and fibroblasts. Exp. Cell Res. **13**:276–291.

Baker, P. C. 1965. Fine structure and morphogenetic movements in the gastrula of the tree frog, *Hyla regilla*. J. Cell Biol. **24**:95–116.

Baker, P. C., and T. Schroeder. 1967. Cytoplasmic filaments and morphogenetic movement in the amphibian neural tube. Devel. Biol. **15**:432–450.

Boell, E. J. 1955. Energy exchange and enzyme development during embryogenesis. In B. H. Willier, P. Weiss, and V. Hamburger, eds., Analysis of Development, pp. 520–555. W. B. Saunders, Philadelphia.

Braun, A. C. 1963. Biochemical changes of a heritable type that result in cellular autonomy. In D. Mazia and A. Tyler, eds., The General Physiology of Cell Specialization, pp. 73–79. McGraw-Hill Book Company, New York.

Braun, A. C. 1968. The multipotential cell and the tumor problem. In H. Ursprung, ed., The Stability of the Differentiated State, pp. 128–135. Springer-Verlag, New York.

Burnett, A. L. 1968. The acquisition, maintenance, and lability of the differentiated state in *Hydra*. In H. Ursprung, ed., The Stability of the Differentiated State, pp. 109–127. Springer-Verlag, New York.

Burnside, M. B., and A. G. Jacobson, 1968. Analysis of morphogenetic movements in the neural plate of the newt *Taricha torosa*. Devel. Biol. **18:**537–552.

Cahn, R. D. 1968. Factors affecting inheritance and expression of differentiation: some methods of analysis. In H. Ursprung, ed., The Stability of the Differentiated State, pp. 58–84. Springer-Verlag, New York.

Carter, S. B. 1967. Haptotaxis and the mechanism of cell motility. Nature **213:**256–260.

Clowes, F. A. L. 1961. Apical Meristems. Blackwell, Oxford.

Coon, H. G. 1966. Clonal stability and phenotypic expression of chick cartilage cells in vitro. Proc. Nat. Acad. Sci. U.S. **55:**66–73.

Dan, K., and K. Okazaki. 1956. Cyto-embryological studies of sea urchins: III. Role of the secondary mesenchyme cells in the formation of the primitive gut in sea urchin larvae. Biol. Bull. **110:**29–42.

Davies, L., J. Priest, and R. Priest. 1968. Collagen synthesis by cells synchronously replicating DNA. Science **159:**91–93.

DeHaan, R. L. 1963. Migration patterns of the precardiac mesoderm in the early chick embryo. Exp. Cell. Res. **29:**544–560.

Dubois, R., and Y. Croisille. 1970. Germ-cell line and sexual differentiation in birds. Phil. Trans. Roy. Soc. Lond. Ser. B. **259:**73–89.

Ebert, J. 1968. Levels of control: a useful frame of perception. In A. A. Moscona and A. Monroy, eds., Current Topics in Developmental Biology, pp. xv-xxv. Academic Press, New York.

Ebert, J. D., and M. E. Kaighn. 1966. The keys to change: factors regulating differentiation. In M. Locke, ed., Major Problems in Developmental Biology, pp. 29–84. Academic Press, New York.

Ephrussi, B. 1953. Nucleo-cytoplasmic Relations in Microorganisms. Oxford University Press, London.

Gehring, W. 1967. Clonal analysis of determination dynamics in cultures of imaginal disks in *Drosophila melanogaster*. Devel. Biol. **16:**438–456.

Gehring, W. 1968. The stability of the determined state in cultures of imaginal disks in *Drosophila*. In H. Ursprung, ed., The Stability of the Differentiated State, pp. 136–154. Springer-Verlag, New York.

Gibbins, J. R., L. G. Tilney, and K. Porter. 1969. Microtubules in the formation and development of primary mesenchyme in *Arbacia punctulata:* I. The distribution of microtubules. J. Cell Biol. **41:**201–226.

Goldschmidt, R. B. 1955. Theoretical Genetics. University of California Press, Berkeley.

Goss, R. J. 1964. Adaptive Growth. Logos Press, London.

Grobstein, C. 1959. Differentiation of vertebrate cells. In J. Brachet and A. E. Mirsky, eds., The Cell, pp. 437–496. Academic Press, New York.

Gustafson, T., and H. Kinnander. 1956. Microaquaria for time-lapse cinematographic studies of morphogenesis in swimming larvae and observations on sea urchin gastrulation. Exp. Cell Res. **11:**36–51.

Gustafson, T., and L. Wolpert. 1963. The cellular basis of morphogenesis and sea urchin development. Intern. Rev. Cytol. **15:**139–214.

Gustafson, T., and L. Wolpert. 1967. Cellular movement and contact in sea urchin morphogenesis. Biol. Rev. Camb. **42:**132–179.

Haaland, J. 1968. Interaction of ionic environment and stress on a bone forming system. Doctoral thesis, University of Minnesota, Minneapolis.

Haas, H., and N. T. Spratt, Jr. 1968. Studies on growth polarity in the ring blastema (germ wall ring) of young chick embryos. Physiol. Zool. **41:**129–148.

Hadorn, E. 1966. Dynamics of determination. In M. Locke, ed., Major Problems in Developmental Biology, pp. 85–104. Academic Press, New York.

Halvorson, H. O. 1964. Genetic control of enzyme synthesis. In Differentiation and Development, pp. 63–76. New York Heart Assoc., Little, Brown, Boston.

Holtfreter, J. 1943. A study of the mechanics of gastrulation: Part I. J. Exp. Zool. **94:**261–318.

Jensen, C. 1966. Interrelationships of the early avian embryo with the vitelline membrane and other substrata: an electron microscope study. Doctoral thesis, University of Minnesota, Minneapolis.

Jensen, C. 1969. Ultrastructural changes in the avian vitelline membrane during embryonic development. J. Embryol. Exp. Morph. **21:**467–484.

Karlson, P. 1965. Mechanisms of Hormone Action. Academic Press, New York.

Karlson, P., and C. E. Sekeris. 1966. Ecdysone, an insect steroid hormone and its mode of action. Recent Prog. Hormone Res. **22:**473–502.

Levi-Montalcini, R. 1958. Chemical stimulation of nerve growth. In W. D. McElroy and B. Glass, eds., The Chemical Basis of Development, pp. 646–664. Johns Hopkins Press, Baltimore.

Levi-Montalcini, R. 1964. Growth control of nerve cells by a protein factor and its antiserum. Science **143:**105–110.

Lewis, W. H. 1970. Chromosomal drift, a new phenomenon in plants. Science **168:**1115–1116.

Needham, J. 1950. Biochemistry and Morphogenesis. Cambridge University Press, London.

Pierce, G. B. 1967. Teratocarcinoma: a model for a developmental concept of cancer. In A. A. Moscona and A. Monroy, eds., Current Topics in Developmental Biology, pp. 223–246. Academic Press, New York.

Rhumbler, L. 1902. Zur Mechanik des Gastrulationsvorganes, insbesondere der Invagination. Arch. Entw. Mech. **14:**401–476.

Roux, W. 1883. Über die Bedeutung der Kernteilungsfiguren, Wilhelm Englemann, Leipzig.

Roux, W. 1905. Die Entwicklungsmechanik, Wilhelm Englemann, Leipzig.

Saunders, J. W., and J. F. Fallon. 1966. Cell death in morphogenesis. In M. Locke, ed., Major Problems in Developmental Biology, pp. 289–314. Academic Press, New York.

Schultz, J. 1965. Genes, differentiation and animal development. In Genetic Control of Differentiation, pp. 116–147. Brookhaven Symp. Biol., Upton, New York.

Shostak, S., and M. Globus. 1966. Migration of epitheliomuscular cells in *Hydra.* Nature **201:**218–219.

Sinnott, E. W. 1960. Plant Morphogenesis. McGraw-Hill Book Company, New York.

Skoog, F., and C. O. Miller. 1957. Chemical regulation of growth and organ formation in plant tissues cultured *in vitro.* Symp. Soc. Exp. Biol. **11:**118–131.

Spemann, H. 1918. Über die Determination der ersten Organanlagen des Amphibienembryo: I-VI. Arch. Entw. Mech. **43:**448–555.

Spratt, N. T., Jr. 1950. Nutritional requirements of the early chick embryo: II. Differential nutrient requirements for morphogenesis and differentiation of the heart and brain. J. Exp. Zool. **114:**375–402.

Spratt, N. T., Jr. 1963. Role of the substratum, supracellular continuity and differential growth in morphogenetic cell movements. Devel. Biol. **7:**51–63.

Spratt, N. T., Jr. 1964. Introduction to Cell Differentiation. Reinhold, New York.

Spratt, N. T., Jr., and H. Haas. 1960. Importance of morphogenetic movements in the lower surface of the young chick blastoderm. J. Exp. Zool. **144:**257–275.

Spratt, N. T., Jr., and H. Haas. 1967. Nutritional requirements for the realization of regulative (repair) capacities of young chick blastoderms. J. Exp. Zool. **164:**31–46.

Steinbach, H. B., and F. Moog. 1955. Cellular metabolism. In B. H. Willier, P. Weiss, and V. Hamburger, eds., Analysis of Development, pp. 70–90. W. B. Saunders, Philadelphia.

Steinberg, M. S. 1964. The problem of adhesive selectivity in cellular interactions. In M. Locke, ed., Cellular Membranes in Development, pp. 321–366. Academic Press, New York.

Stern, C. 1954. Two or three bristles. Am. Scientist **42:**213–247.

Stoker, M. 1967. Contact and short-range interactions affecting growth of animal cells in culture. In A. A. Moscona and A. Monroy, eds., Current Topics in Developmental Biology, Vol. 2, pp. 107–128. Academic Press, New York.

Stone, L. S. 1950. Neural retina degeneration followed by regeneration from surviving retinal pigment cells in grafted adult salamander eyes. Anat. Rec. **106:**89–109.

Thompson, D. W. 1942. On Growth and Form. Cambridge University Press, London.

Torrey, J. G. 1967. Development in Flowering Plants. Macmillan, New York.

Trinkaus, J. P. 1965. Mechanisms of morphogenetic movements. In R. DeHaan and H. Ursprung, eds., Organogenesis, pp. 55–104. Holt, Rinehart and Winston, New York.

Waddington, C. H. 1962. New Patterns in Genetics and Development. Columbia University Press, New York.

Weismann, A. 1892. Das Keimplasm. Eine Theorie der Vererbung. Jena. English translation, 1915. Charles Scribner's Sons, New York.

Weismann, A. 1904. Vorträge öber Descendenztheorie. Vols. 1 and 2. Jena.

Weiss, P. 1941. Nerve Patterns: The mechanics of nerve growth. Growth Suppl. 3, **5:**163–203.

Weiss, P. 1961. Guiding principles in cell locomotion and cell aggregation. Exp. Cell Res. Suppl. **8:**260–281.

Weiss, P., and R. James. 1955. Skin metaplasia *in vitro* induced by brief exposure to vitamin A. Exp. Cell Res. Suppl. **3:**381–394.

Weiss, P., and J. L. Kavanau. 1957. A model of growth and growth control in mathematical terms. J. Gen. Physiol. **41:**1–47.

Wessells, N. K., B. S. Spooner, J. F. Ash, M. O. Bradley, M. A. Luduena, E. L. Taylor, J. T. Wrenn, and K. M. Yamada. 1971. Microfilaments in cellular and developmental processes. Science **171:**135–143.

Wilde, C. E. 1958. Differentiation in response to the biochemical environment. In D. Rudnick, ed., Cell, Organism and Milieu, pp. 3–43. Ronald Press, New York.

Wrenn, J. T., and N. K. Wessells. 1969. An ultrastructural study of lens invagination in the mouse. J. Exp. Zool. **171:**359–367.

Zeevaart, J. A. D. 1966. Hormonal regulation of plant development. In Cell Differentiation and Morphogenesis, pp. 144–179. North Holland Publishing Co., Amsterdam.

17 *Control of Gene Activity and the Program of Nucleoprotein Synthesis*

Since development is the sequential expression of inherited capacities, understanding development as a whole requires understanding the mechanisms that control this expression. Jacob and Monod (1961, 1963) have proposed a model which explains how gene action might be controlled. The model derives from genetic evidence in which exogenous substances control the synthesis of specific enzymes in microorganisms such as *E. Coli.* This process is called *enzyme induction.* The major features of the model are illustrated in Figure 17-1.

According to the model *structural* genes for specific enzyme synthesis depend for their activity on an adjacent *operator* gene on the chromosome. The operator and structural genes involved in a synthetic pathway constitute the *operon.* A *repressor* substance produced by the *regulator* gene normally blocks the operator gene. Some repressor substances are inactivated by specific substances (metabolites); the others are activated. An activated repressor prevents the function of a particular operator gene. The repressor substance may be inactivated by combination with some substance in the cell. In this case the operator and its associated structural genes will function, and messenger RNA molecules will be transcribed. The latter will then participate in the synthesis of correspondingly specific proteins. Under other conditions, an inactive repressor is produced by the regulator gene, and its corresponding operon will function if the substance that activates the repressor is not present. When this substance is present it activates the repressor, which then blocks the activity of the operator gene. The nature of the repressor is unknown, but it might be a protein, possibly a nucleoprotein.

In the model, gene action is controlled by specific substrates present in the cultures of microorganisms. It is tempting to speculate that similar mechanisms might also control gene activity patterns in developing multicellular organisms. Indeed, there is some evidence, which we shall outline below, that such mechanisms operate in higher organisms, but whether they operate in strictly developmental processes like cellular differentiation has not yet been convincingly demonstrated. Perhaps hormones, inductor substances, and the molecules constituting the microenvironment of cells influence gene action much as substrates (metabolites) influence gene action in bacteria. But again, we know only that the environment of cells in developing organisms does influence their differentiation. Nevertheless, the manner in which genes control developmental processes seems unlikely to differ completely from the manner in which they control the day-to-day cellular activities involved in maintenance. As noted in Chapter 2, metabolic processes underlying developmental processes probably do not differ significantly from those operating in the adult organism.

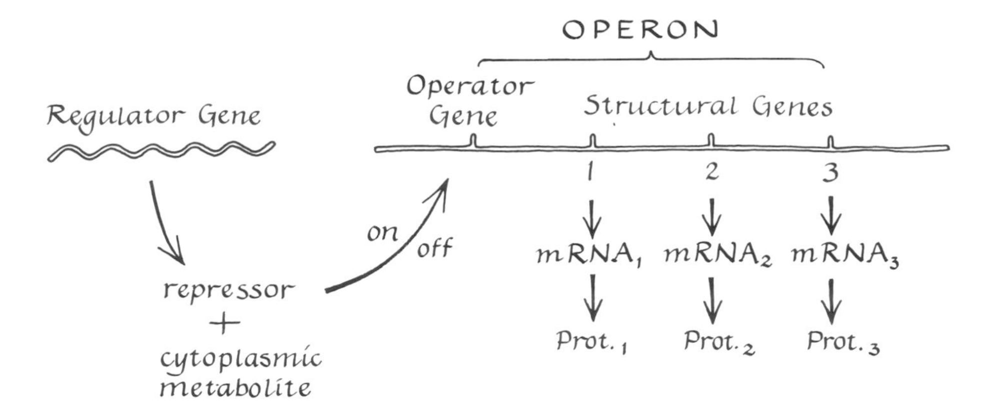

Figure 17-1. *Simplified diagram of the model proposed by Jacob and Monod (1963) for the control of gene action in bacteria.*

Special features of the bacterial operator-regulator model are that control of gene activity is at the transcription level and that the mRNAs are short-lived. Application of the model to multicellular organisms would involve not only a longer-lived mRNA but also the recognition that control could occur at any or all of the following levels: (1) regulator gene, (2) operator gene, (3) structural gene (transcription), (4) mRNA-rRNA-tRNA complex (translation), (5) tRNA synthesis, (6) polypeptide assembly, (7) polysome formation, and (8) possibly other levels (Gross, 1967). In any case, knowing the level or levels at which regulation might occur is of primary importance in applying the scheme to developmental processes.

The mechanism for control of protein synthesis suggested by Jacob and Monod probably does operate in multicellular organisms but by itself seems inadequate for explaining cellular differentiation. As pointed out by Davis (1964), "The regulatory mechanisms seen in bacteria do not furnish a direct model since they are homeostatic in their overall effect: a rise in the level of a metabolite exerts a *negative* feedback action on its own formation, tending to reverse any difference starting to develop in the cell. Differentiation, however, is the opposite of homeostasis: it seeks to perpetuate differences.... Differentiation could in principle be accomplished by *positive* feedback loops, in which a difference in the level of a substance tends to perpetuate itself."

A greatly simplified model of the progressive control of gene action required by the programmatic nature of development is the following:
For further discussion of this problem, see Ebert and Kaighn (1966).

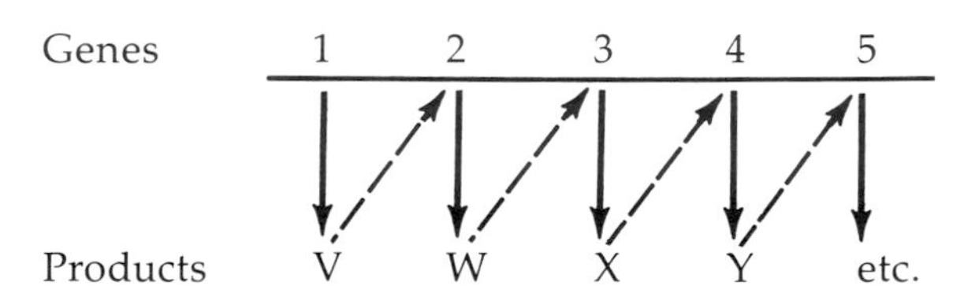

In this model, the products of one gene's action initiate the action of another gene and so forth, thus providing for sequential use of the genetic information. Recall the sequence in puffing of loci in the giant chromosomes of insects (Chapter 3).

In this chapter we shall cite some evidence for the role of nucleoproteins in the development of multicellular organisms and also evidence for the mechanism controlling gene activity. Finally we shall briefly examine evidence for a program of nucleoprotein synthesis underlying early stages in development.

Operation of the Protein Synthesis Mechanism in Multicellular Organisms

The studies of Huang et al. (1960), Huang and Bonner (1962), and Bonner et al. (1963, 1968) suggest that transcription and translation of genetic information (Chapter 3) follow a pathway in higher organisms similar to that in bacteria. Their evidence is briefly summarized below.

1. An enzyme system capable of incorporating ^{14}C-ATP into RNA in the presence of RNA precursors was extracted from pea seedlings. The reaction was DNA-dependent.
2. The DNA-dependent mRNA synthesis was coupled with mRNA-dependent protein synthesis, and DNA-primed proteins were obtained as follows:
3. Chromatin (DNA) from seedlings mediated the synthesis of mRNA. Addition of *E. coli* RNA polymerase increased the mRNA synthesis. Addition of *E.coli* ribosomes resulted in synthesis of protein (pea seed globulins).

Thus, in this *in vitro* system, the information in the pea mRNA was translated on *E. coli* ribosomes into pea seed globulin protein.

A number of other studies with animals and plants point to the universality of the protein synthesis mechanism. Thus, Rich et al. (1963), in work on rabbit reticulocytes, discovered that protein synthesis occurs on clusters of ribosomes, as noted in Chapter 3.

Control of Gene Activity in Multicellular Organisms

For a number of reasons, application of the model for controlling gene action in microorganisms cannot be directly applied to developmental processes in multicellular organisms. For example, (1) cellular differentiation probably requires the coordinated action of many nonadjacent genes, (2) the genome in multicellular organisms is very large compared to that in bacteria, and (3) a large amount of redundancy (repetitive nucleotide sequences) is scattered throughout the genome. For these and other reasons Britten and Davidson (1969) have proposed a modified model involving, in addition to structural and operator genes, *sensor* and *integrator* genes that in effect substitute for the regulator gene of bacteria. In the model, the sensor gene serves as a binding site for inductive agents such as hormones, and this results in activation of the integrator gene. The integrator genes synthesize activator RNA molecules that control the coordinated activity of the *receptor-producer* (operon) complex. The model, as the authors emphasize, is purely tentative and major modifications are to be expected. It does, however, show how an integrator gene can induce transcription of many structural genes in response to a single molecular event.

Histones and Other Proteins in the Regulation of Gene Activity

Masking of DNA by proteins might prevent transcription of portions of the genome. A large fraction of the proteins in chromosomes are histones, so it is not surprising that these particular proteins have been implicated in the control of gene action (Bloch, 1964; Butler, 1965; Bonner et al., 1968; Paul and Gilmour, 1968). Work by Bonner and his group (Huang et al., 1960; Bonner et al., 1963) on the pea seedling system indicates that histones might repress gene action. In an *in vitro* system, chromatin isolated from the cotyledons functioned as a template for the synthesis of seed globulins, but chromatin from the apical meristem (bud) did not direct the synthesis of seed globulins. However, when histones were removed from the pea bud chromatin, seed globulin synthesis occurred. This indicates that the chromatin for the direction of seed globulin synthesis is present in the apical meristem cells but is prevented from functioning by masking histones.

Additional evidence for the role of histones is provided by studies of Allfrey and Mirsky (1962, 1963) and Mirsky (1964) on a DNA-dependent RNA synthesis in isolated calf thymus nuclei. When histones were added to the nuclear preparation, RNA synthesis was completely suppressed. Tryptic removal of the histones caused derepression. Sherbet (1966) reported that histones from the calf thymus inhibited brain development when applied to early chick embryos. Labeling showed that the histones entered the

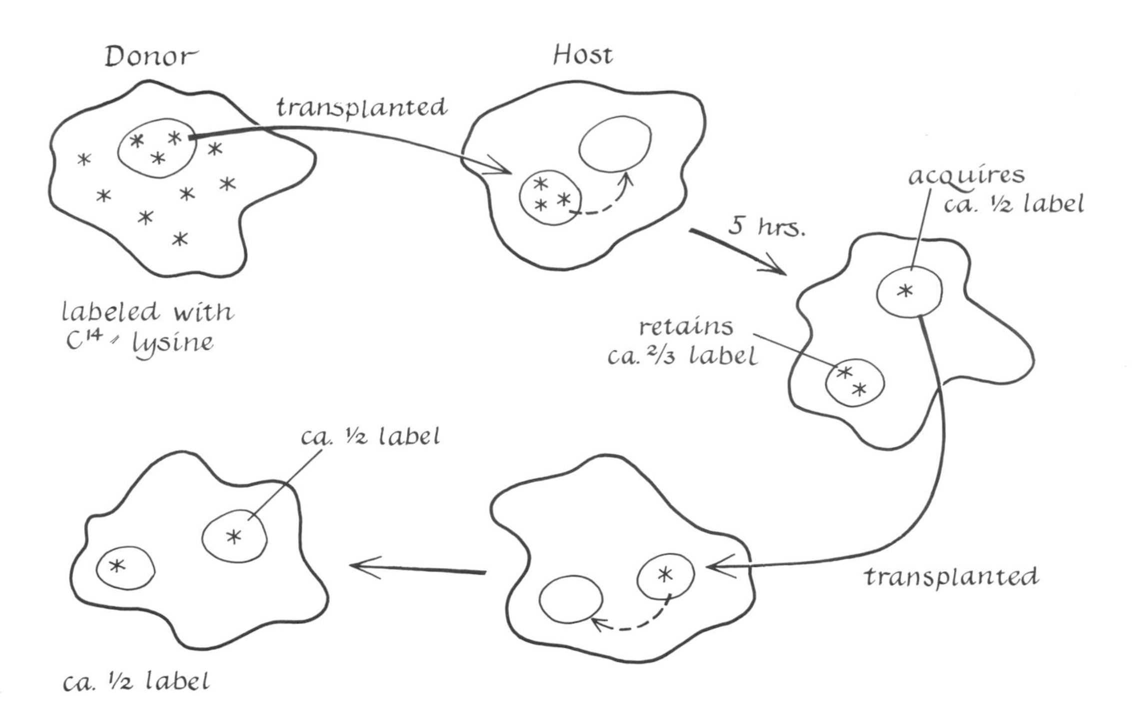

Figure 17-2. *Diagram of the experiments of Goldstein (1964) indicating the movement of labeled proteins from a transplanted nucleus into the cytoplasm and into the host cell nucleus.*

cell nuclei. For further discussion of the role of histones see reviews of Paul (1967), Georgiev (1967), and Bonner et al. (1968).

These studies indicate that histones might repress (regulate) gene action. However, there is doubt whether histones play a gene locus specific role. Histones from pea seedling chromatin were found to bind equally well with calf thymus DNA and pea DNA (Bonner et al., 1968).

Nonhistone proteins in chromosomes and in the cytoplasm might also repress gene activity. This is suggested by the fact that when albumin proteins from adult liver nuclei were injected into frog eggs, development was arrested at the late blastula stage (Markert and Ursprung, 1963). Nuclear transplantation revealed that the effect was inherent in the nuclei and stable. However, it is not clear that the arrested development resulted from complexing the injected albumin with DNA. The studies of Goldstein (1963, 1964) outlined in Figure 17-2 indicate that cytoplasmic proteins might move into the nucleus. As shown in the figure, some labeled material (protein) in the transplanted nucleus moves into the cytoplasm of the unlabeled host ameba and then into the latter's nucleus. A similar transfer occurs when the host ameba nucleus is transplanted into another unlabeled ameba. Other experiments (Figure 17-3) indicate that tritiated adenosine moves from the nucleus into the cytoplasm. This is consistent with the concept of RNA moving out of the nucleus into the cytoplasm.

Hormones in the Regulation of Gene Activity

The role of hormones in regulating gene function has been studied by many biologists (see review of Bonner et al., 1968, and Chapter 19). We have already noted in Chapter 3 the role of ecdysone in inducing puffing patterns in giant chromosomes. Hormones

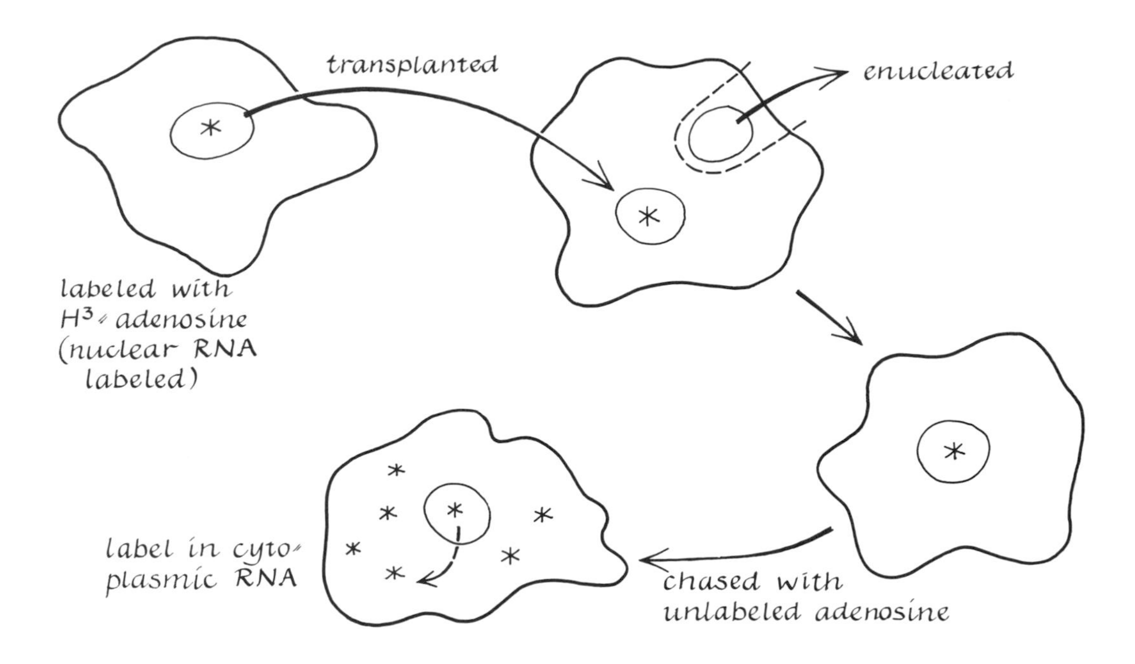

Figure 17-3. *Diagram showing the movement of labeled nuclear RNA from a transplanted nucleus into the cytoplasm of the host ameba.*

might function as derepressors of gene loci. For example, cortisone injected into adrenalectomized rats increased in fifteen minutes the rate of DNA-dependent RNA synthesis (Bonner et al., 1968). Also, dormant potato buds, which show no increased template activity for RNA synthesis in the presence of added RNA polymerase, exhibit a twentyfold increase when treated with *ethylene chlorohydrin*, which mimics the effects of gibberellic acid (Bonner et al., 1968, and Chapter 19). However, hormones appear to have no effect on the template activity of isolated chromatin; this suggests that some intermediate substance, possibly a protein, complexes with the hormone, and that the latter actually regulates the template activity. It also seems probable that hormones may influence, via changing cell permeability, the flux of ions in the cell; the latter may be more directly involved in regulating template activity. Nevertheless, the mechanism of action of hormones remains elusive, as we shall see in Chapter 19.

Role of mRNA and rRNA in Development

As already noted, the Jacob and Monod model for the regulation of gene activity in bacteria does not apply directly to problems of development in multicellular organisms. Relatively stable and long-lived informational units are required to explain satisfactorily many developmental phenomena—for instance, continued development for many hours of enucleated sea urchin and frog eggs, regeneration of a cap by nonnucleated parts of an *Acetabularia* stem, and continued protein synthesis in enucleated eggs (see Davis, 1964; Ebert, 1965). The model must be modified since mRNA in bacteria is extremely unstable (any one molecule exists for only a few minutes). To explain many features of embryonic development which are not explained by the Jacob and Monod model, we can consider the following thesis as a working hypothesis and experimental target.

During oogenesis many RNA copies of parts of the DNA code are made but are not used until later in development. Some of these copies are released into the cytoplasm as stable (or masked) mRNA molecules. Other copies are of the genes for rRNA and are retained in the nucleus of the oocyte (or, in *Xenopus,* in the 100 to 1,000 nucleoli). The genes for the two ribosome components, 28S and 18S (Chapter 3), are closely linked and present in the "nucleolar organizer" of a single autosome. There are about 800 such genes (Brown, 1967). Extensive synthesis of ribosomes released to the cytoplasm is correlated with the excessive copies of rDNA (DNA specifying ribosome synthesis). The many copies of parts of the DNA code (mRNA and rRNA) might constitute an essential part of the functional cytoarchitecture of an egg as transcribed information for later protein synthesis. This differential amplification of the genome is probably of great importance for later development. Thus, the early stages of egg development involve the translation (use) of instructions transcribed during oogenesis (reviews of Spirin, 1966; Wilt, 1966; Tyler, 1967; Brown and Dawid, 1968; Ebert, 1968). Recall (Chapter 5) that cap regeneration in enucleated pieces of the stem of *Acetabularia* indicates the presence of long-lived, stable mRNA in the cytoplasm.

Evidence in support of this thesis is accumulating so rapidly that we can only outline some of the more relevant experimental results. Most of the evidence comes from studies of sea urchin and amphibian (frog and *Xenopus*) eggs and embryos. For details, the following reviews should be consulted: Brown and Littna, 1964; Davidson and Mirsky, 1965; Grant, 1965; Collier, 1966; Brachet, 1967; Brown, 1967; and Davidson, 1968.

Sea Urchin Studies

Oogenesis

Studies with isolated oocytes of *Lytechinus pictus* have shown that the rate of incorporation of RNA precursors (^{14}C-uridine into RNA) and amino acids (^{14}C-valine into proteins) is far greater than the incorporation of these precursors into mature eggs (Piatigorsky et al., 1967). Much rRNA and some tRNA and mRNA are synthesized during oogenesis (Brachet, 1967).

Unfertilized Egg

The unfertilized sea urchin egg loaded with ^{14}C-labeled amino acids fails to incorporate these into proteins. Upon fertilization incorporation is rapid (Gross, 1964, 1967; Tyler, 1966). However, there is apparently no RNA synthesis at fertilization, since actinomycin-D (which blocks RNA synthesis but not protein synthesis, according to Reich et al., 1962) does not block the amino acid incorporation. It was thus concluded that the unfertilized egg contains preformed, stable mRNAs and that no new templates for protein synthesis are necessary at this early stage. Although there is some doubt about the interpretation of this result, since Thaler et al. (1969) have reported that actinomycin does not enter sea urchin eggs until the time of hatching, more recent evidence indicates that early stages are permeable to actinomycin (Greenhouse et al., 1971). Nevertheless there is additional evidence, from experiments not involving actinomycin-D, that protein synthesis is directed by preformed mRNA. Nonnucleated fragments of unfertilized eggs incorporate many kinds of labeled amino acids into protein upon parthenogenetic activation, and the incorporation is as efficient as that in whole fertilized or activated eggs (Denny and Tyler, 1964; Tyler, 1966, 1967). Recall that the anucleate isolated polar lobe of the mud snail *Ilyanassa* can synthesize protein at about one-half the rate of the whole egg (Chapter 13). This constitutes further evidence for a long-lived mRNA. However, there is a possibility (as yet undemonstrated) that DNA of the mitochondria (Chapter 18) might template for mRNA synthesis. It might be instructive to study amino acid incorporation in actinomycin-treated nonnucleated fragments of these eggs. Furthermore, RNA (mRNA?) extracted from unfertilized eggs stimulates the incorporation of labeled amino acids into protein in cell-free ribosomal preparations from fertilized eggs (Maggio et al., 1964). The extracted RNA can hybridize with nuclear DNA (Slater and Spiegelman, 1966).

The above results indicate that activation of protein synthesis at fertilization does not depend upon the production of mRNA by the nucleus. Fertilization is thus an "unmasking" of preformed mRNA molecules (Tyler, 1967). It would be incorrect to assume that no protein synthesis occurs in unfertilized eggs, since there is abundant evidence for protein synthesis (Tyler, 1967). This synthesis probably reflects maintenance processes in the egg, in which synthesis and degradation are in approximate balance. Nevertheless experiments of Bell and Mackintosh (1967) and Mackintosh and Bell (1970) indicate that fertilization does not initiate protein synthesis but increases its rate. Protein synthesis at fertilization follows metabolic events, such as increases in oxygen consumption. Polysomes also form within fifteen minutes following fertilization (Davidson, 1968).

Attempts have been made to explain the seeming inability of mature unfertilized eggs to incorporate labeled amino acids and to synthesize proteins for development. Apparently the block is not due to the absence of the protein synthesis system or the presence of special inhibitors, since unfertilized egg homogenates do not inhibit incorporation of labeled amino acids by ribosome preparations of developing embryos (Hultin, 1961). Nor is the block the result of defective ribosomes, since polyuridylic acid (an exogenous, artificial template) directs the incorporation of labeled phenylalanine into protein when added to unfertilized egg ribosomal preparations (Tyler, 1967). The block appears to be the masking of mRNA by protein. Treating ribosomes from unfertilized eggs with trypsin resulted in apparent unmasking of the mRNA attached to ribosomes and consequent protein synthesis (Monroy et al., 1965; Maggio et al., 1964). Recall that there is evidence for release of a trypsin-like enzyme at fertilization (Chapter 12) which might be responsible for the digestion of the protein that masks mRNA.

Cleavage Stages

Treatment of sea urchin eggs at cleavage stages with actinomycin-D blocks neither cleavage nor protein synthesis. After such treatment, protein synthesis was as intensive as in untreated cleaving eggs (Gross, 1964; Spirin, 1966). It has also been reported that ribosomal RNA is not synthesized during early cleavage stages (Comb et al., 1965; Nemer, 1963), yet at later cleavage stages some synthesis of rRNA occurs (Spirin and Nemer, 1965). Because permeability of eggs in cleavage stages to actinomycin is questionable, interpretation of results following its use is not clear.

Blastula and Gastrula Stages

Labeled amino acids continue to be incorporated into proteins at late blastula and early gastrula stages. However, this incorporation is actinomycin-D sensitive, suggesting a resumption of mRNA synthesis. Although the evidence is not completely convincing, masked mRNAs formed during oogenesis apparently become unmasked during late blastula and early gastrula stages. Newly synthesized mRNAs can also be templates for protein synthesis (Tyler, 1967).

Amphibian Studies

Perhaps the most convincing evidence for a preformed, protein-synthesizing system in mature oocytes of amphibians is the fact that enucleated and activated oocytes of the common frog *Rana pipiens* can synthesize protein as quickly as nucleated controls (Ecker et al., 1968). This suggests that long-lived gene products (presumably RNAs) synthesized during oogenesis are available for use at later stages of development. The studies by Brown and Dawid (1968) Gall (1968), and others of the oocytes of several amphibians show that at least a part of the protein-synthesizing system is preformed for later use. The synthesis of rRNA is intense during oogenesis but ceases in the fertilized egg and is not resumed until gastrulation (Brown and Littner, 1964; Brown and Gurdon, 1964). A particularly significant discovery by Brown's group was that a homozygous mutant of *Xenopus* that has no nucleoli will develop to the swimming tadpole stage (Stage 20, Figure 17-4) before dying. Thus new rRNA is not needed until this stage (Brown and Gurdon, 1964). Normal stages in the development of *Xenopus laevis* are illustrated in Figure 17-4.

Fertilized *Xenopus* eggs, homozygous for the mutant gene locus, which appears to involve a deletion in the nucleolar organizer region of the chromosome, were obtained by crosses between toads heterozygous for the mutation. These heterozygous toads have only one nucleolus (the wild type has two). One-fourth of the fertilized eggs from such a cross are thus homozygous for the mutation. These homozygous fertilized eggs result from the fertilization of mature eggs which have only the mutant gene by sperm containing the mutant gene. Ribosomal RNA

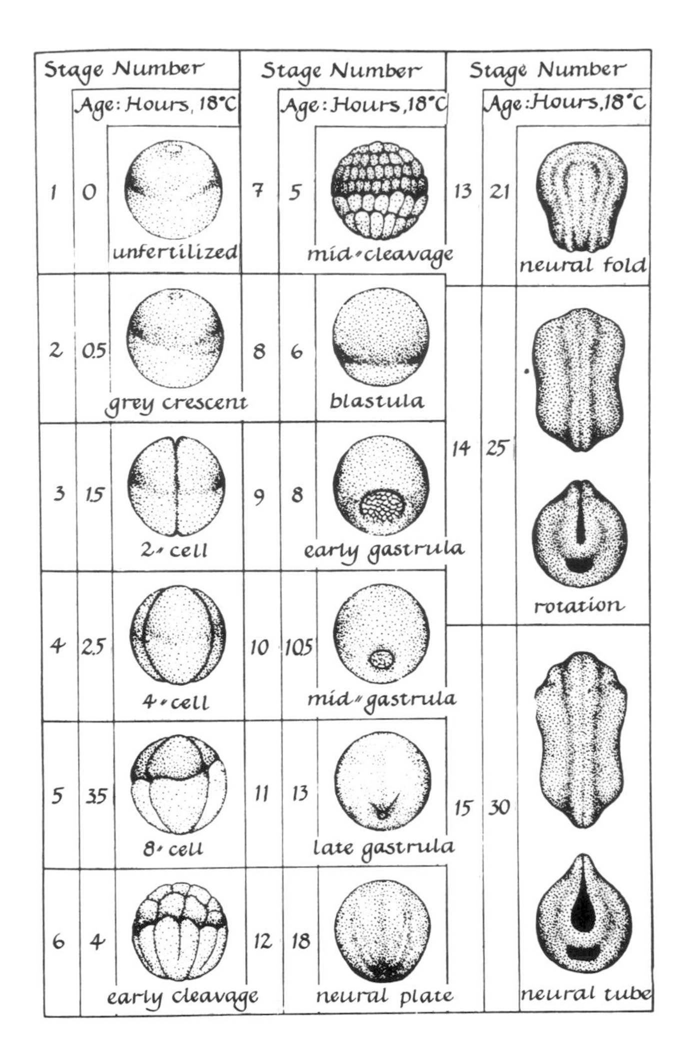

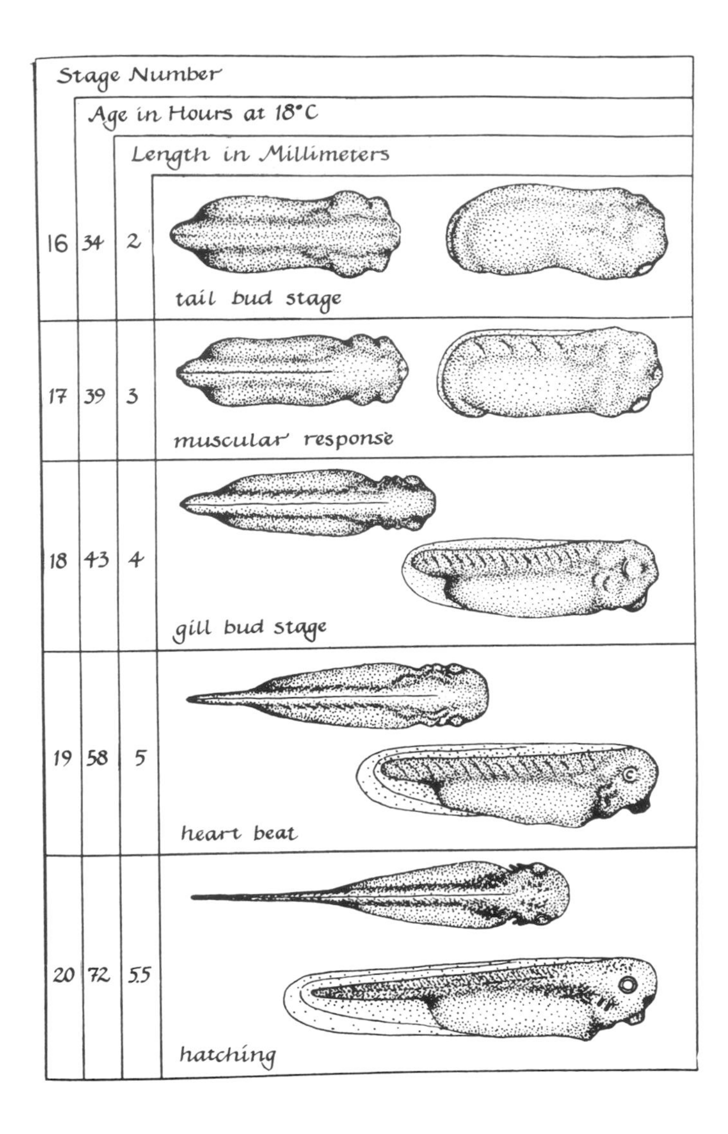

in the mature, unfertilized egg was synthesized under control of the heterozygous nucleus of the oocyte before meiosis and polar body formation occurred; during meiosis the dominant gene was lost from the oocyte to a polar body. As would be expected, Brown's subsequent studies (1967) revealed that, because the rDNA portion of its genome has been deleted, DNA from the anucleolate mutant does not form a hybrid with rRNA.

The problem of whether rRNA is of developmental significance in respect to cell differentiation has been raised by Brown (1967). Although rates of rRNA synthesis vary greatly during embryogenesis and appear to be stage dependent, the ubiquity of rRNA and the lack of evidence that rRNA contributes directly to cell specificity suggests that this RNA functions mainly in the maintenance of cellular activities. However, the marked differences in rRNA synthesis in endodermal and other regions of a *Xenopus* gastrula (Woodland and Gurdon, 1968) could conceivably influence the translation of mRNA.

There is evidence (Brown and Littna, 1964) that the three classes of RNA—mRNA, rRNA and tRNA—are synthesized on a time scale in amphibian development. The program of synthesis, deduced from timed administration of labeled precursors, extraction of total RNA, and gradient centrifugation to determine the size of the RNA molecules, is outlined below for *Xenopus* (review of Davidson, 1968).

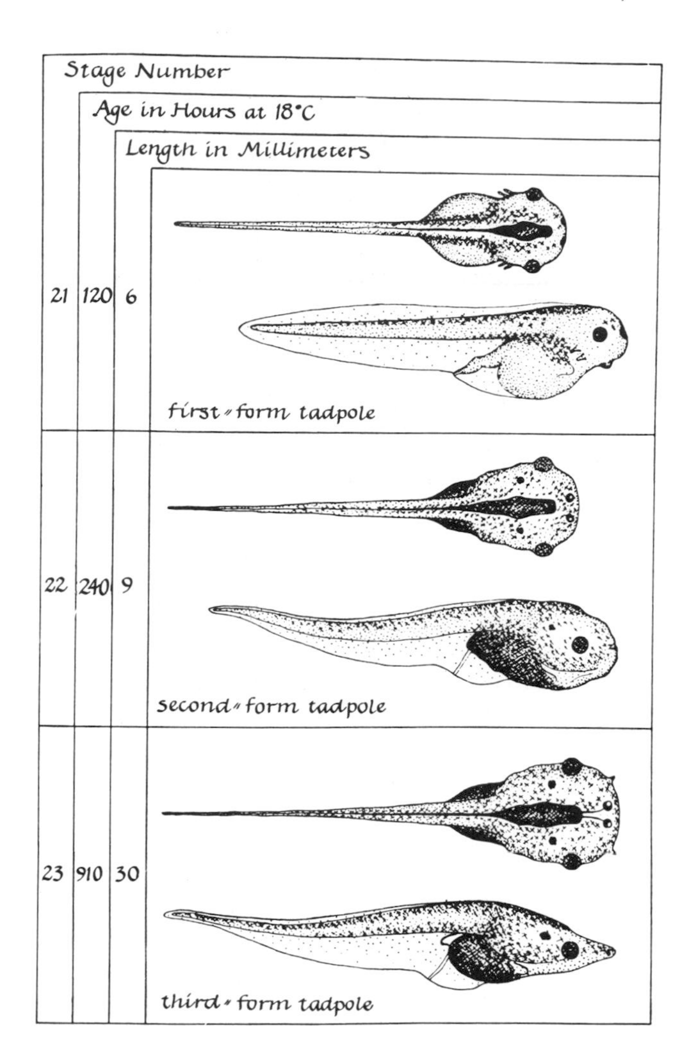

Figure 17-4. *Normal stages in the development of* Xenopus laevis. *From Weisz,* Anat. Record, *Volume 193, p. 165, 1945. Courtesy Wistar Press.*

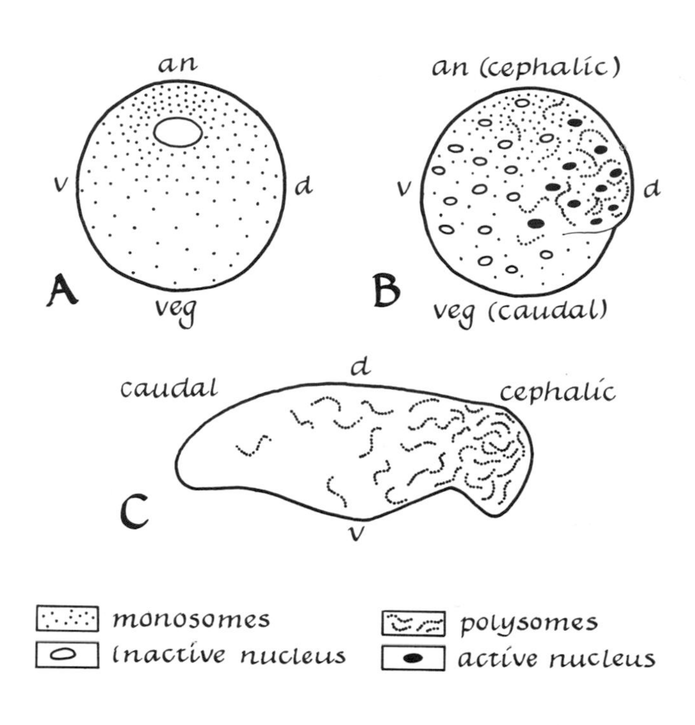

Figure 17-5. *Schematic representation of the distribution and synthesis of RNA during early amphibian development.* A, *unfertilized egg;* B, *gastrula;* C, *young tadpole;* an, *animal pole;* veg, *vegetal pole;* d, *dorsal;* v, *ventral. From Brachet, 1967. Courtesy Little, Brown and Company, Boston.*

oogenesis →	mature egg →	cleavage →	blastula →	gastrulation
rRNA max.	essentially	mRNA min.	mRNA max.	rRNA max.
tRNA min.	none	tRNA min.	tRNA max.	mRNA max.
mRNA min.				tRNA max.

Differences in the distribution and synthesis of ribosomes during early amphibian development are correlated with the program of RNA synthesis. Figure 17-5 summarizes these differences (Brachet, 1967).

Both the unfertilized and the fertilized egg (Figure 17-5A) possess a distinct antero-posterior (A-Vg) gradient in the distribution of ribosomes. At the gastrula stage (Figure 17-5B) RNA synthesis is more active on the dorsal than on the ventral side. This results in a dorso-ventral RNA gradient superimposed on the antero-posterior gradient. Both gradients are also present in the neurula and young tadpole stages. Evidence for these gradients has been reviewed by Brachet (1967), who has also pointed out that they parallel the "morphogenetic gradients" of experimental embryologists. Figure 17-5 also shows an increase in polysome formation during development. In view of the role of polysomes cited above, this indicates an increase in protein synthesis.

Other Studies

Additional evidence which seems to support the concept that preformed, masked mRNA is transcribed prior to its translation on the ribosomes is briefly discussed below.

Seed Germination

Dure and Waters (1965) reported that actinomycin-D inhibits RNA but not protein synthesis in germinating wheat and cotton seeds. However, this is not unequivocal evidence for stable mRNA, since no direct evidence has shown that the concentration of actinomycin used had inhibited all mRNA synthesis. Although studies on wheat and castor bean seeds (Marré, 1967) raise doubt as to the role of masked mRNA in new protein synthesis during germination, studies of germinating cotton seeds reveal that the synthesis of a protease in the cotyledons is catalyzed by mRNA synthesized during seed formation (Ihle and Dure, 1969). Preformed templates for protein synthesis are probably involved in seed germination. Marré found that the early phase of germination in castor bean seeds is characterized by increased protein synthesis and an almost simultaneous appearance of polysomes. Recall that there is also a significant increase in polysomes upon fertilization of sea urchin eggs. The supposition is that some of these sea urchin polysomes are programmed with preexisting mRNA (Tyler, 1967).

Hemoglobin Synthesis

In studies with explanted chick embryos, Wilt (1965, 1966, 1967) found that treatment with actinomycin-D up to eight hours before hemoglobin synthesis begins does not block the synthesis. This shows that the synthesis is independent of the RNA synthesis for eight hours prior to its onset. However, treatment at earlier stages does block hemoglobin synthesis. Thus it appears that a relatively stable mRNA for direction of the hemoglobin synthesis is produced long before it is used. There is also evidence for stable mRNA during blood-cell formation in mammals. Hemoglobin synthesis continues for a prolonged period in the reticulocytes, precursors of erythrocytes, from which the nucleus has been expelled. This was demonstrated by incorporating labeled amino acids into newly synthesized proteins (Marks et al., 1962).

Role of Cytoplasmic DNA

Compared to that in tissue cells, there is a relatively large amount of DNA in the cytoplasm of the eggs of many animal species. Cytoplasmic DNA is about ten times the haploid value in sea urchins and animals with eggs of similar size. Larger eggs, as in amphibians, probably contain much larger amounts of DNA. Most of this DNA is either mitochondrial or is located on yolk spheres in eggs (Tyler, 1967; Dawid, 1968).

Recently Bond et al. (1969) reported evidence for a unique form of DNA associated with the microsomal fraction of adult mouse liver homogenate. This DNA apparently is not mitochondrial and contains potentially a considerable amount of genetic information. We have already considered the possible significance of cytoplasmic DNA as a constituent of the cytoplasmic structure of an egg (Chapter 13); therefore, no further discussion is necessary, although the problem of whether this DNA is active or not remains unsolved.

Summary

A partially hypothetical summary of the program of nucleoprotein synthesis in embryonic development is attempted in the diagrams of Figure 17-6. Data are drawn from the studies of Brown and Littna (1964), Whitely et al. (1966), Gross (1967), Tyler (1967), Denis (1968), and the review of Davidson (1968).

Note in the figure that oogenesis is characterized by synthesis of all classes of RNA – protein and masked or stable RNA. Ribosomal RNA is synthesized in the nucleoli, and large quantities of ribosomes are released to the cytoplasm. The mature egg synthesizes relatively little protein and nucleic acid. Fertilization stimulates the synthesis of more stable

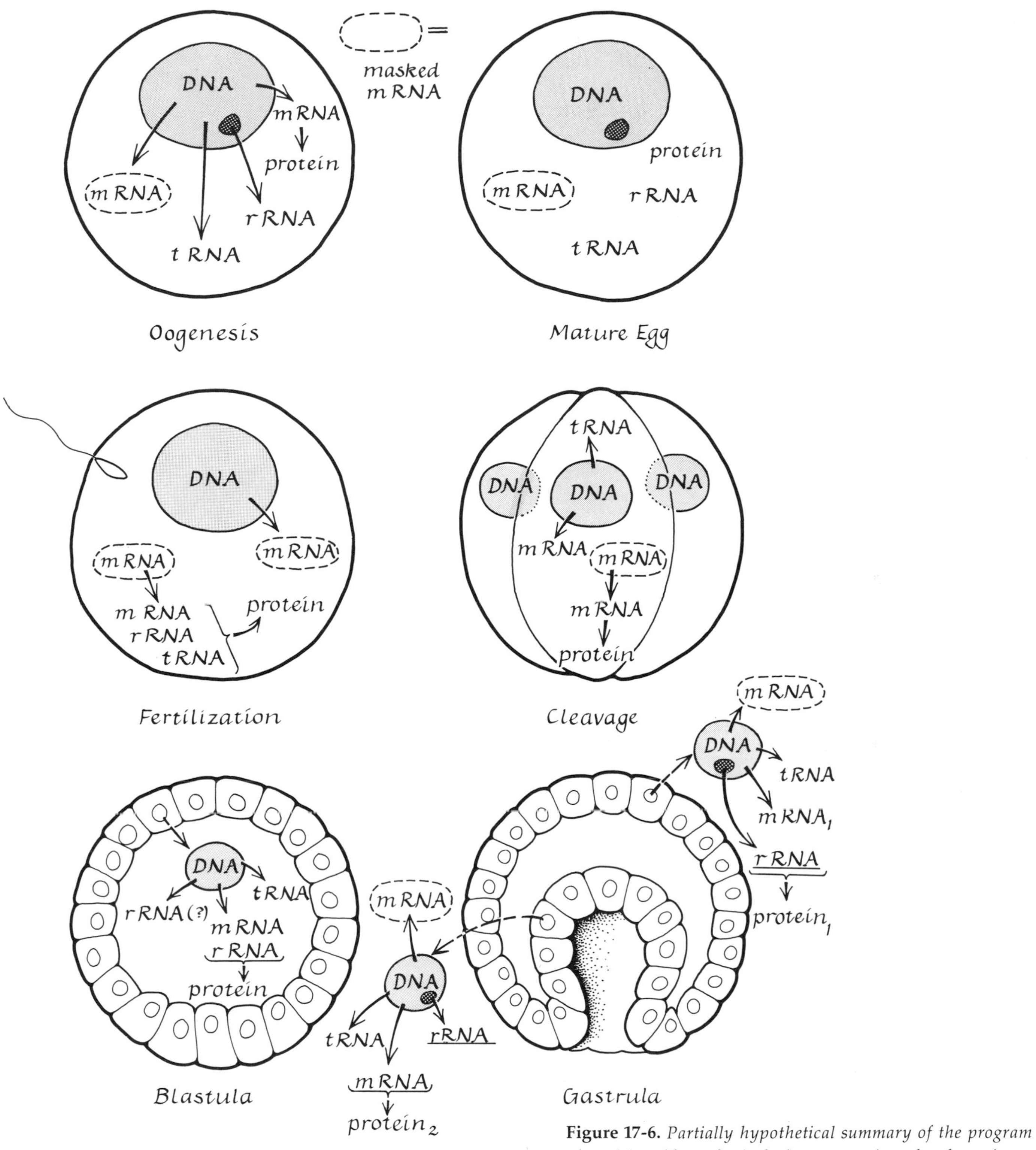

Figure 17-6. *Partially hypothetical summary of the program of nucleic acid synthesis during oogenesis and embryonic development through the gastrula stage.*

mRNA, the unmasking of preformed mRNA, and the synthesis of protein. These activities continue through cleavage and the blastula stage. In the latter stage some rRNA synthesis probably resumes. At the gastrula and later stages, nucleoli reappear and much rRNA is synthesized. Both masked and newly formed mRNA are synthesized, and unmasking of mRNA results in its functioning as a template for protein synthesis. Also indicated in the figure is the presumed synthesis of different kinds of mRNA in ectoderm and endoderm cells.

Conclusion

The mechanism for control of protein synthesis that was deduced from studies with bacteria must be modified, if it is to apply to the control of development in multicellular organisms. Thus, not only is a relatively stable set of instructions (mRNA) needed to explain many developmental phenomena, but also a genetic mechanism which will insure the progressive, programmatic reading of the genetic code over a relatively long period of time. Beyond the apparent fact that orderly reading of the code must itself be programmed in the genome, we know essentially nothing about the molecular mechanisms involved. However, we have cited evidence for sequential and differential gene action during development (Chapter 3) in support of our thesis that each stage of development involves the translation of instructions transcribed at an earlier stage. The messenger RNA hypothesis provides at least a model for such a process.

Much of the genome is redundant, and only a small portion is active at any one time during development. Chromosomal proteins, both histone and nonhistone types, appear to be the immediate regulators of gene activity.

The popular hypothesis of differential gene activity as the basis of cellular specialization is at least partially supported by the experiments and observations we have outlined; nonetheless, direct proof of differential gene activity at the level of messenger RNA remains elusive.

References

Allfrey, V. G., and A. E. Mirsky. 1962. Evidence for the complete DNA-dependence of RNA synthesis in isolated thymus nuclei. Proc. Nat. Acad. Sci. U.S. **48:**1590–1596.

Allfrey, V. G., and A. E. Mirsky. 1963. Mechanisms of synthesis and control of protein and ribonucleic acid synthesis in the cell nucleus. Cold Spring Harbor Symp. Quant. Biol. **28:**247–263.

Bell, E., and F. R. Mackintosh. 1967. Control of synthetic activity during development. In A. V. S. DeReuck and J. Knight, eds., Cell Differentiation, pp. 163–177. Little, Brown, Boston.

Bloch, D. P. 1964. Genetic implication of histone differentiation. In D. Bonner and P.O.P.Ts'O, eds., The Nucleohistones, pp. 335–342. Holden-Day, San Francisco.

Bond, H. E., J. A. Cooper, D. P. Courington, and J. S. Wood. 1969. Microsome-associated DNA. Science **165:**705–706.

Bonner, J., M. Dahmus, D. Fambrough, R. Huang, K. Marushige, and D. Tuan. 1968. The biology of isolated chromatin. Science **159:**47–56.

Bonner, J., R. C. Huang, and R. V. Gilden. 1963. Chromosomally directed protein synthesis. Proc. Nat. Acad. Sci. U.S. **50:**893–900.

Brachet, J. 1967. Behavior of nucleic acids during early development. In M. Florkin and E. Stotz, eds., Comprehensive Biochemistry, pp. 23–54. Elsevier, New York.

Britten, R. J., and E. H. Davidson. 1969. Gene regulation for higher cells: a theory. Science **165:**349–357.

Brown, D. D. 1967. The genes for ribosomal RNA and their transcription during amphibian development. In A. A. Moscona and A. Monroy, eds., Current Topics in Developmental Biology, Vol. 2, pp. 47–73. Academic Press, New York.

Brown, D. D., and I. B. Dawid. 1968. Specific gene amplification in oocytes. Science **160:**272–280.

Brown, D. D., and J. Gurdon. 1964. Absence of ribosomal RNA synthesis in the anucleolate mutant of *Xenopus laevis*. Proc. Nat. Acad. Sci. U.S. **51:**139–146.

Brown, D. D., and E. Littna. 1964. RNA synthesis during the development of *Xenopus laevis*, the South African clawed toad. J. Mol. Biol. **8:**669–687.

Butler, J. A. V. 1965. Gene control by histones. New Scientist. **25:**712–714.

Collier, J. R. 1966. The transcription of genetic information in the spiralian embryo. In A. A. Moscona and A. Monroy, eds., Current Topics in Developmental Biology, Vol. 1, pp. 39–59. Academic Press, New York.

Comb, D. G., S. Katz, R. Brenda, and C. Pinzino. 1965. Characterization of RNA species synthesized during early development of sea urchins. J. Mol. Biol. **14:**195–213.

Davidson, E. H. 1968. Gene Activity in Early Development. Academic Press, New York.

Davidson, E. H., and A. E. Mirsky. 1965. Gene activity in oogenesis. In Genetic Control of Differentiation, pp. 77–98. Brookhaven Symp. Biol., Upton, New York.

Davis, B. D. 1964. Theoretical mechanisms of differentiation. Medicine **43:**639–649.

Dawid, I. B. 1968. Studies on frog oocyte mitochondrial DNA. Carnegie Inst. Year Book **66:**20–25.

Denis, H. 1968. Role of messenger ribonucleic acid in embryonic development. Adv. Morph. **7:**115–150.

Denny, P. C., and A. Tyler. 1964. Activation of protein biosynthesis in non-nucleated fragments of sea urchin eggs. Biochem. Biophys. Res. Comm. **14:**245–249.

Dure, L., and L. Waters. 1965. Long-lived messenger RNA: evidence from cotton seed germination. Science **147:**410–412.

Ebert, J. D. 1965. Interacting Systems in Development. Holt, Rinehart and Winston, New York.

Ebert, J. D. 1968. Levels of control: a useful frame of perception. In A. A. Moscona and A. Monroy, eds., Current Topics in Developmental Biology, pp. xv-xxv. Academic Press, New York.

Ebert, J. D., and M. E. Kaighn. 1966. The keys to change: factors regulating differentiation. In M. Locke, ed., Major Problems in Developmental Biology, pp. 29–84. Academic Press, New York.

Ecker, R. E., L. D. Smith, and S. Subtelny. 1968. Kinetics of protein synthesis in enucleated frog oocytes. Science **160:**1115–1117.

Gall, J. G. 1968. Differential synthesis of the genes for ribosomal RNA during amphibian oogenesis. Proc. Nat. Acad. Sci. U.S. **60:**553–560.

Georgiev, G. P. 1967. Some aspects of the regulation of gene expression in the animal cell. In A. V. S. DeReuck and J. Knight, eds., Cell Differentiation, pp. 148–162. Ciba Found. Symp., Little, Brown, Boston.

Goldstein, L. 1963. RNA and protein in nucleo-cytoplasmic interactions. In R. J. C. Harris, ed., Cell Growth and Cell Division, pp. 129–149. Academic Press, New York.

Goldstein, L. 1964. Nuclear transplantation in *Ameba*. In D. M. Prescott, ed., Methods in Cell Physiology, pp. 97–108. Academic Press, New York.

Grant, P. R. 1965. Informational molecules and embryonic development. In R. Weber, ed., The Biochemistry of Animal Development, pp. 483–593. Academic Press, New York.

Greenhouse, G. A., R. O. Hynes, and P. R. Gross. 1971. Sea urchin embryos are permeable to Actinomycin. Science **171:**686–689.

Gross, P. R. 1964. The immediacy of genomic control during early development. J. Exp. Zool. **157:**21–38.

Gross, P. R. 1967. The control of protein synthesis in embryonic development and differentiation. In A. A. Moscona and A. Monroy, eds., Current Topics in Developmental Biology, Vol. 2, pp. 1–46. Academic Press, New York.

Huang, R. C., and J. Bonner. 1962. Histone, a suppressor of chromosomal RNA synthesis. Proc. Nat. Acad. Sci. U.S. **48:**1216–1222.

Huang, R. C., N. Maheshwari, and J. Bonner. 1960. Enzymatic synthesis of RNA. Biochem. Biophys. Res. Comm. **3:**689–694.

Hultin, T. 1961. Activation of ribosomes in sea urchin eggs in response to fertilization. Exp. Cell Res. **25:**405–417.

Ihle, J. N., and L. Dure III. 1969. Synthesis of a protease in germinating cotton cotyledons catalyzed by mRNA synthesized during embryogenesis. Biochem. Biophys. Res. Comm. **36:**705–710.

Jacob, F., and J. Monod. 1961. On the regulation of gene activity. Cold Spring Harbor Symp. Quant. Biol. **26:**193–211.

Jacob, F., and J. Monod. 1963. Genetic repression, allosteric inhibition and cellular differentiation. In M. Locke, ed., Cytodifferentiation and Macromolecular Synthesis, pp. 30–64. Academic Press, New York.

Mackintosh, F. R., and E. Bell. 1970. Reversible response to puromycin and some characteristics of the uptake and use of amino acids by unfertilized sea urchin eggs. Biol. Bull. **139:**296–303.

Maggio, R., M. L. Vittorelli, A. M. Rinaldi, and A. Monroy. 1964. *In vitro* incorporation of amino acids into proteins stimulated by RNA from unfertilized sea urchin eggs. Biochem. Biophys. Res. Comm. **15:**436–441.

Market, C. L., and H. Ursprung. 1963. Production of replicable persistent changes in zygote chromosomes of *Rana pipiens* by injected proteins from adult liver nuclei. Devel. Biol. **7:**560–577.

Marks, P. A., E. Burka, and D. Schlessinger. 1962. Protein synthesis in erythroid cells: I. Reticulocyte ribosomes active in stimulating amino acid incorporation. Proc. Nat. Acad. Sci. U.S. **48:**2163–2171.

Marré, E. 1967. Ribosome and enzyme changes during maturation and germination of the castor bean seed. In A. A. Moscona and A. Monroy, eds., Current Topics in Cell Differentiation, pp. 75–105. Academic Press, New York.

Mirsky, A. E. 1964. Regulation of genetic expression—general survey. J. Exp. Zool. **157:**45–48.

Monroy, A., R. Maggio, and A. Rinaldi. 1965. Experimentally induced activation of the ribosomes of the unfertilized sea urchin egg. Proc. Nat. Acad. Sci. U.S. **54:**107–111.

Nemer, M. 1963. Old and new RNA in the embryogenesis of the purple sea urchin. Proc. Nat. Acad. Sci. U.S. **50:**230–235.

Paul, J. 1967. Masking of genes in cytodifferentiation and carcinogenesis. In A. V. S. DeReuck and J. Knight, eds., Cell Differentiation, pp. 198–207. Ciba Found. Symp., Little, Brown, Boston.

Paul, J., and R. S. Gilmour. 1968. Organ-specific restriction of transcription in mammalian chromatin. J. Mol. Biol. **34:**305–316.

Piatigorsky, J., H. Ozaki, and A. Tyler. 1967. RNA and protein-synthesizing capacity of isolated oocytes of the sea urchin *Lytechinus pictus.* Devel. Biol. **15:**1–22.

Reich, E., R. M. Franklin, A. J. Shatkin, and E. L. Tatum. 1962. Action of actinomycin-D on animal cells and virus. Proc. Nat. Acad. Sci. U.S. **48:**1238–1245.

Rich, A., J. B. Warner, and H. M. Goodman. 1963. The structure and function of polyribosomes. Cold Spring Harbor Symp. Quant. Biol. **28:**269–285.

Sherbet, G. V. 1966. Effects of histones and other inhibitors on embryonic development. In The Histones, pp. 81–92. Ciba Found. Study Group No. 24, Little, Brown, Boston.

Slater, D. W., and S. Spiegelman. 1966. An estimation of genetic messages in the unfertilized echinoid egg. Proc. Nat. Acad. Sci. U.S. **56:**164–170.

Spirin, A. S. 1966. On "masked" forms of messenger RNA in early embryogenesis and in other differentiating systems. In A. A. Moscona and A. Monroy, eds., Current Topics in Developmental Biology, pp. 1–38. Academic Press, New York.

Spirin, A. S., and M. Nemer. 1965. Messenger RNA in early sea-urchin embryos: cytoplasmic particles. Science **150:**214–217.

Thaler, M. M., M. C. Cox, and C. A. Villee. 1969. Actinomycin D: uptake by sea urchin eggs and embryos. Science **164:**832–834.

Tyler, A. 1966. Incorporation of amino acids into protein by artificially activated non-nucleate fragments of sea urchin eggs. Biol. Bull. **130:**450–461.

Tyler, A. 1967. Masked messenger RNA and cytoplasmic DNA in relation to protein synthesis and processes of fertilization and determination in embryonic development. In M. Locke, ed., Control Mechanisms in Developmental Processes, pp. 170–226. Academic Press, New York.

Weisz, P. B. 1945. The normal stages in the development of the South African clawed toad, *Xenopus laevis.* Anat. Rec. **93:**161–169.

Whitely, A. H., B. J. McCarthy, and H. R. Whitely. 1966. Changing populations of messenger RNA during sea urchin development. Proc. Nat. Acad. Sci. U.S. **55:**519–526.

Wilt, F. H. 1965. Regulation of the initiation of chick embryo hemoglobin synthesis. J. Mol. Biol. **12:**331–341.

Wilt, F. H. 1966. The concept of messenger RNA cytodifferentiation. Am. Zool. **6:**67–74.

Wilt, F. H. 1967. The control of embryonic hemoglobin synthesis. Adv. Morph. **6:**89–125.

Woodland, H. R., and J. B. Gurdon. 1968. The relative rates of synthesis of DNA, sRNA and rRNA in the endodermal region and other parts of *Xenopus laevis* embryos. J. Embryol. Exp. Morph. **19:**363–385.

18 *Extranuclear Genetic Control: Cytoplasmic Inheritance*

There is little doubt that development is ultimately controlled by nuclear genes. As far as we know, information for development is primarily packaged in the nuclear DNA code. However, it has been known for many years that some inheritance patterns cannot be explained in terms of the sole participation of the nuclear genome (reviews of Jinks, 1964, and Wilkie, 1964). Indeed, information (for instance, the cortical structure of *Paramecium*) sometimes is passed from generation to generation quite independently of that contained in the nucleus. In this chapter we shall examine examples of extranuclear or cytoplasmic inheritance. First, it is necessary to distinguish between nuclear and extranuclear inheritance and the very different problems of nuclear and extranuclear control of development.

Does the extranuclear inheritance of specific characteristics in some organisms mean that the nucleus participates only in a supportive, metabolic fashion, in the control of growth and cellular differentiation? Has the nuclear genome in any instance abrogated any of its control of development? Perhaps the nuclear genome has delegated immediate or direct control of inheritance and development to other components of the cell in some organisms. We have emphasized in earlier chapters that genes seem to establish through prior activity the conditions for control of subsequent gene action. Clear examples of this, as we saw in Chapter 13, are found in egg development. Could not the genome have, in the course of evolution, delegated some hereditary and developmental control functions to other cellular organelles? The problem then becomes what form the delegated information takes. Is it exclusively nonnuclear DNA or RNA? As we saw in Chapter 17, there is DNA outside the nucleus, especially in eggs. To what extent is the DNA a copy of part of the nuclear DNA? DNA differing in nucleotide composition from nuclear DNA has been found in chloroplastids, mitochondria, centrioles, kinetosomes of ciliates, and the basal bodies of cilia and flagella. Are DNA and RNA molecules the only ones carrying hereditary or developmental information? Some possible answers to these questions will be considered, but it is significant that there is no evidence that the nuclear genome has abrogated forever its master control of inheritance and development.

Evolution of Nonnuclear (Satellite) Control

Perhaps the evolution of nonnuclear hereditary mechanisms proceeded through the following steps:

1. Continuous nuclear gene action ⟶ products
2. Replication of nuclear gene product dependent upon presence of nuclear gene
3. Replication of nuclear gene product independent of originating gene

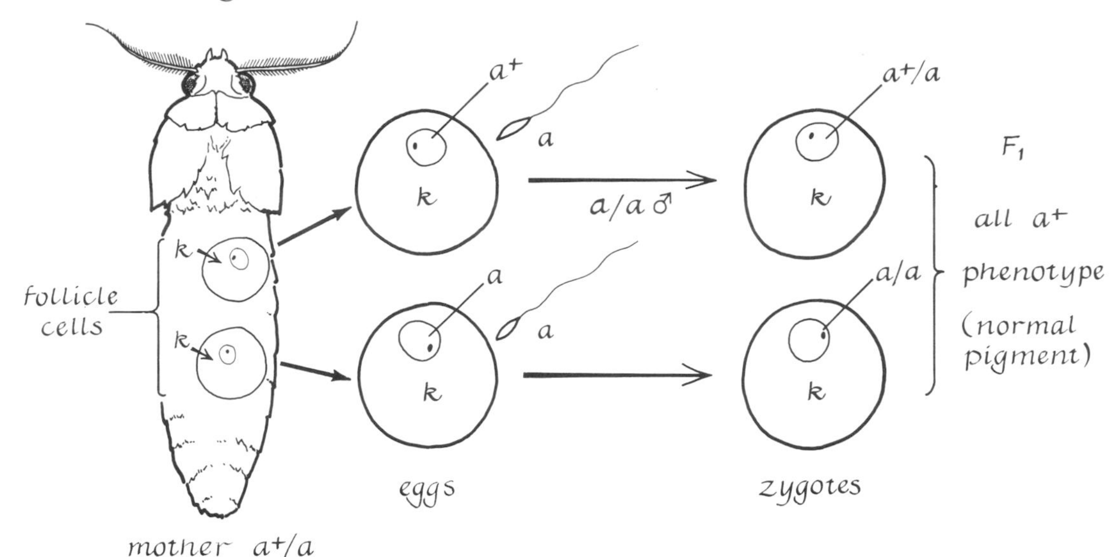

Figure 18-1. *Maternal inheritance involving the transmission of kynurenin in the cytoplasm of the eggs formed by a heterozygous female* Ephestia *(a^+/a) and fertilized by a homozygous recessive male (a/a).*

Probably no replicating molecules arose completely independent of any prior gene action. If such molecules existed, they would constitute a hereditary mechanism that was strictly nonnuclear in its origin.

Maternal Inheritance

Maternal inheritance is distinct from nonnuclear inheritance but is usually considered cytoplasmic. In maternal inheritance, the products of prior gene action in the mother are transmitted to the offspring. These products determine the new individual's phenotype irrespective of its genotype. Three examples of maternal inheritance are outlined below.

Kynurenin Synthesis

In the moth, *Ephestia Kuhniella,* the wild-type gene a^+ controls the synthesis of kynurenin, an intermediate in the formation of *ommochrome,* a pigment of the eye and hypodermis. The recessive gene *a* fails to support the synthesis of kynurenin from tryptophan. When eggs produced by a heterozygous female are fertilized by sperm of a homozygous, recessive male, all the F_1 offspring are phenotypically normal, that is, possess pigmented eyes and hypodermis (Figure 18-1).

As indicated in the figure, even the homozygous recessive zygote, a/a, develops into a normally pigmented moth since the kynurenin in the female's body is transmitted in the cytoplasm of all of her eggs. Caspari (1948) proved this by implanting a normal testis, a^+/a^+, into an a/a female. This is illustrated in Figure 18-2.

Kynurenin synthesized under control of the wild-type genome of the testis cells diffuses into the eggs of the a/a female. When the eggs are fertilized by the

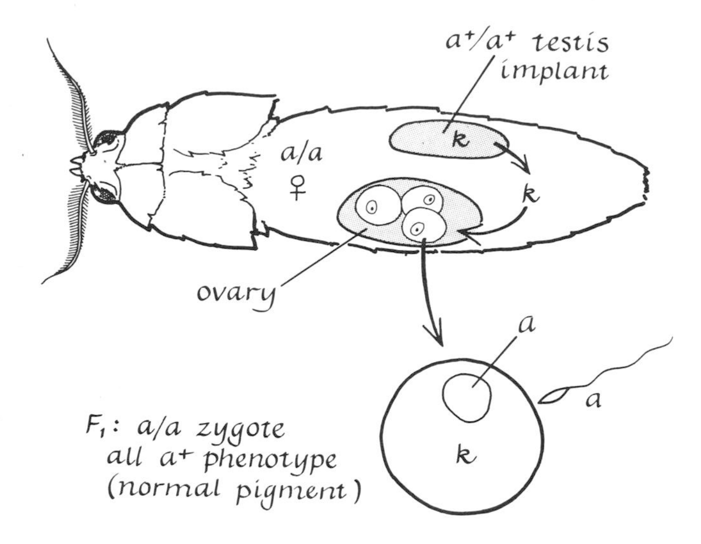

Figure 18-2. *Implantation of the testis of a homozygous wild-type male* Ephestia *(a^+/a^+) into a homozygous recessive female (a/a). Note that eggs of this female fertilized by sperm carrying the recessive gene (a) develop into normally pigmented moths.*

sperm of an a/a male, the a/a zygote develops into a moth with normal pigmentation.

Coiling of the Shell in Limnaea

Another well-known example of maternal inheritance is the control of the spiral cleavage pattern and direction of shell coiling in the snail *Limnaea peregra* (Chapter 13). The coiling, whether dextral or sinistral, is controlled in a Mendelian fashion by a pair of genes; the dextral one is dominant. Inheritance in this case is interesting because the F_1 offspring, regardless of genotype, are always of the same phenotype as the mother. Thus, when the eggs of a homozygous, sinistral female are fertilized by the sperm of a homozygous, dextral male, all the F_1 snails are *sinistral.* The pattern of inheritance is shown below (Boycott et al., 1931).

a. dextral ♂ DD × sinistral ♀ dd
   F_1 Dd all sinistral
b. sinistral ♂ dd × dextral ♀ DD
   F_1 Dd all dextral
c. dextral ♂ Dd × dextral ♀ Dd
   F_1 DD + 2Dd + dd all dextral
d. sinistral ♂ dd × dextral ♀ Dd
   F_1 Dd + dd all dextral

The cytoarchitecture of the egg, formed under control of the female genome, DD, dd, or Dd, is respectively dextral, sinistral, or dextral and is transmitted to the offspring regardless of the male genome.

Inheritance of Cytoplasmic Guidelines

Among numerous examples of how early development is controlled by the cytoplasmic composition and structure of the egg (Chapter 13), one interesting way of demonstrating the role of the female genome is provided by nuclear transplantation experiments (Chapter 20). When the nucleus of a specialized tissue cell from the wall of the intestine of *Xenopus* is transplanted into an enucleated *Xenopus* egg, the egg sometimes develops normally. This means that the program for early development resides primarily in the egg cytoplasm and is thus inherited from the mother.

Following are a few instances in which cytoplasmic organelles that are not products of gene action are transmitted through the cytoplasm.

Kappa in Paramecium

In 1938 Sonneborn found that paramecia of some stocks of *P. aurelia,* later called *killers,* could kill animals of other stocks, called *sensitives.* The ability to kill was hereditary but was not transmitted in a typical Mendelian fashion (Sonneborn, 1943). Transmission of the killing ability was similar to that described in 1938 for the inheritance of killing by carbon dioxide in *Drosophila* (see L' Héritier, 1948, 1951).

It was found that to be killers, paramecia need to contain both nuclear and cytoplasmic hereditary units. The cytoplasmic unit, called *kappa,* may be released by a killer to the culture medium, thereby destroying sensitive paramecia. It also may be transmitted in the cytoplasm during conjugation between a killer and a sensitive animal, or during fission. There is transmission during conjugation when cytoplasmic in addition to micronuclear exchange occurs. This is illustrated in Figure 18-3.

A killer must have a dominant gene, *K*, in order to maintain replication of kappa units, but kappa units are not products of this gene. The kappa unit was found to be a particle, about one micron in diameter, containing DNA (Preer and Stack, 1953). Doubts have been raised as to whether the kappa particles are cytoplasmic organelles that constitute hereditary determinants of a normal paramecium, or whether they are infectious parasites (see reviews of Beale, 1954; Sonneborn, 1964). Beale concludes that the kappa system, although interesting as an example of nucleo-cytoplasmic interaction, cannot be directly applied to the problem of cellular differentiation. A similar conclusion probably applies to the inheritance of CO_2 sensitivity in *Drosophila*. The sensitivity in this case is transmitted by "genoids" or "sigma viruses" through the cytoplasm of the egg, and by infection.

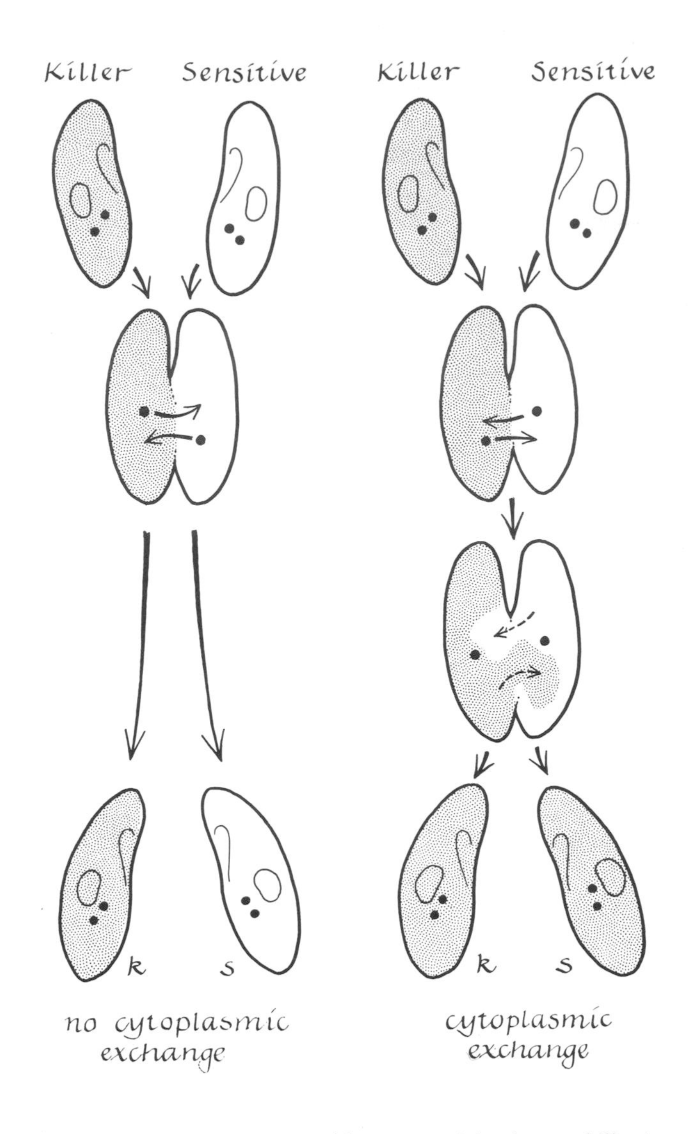

Figure 18-3. *Transmission of kappa particles from a killer to a sensitive paramecium when cytoplasmic exchange accompanies conjugation (right) but not when no cytoplasmic exchange occurs (left).*

Chloroplastid Inheritance

Before discussing the applicability of plastid inheritance to the problem of cell differentiation in plants, we will describe briefly the structure, development, and origin of the various plastid types.

Structure of the Chloroplast

The generalized structure of a chloroplast is illustrated in Figure 18-4A.

The structure illustrated occurs in many kinds of plants from algae to angiosperms (see reviews of Granick, 1963; Schiff and Epstein, 1965; Bogorad, 1967a). An electron micrograph of a chloroplast is shown in Figure 18-4B.

Development of a Chloroplast

Stages in chloroplast development under conditions of light and dark (von Wettstein, 1959) are shown in Figure 18-5A.

In plants such as corn or bean, the proplastids are about one micron in diameter, contain starch, and thus are considered to be amyloplasts. As indicated in Figure 18-5A, plastids must have light to develop normally. The mechanism of this effect is not well known (Bogorad, 1967).

Origin of Plastid Types

Plastids other than chloroplasts are found in higher plants. Until recently, plastid types were thought to arise from a common precursor proplastid, and their specialization to be controlled by local cytoplasmic influences. The formation of several types of plastids from proplastids is shown in the following diagram (Granick, 1963).

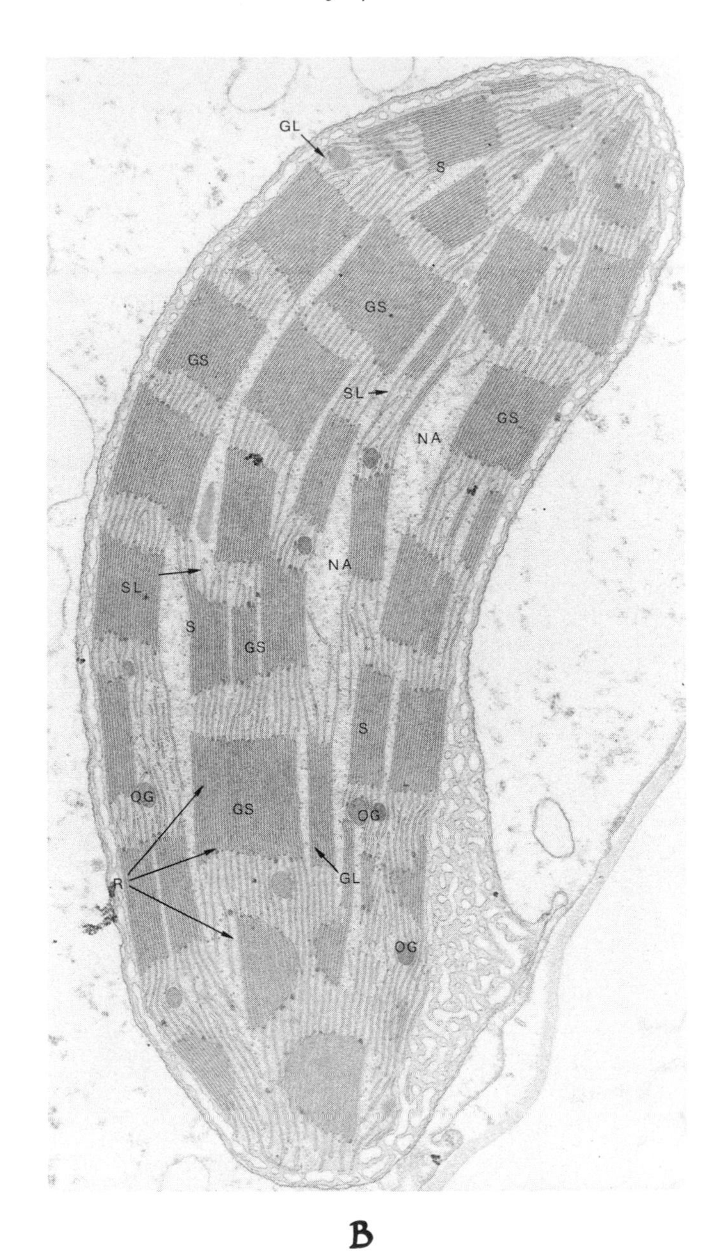

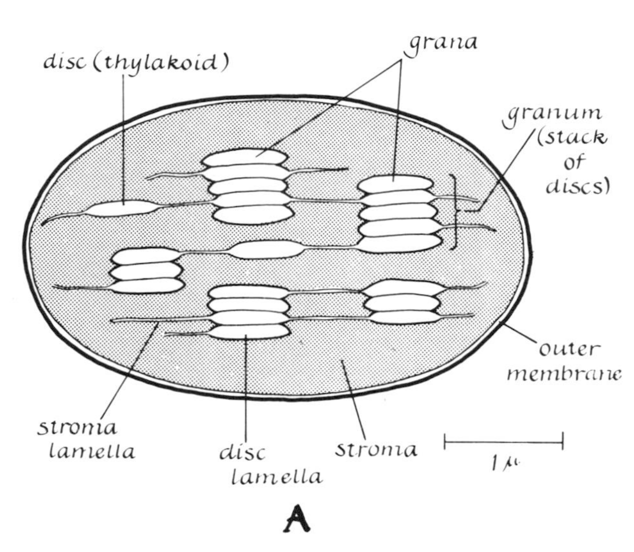

Figure 18-4. *A, diagram illustrating the generalized structure of a chloroplast; B, electron micrograph of a chloroplast in a cell of a corn leaf.* GL, *grana lamellae (thylakoids);* SL, *stroma lamellae;* S, *stroma containing enzymes, ribosomes and DNA;* R, *ribosomes;* OG, *osmiophyllic granules (× 80,000). Photograph courtesy L. K. Shumway. From Jensen and Park, 1967. Courtesy of Wadsworth Publishing Company, Inc.*

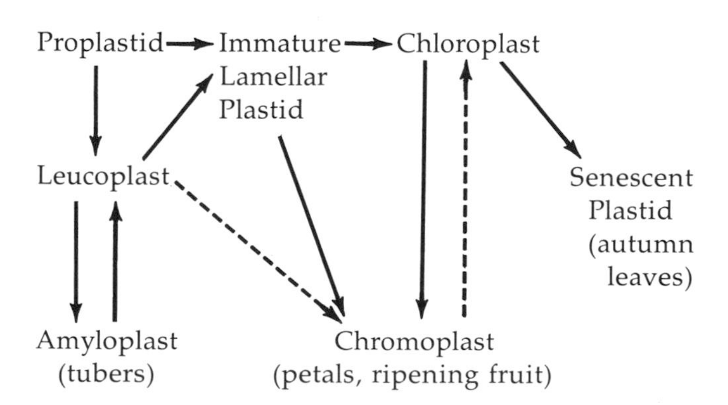

The scheme illustrated is only provisional. Since there is increasing evidence for the interconvertibility of plastid types and for division of mature plastids, the concept of the proplastid as the source of all plastid types is now doubtful.

Continuity and Stability of Plastids

There is good evidence that plastids arise either by division of proplastids or of mature plastids. A

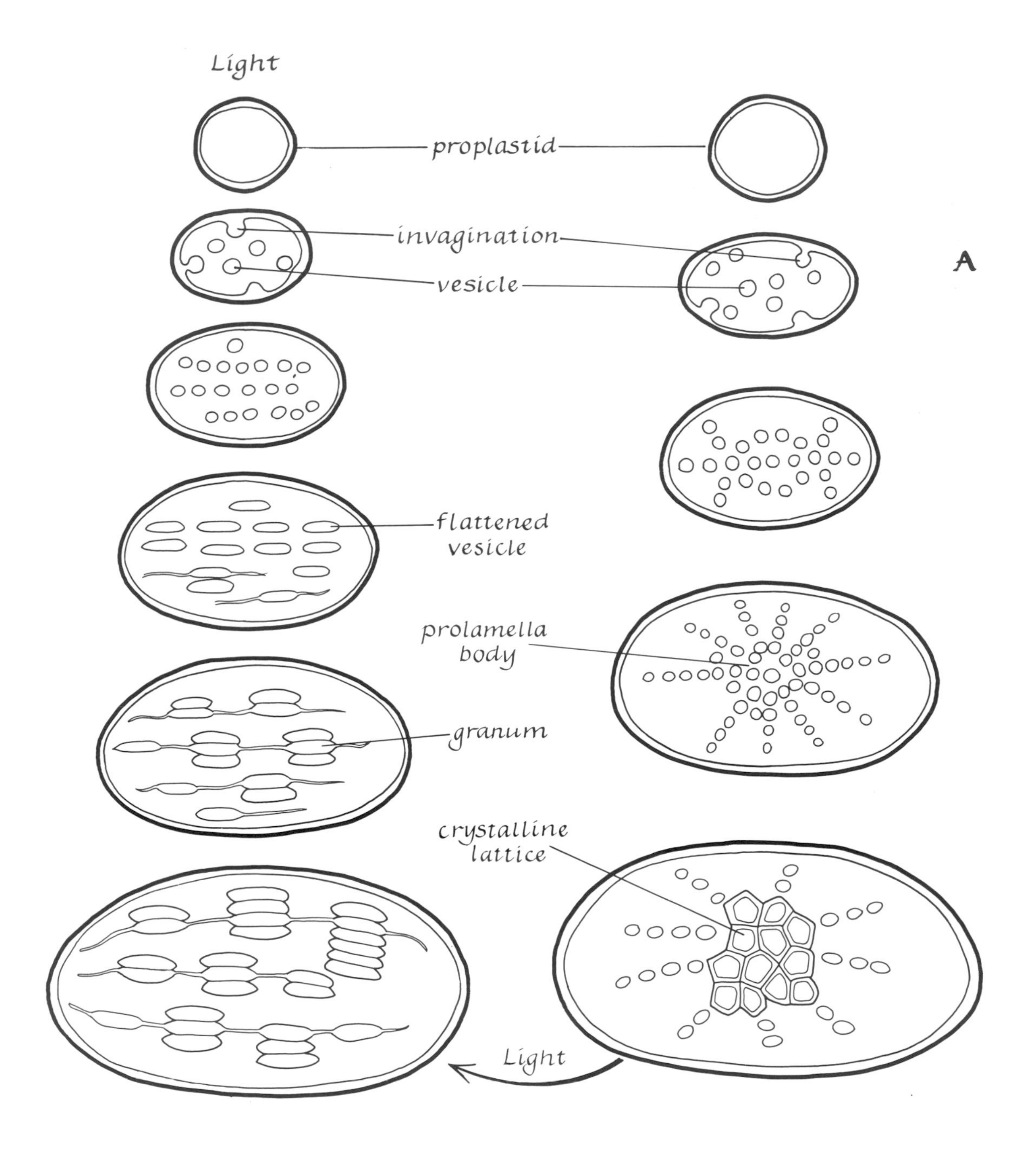
Light
proplastid
invagination
vesicle
A
flattened vesicle
prolamella body
granum
crystalline lattice
Light

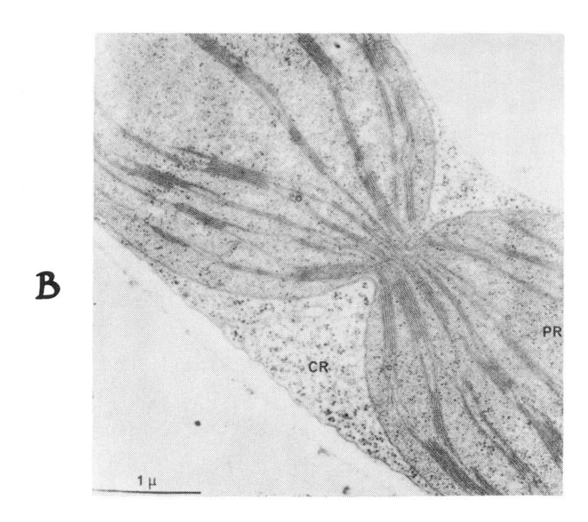

Figure 18-5. *A, stages in development of a chloroplast under light and dark conditions; B, electron micrograph of a dividing chloroplast from a* Spinacae *(spinach) leaf.* CR, *cytoplasmic ribosomes;* PR, *plastid ribosomes. Micrograph by A. D. Greenwood. B from Clowes and Juniper:* Plant Cells. *Copyright 1968 by Blackwell Scientific Publications, Ltd., Oxford.*

remarkable example of the latter is found in the unicellular marine alga *Chromulina psammobia* which has one chloroplast and one mitochondrion, each of which divides by fission when the cell divides. Chloroplast division has also been observed in *Euglena gracilis* (Schiff and Epstein, 1965), in the moss *Splachnum,* in ferns, and in angiosperms (see review of Kirk and Tilney-Bassett, 1967). A dividing chloroplast is illustrated in Figure 18-5B. The division and transmission of an abnormal plastid (having no pyrenoids) in *Spirogyra maxima* (Van Wisselingh, 1920) is particularly significant, since it is an example of plastid autonomy.

The inheritance of mutant plastids constitutes further evidence for continuity and stability of plastid types (Baur, 1909; Kirk and Tilney-Bassett, 1967). For example, in variegated plants (with white and green leaves or leaves with white and green areas), the green or white plastids are inherited cytoplasmically and usually maternally. When the egg of a flower that had formed on a green branch of *Pelargonium zonale* (geranium) was fertilized by pollen from a flower formed on a white branch, all the F_1 plants were green. In the reciprocal cross (flower on a white branch fertilized by pollen from a flower on a green branch), the F_1 plants were variegated, indicating cytoplasmic transmission through the pollen tube (Baur, 1909). By contrast, a similar cross in *Mirabilis jalapa* (four-o'clock) produced F_1 plants all of which were white, indicating transmission only through the egg cytoplasm.

There seems to be no orderly mechanism for the segregation of white and green plastids (or other types of mutant plastids) during mitosis in variegated plants. Depending upon the time in development when segregation occurs, all or only part of the plant or leaf may be variegated. This is illustrated in Figure 18-6.

The leaves of a species or individual often have basically regular patterns of variegation; these patterns could result either from the segregation of plastid types as shown in Figure 18-6 or from a response of the leaf-forming cells to local and different microenvironments (see Chapter 22). The pattern of variegation is also influenced by how the leaf develops (Chapters 4 and 26). Nevertheless, the formation of diverse patterns of variegation in plant leaves remains a challenging problem in developmental biology.

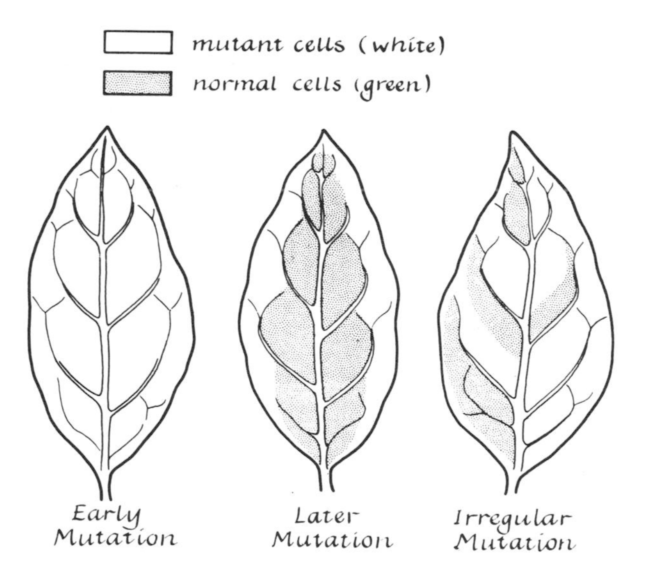

Figure 18-6. *Patterns of variegation resulting from segregation of mutant and normal cells.*

Evidence for Extranuclear Inheritance

The evidence for extranuclear inheritance of plastid types has been summarized by Kirk and Tilney-Bassett (1967) as follows:

1. The inheritance is non-Mendelian.
2. Reciprocal differences occur after reciprocal crosses.
3. Genetically different plastids are sorted out during development.
4. Several genera of plants contain single cells with two morphologically different plastids.

Nuclear and Chloroplast Interrelations

The problem of whether the chloroplast is an independent hereditary unit has been reviewed by Schiff and Epstein (1965). Three possible relationships between the nucleus and chloroplast are shown in Figure 18-7.

As indicated in Figure 18-7A, the plastid is formed anew in each generation under complete and direct control of the nuclear genome. Many genes affect chloroplast development (Bogorad, 1967b). In Figure 18-7C, the nucleus and plastid are shown as independent genetic units. In this case the plastid codes for its own proteins and is inherited independently of information in the nucleus. As indicated in Figure 18-7B, a mutant plastid continues to replicate itself independently of the initiating nuclear gene. An example of this is the gene *iojap* in corn, which produces plants with striped leaves in plants homozygous for this gene (Rhodes, 1946). The dependence of the plastid upon nutrients (for instance, amino acids) produced by the nuclear genome is also indicated in Figure 18-7B.

In any case, there is evidence for separate genomes in nucleus and plastid. For example, DNA and RNA net synthesis in anucleate pieces of *Acetabularia* (Chapter 5) apparently results from the great (but not complete) autonomy of the chloroplasts in these pieces (Brachet, 1968).

Genetic Material of the Plastid

Evidence is accumulating for the presence of a protein-synthesizing system in plastids (Granick, 1963; Schiff and Epstein, 1965; Kirk and Tilney-Bassett, 1967). Since Ris and Plaut (1962) discovered DNA fibrils in chloroplasts of *Chlamydomonas,* chloroplast DNAs have been detected in many chloroplast-containing plants. Biggins and Park (1962) found that a small percentage of the DNA in spinach chloroplasts differs from nuclear DNA in base composition. Template, transfer, and ribosomal RNA as well as RNA polymerase are also present in chloroplasts (reviews of Bogorad, 1967). Evidence for the presence of RNA comes from radioautography and electron microscopy as well as from biochemical studies of isolated chloroplasts. Furthermore, the presence of distinctive plastid ribosomes and some independent RNA metabolism in isolated plastids are both established. Nevertheless, the synthesis of chloroplast enzymes is, in part, controlled by nuclear genes.

Mitochondria

Mitochondria, like chloroplasts, are self-replicating organelles in both plant and animal cells. We saw in Chapter 13 that mitochondrial segregation during early cleavage of animal eggs could be a significant mechanism of differentiation. This significance becomes increasingly apparent as evidence for a separate protein-synthesizing system in mitochondria continues to increase (reviews of Lehninger, 1967; Nass, 1969). Isolated mitochondria incorporate labeled amino acids into insoluble membrane proteins. This synthesis is inhibited by puromycin and exhibits other features of ribosomal protein synthesis. Actinomycin D also inhibits protein synthesis and implies that mitochondrial DNA directs the synthesis of a mitochondrial form of mRNA. Ribosomal and tRNA are also present. Mitochondria possess an actinomycin-sensitive DNA-dependent RNA-polymerase. Furthermore, the DNA of mitochondria differs in molecular weight and base composition from nuclear DNA of the same cell. Dawid (1966) has demonstrated that the mitochondrial DNA in frog eggs is a double-stranded, circular molecule of relatively high molecular weight. However, in higher plants the mitochondrial DNA apparently is not circular. There are apparently no homologies between nuclear and mitochondrial DNA in *Xenopus* or *Rana* (Dawid and Wolstenholme, 1968). Replication of mitochondrial DNA has also been demonstrated (Reich and Luck, 1966; Karol and Simpson,

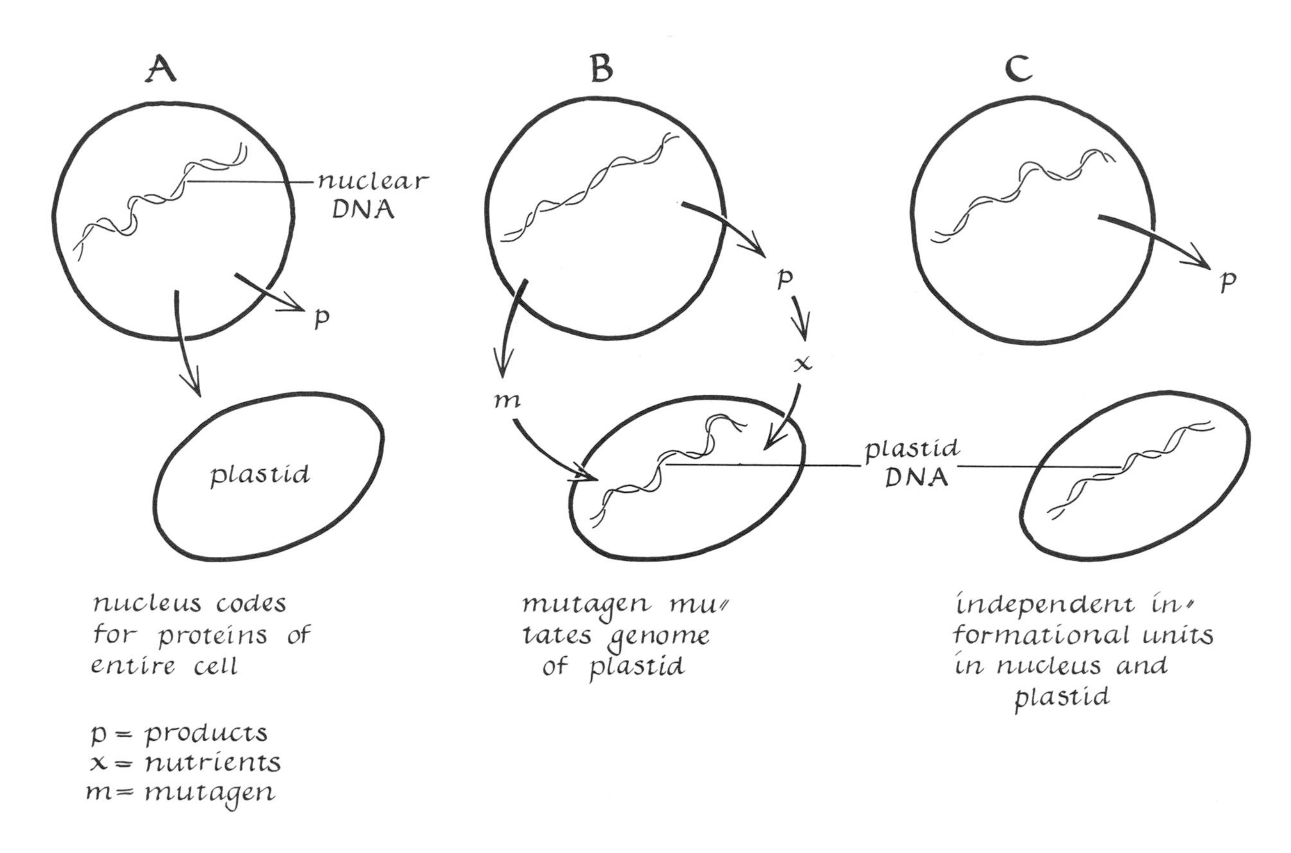

Figure 18-7. *Diagrams of the possible relationships between the nucleus and the chloroplast.*

1968); in *Neurospora* it proceeds at a rapid rate (complete in less than an hour).

Xenopus and *Rana* eggs contain 300 to 500 times more DNA than tissue cells and the bulk of this is mitochondrial DNA (Dawid, 1965). Sea urchin eggs also contain large amounts of mitochondrial DNA (Piko et al., 1967). The possibly independent role of mitochondria in cellular differentiation and inheritance in animals remains to be demonstrated, but in plants (for example, *Neurospora*), there is some evidence for the maternal inheritance of mutant mitochondria which determine the character "slow-growing and poor conidia spore formation." When cytoplasm of a slow-growing strain of *Neurospora* is injected into the wild-type mold, the wild-type becomes slow growing. Furthermore, the extranuclear determinant of slow growth is in the mitochondria (Tatum and Luck, 1967). These studies suggest that mitochondrial DNA replicates independently of nuclear DNA. Inheritance of the mutant character "little" in yeast also involves defective mitochondria (Ephrussi, 1953). The complete or partial loss of genetic information in the cytoplasmically inherited "little" mutants in yeast, coupled with the abnormal structure of their mitochondria, suggests that mutation has occurred in these mitochondria (Nass, 1969).

The evidence cited seems to establish that mitochondrial DNA can code for mitochondrial RNA, but not for all mitochondrial components. Thus an interaction of the nuclear and mitochondrial genetic systems appears to be necessary for organelle biosynthesis (Nass, 1969). Whether mitochondrial DNA can also code for cytoplasmic RNA has not been demonstrated (Fukuhara, 1967). Recall the suggested possibility of this in enucleated eggs (Chapter 17). There is cytoplasmic RNA which hybridizes very efficiently with circular DNA from mitochondria (Attardi and Attardi, 1967), but this cytoplasmic

RNA may be of nuclear origin (Fan and Penman, 1970). Nevertheless, it is highly probable that mitochondrial DNA represents a separate genetic system of the cell.

Cortical Inheritance in Ciliates

The pattern of *kinetosomes* and related structures in the cortex of ciliates such as *Paramecium aurelia* is reproduced faithfully through regular cycles of growth and fission. When ciliates with different types of cortical patterns conjugate, the exconjugants remain stable as to cortical type, since the genome does not influence the types of patterns (Nanney, 1968). Abnormal cortical patterns may be produced by "grafting" at the time of conjugation (Figure 18-8).

The experiment diagramed in Figure 18-8 resulted in formation of a doublet paramecium with extra cortex. The abnormal cortical pattern behaved like a mutation that was hereditary through up to 700 generations even though the animals were genotypically wild type (Beisson and Sonneborn, 1965). The cortical pattern in *Stentor coeruleus* is also inherited independently of the nuclear genome (Tartar, 1956). It might be significant that kinetosomes are organelles similar to centrioles and that they contain DNA (Steinert et al., 1958; Smith-Sonneborn and Plaut, 1967). It appears that the cortical pattern itself is inherited as a unit (see Lwoff, 1950).

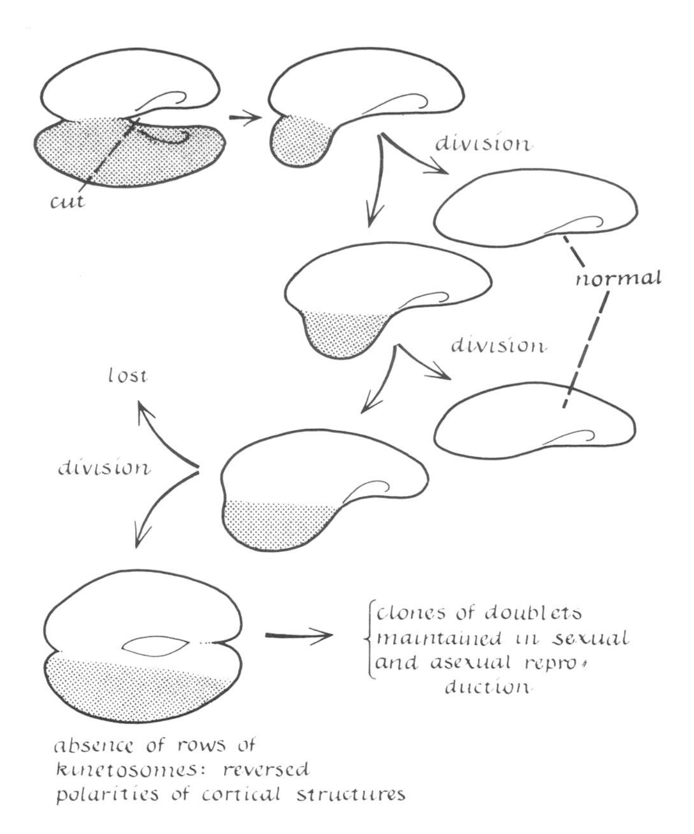

Figure 18-8. *Diagram illustrating cortical inheritance in paramecium.*

Some Additional Examples of Cytoplasmic Inheritance

Micrasterias

Recall that abnormal forms of the desmid *Micrasterias* are inherited through many generations, apparently independently of the nuclear genome, providing another example of extranuclear, cytoplasmic inheritance (Chapter 9).

Slow-Growing Strain in Neurospora

The inheritance of the character "slow growing" in the mold *Neurospora crassa* is apparently not chromosomally determined (Srb, 1965). The character is expressed phenotypically as an appreciably slower growth of conidia and ascospores than occurs in the wild-type mold. The cross between a normal and a slow-growing plant and the pattern of inheritance is shown in Figure 18-9. As shown in the figure, reciprocal crosses of opposite types of parents give different results, the usual test for maternal or cytoplasmic inheritance. By repeated backcrosses in which the slow-growing strain is used as the maternal parent and a different species, *N. sitophila*, is used as a recurrent paternal parent, the slow-growing character has been introduced into *sitophila*.

Red and White Strains in Aspergillus

Evidence for extrachromosomal heredity may be obtained in the absence of sexual reproduction and meiotic segregation. The red variant of the mold *Aspergillus nidulans* segregates persistently through

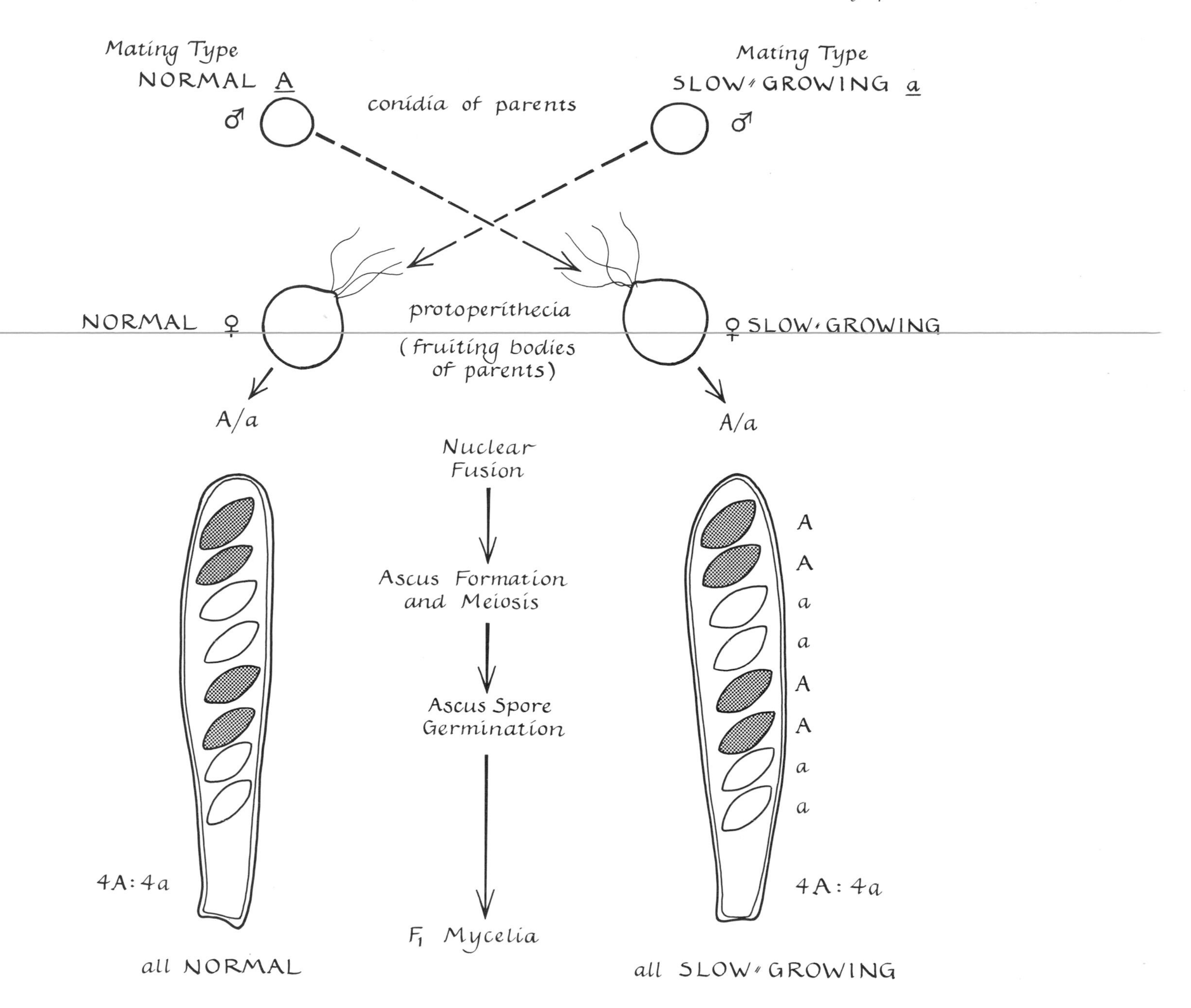

Figure 18-9. *Cytoplasmic inheritance of the character "slow growing" in* Neurospora crassa.

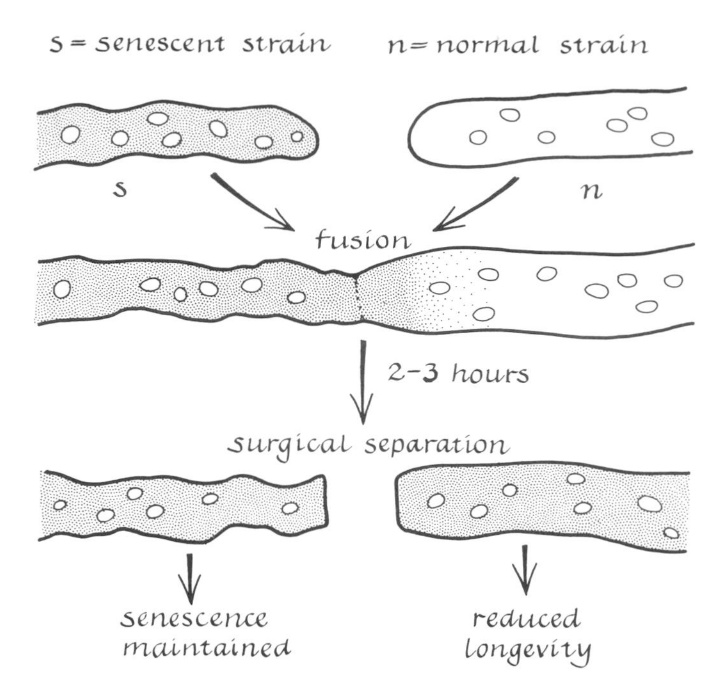

Figure 18-10. *Transmission of the character "senescence" during fusion of hyphae of the senescent,* s, *and normal* n, *strains of* Podospora anserina.

extended cycles of reproduction in which the character is transmitted through uninucleate conidia. Cells exhibiting segregation are presumed to contain homologous red and white (normal) determinants. The pattern of inheritance is indicated below (Srb, 1965).

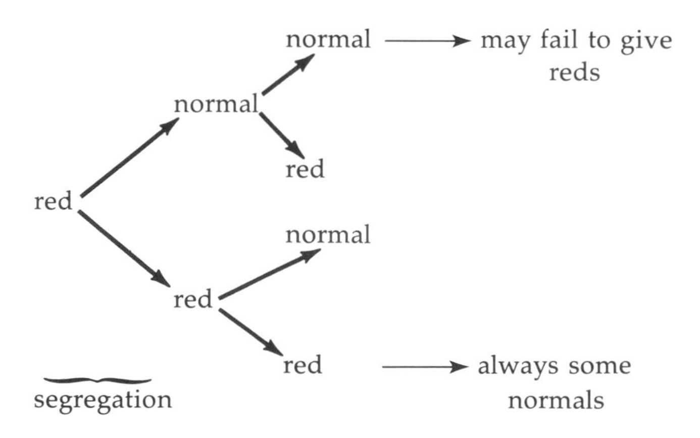

Since no red strains produce only reds, it appears that mycelia containing only red determinants are lethal. Those normal mycelia that fail to produce red strains are presumably derived from conidia that contained no red determinants.

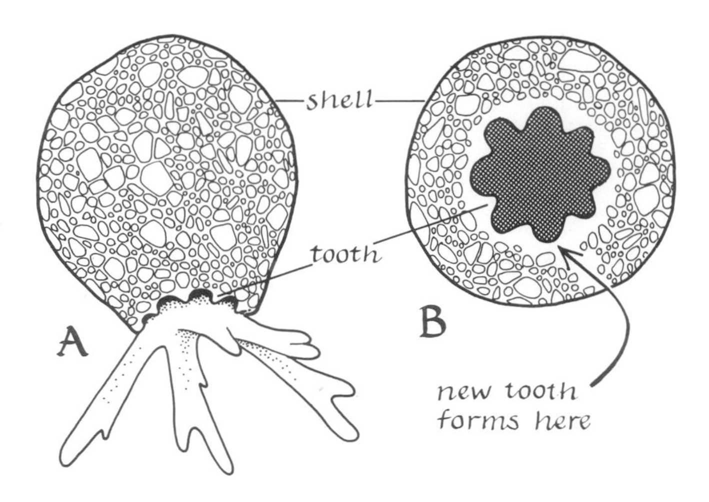

Figure 18-11. *Formation of new teeth of the daughter cell between teeth of the parent cell in* Difflugia corona.

Senescence in Podospora

After a period of vegetative multiplication, strains of the fungus *Podospora anserina* degenerate and finally cease to grow. The character *senescence* is transmitted either maternally, or (as is shown in Figure 18-10) by the fusion of a senescent strain hypha with a normal strain hypha (Srb, 1965).

The use of different genetic markers in the normal and senescent strains demonstrates that nuclear migration does not occur and thus cannot account for the transfer of the senescence character. Some unknown cytoplasmic determinant thus transmits the character.

Number of Teeth in Difflugia

A particularly significant example of extranuclear inheritance in the rhizopod *Difflugia corona* has been described by Jennings (1937). This protozoan builds a shell of sand grains at the opening of which are a definite number of "teeth." During cell division the new daughter cell constructs a shell while in contact with the parent, the new teeth being formed in the spaces between the parent's teeth (Figure 18-11).

Jennings established that the number of teeth in a clone remains constant. When he removed, for example, two teeth from a parent with eight teeth, the daughter cell formed only six teeth, and the descendants of this six-toothed animal in succeeding generations all had only six teeth. After the first generation the six teeth were radially and symmetrically arranged. This example of the direct transmission of a pattern independent of information encoded in the nuclear genome is similar in many respects to the inheritance of pattern in *Micrasterias* during fission (Kallio, 1960).

Conclusion

At the beginning of this chapter we asked whether the nuclear genome ever delegates its control of development to other organelles, including those in the cytoplasm. In some cases of maternal inheritance, products of prior gene action (such as the cytoplasmic organization of the egg) are transmitted directly through the egg cytoplasm and control the phenotype of the offspring irrespective of its genotype. In other types of maternal inheritance, cytoplasmic determinants that are not direct gene products (for instance, mutant mitochondria, mutant plastids, and slow growth in *Neurospora*) are transmitted and control the offspring's phenotype. Although in these latter cases the characteristic is not chromosomally determined, the replication of the cytoplasmic determinants may not be completely independent of the nuclear genome. This is clearly true of the kappa particle, whose replication depends on the K gene.

There are probably protein synthesizing systems (including DNA, rRNA, mRNA, and tRNA) in mitochondria and chloroplasts; and kinetosomes, basal bodies of cilia, and centrioles definitely contain DNA. This extranuclear DNA apparently plays a role in replication of these organelles. What overall role this DNA might play in cellular differentiation is at present unknown. The segregation of mutant mitochondria and chloroplasts during mitosis leads to phenotypically different daughter cells and constitutes a model for cellular differentiation, especially in plants. As we noted in Chapter 13, the quantitative segregation of mitochondria might be a mechanism of cellular differentiation in early cleavage of animal eggs.

The inheritance of cortical patterns in ciliates and of tooth number in *Difflugia* suggests that biological specificity may be maintained and transmitted by supramolecular mechanisms other than those encoded in linear molecular sequences such as DNA or RNA.

Extranuclear control of cell properties can no longer be considered inconsequential, at least in microorganisms, fungi, and higher plants, but whether such mechanisms operate in development of multicellular animals is not established. The control of cell differentiation by segregation of cytoplasmically different regions of the egg and the maternal inheritance of products of maternal gene action are simply examples of delayed phenotypic expression of nuclear gene action.

Finally, it is reasonable to assume that extranuclear genetic systems that appear to be completely independent of nuclear gene action evolved under nuclear gene control. As Nanney (1968) has emphasized: "We do not yet understand sufficiently well the mechanisms whereby the qualities and quantities of gene products are controlled, but the fact of such control is compellingly established and has become the cornerstone of any synthetic edifice in developmental biology."

References

Attardi, G., and B. Attardi. 1967. A membrane-associated RNA of cytoplasmic origin in Hela cells. Proc. Nat. Acad. Sci. U.S. **58**:1051–1058.

Baur, E. 1909. Das Wesen und die Erblichkeitsverhältnisse dä "Varietates albomarginatae hort." von *Pelargonium zonale*. Zeit. Indukt. Abstom. u. Vererb. Lehre. **1**:330–351.

Beale, G. H. 1954. The Genetics of *Paramecium aurelia*. Cambridge University Press, London.

Beisson, J., and T. M. Sonneborn. 1965. Cytoplasmic inheritance of the organization of the cell cortex in *Paramecium aurelia*. Proc. Nat. Acad. Sci. U.S. **53**:275–282.

Biggins, J., and R. B. Park. 1962. Nucleic acid content of the chloroplasts of *Spinacae oleracea*. Bio-Organic Chem. Quart. Rep. u. CRL, p. 39.

Bogorad, L. 1967a. The organization and development of chloroplasts. In J. M. Allen, ed., Molecular Organization and Biological Function, pp. 134–185. Harper & Row, New York.

Bogorad, L. 1967b. Control mechanisms in plastid development. In M. Locke, ed., Control Mechanisms in Developmental Processes, pp. 1–31. Academic Press, New York.

Boycott, A. W., C. Diver, S. L. Garstang, and F. M. Turner. 1931. The inheritance of sinistrality in *Limnaea peregra*. Phil. Trans. Roy. Soc. Lond. Ser. B. **219:**51–130.

Brachet, J. 1968. Synthesis of macromolecules and morphogenesis in *Acetabularia*. In A. A. Moscona and A. Monroy, eds., Current Topics in Developmental Biology, pp. 1–36. Academic Press, New York.

Caspari, E. 1948. Cytoplasmic inheritance. Adv. Genetics **2:**1–66.

Dawid, I. B. 1965. Deoxyribonucleic acid in amphibian eggs. J. Mol. Biol. **12:**581–599.

Dawid, I. B. 1966. Evidence for the mitochondrial origin of frog egg cytoplasmic DNA. Proc. Nat. Acad. Sci. U.S. **56:**269–276.

Dawid, I.B., and R. Wolstenholme. 1968. Renaturation and hybridization studies of mitochondrial DNA. Biophys. Journ. **8:**65–81.

Ephrussi, B. 1953. Nucleo-cytoplasmic Relations in Microorganisms. Oxford University Press, London.

Fan, H., and S. Penman. 1970. Mitochondrial RNA synthesis during mitosis. Science **168:**135–138.

Fukuhara, H. 1967. Informational role of mitochondrial DNA studied by hybridization with different classes of RNA in yeast. Proc. Nat. Acad. Sci. U.S. **58:**1065–1072.

Granick, S. 1963. The plastids: their morphological and chemical differentiation. In M. Locke, ed., Cytodifferentiation and Macromolecular Synthesis, pp. 144–174. Academic Press, New York.

Jennings, H. S. 1937. Formation, inheritance and variation of the teeth in *Difflugia corona:* a study of the morphogenetic activities of rhizopod protoplasm. J. Exp. Zool. **77:**287–336.

Jinks, J. L. 1964. Extrachromosomal Inheritance. Prentice-Hall, Englewood Cliffs, N.J.

Kallio, P. 1960. Morphogenesis of *Micrasterias americana* in clone culture. Nature **187:**164–166.

Karol, M. H., and M. V. Simpson. 1968. DNA biosynthesis by isolated mitochondria: a replicative rather than a repair process. Science **162:**470–473.

Kirk, J. T. O., and R. A. E. Tilney–Bassett. 1967. The Plastids. W. H. Freeman, San Francisco.

Lehninger, A. L. 1967. Molecular basis of mitochondrial structure and function. In J. M. Allen, ed., Molecular Organization and Biological Function, pp. 107–133. Harper & Row, New York.

L'Héritier, P. 1948. Sensitivity to CO_2 in *Drosophila*—a review. Heredity **2:**325–348.

L'Héritier, P. 1951. The CO_2 sensitivity problem in *Drosophila*. Cold Spring Harbor Symp. Quant. Biol. **16:**99–112.

Lwoff, A. 1950. Problems of Morphogenesis in Ciliates. John Wiley, New York.

Nanney, D. L. 1968. Cortical patterns in cellular morphogenesis. Science **160:**496–502.

Nass, M. M. K. 1969. Mitochondrial DNA: advances, problems and goals. Science **165:**25–35.

Pikó, L., A. Tyler, and J. Vinograd. 1967. Amount, location, priming capacity, circularity and other properties of cytoplasmic DNA in sea urchin eggs. Biol. Bull. **132:**68–90.

Preer, J. R., and P. Stack. 1953. Cytological observations on the cytoplasmic factor kappa in *P. aurelia*. Exp. Cell Res. **5:**478–499.

Reich, E., and D. Luck. 1966. Replication and inheritance of mitochondrial DNA. Proc. Nat. Acad. Sci. U.S. **55**:1600–1608.

Rhoades, M. M. 1946. Plastid mutations. Cold Spring Harbor Symp. Quant. Biol. **11**:202–207.

Ris, H., and W. Plaut. 1962. Ultrastructure of DNA-containing areas in the chloroplast of *Chlamydomonas.* J. Cell Biol. **13**:383–391.

Schiff, J. A., and H. T. Epstein. 1965. The continuity of the chloroplast in *Euglena.* In M. Locke, ed., Reproduction: Molecular, Subcellular, and Cellular, pp. 131–189. Academic Press, New York.

Smith-Sonneborn, J., and W. Plaut. 1967. Evidence for the presence of DNA in the pellicle of *Paramecium.* J. Cell Sci. **2**:225–234.

Sonneborn, T. M. 1943. Gene and cytoplasm: I. The determination and inheritance of the killer character in variety 4 of *P. aurelia.* II. The bearing of determination and inheritance of characters in *P. aurelia* on problems of cytoplasmic inheritance, pneumococcus transformations, mutations and development. Proc. Nat. Acad. Sci. U.S. **29**:329–343.

Sonneborn, T. M. 1964. The differentiation of cells. Proc. Nat. Acad. Sci. U.S. **51**:915–929.

Srb, A. M. 1965. Extrachromosomal heredity in fungi. In M. Locke, ed., Reproduction: Molecular, Subcellular, and Cellular, pp. 191–211. Academic Press, New York.

Steinert, G., H. Firket, and M. Steinert. 1958. Cynthese d'acide desoxyribonucleique dans le corps parabasal de *Trypanosoma mega.* Exp. Cell Res. **15**:632–635.

Tartar, V. 1956. Pattern and substance in *Stentor.* In D. Rudnick, ed., Cellular Mechanisms in Differentiation and Growth, pp. 73–100. Princeton University Press, Princeton, N.J.

Tatum, E. L., and D. Luck. 1967. Nuclear and cytoplasmic control of morphology in *Neurospora.* In M. Locke, ed., Control Mechanisms in Developmental Processes, pp. 32–42. Academic Press, New York.

VanWisselingh, C. 1920. Über variabilitat und erblichkeit. Zeit. Ind. Abstam. Vererb. **22**:65–126.

VonWettstein, D. 1959. The effect of genetic factors on the submicroscopic structures of the chloroplast. J. Ultrastruct. Res. **3**:234–235.

Wilkie, D. 1964. The Cytoplasm in Heredity. Methuen, London.

19 *Hormonal Control in Animal and Plant Development*

Hormones help integrate the functional and developmental activities of cells in developing organisms. They are specific chemical substances produced by one group of cells which, in very small concentrations, influence the activity of another, usually nonadjacent, group of cells. Thus hormones constitute a chemical communication between different parts of an organism. The idea of chemical communication between different parts of a plant was described as early as 1880 in an article by Sachs, in which he suggested that substances produced by one organ of a plant are required for development of another organ. However, it was not until 1928 that Went succeeded in isolating the first plant hormone, IAA. The gibberellin hormones of plants were also being discovered around this time in Japan.

The idea of chemical communication between different organs in animals had been suggested by the ancient Greek physicians and philosophers, but only in 1849 did Berthold demonstrate by testicular grafting that the fowl testis produced not only spermatozoa but also one or more blood-borne substances controlling sexual characteristics. In 1895, Oliver and Schäfer discovered that an extract of the adrenal medulla caused a pronounced rise in blood pressure when injected into the circulation. A few years later they isolated and identified the active substance in the extract and called it *adrenalin.* It was not until 1922 that another important animal hormone was isolated. Banting and Best (1922) and Macleod (1922) reported the isolation of *insulin,* a pancreatic hormone that helps control the level of sugar in the blood. The role of hormones in integrating the physiology of adult organisms is well established but is beyond the scope of this book. For information about animal hormones, see the text by Turner (1966); Wilkins (1969) discusses plant hormones. This text deals only with the role of hormones in the control of developmental processes. A vast literature deals with the role of hormones in plant and animal development; space permits us to survey briefly only a few selected examples.

Hormonal Mechanisms in Animals

Insect Metamorphosis

One of the most striking examples of how hormones control developmental processes is found in the metamorphoses of insects (Gilbert and Schneiderman, 1961; Williams, 1963; Gilbert, 1964; Wiggles-

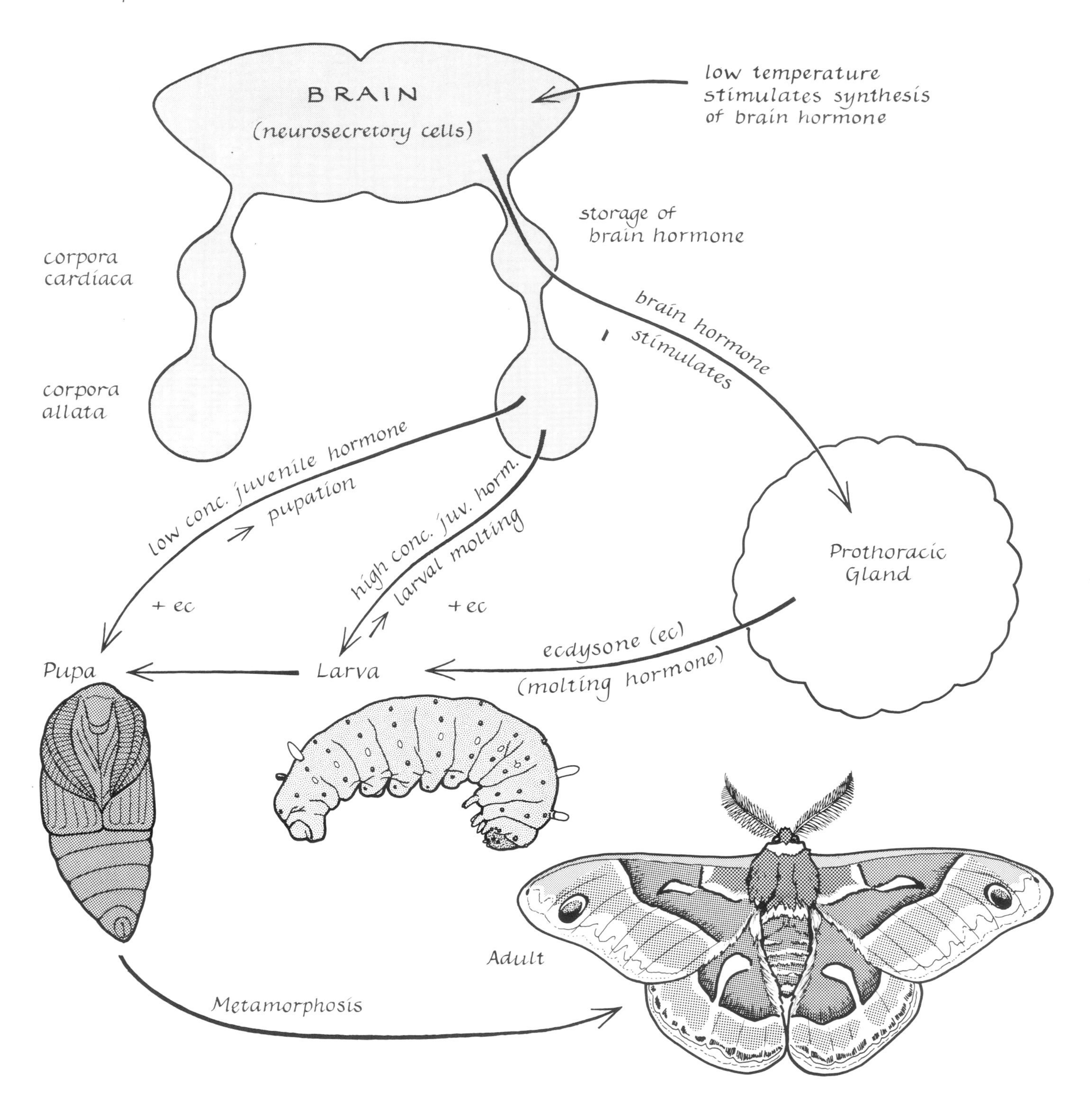

Figure 19-1. *Relationship of three major hormones involved in metamorphosis of the* Cecropia *moth and other insects.*

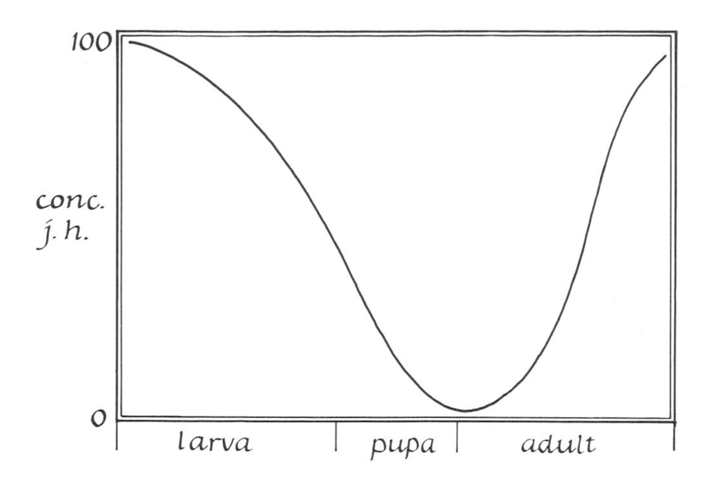

Figure 19-2. *Changes in juvenile hormone concentration during stages of metamorphosis.*

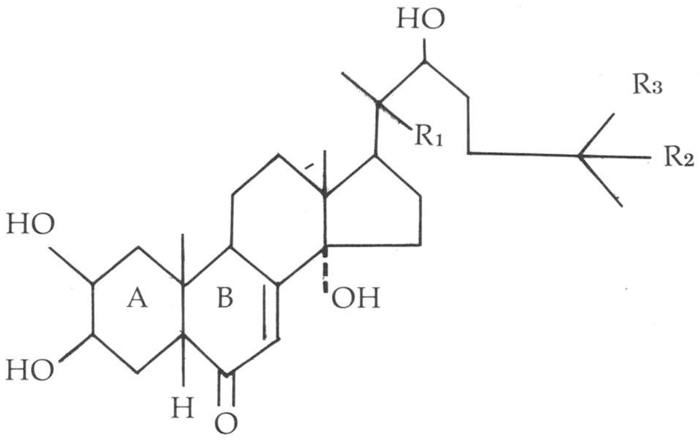

Figure 19-3. *Structural formula of ecdysone.*

worth, 1966). The relationship of the three major hormones involved, *juvenile, brain,* and *molting* hormones, is diagramed in Figure 19-1.

As indicated in the figure, low temperature stimulates the brain of the larva to synthesize the brain hormone. This hormone is stored in the *corpora cardiaca.* When released, the brain hormone stimulates the prothoracic gland to produce and release another hormone, ecdysone. This stimulates epidermal cells to deposit cuticle and thus initiates molting. A third hormone, the juvenile hormone (JH), is secreted by the *corpora allata,* endocrine glands near the brain of the larva. In the presence of ecdysone, a relatively high concentration of juvenile hormone promotes larval development but prevents pupation. A relatively low concentration of juvenile hormone allows pupation but prevents metamorphosis into the adult. In the absence of juvenile hormone, molting results in differentiation of the pupa into an adult. The concentration of juvenile hormone then increases, reaching its highest level in the final week of adult development. The changes in juvenile hormone concentration during the stages of metamorphosis are illustrated in Figure 19-2.

The brain hormone appears to control the entire sequence of larval, pupal, and adult events. Thus, cessation in the production of brain hormone at the time of pupation results in arrested development *(diapause)* of the pupa; no molting occurs. This is because the prothoracic glands stop releasing ecdysone in the absence of stimulation by the brain hormone. Low temperature in nature or artificial chilling of the pupa stimulates the brain to produce its hormone, which in turn stimulates the prothoracic gland to produce ecdysone. This is then followed by metamorphosis of the pupa into the adult.

The brain hormone has been variously reported to be cholesterol, a steroid, or a protein. The exact nature of the juvenile hormone also remains unclear, although it has been reported to be farnesol (a precursor of cholesterol) and other sterols (Williams and Law, 1965). In contrast, the prothoracic gland hormone, ecdysone, has been crystallized, starting with one thousand pounds of male bee pupae (Butenandt and Karlson, 1954). Ecdysone is a steroid with the empirical formula $C_{18}H_{30}O_4$. Its structural formula is shown in Figure 19-3 (Karlson and Sekeris, 1966).

The sequence of metamorphic events just described is undoubtedly the phenotypic expression of sequential gene action. The hormones involved are products of gene action, and they in turn control the action of other genes in directing cell proliferation and differentiation. Hormones are thus agents of the genes, the latter retaining control of the entire program of development.

Hormones controlling sex differentiation in insects (for example, in beetles) have been demonstrated. When testes of a fifth instar male are grafted into a female larva, the host develops into a functional male adult (Naisse, 1963).

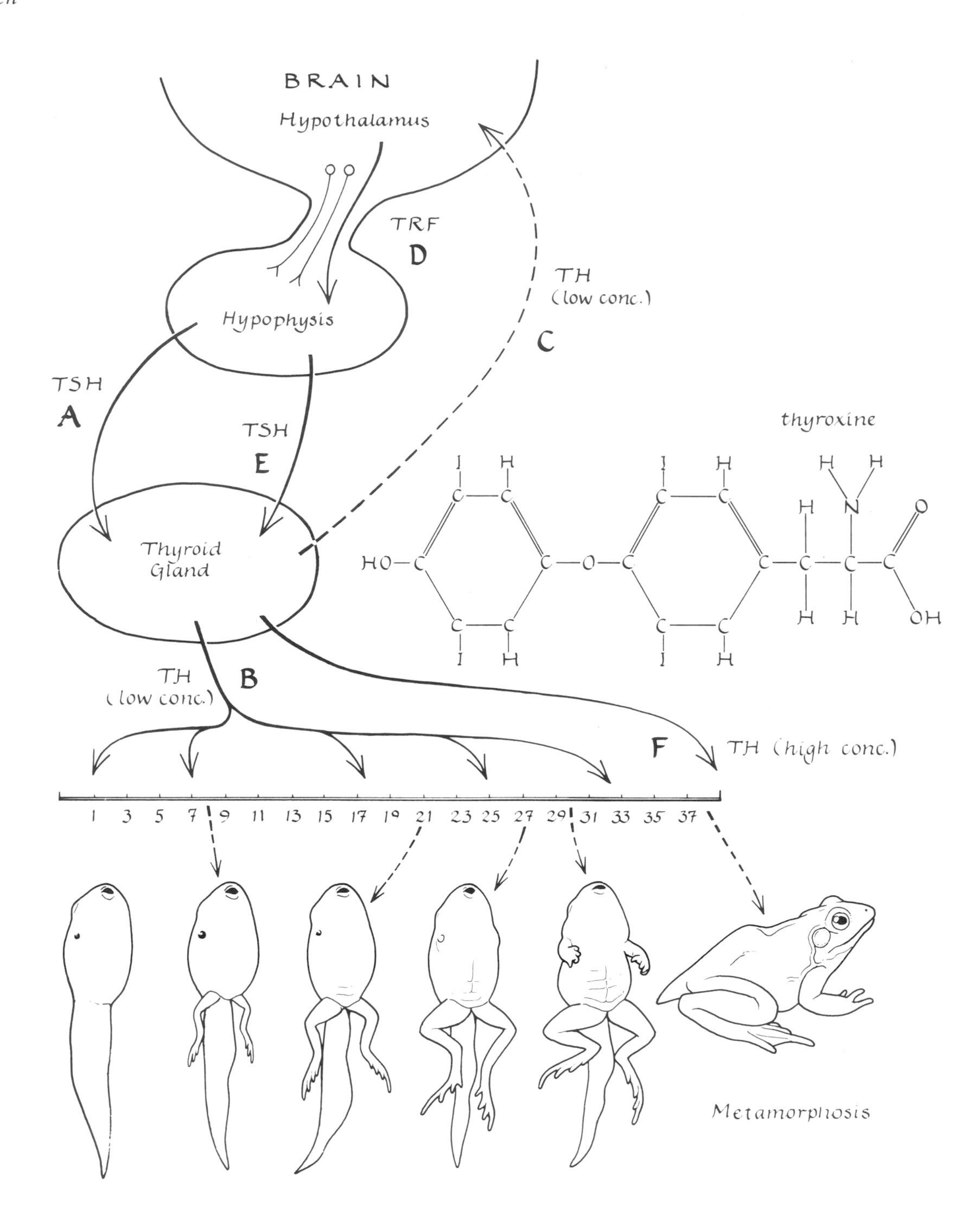

Figure 19-4. *Hormonal interrelations in amphibian metamorphosis.*

Hormones in Other Invertebrates

Since most studies of hormones involved in development of noninsect invertebrates have been done with the class *Crustacea,* we shall deal only with them (Carlisle and Knowles, 1959; Highnam and Hill, 1969; Tombes, 1970). The principal endocrine organs of crustacea are located at the base of the eye stalk. One organ is called the x-organ, the other the y-organ. One hormone liberated by the x-organ in *Carcinus maenas* influences gonadal development (Demsusy, 1962). Another hormone, which controls the molting process, is an ecdysone, similar to that found in insects (Schneiderman and Krishnakumaran, 1968). This hormone is produced by the y-organ, which is regulated in turn by the x-organ (Passano, 1961). Removal of the x-organ results in precipitous molting and thus indicates that the x-organ produces a molt-inhibiting hormone. If the y-organ is also removed, no molting occurs.

Sex hormones also play a role in development of crustacea (Charniaux-Cotton, 1965). For example, in amphipods sex development is controlled by the presence or absence of male hormone alone. In the presence of androgenic hormone, produced by the androgenic gland, the male sex is differentiated regardless of the genetic sex. When androgenic glands of the male *Orchestia gammarellus* are transplanted into a genetic female, the ovaries rapidly transform into testes and after one or two molts large claws characteristic of the male are formed. When the androgenic glands are removed from a young genetic male, the primordial germ cells of the testis develop into oocytes. These germ cells are thus bipotential in respect to their capacity to differentiate into eggs or spermatozoa. Even in hydras, parabiosis demonstrates the existence of a hormone-like substance secreted by males and capable of inducing spermatogenesis in females (Brien, 1964). A neurosecretory substance, the growth hormone, produced by cells of the nerve net in the hypostome of *Hydra,* apparently helps control interstitial cell differentiation. When growth hormone production is relatively great the interstitial cells can differentiate into nematoblasts, nerve cells, or other somatic cell types. When growth hormone production is decreased, the interstitial cells can differentiate into gametes (Burnett and Diehl, 1964).

Control of Amphibian Metamorphosis

The first indication that hormones control metamorphic changes in amphibians was obtained by Gudernatsch (1912); he fed powdered sheep-thyroid gland to tadpoles and observed precocious metamorphosis. Allen (1918) proved the thyroid hormone to be the cause of metamorphosis: he removed the gland from frog embryos in the tail-bud stage and found that these animals grew but failed to metamorphose. Similar results have been obtained with urodeles. The hormones released by the thyroid gland were later shown to be *thyroxine* (TH) and *triiodothyronine.* The activity of triiodothyronine is greater than that of thyroxine, although the latter is produced in much greater quantity by the thyroid gland (Wilkins, 1960).

Metamorphosis includes the growth of hind and forelegs, resorption of the tadpole tail, changes in body proportions and in the internal organs (shortening and uncoiling of the intestine, for instance), and numerous changes in metabolism, particularly in the activities of enzymes involved in urea synthesis (Cohen, 1970) during the change from the excretion of ammonia by the tadpole to that of urea by the adult. These changes occur in sequence, and are controlled by hormones produced by the thyroid gland, the anterior lobe of the hypophysis, and the hypothalamus. These hormonal interrelations are shown in Figure 19-4.

Metamorphic changes begin when the tadpole hypophysis releases the thyroid-stimulating hormone, TSH (Figure 19-4A), which causes the thyroid gland to produce its hormone TH (Figure 19-4B) at a relatively low level. The thyroid hormone initiates growth of the hind limbs and later the forelimbs; during this period it also apparently stimulates the hypothalamus (Figure 19-4C) to produce and release another hormone, thyrotropin-releasing factor (TRF). Vascular connections transmit TRF from the hypothalamus to the hypophysis (Figure 19-4D) and the TRF causes the latter to release more TSH into the circulation (Figure 19-4E). TSH in turn causes the production of more thyroid hormone (Figure 19-4F), which brings about the final transformation of the tadpole into a frog (Etkin, 1966). Thus, under normal conditions the thyrotropic hormone (TSH) stimulates the thyroid to produce and release thyroxine, which in turn causes the metamorphic events.

Various tissues of the tadpole respond differently to the thyroid hormone. Cells of the tail and gills are induced to die, whereas limb cells are stimulated to divide and differentiate. In general, the sensitivity of tissues to the hormone is reflected in the order in which the metamorphic changes occur (Etkin, 1955; Etkin and Gilbert, 1968). Parts with a low threshold (legs, reacting by growth) respond earlier than parts with a higher threshold (tail, reacting by reduction).

If the hypophysis is removed from a frog embryo, no metamorphosis occurs; and if the thyroid gland is removed, no amount of hypophyseal tissue or extract will induce metamorphosis. However, hypophysectomized tadpoles artificially supplied with thyroxine do metamorphose.

Weber (1967) has presented a review of the earlier studies of the biochemistry of amphibian metamorphosis.

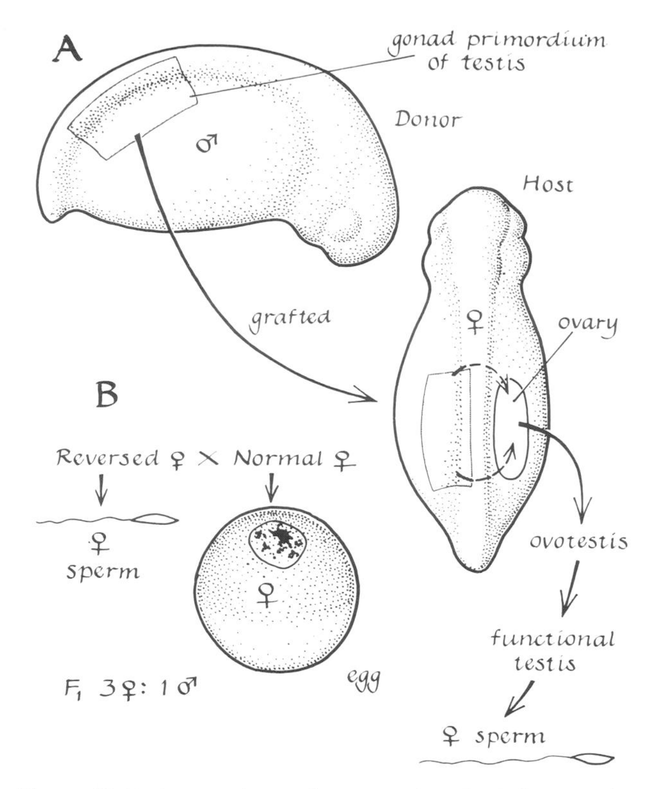

Figure 19-5. *A, experiment demonstrating the influence of sex hormones on the differentiation of the gonads in the axolotl; B, abnormal sex ratio in the cross between a reversed female and a normal female.*

Hormones in Vertebrate Embryology

We know that hormones are synthesized during embryonic development (Willier, 1955) and that they control cellular differentiation in adults. (For example, Hardy, 1953, found that the female sex hormone stimulated keratin synthesis in vaginal wall cells.) Nonetheless, the problem of the role hormones play in differentiation, exclusive of the endocrine system itself, remains challenging. Perhaps more is known about the role of sex hormones during development than that of any other hormones (review of Witschi, 1967).

One of the first examples of the role of hormones or hormone-like substances in the control of sex differentiation was described by Lillie (1917). When vascular connections are established between cattle embryo twins of opposite sex, the gonads and sex ducts of the female partner are modified in the male direction. The modified female is known as a freemartin. Lillie interpreted this result to mean that male sex hormone, which is released relatively early in development from the testis of the male twin, passes through the blood connections and modifies the way the reproductive system of the female twin develops. However, since the male sex hormone testosterone appears unable to modify ovarian development in genetic female cattle, this interpretation is doubtful. An alternative suggestion is that a male inductor substance called medullarin might be involved in the modification of the female reproductive system of the freemartin (see review of Short, 1970).

The injection of sex hormones into chicken eggs (Willier, 1952) and the implantation of sexually differentiated gonad tissue into the coelom of younger host chick embryos (Wolff, 1947) indicate that embryonic gonads produce sex hormones which control the differentiation of the reproductive system. Numerous additional experiments with amphibians, birds, and mammals strongly support the idea that sex differentiation (both primary and secondary sex characters) is controlled by sex hormones released by the embryonic gonads (Willier, 1955; Jost, 1965; Wells, 1965; Haffeh, 1970; and Harris and Edwards, 1970).

An experiment demonstrating the role of sex hormones in controlling the differentiation of the reproductive system and primordial germ cells in the Mexican axolotl *(Siredon mexicanum)* is illustrated in Figure 19-5A. The gonad primordium on one side of

A eggs: w + z (normal ♀)
sperm: w + z (reversed ♀)
F_1 1ww + 2wz + 1zz
3 ♀ = 1 ♂

B ww ♀ × zz ♂
F_2 wz (all ♀'s)

Figure 19-6. *A, results of crossing a normal female (WZ) with a reversed female (WZ); B, results of crossing a WW female with a ZZ male of the* F_1 *generation.*

a sexually undifferentiated donor embryo was excised and grafted into a corresponding position in a host embryo of similar age, one of whose gonad rudiments had been previously removed. When the donor was a genetic male (determined by raising the donor to sexual maturity) and the host was a genetic female, the ovary of the host gradually became a testis under the influence of male sex hormones produced by the graft. Sometimes transformation was complete; the ovary developed into a functional testis in which sperm were formed. Note in Figure 19-5 that the sperm produced by the transformed ovary are female in genotype; normal male sperm are always male in genotype. This is another example of the bipotentiality of the primordial germ cells (Chapter 12). The reversed female axolotl was mated with a normal female and, as shown in Figure 19-5B, offspring were obtained in a ratio of three females to one male when sex at maturity was ascertained (Humphrey, 1931 and 1945). This ratio indicated that the female axolotl is heterozygous for sex chromosomes and the male homozygous. In amphibians, the symbols W and Z are used in place of X and Y. The normal amphibian female *Axolotl* produces two kinds of eggs, half with a W chromosome and half with a Z chromosome (Beatty, 1970). The normal male produces one kind of sperm, each containing one Z chromosome. The reversed female, however, produces two kinds of sperm, half with a W and half with a Z chromosome. This explains the unusual sex ratio in F_1 individuals produced by fertilizing normal eggs with sperm of the reversed female (Figure 19-6A).

Note in Figure 19-6B that when Humphrey crossed an F_1 female of WW genotype with a normal male, all the F_2 individuals were females. These experiments demonstrate that the male sex hormone produced by the grafted embryonic testis controls the differentiation of genotypically female primordial germ cells of the transformed ovary into sperm. It might be thought that the male sex hormone, when produced earlier or in greater quantity, stimulates the activity of normally quiescent male-determining genes present in the genetic female. Probably, however, the male sex hormone simply suppresses the activity of the female-determining genes, as it does in crustacea and other invertebrates noted previously. Functional sex has also been reversed by gonad transplantation in *Xenopus laevis*. When sex-reversed genetic males that produce eggs are crossed with normal males, the resulting offspring are all males (Mikamo and Witschi, 1963).

Evidence for the time at which various hormones are synthesized during embryonic development has been summarized by Willier (1955). A few examples are the following: in the chick embryo, hypophyseal, thyroid, sex, and insulin hormones are secreted between the tenth and the thirteenth days of incubation; in the frog embryo, thyroid hormone is secreted at the 10 mm stage and thyrotropic and growth hormones are released after the tail-bud stage; in the pig embryo, at 40 mm length the adrenal medulla is active, at 90 mm, the thyroid, and at 175 mm the anterior hypophysis.

Mechanism of Action of Animal Hormones

The possible but not mutually exclusive mechanisms by which hormones might act have been listed by Villee (1962, 1967): hormones might control the rate of enzyme synthesis, influence the steric form of enzymes and thus their activities, influence cell permeability, function as coenzymes, or, finally, initiate transcription of DNA. Unfortunately, little is known about exactly how any single hormone works, either at the cellular or the molecular levels (see Karlson, 1965). However, increasing evidence in recent years indicates that the effects of many hormones are the result of their activation of genes (see review of Davidson, 1965). Some hormones enter cell nuclei. For example, *aldosterone* (released by the adrenal cortex), which controls sodium and potassium excretion, was found in the nuclei after being radioactively labeled and administered to a preparation of toad bladder tissue. Other hormones, such as *progesterone* (secreted by the corpus luteum), do not exhibit any specific localizations. Most studies of how hormones affect genes have been concerned with the steroid hormones, *estrogens.* Stimulation of protein synthesis in the uterus by estrogens is blocked by the antibiotic *puromycin,* which specifically inhibits protein synthesis—evidence that estrogens act on the genetic, protein-synthesizing system. Furthermore, about thirty minutes after treatment with estrogens, RNA synthesis increases markedly. When actinomycin is used to block the synthesis of RNA, administration of estrogens has no effect on protein synthesis.

Hormone action is typically specific; that is, hormones act on certain cells but not on others. An excellent example of the selective activation of repressed genes by estrogens is its action on the liver in chickens. During egg formation in the hen, estrogen produced by her ovary stimulates the synthesis of the yolk proteins *lipovitellin* and *phosvitin* in her liver. A rooster, which does not need to synthesize these proteins, will, however, make them in large quantities when treated with estrogen.

Thyroid hormone has been reputed to stimulate protein synthesis in cultured human cells (Siegel and Tobias, 1966) and to stimulate the incorporation of amino acids into proteins in the liver of mammals (Tata, 1966). Villee (1967) has presented evidence that the sex hormone *testosterone* acts on the genetic mechanism leading to the production of specific mRNA, which in turn leads to the production of proteins in the seminal vesicle of the rat. Thus when RNA extracted from the testosterone-treated seminal vesicle of one rat is injected into the lumen of the seminal vesicle of another rat, the result is protein synthesis similar to that induced by injection of testosterone. Mitochondrial biosynthesis of proteins and lipids in male sex accessory gland cells is stimulated by testosterone (Doeg, 1969). Whether or not this results from direct action of the hormone on the mitochondrial DNA (Chapter 18) or on the nucleic DNA is not known at present.

Ecdysone injected into insect *(Chironomus thummi)* larvae induces localized gene action (seen as "puffs") in the giant chromosomes of salivary glands (see Chapter 3). Juvenile hormone appears to have an opposite effect: it partly suppresses puffing in loci that are normally active in prepupal stages (Laufer and Greenwood, 1969). Karlson and Sekeris (1967) have postulated that ecdysone enters the cell nucleus

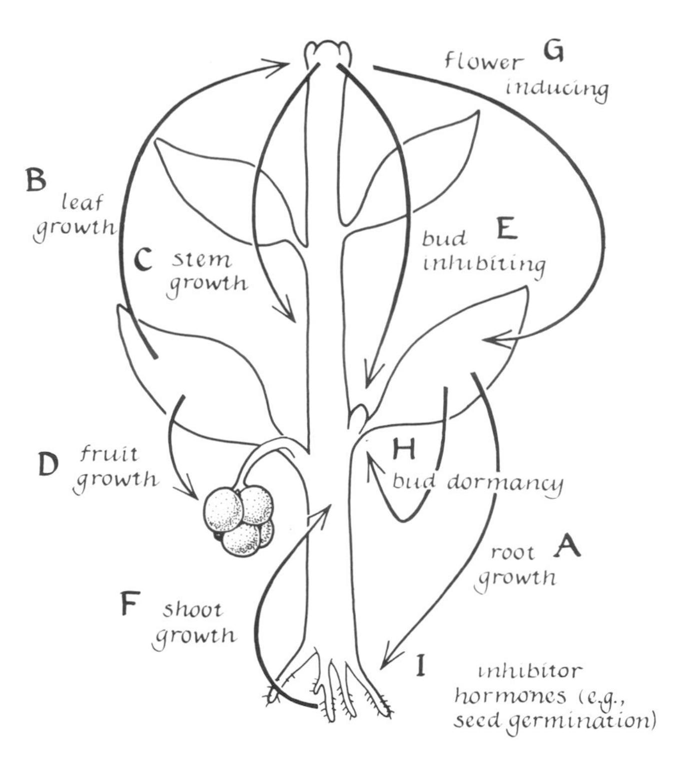

Figure 19-7. *Types of hormone-like control in plants.*

and acts directly on the chromosome. The result is an induction of RNA and protein synthesis. An alternative hypothesis suggested by Kroeger (1963) is that ecdysone causes the nuclear membrane to become more permeable, thereby increasing the influx of sodium ions and the release of potassium ions by the nucleus. This was presumed to produce a change in the relationship of histone protein and DNA, thereby making regions of DNA accessible for transcription (that is, puffing). How this could explain differential and sequential gene action remains a problem.

Possibly a general mechanism of action of steroid hormones might involve forming a hormone-receptor complex in the cell membrane. This complex would then pass into the cytoplasm, altering the permeability of the cell membrane. The hormone-receptor complex would interact with the DNA-histone complex of the chromosomes and allow specific loci to be transcribed.

A hormone may influence one type of tissue but not another, as already noted. There is increasing evidence that many hormones must interact with specific hormone receptors in the cell membranes in

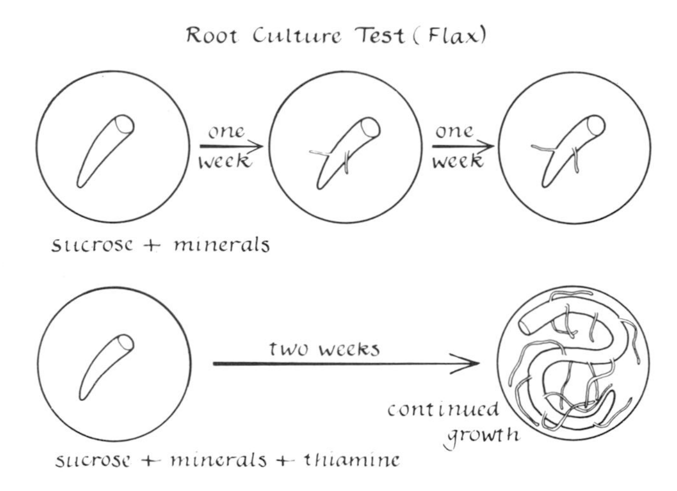

Figure 19-8. *Influence of thiamine on the growth of excised roots.*

order to exert their action. For example, there is evidence for an estrogen-binding macromolecule in uterine tissue (Toft et al., 1967). Binding sites for insulin and thyroid-stimulating hormone have also been described (Macchia and Pastan, 1967).

Finally, much recently accumulated evidence indicates that the nitrogen-containing polypeptide hormones (for example, thyroid-stimulating hormone, adrenocorticotropic hormone, ACTH) act by stimulating or suppressing the formation of cyclic adenosine monophosphate. This substance helps control many enzyme activities (see reviews of Robinson et al., 1968, and Rasmussen, 1970). Thus the polypeptide hormones do not act directly on the genome as steroid hormones such as ecdysone and estrogen apparently do.

Hormonal Mechanisms in Plants

The control of growth and differentiation by hormones, particularly in adult plants, has been studied extensively by botanists (see reviews of Skoog, 1954; Went and Thimann, 1937; Went, 1962; Leopold, 1964; Wetmore et al., 1964; Carr, 1966; Zeevaart, 1966; Galston, 1967; Wilkins, 1969; and others). Although only five types of phytohormones *(auxins, gibberellins, cytokinins, ethylene,* and *abscissic acid)* have been studied carefully and chemically identified, other hormones are probably involved in plant development. In Chapter 10, we noted the apparent role of hormone-like substances in the differentiation of antheridia and archegonia in ferns, and in Chapter 11, their role in the development of sex organs of flowers of some angiosperms. The influence of sex hormones has been demonstrated in some algae and fungi. We shall cite an example of the latter in Chapter 22.

Types of Hormone-Like Control

On the basis of hormone-like control, about eight humoral relationships have been described. These are diagramed in Figure 19-7.

Unlike animal hormones, each of which has specific target organs, plant hormones or growth substances have multiple functions which often overlap. As shown in Figure 19-7, developmental processes in various plant organs or structures are controlled by substances produced in other plant organs. Through mechanisms that control the concentrations and time of production of the hormones, the growth and differentiation of the plant is regulated in an orderly fashion. Some evidence for the role these hormones play is outlined below.

Root Growth Hormone

Went (1938) found that whole pea seedlings placed in a synthetic medium containing salts and sugars grow to normal size, but excised roots placed on the same culture medium do not grow. Some substances, which Went called calines, produced in the remainder of the seedling and translocated to the roots are necessary for their growth. Addition of *thiamine* (vitamin B_1) at a concentration of about 10^{-6} to the medium results in continued growth of the excised roots of some species (Figure 19-8).

Some species also require substances such as *pyridoxine* (Vitamin B_6) and *nicotinic acid* (niacin) for normal growth. Thiamine is produced in the leaves and is transported down to the root through the phloem. If the phloem is interrupted by girdling a leaf petiole, thiamine accumulates above the girdle. The B complex vitamins—thiamine, pyridoxine, and nicotinic acid—are coenzymes for carboxylases, transaminases, and dehydrogenases, respectively. In plants, they function like hormones.

Leaf Growth Hormone

A cytokinin (see below) produced in older leaves of a tobacco plant stimulates the growth of young leaves (Skoog and Tsui, 1948).

Stem Growth Hormones: Auxins

Sections of stems placed in culture do not grow unless *B-indoleacetic acid* (IAA) is in the medium at a concentration of approximately 10^{-4} molal. (B-indoleacetic acid was first isolated in human urine by Kögl et al., 1934, and was later found in corn seeds.) We shall see below that the shoot apex produces relatively large amounts of auxins, and that auxins are involved in the tropistic responses of plant shoots, such as the bending of shoots toward light sources. Auxins also have many other effects on plant development.

Fruit Growth Hormone: Ethylene

Plants placed in proximity to ripening fruits exhibit many growth abnormalities (Elmer, 1932). Apparently this is because fruit growth and ripening are both influenced by *ethylene* (C_2H_4), which is produced in regions of the plant where auxin content is relatively high (Burg, 1968; Mapson, 1970). When IAA itself is applied to the ovaries of some plants, it will stimulate fruit development (Luckwill, 1959), resulting in the development of *parthenocarpic* (seedless) fruits.

Like auxins, ethylene has manifold effects: stimulation of leaf abscission, inhibition of stem elongation, initiation of flowering in the pineapple, initiation of root hair formation and bending downward of leaves *(epinasty)*, to mention a few (see reviews of Burg, 1968; Pratt and Goeschl, 1969). Many plant tissues and organs, including the leaves, produce ethylene gas. Ethylene also appears to be involved in the differentiation and polarity of carrot embryos growing in tissue culture (Halperin, 1964). In general, ethylene helps induce senescence in fruits and flowers, causing the latter to fade and drop their petals.

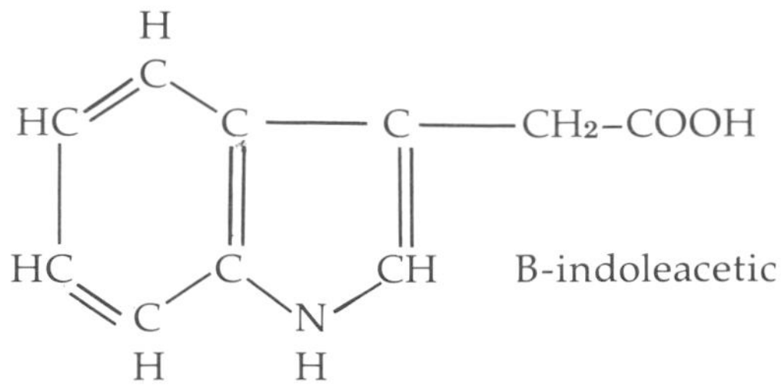

Lateral Bud Inhibition Hormones: Auxins

It has long been known that removing the shoot apex, a major source of auxins, releases lateral, dormant buds from inhibition. This is why pruning trees, shrubs, and other plants increases their branching. The production of auxins in relatively high concentrations by the shoot apex is the basis of *apical dominance* (Chapters 4 and 5).

H C HC C — C — CH_2–COOH HC C CH C N H H

B-indoleacetic acid (IAA)

Figure 19-9. *Structural formula of β-indole acetic acid.*

Shoot Growth Hormones

Went (1938) described a hormone-like relationship between roots and shoot growth. It has long been known that pruning the roots, as in the Bonsai culture procedure, reduces the growth of the remainder of the plant. This could be partly the result of nutritional deprivation. However, it could also be due to loss of cytokinins and gibberellins; these are transported from the roots to the shoots, where they apparently act as growth hormones.

Flower-Inducing Hormone

Grafting a leaf or branch of a flowering plant to a nonflowering plant can induce flower formation (Zeevaart, 1962).

Bud Dormancy and Senescence Hormone

This hormone, called *dormin* or *abscissic acid,* is produced in the leaves by the short day–long night photoperiod of autumn. It induces dormancy in the leaf and flower buds that formed during the summer. The same hormone causes the senescence changes involved in the falling (abscission) of leaves and flowers in the early autumn (Salisbury and Ross, 1969).

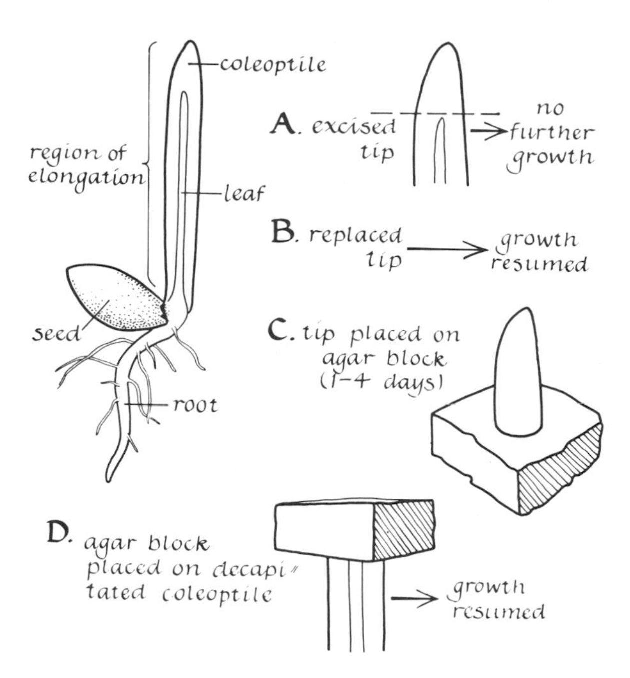

Figure 19-10. *Experiments A, B, C, and D demonstrate the production of auxins by the apex of the choleoptile of an oat seedling.*

Inhibitor Hormones

Among these hormones are the unsaturated *lactones* in seed coats and ripe fruits which inhibit seed germination, and a *flavonoid* which has been isolated from dormant flower buds of the peach tree. A flavonoid has been found to inhibit coleopile growth (Figure 19-10).

Role of Auxins

In addition to stimulating cell elongation, auxins influence tropisms, xylem and root initiation and differentiation, apical dominance, cell division, flower initiation in some plants, pollen tube growth, seed germination, fruit growth, callus formation on a cut stem surface, and abscission of leaves, flowers, and fruit. Almost every growth process in plants seems to be influenced by auxins. At the cell level, auxins increase plasticity (irreversible stretching) and elasticity (reversible stretching) of cell walls, cytoplasmic viscosity, respiration rates, and many other metabolic activities (Leopold, 1964). Space permits only a brief examination of some of these effects.

A wide range of compounds, including B-indoleacetic acid, can act as auxins. The structure of IAA is shown in Figure 19-9.

In general, auxins are synthesized in the shoot and root apices and in developing leaves—that is, in rapidly multiplying cell populations. However, no auxin has yet been crystallized from apical meristems (Salisbury and Ross, 1969).

Cell Elongation

The so-called primary effect of auxin is on the elongation of plant cells (Went, 1928, 1935). Went demonstrated this using the *coleoptile* of an oat *(Avena sativa)* seedling. The basic experiment is illustrated in Figure 19-10.

As indicated in the figure, the apex of the coleoptile, which is normally necessary for shoot growth, can be replaced by an agar block containing what turns out to be an auxin (probably IAA). The auxin produced by the apex diffuses into the agar block; thus, the auxin alone is necessary for growth of the coleoptile.

By modifying the procedure shown in Figure 19-10, Went and Thimann (1937) developed a way to assay the quantity of auxin present in the agar block. In this assay, 5×10^{-10} g of indoleacetic acid in the agar block produced a curvature of 10° when placed on one side of the decapitated coleoptile for 90 minutes. By using different known concentrations of IAA, a graph relating the degree of curvature to the concentration of IAA could be constructed. Then, by measuring the curvature produced by a block containing an unknown amount of IAA, the amount could be determined. This assay, known as the "Avena test," is diagramed in Figure 19-11.

By use of the Avena test the distribution of auxin in a seedling could be determined (Figure 19-12).

As the figure shows, auxin concentration was highest in the apex of the coleoptile and relatively high in the root apex. The coleoptile tip, however, does not contain dividing cells at this stage in the growth of the germinating seed.

Translocation of Auxin

Auxin produced in the plant's apex is transported down the stem at a rate of about 10 to 12 mm per hour

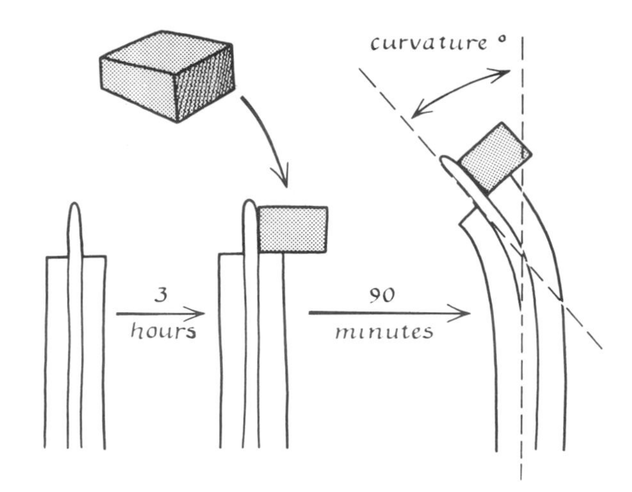

Figure 19-11. *The "Avena test" of Went and Thimann.*

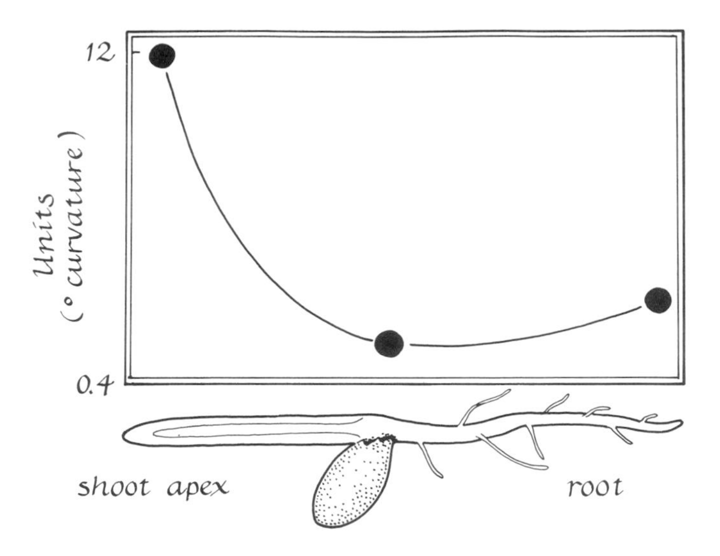

Figure 19-12. *Distribution of auxins in a seedling, determined by the "Avena test."*

and can be transported against a concentration gradient of auxin (Went, 1928, 1935; Leopold, 1963). The direction of translocation is polarized, as shown by the experiment originally devised by Went and illustrated in Figure 19-13.

Transport is primarily *basipetal,* that is, from more apical to more basal regions in the stem. Basipetal transport has also been demonstrated in intact pea seedlings by use of ^{14}C-labeled IAA applied to the shoot apex, followed by scintillation-counter detection of the label. The IAA is transported into the root system, where it accumulates in the lateral root primordia (Morris et al., 1969). The polarized transport is not due to simple diffusion but seems to occur because auxin is secreted only at one pole of the cell (Figure 19-13B). Transport in the stem thus appears to have a basis in the polarity of the individual cells (Leopold, 1963). The polar transplant in stems can be influenced by the nutritional status of the plant and by external environmental agents such as light (Wilkins, 1969).

Transport in roots is primarily from base to apex, that is, acropetal. This was demonstrated by a method similar to that shown in Figure 19-13, except that ^{14}C-IAA was present in one of the agar blocks (Wilkins, 1968).

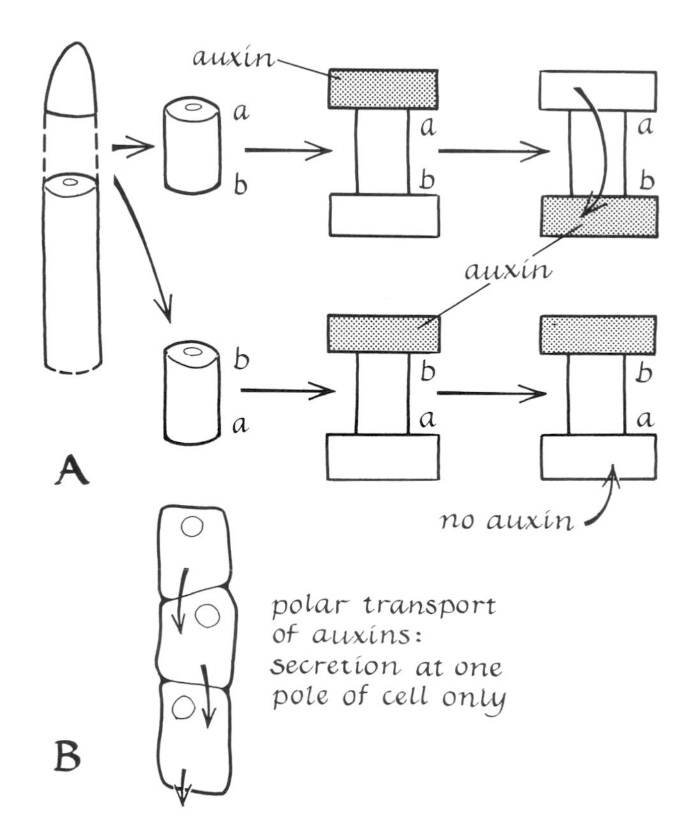

Figure 19-13. *A, experiment demonstrating the polar transport of auxins; B, secretion of auxins at one pole of the cell.*

Phototropism and Geotropism

Plant shoots bend toward a light source. Charles Darwin (1897) discovered that this tropism required

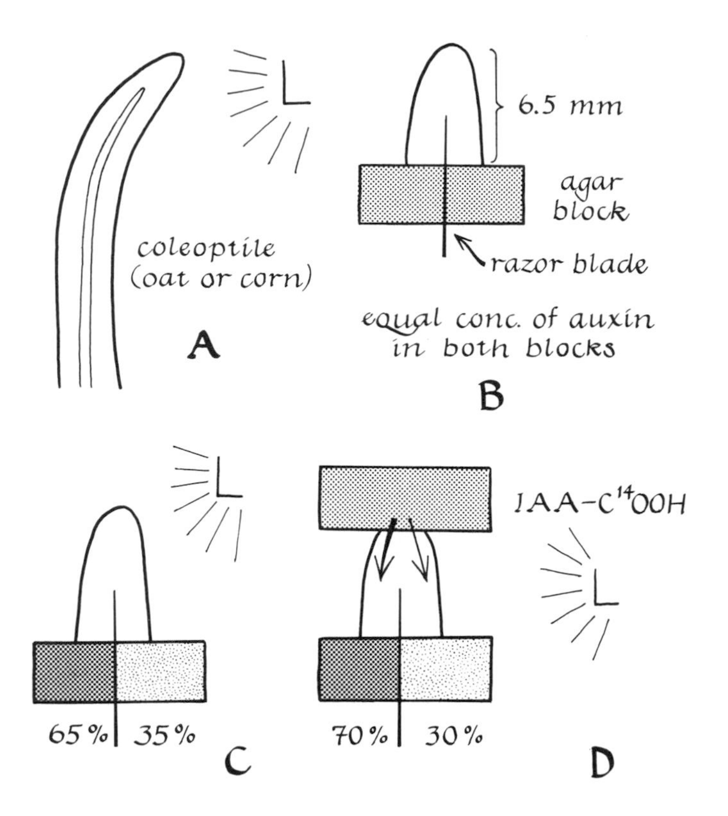

Figure 19-14. *A, coleoptile bending toward source of illumination; B, control for experiment illustrated in C. In C, the tip of the coleoptile was partially divided by insertion of a razor blade and placed on two agar blocks also separated by the razor blade. When the block under the nonilluminated part of the tip was assayed for auxin content, it was found to contain more hormone than the block under the illuminated part of the tip. D, similar result obtained when an agar block containing labeled indoleacetic acid was placed on the coleoptile tip unilaterally illuminated.*

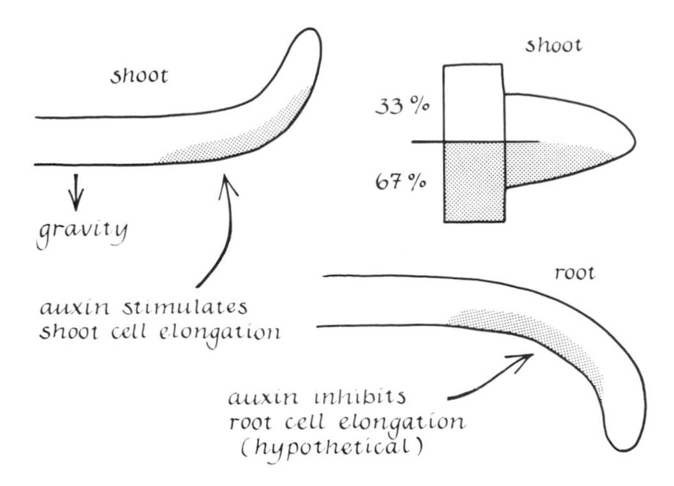

Figure 19-15. *Experiment illustrating the role of auxins in the geotropic response of a plant shoot (left) and hypothetically in the root.*

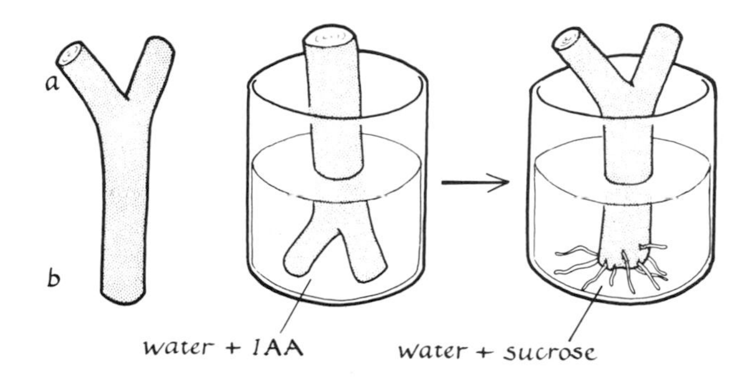

Figure 19-16. *Role of auxins in promoting the adventitious formation of roots on stem cuttings.*

the tips of the coleoptiles of grasses. Studies by other early botanists, including Paal of Hungary and Boysen-Jensen of Denmark (1910–1913), suggested that some substance moved from the tip to more basal regions where the bending occurs (Salisbury and Ross, 1969). It is now known that the bending occurs because more auxin is on the unlighted than on the lighted side of the shoot (Briggs et al., 1957; Pickard and Thimann, 1964). Pickard and Thimann's basic experiments demonstrate that auxin is not destroyed on the illuminated side but is laterally transported to the dark side (Figure 19-14). The higher concentration of auxin in the darker part of the coleoptile causes greater cell elongation on that side and the consequent bending of the shoot toward the light source.

The geotropic responses of plant shoots and perhaps roots is also controlled by auxin distribution. Again no auxin is destroyed (Figure 19-15).

Note in the figure that auxin is translocated to the lower side of a horizontally positioned root or shoot. However, the higher auxin concentration in the lower part of the shoot stimulates cell elongation and causes upward bending, whereas, hypothetically, the higher concentration in the lower part of the root would inhibit cell elongation and cause downward bending. Thus the same hormone might produce diverse responses in different parts of a plant. Different concentrations of the same hormone are known to produce diverse effects on growth of plant parts (Figure 19-17).

Root Initiation in Cuttings

Cuttings from shoots of many plants will, when placed in water, form adventitious roots at the basal end. Auxin (IAA) facilitates such root formation. This is shown in Figure 19-16.

In some plants the cutting must be inverted, since the IAA taken up will be transported to the basal end only in an apical-basal direction.

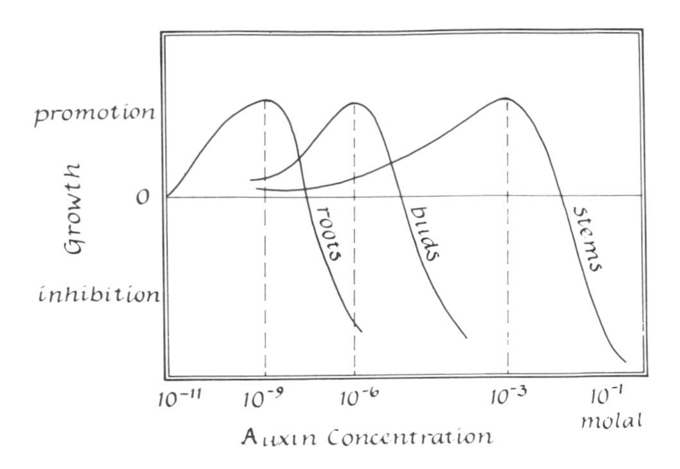

Figure 19-17. *Effects of different concentrations of auxins on root, bud, and stem growth.*

Inhibition of Lateral Buds: Apical Dominance

The synthesis of auxin by the growing apex is the basis of apical dominance (Audus, 1959; Wilkins, 1969). Auxin passing down the shoot inhibits the growth of lateral buds. When the source of inhibition is removed by removing the shoot apex, the lateral buds start to grow. In some plants after removal of the apex, auxin applied to the cut surface will prevent bud growth.

Apical dominance is not mediated solely by auxins; it is also influenced by the balance between auxin concentration and that of other hormones, such as the cytokinins and gibberellins. Apical dominance is further controlled by nutritive and environmental factors (Phillips, 1969).

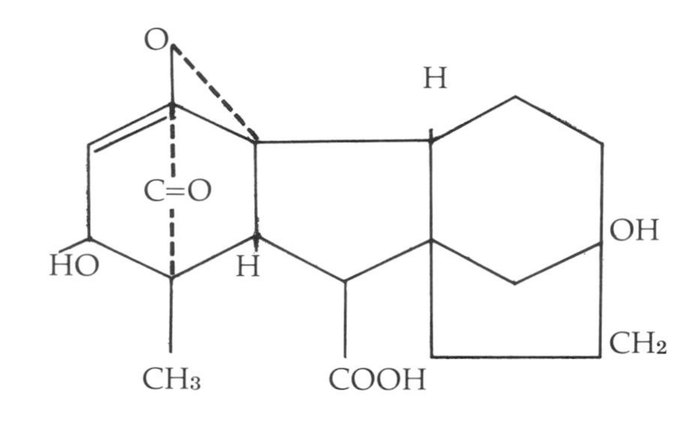

Figure 19-18. *Structural formula of gibberellic acid.*

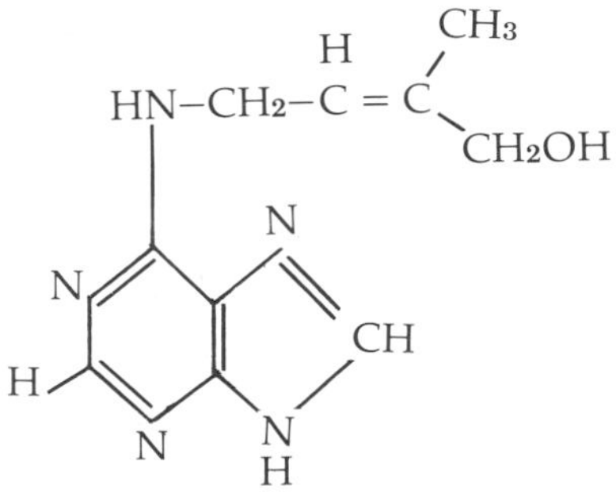

Cell Differentiation

In addition to their many other effects, auxins promote the differentiation of xylem cells in *Coleus* (Jacobs, 1961). Auxins applied to undifferentiated callus tissue growing *in vitro* stimulate complex patterns of cell differentiation (Chapter 22).

Quantitative Effects of Auxin

As noted above, different concentrations of a hormone can produce diverse responses in the same target organ. Also, the same concentration can influence diversely the responses of different target organs. This is true of auxin effects, as Figure 19-17 shows (Thimann, 1937). Note that a relatively high concentration (10^{-3} molal) of auxin stimulates stem

Figure 19-19. *Structural formula of a naturally occurring cytokinin, zeatin.*

growth but inhibits the growth of buds and roots. Conversely, only a relatively low concentration (10^{-9} molal) promotes the growth of roots. Also, an intermediate concentration (10^{-6} molal) stimulates the growth of lateral buds but not that of roots.

Role of Gibberellins

Hormones called gibberellins are natural growth-regulators in plants. Gibberellins were first extracted from the fungus *Gibberella fujikuroi* in the 1930s by the Japanese botanists T. Yabuta and T. Hayashi. They were later found in higher plants (see review of Phinney and West, 1960). About twenty-seven different gibberellins have now been found in fungi and higher plants. The gibberellins act in small quantities, as low as 10^{-6} molal. The structural formula of gibberellic acid, the prototype of the gibberellins, is shown in Figure 19-18.

The gibberellins, like auxins, usually stimulate cell elongation excessively, causing plants to become tall and spindly. In some plants they stimulate cell division accompanied by increase in cell length. They also participate with auxins in inhibiting lateral bud growth (Wilkins, 1969). They inhibit the initiation but not the growth of roots. Unlike auxins, endogenous gibberellins appear to be transported in a nonpolar way, moving both acropetally and basipetally from mature leaves where they are synthesized (Wilkins, 1969). In general, the gibberellins appear to help determine the form of the plant—whether, for instance, it becomes a bush or a vine. They affect stem elongation most strongly, but sometimes they also stimulate leaf, flower, and fruit growth. The gibberellins appear to be involved in regulatory mechanisms responsible for the developmental responses of the plant to environmental influences (Michniewicz and Lang, 1962; review of Leopold, 1964; and Salisbury and Ross, 1969). For example, they participate in the breaking of dormancy in buds and seeds, acting in place of cold treatment or light. They can substitute for long-day photoperiod in the induction of flowering (Chapter 22).

Antigibberellins are known—compounds like 2,4-dichlorobenzyl-tributyl-phosphonium chloride, called *phosphon D.* The antigibberellins interfere with enzymes involved in gibberellin synthesis.

Role of Cytokinins

A class of N^6-substituted adenine derivatives, the cytokinins, have a profound influence on plant growth (Leopold, 1964; Helgeson, 1968; Letham, 1969). Cytokinins were first extracted from autoclaved DNA of herring sperm by Miller in the laboratory of Folke Skoog at the University of Wisconsin (see review of Miller, 1961). Cytokinins have since been synthesized; synthetic cytokinins have effects similar to those of naturally occurring cytokinins. The structural formula of a naturally occurring cytokinin, *zeatin,* is shown in Figure 19-19.

Cytokinins primarily effect cell division in the plant. Their effect on tobacco pith cells is shown in Figure 19-20A.

The cytokinins can alter many growth processes—such as leaf growth, apical dominance, and light responses—independently of the stimulation of cell division. The growth rate of tobacco callus tissue on a medium containing 6-benzylaminopurine, a synthetic cytokinin, varies directly with the concentration of the hormone. The morphology (growth habit) of the cultures is also influenced by cytokinin concentration (Helgeson, 1968).

As Wickson and Thimann (1958) found in studying the pea *Pisum sativum,* the cytokinins interact with auxins in controlling apical dominance. As previously noted, an auxin applied to the cut surfaces of a shoot from which the apex has been removed inhibits growth of lateral buds. Application of kinetin or other cytokinins overcomes the inhibition due to auxin. Kinetin can also activate dormant winter buds in some plants. Further, lettuce seeds, when pretreated with kinin solutions, germinate at levels of light intensity which alone would not induce germination.

Unlike auxins and gibberellins, the cytokinins apparently are not transported up or down through the plant body, except from the roots where they are formed to the shoots via the xylem water-conduction cells. However, by use of ^{14}C-kinetin and ^{14}C-gibberellic acid, lateral transport and interchange of these substances between the xylem and phloem of willow *(Salix viminalis)* stems has been demonstrated (Bowen and Wareing, 1969).

Perhaps the most significant role of cytokinins in plant development is their regulation of hormone action. This concept is supported by the classical ex-

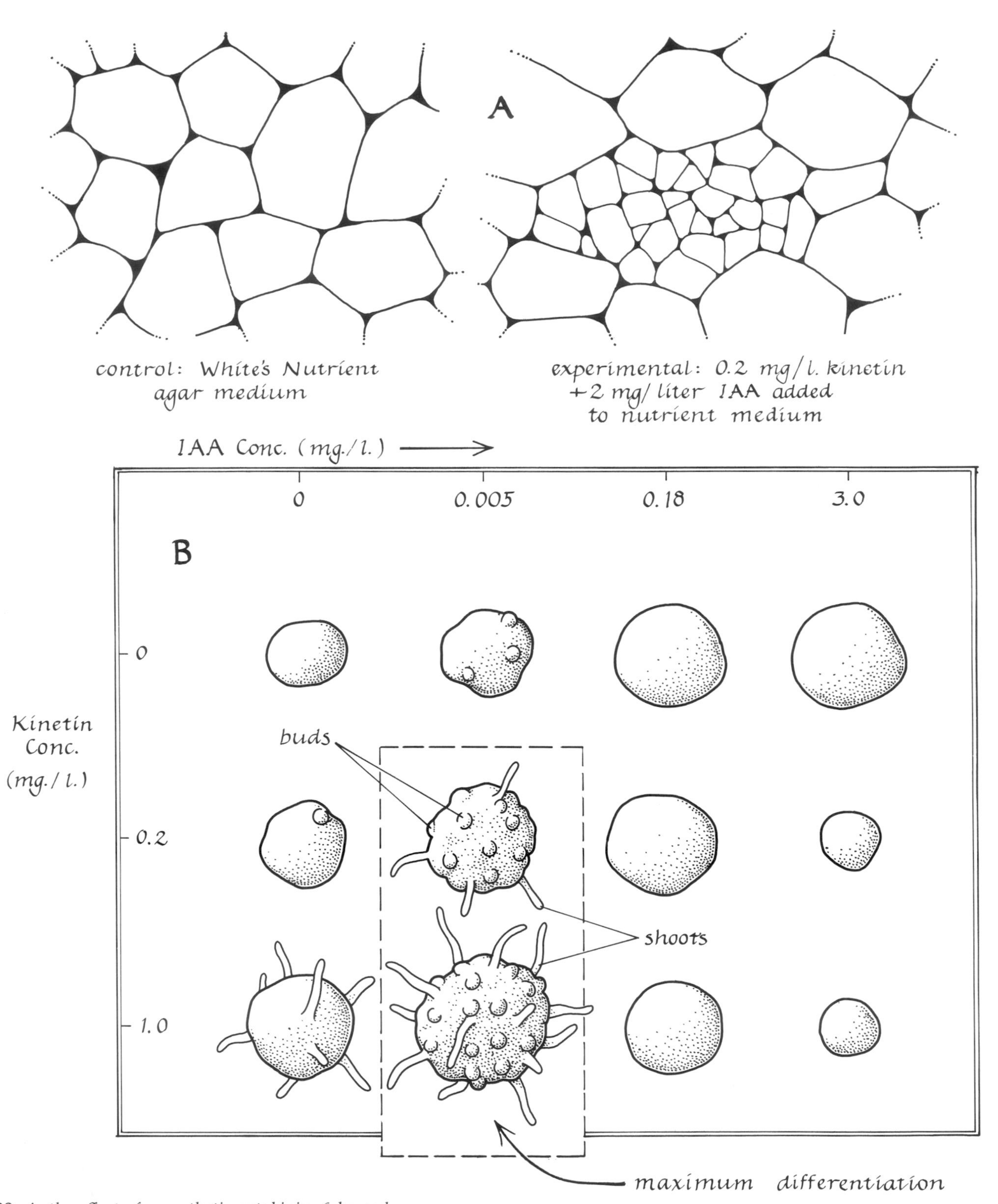

Figure 19-20. *A, the effect of a synthetic cytokinin, 6-benzylaminopurine, on cell division of tobacco pith cells in vitro; B, diagrams illustrating the effects of different concentrations of auxin and kinetin in the culture medium on formation of adventitious buds and shoots by tobacco callus tissue.*

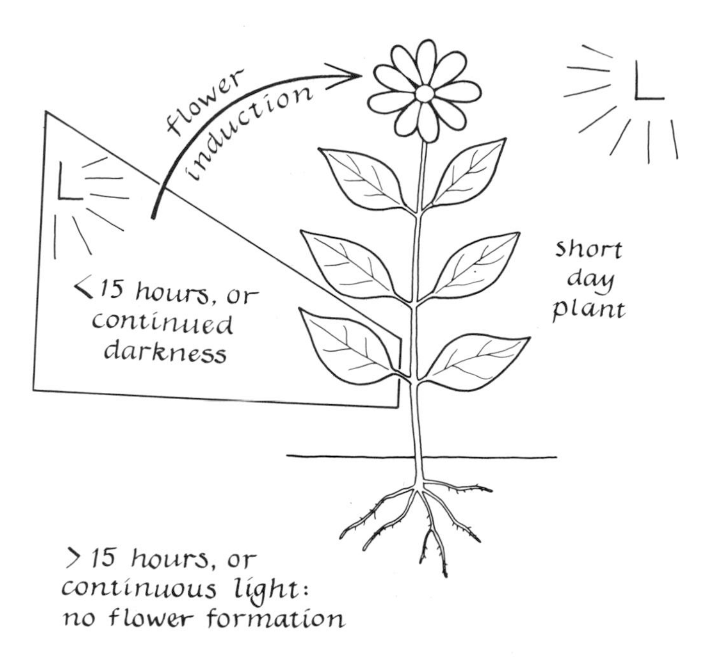

Figure 19-21. *One leaf of the cocklebur, a short day plant, in continuous darkness or less than 15 hours of light per day is sufficient to induce flowering even though the remainder of the plant is exposed to more than 15 hours of light per day or to continuous light.*

periments of Skoog and Miller (1957) on the hormonal control of morphogenesis *in vitro*. These workers demonstrated that the balance between auxins and cytokinins determines the developmental response of tobacco stem callus tissue growing *in vitro*. At low cytokinin-to-IAA ratios, there was essentially no differentiation in the callus (adventitious buds or shoots formed minimally or not at all), whereas increasing the ratio of cytokinin to auxin (1.0 mg kinetin/0.005 mg IAA) enhanced growth and differentiation (Figure 19-20B).

It seems clear that the balance between hormones and metabolites, as well as their distribution within the plant, controls plant development. For example, the concentration of glucose influences the morphology of the water fern *Marsilea* (Chapter 22, Figure 22-13); sucrose concentration, the formation of rhizomes in grass; B-alanine concentration, activation of axillary buds in pea plants; organic nitrogen, the induction of flowering in *Utricularia* (bladderwort).

Flower-Inducing Hormone

There is evidence that a hormone-like substance of unknown composition induces the shoot apex to stop forming leaves and transform into a floral apex (Bonner and Liverman, 1953; Zeevaart, 1962; Salisbury, 1963; see also Chapters 10, 11, and 22). Flower induction is a response of some plants to *photoperiod*, the relative length of light and dark periods in a twenty-four-hour day. Some plants (for example, tulips) flower in the spring or early summer when the days are relatively long; others (such as chrysanthemums) flower in the late summer or fall when the days are relatively short (less than fifteen hours long). The former are known as long-day plants, the latter as short-day plants. If a short-day plant is kept under artificial long-day conditions (more than fifteen hours in some plants), no flowers develop. However, if a leaf of such a plant (for instance, cocklebur) is covered with an opaque, black bag so that it is in continuous darkness or receives less than fifteen hours of illumination each day, the plant will flower. One leaf synthesizes enough flower-inducing hormone during a single long night to influence the shoot apex (Figure 19-21).

Grafting a nonflowering branch onto a flowering plant can cause the grafted branch to flower. One or more substances thus pass from the host plant into the graft and induce flowering (Zeevaart, 1962).

The response of plants to photoperiod is regulated by a bluish billiprotein pigment, phytochrome, but its mechanism of action is not known (Chapter 22). This pigment is not the flower-inducing hormone itself, but probably acts in a manner similar to that of other metabolites in regulating hormone action (see Galston, 1967). Gibberellins stimulate flowering in long-day plants even when they are grown under noninductive conditions of photoperiod. However, this effect might be indirect (Leopold, 1964). Recall (Chapter 11) that auxins also stimulate the development of female structures even in genetically male flowers (see Torrey, 1967).

Mechanism of Plant Hormone Action

Very little is known about how hormones work at the molecular level. Probably they work in several ways. There is some evidence that phytohormones might act at or very near the level of the gene as repressors or derepressors (Zeevaart, 1966). Auxins, gibberellins, and cytokinins reportedly stimulate DNA-dependent RNA synthesis in isolated plant cell nuclei (see Letham, 1969). Varner and Chandra (1964) have reported that seeds first respond to gibberellins by synthesizing *de novo* the enzyme α-amylase, which digests and releases endosperm products during seed germination. Although the multiple effects of plant hormones make the problem of solving their mechanism of action difficult, the fact that small concentrations are greatly amplified suggests that they are involved in enzymatic reactions and possibly in the formation of RNAs involved in enzyme synthesis. Some effects of auxins are so rapid that they do not appear to depend upon prior RNA synthesis. In studies of the primary effect of IAA upon cell elongation in oat coleoptiles, it was found that the lag period before a steady-state rate of elongation was reached could be shortened to zero by increasing the IAA concentration. Since induction of protein synthesis via transcription and translation is a time-consuming process preceded by a lag of about two hours in higher plants, these results indicate that auxins can exert their effects directly and without induction of protein synthesis via action on the genome (Nissl and Zenk, 1969).

Small amounts of IAA (10^{-10} molal) added to mitochondrial preparations from inbred lines of corn can almost double the rate of oxygen uptake by these preparations (Sarkissian and McDaniel, 1966). This indicates that auxins can bypass the nuclear genome in exerting some of their effects. It thus appears that phytohormones, especially auxins, do not necessarily act directly or initially at the level of the gene, although some studies show that ^{14}C-labeled IAA enters and accumulates inside cell nuclei. Recall that some animal hormones also accumulate inside nuclei.

Conclusion

Both animal and plant hormones play a fundamental and indispensable role in controlling the functional and developmental activities of animals and plants. By acting correlatively, hormones are an essential integrating mechanism in both plant and animal development. Plants, which have no coordinating system analogous to the nervous system of animals, appear to be almost entirely dependent on hormone-based mechanisms for coordination of growth and development. In general, individual plant and animal hormones can have manifold effects on development, and different concentrations of the same hormone can have diverse, sometimes opposite effects. As products of gene action, hormones are a way of insuring the orderly expression of the genome during development. Thus, the synthesis and release of hormones under control of the genes at one stage will condition the pattern of subsequent gene action.

References

Allen, B. M. 1918. The results of thyroid removal in larvae of *Rana pipiens*. J. Exp. Zool. **24:**499–519.

Audus, I. J. 1959. Correlations. J. Linn. Soc. Lond. Bot. **56:**177–187.

Banting, F. G., and C. H. Best. 1922. The internal secretion of the pancreas. J. Lab. Clin. Med. **7:**251–266.

Beatty, R. A. 1970. Genetic basis for the determination of sex. Phil. Trans. Roy. Soc. Lond. Ser. B. **259:**3–14.

Berthold, A. A. 1849. Transplantation der Hoden. Arch. f. Anat. Physiol. u. Wissensch. Med., pp. 42–46.

Bonner, J., and J. Liverman. 1953. Hormonal control of flower initiation. In W. E. Loomis, ed., Growth and Differentiation in Plants, pp. 283–303. Iowa State College Press, Ames.

Bowen, M. R., and P. F. Wareing. 1969. The interchange of ^{14}C-kinetin and ^{14}C-gibberellic acid between bark and xylem of willow. Planta **89:**108–125.

Brien, P. 1964. Contribution à l'étude de la biologie sexuelle chez les hydres d'eau douce: induction gamétique et sexuelle par méthode des greffes an parabiose. Bull. Biol. France Belg. **97:**214–283.

Briggs, W. R., R. Tocher, and J. Wilson. 1957. Phototropic auxin redistribution in corn coleoptiles. Science **126:**210–212.

Burg, S. F. 1968. Ethylene, plant senescence and abscission. Plant Physiol. **43:**1503–1511.

Burnett, A., and N. Diehl. 1964. The nervous system of Hydra: III. The initiation of sexuality with special reference to the nervous system. J. Exp. Zool. **157:**237–250.

Butenandt, A., and P. Karlson. 1954. Über die Isolierung eines Metamorphose-Hormons der Insekten in Kristallisierter Form. Z. Naturforsch. 9b. **6:**389–391.

Carlisle, D. B., and F. Knowles. 1959. Endocrine Control in Crustaceans. Cambridge University Press, London.

Carr, D. J. 1966. Metabolic and hormonal regulation of growth and development. In E. Cutter, A. Allsopp, F. Cusick, and I. M. Sussex, eds., Trends in Plant Morphogenesis, pp. 253–283. Longman, London.

Charniaux-Cotton, H. 1965. Hormonal control of sex differentiation in invertebrates. In R. DeHaan and H. Ursprung, eds., Organogenesis, pp. 701–740. Holt, Rinehart and Winston, New York.

Cohen, P. P. 1970. Biochemical differentiation during amphibian metamorphosis. Science **168:**533–543.

Darwin, C. 1897. The Power of Movement in Plants. D. Appleton, New York.

Davidson, E. H. 1965. Hormones and Genes. Sci. Amer. **212:**36–45.

Démsusy, N. 1962. Role de la glande de mue dans l'évolution ovarienne du crabe *Carcinus maenas.* Cahiers Biol. Marine **3:**37–56.

Doeg, K. A. 1969. Control of mitochondrial lipid biosynthesis by testosterone in male sex accessory gland tissue of normal rats. Endocrinol. **85:**974–976.

Elmer, O. H. 1932. Growth inhibition of potato sprouts by the volatile products of apples. Science **75:**193.

Etkin, W. 1955. Metamorphosis. In B. H. Willier, P. Weiss, and V. Hamburger, eds., Analysis of Development, pp. 631–663. W. B. Saunders, Philadelphia.

Etkin, W. 1966. How a tadpole becomes a frog. Sci. Amer. **214:**76–88.

Etkin, W., and L. I. Gilbert, eds. 1968. Metamorphosis: A Problem in Developmental Biology. Appleton-Century-Crofts, New York.

Galston, A. W. 1967. Regulatory systems in higher plants. Am. Scientist **55:**144–160.

Gilbert, L. I. 1964. Hormones regulating insect growth. In G. Pincus, ed., The Hormones, Vol. 4, pp. 67–134. Academic Press, New York.

Gilbert, L. I., and H. A. Schneiderman. 1961. Some biochemical aspects of insect metamorphosis. Am. Zool. **1:**11–51.

Gudernatsch, F. 1912. Feeding experiments on tadpoles. Arch. Entw. Mech. **35:**457–483.

Haffen, K. 1970. Biosynthesis of steroid hormones by the embryonic gonads of vertebrates. Adv. Morph. **8:**285–306.

Halperin, W. 1964. Doctoral thesis, University of Connecticut, Storrs.

Hardy, M. H. 1953. Vaginal cornification of the mouse produced by estrogens *in vitro.* Nature **172:**1196–1197.

Harris, G. W., and R. G. Edwards. 1970. A Discussion on the Determination of Sex. Phil. Trans. Roy. Soc. Lond. Ser. B. **259:**1–206.

Helgeson, J. P. 1968. The cytokinins. Science **161:**974–981.

Highnam, K., and L. Hill. 1969. The Comparative Endocrinology of the Invertebrates. Elsevier, New York.

Humphrey, R. R. 1931. Studies on sex reversal in Amblystoma: III. Transformation of the ovary of *A. tigrinum* into a functional testis through the influence of a testis resident in the same animal. J. Exp. Zool. **58:**333–365.

Humphrey, R. R. 1945. Sex determination in ambystomid salamanders: a study of the progeny of females experimentally converted into males. Am. J. Anat. **76:**33–66.

Jacobs, W. P. 1961. Auxin as a limiting factor in the differentiation of plant tissue. In Recent Advances in Botany, pp. 786–790. University of Toronto Press, Toronto.

Jost, A. 1965. Gonadal hormones in the sex differentiation of the mammalian fetus. In R. DeHaan and H. Ursprung, eds., Organogenesis, pp. 611–628. Holt, Rinehart and Winston, New York.

Karlson, P. 1965. Mechanisms of Hormone Actions. Academic Press, New York.

Karlson, P., and C. E. Sekeris. 1966. Ecdysone, an insect steroid hormone and its mode of action. Recent Prog. Hormone Res. **22**:473–502.

Kögl, F., A. J. Haagen Smit, and H. Erxleben. 1934. Über ein neues Auxin (Heteroauxin) aus Harn. XI Mitt. Zeit. Physiol. Chem. **228**:104–112.

Kroeger, H. 1963. Chemical nature of the system controlling gene activities in insect cells. Nature **200**:1234–1235.

Laufer, H., and H. Greenwood. 1969. The effects of juvenile hormone on larvae of the dipteran, *Chironomus thummi*. Am. Zool. **9**:603 (abstr.).

Leopold, A. C. 1963. The polarity of auxin transport. In Meristems and Differentiation, pp. 218–234. Brookhaven Symp. Biol., Upton, New York.

Leopold, A. C. 1964. Plant Growth and Development. McGraw-Hill Book Company, New York.

Letham, D. S. 1969. Cytokinins and their relation to other phytohormones. Bioscience **19**:309–316.

Lillie, F. R. 1917. The free-martin: a study of the action of sex hormones in the foetal life of cattle. J. Exp. Zool. **23**:371–452.

Luckwill, L. C. 1959. Fruit growth in relation to internal and external chemical stimuli. In D. Rudnick, ed., Cell, Organism and Milieu, pp. 223–251. Ronald Press, New York.

Macchia, V., and I. J. Pastan. 1967. Action of phospholipase C on the thyroid. J. Biol. Chem. **242**:1864–1869.

Macleod, J. J. R. 1922. The source of insulin: a study of the effect produced on blood sugar by extracts of the pancreas and principle islets of fishes. J. Metab. Res. **2**:149–172.

Mapson, L. W. 1970. Biosynthesis of ethylene and the ripening of fruit. Endeavour **29**:29–33.

Michniewicz, M., and A. Lang. 1962. Effect of nine different gibberellins on stem elongation and flower formation in cold-requiring and photo-periodic plants grown under non-inductive conditions. Planta **58**:549–563.

Mikamo, K., and E. Witschi. 1963. Functional sex-reversal in genetic females of *Xenopus laevis*, induced by implanted testes. Genetics **48**:1411–1421.

Miller, C. O. 1961. Kinetin and related compounds in plant growth. Ann. Rev. Plant Physiol. **12**:395–408.

Morris, D. A., R. E. Briant, and P. G. Thomson. 1969. The transport and metabolism of ^{14}C-labeled indoleacetic acid in intact pea seedlings. Planta **89**:178–197.

Naisse, J. 1963. Détermination sexuelle chez *Lampyris noctiluca* L. (Insecte Coléoptère Malacoderme). Compt. Rend. Acad. Sci. **256**:799–800.

Nissl, D., and M. H. Zenk. 1969. Evidence against induction of protein synthesis during auxin-induced initial elongation of *Avena* coleoptiles. Planta **89**:328–341.

Oliver, G., and E. A. Schäfer. 1895. The physiological effects of extracts of the suprarenal capsules. J. Physiol. **18**:230–276.

Passano, L. M. 1961. The regulation of crustacean metamorphosis. Am. Zool. **1**:89–95.

Phillips, I. D. J. 1969. Apical dominance. In M. B. Wilkins, ed., The Physiology of Plant Growth and Development, pp. 163–202. McGraw-Hill Book Company, New York.

Phinney, B. O., and C. A. West. 1960. Gibberellins as native plant growth regulators. Ann. Rev. Plant Physiol. **11**:411–436.

Pickard, B., and K. Thimann. 1964. Transport and distribution of auxin during tropistic responses: II. The lateral migration of auxin in phototropism of coleoptiles. Plant Physiol. **39:**341–349.

Pratt, H. K., and J. D. Geoschl. 1969. The physiological roles of ethylene in plants. Ann. Rev. Plant Physiol. **20:**541–584.

Rasmussen, H. 1970. Cell communication, calcium ion, and cyclic adenosine monophosphate. Science **170:**404–412.

Robinson, G. A., R. W. Butcher, and E. W. Sutherland. 1968. Cyclic AMP. Ann. Rev. Biochem. **37:**149–174.

Sachs, J. 1880. Stoff und Form der Pflanzenorgane. Arb. Bot. Inst. Würzburg **3:**452–488.

Salisbury, F. B. 1963. The Flowering Process. Macmillan, New York.

Salisbury, F. B., and C. Ross. 1969. Plant Physiology. Wadsworth Publishing Company, Inc., Belmont, Calif.

Sarkissian, I. V., and R. G. McDaniel. 1966. Regulation of mitochondrial activity by indoleacetic acid. Biochem. Biophys. Acta. **128:**413–418.

Schneiderman, H. A., and A. Krishnakumaran. 1968. The control of molting in arthropods by steroid hormones. Biol. Bull. **135:**435–436.

Short, R. V. 1970. The bovine freemartin: a new look at an old problem. Phil. Trans. Roy. Soc. Lond. Ser. B. **259:**141–147.

Skoog, F. 1954. Chemical regulation of growth in plants. In E. J. Boell, ed., Dynamics of Growth Processes, pp. 148–182. Princeton University Press, Princeton, N.J.

Skoog, F., and C. O. Miller. 1957. Chemical regulation of growth and organ formation in plant tissues cultured *in vitro*. Symp. Soc. Exp. Biol. **11:**118–131.

Skoog, F., and C. Tsui. 1948. Chemical control of growth and bud formation in tobacco stem segments and callus cultured *in vitro*. Am. J. Bot. **35:**782–787.

Tata, J. R. 1966. *In vitro* synthesis of nuclear protein during growth of the liver induced by hormones. Nature **212:**1312–1314.

Thimann, K. V. 1937. On the nature of inhibitions caused by auxin. Am. J. Bot. **24:**407–412.

Toft, D., G. Shyamala, and J. Gorski. 1967. A receptor molecule for estrogens: studies using a cell-free system. Proc. Nat. Acad. Sci. U.S. **57:**1740–1743.

Tombes, A. S. 1970. An Introduction to Invertebrate Endocrinology. Academic Press, New York.

Torrey, J. G. 1967. Development in Flowering Plants. Macmillan, New York.

Turner, C. D. 1966. General Endocrinology, 4th ed. W. B. Saunders, Philadelphia.

Varner, J. E., and G. R. Chandra. 1964. Hormonal control of enzyme synthesis in barley endosperm. Proc. Nat. Acad. Sci. U.S. **52:**100–106.

Villee, C. A. 1962. The role of steroid hormones in the control of metabolic activity. In J. M. Allen, ed., The Molecular Control of Cellular Activity, pp. 297–318. McGraw-Hill Book Company, New York.

Villee, C. A. 1967. Hormonal expression through genetic mechanisms. Am. Zool. **7:**109–113.

Weber, R. 1967. Biochemistry of amphibian metamorphosis. In R. Weber, ed., Biochemistry of Animal Development, pp. 227–301. Academic Press, New York.

Wells, L. J. 1965. Fetal hormones and their role in organogenesis. In R. DeHaan and H. Ursprung, eds., Organogenesis, pp. 673–680. Holt, Rinehart and Winston, New York.

Went, F. W. 1928. Wuchsstoff und Wachstum. Rec. Trar. Bot. Néerl. **25:**1–116.

Went, F. W. 1935. Auxin, the plant-growing hormone. Bot. Rev. **1:**162–182.

Went, F. W. 1938. Specific factors other than auxin affecting growth and root formation. Plant Physiol. **13:**55–80.

Went, F. W. 1962. Plant growth and plant hormones. In W. H. Johnson and W. C. Steere, eds., This is Life, pp. 213–254. Holt, Rinehart and Winston, New York.

Went, F. W., and K. V. Thimann. 1937. Phytohormones. Macmillan, New York.

Wetmore, R. H., A. E. DeMaggio, and J. P. Rier. 1964. Contemporary outlook on the differentiation of vascular tissues. Phytomorphol. **14:**203–217.

Wickson, M., and K. V. Thimann. 1958. The antagonism of auxin and kinetin in apical dominance. Plant Physiol. **11:**62–74.

Wigglesworth, V. B. 1966. Hormonal regulation of differentiation in insects. In Cell Differentiation and Morphogenesis, pp. 180–209. North Holland Publishing Company, Amsterdam.

Wilkins, L. 1960. The Thyroid Gland. Sci. Amer. **202:**119–129.

Wilkins, M. B. 1968. Auxin transport in roots. Nature **219:**1388–1389.

Wilkins, M. B., ed. 1969. The Physiology of Plant Growth and Development. McGraw-Hill Book Company, New York.

Williams, C. M. 1963. Differentiation and morphogenesis in insects. In J. M. Allen, ed., The Nature of Biological Diversity, pp. 243–260. McGraw-Hill Book Company, New York.

Williams, C. M., and J. H. Law. 1965. The juvenile hormone: IV. Its extraction, assay and purification. J. Insect Physiol. **11:**569–580.

Willier, B. H. 1952. Development of sex-hormone activity of the avian gonad. Ann. N.Y. Acad. Sci. **55:**159–171.

Willier, B. H. 1955. Ontogeny of endocrine correlation. In B. H. Willier, P. Weiss, and V. Hamburger, eds., Analysis of Development, pp. 574–619. W. B. Saunders, Philadelphia.

Witschi, E. 1967. Biochemistry of sex differentiation in vertebrate embryos. In R. Weber, ed., The Biochemistry of Animal Development, Vol. 2, pp. 193–225. Academic Press, New York.

Wolff, E. 1947. Recherches sur l'intersexualité expérimentale produite par la methode des greffes de gonades à l'embryon de poulet. Arch. Anat. Micr. Morph. Exp. **36:**69–90.

Zeevaart, J. A. D. 1962. Physiology of flowering. Science **137:**723–731.

Zeevaart, J. A. D. 1966. Hormonal regulation of plant development. In Cell Differentiation and Morphogenesis, pp. 144–179. North Holland Publishing Company, Amsterdam.

20 *Nucleo-Cytoplasmic Interactions*

The influence of the nucleus on cytoplasmic processes and that of the cytoplasm on nuclear processes constitute basic mechanisms for controlling cellular activities in development (Harris, 1968). This was recognized long ago by Dreisch (1894) and Morgan (1934), who believed that the distribution of equipotential nuclei in chemically heterogeneous cytoplasm leads to differential gene activity. More recently radioactive labeling has shown that proteins and nucleic acids are exchanged between nucleus and cytoplasm in amebas (Goldstein, 1958, 1963). Using ^{3}H-labeling techniques, Goldstein and Prescott (1968) demonstrated that proteins continually shuttle between the nucleus and cytoplasm of *Amoeba proteus* (see also Goldstein, 1967, and Chapter 17). Proteins also transfer from the cytoplasm of cleaving eggs of *Xenopus* to injected nuclei (Arms, 1968). We have defined development as the progressive acting out of an inherited program (Stern, 1954, 1955). This program resides primarily in the nucleus or nuclei of the reproductive units and later in the nuclei of cells derived by divisions of the reproductive unit. As noted in Chapter 13, there is much evidence, especially from the study of eggs, for the presence of a program in the cytoplasm as well as in the nucleus. The cytoplasmic program, in the form of a graded or localized cytodifferentiation, and including perhaps stable mRNA molecules and ribosomes built up under control of the oocyte and follicle cell nuclear genes, constitutes the environment for further expression of the nuclear program.

In this chapter we shall first consider some evidence for the importance of the nucleus and its products and then survey some experiments demonstrating the influence of the nucleus on cytoplasmic properties. Next we shall consider experiments demonstrating the influence of the cytoplasm on nuclear activity. Finally, we shall cite evidence that cleavage-stage nuclei are developmentally equivalent, and that nuclei from progressively older amphibian embryos are increasingly less able to substitute for the egg nucleus after it has been removed. Examples of nuclear transplantation in other organisms will also be briefly considered.

Importance of the Nucleus and/or its Products

Effects of Removing the Nucleus

Ameba

In 1952, Brachet showed that the nonnucleated fragment of an ameba behaves differently from the nucleated fragment. This is illustrated in Figure 20-1. His simple experiment, which any student can perform with a fine glass needle, shows that the nucleus of this cell is necessary for indefinitely continued metabolic activities and for subsequent cell division. However, protein synthesis, measured by incorporation of labeled amino acids, continues for many hours in the enucleated half (Mazia and Prescott, 1955).

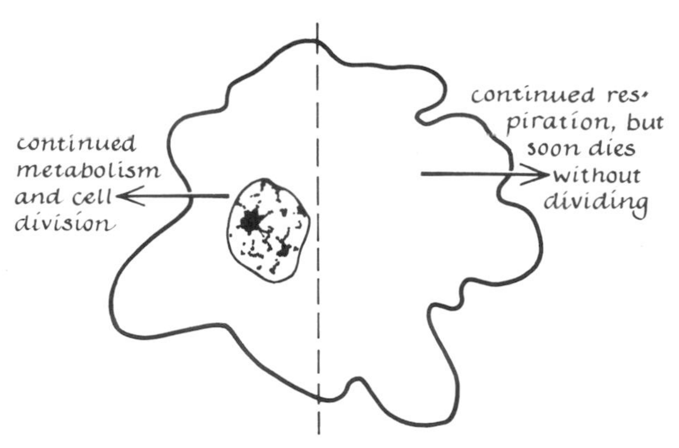

Figure 20-1. *Importance of the nucleus for indefinitely continued metabolic activities.*

Spirogyra

Nonnucleated protoplasmic masses of this green alga produced by treatment with 15 percent sucrose solution may live for up to six weeks. Such pieces did not synthesize cell-wall material unless connected by a protoplasmic strand with a nucleated piece (Morgan, 1901). In more recent studies reviewed by Harris (1968), enucleated cells of *Spirogyra* continued to grow, form new cytoplasm, and synthesize protein for up to about two months. Although the nucleus is essential for continued synthesis of cell components and for cell survival, the cell can live for a relatively long period without its nucleus, presumably because of long-lived mRNA molecules in its cytoplasm. This is even more strikingly demonstrated by the experiments on *Acetabularia*.

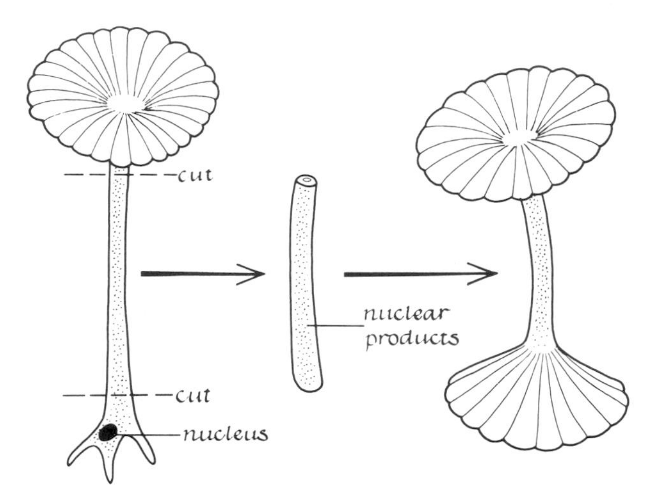

Figure 20-2. *Regeneration of caps by an isolated piece of the stem of* Acetabularia *that contains nuclear products but no nucleus.*

Acetabularia

Recall (Chapter 5) that a nonnucleated piece of the stem of this alga can regenerate a cap at each end. The experiment illustrated in Figure 20-2 demonstrates the importance of nuclear products, which appear to be relatively stable and long-lived mRNA molecules (Farber et al., 1968).

As indicated in Figure 20-2, light is necessary for cap regeneration. When illumination is restricted, stalk elongation occurs but no caps are formed. When the enucleated stalk is illuminated several weeks later, cap formation occurs. This shows that mRNA or something equivalent to it in the cytoplasm is stable for long periods.

After three months the respiration of anucleate parts of *Acetabularia* was not reduced, but the uptake of ^{32}P was (Brachet, 1952). The uptake of $^{14}CO_2$ after five weeks was 70 percent that of nucleated parts. These experiments have been criticized because bacteria in the cultures could have accounted for much of the uptake of labeled compounds. Never-

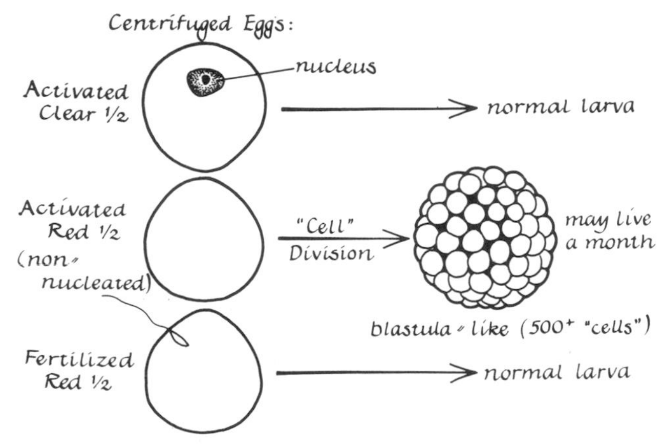

Figure 20-3. *Development of centrifugally separated halves of the sea urchin egg.*

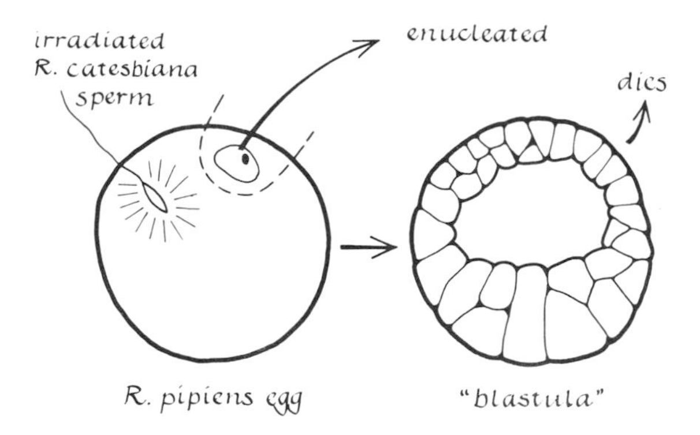

Figure 20-4. *Development of an enucleated egg of* Rana catesbiana *fertilized by an irradiated sperm.*

theless, new plastids, membranes, and other organelles (the formation of which involves protein synthesis) must form when the new caps regenerate. We shall presently see that the species-specific structure of the regenerated caps is controlled by specific gene products in the isolated stem piece. The dry weight of regenerating anucleate parts also increases, but eventually the anucleate part dies (Hammerling, 1966).

Stentor

Balamuth (1940) first demonstrated the importance of the nucleus for regeneration in *Stentor coeruleus.* In fragments lacking the macronucleus, the *kinetosomes* (basal bodies from which the cilia develop) fail to replicate (Tartar, 1961), and the development of an oral primordium is arrested. The micronucleus, however, is not necessary for regeneration (Schwartz, 1935).

Sea Urchin

As we saw in Chapter 13, Figure 13-17, centrifugation can pull the unfertilized sea urchin egg *(Arbacia punctulata)* into a nucleated clear half and a nonnucleated red half (Harvey, 1936, 1951). The nucleated and nonnucleated halves were activated by placement in *hypertonic sea water* (sea water reduced to half its volume by boiling). The development of these halves upon return to normal-strength sea water is shown in Figure 20-3. The figure also shows the development of a fertilized half.

The nonnucleated activated half of the egg, called a parthenogenetic merogone, can undergo orderly early cleavage and hatch out of the vitelline membrane. This indicates that the program of cleavage has been preformed during oogenesis under control of the nucleus, thus constituting further evidence for the significance of egg cytoarchitecture (Chapter 13). As in *Acetabularia,* it may be nuclear products in the cytoplasm that maintain development for a while. Shapiro (1935) reported that the enucleated half of the sea urchin egg consumes less oxygen than the half with the nucleus. Nevertheless, protein synthesis, indicated by the incorporation of glycine into proteins, occurs in the nonnucleated halves (Malkin, 1954). This has been confirmed by later studies, as noted in Chapter 17, and suggests that the cytoplasm of these eggs contains preformed stable mRNA. However, no cellular differentiation occurs, and the pseudoblastula eventually dies.

Frog

Fertilization of the enucleated egg of a frog *(Rana catesbiana)* with an irradiated sperm results in partial development (Briggs et al., 1951). In this case, as shown in Figure 20-4, the irradiated sperm nucleus does not divide, but the sperm aster functions in cleavage.

Removal of the nucleus from a mature frog egg does not depress its oxygen consumption or carbon dioxide production (Brachet, 1960), presumably because synthesis of the enzymes involved continues under the direction of stable mRNA.

A summary of the ways in which nonnucleated eggs or egg fragments can develop is provided in Table 20-1 (Fankhauser, 1955).

Tissue Cultured Cells

Enucleated fragments of cultured human cells survive for up to four days, move about, and incorporate amino acids into protein. Similarly, enucleated fragments of prospective pigment cells (melanoblasts) from urodele embryos can synthesize pigment granules (review of Harris, 1968).

Although the above studies show that the nucleus is necessary for continued, long-term survival of a cell, it is equally clear that cell activities can continue for significantly long periods without the nucleus. This constitutes evidence that templates for protein synthesis can persist in the cytoplasm after removal

Table 20-1. Development of Nonnucleated Eggs or Egg Fragments in Animals

Species	*Method*	*Cleavage and Final Stage Reached*	*Author*
Asterias forbesii	Removal of maturation spindle from unfertilized egg, treatment with CO_2 to induce parthenogenesis (formation of cytasters)	Irregular cleavage to "morula"	McClendon, 1908
Arbacia punctulata	Fragmentation of unfertilized egg by centrifuging, hypertonic treatment of nonnucleated halves or quarters to induce parthenogenesis (formation of cytasters)	More or less regular cleavage to blastula without blastocoele (up to 500 cells)	Harvey, 1936, 1940
Arbacia pustulosa	Same	16 cells	Harvey, 1938
Parechinus microtuberculatus	Same	Irregular cleavage or fragmentation to blastula-like structure	
Paracentrotus lividus	Same	Same	
Sphaerechinus granularis	Same	Eggs fragment irregularly	
Chaetopterus pergamentaceus	Same	2 cells	Harvey, 1939
Amblystoma mexicanum (axolotl)	Removal of maturation spindle by puncture of egg, spontaneous degeneration of single sperm nucleus; sperm aster probably functioned in cleavage	Advanced, normal blastula with blastocoele (single case)	Stauffer, 1945
Rana pipiens	Eggs inseminated with heavily x-rayed sperm of *R. catesbiana,* maturation spindle removed by puncture; sperm nucleus damaged, did not take part in development, sperm aster probably functioned	Partial blastulae, some cells with degenerated remnants of paternal chromatin	Briggs, Green, and King, 1951
Triturus palmatus, T. viridescens	Polyspermy and removal of maturation spindle by division or puncture of egg, abnormal mitotic figures and cytasters appear frequently, both sperm asters and cytasters may function in cleavage	Advanced blastulae with large areas of nonnucleated cells	Fankhauser, 1929, 1934; Fankhauser and Moore, 1941

From Fankhauser in Willier, Weiss, and Hamburger, *Analysis of Development,* W. B. Saunders Company, Philadelphia, 1955.

Table 20-2. Methods for Production of Haploid Embryos

Type of Experiment	*Procedures*	*Origin of Mitotic Apparatus Assuring Normal Cleavage*
Parthenogenesis (activation of unfertilized egg by an agent other than sperm)	Hyper- and hypotonic solutions; puncture with needle; high or low temperatures; etc.	New division center (centrosome) arises in egg cytoplasm
Gynogenesis (development of a fertilized egg with maternal chromosomes alone)	Hybridization; radiation of sperm fluid with x-rays, etc.; treatment of sperm fluid with chemicals such as acriflavine or toluidine blue; low or high temperature acting during fertilization; also occurs spontaneously	Division center contributed by sperm as in normal fertilization
Gyno-merogony (development of fragment of fertilized egg containing egg nucleus alone)	Fragmentation of fertilized egg isolating egg nucleus in one half	New division center may arise in cytoplasm
Androgenesis (development of fertilized egg with sperm nucleus alone)	Radiation of egg before fertilization; mechanical removal of egg nucleus; low or high temperature acting during fertilization; also occurs spontaneously	Division center contributed by sperm as in normal fertilization
Andro-merogony (development of an egg fragment with the sperm nucleus alone)	Fragmentation of unfertilized egg and fertilization of nonnucleated half, or division of fertilized egg isolating sperm nucleus in one half	Same

From Fankhauser in Willier, Weiss, and Hamburger, *Analysis of Development*, W. B. Saunders Company, Philadelphia, 1955.

of the nucleus and that synthesis of proteins on these templates is regulated by mechanisms that operate in the cytoplasm.

Effects of a Haploid Nucleus in Normally Diploid Organisms

Haploid embryos have been artificially produced in the ways summarized in Table 20-2. Types of haploid larvae of the salamander *(Triturus)* are illustrated in Figure 20-5. Typically, the haploid exhibits stunted growth and edema. The expression of lethal genes in haploids may be a reason for their poor development and frequent death. The most normal haploid so far produced came from an egg fragment. This suggests that the nucleo-cytoplasmic volume ratio may be important for normal development.

Effects of a Polyploid Nucleus in Normally Diploid Cells

Although different tissues in mammals have different chromosome numbers, this does not cause differences in cell types. However, artificially produced extra sets of chromosomes (artificially induced *polyploidy*) do influence development in many plants (Sinnott, 1960) and animals (see Rudnick, 1958, for a review). An example of the relationship between nuclear, cell, and body or organ size is revealed in artificially produced polyploid urodeles (Fankhauser, 1955) and is illustrated in Figures 20-6 and 20-7.

The figures show that an increase in chromosome number results in an increase in both nuclear and cell volume, but a decrease in cell division activity. The maintenance of normal body and organ size occurs by proportionate decrease (in polyploids) and an increase (in haploids) in cell number and a change in cell shape. The amphibians *Pleurodeles waltlii* and *Triturus viridescens* provide examples of how chromosome number influences cell shape (Davison, 1959). In triploids the capillary diameters remain approximately the same as in diploids, but the erythrocytes, which are larger, become disproportionately elongated and can thus pass through the capillaries. Other examples of the influence of chromosome number on cell shape are known. In plants as well as animals, polyploidy results in an increase in cell and nuclear size and in *Micrasterias*, a change in cell shape (Kallio, 1953).

How the proportionality between nuclear, cell, organ, and body size is regulated is an important but unsolved problem. Apparently, the guidelines controlling cell division are supracellular, and the unit of structure is the body or organ.

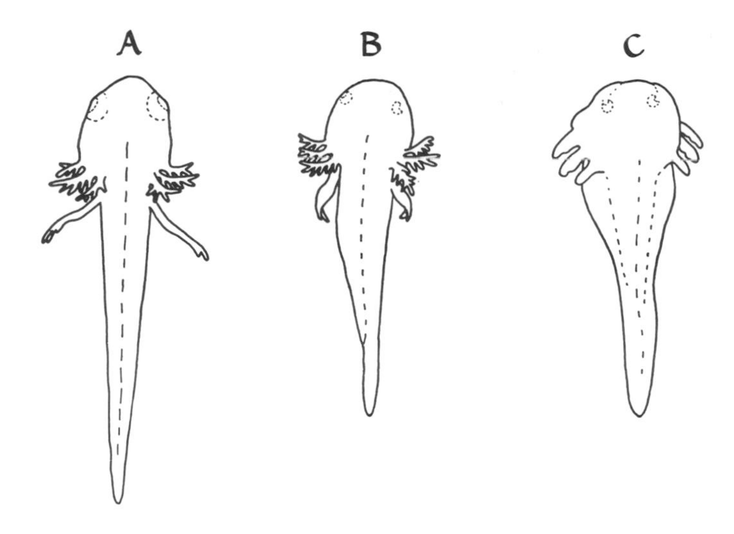

Figure 20-5. *A and B, types of haploid larvae of* Triturus pyrrhogaster *from heat-treated eggs; C, haploid larva of axolotl from cold-treated egg. Note the stunted growth of the larvae in B and C. From Fankhauser in Willier, Weiss, and Hamburger,* Analysis of Development, *W. B. Saunders Company, Philadelphia, 1955.*

Effects of Polyspermy on Egg Development

Normal development usually depends upon the joint participation of one female and one male haploid genome. When sea urchin eggs are fertilized by two spermatozoa, both sperm nuclei often fuse with the egg nucleus. In most of these zygotes, both sperm asters divide to form a mitotic figure with four (instead of two) poles connected by spindles among which the chromosomes are distributed at random. This results in four nuclei at anaphase, and four cells after cleavage, each with different numbers of chromosomes (Figure 20-8).

The abnormal chromosome sets are reproduced during later cleavages; this results in a blastula with cells containing different numbers of chromosomes. Such embryos develop abnormally and soon disintegrate. In a careful analysis of these dispermic eggs, Boveri (1907) showed that the abnormal development was the result not of the quantity of chromosome material but of abnormal combinations of individual chromosomes. Thus, only embryos in which the cells contain at least one of each kind of chromosome can

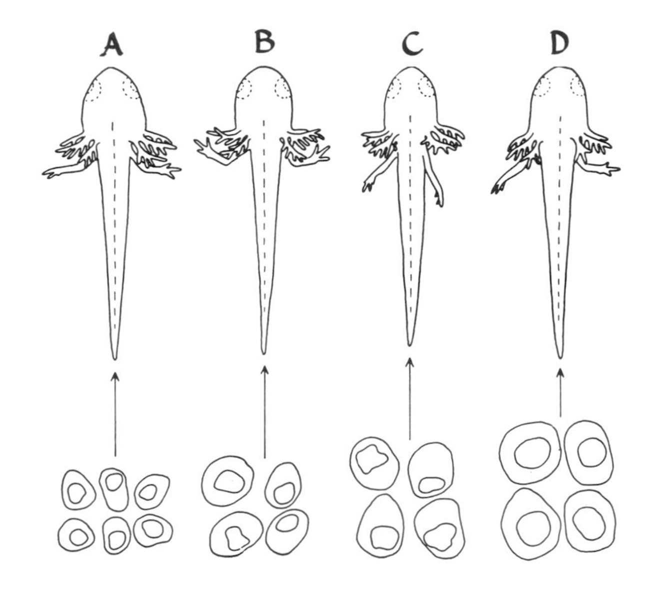

Figure 20-6. *Relationship between nuclear, cell, and body size in A, diploid; B, triploid; C, tetraploid; and D, pentaploid* Triturus viridescens *larvae. Below each larva are groups of gland cells from the epidermis of the tailfin. From Fankhauser in Willier, Weiss, and Hamburger,* Analysis of Development, *W. B. Saunders Company, Philadelphia, 1955.*

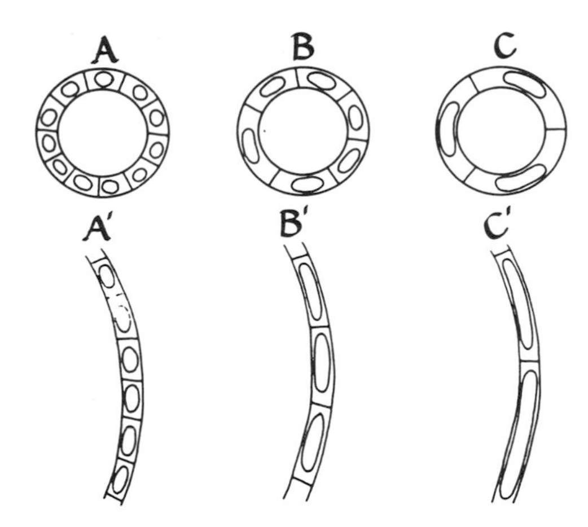

Figure 20-7. *Diagrams of cross-sections through a pronephric tubule (above) and the lens epithelium (below) in A, a haploid larva; B, a diploid larva; and C, a pentaploid larva of* Triturus viridescens. *Note that normal size and structure are maintained with cells of different sizes by adjustment of number and shape of individual cells. From Fankhauser in Willier, Weiss, and Hamburger,* Analysis of Development, *W. B. Saunders Company, Philadelphia, 1955.*

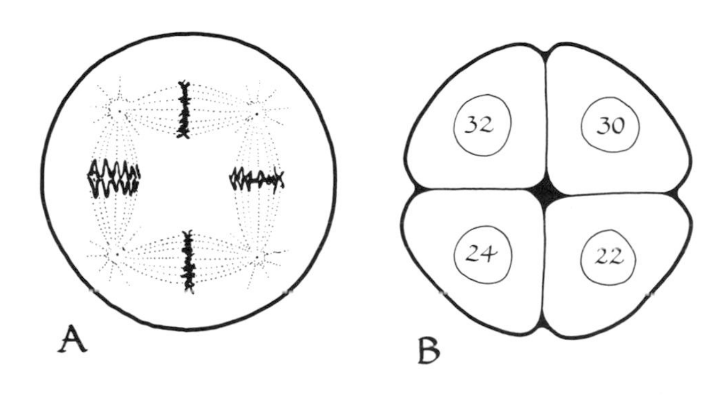

Figure 20-8. *Polyspermy (dispermy) in the sea urchin egg results in abnormal combinations of chromosomes in the blastomeres.*

develop normally. These results demonstrate that as a prerequisite for normal development every cell must have a normal complement of chromosomes. This constitutes conclusive evidence for the indispensability of a nucleus with all the information characteristic of the species.

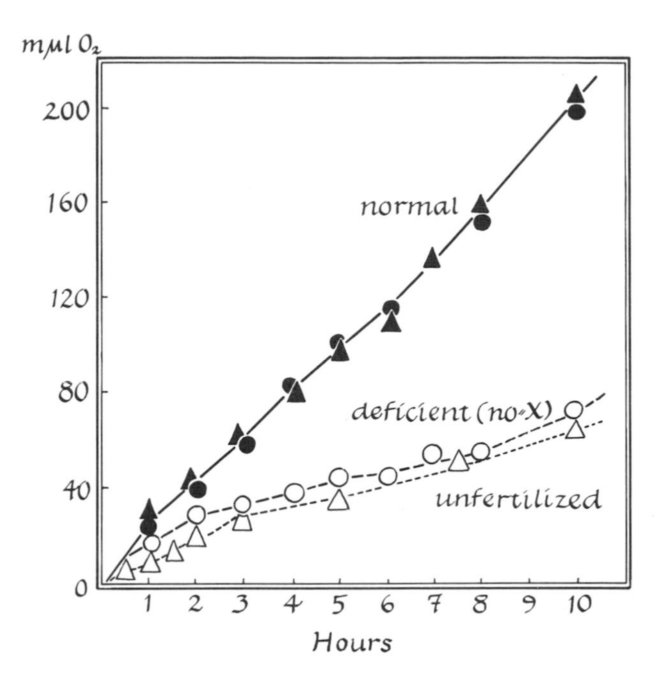

Figure 20-9. *Oxygen consumption of normal, unfertilized, and deficient (no X-chromosome) single eggs of* Drosophila melanogaster. *From Poulson, Chromosomal control of embryogenesis in* Drosophila. Amer. Naturalist, *Volume 79, 1945, courtesy University of Chicago Press.*

Effects of Lethal Genes and Missing Chromosomes

Defective nuclei—those containing lethal genes—also profoundly influence development (see review of Hadorn, 1955). Even small chromosome deficiencies or gene mutations can lead to abnormal or arrested development at early embryonic stages. When an entire chromosome, such as the X chromosome of *Drosophila,* is absent from the zygote (in one-fourth of the eggs of experimentally produced attached-X females), cleavage is abnormal and the nuclei remain mostly in the anterior half of the egg. Meanwhile, the cortical layer in the posterior end of the egg breaks down, suggesting that the integrity of the cytoplasm depends upon the presence of nuclei. The oxygen consumption of such nullo-X eggs is greatly depressed and the embryos die at an early stage (Poulson, 1945). This is illustrated in Figure 20-9.

Note in the figure that nullo-X zygotes consume about as much oxygen as unfertilized eggs, but that for the first two-thirds of an hour after fertilization they consume almost as much as normal eggs. One would expect that a deficiency in the genome would affect both developmental processes and their accompanying metabolic processes.

Lethal Hybrids

The abnormal development of hybrids produced by crossing varieties of animals not closely related illustrates the importance of nucleo-cytoplasmic interactions in development. The crossing of two sea urchin genera—*Sphaerechinus* and *Echinus*—provided an early example of the failure of a hybrid genome to support development beyond gastrulation (Boveri, 1893 and 1918). In crosses of this type, the incompatibility of the paternal chromosomes and the egg cytoplasm can result in elimination of the former (Baltzer, 1910). Hybridization between various species of amphibians has been extensively studied. This frequently results in arrest of development at gastrulation (review of Fankhauser, 1955) and indicates that gastrulation is the time when the paternal genome begins to play a role in development (review of Davidson, 1968). Further discussion of the many examples of this kind of nucleo-cytoplasmic incompatibility can be found in the reviews of Fankhauser (1955), Brachet (1960), and Davidson (1968). One example of how the hybrid produced by crossing *R. pipiens* with *R. sylvatica* synthesizes abnormally little DNA is shown in Figure 20-10 (Gregg and Løvtrup, 1955).

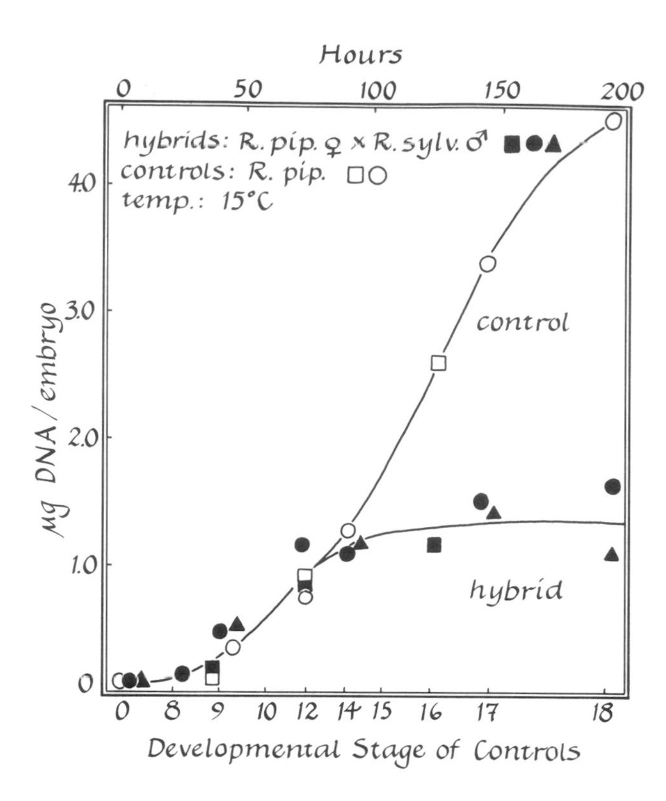

Figure 20-10. *DNA synthesis in normal and hybrid embryos. From Gregg and Lovtrup,* Biological Bull., *Volume 108, 1955, courtesy Marine Biological Laboratory, Woods Hole, Mass.*

Influence of the Nucleus on Cytoplasmic Properties

Nuclear Transplantation in Ameba

When an *Amoeba proteus* nucleus is transplanted into the cytoplasm of an enucleated *Amoeba discoides*, the nucleo-cytoplasmic hybrid becomes active and may undergo mitosis (Lorch and Danielli, 1950, 1953). The hybrid is intermediate in type in terms of cell shape and movement. However, the hybrid's antigens are of the *proteus* (nucleus) type, whereas nuclear size and division rate are of the *discoides* (cytoplasmic) type. This is a clear example of nucleo-cytoplasmic interaction (Danielli et al., 1955).

Nuclear Influence in Acetabularia

Parts of two different species of this unicellular alga may be grafted together in various ways, as diagramed in Figure 20-11. The two species most frequently used are *A. mediterranea* and *A. crenulata*. The caps of the two species look quite different, so that one can easily assign to one species or the other the type of cap formed during regeneration of the chimaera. In the grafting experiments, the single nucleus (in the rhizoid) of one species is combined with cytoplasm (in an isolated segment of the stalk) of the other species. The following paragraph represents Werz (1965) and Hammerling's (1966) interpretation of the results illustrated in Figure 20-11.

The nucleus produces and releases specific products into the cytoplasm. These products, in conjunction with cytoplasmic products, direct the synthesis of cytoplasmic structures constituting the architecture of the cap. The type of cap formed is controlled

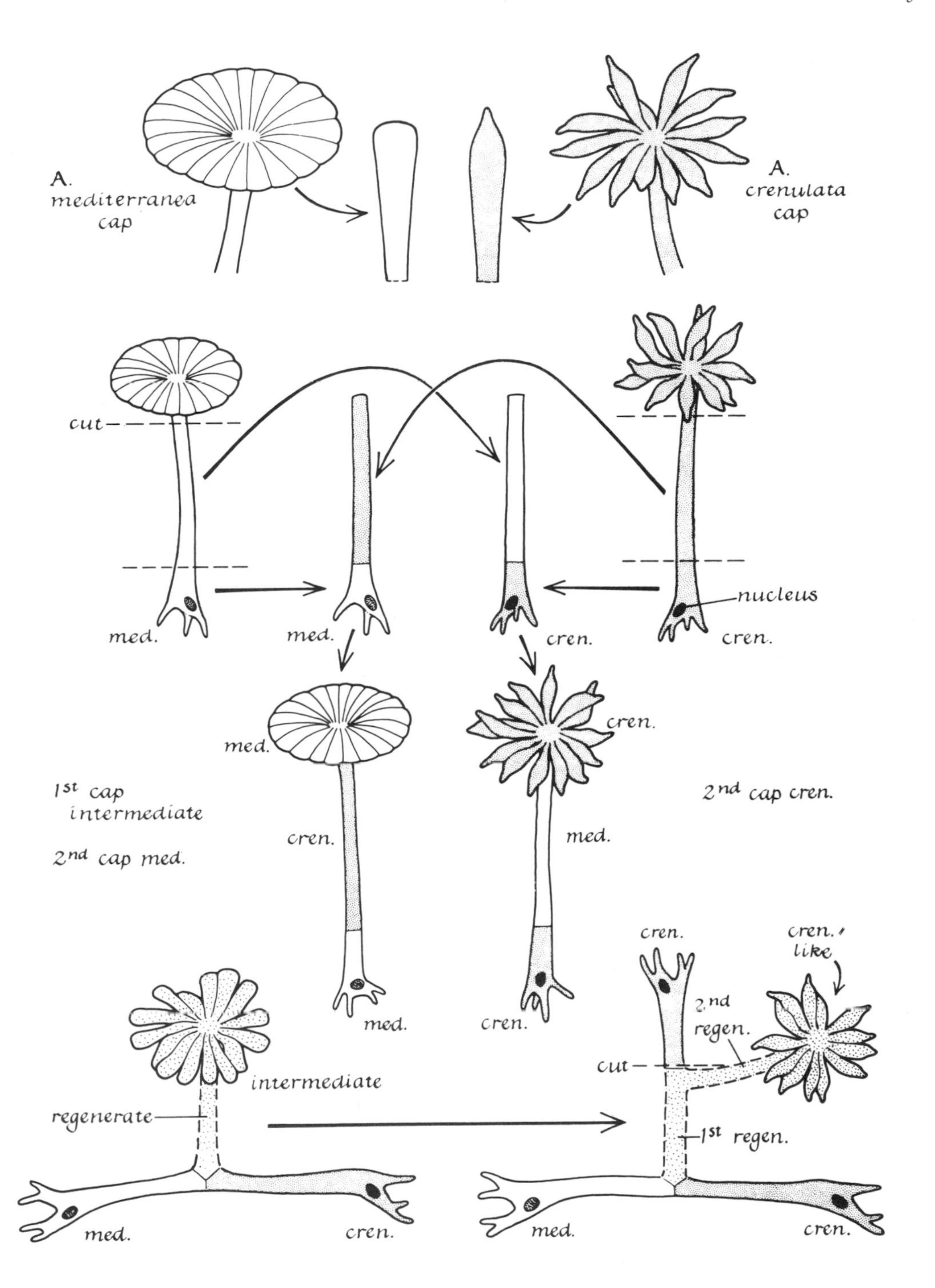

Figure 20-11. *Results of grafting the stalk of one species to the rhizoid of another in* Acetabularia. *Bottom, grafts involving both* A. crenulata *and* A. mediterranea *stalks and nuclei.*

by specific nuclear products of the species furnishing the nucleus in either *mediterranea* nucleus × *crenulata* cytoplasm or the reciprocal. However, in more complex combinations involving both a *crenulata* and a *mediterranea* nucleus, the type of cap is intermediate. Furthermore, when one *mediterranea* and two *crenulata* nuclei are grafted so as to furnish products to the cytoplasm, the type of cap regenerated is more *crenulata*-like than intermediate. These experiments show that nuclear control can depend on the amounts of specific gene products released into the cytoplasm.

Keck's experiments (1961) provide an example of how the nucleus of one species of *Acetabularia* dominates the nucleus of a species of a related genus. In work with *A. mediterranea* and *Acicularia schenckii*, Keck demonstrated the presence of an esterase isozyme enzyme specific to each, the synthesis of which is directed by the nucleus. However, in grafts between the two species, the enzyme synthesized by a *mediterranea* nucleus persists in the cytoplasm of the enucleated stalk even when the stalk is grafted to a rhizoid containing an *Acicularia* nucleus. This is an unusual situation, since most plant and animal enzymes are continually being degraded and synthesized. Some of the experiments are diagramed in Figure 20-12.

The experiments illustrated in Figure 20-12 show the apparent conversion of the *Acicularia* enzyme into the *mediterranea* enzyme as a result of the dominance of the *mediterranea* nucleus over the *Acicularia* nucleus (or even over two *Acicularia* nuclei).

The nucleus of the mature *Acetabularia* with a full-size cap is about 80 μ in diameter compared with a gamete nucleus 4 μ long. This large nucleus has a large lobulated and vacuolated nucleolus with a high RNA content. Presumably the nucleus synthesizes a large amount of rRNA. Nuclear products in the cytoplasm appear to be complexes of template RNA and rRNA. In any case, these products are extremely stable.

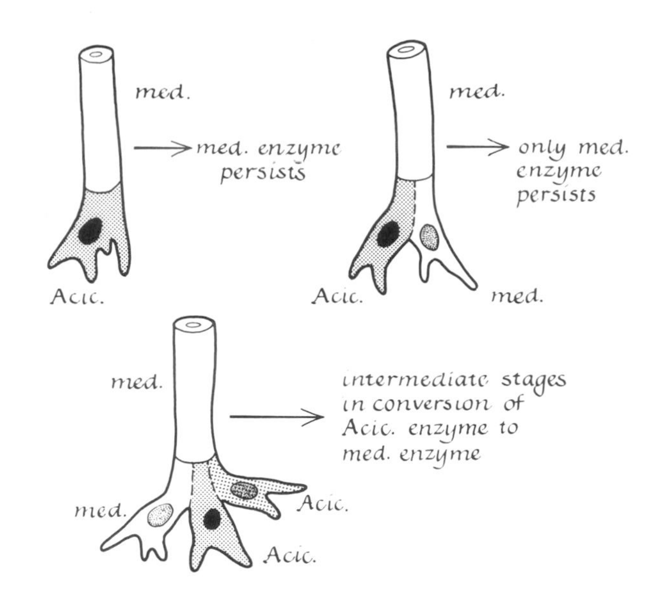

Figure 20-12. *Diagrams illustrating the stability of the* A. mediterranea *enzyme and dominance of the* mediterranea *nucleus over the* Acicularia *nucleus in graft combinations.*

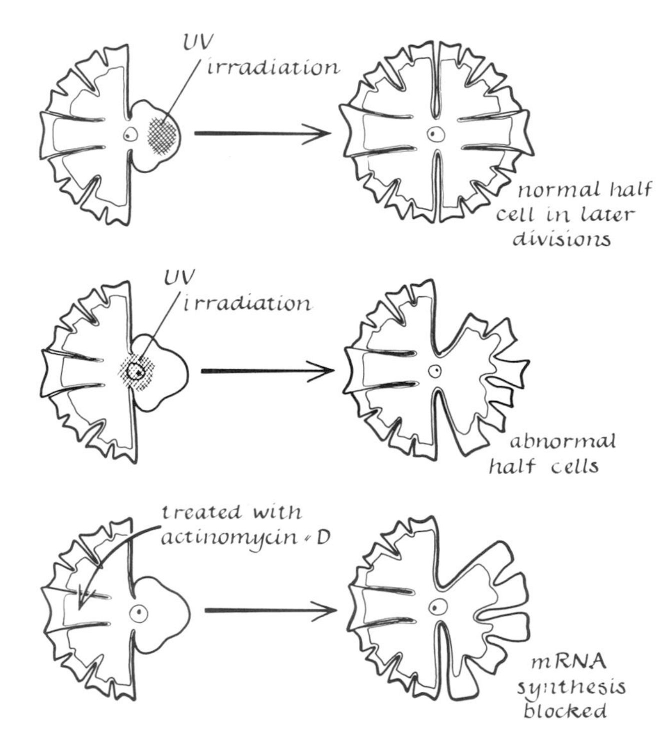

Figure 20-13. *Influence of the nucleus on formation of the new daughter half-cell in* Micrasterias.

Nuclear Influence in Micrasterias

Ultraviolet irradiation of the nucleus or treatment with actinomycin-D (which blocks DNA transcription) at an early stage of asexual reproduction in the alga *Micrasterias* causes abnormality in the new

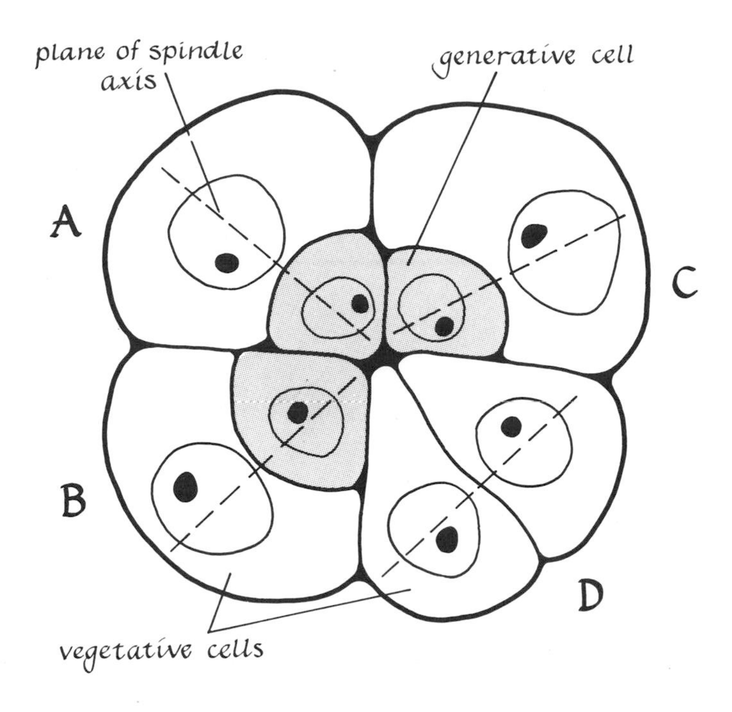

Figure 20-14. *Influence of the kind of cytoplasm surrounding the nuclei of the pollen grain of* Scilla sibirica *on their differentiation into generative and vegetative nuclei. In* D, *the abnormal plane of division results in two vegetative cells instead of one vegetative and one generative cell as in* A, B, *and* C.

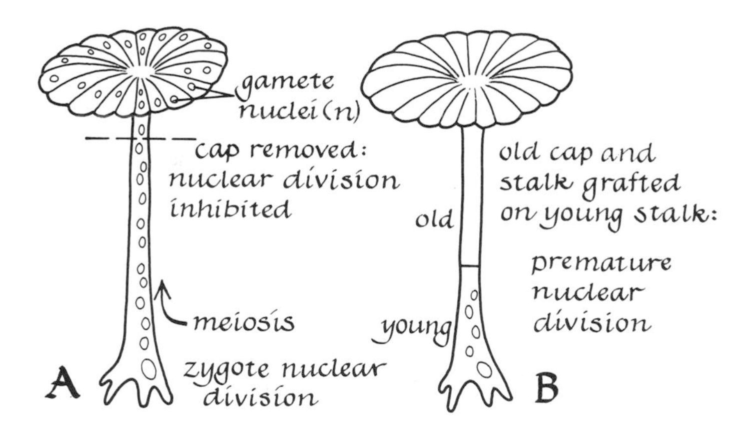

Figure 20-15. *Influence of the cytoplasm on nuclear division demonstrated by removal of the cap, A, and by grafting an old cap to a younger stalk, B.*

daughter half-cell (Selman, 1966). This is illustrated in Figure 20-13.

Influence of the Cytoplasm on Nuclear Activity

Previous examples cited illustrate nuclear influence on cytoplasmic properties. Now we shall cite examples of cytoplasmic influence on nuclear activities in several different organisms.

Tradescantia and Scilla

In Chapter 11 we noted how the two nuclei produced by the first division in the pollen grain of *Tradescantia* can be affected by the composition of the surrounding cytoplasm. If the normal radial position of the mitotic spindle in *Scilla sibirica* (squill) is changed by accident or by temperature shock to a tangential position, differentiation into generative and vegetative nuclei does not occur (La Cour, 1949). This is shown in Figure 20-14D.

Note in Figure 20-14 that pollen grains *A, B,* and *C* have divided normally but that the tangential division of *D* results in both nuclei being surrounded by RNA-rich cytoplasm. Both become vegetative and undergo no further division. Failure of normal nuclear differentiation also occurs in dwarf pollen grains; the first division produces two "vegetative" nuclei which abnormally divide again. Apparently the cytoplasm in the small grains is uniformly deficient in RNA.

Acetabularia

Figure 20-15 illustrates the influence of the cytoplasm on nuclear activity in *Acetabularia* (Hammerling, 1966). Two examples of how the cytoplasm influences nuclear activity in *A. mediterranea* are shown in the figure. When nuclear divisions preparatory to gamete formation are underway, removing the cap suppresses the division (Figure 20-15A). When a new cap of maximum size regenerates, the divisions are resumed. Grafting an old cap and stalk to a young rhizoid induces premature nuclear division (Figure 20-15B).

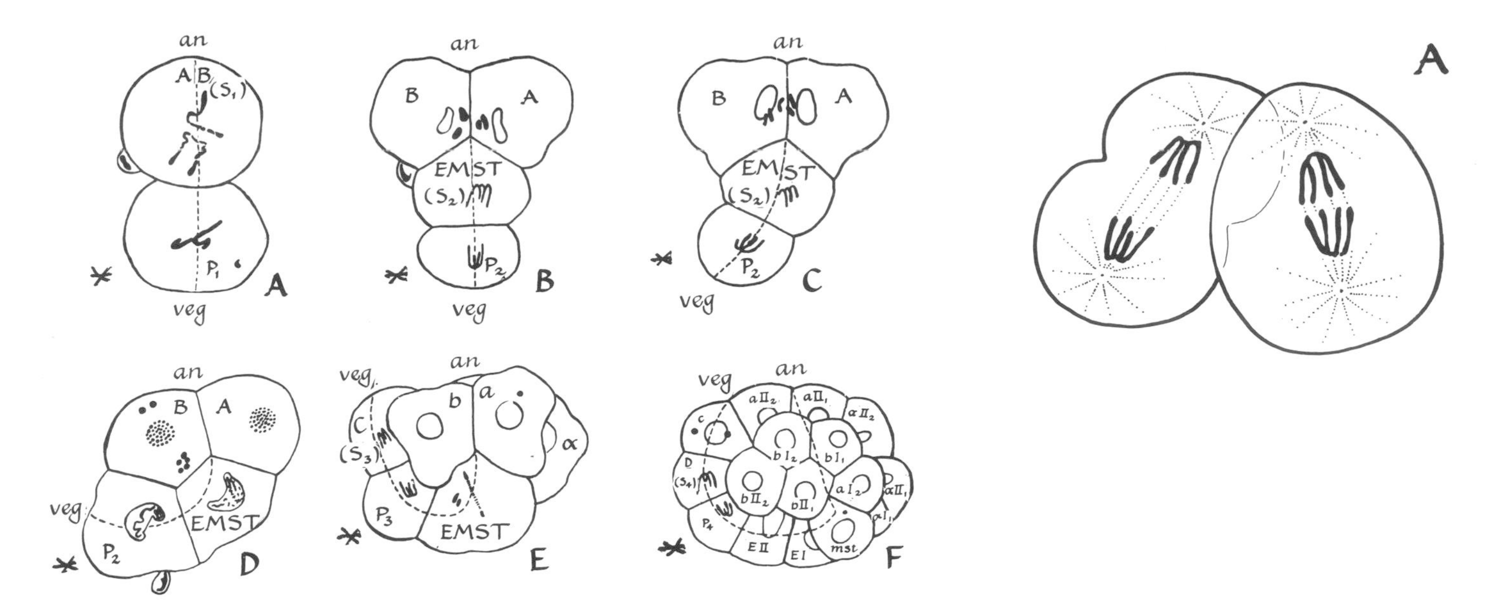

Figure 20-16. *Cleavage of the egg of* Ascaris megalocephala, *illustrating location of the stem cells, P_1 and P_2, which do not undergo chromatin diminution and from which the germ cells are derived. After von Ubisch in Willier, Weiss, and Hamburger,* Analysis of Development, *W. B. Saunders Company, Philadelphia, 1955.*

Stentor

The macronucleus of *Stentor* responds to cytoplasmic influences during regeneration. This is indicated by the condensation of the macronucleus, which regularly follows formation of a new oral primordium in a posterior piece. A *Stentor* with two oral regions, produced by parabiosing a large and a small *Stentor*, has double macronuclear chains. When the oral region of the smaller *Stentor* is later resorbed, a single macronuclear chain is formed (Tartar, 1961). Weisz (1954) describes other examples of nucleocytoplasmic influences in *Stentor*.

Ascaris megalocephala

Chromosome diminution during early cleavage in *Ascaris* was first described by Boveri (1887, 1899). Its very small chromosome number has made this parasitic worm the object of numerous cytological studies. Figure 20-16 shows that during chromatin diminution, the midportion of each chromosome (there are only two in *A. megalocephala univalens*) breaks up into about ten fragments, leaving the two large terminal pieces. The terminal pieces remain in the cytoplasm and gradually disappear, but the small fragments become incorporated in the nucleus and constitute the persistent and functional chromosomes.

The asterisks in Figures 20-16 and 20-17 denote the cells P_1 through P_4, which do not undergo chromatin diminution. Boveri surmised that cytoplasmic differences in different regions of the egg were responsible for whether or not diminution occurred. These differences also cause the differences between somatic and germ-line cells.

Experiments with centrifuged eggs by Boveri (1910) and Hogue (1910) supported Boveri's supposition. In many of the centrifuged eggs, two germ-line cells (Figure 20-17, P_2) were formed in which no diminution occurred. This was interpreted as resulting from the first cleavage spindle's changing from a vertical to a horizontal position, so that the first two cells had essentially the same cytoplasmic composition. This is illustrated in Figure 20-17D.

Miastor (Gall Midge)

Another instance of cytoplasmic composition influencing nuclear activity is found in chromosome elimination in the gall midge (White, 1946, 1947).

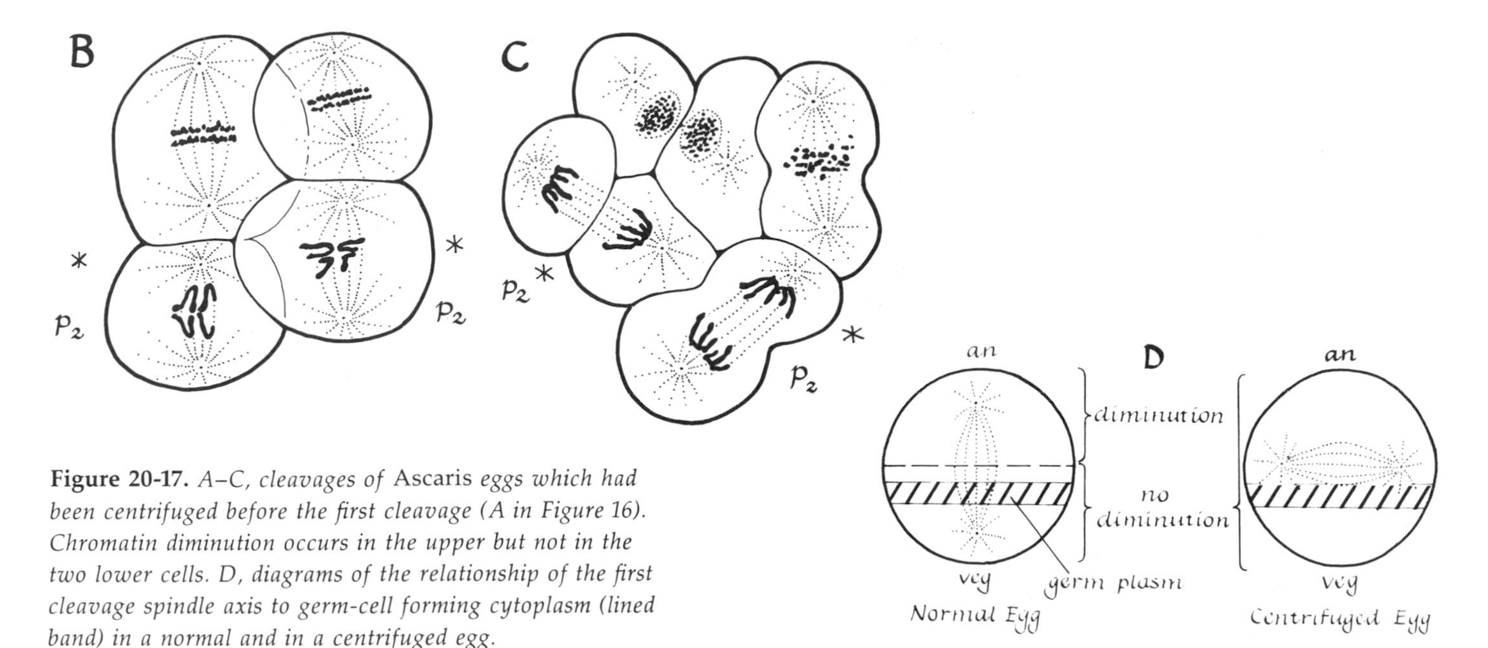

Figure 20-17. *A–C, cleavages of* Ascaris *eggs which had been centrifuged before the first cleavage (A in Figure 16). Chromatin diminution occurs in the upper but not in the two lower cells. D, diagrams of the relationship of the first cleavage spindle axis to germ-cell forming cytoplasm (lined band) in a normal and in a centrifuged egg.*

The germ cells that are cleaved out of the polar germ-plasm region of the egg contain forty-eight chromosomes in their nuclei, the original number in the

These experiments were later repeated by King and Beams (1938), who used high-speed centrifugation (150,000 g). In this case cytokinesis was blocked but nuclear division continued. Diminution occurred in all nuclei at the second or third mitosis.

According to Boveri (1899) and von Ubisch (1943), the cytoplasmic region that becomes incorporated into the germ line (Figure 20-16F, P_4) is located halfway between the egg's equator and its vegetal pole (Figure 20-18).

There are more radical examples of chromatin diminution, involving elimination of whole chromosomes in the somatic cells of the fungus fly *Sciara* and other insects.

Sea Urchin

When an enucleated egg of one sea urchin species (for example, *Sphaerechinus granularis*) is fertilized by the sperm of another species *(Paracentrotus lividus)*, normal development to the pluteus larval stage occurs. Later development is abnormal and the larva dies (Boveri, 1918). This indicates that there are two distinguishable periods in development: one in which the egg cytoplasm controls development, followed by one in which specific chromosomal properties of the sperm nucleus are expressed and result in injury of the sperm nucleus or cause abnormal nucleo-cytoplasmic interaction.

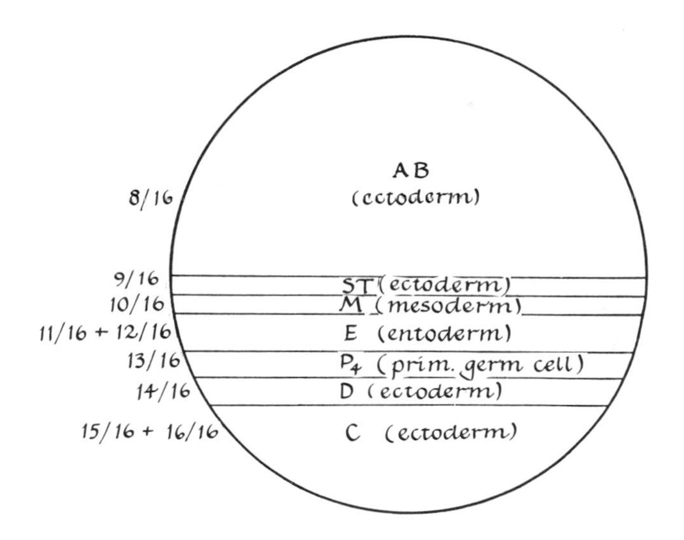

Figure 20-18. *Diagram illustrating the location of the cytoplasmic region (P_4), which becomes incorporated into the germ line cells. After von Ubisch in Willier, Weiss, and Hamburger,* Analysis of Development, *W. B. Saunders Company, Philadelphia, 1955.*

zygote. Nuclei of the female's somatic cells have only twelve chromosomes; nuclei of male somatic cells, only six (Figure 20-19).

Wachtliella persicariae, a related species of insect, exhibits a similar type of chromosome elimination during early cleavage. Centrifugation of this insect's egg displaces the polar germ plasm into the somatic region of the egg and arrests chromosome elimination in nuclei surrounded by the pole plasm (Geyer-Duszynska, 1959).

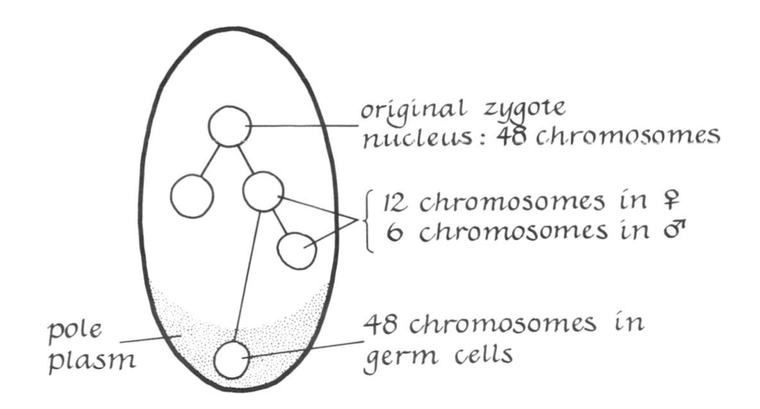

Figure 20-19. *Diagram of the egg of the gall midge* (Miastor), *illustrating chromosome elimination in nuclei surrounded by soma plasm but not in nuclei surrounded by the polar germ plasm.*

Amphibians

Numerous studies of developing amphibian embryos demonstrate the influence of the cytoplasm on nuclear activities. The methods used include hybridization between different species, transplantation of nuclei from cells of embryos into enucleated eggs, and study of a gene involved in oogenesis.

Damage to a Nucleus by the Cytoplasm

When a *Rana sylvatica* egg is fertilized by an *R. pipiens* sperm and the egg is then enucleated, development mediated by the *pipiens* sperm nucleus proceeds only to the blastula stage. *R. pipiens* nuclei of this blastula which were "damaged" by residence in cytoplasm of another species *(sylvatica)* fail to recover even when transplanted back into the cytoplasm of its own species of egg *(pipiens).* These basic experiments of Moore (1958, 1960), illustrated in Figure 20-20, demonstrate a type of permanent cytoplasmic influence on nuclear capacity.

Hennen's studies (1963) show that damage to a nucleus in foreign cytoplasm can result in *aneuploid* chromosome sets, in which one or more kinds of chromosomes are absent and/or extra chromosomes are present. The more severe the chromosome anomaly, the earlier the arrested development.

Control of Nucleoprotein Synthesis

Implantation of adult *Xenopus* liver or brain nuclei into enucleated, unfertilized eggs of *Xenopus* results within thirty minutes in the incorporation by the adult nuclei of thymidine into their DNA. These adult liver or brain nuclei never synthesize DNA under normal conditions (Graham et al., 1966); thus the cytoplasm of the egg can stimulate DNA synthesis. These adult nuclei do not support continued normal development as do implanted nuclei from embryonic *Xenopus* cells.

There are similar examples of nuclear activity being regulated by the egg cytoplasm. For instance, the nucleus of a blastula, gastrula, or neural fold stage of *Xenopus* stops synthesizing RNA when transplanted into an enucleated *Xenopus* egg (Gurdon, 1967). Recall (Chapter 17) that when *Xenopus* develops normally, RNA synthesis, particularly that of rRNA, occurs during oogenesis but ceases in the mature egg and is not resumed until the late blastula or gastrula stage. Thus when the transplant egg reaches the blastula and gastrula stages, the implanted nucleus resumes RNA synthesis, as would be expected (Gurdon and Brown, 1965). The nucleoli are involved in rRNA synthesis and are present in the nuclei of blastula and later stages but not in mature eggs; they disappear within forty minutes after the nucleus of such a cell is transplanted into the egg. This effect of the egg cytoplasm upon nuclear activity is not species specific; for instance, mouse-liver nuclei, which do not normally synthesize DNA, begin to do so within one or two hours after implantation into frog eggs.

Cytoplasm of the fertilized *Xenopus* egg also controls the rate of nuclear division. Injecting accessory sperm nuclei into eggs at various times during their

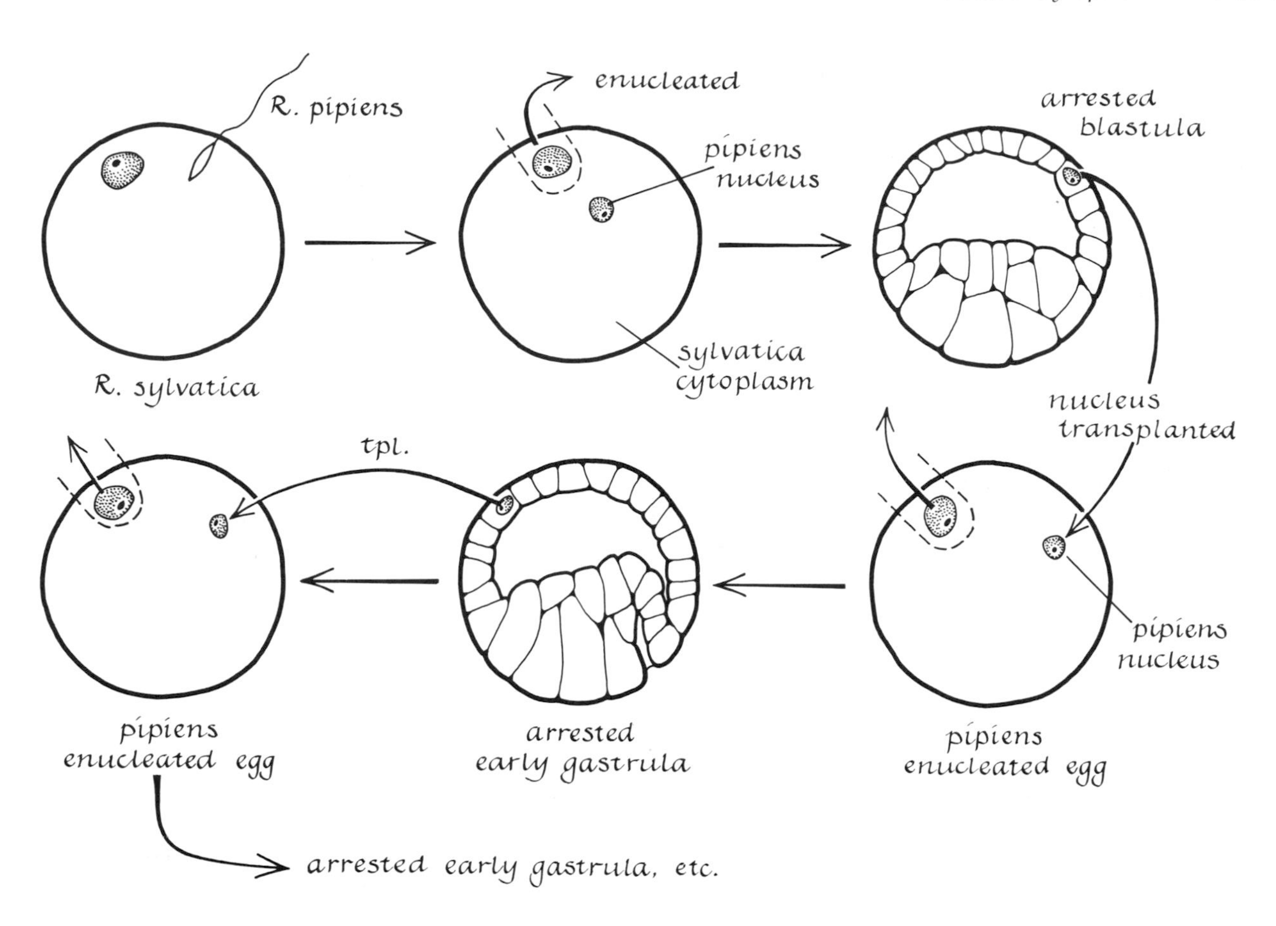

Figure 20-20. *Diagrams illustrating damage to the sperm nucleus of one species* (R. pipiens) *resident in the cytoplasm of an enucleated egg of another species* (R. sylvatica). *Note that transplantation of the damaged* pipiens *nucleus of the arrested blastula back into an enucleated* pipiens *egg does not repair the damage.*

cell cycles causes the sperm nuclei to enter mitosis in synchrony with the egg zygote nucleus. However, nuclei injected into unfertilized eggs synthesize DNA but do not divide. Thus the controls of DNA synthesis and mitosis apparently are different (Graham, 1966).

A Gene Involved in Oogenesis

Eggs of *Ambystoma mexicanum* homozygous for the gene *o* develop only to the gastrula stage. However, *o/o* eggs injected with cytoplasm from +/+ eggs develop fairly normally to larval stages (Briggs and Justus, 1968). Apparently the normal allel (+) of the gene *o* controls the synthesis during oogenesis of some protein necessary for gastrulation. Thus, in the cross +/*o* ♂ × +/+ ♀ all the F_1 individuals develop normally, but in the cross +/*o* ♂ × *o*/*o* ♀ the F_1 individuals develop only to the gastrula stage.

Cytoplasmic Control of Skin Structure

Triturus palmatus eggs fertilized by *T. cristatus* sperm and then enucleated are not viable for long after gastrulation, but if a piece of gastrular epidermis is transplanted to a normal gastrula of *T. alpestris,* normal skin differentiation occurs in the graft. The species type of skin which results is that of the egg *(palmatus)*, not the sperm; this demonstrates that the cytoplasm determines the character of the skin (Hadorn, 1936). This experiment is shown in Figure 20-21. The experiment illustrates an important aspect of our thesis that the cytoplasm as well as the nucleus contains a program for development and that

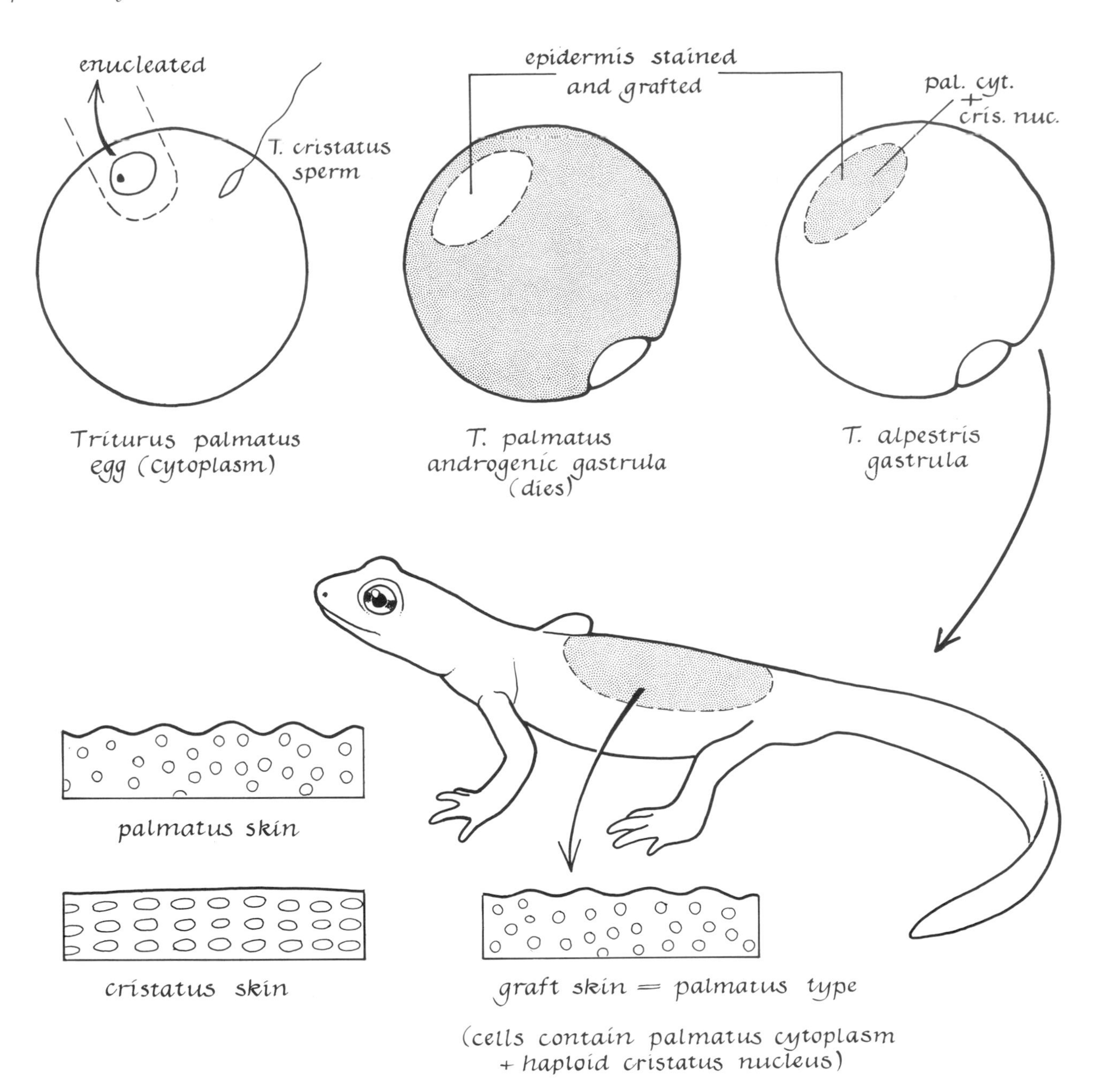

Figure 20-21. *Diagrams illustrating the control of skin character by the species* (T. palmatus) *furnishing the cytoplasm, not the species* (T. cristatus) *furnishing the nuclei in the skin cells of the graft.*

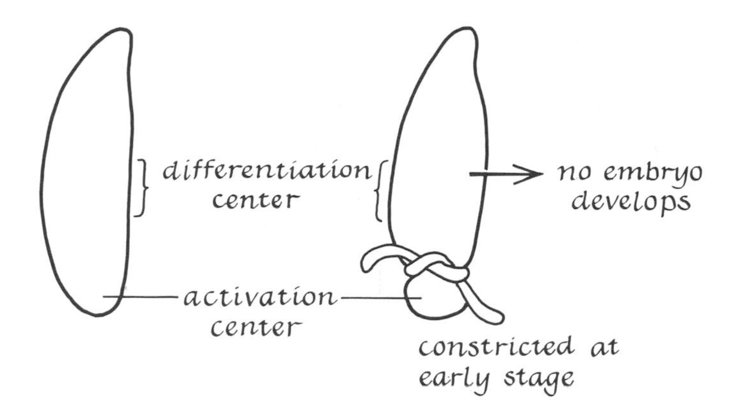

Figure 20-22. *Separation of the activation center from the differentiation center in the egg of* Platychemis pennipes *before the nuclei have entered the activation center prevents embryo formation.*

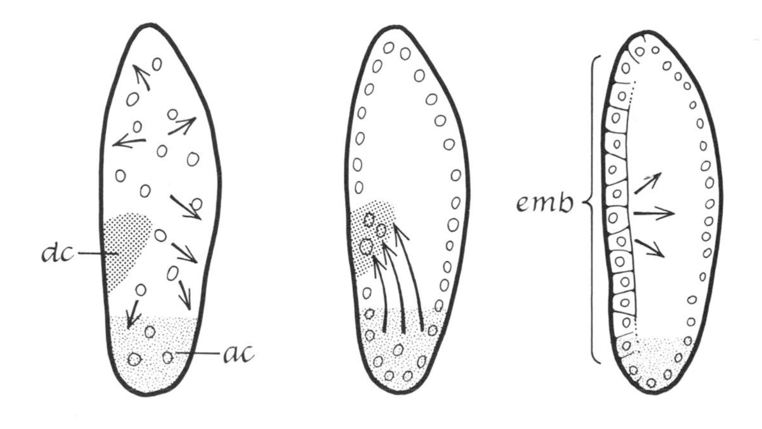

Figure 20-23. *Diagrams of the* Platychemis *egg, illustrating Seidel's interpretation of the result of the experiment shown in Figure 20-22.*

this program is built up during oogenesis (see Briggs and Justus, 1968). However, the cytoplasm of an amphibian egg does not control all skin properties (see Figure 20-26B).

Dragonfly

The egg of the dragonfly *Platychemis pennipes* furnishes a clear example of the importance of nucleo-cytoplasmic interactions in development (Seidel, 1936, 1960). This egg contains a special cytoplasmic region at the posterior pole called the activation center. Another special cytoplasmic region near the middle of the egg, the differentiation center, initiates embryo and organ formation (Figure 20-22).

As indicated in Figure 20-22, if the egg is constricted with a fine hair so as to separate the activation center plasm from the rest of the cytoplasm before the dividing nuclei have entered this region (Figure 20-23), no embryo parts develop. If constriction is later—for example, after organogenesis in the differentiation center has begun—normal development ensues. These results indicate that interaction between the nuclei and cytoplasm in the activation center releases a product which diffuses anteriorly and initiates differentiation in the differentiation center (Figure 20-23).

Chicken and Mammal Somatic Cell Hybrids

Interspecific cell hybrids produced by fusing tissue-cultured cells is a relatively new method for studying nucleo-cytoplasmic interactions (see Harris, 1968, for method). This remarkable procedure promises to contribute significantly to our understanding of development (Ephrussi and Weiss, 1968). A few examples will illustrate this significance.

When a nucleated erythrocyte of the domestic fowl was fused with a HeLa cell (a human tumor cell), a *heterokaryon* (a cell containing two genetically different nuclei) formed. The erythrocyte nucleus synthesized RNA; this is significant, since RNA synthesis is minimal in the nucleus of a normal erythrocyte. Thus, a normally quiescent gene in the erythrocyte can be activated by an unfamiliar stimulus, namely, the HeLa cell cytoplasm (Harris, 1965). Division of the erythrocyte nucleus and hemoglobin synthesis have also occurred in this system (Bolund et al., 1969).

Differentiated neurons normally do not synthesize DNA or divide. However, when a neuron of the three-week-old mouse was fused with a monkey fibroblast in the presence of *Sendai virus* (using Harris's technique), uptake of ^{3}H-thymidine occurred (Jacobson, 1969). Thus, the relatively undifferentiated cytoplasm in the fibroblast stimulated DNA synthesis. Heterokaryons of neurons and epithelial cells of other mammals and larval salamanders have also been experimentally produced by Jacobson (1969).

Developmental Equivalence of Cleavage Nuclei

As noted in Chapter 13, isolating the blastomeres of a sea urchin egg up to the four-cell stage and separating them into halves up to the blastula stage along the animal-vegetal axis causes complete but small larvae to develop. These experiments demonstrate that the nuclei of the first few blastomeres are developmentally equivalent. An experiment performed by Driesch (1894) that demonstrates this even more clearly is illustrated in Figure 20-24.

When developing eggs were slightly flattened by being confined between two sheets of glass, the direction of the planes of division was changed. This resulted in abnormal positions of the blastomeres relative to one another. The cleavage nuclei thus became located in "wrong" cells (compare Figure 20-24A with 20-24B). Nevertheless, the compressed eggs developed normally after release of pressure. Thus, the normal relationship of nuclei and cytoplasmic regions is not necessary for normal development. Consequently, the nuclei are equivalent during early development of the eggs.

Spemann (1914) reported a similar demonstration of the developmental equivalence of cleavage nuclei. He partially constricted eggs of *Triton* (newt) with a hair ligature (Figure 20-25). At about the sixteen- to thirty-two-cell stage, a nucleus in the cleaving half which contained the original egg nucleus migrated into the uncleaved half and initiated cleavage in it (Figure 20-25C). Eventually, two larvae formed, showing that a fifth- or sixth-generation nucleus was equivalent to an egg nucleus (see also Spemann, 1938).

These classical experiments of Driesch and Spemann demonstrated the developmental equivalence of nuclei at early stages, but the problem of whether this developmental equivalence is maintained throughout development still remained. The many nuclear transplantation experiments performed since Briggs and King (1952) perfected the method provide evidence for an answer to this problem (reviews of Gurdon, 1964, 1966; King, 1966; Lehman, 1967). The following section describes some of these experiments.

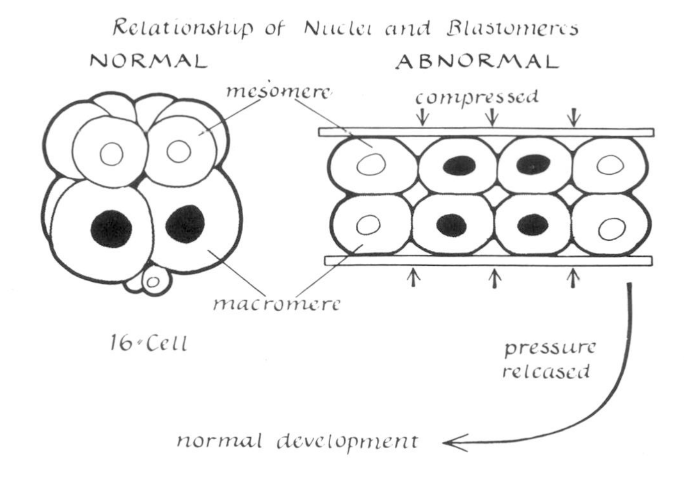

Figure 20-24. *Driesch's experiment demonstrating the equivalence of the cleavage nuclei at the 16-cell stage of the sea urchin embryo.*

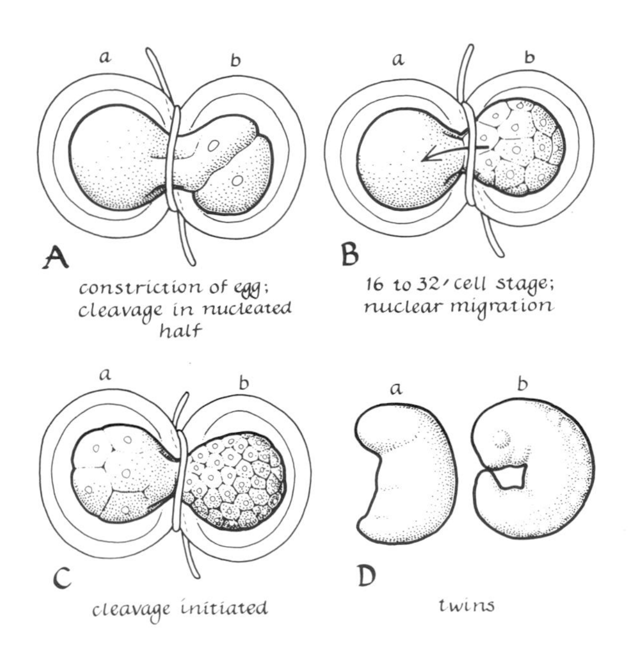

Figure 20-25. *Spemann's experiment demonstrating the developmental equivalence of the cleavage nuclei in the egg of* Triton. *A and B, partial constriction of the fertilized egg and cleavage in the half containing the egg nucleus; C, migration of a nucleus into the uncleaved half; D, larva (left) developed by the half with delayed nucleation and larva (right) developed by the half containing the original egg nucleus.*

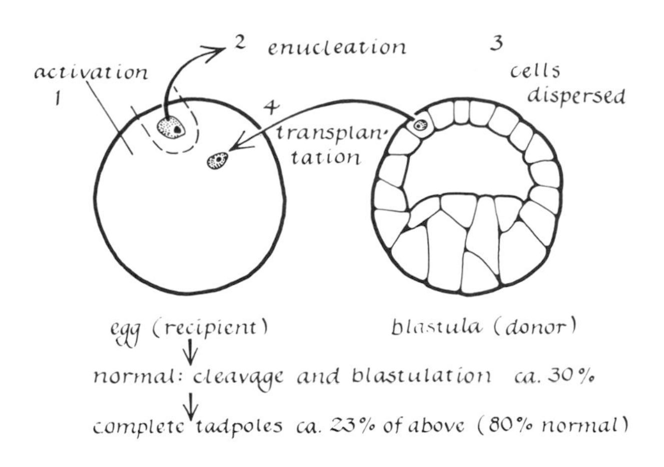

Figure 20-26. *A, method of nuclear transplantation devised by Briggs and King. The unfertilized frog egg is first activated and then enucleated. Cells of the donor embryo are dispersed and the nucleus of one of these cells is picked up in a fine bore pipette and injected into the enucleated egg. The percentages of normal and abnormal development of the transplant eggs are only approximate. B, sexually mature male nuclear-transplant* R. pipiens *(kandiyohi dominant mutant) with two metamorphosed progeny, one of which was wild-type, the other kandiyohi in skin coloration. In this case, the transplanted nucleus was from the animal hemisphere of the blastula of a kandiyohi frog and the enucleated egg was from a wild-type frog. Photograph, courtesy of Dr. R. G. McKinnell, from* Intraspecific nuclear transplantation in frogs, J. Heredity, *Volume 53, 1962, copyright American Genetic Association.*

B

Nuclear Transplantation

Amphibians

Blastula Stage Ectodermal Nuclei

The method of transplanting the nucleus of an ectodermal cell from the roof of the blastula of the frog *Rana pipiens* into an activated and enucleated egg is diagramed in Figure 20-26A (Briggs and King, 1952). Since a small percentage of the eggs receiving the transplant nucleus do develop, it can be concluded that some ectodermal cell nuclei at the blastula stage can substitute for an egg nucleus. Competence of a transplant ectodermal nucleus is further shown by the fact that it can support development of the egg into a fertile adult frog. An adult nuclear transplant male produced by transplanting an animal hemisphere nucleus of a heterozygous kandiyohi (skin pigmentation) dominant mutant of *Rana pipiens* into an enucleated wild-type *R. pipiens* egg is shown in Figure 20-26B. Note that the frog has the kandiyohi-type skin pigmentation. Also shown in the figure are two offspring resulting from fertilization of eggs of a wild-type female with sperm from this frog. Since one of these offspring is wild-type and the other kandiyohi, the transplanted nucleus must have been heterozygous with respect to the kandiyohi character. Further crosses confirmed this (McKinnell, 1962).

Blastula Stage Ectodermal, Mesodermal, and Endodermal Nuclei

The results of substituting mesodermal and endodermal as well as ectodermal cell nuclei for the egg nucleus are summarized in Figure 20-27 (Briggs and King, 1952).

These experiments show that nuclei at the blastula stage, regardless of germ layer origin, can function like the egg nucleus in supporting apparently normal development. However, these results do not show that all nuclei of a blastula are developmentally competent, nor that some are and some are not developmentally competent. For example, it is not known why most injected nuclei in the first experiment (Figure 20-26) failed to support normal development. Possibly these nuclei were injured by the manipulation (Hennen, 1963).

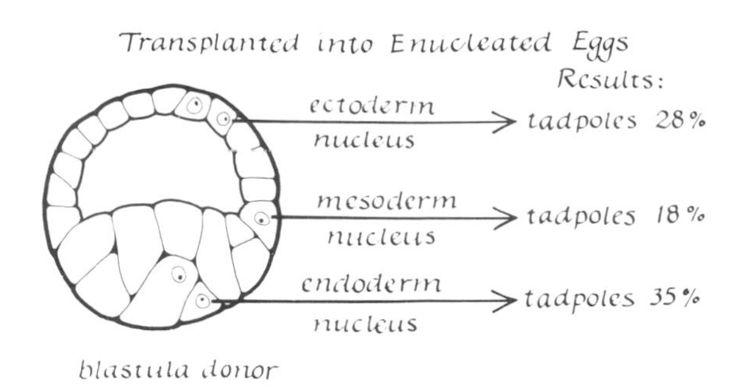

Figure 20-27. *Results of transplanting ectodermal, mesodermal, and endodermal cell nuclei of a frog* (R. pipiens) *blastula stage into an enucleated frog* (R. pipiens) *egg.*

Serial Transplantation of Late Gastrula Endoderm Nuclei

The method of serial transplantation and the generalized results it yields are diagramed in Figure 20-28 (King and Briggs, 1956).

Note that nuclei derived from a single nucleus (a nuclear clone) of the original gastrula (D in the figure) are much less variable in supporting development than are nuclei derived from different endodermal cells. Some nuclei (G in the figure) can support normal development of the transplant egg, resulting in development of a normal tadpole that can metamorphose into an adult fertile frog.

The nuclear transplantation experiments of Briggs and King demonstrate that the capacity of some endodermal nuclei to support normal development decreases with increase in developmental age, and that the nuclear condition responsible for abnormal development is highly stabilized and heritable. However, other endoderm nuclear transplant embryos were arrested at various stages and failed to display any consistent pattern of deficiencies. These latter embryos had abnormal chromosome constitutions, whereas abnormal embryos exhibiting a regular pattern of deficiencies (consistent with the endodermal origin of their nuclei) were *euploid* (normal) in chromosome composition (Briggs et al., 1960). One reason for the irregularity of development of endoderm nuclear-transplant embryos during serial transplantation might be inadvertent injury to the nucleus and consequent chromosome loss or damage during manipulation. Thus only embryos in which the chromosome complement remains euploid would provide a valid test of the properties of the nuclei as they exist in the donor cell prior to transplantation. This is an important point, since Briggs and King's conclusion that their results demonstrate nuclear differentiation has been criticized by Gurdon

Figure 20-28. *Serial transplantation of late gastrula endoderm nuclei. Note that there is relatively small variation in the type of development within a nuclear clone (eggs receiving nuclei derived from endoderm nucleus, D). In contrast, there is much greater variation in the type of development in transplant eggs receiving different endoderm nuclei of the original donor gastrula (top row of embryos). Note also the similarity in type of development of the first and second generation embryos receiving clone D nuclei.*

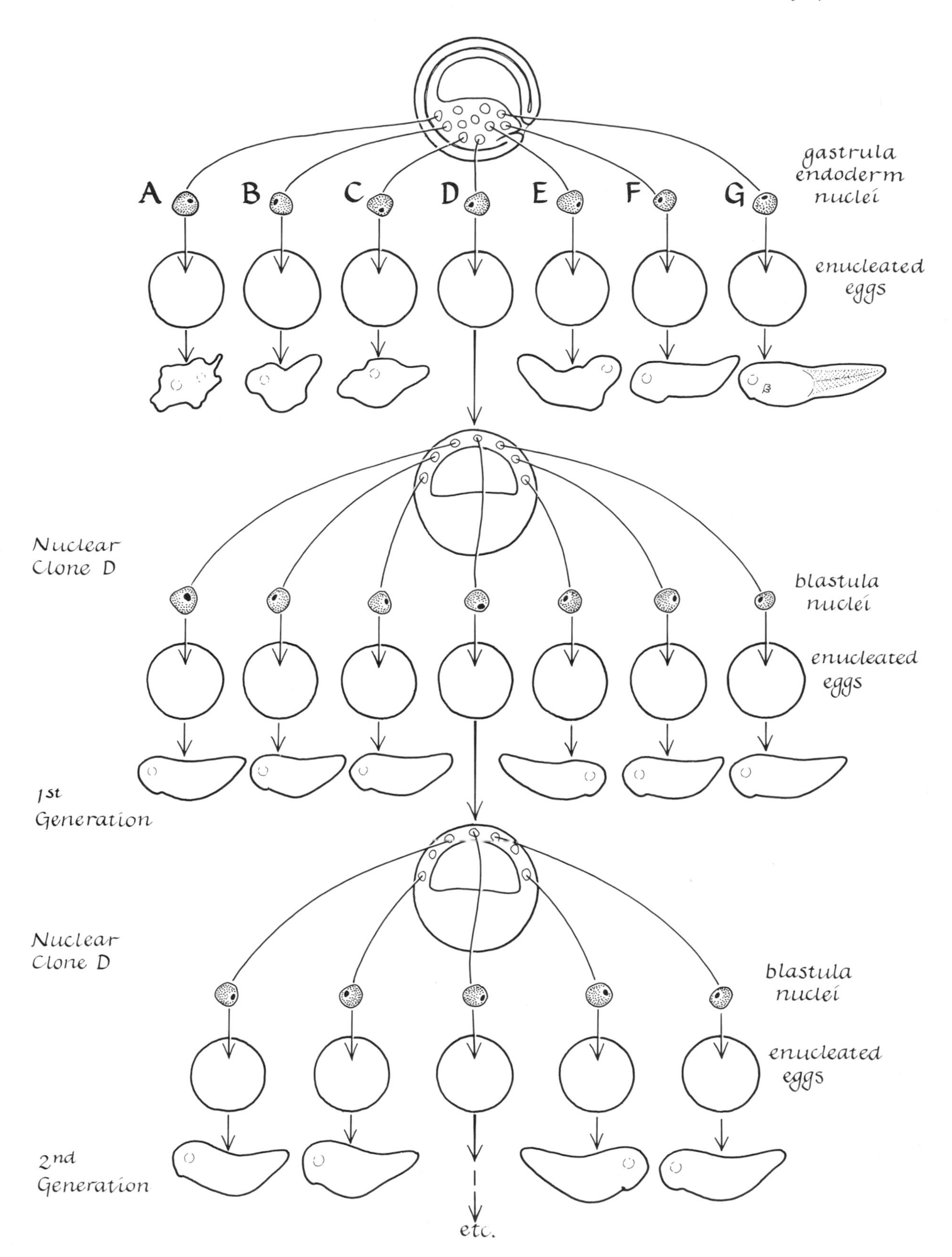
gastrula
endoderm
nuclei
A
B
C
D
E
F
G
enucleated
eggs
Nuclear
Clone D
blastula
nuclei
enucleated
eggs
1st
Generation
Nuclear
Clone D
blastula
nuclei
enucleated
eggs
2nd
Generation
etc.

(1962, 1966) on the grounds that the inability of a nucleus to support normal development is the result of injury during transplantation, not the result of a change in nuclear potentiality prior to its transplantation. However, DiBerardino and Hoffner (1970) demonstrated that the chromosomes of many endodermal nuclei have acquired restrictions in the form of condensed chromatin (heterochromatin); thus their inability to support normal development is not the result of injury during transplantation.

Transplantation of Nuclei of Neural Plate Cells at Progressively Older Stages

Tests of neural nuclei in *Rana, Xenopus,* and *Axolotl* (cited by DiBerardino and King, 1967) have shown that the majority of these nuclei from embryos of increasing developmental age increasingly fail to support normal egg development. In the experiments of DiBerardino and King, neural nuclei from late gastrula, early neurula, and midneurula stages of *Rana* were transplanted into enucleated eggs. Figure 20-29 summarizes the results.

Stage of Donor			*Percentage of Normal Larvae*
Stage 8	Blastula (animal pole)	⟶	45
Stage 12	Early Gastrula (neural plate)	⟶	23
Stage 13	Early Neurula (neural plate)	⟶	11
Stage 14	Late Neurula (neural fold)	⟶	3

Figure 20-29. *Transplant nuclei from increasingly older donor embryos increasingly fail to support normal development.*

Larval abnormality was not always correlated with chromosome abnormalities; this increases the probability that at least some neural nuclei gradually lose the capacity to support normal development of enucleated eggs prior to their transplantation. Presumably, progressive changes in the composition of the cytoplasm accompanying the early stages in the differentiation of some of the neural cells influence nuclear capacity to support normal egg development. Serial transplantation of neural nuclei from the open neural-fold stage demonstrated that the nuclear condition responsible for the restricted development is stable and cannot be reversed by exposure to test-egg cytoplasm.

Transplantation of Endodermal Nuclei from the Tadpole Stage in Xenopus

In these experiments, nuclei of intestinal epithelial cells of feeding tadpoles were transplanted into enucleated eggs (Gurdon, 1962, 1966; Fischberg et al., 1963). Unlike the donor cells in the experiments just described, the donor cells in these experiments are essentially fully differentiated, as indicated in Figure 20-30.

Gurdon's results are summarized below.

1. 1.5 percent (10 out of 726) became normal tadpoles (first transfer).
2. Serial transfer of nuclei from abnormal embryos into enucleated eggs increased the number of normal tadpoles to 7 percent.
3. Excluding injured nuclei and those in unsuitable mitotic stages increased the percentage of normal tadpoles to 24 percent.

In order to explain the increased percentage of normal transplant embryos during serial transplantation, Gurdon presumed that the poor quality of the cytoplasm in some of the eggs prevents the first transfer nuclei from expressing their capacities, but that this is rectified when they are transferred to more normal eggs. Gurdon's experiments show that a nucleus can support the formation of a differentiated intestinal cell and at the same time contain the genetic information needed to form all other types of differentiated cells. This seems reasonable and compatible with current dogma that all cells contain the same genetic information. However, genetic information content and genetic information use are not equivalent. Nuclear differentiation probably occurs as a normal developmental process in response to cytoplasmic influences. Furthermore, the differentiated state of the nucleus might be stable. This does not imply a loss of genetic competence, but only an inhibition of the expression of certain genetic capacities. The inhibition appears to be permanent in some animal cells, but not in many plant cells, where a single tissue cell can develop into a whole plant. We may tentatively conclude that because of their cytoplasmic environments, some nuclei can express full genetic capacity under appropriate influences whereas others cannot.

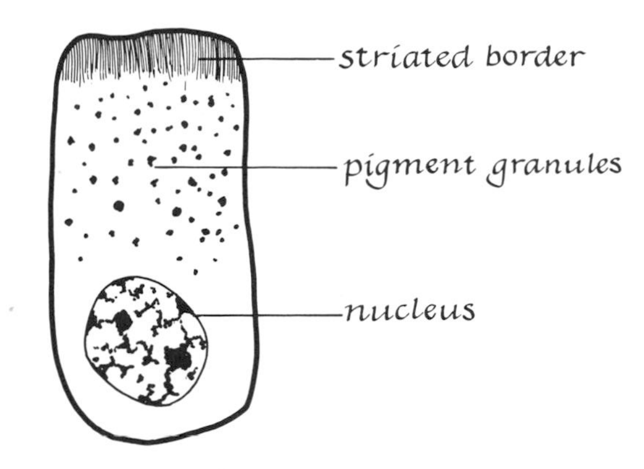

Figure 20-30. *Diagram of a relatively differentiated intestinal epithelial cell of the feeding tadpole of* Xenopus. *Nuclei of cells of this type were used as transplant nuclei by Gurdon.*

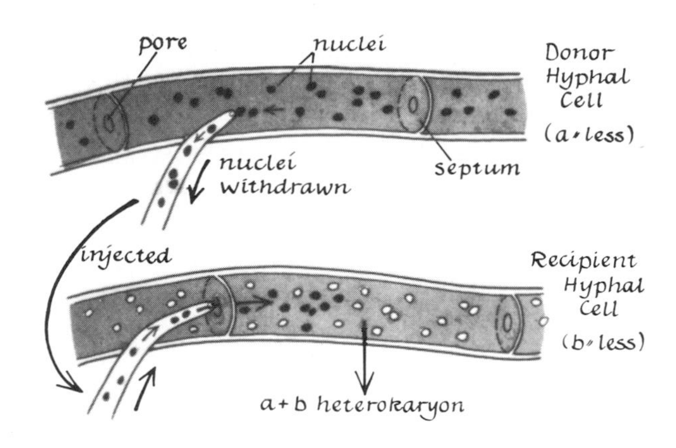

Figure 20-31. *Method of nuclear transplantation in* Neurospora.

Lucke Tumor Cell Nucleus

When an adenocarcinoma cell from a renal tumor of *Rana pipiens* was transplanted into an enucleated *R. pipiens* egg, its nucleus supported the egg's development to the swimming tadpole stage, Shumway stage 25 (see Chapter 12; see also King and McKinnell, 1960; McKinnell, 1962a; DiBerardino and King, 1965). In more recent experiments, McKinnell and co-workers (1969) dissociated and transplanted triploid renal tumor cells into activated and enucleated *R. pipiens* eggs. Pluripotency of the implanted nuclei was evidenced by the formation of swimming triploid tadpoles.

Primordial Germ Cell Nucleus

Primordial germ cells of amphibians such as *R. pipiens* are cleaved out of a distinctly staining, ultraviolet-sensitive cytoplasmic region of the egg. These cells probably give rise to all the definitive gametes (Smith, 1965). As noted earlier, somatic endoderm nuclei become restricted in their capacity to support development when transplanted into enucleated eggs. Since primordial germ cells would be expected to exhibit unrestricted capacities, their nuclei might be expected to be more competent than endodermal cell nuclei when transplanted into enucleated eggs. Experimental results (Smith, 1965) support this expectation. Nuclei from primordial germ cells promoted cleavage in 43 percent of the transplant eggs, and 40 percent of the cleaving eggs developed into normal tadpoles. Only 18 percent of somatic endodermal cell nuclei promoted cleavage, and none of the transplant eggs developed normally. Perhaps the cytoplasmic environment of the germ cell nuclei, derived from the germinal cytoplasmic region of the egg, maintains full developmental capacity in these nuclei, as the germinal plasm in *Ascaris* and insect eggs prevents a reduction in chromatin material. Fertile adult male and female *Xenopus* developed from enucleated eggs in which an endodermal cell nucleus of the blastula or gastrula stage had been transplanted (Gurdon, 1962b). However, these nuclei may have been derived from the prospective primordial germ cells which are present in the endodermal region of the blastula (Chapter 12).

Neurospora

Nuclear transplantation has been accomplished in the fungus *Neurospora crasa* (Wilson, 1963). The diagram in Figure 20-31 illustrates the method.

When nuclei from a strain of *Neurospora* which is unable to synthesize a particular amino acid *(a)* are transplanted into another strain which can synthesize that amino acid but is unable to synthesize another amino acid *(b)*, the heterokaryon (*a*-less and *b*-less nuclei) grows on a minimal medium lacking both *a* and *b* amino acids. Both *a*-less and *b*-less spores are formed during fruiting.

Stentor

Three macronuclear nodes of *Stentor polymorphus* were grafted into an enucleated *S. coeruleus*. The results revealed that the *polymorphus* nucleus could support initial but not final stages of regeneration. The animal died after three days. When transplantation is between animals of the same species, the transplanted macronucleus can support regeneration completely (Tartar, 1953, 1961). The method of transplantation is illustrated in Figure 20-32.

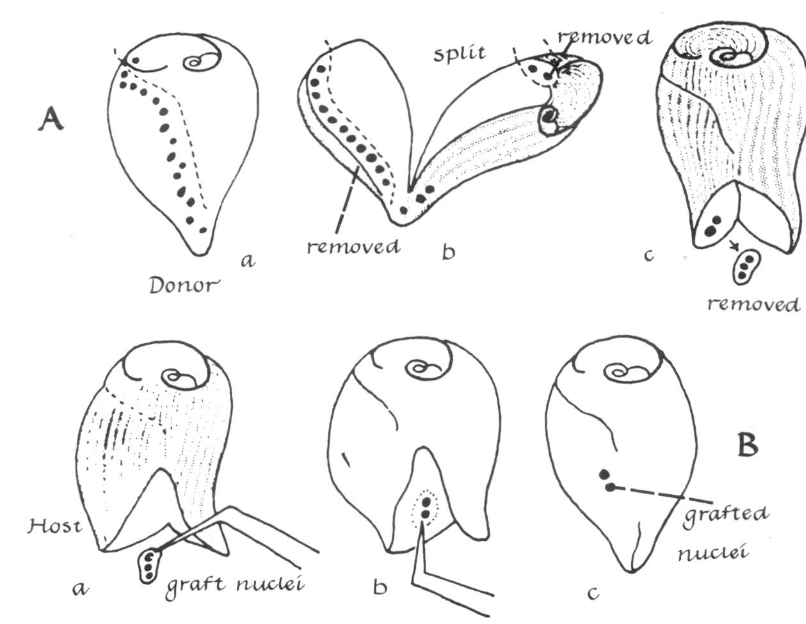

Figure 20-32. *Tartar's method of transplanting nuclei in* Stentor. *Top row: method of enucleating the host animal; bottom row: method of grafting macronuclear nodes into the enucleated host. Reprinted with permission of Dr. Tartar:* The Biology of Stentor, *copyright 1961 by Pergammon Press.*

Honey Bee

Figure 20-33 shows how preblastoderm nuclei were substituted for the egg nucleus (Du Praw, 1960, 1967). A dwarf but "normal" larva forms which, however, does not metamorphose. The reason for this is not known.

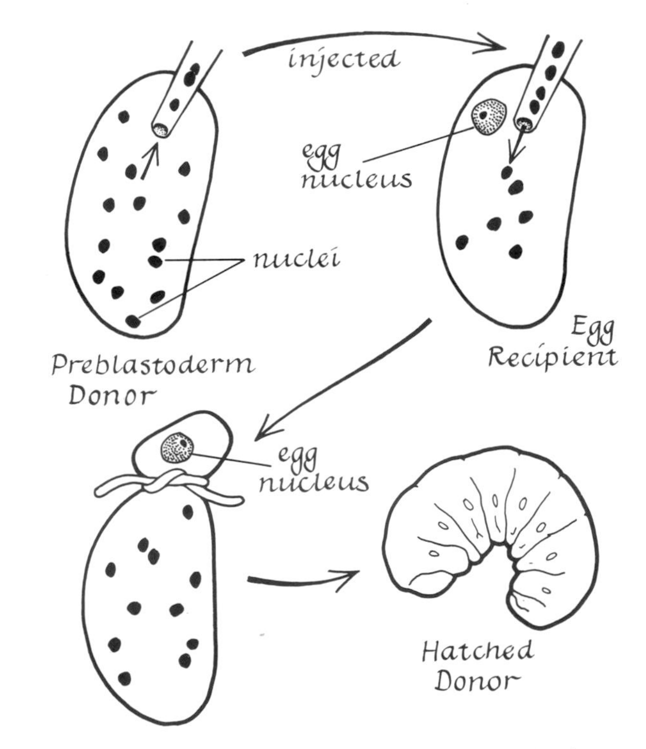

Figure 20-33. *Method of nuclear transplantation in the egg of the honey bee.*

Summary

Table 20-3 summarizes successful nuclear transplantation experiments performed up to 1964. Undoubtedly, further experiments will provide more information and will perhaps help us to understand why nuclei from some highly differentiated cells seem to lose their developmental potency when transplanted into eggs.

Conclusion

The experiments described in this chapter attest to the important role of nucleo-cytoplasmic interactions in development of all kinds of organisms. Attempting to assess the relative importance of nucleus and cytoplasm in development is futile, since the interdependence of the two cell components is the important factor. Although the nucleus is the primary locus of inherited instructions for continued developmental potentiality, the use of these instructions depends entirely upon an appropriate cytoplasmic environment (see review of Chen, 1967). The nuclear transplantation experiments show this clearly. Nevertheless, even in patterns of egg development in which

Table 20-3. Species in which living cell nuclei have been successfully transplanted, through 1964.

Species	Total Transfers	Late Blastulae (% of Total)	Tail-bud Embryos (% of Total)	Feeding Tadpoles (% of Total)	References
Rana pipiens	92	42	40	36	Briggs and King
			(usual range 40-60%)		(1957)
Rana sylvatica	24	13	?	8	Hennen (1963)
Rana nigromaculata	387	14	2	1	Sambuichi (1957)
Rana temporaria	416	7	0.5	–	Stroeva and
Rana arvalis	144	10	2	–	Nikitina (1960)
Xenopus laevis	279	62	38	36	Gurdon (1962c)
Ambystoma mexicanum	138	78	44	11	Signoret et al. (1962)
Pleurodeles waltlii	207	30	16	9	Picheral (1962)
Amoeba proteus					Danielli et al.
Amoeba discoides					(1955)
Apis mellifera	Many nuclei				DuPraw (1960)
Neurospora crassa	injected				Wilson (1963)
Stentor coeruleus	Macronuclear grafts		High percentage of success		Tartar (1953)
Acetabularia mediterranea					Hämmerling
Acetabularia crenulata	Rhizoid grafts				(1953)
Acicularia schenckii					
Acetabularia species—transfer of isolated nuclei					Werz (1962)

From Gurdon in *Advances in Morphogenesis*. Copyright 1964 by Academic Press.

cytoplasmic composition is an important guideline for nuclear expression, that composition has been established by nuclear activity during egg formation (Chapter 12). The relationship between nucleus and cytoplasm appears to be a nucleo-cytoplasmic interaction system set up in normally developing organisms under rigid genetic control. Understanding development thus becomes a matter of understanding the controlled, selective use of genetic information within the framework of a genetically established cytoplasmic and extracellular environment.

References

Arms, K. 1968. Cytonucleoproteins in cleaving eggs of *Xenopus laevis*. J. Embryol. Exp. Morph. **20**:367–374.

Balamuth, W. 1940. Regeneration in protozoa: a problem of morphogenesis. Quart. Rev. Biol. **15**:290–337.

Baltzer, F. 1910. Ueber die Beziehung zwischen dem Chromatin und der Entwicklung und Vererbungsrichtung bei Echinodermen-Bastarden. Arch. f. Zellforsch. **5**:497–621.

Bolund, L., N. Ringerta, and H. Harris. 1969. Changes in the cytochemical properties of erythrocyte nuclei reactivated by cell fusion. J. Cell Sci. **4**:71–87.

Boveri, T. 1887. Über Differenzierung der Zellkerne während den Furschung des Eies von *Ascaris megalocephala*. Anat. Anz. **2**:688–693.

Boveri, T. 1893. An organism produced sexually without characteristics of the mother. Am. Nat. **27**:222–232.

Boveri, T. 1899. Die Entwickelung von *Ascaris megalocephala* mit besonderer Rücksicht auf die kernverhältnisse. Zeit. Gustav Fischer, Jena.

Boveri, T. 1907. Zellen-Studien, VI. Entwicklung dispermer Seeigeleier. Ein Beitrag zur Befruchtungslehre und zur Theorie des Kernes. Zeit. Gustav Fischer, Jena.

Boveri, T. 1910. Ueber die Teilung centrifugierter Eier von *Ascaris megalocephala*. Arch. Entw. Mech. **30**:101–125.

Boveri, T. 1918. Zwei Fehlerquellen bei Merogonieversuchen und die Entwicklungs-fahigkeit merogonischer und partiell merogonischer Seeigelbastarde. Arch. Entw. Mech. **44**:417–471.

Brachet, J. 1952. The role of the nucleus and the cytoplasm in synthesis and morphogenesis. Symp. Soc. Exp. Biol. **6**:173–200.

Brachet, J. 1960. The Biochemistry of Development. Pergamon Press, London.

Briggs, R., E. Green, and T. King. 1951. An investigation of the capacity for cleavage and differentiation in *Rana pipiens* eggs lacking "functional" chromosomes. J. Exp. Zool. **116**:455–499.

Briggs, R., and J. T. Justus. 1968. Partial characterization of the component from normal eggs which corrects the maternal effect of gene *o* in the Mexican axolotl (*Ambystoma mexicanum*). J. Exp. Zool. **167**:105–116.

Briggs, R., and T. King. 1952. Transplantation of living nuclei from blastula cells into enucleated frogs' eggs. Proc. Nat. Acad. Sci. U.S. **38**:455–463.

Briggs, R., T. King, and M. DiBerardino. 1960. Development of nuclear-transplant embryos of known chromosome complement following parabiosis with normal embryos. In Symposium on Germ Cells and Development, pp. 441–477. Inst. Int. d'Emb.

Chen, P. S. 1967. Biochemistry of nucleo-cytoplasmic interactions in morphogenesis. In R. Weber, ed., The Biochemistry of Animal Development, pp. 115–191. Academic Press, New York.

Danielli, J. F., I. J. Lorch, M. J. Ord, and E. G. Wilson. 1955. Nucleus and cytoplasm in cellular inheritance. Nature **176**:1114–1115.

Davidson, E. H. 1968. Gene Activity in Early Development. Academic Press, New York.

Davison, J. 1959. Studies on the form of the amphibian red blood-cell. Biol. Bull. **116**:397–405.

DiBerardino, M., and N. Hoffner. 1970. Origin of chromosomal abnormalities in nuclear transplants—a reevaluation of nuclear differentiation and nuclear equivalence in amphibians. Devel. Biol. **23**:185–209.

DiBerardino, M., and T. King. 1965. Transplantation of nuclei from the frog renal adenocarcinoma: II. Chromosomal and histologic analysis of tumor nuclear-transplant embryos. Devel. Biol. **11**:217–242.

DiBerardino, M., and T. King. 1967. Development and cellular differentiation of neural nuclear-transplants of known karyotype. Devel. Biol. **15**:102–128.

Driesch, H. 1894. Analytische Theorie der Organischen Entwicklung. W. Engelmann, Leipzig.

DuPraw, E. J. 1960. Further developments in research on the honeybee egg. Glean. Bee Cult. **88**:104.

DuPraw, E. J. 1967. The honeybee embryo. In F. H. Wilt and N. K. Wessells, eds., Methods in Developmental Biology, pp. 183–217. Thomas Y. Crowell, New York.

Ephrussi, B., and M. Weiss. 1968. Regulation of the cell cycle in mammalian cells: inferences and speculations based on observations of interspecific somatic hybrids. In M. Locke, ed., Control Mechanisms in Developmental Processes, pp. 32–42. Academic Press, New York.

Fankhauser, G. 1955. The role of the nucleus and cytoplasm. In B. H. Willier, P. Weiss, and V. Hamburger, eds., Analysis of Development, pp. 126–150. W. B. Saunders, Philadelphia.

Farber, F. E., N. Cape, M. Decroly, and J. Brachet. 1968. The *in vitro* translation of *Acetabularia mediterranea* RNA. Proc. Nat. Acad. Sci. U.S. **61**:843–846.

Fischberg, M., A. W. Blackler, V. Uehlinger, J. Reynaud, A. Droin, and J. Stock. 1963. Nucleo-cytoplasmic control of development. In Proc. Int. Congr. Genetics, pp. 187–198.

Geyer-Duszynska, I. 1959. Experimental research on chromosome elimination in Cecidomyidae (Diptera). J. Exp. Zool. **141:**391–447.

Goldstein, L. 1958. Localization of nucleus-specific protein as shown by transplantation experiments in *Amoeba proteus.* Exp. Cell Res. **15:**635–637.

Goldstein, L. 1963. RNA and protein in nucleo-cytoplasmic interactions. In R. J. C. Harris, ed., Cell Growth and Cell Division, pp. 129–149. Academic Press, New York.

Goldstein, L., ed. 1967. The Control of Nuclear Activity. Prentice-Hall, Englewood Cliffs, N.J.

Goldstein, L., and D. Prescott. 1968. Proteins in nucleocytoplasmic interactions. J. Cell Biol. **36:**53–61.

Graham, C. F. 1966. The regulation of DNA synthesis and mitosis in multinucleate frog eggs. J. Cell Sci. **1:**363–374.

Graham, C. F., K. Arms, and J. B. Gurdon. 1966. The induction of DNA synthesis by frog egg cytoplasm. Devel. Biol. **14:**439–481.

Gregg, J., and S. Løvtrup. 1955. Synthesis of desoxyribonucleic acid in lethal amphibian hybrids. Biol. Bull. **108:**29–34.

Gurdon, J. B. 1962a. The developmental capacity of nuclei taken from intestinal epithelium cells of feeding tadpoles. J. Embryol. Exp. Morph. **10:**622–640.

Gurdon, J. B. 1962b. Adult frogs derived from the nuclei of single somatic cells. Devel. Biol. **4:**256–273.

Gurdon, J. B. 1964. The transplantation of living cell nuclei. Adv. Morph. **4:**1–43.

Gurdon, J. B. 1966. The cytoplasmic control of gene activity. Endeavour **25:**95–99.

Gurdon, J. B. 1967. Nuclear transplantation and cell differentiation. In A. V. S. DeReuck and J. Knight, eds., Cell Differentiation, pp. 65–78. Ciba Found. Symp., Little, Brown, Boston.

Gurdon, J. B., and D. D. Brown. 1965. Cytoplasmic regulation of RNA synthesis and nucleolus formation in developing embryos of *Xenopus laevis.* J. Mol. Biol. **12:**27–35.

Hadorn, E. 1936. Uebertragung von Artmerkmalen durch das entkernte Eiplasma beim merogonischen Tritonbastard, *palmatus* Plasma × *cristatus* kern, Verh. Deutsch Zool. Ges. Freiburg, pp. 97–104.

Hadorn, E. 1955. Letalfaktoren in ihrer Bedeutung für Erbpathologie und Genphysiologie der Entwicklung. G. Thieme, Stuttgart.

Hammerling, J. 1966. Nucleo-cytoplasmic relationships in the development of *Acetabularia.* In R. A. Flickinger, ed., Developmental Biology, pp. 23–47. William C. Brown, Dubuque, Iowa.

Harris, H. 1965. Behavior of differentiated nuclei in heterokaryons of animal cells from different species. Nature **206:**583–588.

Harris, H. 1968. Nucleus and Cytoplasm. Clarendon Press, Oxford.

Harvey, E. B. 1936. Parthenogenetic merogony or cleavage without nuclei in *Arbacia punctulata.* Biol. Bull. **71:**101–121.

Harvey, E. B. 1951. Cleavage in centrifuged eggs, and in parthenogenetic merogones. Ann. N.Y. Acad. Sci. **51:**1336–1348.

Hennen, S. 1963. Chromosomal and embryological analysis of nuclear changes occurring in embryos derived from transfers of nuclei between *Rana pipiens* and *Rana sylvatica.* Devel. Biol. **6:**133–183.

Hogue, M. J. 1910. Über die Wirkung der Centrifugalkraft auf die Eier von *Ascaris megalocephala.* Arch. Entw. Mech. **29:**109–145.

Jacobson, C. O. 1968. Reactivation of DNA synthesis in mammalian neuron nuclei after fusion with cells of an undifferentiated fibroblast line. Exp. Cell Res. **53**:316–318.

Jacobson, C. O. 1969. Production of artificial heterokaryons from mammalian neurons and various undifferentiated cells. Zool. Bidrag Fran Uppsala **38**:241–247.

Kallio, P. 1953. The effect of continued illumination on desmids. Arch. Soc. Zool. Bot. Fenn. "Vanamo" **8**:58–74.

Keck, K. 1961. Nuclear and cytoplasmic factors determining species specificity of enzyme proteins in *Acetabularia.* Ann. N.Y. Acad. Sci. **94**:741–752.

King, R. L., and H. W. Beams. 1938. An experimental study of chromatin diminution in *Ascaris.* J. Exp. Zool. **77**:425–443.

King, T. 1966. Nuclear transplantation in amphibia. In D. M. Prescott, ed., Methods in Cell Physiology, Vol. 2, pp. 1–36. Academic Press, New York.

King, T., and R. Briggs. 1956. Serial transplantation of embryonic nuclei. Cold Spring Harbor Symp. Quant. Biol. **21**:271–290.

King, T., and R. G. McKinnell. 1960. An attempt to determine the developmental potentialities of the cancer cell nucleus by means of transplantation. Cell Physiol. Neoplasia, pp. 591–617. University of Texas Press, Austin.

LaCour, L. F. 1949. Nuclear differentiation in the pollen grain. Heredity **3**:319–337.

Lehman, H. E. 1967. Nuclear transplantation, a tool for the study of nuclear differentiation. In A. Tyler, R. C. von Borstel, and C. B. Metz, eds., The Beginnings of Embryonic Development, pp. 201–230. Am. Assoc. Adv. Sci., Washington, D.C.

Lorch, I. J., and J. F. Danielli. 1950. Transplantation of nuclei from cell to cell. Nature **166**:329–330.

Lorch, I. J., and J. F. Danielli. 1953. Nuclear transplantation in Amoeba: II. The immediate results of transfer of nuclei between *Amoeba proteus* and *Amoeba discoides.* Quart. J. Micr. Sci. **94**:461–480.

Malkin, H. M. 1954. Synthesis of ribonucleic acid purines and protein in enucleated and nucleated sea urchin eggs. J. Cell Comp. Physiol. **44**:105–112.

Mazia, D., and D. M. Prescott. 1955. Role of the nucleus in protein synthesis in *Amoeba.* Biochem. Biophys. Acta. **17**:23–34.

McKinnell, R. G. 1962a. Development of *Rana pipiens* eggs transplanted with Lucké tumor cells. Am. Zool. **2**:430–431 (abstr.).

McKinnell, R. G. 1962b. Intraspecific nuclear transplantation in frogs. J. Heredity **53**:199–207.

McKinnell, R. G., B. A. Deggins, and D. D. Labat. 1969. Transplantation of pluripotential nuclei from triploid frog tumors. Science **165**:394–396.

Moore, J. A. 1958. The transfer of haploid nuclei between *Rana pipiens* and *Rana sylvatica.* Exp. Cell Res. Suppl. **6**:179–191.

Moore, J. A. 1960. Serial back-transfers of nuclei in experiments involving two species of frogs. Devel. Biol. **2**:535–550.

Morgan, T. H. 1901. Regeneration. Macmillan, New York.

Morgan, T. H. 1934. Embryology and Genetics. Columbia University Press, New York.

Poulson, D. F. 1945. Chromosome control of embryogenesis in Drosophila. Am. Nat. **79**:340–363.

Rudnick, D., ed. 1958. Cytodifferentiation. University of Chicago Press, Chicago.

Schwartz, V. 1935. Versuche über Regeneration und Kerndimorphismus bei *Stentor coeruleus* Ehrbg. Arch. Protistenk. **85**:100–139.

Seidel, F. 1936. Entwicklungsphysiologie des Insekten-Keims. Verh. Deutsche Zool. Ges. **38**:291–336.

Seidel, F. 1960. Körpergrundgestalt und Keimstruktur eine Erörterung über die Grundlagen der Vergleichenden und experimentellen Embryologie und deren Gültigkeit bei phylogenetischen Überlegungen. Zool. Anz. **164**:245–305.

Selman, G. G. 1966. Experimental evidence for the nuclear control of differentiation in *Micrasterias.* J. Embryol. Exp. Morph. **16:**469–485.

Shapiro, H. 1935. The respiration of fragments obtained by centrifuging the egg of the sea urchin, *Arbacia punctulata.* J. Cell Comp. Physiol. **6:**101–116.

Sinnott, E. W. 1960. Plant Morphogenesis. McGraw-Hill Book Company, New York.

Smith, L. D. 1965. Transplantation of the nuclei of primordial germ cells into enucleated eggs of *Rana pipiens.* Proc. Nat. Acad. Sci. U.S. **54:**101–107.

Spemann, H. 1914. Über verzöegerte Kernversorgung von Keimteilen. Verh. Deutsche Zool. Ges. pp. 216–221.

Spemann, H. 1938. Embryonic Development and Induction. Yale University Press, New Haven, Conn.

Stern, C. 1954. Two or three bristles. Am. Scientist **42:**212–247.

Stern, C. 1955. Gene action. In B. H. Willier, P. Weiss, and V. Hamburger, eds., Analysis of Development, pp. 151–169. W. B. Saunders, Philadelphia.

Tartar, V. 1953. Chimeras and nuclear transplantation in ciliates. *Stentor coeruleus* × *S. polymorphus.* J. Exp. Zool. **124:**63–104.

Tartar, V. 1961. The Biology of Stentor. Pergamon Press, New York.

Ubisch, L. von. 1943. Über die Bedeutung der Diminution von *Ascaris megalocephala.* Acta. Biotheoretica **7:**163–182.

Weisz, P. B. 1954. Morphogenesis in protozoa. Quart. Rev. Biol. **29:**207–229.

Werz, G. 1965. Determination and realization of morphogenesis in *Acetabularia.* In Genetic Control of Differentiation, pp. 185–203. Brookhaven Symp. Biol., Upton, New York.

White, M. J. D. 1946. The cytology of the cecidomyidae (Diptera): II. The chromosome cycle and anomalous spermatogenesis of *Miastor.* J. Morph. **79:**323–369.

White, M. J. D. 1947. Chromosome studies on gall midges. Carnegie Inst. Year Book **46:**165–169.

Wilson, J. F. 1963. Transplantation of nuclei in *Neurospora crasa.* Am. J. Bot. **50:**780–786.

21 *Induction: Cellular Interactions*

In multicellular organisms every cell is associated with neighboring cells. No cell develops independently of its surrounding cells; consequently, one cell's influence upon another is an important aspect of development. For example, products of one group of cells can influence or control the differentiation of other cells, either like or different from the first group, since these products partly constitute the microenvironment of the other cells. The two cell groups might be adjacent, as in the many interactions called *inductions*, or they might be far removed, as in endocrine control of cellular behavior (Chapter 19). Inductive cellular interaction will be our major concern in this chapter.

The term *induction* is also used to describe other biological processes. We speak, for example, of the induction of enzyme synthesis by a substrate, transduction in bacteria, and the induction of flowering in plants by photoperiod (Chapter 22). In this chapter the term will be used mostly to refer to cellular interactions whose results can continue, in some cases, without the inducing stimulus. In this sense the term induction usually describes how one type of cell, the *inductor*, influences a different type, the *responding cell* (or *induced cell*).

The significance of inductive cellular associations is demonstrated by removing a small group of cells from its normal association and placing it *in vitro* on a culture medium, or by grafting the cell group to a different region of the developing organism. For example, if a group of cells which normally forms nerve cells is removed from the brain-forming area of an amphibian blastula and explanted *in vitro* (Figure 21-1A), the cells usually will not become nerve cells. If such a group of cells is grafted into the prospective skin-forming region of the blastula (Figure 21-1B) they will become skin cells (Spemann, 1938). If prospective skin-forming cells are grafted into the brain-forming region, they will become brain cells (Figure 21-1C). In short, the realization of a cell's fate depends upon its association with surrounding cells.

The cells that influence development of ectoderm cells into brain cells are the chorda-mesoderm cells which underlie them. Recall (Chapter 12) that gastrulation movements bring the chorda-mesoderm cells into position beneath the neural plate area of the ectoderm. The chorda-mesoderm then induces the ectoderm cells to specialize as nerve cells. The induction of the *neural plate* (prospective brain and spinal cord) is called "primary induction" (Spemann and Mangold, 1924; Saxen and Toivonen, 1962) because it is a very early manifestation of one group of cells influencing another, and because it results

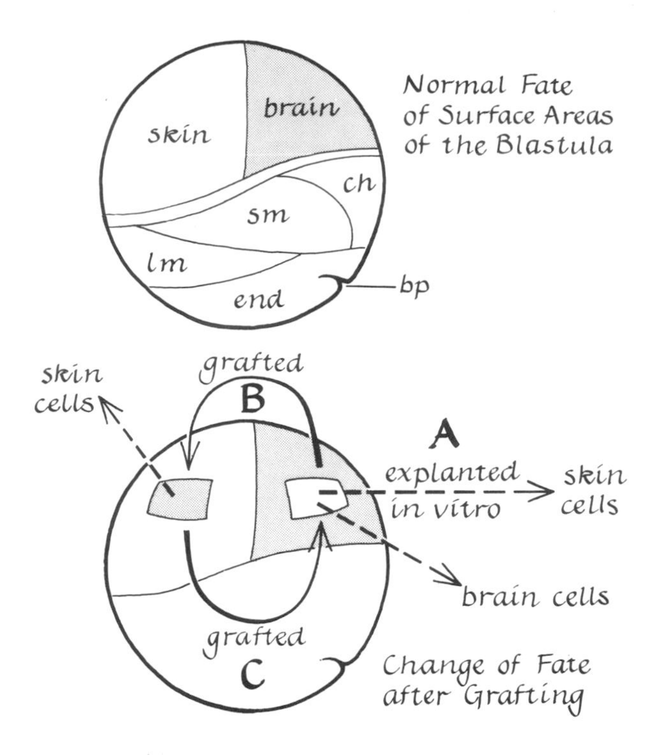

Figure 21-1. *Top, diagram illustrating the normal fate of areas on the surface of the early gastrula of a urodele amphibian. Bottom, the change in fate of a group of prospective skin cells when grafted into the brain-forming area and change in fate of prospective brain cells when grafted into the skin-forming area. Also shown is the change in fate of prospective brain cells when explanted in vitro. Bp, blastopore; sm, somatic mesoderm; lm, lateral mesoderm; end, endoderm; ch, chorda.*

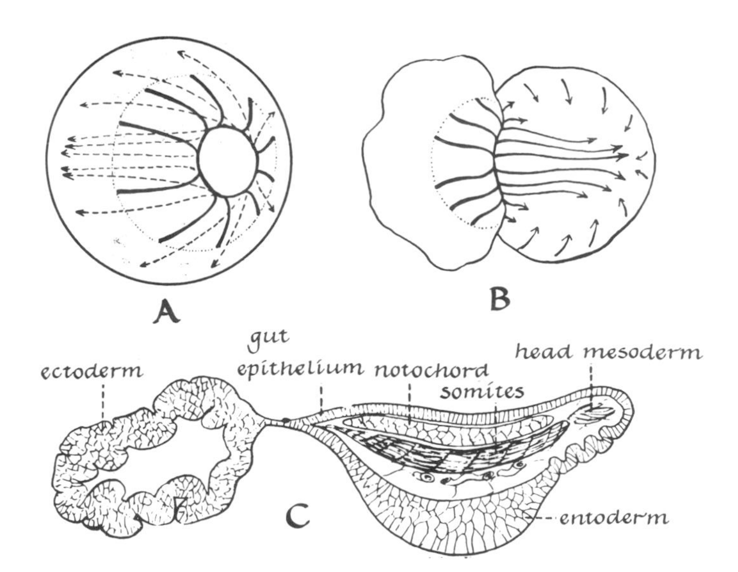

Figure 21-2. *Diagram illustrating the morphogenetic movements, A, in normal gastrulation and, B, in exogastrulation. C, differentiation in an exogastrulated embryo. After J. Holtfreter and V. Hamburger in Willier, Weiss, and Hamburger,* Analysis of Development, *W. B. Saunders Company, Philadelphia, 1955.*

in establishment of the basic body plan of the embryo. Other types of induction, particularly in animal embryos, are usually described as nonprimary or secondary.

If invagination of the chorda-mesoderm during gastrulation in amphibian species is blocked by treating the gastrula with LiCl, *exogastrulation* occurs and no neural plate develops (Holtfreter and Hamburger, 1955). This is illustrated in Figure 21-2.

Primary Induction

Spemann and Mangold (1924) reported the first clear demonstration of primary induction. This significant experiment is diagramed in Figure 21-3. A group of chorda-mesoderm cells just anterior to the dorsal lip of the blastopore in an early *Triturus (Triton) cristatus* gastrula was cut out and grafted into the flank or belly region of the gastrula of *Triturus taeniatus*. The grafted cells invaginated and disappeared from the surface by rolling under the belly ectoderm. Donor and host gastrulae were from darkly and lightly pigmented species in order to trace the fate of the graft cells.

Two significant events then occurred: First, the graft (darkly pigmented cells) self-differentiated into a rather complete mesodermal axis of the body: axial notochord, somites, pronephros, and lateral plate (Chapter 12). This is the normal fate of this gastrula region. Host (lightly pigmented) mesoderm cells were partially incorporated in the mesodermal axial system; for instance, a somite might contain both dark and light (donor and host) cells, constituting a chimaeric organ. Secondly, the mesodermal axis induced the overlying lightly pigmented host cells in the ectoderm to form a complete neural plate matching the underlying mesodermal structures. Finally an almost complete "secondary embryo" was formed. Spemann (1938) called the chorda-mesodermal cell

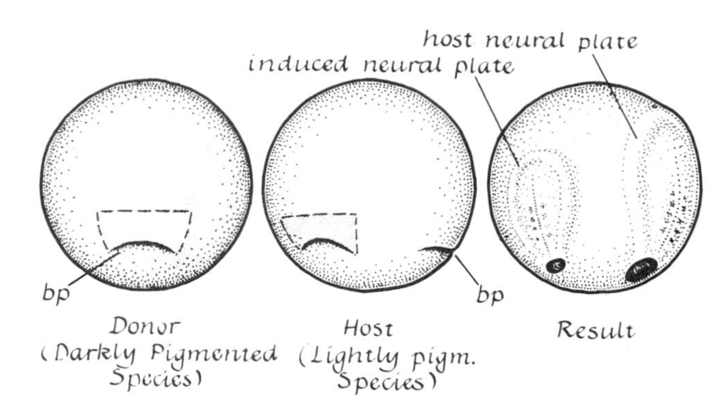

Figure 21-3. *Diagram and result of Spemann and Mangold's experiment illustrating primary induction.*

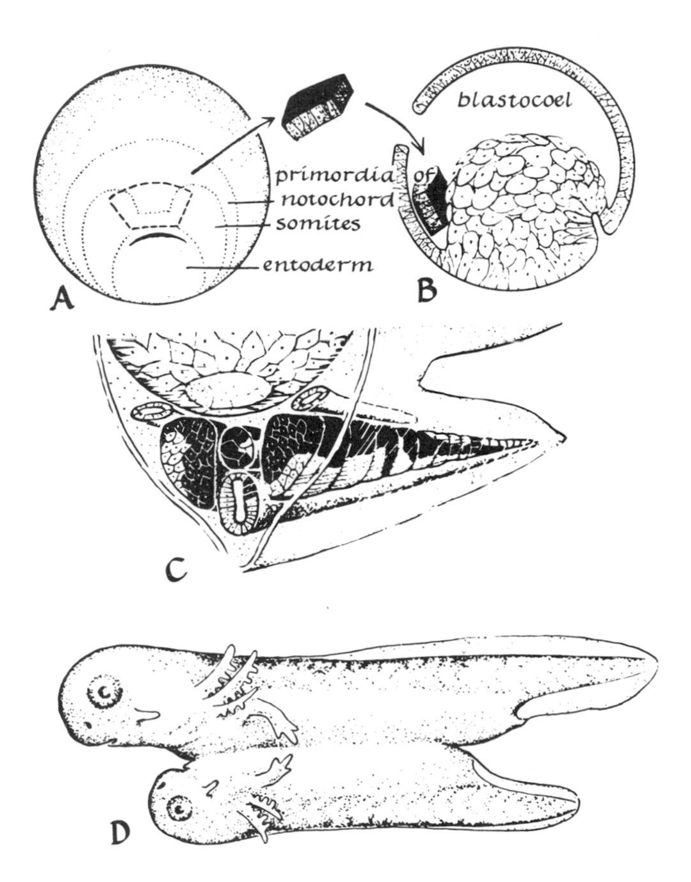

Figure 21-4. *A and B, diagram of the transplantation of a piece of the dorsal blastoporal lip into the blastocoel of another gastrula. C and D, self-differentiation and induction by the graft. In C, the tissues derived from the graft are shown in black, the induced tissues of the host in white. After J. Holtfreter and V. Hamburger in Willier, Weiss, and Hamburger,* Analysis of Development, *W. B. Saunders Company, Philadelphia, 1955.*

group an "organizer" or "organization center" because of its precocious self-differentiating capacity and because of its ability to induce an organized (regionalized and patterned) central nervous system.

Little is known about the chemical basis of primary induction, despite many attempts to identify the substance or substances presumably released by the chorda-mesoderm cells. Saxen and Toivonen (1962) and Tiedeman (1967) discuss a chemical approach to the problem of the organizer. Spemann and Mangold found that only grafts from the dorsal lip and adjacent prospective mesodermal regions induced a response in the host gastrula.

The experiment illustrated in Figure 21-3 has been repeated many times with slight modifications such as the one shown in Figure 21-4. Note in the figure that a group of cells of the dorsal lip is inserted into the host gastrula through an opening on its ventral side. In this region of the host, the graft induces a secondary embryo (Holtfreter and Hamburger, 1955). Primary induction has been demonstrated in the chick as well as in amphibians. In birds the node and adjacent anterior portion of the primitive streak (comparable in function to an amphibian blastopore) functions like an organizer by inducing a secondary neural plate when implanted below prospective skin ectoderm (Waddington, 1952).

Types of Induction and Competence

There are two general types of induction: *homoiogenetic (homotypic)* and *heterogenetic (heterotypic).* An example of homotypic induction is the induction by grafted nerve cells of more nerve cells out of responsive epidermis cells. Another example is the induction by somite cells of more somite cells out of lateral plate mesoderm cells. The inducer thus influences surrounding cells to develop into cells like the inducer. An example of heterotypic induction is the induction by chorda-mesoderm cells of nerve cells out of responsive epidermis cells. Here the inductor influences adjacent cells to develop into a third cell type. Both types of induction occur in primary induction.

Induction may involve an influence of one cell type upon another, the inductor not being significantly influenced by the responding cell, as in primary inductions. Alternatively, it may involve cellular interaction, as in secondary induction. In both types of induction (homo- or heterotypic, and whether interactive or not), not only must the inductor be competent to induce but the cells which it influences must be able to respond. In general, as cells' developmental age increases, their ability to respond to fully adequate inductors decreases (Saxen and Toivonen, 1962). Further, as inductors get older, their potency decreases. Finally, no inductor can influence the differentiation of originally competent cells.

Selected Examples of Induction

Induction of the neural plate, and in some cases a secondary embryo, by grafts of underlying chordamesoderm is apparently one of the more complex examples of induction. Other induction systems might be more amenable to experimental analysis (Wolff, 1966). We shall now survey a few apparently simpler induction systems.

Vascular Tissue of Plants

When a bud of *Syringa vulgaris* was grafted to an undifferentiated mass of callus tissue growing in culture, groups of vascular tissues were induced at a distance from the graft in the tissue (Wetmore and Sorokin, 1955). In a similar experiment, a bud grafted to a piece of chicory root also induced vascular tissue. Induction occurred even when porous cellophane was inserted between the grafted chicory bud and the root tissue (Camus, 1949). Thus some inductions require no cellular continuity between the inductor (the bud) and the responding tissue (callus or root). This is illustrated in Figure 21-5.

In the system illustrated in Figure 21-5, at least one inductive substance produced by the grafted shoot apex is an auxin, since a plug of agar impregnated with auxin and sucrose and inserted into a notch in the callus tissue induces xylem differentiation (Wetmore and Sorokin, 1955). At certain concentrations of sucrose, phloem and cambial tissues can also be induced, and the spatial arrangement of these tissues in relation to the xylem closely simulates that in a section through a normal stem (Chapters 4 and 22).

Another experiment which demonstrates the induction of vascular tissue in plants is diagramed in

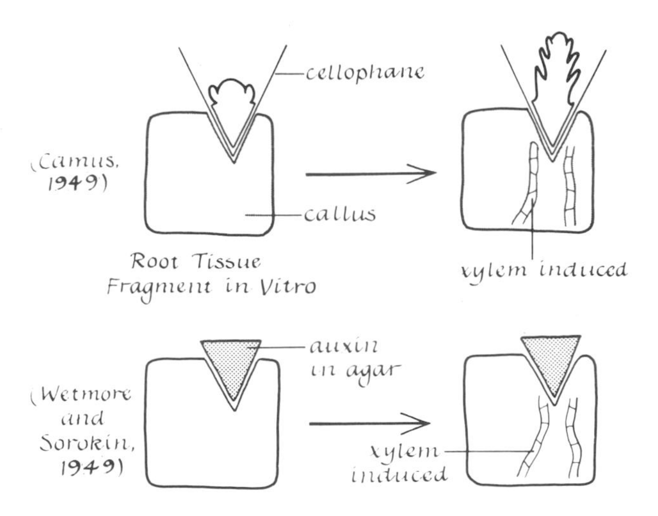

Figure 21-5. *Induction of vascular tissue in callus tissue by a grafted shoot apex (upper) and by a plug of agar containing auxin (lower).*

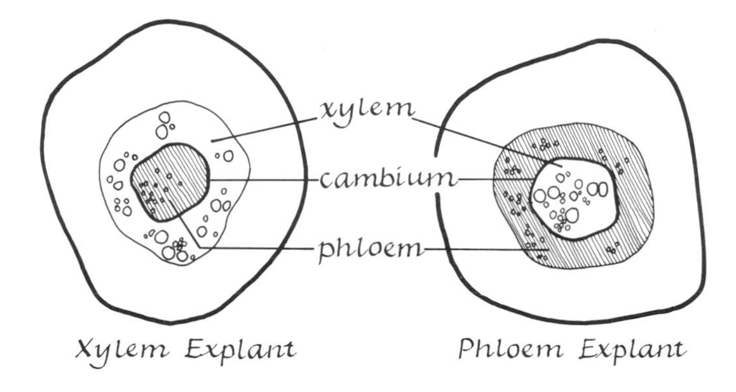

Figure 21-6. *Homoiogenetic induction of xylem in a callus derived from xylem tissue and of phloem in a callus derived from phloem tissue. Note that in both cases a complete vascular bundle forms with phloem and xylem separated by a ring of cambial cells.*

Figure 21-6. In the differentiation of vascular bundles inside pieces of explanted callus tissue derived from xylem or phloem tissue of Jerusalem artichoke (Figure 21-6), the new xylem is formed adjacent to the xylem explant and the new phloem adjacent to the phloem explant (Gautheret, 1965). This seems to be an example of homotypic induction in which the position of the new xylem or phloem in the vascular bundles is controlled by the nature of the explanted tissue. Of special interest, however, is the

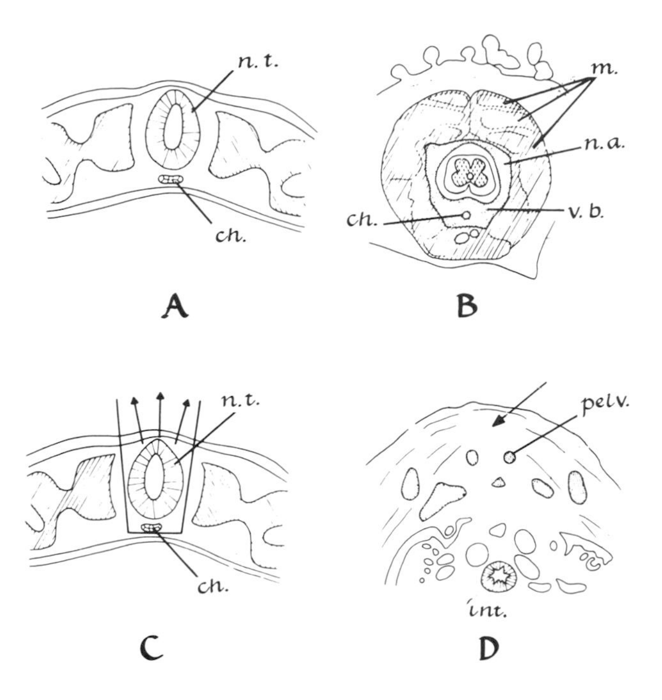

Figure 21-7. *The inducing role of chorda and spinal cord in the formation of vertebrae. A and B, cross section of normal chick embryo at 40 hours and 7 days of incubation; C, removal of chorda and spinal cord at 40 hours; D, the result at 7 days: no formation of vertebrae.* N.T., *neural tube;* Ch, *chorda;* M, *muscles;* n.a., *neural arch;* v.b., *vertebral body;* Int., *intestine;* pelv., *pelvic girdle. After Strudel, from Wolff, 1966, courtesy John Wiley and Sons, Inc.*

fact that xylem, always separated by cambium from phloem, develops in the callus regardless of its origin. Thus a complete bundle, not just xylem or phloem alone, is formed. The mechanisms controlling this kind of patterned differentiation are not known. Induction in this case probably involves establishing an internal environmental gradient system to which the cells respond in accord with their positions (Wardlaw, 1966). We shall examine this possibility in Chapter 22.

Vertebrae

The formation of the vertebrae in vertebrates results from an inductive interaction between the notochord, spinal cord, and somite mesoderm (Strudel, 1955). This is illustrated in Figure 21-7.

Epithelio-Mesenchymal Inductions

The metanephric kidney, pancreas, salivary glands, thyroid, epidermal skin derivatives, and the like, develop from an interaction of epithelial and mesenchymal cells (see reviews of Wessells, 1964, 1970 and Fleischmajer, 1968). For example, if mouse kidney mesenchyme and epithelium are separated and grown *in vitro,* neither tissue develops the tubule systems characteristic of the secreting and collecting systems of the normally developing kidney. If reassociated, the epithelium of the ureteric bud component is induced by the mesenchyme to branch. The mesenchyme is also induced (interactively) by the epithelium to form secreting tubules and glomeruli (Grobstein, 1955). The dorsal half of the embryonic mouse spinal cord also will induce tubule formation in embryonic mouse kidney mesenchyme. Tubule-inducing capacity thus is not confined to the epithelium of the ureteric bud.

In general, epithelial cells, including skin epidermis cells, cease dividing when separated from mesenchyme cells. Normally, epidermis remains unkeratinized on the cornea, secretes mucus in the nasal passages, forms keratin on the body surfaces, and becomes a single squamous sheet in the extraembryonic membranes. When undifferentiated skin epidermis of the chick is combined *in vitro* with different kinds of mesenchyme, each kind of mesenchyme induces a characteristic response in the epidermis cells (McLoughlin, 1963). Figure 21-8 shows some types of epithelio-mesenchymal inductions.

A striking demonstration of the role of epithelio-mesenchymal inductions in the development of epidermal derivatives in birds has been reported by Rawles (1963). Chick epidermal cells can form feathers, beaks, or scales. When prospective feather-forming epidermis from the back is grafted to dermis of the leg, it develops scales instead of feathers. When back epidermis is grafted to the dermis of the beak region, the cells form a beak. This is an example of specific induction. The specificity of induction is considered in a separate section later in this chapter. It is clear that the epidermis cells are uncommitted until induced to specialize in directions provided by the dermis of the chick body.

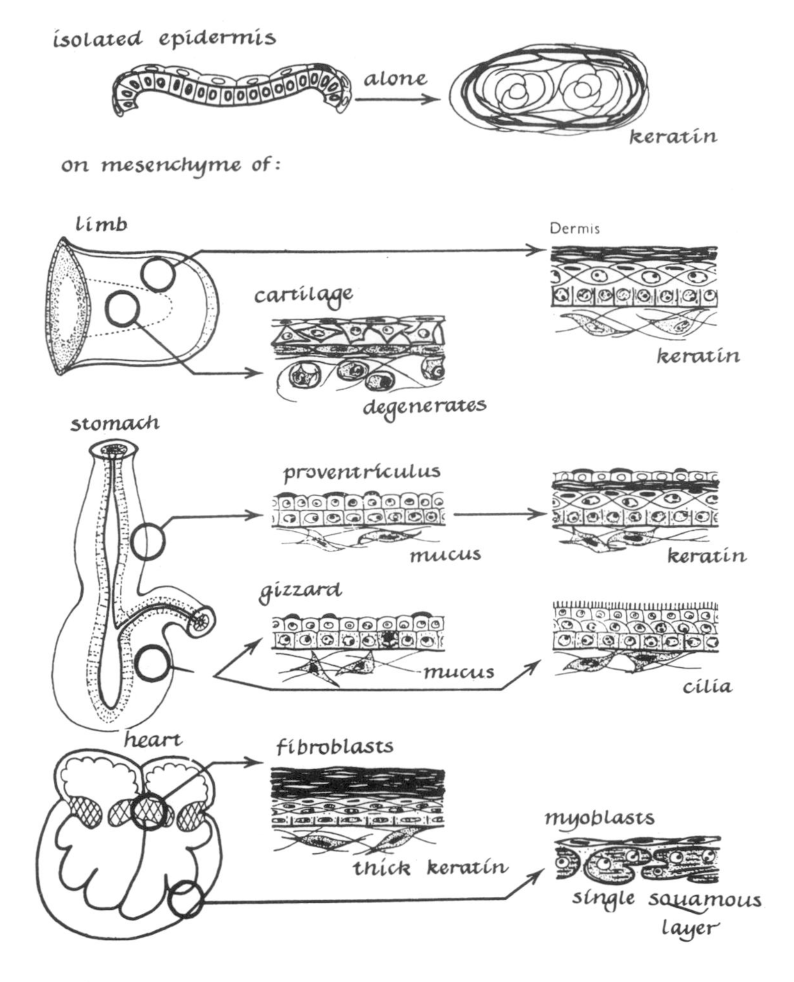

Figure 21-8. *The result of explanting epidermis on different types of connective tissue of the chick embryo. Epidermis explanted on limb bud mesenchyme becomes keratinized but degenerates or completely keratinizes when explanted on limb cartilage. Epidermis explanted on mesenchyme of the proventriculus of the stomach secretes mucus and later keratinizes. But when explanted on the mesenchyme of the gizzard, the epidermal cells secrete mucus and form cilia, and no keratinization occurs. Epidermis on heart myoblasts develops into a single layer of squamous cells, and no keratinization occurs. But when explanted on heart fibroblasts, the epidermis keratinizes. From McLoughlin, 1963. Courtesy Cambridge University Press.*

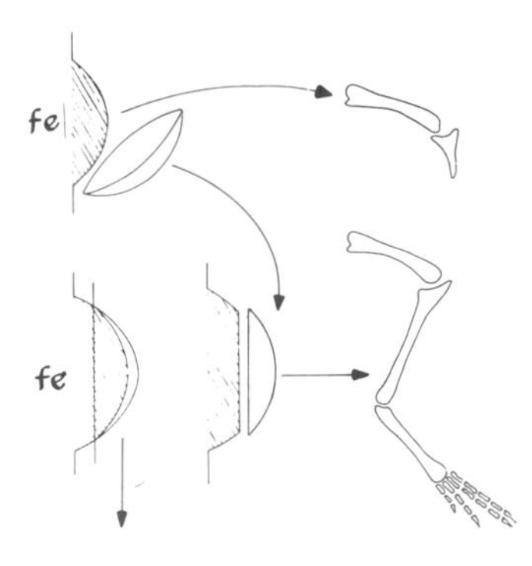

Figure 21-9. *Induction by the ectodermal apical cap of the chick leg bud. Removal of the cap results in a limb with only the femur and proximal part of the tibia present. Grafting an apical cap to the basal part of the leg bud (after removal of the distal part) results in formation of a complete leg skeletal apparatus. In this latter case, the apical cap induces the basal mesoderm of the limb bud to form the distal parts of the leg skeleton. After Hampé, from Wolff, 1966, courtesy John Wiley and Sons, Inc.*

Lens

In many vertebrates, removing the optic cup before the lens has formed blocks its formation. Conversely, a grafted optic cup below head and trunk ectoderm will induce lens formation in prospective skin cells (see Weiss, 1939, for review).

Limb Skeleton

The apical (ectodermal) cap that covers the developing limb bud in the chick embryo induces formation of a complete apendicular skeleton (Saunders, 1948; Hampe, 1960; Saunders and Gasseling, 1968). Experiments demonstrating this are shown in Figure 21-9.

Endo-mesodermal Inductions

Development of components of the circulatory system of vertebrates depends upon interaction between the endoderm and mesodermal germ layers. For example, normal development of the salamander heart seems to depend upon a specific heart inductor located in anterior endoderm (Jacobson and Duncan, 1968). In the chick embryo endoderm appears to pro-

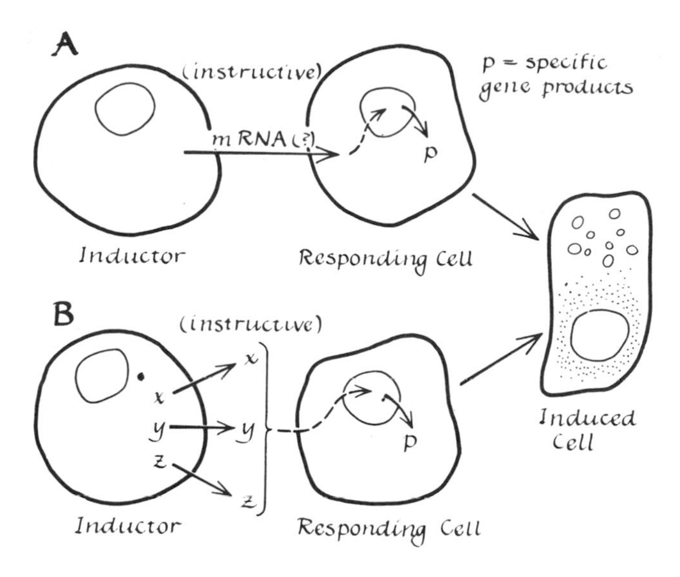

Figure 21-10. *Two possible mechanisms of induction.*

vide a stimulus for blood island development (Miuria and Wilt, 1969).

Mechanisms of Induction

Very little is known about the mechanisms of inductive actions or interactions (Grobstein, 1966; Tiedeman, 1967). This was evident at a workshop on morphogenetic tissue interactions held in Nokkala, Finland, in 1969. In a review of this workshop, Ebert (1969) reported much discussion on whether inductive interactions are permissive or instructive. Perhaps no mechanism is common to all types of induction. Possibly the inductor transfers informational molecules to the responsive cells, but this has yet to be proved in inductive systems of normally developing organisms. Perhaps induction results from permissive environmental conditions that elicit a specific response in the competent, responsive cells. These two possible mechanisms are diagramed in Figure 21-10.

As indicated in the figure, the induced cell characteristically responds by synthesizing specific types of proteins. This synthesis does not occur in the absence of the inductor cell; thus, in isolated ectoderm of *Triturus*, labeled precursors of RNA indicate that rRNA and tRNA synthesis occurs but that mRNA synthesis is less than in ectoderm combined with inductive mesoderm (Tiedeman, 1967).

Care must be exercised in using the expressions *instructive* and *permissive*. Whether mediated by informational molecules imparted to the responding cells or by the environmental composition surrounding these cells, the inductor cells are still instructing or directing the kind of response. Clearly this is the case in inductive systems of normally developing organisms.

Whatever the nature of the inductor's influence on the responsive cells might be, it is obvious that some mechanism of cell-cell communication is involved. One or more of at least eight different methods of cellular communication appear to be used. These are diagramed in Figure 21-11.

Note in the figure that desmosomata (or plasmodesmata) constitute the most intimate intercellular connections. In plants, there is contact via plasmodesmata between the endoplasmic reticulum of two adjacent cells (Clowes and Juniper, 1968). Such intimate contact is rare in animal cells.

Biologists are beginning to give methods of cellular communication much-needed attention. In addition to the hormonal (a), inductor substance (b), and morphological types of communication (c, d, e, f, g, and h) illustrated in Figure 21-11, certain cells, including those of early embryos, may be electrically (ionically) coupled (Sheridan, 1968; Ito and Lowenstein, 1969; Bennet and Trinkaus, 1970). It will be interesting if this coupling can be correlated with inductive systems. Tight junctions (Figure 21-11F), which preferentially allow small ions to pass, might be instrumental in the flow of signals controlling gene activity concerned with development (Lowenstein, 1969).

The inductor's influence in most systems appears to be exerted over a period of time. Induction may be a gradual, stepwise process. Completion of the induction requires a series of influences passing from the inducer cells. In induction of the neural plate, there is a primary, activating principle (neural factor or evocator) which seems to bring about a "low grade" of neuralization (for instance, induction of a neural plate). This is followed by a superimposed transforming (regionalizing) principle (mesodermal factor) which specifies the antero-posterior pattern of cell specialization within the neural plate (Toivonen and Saxen, 1968). The mechanics of the gastrula-

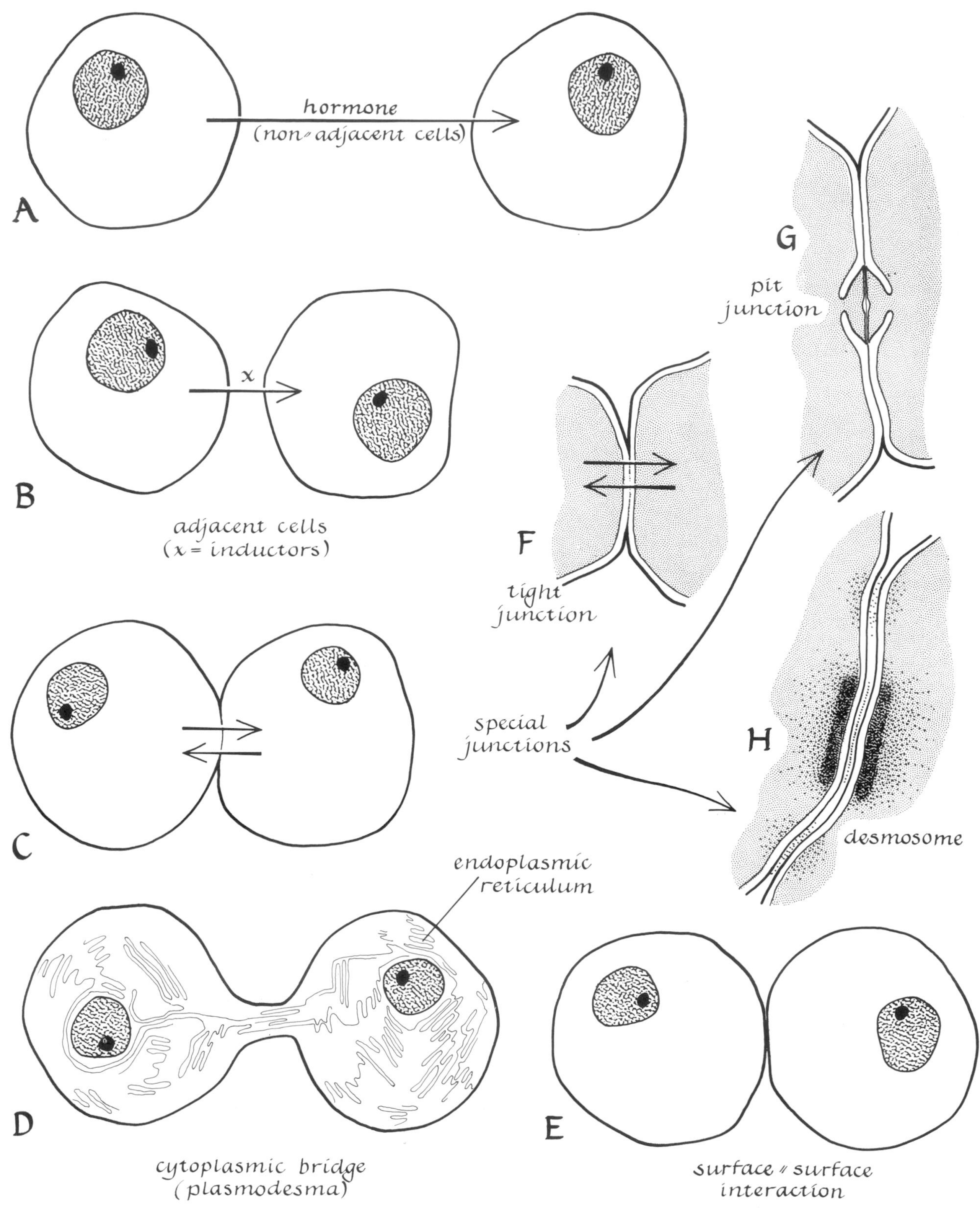

Figure 21-11. *Types of cell-cell communication that might be used in induction.*

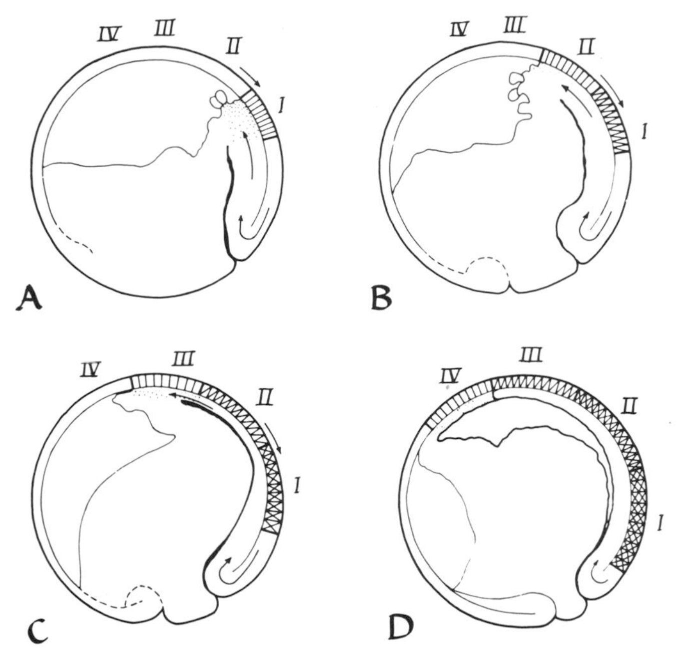

Figure 21-12. *Diagrams illustrating the oppositely directed movements of invaginating chorda-mesoderm and epibolically spreading neural ectoderm during amphibian gastrulation. From Nieukoop, 1966, courtesy John Wiley and Sons, Inc.*

tion process in amphibians illustrates how the two steps operate on a time scale (Nieukoop, 1966) as shown in Figure 21-12.

Note in Figure 21-12 that regions I through IV of the neural plate area touch the underlying chorda-mesoderm, cm, for an increasingly long time as gastrulation proceeds. Note also that the length of contact time is maximal in region I and minimal in region IV. Length of time of contact between chorda-mesoderm and the ectoderm, although probably involved in neural plate induction, is apparently not the only mechanism operating in induction in amphibians. It has been suggested that the chorda-mesoderm is regionalized, its anterior portion releasing the activating principle and its posterior portion releasing the transforming principle (Nieukoop, 1967). The chorda-mesoderm thus functions in some respects like a preformed template, stamping out the regional pattern in the neural plate, but doing this progressively as the chorda-mesoderm moves inward and forward below the ectoderm during gastrulation. It has long been known that regional differences in the chorda-mesoderm are reflected in differences in inductive capacity. Mesoderm anterior

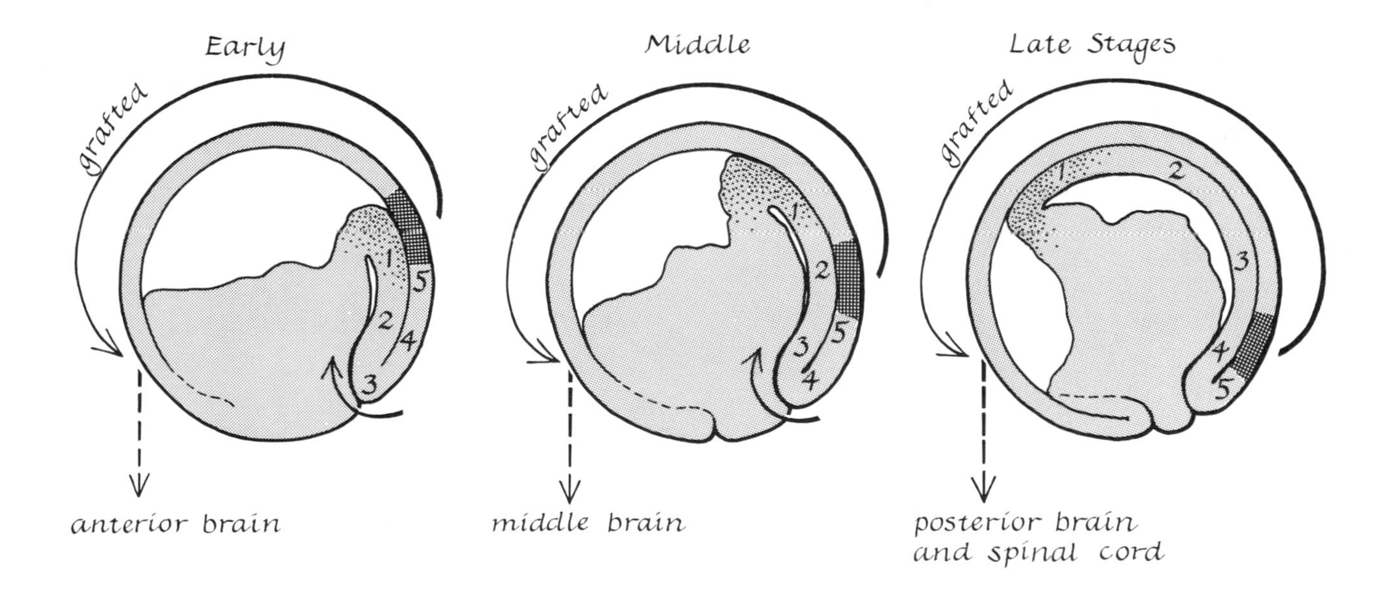

Figure 21-13. *Diagrams illustrating the experiments of Eyal-Giladi.*

to the end of the notochord induces forebrain structures whereas the chordal and parachordal mesoderm induces more posterior brain structures (Lehmann, 1945).

The length of contact between chorda-mesoderm and ectoderm is important, at least in some cases, in determining which part of the neural axis is induced. This is demonstrated by Eyal-Giladi's experiments (1954), in which pieces of the neural-plate area of the *Axolotl* and *Pleurodeles* were grafted to the ventral side of the gastrula at different stages (Figure 21-13), and by experiments of Johnen (1961), in which a "sandwich" method was used (Figure 21-14).

As illustrated in Figure 21-13, the same region of the posterior neural plate (lined area) was grafted to the ventral side of the gastrula. Induction of progressively more posterior parts of the brain with increasing age of the gastrula was observed. Thus, early contact between the chorda-mesoderm resulted in induction of anterior brain whereas later and longer contact resulted in induction of more posterior parts of the brain in the graft.

In the sandwich method, a piece of ectoderm from a middle-stage gastrula of an *Axolotl* was folded around a piece of chorda-mesoderm (Figure 21-14A). After varying periods of time (one-half to six hours), the chorda-mesoderm was removed and discarded and the ectoderm allowed to continue to develop. The results, shown in Figure 21-14B, indicate an increase in the number of explants in which more posterior neural regions were induced when the period of contact was made longer. Other experiments have produced similar results (Saxen and Toivonen, 1962).

Many investigators have tried to discover exactly how the inductor cells influence the differentiation of adjacent cells. Some of their attempts are described below.

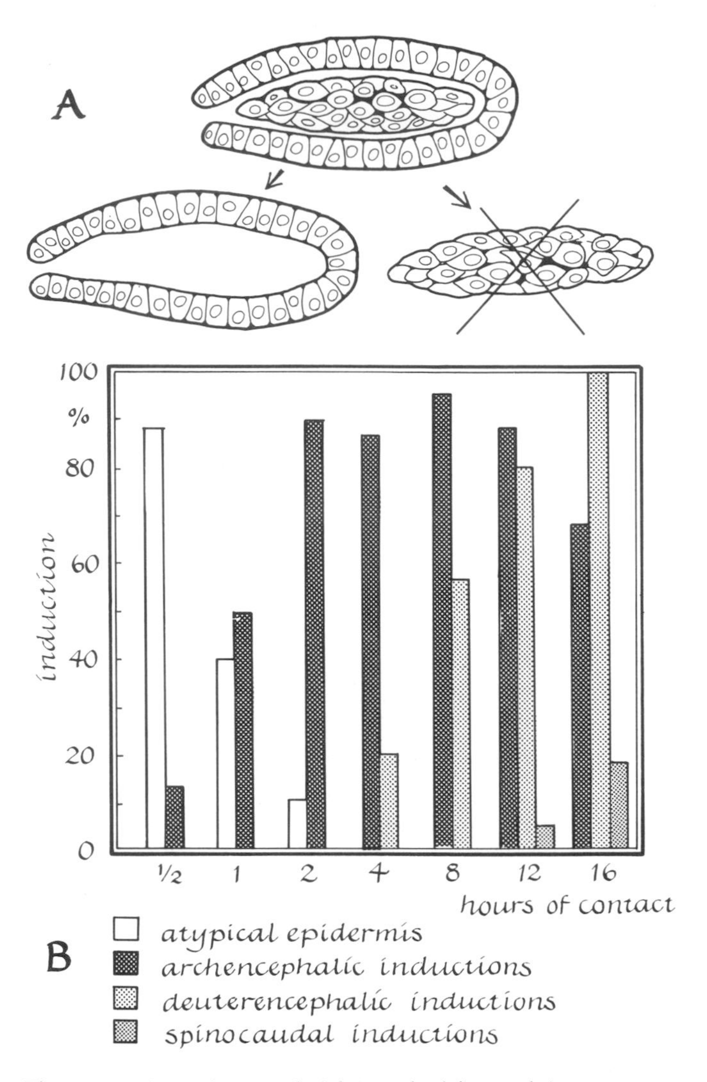

Figure 21-14. *A, the "sandwich" method for studying induction; B, graph illustrating the percentage of archencephalic (forebrain), deuterencephalic (midbrain and hindbrain), and spinocaudal inductions in which archenteron roof was the inductor and the length of its contact with competent epidermis varied from ½ to 16 hours. After Johnen in Saxen and Toivonen. Copyright 1962 by Academic Press.*

Transfilter Induction

Early studies suggested that induction required physical contact between inductor cells and responsive cells. Grobstein's more recent studies (1955, 1956, 1957, 1967) indicate that such direct cell-cell contact is not required. When a group of explanted mouse kidney mesenchyme cells and dorsal spinal cord cells were separated by a millipore filter (20 μ thick) whose pores were apparently too small (0.45 μ)

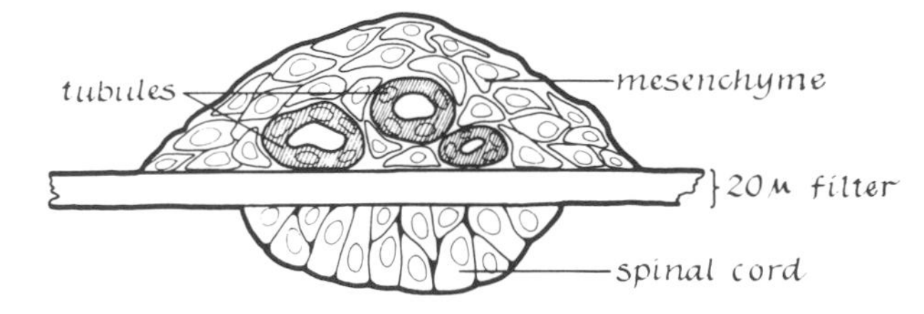

Figure 21-15. *Diagram of the transfilter induction of tubules in mouse embryo mesenchyme by mouse spinal cord cells.*

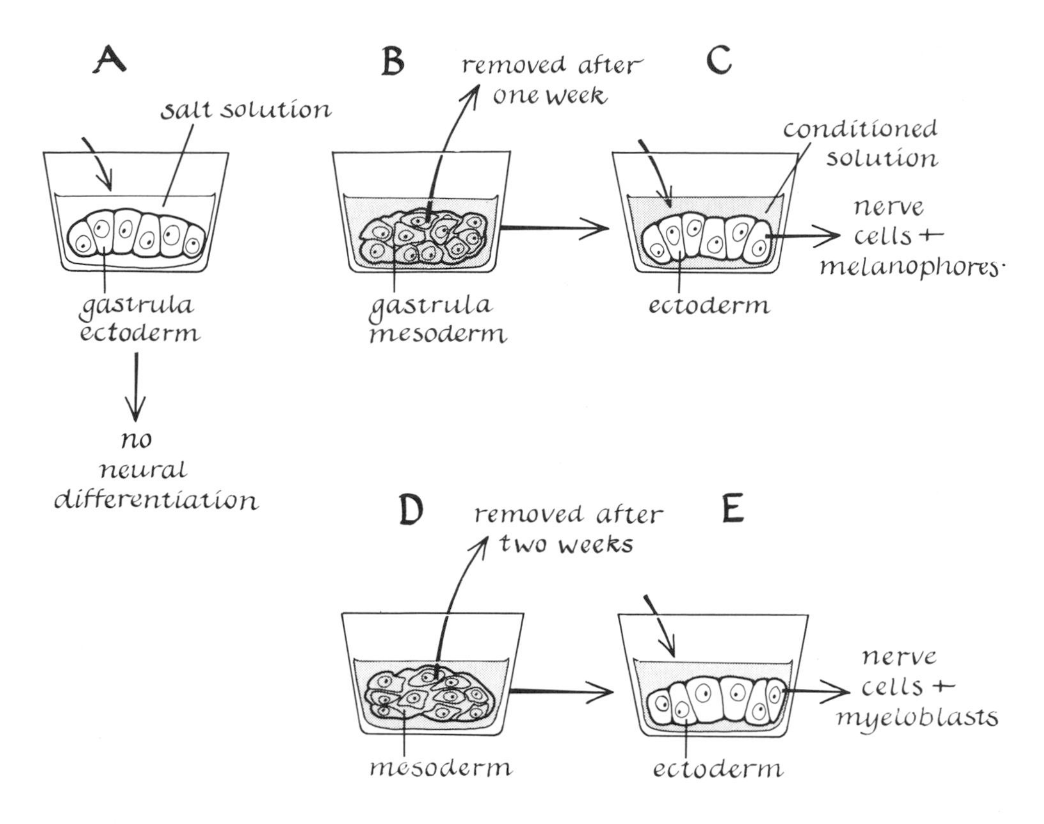

Figure 21-16. *Experiments by Niu and Twitty suggesting the diffusion of an inductor from explanted mesoderm into the culture medium and its later action on competent ectoderm.*

to allow any cell-cell contact, induction of tubules in the mesenchyme occurred (Figure 21-15).

Grobstein (1957, 1961) also made electron micrographs of preparations like the one illustrated in Figure 21-15 and found no evidence of direct cell-cell contact—no complete penetration of the filter by cellular processes such as filapodia. Note that induction in this system occurs vertically; that is, tubules develop only in the mesenchyme above the spinal cord cells. The pores run vertically through the filter.

Diffusion of the Inductor

Using *Triturus torosus* and other amphibians, Niu and Twitty (1953) performed the experiment illustrated in Figure 21-16.

Niu (1956) later reported that the conditioned medium (Figure 21-16C) had an adsorption spectrum similar to that of nucleoprotein. This suggests that the mesoderm releases an information-carrying inductor molecule into the culture medium, but much work must be done before the transfer of specific informational molecules can be proved to be a mechanism of induction.

Transmission Problem

A number of investigators have demonstrated the transfer of radioactively labeled compounds from the inducing cell group to the responding group (Vainio et al., 1962; Koch and Grobstein, 1963). Combining ectoderm of *Triturus* with guinea pig bone marrow labeled with antigen, Vainio et al. have demonstrated the transfer of large molecular antigenic material by the fluorescent antibody technique. However, no studies have demonstrated that any known inducing molecules are transferred during induction in animal embryos (Grobstein, 1967).

However, nonembryonic types of induction, such as the induction of crown galls in plants (Chapter 16) appear to involve specific informational molecules.

Simple Molecules as Inductors

Ions can induce various cell types from prospective epidermis of *Rana pipiens* gastrula (Barth and Barth, 1963, 1964, 1966, 1967). A graded series of concentrations of an inorganic ion (for example, Li^+, Na^+, K^+, Ca^{++}, Mg^{++}, HCO_3) can induce cell types ranging from nerve cells to pigment cells (Figure 21-17).

A constant concentration of the same ion applied for different durations of treatment gives various cell types. It has been suggested that the ions act directly on DNA complexes, but this appears unlikely. These experiments suggest that the composition of the environment, not a specific information-carrying molecule, mediates induction of nerve or pigment cell types. Recall (Chapter 16) that the proportion of Na^+ to Ca^{++} ions influences cellular differentiation in *Hydra*.

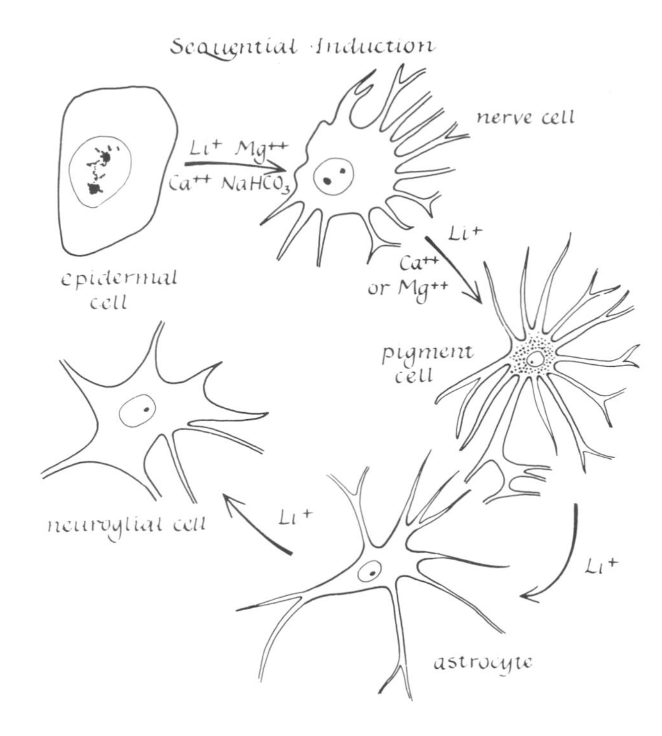

Figure 21-17. *Simple molecules producing sequential inductions in explanted epidermis of* Rana pipiens.

Release of Inductors

It has been known for many years that living tissues which do not exhibit inductive activity may do so after being killed or treated in various ways. An experiment involving treatment with glutathione, which presumably releases latent inductive capacities, is illustrated in Figure 21-18 (Waheed and Mulhekar, 1967).

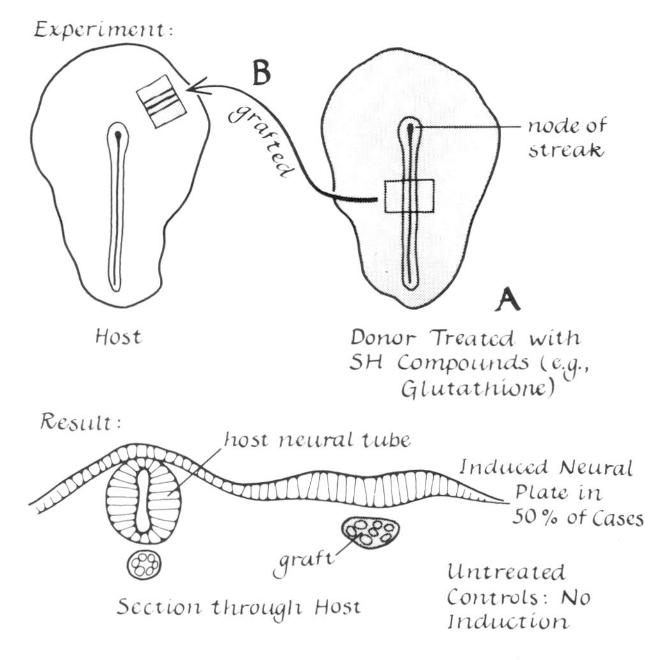

Figure 21-18. *Diagram illustrating the release of latent inductive capacity in a graft of the primitive streak of the chick embryo by treatment with glutathione.*

Proportion of Inductor to Responder Cells

In experiments on primary induction described by Toivonen and Saxen (1968), cells of the prospective forebrain region and axial mesoderm of *Triturus vulgaris* neurulae were disaggregated and combined in different ratios. As the ratio of mesodermal (inductor) cells to prospective forebrain (responsive) cells increased, the induction of more posterior regions of the central nervous system increased correspondingly (Figure 21-19).

The results shown in Figure 21-19 suggest that, in normal development, the quantity of inductor substances emanating from the mesodermal cells is a

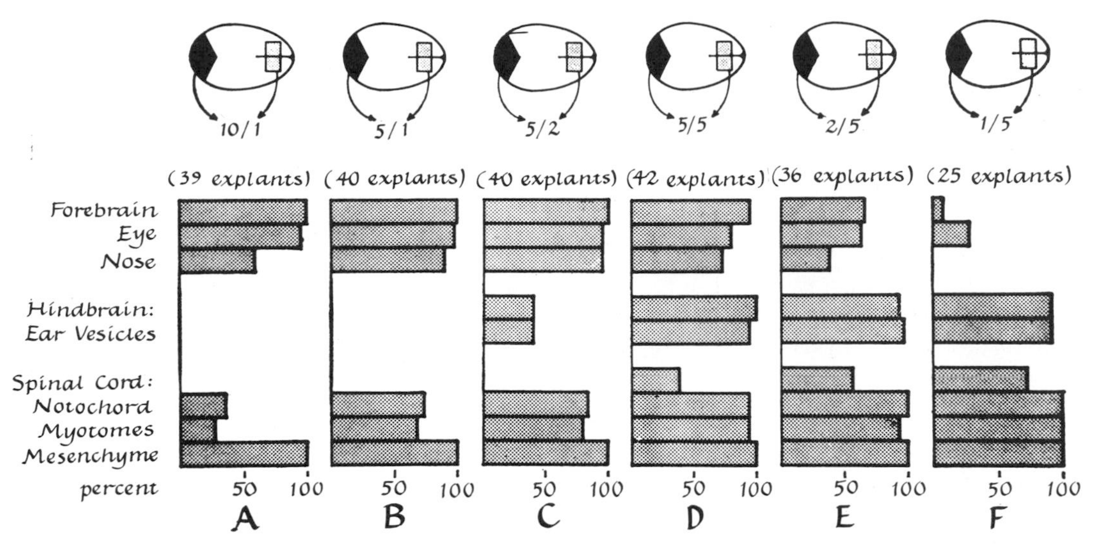

Figure 21-19. *Percentages of different structures in reaggregates of prospective forebrain and chorda-mesoderm cells cultivated for 14 days in vitro. The schemes above the bar graph indicate the ratios of responsive ectoderm cells to mesodermal (inductor) cells in reaggregates of types A to F. From Toivonen and Saxen,* Science, *Volume 159, pp. 539–540, 1963. Copyright by the American Association for the Advancement of Science.*

factor in regionalization (antero-posterior) of the neural plate. In normal development the length of time the prospective neural plate is exposed to inductive influences varies from a minimal time in the anterior part to a maximal time in the posterior region (see Figures 21-12 and 21-13). Thus the more posterior portions of the plate are exposed to more inductive agents than the anterior parts. Spratt and Haas (1960, 1961) and Spratt (1966) suggested a similar hypothesis for epigenetic induction of the neural axis in the chick blastoderm.

Specificity versus Nonspecificity of Induction

In normal development inductive actions are specific and orderly. However, nonspecific inductions have been demonstrated in artificial induction systems (Goetinek, 1966). Recall that mouse spinal cord as well as mouse kidney epithelium (ureteric bud) induces tubule formation in kidney mesenchyme (Grobstein, 1955). Also, in the pancreas system, any kind of mesenchyme induces development of the duct system in the epithelial component. However, only salivary gland mesenchyme induces branching of the epithelial component in the salivary gland.

In general, epithelio-mesenchymal inductions are highly specific. These include induction of feathers, scales, claws, and the uropygial gland of birds. In these and other types of epithelio-mesenchymal inductions, collagen secreted as an extracellular product by the mesenchyme cells is organized into a fabric within the basement membrane that separates the two types of tissue (Chapter 23). Whether this collagen fabric has any effect in the specificity of induction is an unsolved problem, but its position suggests that it might be responsible for orienting epithelial cells (McLoughlin, 1963).

A type of genetic specificity resident in the mesoderm is illustrated in combinations of chick limb-bud ectoderm with duck limb-bud mesoderm and the reverse (Hampe, 1960). This is illustrated in Figure 21-20. Note that the limb's webbed character is determined by the genetic composition of the mesoderm, not the ectoderm. In a mutant that produces webbing in chick legs, combining the mutant mesoderm with duck ectoderm results in a webbed leg (Goetinck, 1966). Note also that duck ectoderm plus normal chick mesoderm produces an unwebbed leg.

Ebert (1966) describes an experiment indicating that specific information-carrying molecules might mediate some types of induction. An antigen (type A) was mixed with macrophages. After a suitable interval the macrophages were homogenized. Their RNA was extracted and was added to lymphocytes growing in culture. Lymphocytes do not ordinarily make antibodies unless primed with antigen. Thus control cultures produce no antibody. However, the lymphocytes to which the RNA was added made antibody specific for the antigen (type A) that had been added to the macrophages.

Specific types of induction involving the transfer of genetic material occur in bacteria. One type of transfer called *transformation* was discovered by Griffith (1928) in pneumoccocal types. However, only in 1944 did Avery and co-workers demonstrate that the transforming substance released by one type of bacterium and picked up by another was DNA. Part of the DNA released to the culture medium is integrated with the DNA of the transformable bacteria. Zinder and Lederberg (1952) discovered another type of transfer called *transduction*. This involves transferring a small part of the DNA from one bacterium to another by a bacterial virus called a *bacteriophage*.

Transfer of genetic material from a bacterium to a multicellular organism also occurs under experimental conditions. One example is the transfer of ^{32}P-labeled DNA from the bacterium *M. lysodeikticus* to the DNA of root-cell nuclei in germinating barley seeds. In this experiment, the barley seeds were placed in solutions of labeled bacterial DNA; this DNA was integrated with the barley DNA within about two hours. The bacterial DNA is replicated in the barley nuclei (Ledoux and Huart, 1968).

A strikingly specific kind of induction occurs in the development of insect galls in plants. These are abnormal but organized growths that are often complex in internal and external morphology (Figure 21-21); they are thus distinct from crown galls (Figure 16-11).

Although the chemical nature of the substance injected by the insect into the plant tissue when it stings or lays its egg in the plant leaf or stem is not known, the type of gall produced is specific to the species of insect. Different species of insects can induce different types of galls on the same species of plant. Thus, the inductive agent is highly specific (Braun, 1969), but how it directs the response of the plant cells is unknown.

In some of these latter types of specific induction, the participation of informational molecules has been demonstrated, but how general such a mechanism might be and whether it exists in embryonic induction systems remains to be demonstrated.

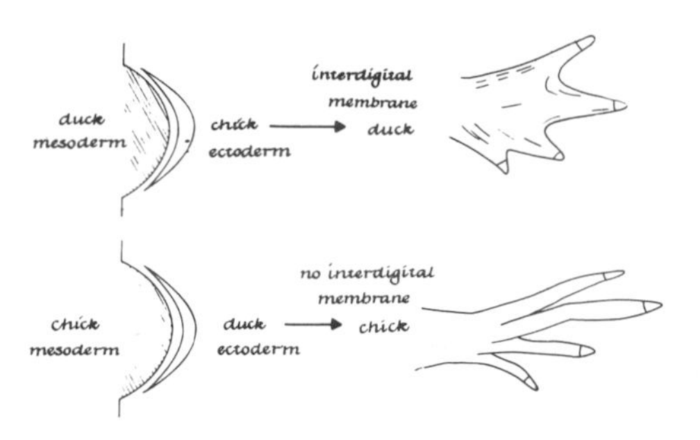

Figure 21-20. *Specific induction of the webbing character (interdigital membrane) by duck mesoderm. After Hampé, from Wolff, 1966, courtesy John Wiley and Sons, Inc.*

Conclusion

Cellular interactions in development are extremely important. Many kinds of inductive interactions are known, and apparently no mechanism applies to all of them. It is unlikely that any one kind of molecule is the inductive agent in any kind of induction in animals (Grobstein, 1966). Probably many agents (both large and small molecules) released by inductor cells constitute a specific microenvironment to which competent cells respond. There is no reason to assume that the composition of this inductive microenvironment in normal development is less specific and regulated than that of the cells whose products contribute to it.

References

Avery, O. T., C. M. MacLeod, and M. McCarty. 1944. Studies on the chemical nature of the substance inducing transformation of pneumococcus types: induction of transformation by a desoxyribonucleic acid fraction isolated from pneumococcus Type III. J. Exp. Med. **79:**137–158.

Barth, L. G., and L. J. Barth. 1963. The relation between intensity of inductor and type of cellular differentiation of *Rana pipiens* presumptive epidermis. Biol. Bull. **124:**125–140.

Barth, L. G., and L. J. Barth. 1964. Sequential induction of the presumptive epidermis of the *Rana pipiens* gastrula. Biol. Bull. **127:**413–427.

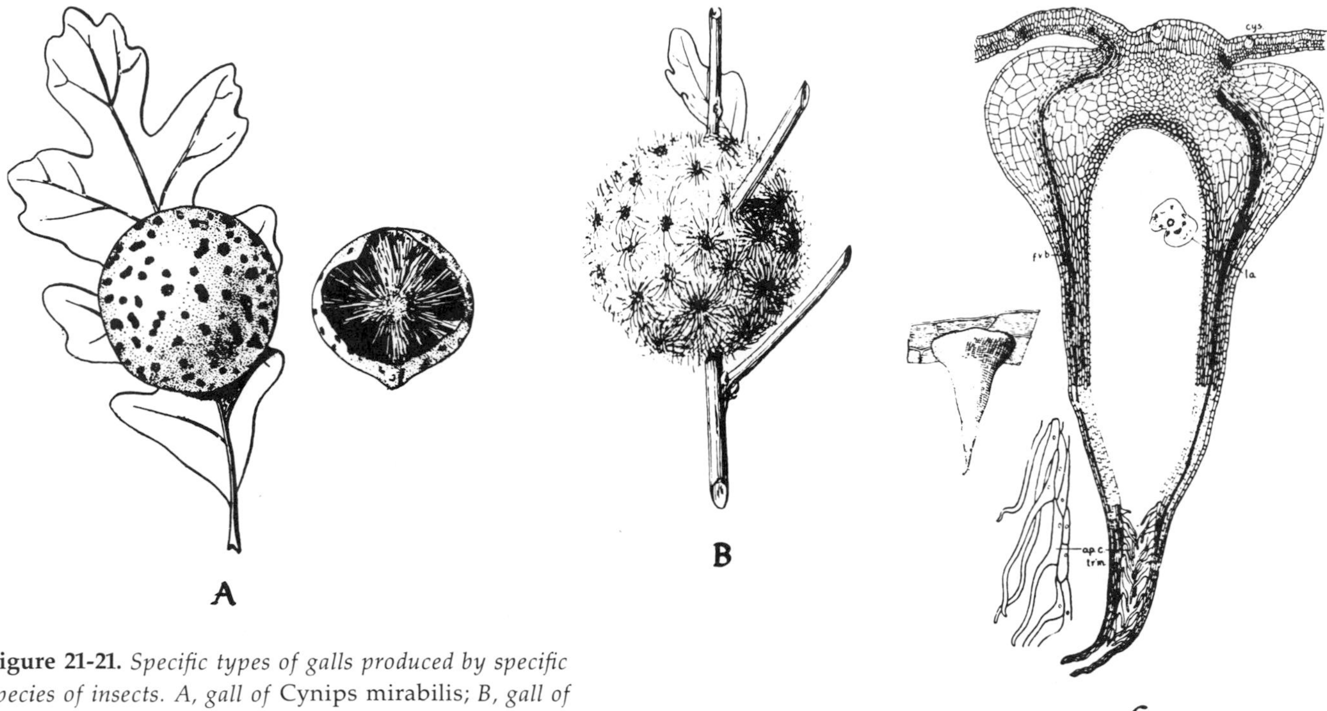

Figure 21-21. *Specific types of galls produced by specific species of insects. A, gall of* Cynips mirabilis; *B, gall of wool sower,* Callirhytis seminator; *C, gall of* Cecidomyia unquicola. *Note the enormous increase in number of plant cells and the differentiation of layers of cells in the gall as shown in the enlarged section in C. Reprinted from E. P. Felt:* Plant Galls and Gall Makers, *copyright 1940 by Comstock Publishing Company, used by permission of Cornell University Press.*

Barth, L. G., and L. J. Barth. 1966. The role of sodium chloride in sequential induction of the presumptive epidermis of *Rana pipiens* gastrulae. Biol. Bull **131:**415–426.

Barth, L. G., and L. J. Barth. 1967. The uptake of Na-22 during induction in presumptive epidermis cells of *Rana pipiens* gastrula. Biol. Bull. **133:**495–501.

Bennett, M. V. L., and J. P. Trinkaus. 1970. Electrical coupling between embryonic cells by way of extracellular space and specialized junctions. J. Cell Biol. **44:**592–610.

Braun, A. C. 1969. Abnormal growth in plants. In F. C. Steward, ed., Plant Physiology, Vol. 5B, pp. 379–420. Academic Press, New York.

Camus, G. 1949. Recherches sur le rôle des bourgeons dans les phénomènes de morphogénèse. Rev. Cytol. Biol. Veg. **9:**1–199.

Clowes, F. A. L., and B. E. Juniper. 1968. Plant Cells. Blackwell, Oxford.

Ebert, J. D. 1966. Interacting Systems in Development. Holt, Rinehart and Winston, New York.

Ebert, J. D. 1969. Morphogenetic tissue interactions. Science **166:**1314–1316.

Eyal-Giladi, H. 1954. Dynamic aspects of neural induction in Amphibia. Arch. Biol. **65:**179–259.

Fleischmajer, P., ed. 1968. Epithelio-Mesenchymal Interactions. Williams & Wilkins, Baltimore.

Gautheret, R. J. 1965. Factors affecting differentiation in plant tissues grown *in vitro*. In Cell Differentiation and Morphogenesis, pp. 55–95. North Holland Publishing Company, Amsterdam.

Goetinek, P. F. 1966. Genetic aspects of skin and limb development. In A. A. Moscona and A. Monroy, eds., Current Topics in Developmental Biology, pp. 253–283. Academic Press, New York.

Griffith, F. 1928. The significance of pneumococcal types. J. Hygiene **27**:113–159.

Grobstein, C. 1955. Inductive interaction in the development of the mouse metanephros. J. Exp. Zool. **130**:319–340.

Grobstein, C. 1956. Transfilter induction of tubules in mouse metanephrogenic mesenchyme. Exp. Cell Res. **10**:424–440.

Grobstein, C. 1957. Kidney tubule induction in mouse metanephrogenic mesenchyme without cytoplasmic contact. J. Exp. Zool. **135**:57–73.

Grobstein, C. 1961. Cell contact in relation to embryonic induction. Exp. Cell Res. Suppl. **8**:234–245.

Grobstein, C. 1966. Mechanisms of organogenetic tissue interaction. In Cell, Tissue and Organ Culture, pp. 279–299. 2nd Decennial Rev. Conf., Bedford, Pa.

Grobstein, C. 1967. The problem of the chemical nature of embryonic inducers. In A. V. S. DeReuck and J. Knight, eds., Cell Differentiation, pp. 131–138. Ciba Found. Symp., Little, Brown, Boston.

Hampé, A. 1960. Sur l'induction et la compétence dans les relations entre l'epiblaste et le mésenchyme de la patte de Poulet. J. Embryol. Exp. Morph. **8**:246–250.

Holtfreter, J., and V. Hamburger. 1955. Amphibians. In B. H. Willier, P. Weiss, and V. Hamburger, eds., Analysis of Development, pp. 230–296. W. B. Saunders, Philadelphia.

Ito, S., and W. R. Loewenstein. 1969. Ionic communication between early embryonic cells. Devel. Biol. **19**:228–243.

Jacobson, A. G., and J. T. Duncan. 1968. Heart induction in salamanders. J. Exp. Zool. **167**:79–103.

Johnen, A. G. 1961. Experimentelle Untersuchungen über die Bedeutung des Zeitfactors beim Vorgang der Neuralen Induktion. Arch. Entw. Mech. **153**:1–13.

Koch, W. E., and C. Grobstein. 1963. Transmission of radioisotopically labeled materials during embryonic induction *in vitro*. Devel. Biol. **7**:303–323.

Ledoux, L., and R. Huart. 1968. Integration and replication of DNA of *M. lysodeikticus* in DNA of germinating barley. Nature **218**:1256–1259.

Lehman, F. E. 1945. Einführung in die Physiologische Embryologie. Birkhäuser, Basel.

Lowenstein, W. R. 1969. Emergence of order in tissues and organs. Communication through cell junctions. Implications in growth control and differentiation. In M. Locke, ed., The Emergence of Order in Developing Systems, pp. 151–183. Academic Press, New York.

McLoughlin, C. B. 1963. Mesenchymal influences on epithelial differentiation. Symp. Soc. Exp. Biol. **17**:359–388.

Miuria, Y., and F. H. Wilt. 1969. Tissue interactions and the formation of the first erythroblasts of the chick embryo. Devel. Biol. **19**:201–211.

Nieukoop, P. D. 1966. Induction and pattern formation as primary mechanisms in early embryonic differentiation. In Cell Differentiation and Morphogenesis, pp. 120–143. John Wiley, New York.

Nieukoop, P. D. 1967. The organization centre: II. Field phenomena, their origin and significance. Acta. Biotheoretica **17**:151–177.

Niu, M. C. 1956. New approaches to the problem of embryonic induction. In D. Rudnick, ed., Cellular Mechanisms in Differentiation and Growth, pp. 155–171. Princeton University Press, Princeton, N.J.

Niu, M. C., and V. C. Twitty. 1953. The differentiation of gastrula ectoderm in medium conditioned by axial mesoderm. Proc. Nat. Acad. Sci. U.S. **39**:985–989.

Rawles, M. 1963. Tissue interactions in scale and feather development as studied in dermal-epidermal recombinations. J. Embryol. Exp. Morph. **11**:765–789.

Saunders, J. W. 1948. The proximo-distal sequence of origin of the parts of the chick wing and role of the ectoderm. J. Exp. Zool. **108**:363–403.

Saunders, J. W., and M. T. Gasseling. 1968. Ectodermal-mesenchymal interactions in the origin of limb symmetry. In P. Fleischmajer, ed., Epithelio-Mesenchymal Interactions, pp. 78–97. Williams & Wilkins, Baltimore.

Saxen, L., and S. Toivonen. 1962. Primary Embryonic Induction. Logos Press, London.

Sheridan, J. D. 1968. Electrophysiological evidence for low resistance in intercellular junctions in the early chick embryo. J. Cell Biol. **37**:650–659.

Spemann, H. 1938. Embryonic Development and Induction. Yale University Press, New Haven, Conn.

Spemann, H., and H. Mangold. 1924. Über Induktion von Embryonalanlagen durch Implantation artfremde Organisatoren. Arch. f. Mik. Anat. u. Entw. Mech. **100**:599–638.

Spratt, N. T., Jr. 1966. Some problems and principles of development. Am. Zool. **6**:9–19.

Spratt, N. T., Jr., and H. Haas. 1960. Integrative mechanisms in development of the early chick blastoderm: I. Regulative potentiality of separated parts. J. Exp. Zool. **145**:97–137.

Spratt, N. T., Jr., and H. Haas. 1961. *Ibid.*: II. Role of morphogenetic movements and regenerative growth in synthetic and topographically disarranged blastoderms. J. Exp. Zool. **147**:57–94.

Strudel, G. 1955. L'action morphogène du tube nerveux et de la chorde sur le différenciation des vertébres et des muscles vertébraux chez l'embryon de poulet. Arch. Anat. Micr. Morph. Exp. **44**:209–235.

Tiedeman, H. 1967. Biochemical aspects of primary induction and determination. In R. Weber, ed., The Biochemistry of Animal Development, Vol. 2, pp. 3–55. Academic Press, New York.

Toivonen, S., and L. Saxen. 1968. Morphogenetic interaction of presumptive neural and mesodermal cells mixed in different ratios. Science **159**:539–540.

Vainio, T., L. Saxen, S. Toivonen, and J. Rapola. 1962. The transmission problem in primary embryonic induction. Exp. Cell Res. **27**:527–538.

Waddington, C. H. 1952. The Epigenetics of Birds. Cambridge University Press, London.

Waheed, M. A., and L. Mulhekar. 1967. Studies on induction by substances containing sulphydryl groups in post-nodal pieces of chick blastoderm. J. Embryol. Exp. Morph. **17**:161–169.

Wardlaw, C. W. 1966. Leaves and buds: mechanisms of local induction in plant growth. In Cell Differentiation and Morphogenesis, pp. 96–119. North Holland Publishing Company, Amsterdam.

Weiss, P. 1939. Principles of Development. Henry Holt, New York.

Wessells, N. K. 1964. Tissue interactions and cytodifferentiation. In Differentiation and Development, pp. 139–155. Little, Brown, Boston.

Wessells, N. K. 1970. Some thoughts on embryonic induction in relation to determination. J. Investigative Dermatology **55**:221–225.

Wetmore, R. H., and S. Sorokin. 1955. On the differentiation of xylem. J. Arnold Arboret. **36**:305–317.

Wolff, E. 1966. General Introduction: General factors in embryonic differentiation. In Cell Differentiation and Morphogenesis, pp. 1–23. John Wiley, New York.

Zinder, N. D., and J. Lederberg. 1952. Genetic exchange in *Salmonella*. J. Bact. **64**:679–699.

22 *Environmental Control of Development*

The influence of the immediate environment on the activities of genes, nuclei, cells, and organisms is important enough to merit a separate chapter. We have defined development as the action of genes in creating a new organism, repairing defects, and adding to an adult organism. As we have seen in preceding chapters, there is much evidence that an organism's development is the result of interaction between the genome and its external environment. This raises the old question of the relative roles of heredity and environment in development. Both are essential, but in some instances one is more important than the other; thus, no general statement can be made in answer to the question.

Genes, cells, even whole organisms do not make all of their own decisions during their development. Many of these are made for them by their surroundings. Almost all of the preceding chapters provide examples of the role of the environment in development. This chapter presents the concept of an environmental continuum which controls development at all levels of biological organization and shows how the internal environment of an organism arises and increases in complexity during development. The environment of the genome is apparently the most important controlling influence in development. There is, of course, no such entity as the environment. There is a continuum of environments beginning with the immediate surroundings of the genes and ending with the outside world (Figure 22-1).

The composition of the organism's internal (intercellular) environment is largely a product of gene action. By the action of certain genes at one time (during oogenesis, for example) the genome provides for the control of the action of other genes later in development. This provides the orderly program of sequential gene action that is the cardinal feature of development.

We shall first consider some of the ways in which the developing organism builds up the internal environment surrounding the genes, cells, tissues, and organs, and how this internal environment functions in controlling gene action. Since we have already discussed cellular interactions in which the products released by one type of cell can influence the development of another type (Chapter 21), we shall not be concerned with this aspect of environmental influence in this chapter. However, we will discuss interactions between originally similar cell types.

This chapter will also examine instances in which the external environment influences patterns of gene expression. Changes in the natural external environment (such as seasonal changes) and changes in the location of organisms or parts of them in the heterogeneous and complex external environment constitute natural influences on development. We will also consider the effects of artificial, well-defined external environments, such as those obtained in cell, tissue, and organ culture.

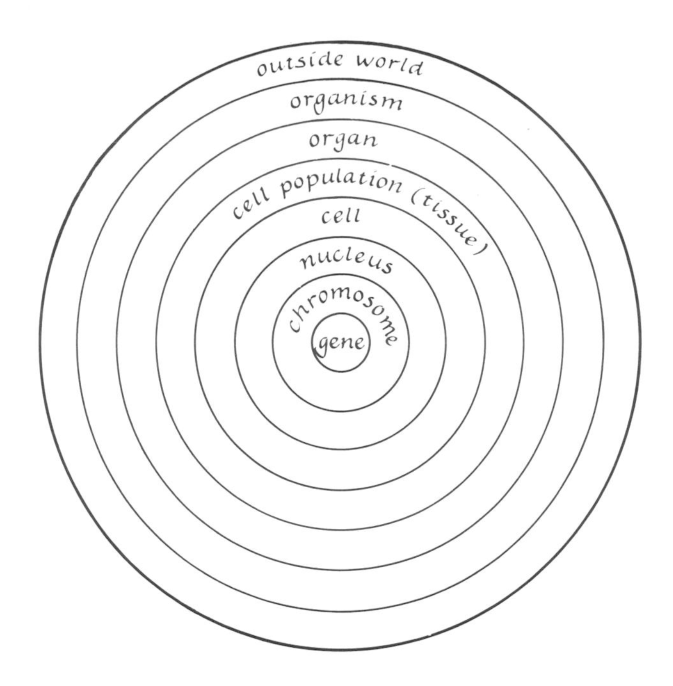

Figure 22-1. *The continuum of environments.*

Genesis and Role of the Internal Environment

Reproductive units are either single cells or collections of cells (Chapters 9 through 14). In multicellular units the internal microenvironment is largely inherited from the parent. On the other hand, the internal microenvironment of unicellular units develops progressively as the number of cells increases. This is diagramed in Figure 22-2.

As indicated in the figure, the blastocoele constitutes a new environment for the cells—an environment whose composition, because of the concentration of cell products released into it, is quite different from the external environment. Furthermore, the cells of the blastula, positioned between the external and internal environments, undergo polarized differentiation; for example, in a sea urchin, cilia form only on the outer cell surfaces. Recall that flagella form on the inner surface of the cells during asexual daughter colony formation from a gonidial cell, and on the inner surface of the cells of the young daughter colony developing from the zygote in *Volvox* (Chapters 9 and 10). The movement of some of the surface cells into the blastocoele (for instance, in gastrulation) establishes a still different environmental domain. Localized aggregations of cells arise as development progresses; as they increase in size, they contribute to the increasing complexity of the internal environment. A similar method of developing an internal environment occurs in a germinating spore of a fern or moss as it develops into a multicellular prothallium or protonema.

The hypothetical origin of the internal environment is illustrated in Figure 22-3. As shown in the figure, the aggregate formed by the descendants of a single cell might be spherical in the absence of external restraints; it might be a flat disk if development occurs on a solid substratum or if the mitotic

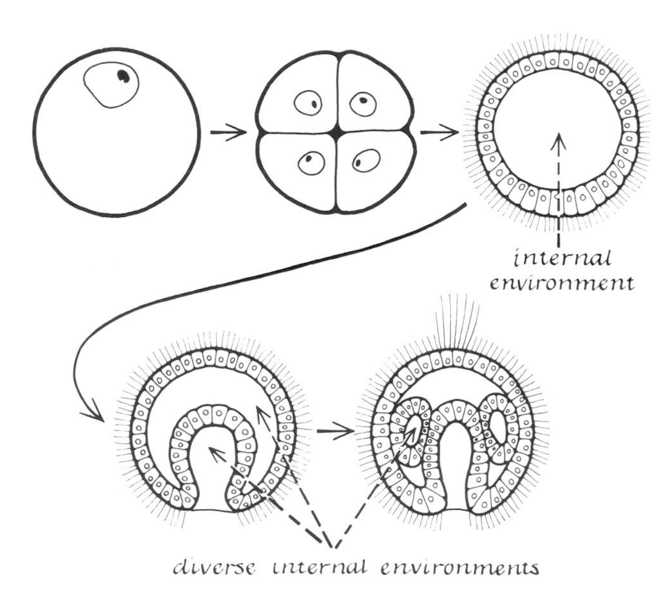

Figure 22-2. *Diagram of the genesis of the internal environment during development of a generalized animal embryo.*

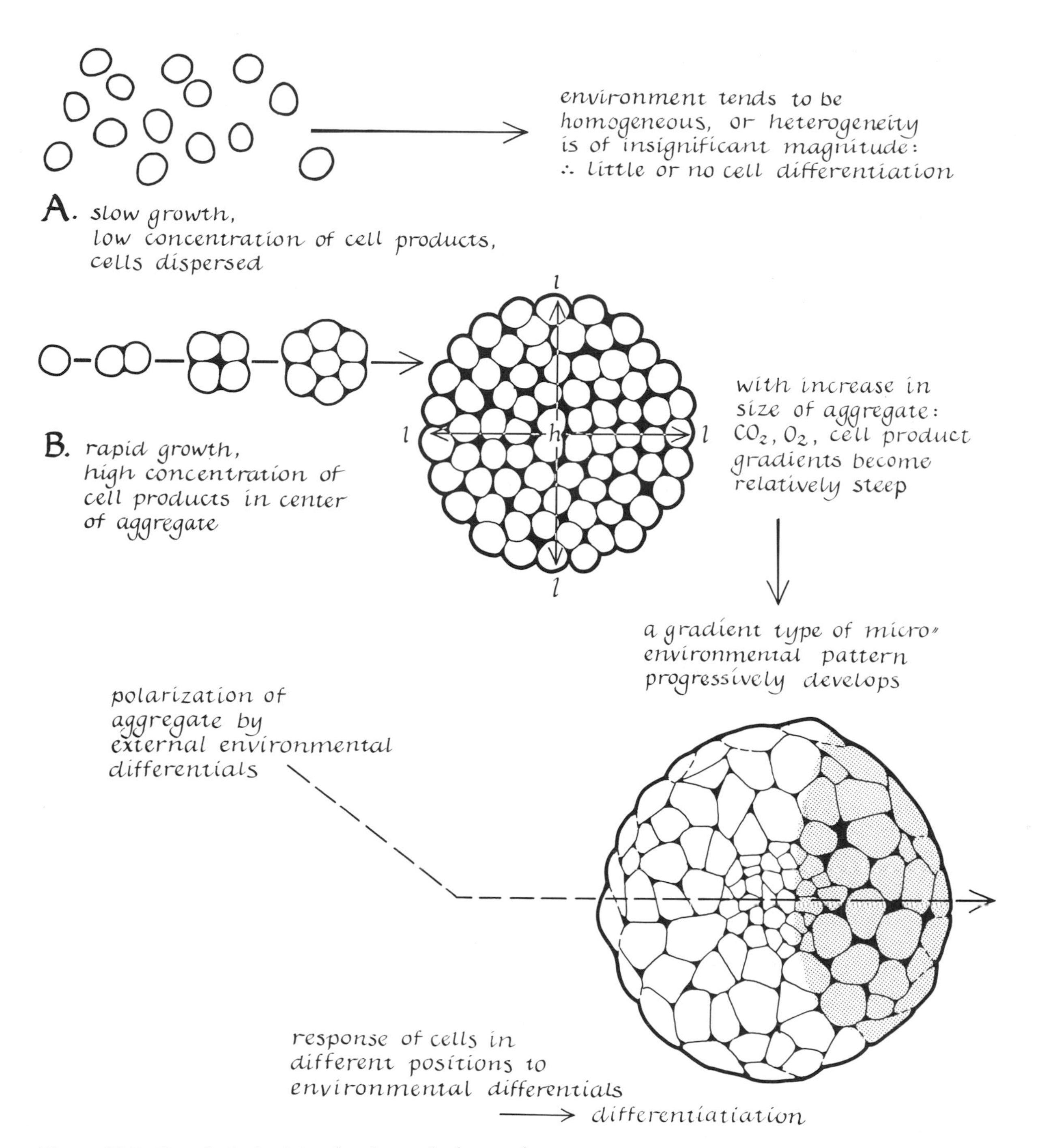

Figure 22-3. *Hypothetical origin of an increasingly complex internal environment inside a cellular aggregate.*

spindles all lie in one plane. In any case, a graded difference in composition of the intercellular environment from outside (or edge) to inside probably would result. As the aggregate increases in mass, the environment surrounding interior cells would increasingly differ from that surrounding more externally located cells. These differences might involve concentrations of cell products. At least initially there would be quantitatively graded differences. Certain components (for example, CO_2 and cell products) would reach their highest concentrations in the center of the mass. Oxygen tension would be maximal at the surface and minimal in the center. The capacities of the cells to use oxygen and carbon dioxide would in turn influence the distribution of these components. In aggregations of photosynthesizing plant cells the distribution might be the reverse of that just described. Differences in pH between the surface and interior would also arise.

Since the aggregate develops in a nonhomogeneous environment external to it, the initially spherically or radially symmetrical aggregate tends to become polarized by these external influences. The external differentials are probably gradient in nature. This would be the case either in a whole organism or in an internal part of an organism whose external environment would be the internal environment of the organism. In a graded external environment, one side of the aggregate would be exposed to a higher concentration than the other. Differences between the concentrations of the graded component (or components) on the two sides would probably depend either upon the "steepness" or slope of the gradient or upon the size (diameter or thickness) of the aggregate, or both. This is illustrated in Figure 22-4A.

The diagram shows how cells on opposite sides of an aggregate become exposed to increasingly different microenvironments from d_s (small differences) to d_l (larger differences), as the aggregate's size increases. When the differences become great enough, cells on one side but not on the other would respond, depending upon their threshold of sensitivity, and begin to become different. The originally spherical aggregate would then be polarized and exhibit radial symmetry around an axis of polarity, a–p or a–b (Figures 22-3 and 22-4). A second gradient, d′–v′ (Figure 22-4B), in the external environment could lead to establishment of another axis of polarity. If this axis were at right angles to the first, the aggregate would become bilaterally symmetrical. For example, cellular aggregates in a hormonal gradient within an organism might become polarized. The above model is highly simplified, since a complex of graded differentials is probably present in developing organisms (Prat, 1948; Child, 1941).

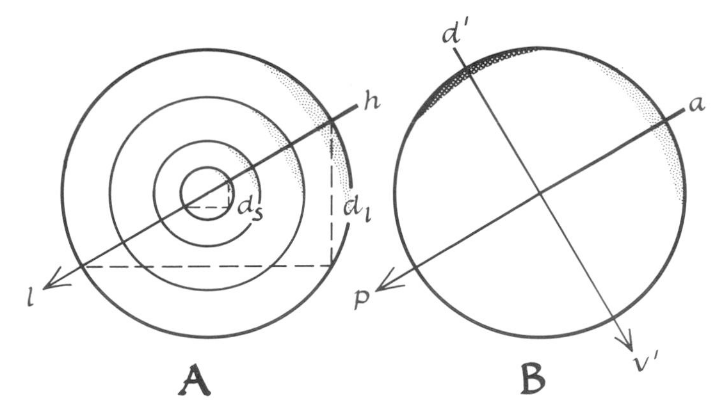

Figure 22-4. *A, origin of polarity in an originally spherical symmetrical cellular aggregate. B, origin of bilateral symmetry in response to two external environmental gradients,* a-p *and* d′-v′. *From* Introduction to Cell Differentiation *by Nelson T. Spratt, Jr. Copyright © 1964 by Reinhold Publishing Corporation, by permission of Van Nostrand Reinhold Company.*

The major point is that the development of an internal environmental pattern is as important as the development of the cells themselves. This is a well-established principle, since there is extensive evidence that the environment surrounding a cell is the crucial factor in determining which of its inherited capacities is to be expressed. Furthermore, these environmental differentials can arise in an aggregate of initially equivalent cells; once the cells respond differentially, the two or more resulting cell types can influence one another via the different products they release. The latter interactions might lead to further cellular differentiation (Chapter 21).

Now we will examine a few examples of how a graded internal environment might control not only cell differentiation but also the spatial arrangement of the specialized cell types. As described in Chapter 19, different auxin concentrations may affect cellular specialization very differently. Different concentrations of sucrose in combination with auxin when applied to plant callus tissue *in vitro* also have diverse influences on the differentiation of the cells

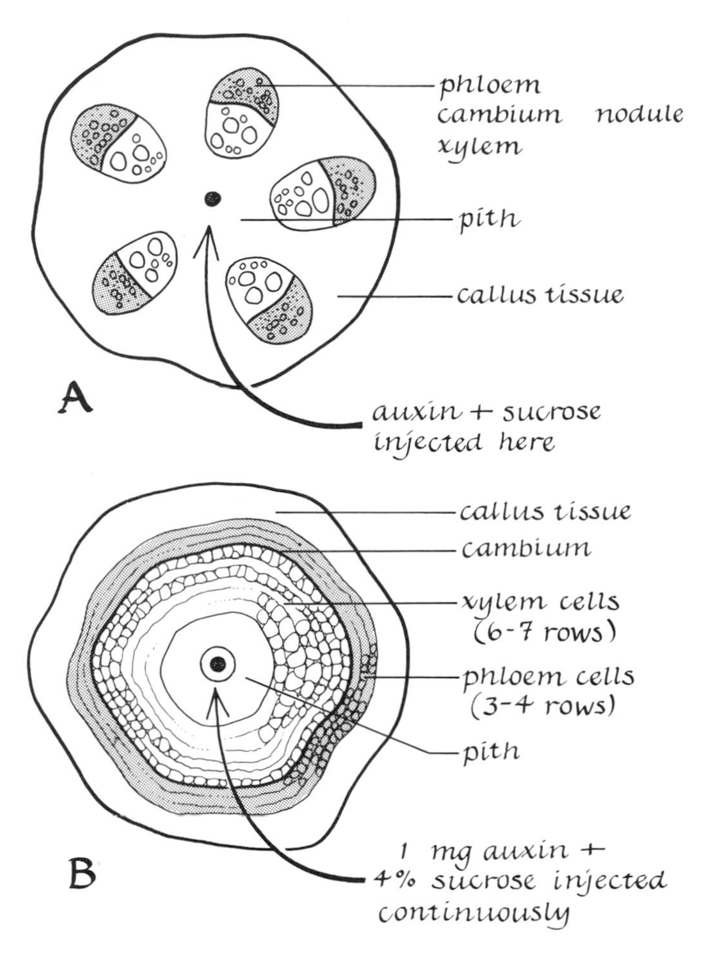

Figure 22-5. *Diagrams illustrating the response of cells of a* Syringa *callus to experimentally produced auxin gradients. A, auxin plus sucrose in agar applied to center of the callus; B, auxin plus sucrose injected continuously into the center of the callus.*

(Wetmore and Rier, 1963; Wardlaw, 1965). A low concentration of sucrose (1.5 to 2.5 percent) results in xylem differentiation; a high sucrose concentration (3 to 4 percent) causes phloem differentiation; and an intermediate concentration (2.5 to 3.5 percent) results in differentiation of xylem, cambium, and phloem.

Wetmore and Rier applied auxin (50 mg/liter) plus 4 percent sucrose in agar to the center of a mass of callus tissue of *Syringa vulgaris* (Figure 22-5).

In Figure 22-5A, nodules of phloem, xylem, and cambium developed and were spaced in a ring. The ring's diameter varied directly with the auxin concentration over a range from 0.5 to 1.0 mg/liter. In the experiment shown in Figure 22-5B, a complete ring of xylem and phloem developed around the pipette containing auxin (1 mg/liter) and sucrose (4 percent) inserted into the center of the callus. The continuous flow out of the pipette could explain the complete ring formation.

A striking example of the origin of differentials in a cell population derived from a single, isolated cell (that is, a clone population) is the development of a whole plant from a single tissue cell. This has been accomplished with single tissue cells of moss and fern gametophytes (Ito, 1962; Ward, 1963), carrot roots, stems, and leaves (Steward et al., 1964; Steward et al., 1966), and other plant species. The development of isolated carrot cells is illustrated in Figure 22-6.

The free cells placed in culture divide and form cell populations; the cell differentiation in these populations is apparently a consequence of their positions. Internal environmental guidelines are progressively built up as the cell mass grows and the cells thus come to lie in different microenvironments to which they respond differentially. How the mass of callus tissue develops growth centers and thus acquires polarity is not well understood but gravity appears to be involved (see Chapter 26).

Another example of the importance of the position of cells in controlling their differentiation is provided by Wardlaw's studies (1965) of how the shape of the fern shoot apex affects the differentiation of vascular tissues in *Dryopteris dilatata.* The vascular tissues collectively constitute the *stele;* it was found that the shape of the stele in cross section conforms closely with that of the shoot formed below the apex. By making various radial or tangential incisions through the apical region, shoots of different cross-sectional shapes were produced and the steles reflected these shapes (Figure 22-7). Similar results have been obtained in the lupine *(Lupinus albus)* by Ball (1946).

The animal kingdom provides numerous examples of the emergence of differences in populations of initially equivalent cells. One is the differentiation of a *blood island.* In birds and other vertebrates blood islands are aggregates of mesoderm cells in the wall of the yolk sac. Changes that occur in a blood island are illustrated in Figure 22-8. The outer cells become endothelial cells constituting a capillary wall (formed by prior fusion of many blood islands). Some inner cells become hemoglobin-containing

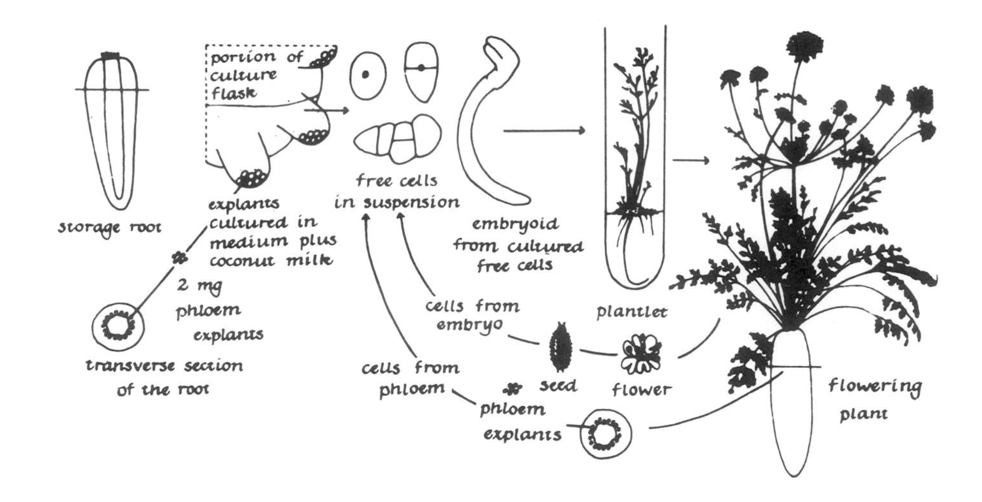

Figure 22-6. *Morphogenetic responses of free cells of* Daucus carota *determined by their origin and culture. From Steward et al., 1963.*

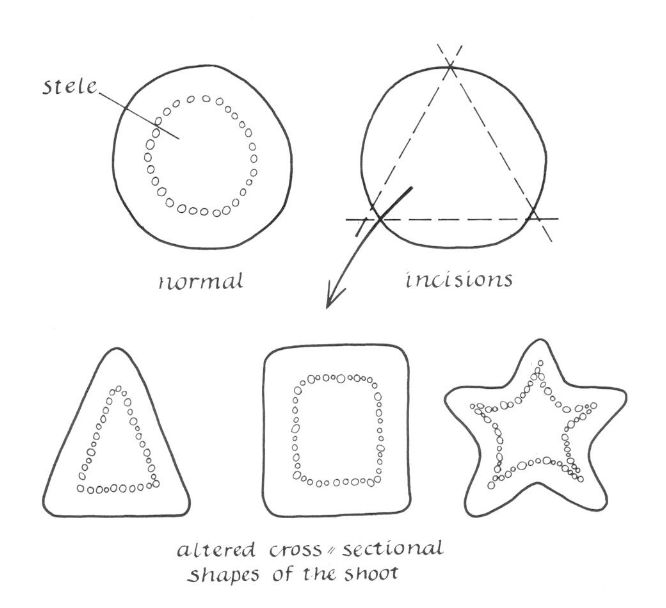

Figure 22-7. *Influence of the experimentally altered cross-sectional shape of the fern apex on the shape of the stele, as seen in cross sections of the stem below the apex.*

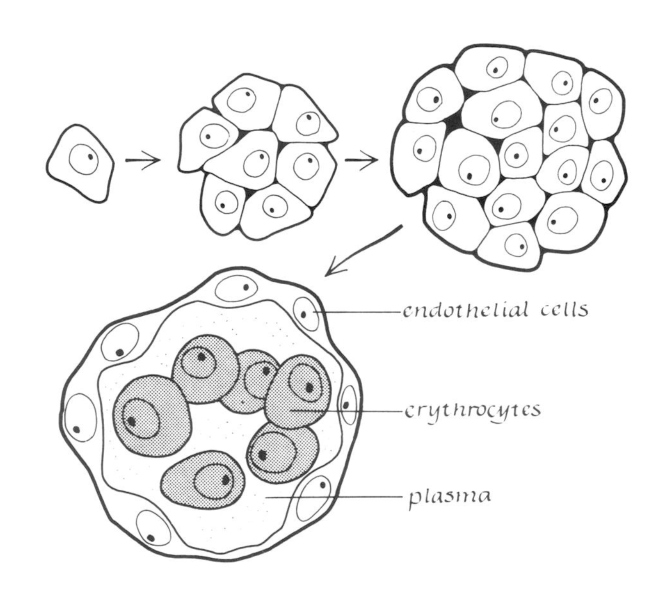

Figure 22-8. *Differentiation in a blood island.*

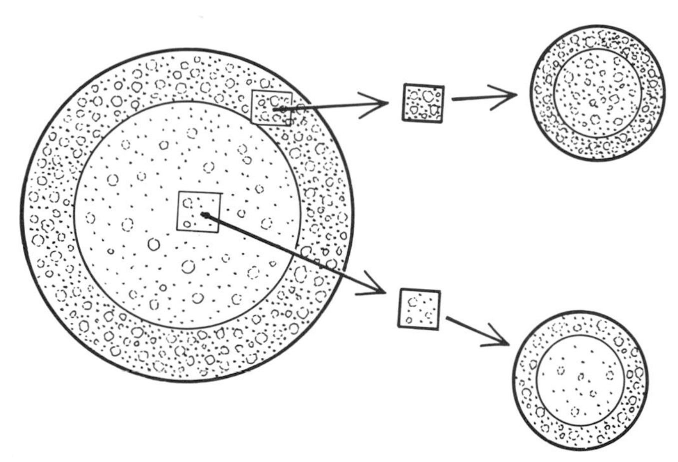

Figure 22-9. *Influence of the environment on the pattern of intracellular yolk use in isolated and cultured pieces of a young chick blastoderm. From* Introduction to Cell Differentiation *by Nelson T. Spratt, Jr. Copyright © 1964 by Reinhold Publishing Corporation, by permission of Van Nostrand Reinhold Company.*

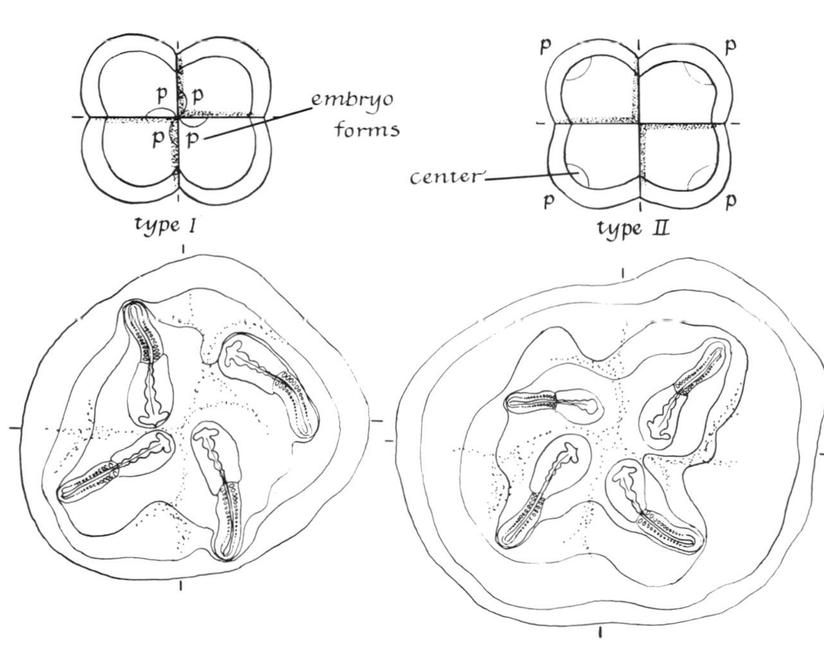

Figure 22-10. *Diagrams illustrating the composition of two types of systems produced by fusing four unincubated chick blastoderms explanted in vitro. Lower, drawings of representative results. From Spratt and Haas,* J. Exp. Zool., *Volume 147, p. 283, 1961. Courtesy Wistar Press.*

erythrocytes; others apparently break down and contribute to the blood plasma.

Another example is the role of the internal environment of an unincubated chick blastoderm in controlling the differentiation of the outer, *opaque area* (relatively yolk-rich cells) and the inner, *pellucid area* (relatively yolk-poor cells). When small groups of inner and outer cells are excised from the flat, disk-shaped embryo and explanted to the surface of a nutrient agar culture medium, each group rapidly becomes circular in outline (Figure 22-9).

Groups of cells above a minimum size (about 500 cells compared with the 30,000 to 60,000 cells of a whole embryo) increase in size and develop inner and outer areas, copying the radially symmetrical pattern of the whole embryo. Furthermore, yolk-poor and yolk-rich areas are present in about the same proportion as in a whole embryo (Spratt, 1964). Differentiation in these isolated groups of cells is epigenetic, since they contain no preformed guidelines.

The apparent role played by the internal environment of the young chick embryo in directing a complex pattern of cell differentiation is illustrated in Figure 22-10. When four unincubated, explanted chick blastoderms are fused so that the embryo-body-initiating centers are in the geometric center (Figure 22-10, type I), the cells of the four centers no longer form embryo bodies. Instead, embryo bodies arise from the peripheral areas as they do in the control experiment (type II). Even in systems this small (about 4 mm in diameter), the positions of the cells determine their differentiation (Spratt and Haas, 1961).

There are numerous other examples of the internal environment's control of gene action. One striking example is found in hermaphroditic amphibians, in which the location of primordial germ cells within the gonad controls their differentiation. According to Witshi (1934, 1967), follicle cells of the cortex of the gonad produce a female inductor substance, and interstitial cells of the medulla produce a male inductor. Primordial germ cells of a single animal, whether genetically male or female, that enter and remain in the cortex become eggs; those remaining in the medulla become sperm. Recall (Chapter 19) that differentiation of the primordial germ cells of the *Axolotl* is controlled by sex hormones or hormone-like substances produced by the developing gonad. Thus the primordial germ cells are apparently capable of developing into either eggs or spermatozoa.

As noted earlier in this chapter, as size and density in a cellular aggregate increase so does complexity in the internal (intercellular) environment. As a consequence of this, the number of kinds of specialized cells increases. In some instances complexity of the morphological pattern which develops also increases. Conversely, decrease in size and cell population density of the aggregate below a minimum number results in failure of differentiation of the cells, or failure in further differentiation of already committed cells. This has been demonstrated in plants, slime molds, coelenterates, insects, ascidians, fish, amphibians, birds, and mammals (Grobstein, 1955). Some of these examples, generally referred to as examples of *cell population density* or *mass effect,* are reviewed by Saxen and Toivonen (1962). We shall cite only a few.

The number of branches developing on the chloroplasts of different species of desmids increases with increase in diameter of the chloroplast. One species in which the mean diameter of the plastid is 1.2 μ has six branches; another in which the mean diameter is 6.4 has fifteen (Bower, 1930). The complexity of the patterns of xylem differentiation in the steles of roots increases with increase in root diameter (Bower, 1930). This is illustrated in Figure 22-11.

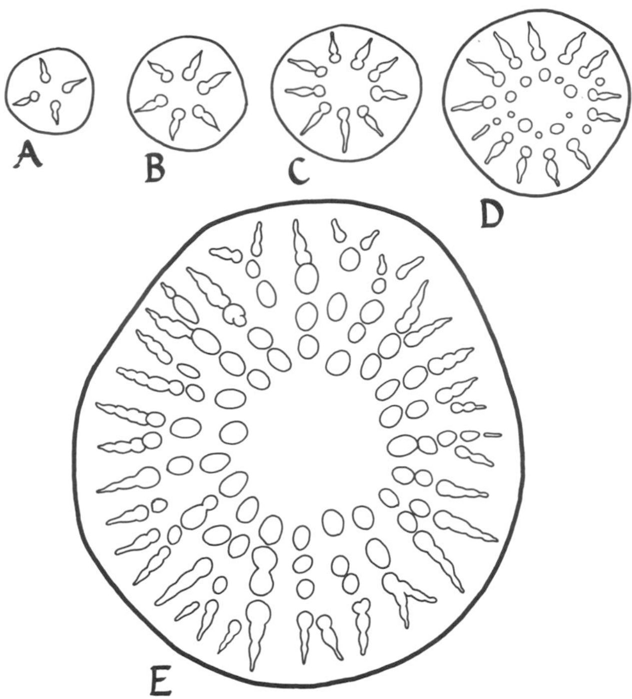

Figure 22-11. *Complexity of the pattern of xylem differentiation in the stele of a root of* Colocasia *(Arum family) with increase in diameter of the root. This also occurs in the steles of roots of other species of plants.*

Very small pieces of *Tubularia* regenerate hydranths with fewer tentacles than normal (Morgan, 1901). Recall (Chapter 5) that regions of higher cell population density are dominant in regeneration in pieces of the stem of *Tubularia.* In *Hydra,* the number of regeneration centers arising in minced body sections increases with increase in tissue volume (Chalkley, 1945). Recall (Chapter 9) that the size of the bud disc in tunicates is directly related to the the presence of sexual differentiation in the new individuals.

As the number of combined explants of the dorsal blastoporal lip of urodele amphibian gastrulae increases so do the kinds of specialized cells which develop (Lopashov, 1935). The results are summarized as follows:

one dorsal lip	→	muscle cells
two to four dorsal lips	→	muscle and notochord cells
four to five dorsal lips	→	muscle, notochord, and epidermis cells
six to ten dorsal lips	→	muscle, notochord, epidermis, and nerve cells

In an opposite kind of experiment, Grobstein and Zwilling (1953) explanted *in vitro* increasingly small fragments of the neural plate area of young chick blastoderms. After a short period of explantation, the fragments were grafted to the chorio-allantoic membrane of an eight-day chick embryo. In fragments about 0.1 mm square, the differentiation of nerve tissue was significantly less than in larger fragments. Thus the number of cells is important for their differentiation. Products released by the cells during explantation and in the grafts are necessary for their further differentiation. Cell products would be more concentrated in aggregates of closely packed cells than in more dispersed aggregates where diffusion of the necessary products would be greater.

Evidence for the importance of cell products in differentiation is provided by some experiments of Wilde (1961) in which single cells of *Ambystoma* neuroepithelium isolated in microdrop cultures failed to differentiate, though in larger cultures containing twenty-five to one hundred cells, differentiation was observed. Washing the cells before cultivation in-

Figure 22-12. *Differential influence of an aqueous and an air environment on the morphology of leaves of the water buttercup. Reprinted with permission of The Macmillan Company from* Development in Flowering Plants *by Torrey. Copyright © by John G. Torrey, 1967.*

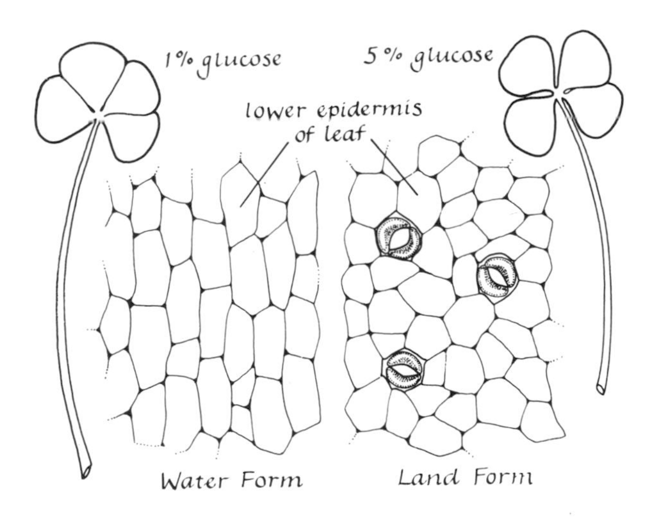

Figure 22-13. *Experimentally increasing the concentration of glucose in the culture medium favors development of the land form of the fern* Marsilea Drummondii.

hibited their differentiation; thus, an increase in cell population density apparently leads to an increase in the concentration of cell products, which are important components of the intercellular microenvironment. Further aspects of the importance of cell products will be discussed in Chapter 23.

Role of the External Environment

The pattern of development of many organisms (especially plants) is influenced by their external environment (see reviews of Went, 1954; Leopold, 1964; Galston, 1967; Salisbury and Ross, 1969). The environmental stimuli appear to be transduced into chemical effects that control development. Growth-regulating substances—auxins, gibberellins, cytokinins and probably others (Chapter 19)—are involved, but how most environmental agents actually work is unknown. What follows is a description of properties of the external environment known to control development.

Physical Properties

Water and Air

Those leaves of the water buttercup *(Ranunculus aquatalis)* that develop below the water's surface are finely divided; those developing in the air are lobed (Ashby, 1949). When a leaf develops half in and half out of the water, morphogenesis of the halves is different (Figure 22-12).

Other aquatic or amphibious plants respond similarly. For example, in the fern *Marsilea,* experimentally increasing the concentration of glucose in the liquid culture medium favors the development of the land form (Figure 22-13). According to Allsopp (1953, 1955, 1967) the effect is osmotic and is related to water balance in the plant. There are numerous examples of humidity influencing plant development. In general, dry air promotes xylem differentiation, and humid air has the opposite effect (Sinnott, 1960). Recall that the transition from an aqueous to an air environment causes fruiting bodies to develop in cellular slime molds (Chapter 6).

Gravity

Gravity has long been known to influence plant development. Through control of auxin distribution, the shoot grows upward and the root downward (Chapter 19). Gravity also influences cellular differentiation in leaves and stems. The upper and lower surfaces of leaves are morphologically different; when leaves of some plants have been artificially positioned, these differences have been reversed (Sinnott, 1960). The formation of "reaction wood" (wood with a special xylem structure) along the lower but not the upper part of horizontal branches is another example of the influence of gravity (Sinnott, 1960). How the plant organs detect the gravitational force is unknown. It has been suggested that particles in the cells act as gravity sensors. These particles are called *statoliths.* Gravity displaces cytoplasmic inclusions such as mitochondria, chloroplasts, starch grains, and the nucleus, but whether any of these are the sensors remains to be demonstrated. The distribution of these inclusions might influence the distribution of auxins. When the force of gravity acting on young fern prothallia *(Onoclea sensibilis)* is equalized in approximately all directions by growing them in a liquid culture medium that is constantly shaken, the prothallia develop as spherical, tumor-like masses of tissue with no polar organization. When these masses are later transferred to a solid culture medium without any agitation, they become approximately normally polarized prothallia (Näf, 1953).

Gravity also influences the distribution of cytoplasmic inclusions in eggs and probably influences development. It would be interesting to discover how so-called zero gravity, attainable in orbiting, man-made satellites, affects developing organisms.

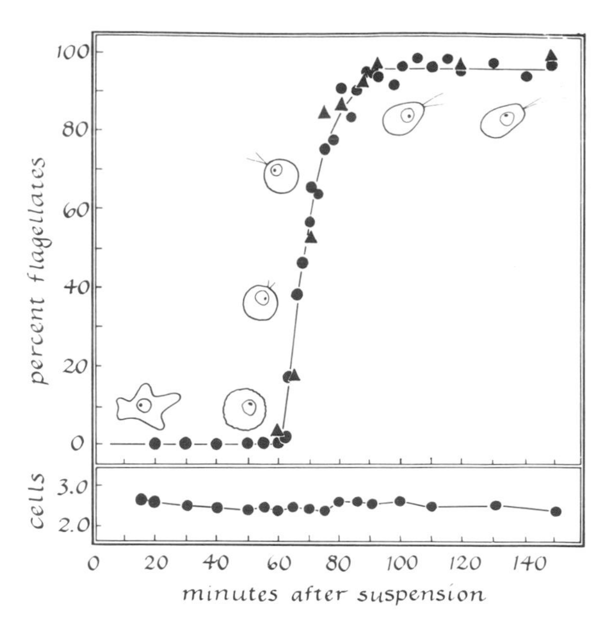

Figure 22-14. *Appearance of flagellates in a population of* Naegleria. *The upper curve gives the percentage of cells with flagella when the amebae were transferred from a solid to a liquid medium (circles and triangles indicate separate experiments). The lower curve gives the number of cells* $\times 10^5$ *per ml. From Fulton and Dingle in* Developmental Biology, *Volume 15. Copyright 1967 by Academic Press.*

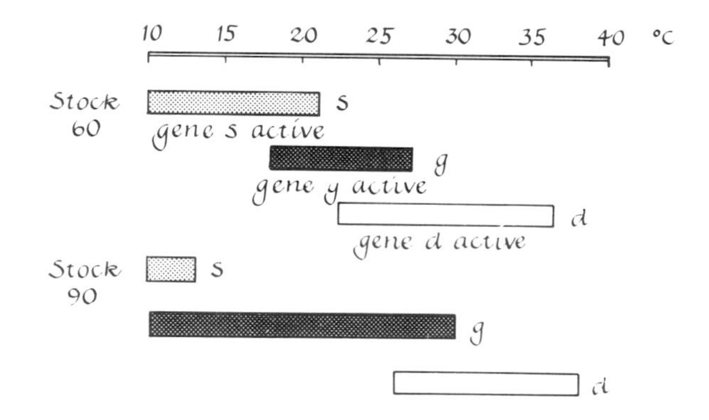

Figure 22-15. *Formation of antigens in* Paramecia *cultured at different temperatures.*

Mechanical Factors

The movement of plant parts such as tendrils of the pea results from their contact with a physical support followed by differential growth on two sides of the tendril (Darwin and Darwin, 1881). This results in the familiar winding of the tendril around the support. Recent studies (Galston, 1967) indicate that a contractile ATPase might be involved, but there is no direct evidence for the presence of a contractile protein in the tendril cells; there is, however, some evidence for an ATP-ATPase system underlying several kinds of plant movements (Sibaoka, 1969).

The forced bending of a stem or root results in increased tension on the convex side and the development of small, thick-walled cells (Bücher, 1906).

Tension in the root of a birch tree has a similar effect (Jaccard, 1914). It is a common observation that repeated bending and shaking of growing plants by wind may cause radical changes in their growth patterns.

A particularly obvious example of the influence of a solid as opposed to a liquid environment on the response of a cell is found in cultures of the ameba *Naegleria gruberi.* On a solid medium the organism is a typical ameba, but when transferred to a liquid medium it transforms within about seventy-five minutes into a flagellate with two flagella (Willmer, 1956). This is shown in Figure 22-14. Details of this transformation have been studied by Dingle and Fulton (1966) and Fulton and Dingle (1967) and will be described in Chapter 25.

The true slime mold *Didymium nigripes* also transforms from an ameboid to a flagellated state when transferred from a solid to a liquid medium. For unknown reasons, bacteria in the liquid medium will inhibit flagella formation (S. Kerr, personal communication).

Temperature

In general, within the tolerance limits of the organism, the higher the temperature, the more rapid is growth and development, but sometimes colder temperatures stimulate more rapid development. Colder temperatures stimulate the growth of hair in temperate and arctic mammals. Lower night temperatures sometimes stimulate plant growth more than day temperatures; this is called *thermoperiodism* (Sinnott, 1960). Plants such as the buttercup, when grown at a relatively high mean temperature (17° C), form undivided, juvenile-type leaves; at a lower temperature (7.5° C), they form adult-type lobed or divided leaves. Temperature also influences cell differentiation in animals. At 8° C, *Hydra pirardi* may develop testes; when the temperature is raised to 19° C, the testis involutes and bizarre bud-like structures develop in place of it (Brien, 1961).

In some plants, such as rye, exposure of the seeds or seedlings to relatively low temperatures (1° to 7° C) for several weeks results in more rapid flower and fruit development when the temperature is raised. This is called *vernalization* (Wilkins, 1969). Low temperature alone, however, is not sufficient, since the vernalized plants usually have a long day requirement (Salisbury, 1963). The low temperature is perceived by the shoot apex; exposure of other plant parts, such as roots or leaves, to low temperature does not vernalize the plant (Torrey, 1967). Lang (1953) has suggested that a substance, *vernalin,* is produced during the cold period, since nonvernalized plants become vernalized when grafted to vernalized plants. Vernalin may be a gibberellin-type substance, since treatment with gibberellins can often replace vernalization. It has been suggested that vernalization demonstrates that each stage of development requires certain temperature and light conditions before the next stage can begin (see Sinnott, 1960, for references).

Low temperature also breaks dormancy in various seeds and buds and induces the formation of tubers and other underground storage organs in some plants. The influence of low temperature can persist long after the stimulus is removed, and thus is a type of inductive effect (Chapter 21).

At different environmental temperature ranges of cultures of *Paramecia,* different antigen-determining genes are active (Beale, 1954). Among the many kinds of antigens found in the *Paramecia* cytoplasm, formation of those described as s, g, and d antigens are shown in Figure 22-15.

Note that the s antigen is formed at temperatures ranging from 10° to 18° C, the g type between 18° and 27° C, and the d type from 25° to 35° C. These ranges differ from one stock to another. Thus, in Stock 90 the g type is stable over a wider range (10° to 30° C) than in Stock 60. The opposite is true of the s type. The mechanism of action of changes in temperature is not known, but it has been suggested that different temperatures change the composition of the cytoplasm and that the cytoplasmic state controls the pattern of gene activity (Beale, 1954).

Light

Perhaps of all environmental influences, that of light is most important in controlling animal and plant development. The intensity, quality (wavelength), and duration (photoperiod) of light are all important. Plants, especially dicotyledons, grown in the dark or very dim light are tall and spindly and have small leaves. There is maximum cell elongation, poor differentiation, and little or no chlorophyll synthesis.

The responses of plants to directional illumination—for instance, the bending of the growing shoot toward the light source, or the turning of the sunflower as it follows the sun from east to west—are well-known examples of the influence of light. A gradient in light intensity also seems to influence the direction of polarity in reproductive units such as the egg of *Fucus* (Chapter 13) and moss and other spores. In the germination of a moss spore, the protonema grows out of the illuminated side, whereas the rhizoid emerges at the opposite side (see review of Sinnott, 1960). In *Equisetum,* the horsetail or scouring rush, light influences the position of the mitotic spindle of the germinating spore's first division (Figure 22-16).

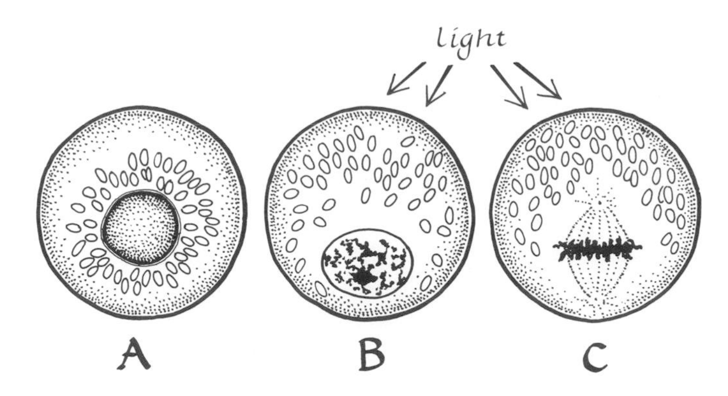

Figure 22-16. *Influence of light on the position of the mitotic spindle in a spore of* Equisetum. *The beginning of polarization is shown by the changed positions of the plastids and the nucleus.*

Growth of *Acetabularia* stops when plants are placed in the dark. Activity of the nucleolus ceases, as indicated by a decrease in its size and the assumption of a spherical compact appearance. Upon return to light, the nucleolus assumes its normal, lobulated structure, and growth is resumed within one or two months (Hammerling, 1966). Light-dependent synthetic processes in the cytoplasm apparently influence nuclear activity (Chapter 20).

In some cases the quality of light influences the response. A fern prothallus of *Dryopterus felix-mas* grown in red light (about 6,000 to 7,000 Å) is filamentous in form, whereas blue light (about 5,000 Å) produces the typical heart-shaped thallus (Ohlenroth and Mohr, 1964). On the other hand, continuous blue light inhibits the transformation of the filamentous moss protonema of *Pohlia nutans* into the leafy upright gametophyte (Mitra et al., 1965). The quality of light also influences the germination of seeds such as lettuce and the growth of seedlings such as *Catalpa* (Indian bean); red light is optimal in both cases (Parker and Borthwick, 1950). Leaf shape in some plants is influenced by the wavelength of light. Additional examples of light's influence on plant growth and development are reviewed by Mohr (1969).

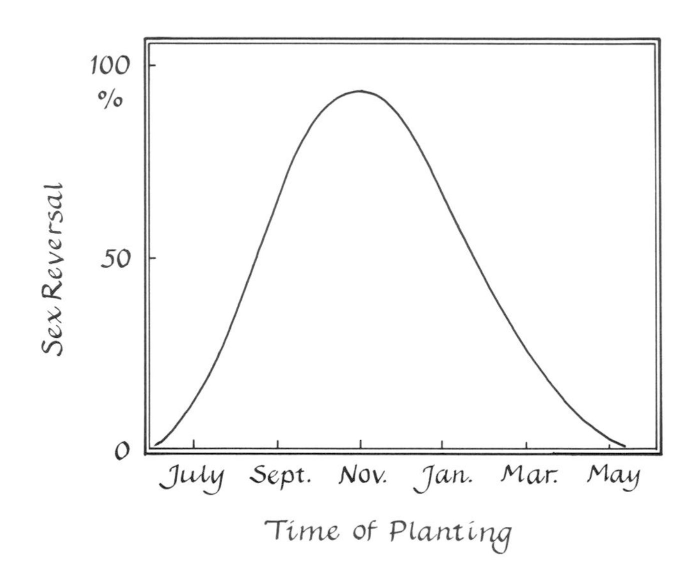

Figure 22-17. *Influence of photoperiod on the type of flower (staminate vs. pistillate) developing in hemp plants from which marijuana is obtained.*

Aggregation in cellular slime molds (Chapter 6) appears to be stimulated by blue light (Reinhardt and Mancinelli, 1968). When the migrating slug of *Acrasis rosea* is illuminated by overhead light it stops and fruiting body formation begins. The light acts as a trigger, since slugs illuminated for fifteen to thirty minutes and then returned to darkness stop moving and construct fruits (Nowell et al., 1969).

What molecule functions as the photoreceptor in the above examples is not known with certainty, but there is evidence that flavoproteins may be involved in responses to blue light (Davis, 1968), and that the pigment *phytochrome* may be involved in responses to red (and perhaps sometimes blue) light. These substances are present in algae *(Mesotaenium)* and liverworts *(Sphaerocarpus)* (Taylor and Bonner, 1967), as well as in higher plants.

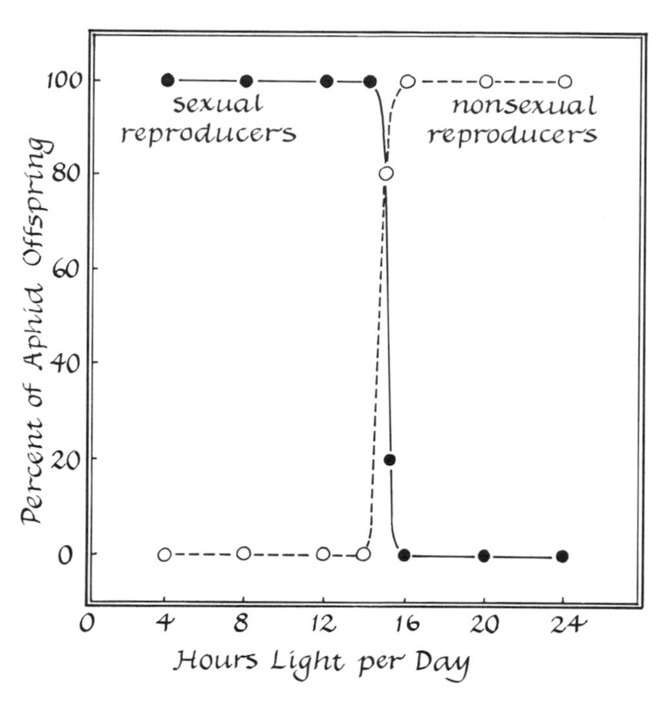

Figure 22-18. *Effect of photoperiod on reproductive type in the aphid* Megoura viciae. *From Beck in* Animal Photoperiodism, *copyright Henry Holt and Company, 1963. Courtesy Holt, Rinehart and Winston.*

Photoperiod has a profound effect upon many developmental processes in plants and animals, particularly sex development. In general, long days stimulate sex development, which is clearly of ecological significance. Kjellman (1885) performed some of the first controlled light-duration experiments as early as 1878. Many studies have followed. See Naylor (1952) for early bibliography. The influence of photoperiod on flowering plants was discovered in 1920 by Garner and Allard, two U.S. Department of Agriculture scientists. According to the length of the dark period required for inducing flowering, some plants are classified as either long- or short-day plants. This is also of ecological significance, allowing seeds to be formed at an appropriate time of year. Day-neutral plants are also known (Salisbury, 1963; Torrey, 1967). It is believed that the photoperiod is perceived by the plant leaves and that under appropriate day length, a hormone, *florigen,* is produced in the leaves and travels via the phloem to the shoot apex (Chapter 19).

In plants such as hemp, photoperiod induces not only flowering but also the type of flower, whether staminate or pistillate (Shaffner, 1931). Genetically staminate plants begin to bear pistillate flowers as the length of day decreases, maximal reversal occurring in November and December, as shown in Figure 22-17. Presumably this effect is mediated by an increase in the concentrations of auxins (Chapters 11 and 19)

The response of plants to photoperiod also is regulated by the billiprotein pigment phytochrome, which exists in two forms that are interconvertible by light (see Bonner, 1966, and Wilkins, 1969, for reviews). One has an absorption peak at 6,500 Å in the red, the other at 7,300 Å in the far red. In general, the red-absorbing form is inactive or inhibitory, while the far-red absorbing form favors or stimulates flower development, leaf growth, and seed germination in different plants. How phytochrome and other light-absorbing pigments work is not known. Mohr (1966) suggested that the pigment acts at the gene level, but action on the cell membrane is more likely (Galston, 1967). However, it is tempting to postulate that hormone-like substances are produced in cells when light is absorbed and that these substances activate genes which then code for proteins involved in the particular response.

Photoperiod influences the reproductive cycles of many animals, both invertebrate and vertebrate (review of Beck, 1963). In the water flea *Daphnia pulex,* the encapsulated eggs require a period of darkness followed by a light period for initiation of development (Davison, 1969). Changes in photoperiod are instrumental in inducing sexual reproduction in the planarian *Dugesia tigrina* (Vowinckel, 1970). In aphids (plant lice), the alternation of parthenogenetic production of offspring with sexual reproduction from fertilized eggs is controlled by day length. In one aphid species which lives on bean plants, sexually reproducing aphids are born when the daylight component of the photoperiod is less than fifteen hours. When day lengths are more than fifteen hours, nonsexual forms are produced (Lees, 1959). This is shown in Figure 22-18.

The photoperiod exerts its effect by influencing hormone production of the mother aphid. These hormones control the development of the embryos, providing an example of the interplay of external and internal environmental influences on development.

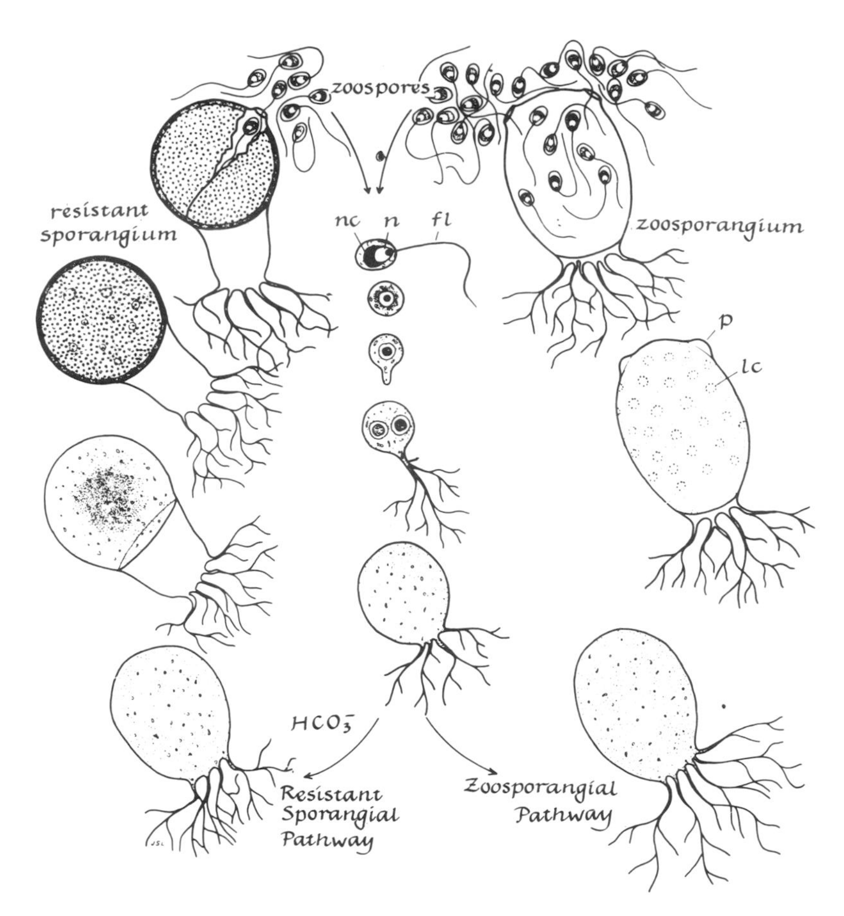

Figure 22-19. *Life cycle and development of* Blastocladiella emersonii. *Stages illustrating zoospore germination and development have been drawn on a larger scale than the rest of the cycle. From Lovett, 1967, copyright 1967 by the Thomas Y. Crowell Company.*

Asexual reproduction by budding in both the green *Chlorohydra* (which contains the symbiotic alga *Chlorella*) and the albino strain (which lacks the symbionts) is greater in both constant light and intermittent light (twelve hours' light followed by twelve hours' darkness) than under constant darkness (Epp and Lytle, 1969). Budding appears to be regulated by an unknown metabolite derived from food and the symbiotic algae.

Photoperiod also controls the growth and periodic shedding of deer antlers *(Cervus elaphus)* (Chapter 2). Short days stimulate antler growth. Under long-day conditions, the antlers are shed (Jaczewski, 1954).

Chemical Properties

Chemical components of the external environment are known to influence development by selecting the program of gene action. However, the mechanism of action of these components remains mostly unknown and thus is a central problem in development. A few examples of the influence of chemical components in the environment are discussed in what follows.

Inorganic Components

In the water mold *Blastocladiella emersonii* (Figure 22-19), the presence of bicarbonate ions influences the sporophyte to form resistant sporangia involving the synthesis of melanin and chitin. In the absence of such ions, a *mitosporangium* (zoosporangium), which releases motile spores, is formed (Emerson, 1954; Cantino and Lovett, 1964; Lovett, 1967). Emerson (1954) suggested that this might be a pH effect.

The bicarbonate ion also induces the interstitial cells of *Hydra pseudoligactis* to transform into cnidoblasts and the gland cells to become mucous cells when the animal is cultured in a medium containing 0.2 percent $NaHCO_3$ and 0.2 percent $CaCl_2$ (Macklin and Burnette, 1966). The transformation of the ameboid into the flagellated state in the ameba *Nae-*

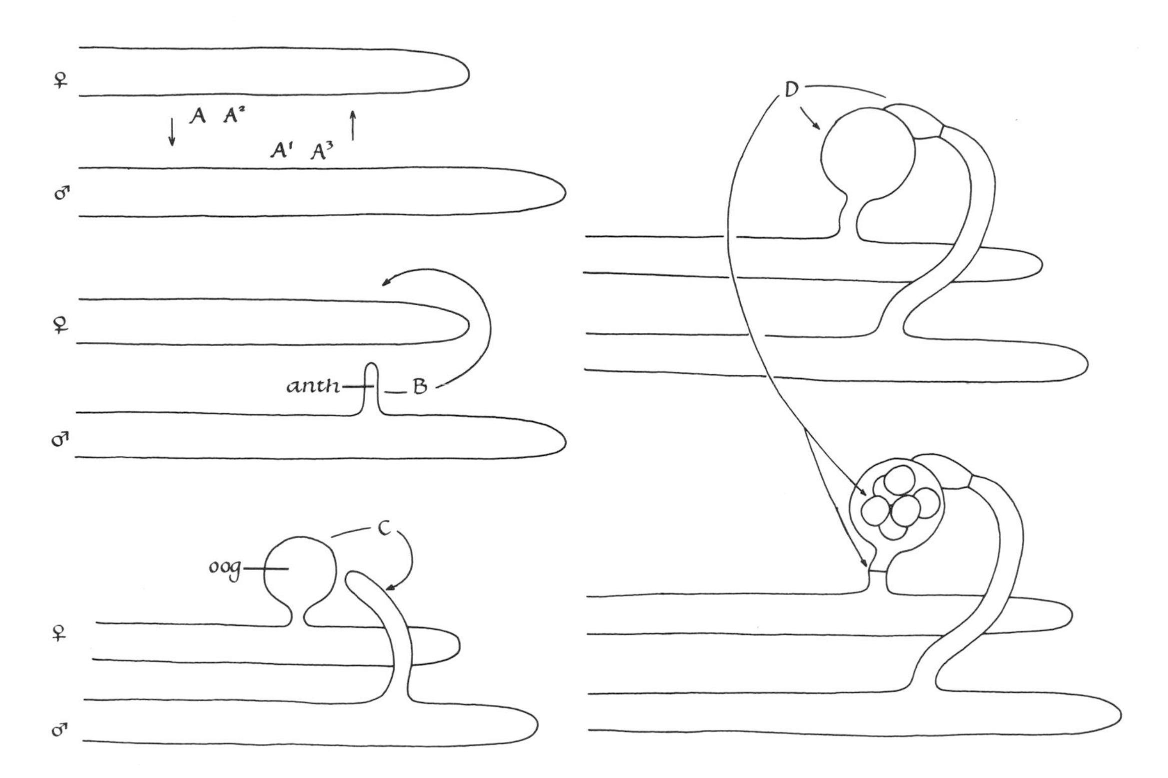

Figure 22-20. *Sequence showing regulation of development in* Achlya bisexualis. *Starting at left, interaction of hormones A and A^2 produced by female thallus and A^1 and A^3 produced by male thallus. This initiates development of the antheridial filament,* anth. *Production of hormone B by the male filament initiates development of the oogonial initial,* oog. *Hormone C produced by the oogonial initial controls directional growth of the antheridial filament and delimitation of the antheridium. Hormone D produced by the antheridium controls delimitation of the oogonium and development of oospheres. From Scagel et al., 1966. Courtesy of Wadsworth Publishing Company, Inc.*

gleria is stimulated by HCO_3 and $NaHPO_4$, but inhibited by Li^+, Mg^{++}, and SO_4^{--} ions (Dingle and Fulton, 1966).

Organic Components

Several organic compounds influence the development of cultured fungi, fern gametophyte, and sporophyte tissue. In some *Mucors* among the fungi, the mycelial (filamentous) type can be transformed into the yeast-like type with fermentative capacities by eliminating potassium from and adding 10 percent sucrose to the culture medium (Nickerson, 1954). Another type of transformation studied by Nickerson is the formation of filaments by the yeast *Candida albicans* when grown on a medium containing very little sucrose. However, in the presence of 0.001 M cysteine all growth is of the yeast (nonfilamentous) type. These transformations involve metabolic as well as morphological changes.

As noted in Chapter 14, the presence or absence of sucrose in the culture medium on which fern sporophyte callus tissue is growing influences the path of differentiation of the callus cells. When an auxin is present in the medium no cell differentiation occurs (Bristow, 1962). The influence of sucrose on the development of fern gametophytes in culture has been investigated by Whittier and Steeves (1960, 1962) and Wetmore et al. (1963). In the presence of 1 percent sucrose, sporophyte tissue (xylem) and even a sporophyte may develop from a cultivated gametophyte; this is an example of apogamy. How the sucrose works is not known, but an osmotic effect similar to that claimed for the influence of glucose on *Marsilea* is a possibility. Indeed, the specificity of developmentally significant environmental agents has rarely been determined. Earlier we mentioned one exception—the specific response of fern spermatozoa to organic acids.

Hormones

The function of hormones as constituents of the internal environment has been discussed in Chapter 19. Raper (1955, 1960, 1966) describes how external hormones help control the sexual process in the water mold *Achlya bisexualis.* In this mold, the sex organs are borne on separate individuals, and the development of the organs requires cooperation of the two individuals. The sequence in formation of the sex organs in male and female thalli and the regulation of these changes by a series of hormones are illustrated in Figure 22-20.

Hormone A, called *antheridiol,* has been isolated and purified; it is a sterol with the carbon skeleton of *stigmasterol.* Hormone B, which initiates formation of oogonia, has not yet been isolated or identified, nor have hormones C and D (Barksdale, 1969). In the species of *Achlya* which bears antheridia and oogonia on the same thallus, a similar hormonal relay mechanism controls the development of the sex structures.

Sex hormones are also involved in forming the gametangia of the fungus *Mucor,* a relative of the black bread mold *Rhizopus* (Chapter 10). The hormones in *Mucor* and related genera are terpenoid carboxylic acids (Barksdale, 1969). A hormone produced by the female controls sexual differentiation in *Bonellia viridis,* an echiuroid worm (Baltzer, 1928). If an intersexual larva is free living, it develops into the relatively large female, but if it attaches to the proboscis of an adult female it becomes a microscopic male. The male migrates into the nephridia of the female and lives there as a parasite.

The Controlled Environment

Perhaps the most fruitful approach to the problem of understanding how environmental constituents influence the development of organisms is by studying their growth under artificial and known environmental conditions. Culture procedures permit the investigator to manipulate the environment and to observe the responses of the developing material. There are three basic methods of studying development under controlled conditions. Studying the intact organism in a controlled environment permits one to assess the effects of external environmental variables such as light or temperature on the development of the whole organism. In the analytical method, one studies separated parts of organisms grown in culture. This permits more rigid control of variables in the internal environment than is possible in study of cultured intact organisms. The synthetic method consists of studying separated parts combined in normal or abnormal relations. This reveals insights into the interactions of the parts that would otherwise be obscure. Examples of the dissociation and reaggregation of tissue cells and of induction were discussed in Chapters 8 and 21. Unfortunately, in many studies in which culture methods have been used, the composition of the culture medium is not well defined. Consequently, conclusions have been drawn regarding the behavior of the tissues or cells without considering the influence of environmental variables. We have already noted that the structure and composition of the internal environment undergo developmental changes concomitantly with the changes in the cells residing within this environment. Advances in our understanding of development and its control will probably depend just as much upon an analysis of the microenvironment of the cells as upon the analysis of the genetic material itself.

The first published studies of the culture of plant cells and tissues were Haberlandt's (1902); Harrison (1907) published the first such work dealing with animal tissues. Since then, extensive use of culture methods to study development has resulted in thousands of published reports. However, only recently have culture media of known chemical composition that permit essentially normal development been prepared. No study of development under conditions where the composition and influence of the culture medium are unknown can reveal meaningful conclusions regarding the influence of the environment.

Summary and Conclusions

Any study of environmental control of an organism must take into account an environmental continuum extending from the gene locus, or even part of a gene, the *site,* to the outside world. This continuum is illustrated as follows:

Unit Levels	*Environmental Levels*
site	gene (environment of the site)
gene	DNA molecule (environment of the gene)
DNA molecule	chromosome (environment of the DNA molecule)
chromosome	nuclear plasm
nucleus	cytoplasm
cell	tissue (cell population)
tissue	organ
organ	organism
organism	outside world

Note that a particular level of structure (for example, the chromosome) constitutes the environment of the level below it (in this case, the DNA molecule). Each level of complexity operates in the context of (or, is restrained by) the levels above it but is limited by the properties of the levels below it and harnesses these lower levels (Polanyi, 1968). The mechanisms of developmental processes can only be discovered by studying the processes in their contexts, not alone by analyzing the processes themselves. As Polanyi has further emphasized, studying the disassembled parts of a machine will never reveal their coordinated working to produce a functional machine. This is why so much emphasis is placed on the function of the environmental continuum as a mechanism of development.

Although the environment external to a developing organism may select among several potential programs of development, there is no evidence that it ever creates a program. The internal environmental continuum is itself a product of sequential gene action. In this way an organism inherits not only information or potentialities but also the method for orderly use of the information or realization of potentialities.

References

Allsopp, A. 1953. Experimental and analytical studies of pteridophytes: 19. Investigations on *Marsilea*, 2. Induced reversion to juvenile stages. Ann. Bot. n.s. **17:**37–55.

Allsopp, A. 1955. Investigations on *Marsilea*, 5. Culture conditions and morphogenesis, with special reference to the origin of land and water forms. Ann. Bot. n.s. **19:**247–264.

Allsopp, A. 1967. Heteroblastic development in vascular plants. Adv. Morph. **6:**127–171.

Ashby, E. 1949. Leaf shape and physiological age. Endeavour **8:**18–25.

Ball, E. 1946. Development in sterile culture of stem tips and subjacent regions of *Tropaeolum majus* L. and *Lupinus albus* L. Am. J. Bot. **33:**301–318.

Baltzer, F. 1928. Neue Versuche über die Bestimmung des Geschlechts bei *Bonellia viridis*. Rev. Suisse Zool. **35:**225–231.

Barksdale, A. W. 1969. Sexual hormones of *Achlya* and other fungi. Science **166:**831–837.

Beale, G. H. 1954. The Genetics of *Paramecium aurelia*. Cambridge University Press, London.

Beck, S. D. 1963. Animal Photoperiodism. Holt, Rinehart and Winston, New York.

Bonner, B. 1966. Phytochrome and the red–far red system. In W. A. Jensen and L. C. Kavaljian, eds., Plant Biology Today, pp. 185–208. Wadsworth Publishing Company, Inc., Belmont, Calif.

Bower, F. O. 1930. Size and Form in Plants. Macmillan, New York.

Brien, P. 1961. Etude d'*Hydra pirardi*. Origine et repartition des nematocystes. Gametogenese. Involution postgametique. Evolution reversible des cellules interstitielles. Bull. Biol. France Belg. **95:**301–363.

Bristow, J. M. 1962. The controlled *in vitro* differentiation of callus derived from a fern, *Pteris cretica* L., into gametophytic or sporophytic tissues. Devel. Biol. **4:**361–375.

Bücher, H. 1906. Anatomische Veranderungen bei gewaltsamer Krümmung und geotropischer Induktion. Jahrb. Wiss. Bot. **43:**271–360.

Cantino, E. C., and J. S. Lovett. 1964. Non-filamentous aquatic fungi: model systems for biochemical studies of morphological differentiation. Adv. Morph. **3**:33–93.

Chalkley, H. W. 1945. Quantitative relation between the number of organization centres and tissue volume in regenerating masses of minced body sections of *Hydra*. J. Nat. Cancer Inst. **6**:191–195.

Child, C. M. 1941. Patterns and Problems of Development. University of Chicago Press, Chicago.

Darwin, C., and F. Darwin. 1881. The Power of Movement in Plants. Appleton-Century, New York.

Davis, B. 1968. Is riboflavin the photoreceptor in the induction of two-dimensional growth in fern gametophytes? Plant Physiol. **43**(7):1165–1167.

Davison, J. 1969. Activation of the ephippial egg of *Daphnia pulex*. J. Gen. Physiol. **53**:562–575.

Dingle, A. D., and C. Fulton. 1966. Development of the flagellar apparatus of *Naegleria*. J. Cell Biol. **31**:43–54.

Emerson, R. 1954. The biology of water molds. In D. Rudnick, ed., Aspects of Synthesis and Order in Growth, pp. 171–208. Princeton University Press, Princeton, N.J.

Epp, L. G., and C. F. Lytle. 1969. The influence of light on asexual reproduction in green and aposymbiotic hydra. Biol. Bull. **137**:79–94.

Fulton, C., and A. D. Dingle. 1967. Appearance of the flagellate phenotype in populations of *Naegleria* amebae. Devel. Biol. **15**:165–191.

Galston, A. W. 1967. Regulatory systems in higher plants. Am. Scientist **55**:144–160.

Grobstein, C. 1955. Tissue disaggregation in relation to determination and stability of cell type. Ann. N.Y. Acad. Sci. **60**:1095–1106.

Grobstein, C., and E. Zwilling. 1953. Modification of growth and differentiation of chorio-allantoic grafts of chick blastoderm pieces after cultivation at a glass-clot interface. J. Exp. Zool. **122**:259–284.

Haberlandt, G. 1902. Kulturversuche mit isolierten Pflanzenzellen. Sitzungsber. Akad. Wissen. Wien, Math. Naturwiss. Kl. **111**:69–92.

Hammerling, J. 1966. Nucleo-cytoplasmic relationships in the development of *Acetabularia*. In R. A. Flickinger, ed., Developmental Biology, pp. 23–47. William C. Brown, Dubuque, Iowa.

Harrison, R. G. 1907. Observations on the living developing nerve fiber. Anat. Rec. **1**:116–118.

Ito, M. 1962. Studies on the differentiation of fern gametophytes: I. Regeneration of single cells isolated from cordate gametophytes of *Pteris vittata*. Bot. Mag. Tokyo **75**:19–27.

Jaccard, P. 1914. Structure anatomique de racines hypertendues. Rev. Gén. Bot. **25**:359–372.

Jaczewski, Z. 1954. The effect of changes in length of daylight on growth of antlers in the deer (*Cervus elaphus* L.). Folia Biol. Warsaw **9**:47–99.

Kjellmann, F. R. 1885. Studien und Forchungen Veranlasst durch meine Reisen im hohen Norden, pp. 443–521. A. E. Nordenskiöld, ed., Leipzig.

Lang, A. 1953. Physiology of Flowering. Ann. Rev. Plant Physiol. **3**:265–306.

Lees, A. D. 1959. The role of photoperiod and temperature in the determination of parthenogenetic and sexual forms in the aphid, *Megoura viciae*. J. Insect Physiol. **3**:92–117.

Leopold, A. C. 1964. Plant Growth and Development. McGraw-Hill Book Company, New York.

Lopashov, G. V. 1935. Die Entwicklungsleistungen des Gastrulaektoderms in Abhängigkeit von Veränderungen der Masse. Biol. Zentrbl. **55**:606–615.

Lovett, J. S. 1967. Aquatic fungi. In F. H. Wilt and N. K. Wessells, eds., Methods in Developmental Biology, pp. 341–358. Thomas Y. Crowell, New York.

Macklin, M. and A. L. Burnett. 1966. Control of differentiation by calcium and sodium ions in *Hydra pseudoligactis*. Exp. Cell Res. **44**:665–668.

Mitra, G. C., L. P. Misra, and C. Prabha. 1965. Interaction of red and blue light on the development of the protonema and bud formation in *Pohlia nutans.* Planta **65:**42–48.

Mohr, H. 1966. Differential gene activation as a mode of action of phytochrome 730. Photochem. Photobiol. **5:**469–483.

Mohr, H. 1969. Photomorphogenesis. In M. B. Wilkins, ed., The Physiology of Plant Growth and Development, pp. 507–556. McGraw-Hill Book Company, New York.

Morgan, T. H. 1901. Regeneration. Macmillan, New York.

Näf, U. 1953. Some contributions to the development of the gametophytic phase of the fern, *Onoclea sensibilis* L. Doctoral thesis, Yale University, New Haven, Conn.

Naylor, A. W. 1952. Reactions of plants to photoperiod. In W. E. Loomis, ed., Growth and Differentiation in Plants, pp. 149–178. Iowa State College Press, Ames.

Newell, P. C., A. Telser, and M. Sussman. 1969. Alternative developmental pathways determined by environmental conditions in the cellular slime mold *Dictyostelium discoideum.* J. Bact. **100:**763–768.

Nickerson, W. J. 1954. Experimental control of morphogenesis in microorganisms. Ann. N.Y. Acad. Sci. **60:**50–57.

Ohlenroth, K., and H. Mohr. 1964. Die steuerung der Proteinsynthese durch Blaulicht und Hellrot in den Vorkeimen von *Dryopteris felix-mas* L. Schott. Planta **62:**160–170.

Parker, M. W., and H. A. Borthwick. 1950. Influence of light on plant growth. Ann. Rev. Plant Physiol. **1:**43–58.

Polanyi, M. 1968. Life's irreducible structure. Science **160:**1308–1312.

Prat, H. 1948. Histophysiological gradients and plant organogenesis. Bot. Rev. **14:**603–643.

Raper, J. R. 1955. Some problems of specificity in the sexuality of plants. In E. G. Butler, ed., Biological Specificity and Growth, pp. 119–140. Princeton University Press, Princeton, N.J.

Raper, J. R. 1960. The control of sex in fungi. Am. J. Bot. **47:**794–808.

Raper, J. R. 1966. Genetics of Sexuality in Higher Fungi. Ronald Press, New York.

Reinhardt, D., and A. Mancinelli. 1968. Developmental responses of *Acrasis rosea* to the visible light spectrum. Devel. Biol. **18:**30–41.

Salisbury, F. B. 1963. The Flowering Process. Macmillan, New York.

Salisbury, F. B., and C. Ross. 1969. Plant Physiology. Wadsworth Publishing Company, Inc., Belmont, Calif.

Saxen, L., and S. Toivonen. 1962. Primary Embryonic Induction. Logos Press, London.

Schaffner, J. H. 1931. The fluctuation curve of sex reversed in staminate hemp plants induced by photoperiodicity. Am. J. Bot. **18:**424–430.

Sibaoka, T. 1969. Physiology of rapid movements in higher plants. Ann. Rev. Plant Physiol. **20:**165–184.

Sinnott, E. W. 1960. Plant Morphogenesis. McGraw-Hill Book Company, New York.

Spratt, N. T., Jr. 1964. Introduction to Cell Differentiation. Reinhold, New York.

Spratt, N. T., Jr., and H. Haas. 1961. Integrative mechanisms in development of the early chick blastoderm: III. Role of cell population size and growth potentiality in synthetic systems larger than normal. J. Exp. Zool. **147:**271–293.

Steward, F. C., A. E. Kent, and M. O. Mapes. 1966. The culture of free plant cells and its significance for embryology and morphogenesis. In A. A. Moscona and A. Monroy, eds., Current Topics in Developmental Biology, pp. 113–154. Academic Press, New York.

Steward, F. C., M. O. Mapes, A. E. Kent, and R. D. Holsten. 1964. Growth and development of cultured plant cells. Science **143:**20–27.

Taylor, A. O., and B. A. Bonner. 1967. Isolation of phytochrome from the alga, *Mesotaenium,* and the liverwort, *Sphaerocarpus.* Plant Physiol. **42:**762–766.

Torrey, J. G. 1967. Development in Flowering Plants. Macmillan, New York.

Vowinckel, C. 1970. The role of illumination and temperature in the control of sexual reproduction in the planarian *Dugesia tigrina* (Girard). Biol. Bull. **138:**77–87.

Ward, M. 1963. Developmental patterns of adventitious sporophytes in *Phlebodium aureum.* J. Sm. J. Linn. Soc. Bot. **58:**377–380.

Wardlaw, C. W. 1965. Leaves and buds: mechanisms of local induction in plant growth. In Cell Differentiation and Morphogenesis, pp. 96–119. North Holland Publishing Company, Amsterdam.

Went, F. W. 1954. Physical factors affecting growth in plants. In E. J. Boell, ed., Dynamics of Growth Processes, pp. 130–147. Princeton University Press, Princeton, N.J.

Wetmore, R. H., A. E. Demaggio, and G. Morel. 1963. A morphogenetic look at the alternation of generations. J. Indian Bot. Soc. **42A:**306–320.

Wetmore, R. H., and J. P. Rier. 1963. Experimental induction of vascular tissues in callus of angiosperms. Am. J. Bot. **50:**418–430.

Whitier, D. P., and T. A. Steeves. 1960. The induction of apogamy in the bracken fern. Canad. J. Bot. **38:**925–930.

Whitier, D. P., and T. A. Steeves. 1962. Further studies on induced apogamy in ferns. Canad. J. Bot. **40:**1525–1531.

Wilde, C. 1961. Factors concerning the degree of cellular differentiation in organotypic and disaggregated tissue cultures. In M. E. Wolff, ed., Actes du colloque international sur "La culture organotypique: associations et dissociations d'organes en culture *in vitro,*" pp. 183–198. Paris.

Wilkins, M. B. 1969. The Physiology of Plant Growth and Development. McGraw-Hill Book Company, New York.

Willmer, E. N. 1956. Factors which influence the acquisition of flagella by the amoeba, *Naegleria gruberi.* J. Exp. Biol. **33:**583–603.

Witschi, E. 1934. Genes and inductors of sex differentiation in amphibians. Biol. Rev. Camb. **9:**460–488.

Witschi, E. 1967. Biochemistry of sex differentiation in vertebrate embryos. In R. Weber, ed., The Biochemistry of Animal Development, pp. 193–225. Academic Press, New York.

Part Four:

Special Aspects of Development

23 The Role of Extracellular Products

In addition to growth, cell movement, morphological specialization of cells, and cell death, development involves the production and release of specific cell products. This chapter discusses the role of extracellular products in development. It is difficult to draw a sharp line separating morphological from functional aspects of cellular differentiation, but it is possible to distinguish products of gene action that remain inside a cell and constitute its morphology from products secreted by the cell. Through various, often intricate patterns of association, these products characterize to a large extent the development and structure of many organisms. The extracellular products acquire their definitive form after release from the cell, and the cell type is primarily identified by its product rather than by its morphology, which may remain rather mundane. This is particularly true of cellular differentiation in plants, where the organization of the cell wall surrounding the cell constitutes its distinctive feature. However, it is true of many animal organisms as well.

There are many examples of products remaining inside cells that are not components of the cell's structure and that are not necessary for the life of the cell. This category includes crystalline substances such as the complex crystoliths in the leaf cells of rubber and fig plants, the silica bodies in the leaves of grasses and palms, calcium oxalate crystals in many plants, and starch grains. Many invertebrate cells and vertebrate bones also contain crystalline cell products (see review of King, 1969). A particularly complex intracellular product is the *nematocyst* of cnidaria. According to Lenhoff (1968), the nematocyst is composed of a protein resembling collagen, of which 12 to 20 percent is the amino acid *hydroxyproline,* a common constituent of vertebrate collagen. Various crystal types such as silicate dendrites and spherulites occur intracellularly in both plants and animals. These progressively develop from a primary form, a *microlite* (Liang-Ho, 1965). Although they do distinguish cell types, these intracellular inclusions do not appear to be significant in development of the organism as a whole.

Extracellular products constitute the internal microenvironment of the cells of multicellular organisms. These include a multitude of substances, such as inductors and hormones, which were discussed in Chapters 19 and 21. The primary concern of this chapter is those products that seem to have a direct influence in the control of cell movements and cell distribution, and an overall function in shaping the organism. Extracellular materials control the orderly development of cells by constraining and sustaining cellular assemblies (Picken, 1960). Furthermore, in many organisms development eventually involves not the production of more cells but of more extra-

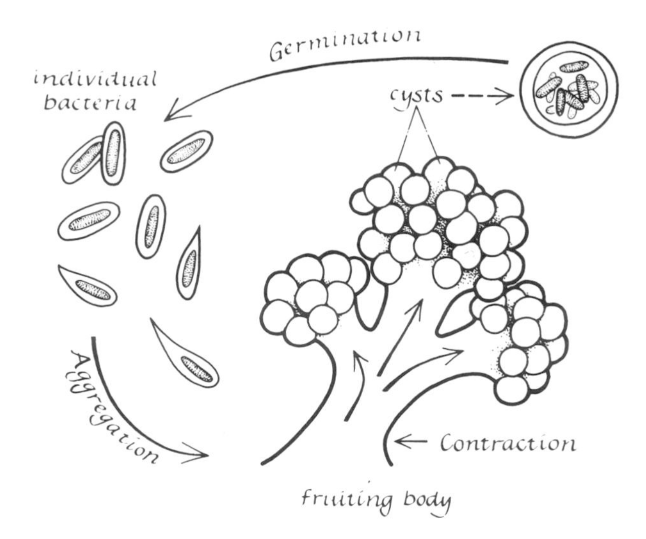

Figure 23-1. *Diagrams of individual slime bacteria and the fruiting body formed by an aggregate of the bacteria.*

cellular substances, as for example, mollusk shells. The cells sometimes employ extracellular minerals from their environment to construct complex exoskeletons, such as the silicaceous shell of *Difflugia* (Chapter 18). In these cases the skeletons are not direct gene products, but their assembly is a product of gene action.

The following examples of organisms in which cell products are significant factors in development illustrate the function of these products.

Myxobacteria

In the so-called slime bacteria (for example in *Chondromyces crocatus*), the slime extruded by the rod-shaped bacteria appears to play an important role in aggregation and fruiting body formation. As the rods move they excrete slime tracks, which appear to guide the movement of other rods. Finally, a complex fruiting body is formed (Figure 7-26). This process has been compared to a funnel the base of whose flaring portion progressively constricts, forcing the flaring part upwards. Apparently the contraction of the slime accomplishes this type of morphogenesis (Figure 23-1), but many questions remain (Bonner, 1952).

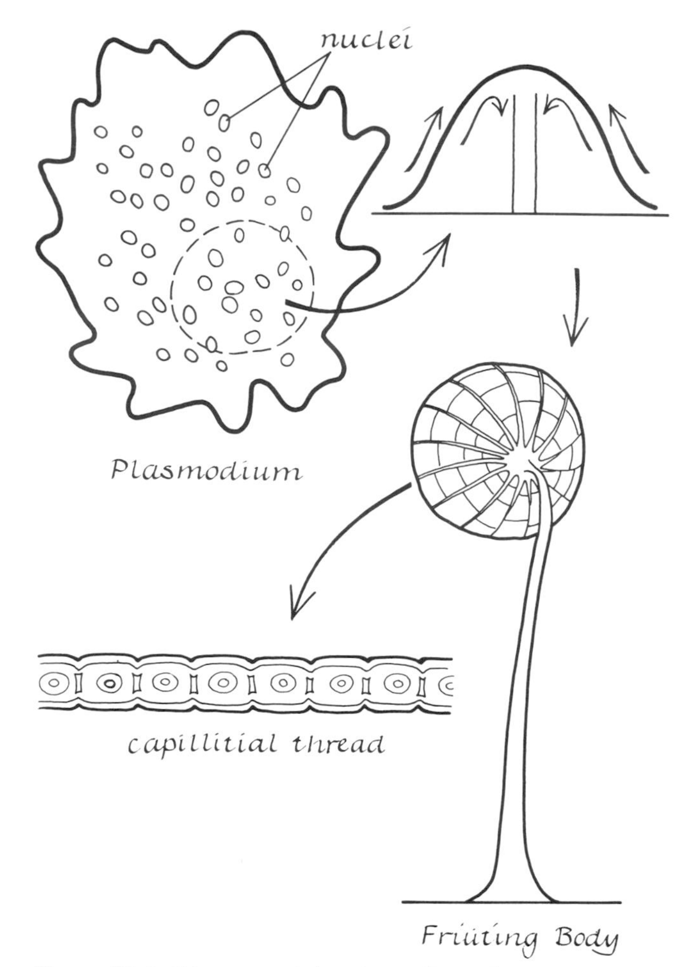

Figure 23-2. *Diagrams of the plasmodium of a myxomycete, the formation of the base of the stalk and its elongation by addition of exudate to its tip, a mature fruiting body of* Dictydium, *and part of a sculptured capillitial thread.*

True Slime Molds (Myxomycetes)

The complexly architectured fruiting bodies of the true slime molds, exclusive of the spores, are largely the result of a modeling of a noncellular exudate. According to Howard (1931), a short stalk, which is deposited in the center of each fruiting body region of the multinucleate plasmodium, is elongated by addition of exudate to its tip. The protoplasm streams up the stalk until all of it is at the tip. Here more exudate is produced, and through its condensation the complex *capillitium* is formed. In *Dictydium cancellatum* an amazing, birdcage-like structure is formed (Figure 23-2). The capillitium is a network

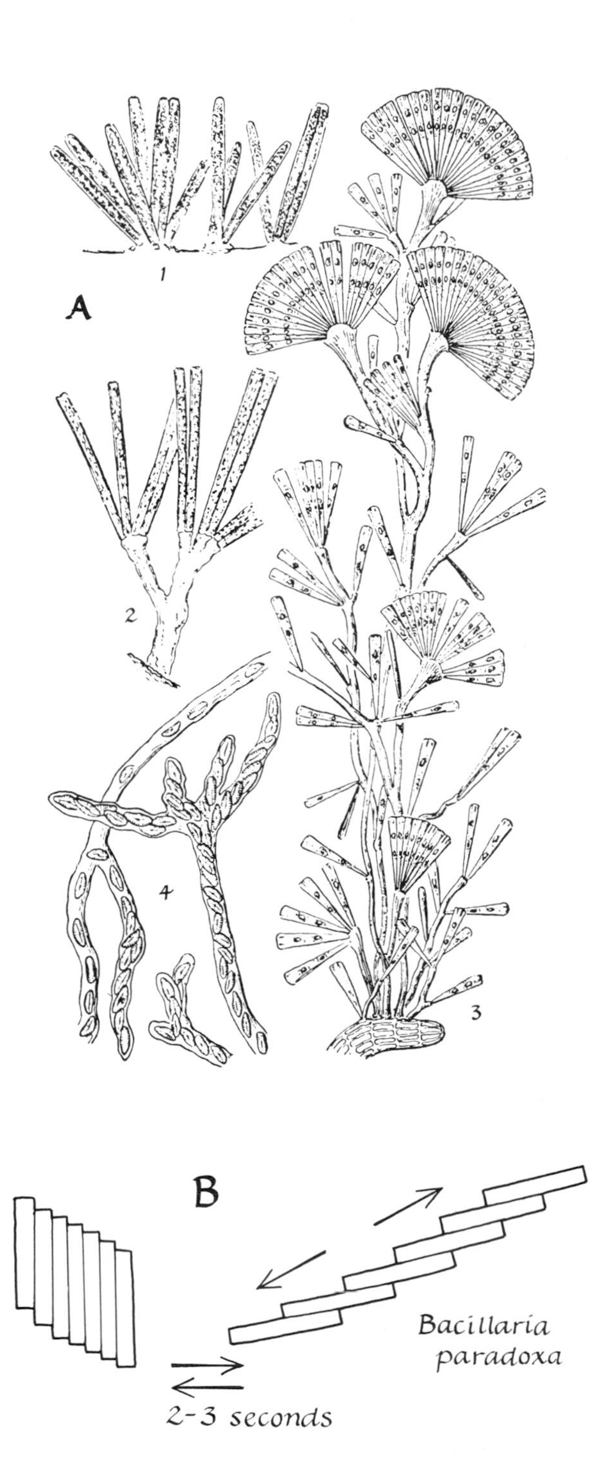

Figure 23-3. *A, various colonial diatoms:* 1 *and* 2, *two species of* Synedra; 3, Licomorpha flagellata; 4, Cymbella. *B, diagram illustrating movement in colonial diatom* Bacillaria paradoxa. *A, from Oltmans, 1904, Gustav Fischer, Stuttgart.*

of threads, each of which is intricately sculptured. Whether the cytoplasm of the plasmodium is a mold guiding the assembly of cell molecules to form the capillitium or whether the cytoplasm is an environment for a self-assembly of the molecules is not known.

Colonial Algae

In the colonial blue green and green algae the gelatinous extracellular product is important in holding the cells together, especially in those forms that appear to have no intercellular cytoplasmic strands. The definite orientation of the cells in relation to the polarity of the colony, as we saw in Chapter 7, might be controlled by the structure of the matrix in which they are embedded.

Colonization in diatoms such as *Navicula rhombica* is achieved largely through the synthesis of cell products which bind the cells together in remarkably orderly patterns (Wilson, 1929; Smith, 1933). In some diatoms the hard silica valves surrounding the protoplasm of each individual are stuck together in intricate ways by interlocking spines or adhesive disks. Some colonial diatoms are illustrated in Figure 23-3A.

One of the most remarkable colonial diatoms is *Bacillaria paradoxa* (Figure 23-3B), whose cells can slide along one another. At one instant the spindle-shaped cells are aligned like pickets in a fence; at the next they stretch out but remain partially attached to one another (Smith, 1933).

How the protoplast of a diatom constructs the complex valves is not known. In some fashion the protoplast seems to function like a template for assembly of silicon. For example, Hendy (1945) kept protoplasts ejected from the shells of marine diatoms in silica-free water for nine months, during which time the elliptical globs of protoplasm divided many times. Upon return to water containing silica, the diatoms rebuilt shells identical to the original ones.

Radiolarians and Foraminiferans

Remarkably complex silica or calcium carbonate shells secreted by these protozoa constitute the diagnostic features of the adults (Figure 23-4).

Again the protoplast apparently functions like a template, but the mechanism is unknown (Thompson, 1942). Some self-assembly (Chapter 25) probably occurs, since the shape of the shell rarely reflects the shape of the protoplasmic mass.

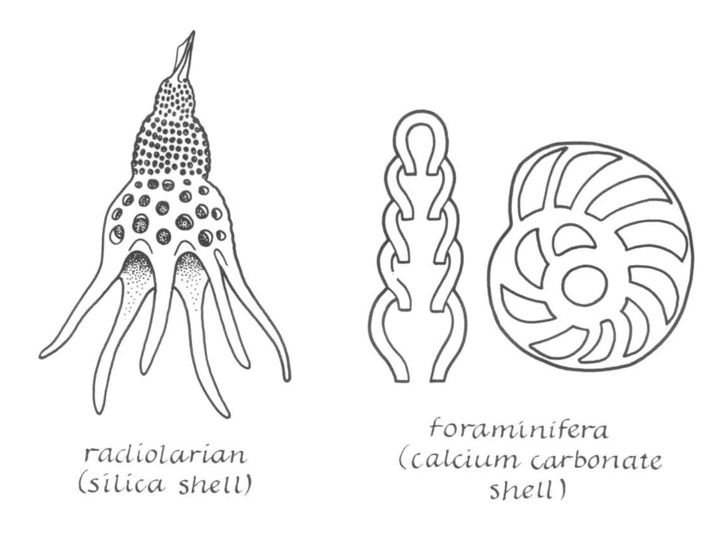

Figure 23-4. *Diagrams of the shells of a Radiolarian and two species of Foraminifera.*

Sponges

The body shape of many sponges results from the arrangement of the silicaceous or calcium carbonate spicules that form the sponge skeleton. A remarkable example of this is the silicaceous Venus flower basket of the sponge *Euplectella suberea* (Figure 23-5).

A monaxon spicule of a sponge is formed initially inside a binucleate scleroblast cell. As the calcium carbonate spicule enlarges, the scleroblast divides. When the spicule reaches its maximum size, the scleroblasts wander into the mesoglea. Triradiate spicules are secreted by three scleroblasts, each of which secretes a monaxon spicule. The three monaxon spicules then fuse to form the triradiate form (Figure 23-6).

Silicaceous spicules are also formed intracellularly and outgrow the cell. The hexactinellid sponge *Monorhaphis* builds a spicule two to three meters long and about one centimeter in diameter (Needham, 1964). Young spicules of the sponge *Leucosolenia complicata* placed in a calcium carbonate solution continue to grow in normal fashion by deposition of calcite. This appears to be a crystallization process (Jones, 1954, 1955). Some of the types of spicules found in sponges are shown in Figure 23-7.

The spicule in Figure 23-7D is constructed with silica; an almost identical spicule is constructed with calcium carbonate in the sea-cucumber, an echinoderm. In this case, unlike the one cited above, the spicule shape is apparently not determined by the properties of the molecules, since they are radically different (see Thompson, 1942). A template mechanism might be involved here, but how unrelated organisms (sponges and echinoderms) could have similar cytoplasmic templates (if they do) is strange, although it could be the result of convergent evolution. How spicules are assembled in designs characterizing the shape of each species of sponge, as well as how individual spicules form, is not known.

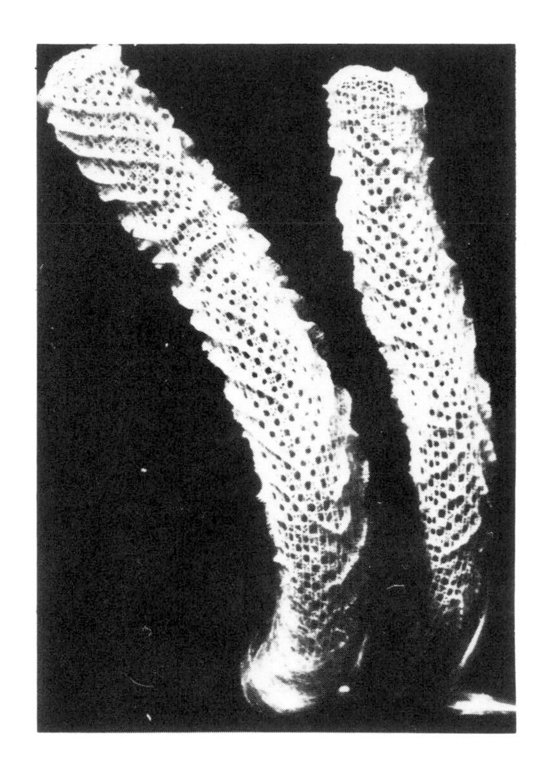

Figure 23-5. *Photograph of the silicaceous skeleton of the sponge* Euplectella suberea. *American Museum of Natural History, reproduced by permission.*

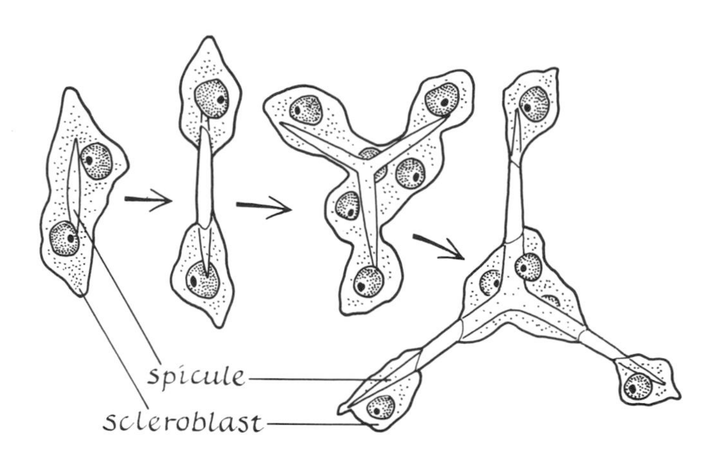

Figure 23-6. *Stages in formation of a triradiate spicule of a sponge.*

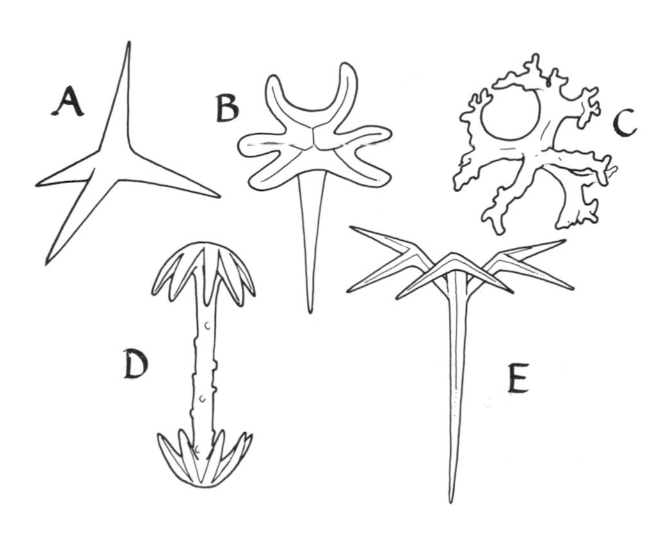

Figure 23-7. *Some types of spicules found in sponges.*

Coelenterates

The mesoglea of the sea anemone (a coelenterate) is a complex system of radial, longitudinal, and diagonal fibers (see Picken, 1960). Hydra contains a similarly organized mesoglea. This extracellular matrix might account for the shape of these organisms.

Figure 23-8. *The intricate design of a mollusc shell.*

Molluscs

The great diversity of the 100,000 mollusc species is largely a diversity in extracellular products and the way they assemble to form shells of remarkably complex design. As the mollusc develops, the *mantle* (shell-forming organ) deposits a horny acid-resistant cuticle and an underlying layer of $CaCO_3$ crystals. In gastropods, the *nacre* (mother-of-pearl) of the shell consists of aragonite that is deposited in the form of crystal stacks. This crystallization involves a complex system of organic membranes (Wise, 1970). However, very little is known about the mechanisms that control the intricate shell designs (Figure 23-8).

Echinoderms

The echinoderm skeleton is composed of plates and spines, each of which is a single crystal of magnesium-rich calcite (Donnay and Pawson, 1969; Nissen, 1969). However, the cells secreting the crystal-forming substances appear to direct the crystallization.

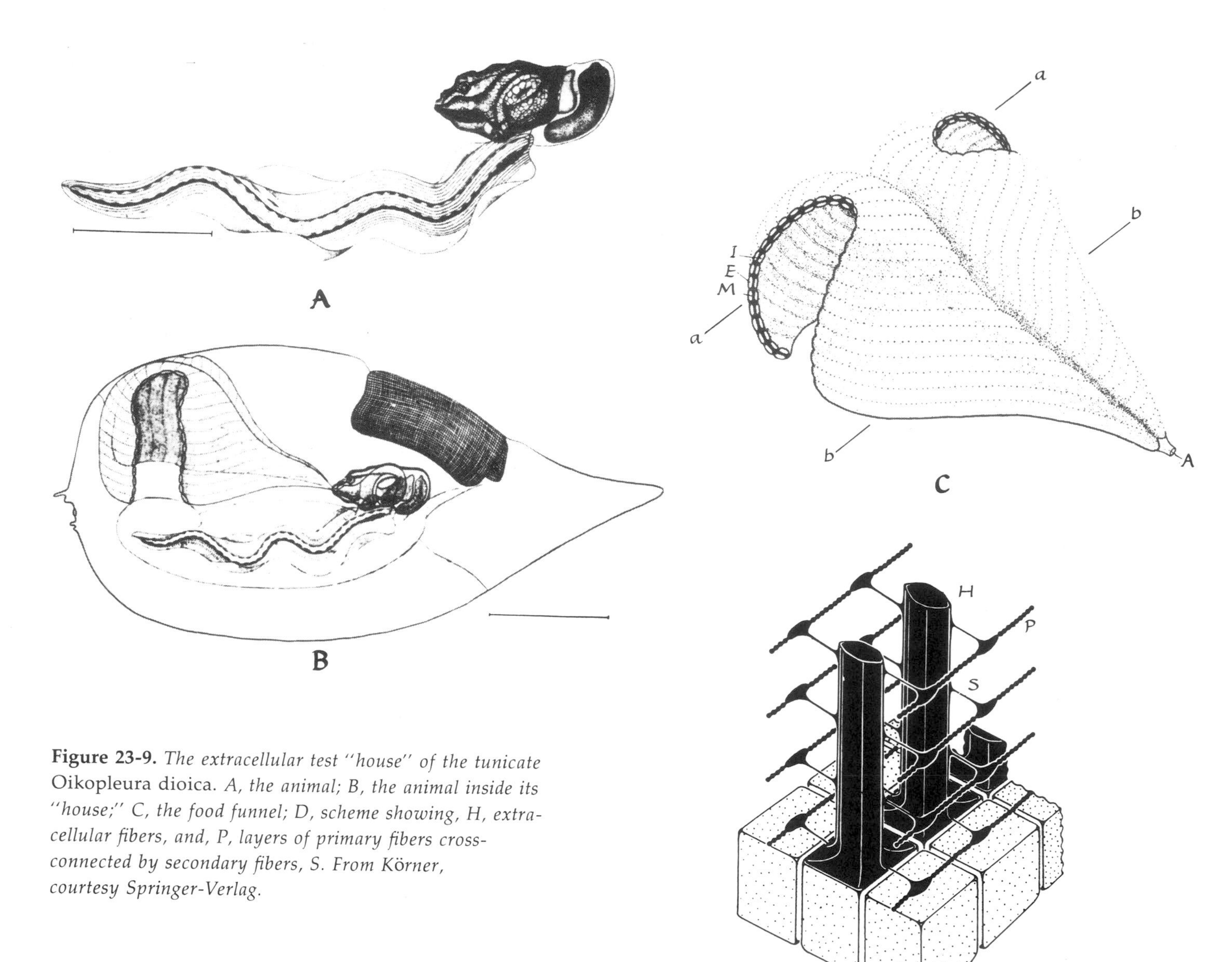

Figure 23-9. *The extracellular test "house" of the tunicate* Oikopleura dioica. *A, the animal; B, the animal inside its "house;" C, the food funnel; D, scheme showing, H, extracellular fibers, and, P, layers of primary fibers cross-connected by secondary fibers, S. From Körner, courtesy Springer-Verlag.*

Arthropods

The complex exoskeletons of arthropods consist of *chitin,* a linear polymer of acetyl-B-glucosamine. This extracellular product becomes organized into bristles, scales, and other structures. The subunits of chitin appear to be largely self-assembled (see Richards, 1951; Picken, 1960). Whether an arthropod's shape is largely a consequence of the shape of its exoskeleton, which is a single entity, is possible but unproved.

Tunicates

The *test* (coat) of the larvacean tunicate *Oikopleura* is a spectacular extracellular structure. The test is extraordinarily complex; it includes feeding funnel and filters, although it is a purely epidermal secretion. The fibrous material of the test seems to be

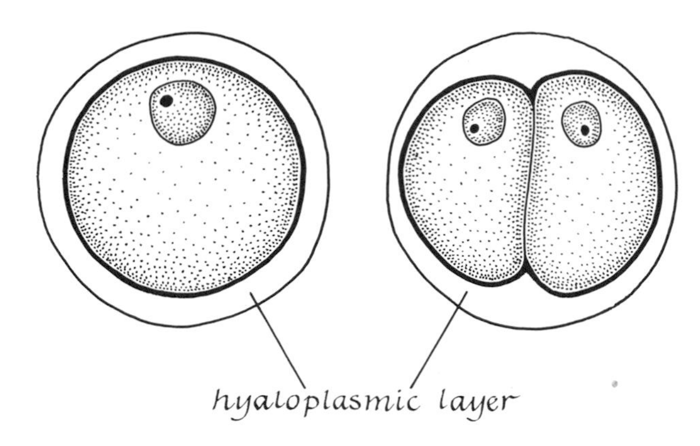

Figure 23-10. *Diagram of the hyaloplasmic layer of a sea urchin egg.*

mainly chondroid and mucoid (Körner, 1952). The test is illustrated in Figure 23-9.

Vertebrates

The scales, horns, hair, and feathers of vertebrates are assemblies of keratin, a complex protein synthesized by epidermal cells. The completed structures are entirely extracellular and in feathers are very complex. Although the general features of development of a feather are known (Chapter 4), the details remain unknown.

Collagen (Chapter 25) which comprises about 25 to 35 percent of the total body protein of higher animals is an extremely important extracellular product (Picken, 1960). Many invertebrates also contain collagen or collagen-like substances (Rudall, 1955). Evidence for the importance of collagen in the differentiation of myoblasts *in vitro* has been presented by Konigsberg and Hauschka (1965) and Hauschka (1968). Collagen appears to be the principal substance in culture media conditioned by the growth of fibroblasts. Fibroblasts deposit collagen on the surface of petri plates, and this collagen permits muscle clones to develop from a single myoblast. These results, however, remain to be confirmed. Collagen also appears to be necessary for normal morphogenesis *in vitro* of lung and ureteric bud (kidney) primordia (Wessels and Cohen, 1968) and salivary gland primordia (Grobstein and Cohen, 1965) in the mouse embryo. Collagenase treatment of the primordia results in loss of normal morphology. Upon addition of mesoderm cells (the source of collagen) normal morphogenesis resumes.

The fibrillar net of collagen in the skin of sharks and teleosts is oriented in a definite relation to the structural axis of the embryo. The developing body surface of these fishes is thus contained by a system of geodetics consisting of open helices and elliptical annuli (Picken, 1960).

The collagenous lattice in the skin of the chick embryo appears to affect the spacing of dermal condensations which precede the development of scales. These condensations form where the collagen fibers intersect. Scaleless dorsal skin has no such lattice (Goetinck, 1970).

Collagen is the chief component of the *basement lamella,* a discrete extracellular structure interposed between the ectodermal and mesodermal layers of the skin of vertebrates. During its development the lamella becomes composed of fibrils regularly arranged in an orthogonal grid of near-crystalline regularity. The epidermis of frog tadpoles can be stripped from the lamella by ultrasonic radiation (Bell, 1959), and much of the lamella can be peeled off the underlying mesoderm (Edds and Sweeny, 1961). The coarseness of the fibrous grid varies and is related to the general contours of the body surface. Edds and Sweeny suggest that the pattern in the lamella is determined not only by the epidermal cells but perhaps by the "organism as a whole." This remains to be explained more comprehensibly.

Collagen also appears to be involved in the aging of animals. Collagen tends to increase in amount and to change in structure with increasing age. The structural changes include an increase in cross-linkage of collagen fibers; this leads to tissue shrinkage around capillaries, which decreases the supply of nutrients and oxygen to local regions of the body (Bjorksten, 1968; Verzan, 1963).

Both the film or coat called the *surface coat* which covers the amphibian egg and its later stages of development, and the hyaloplasmic layer surrounding the cleaving sea urchin egg (Gray, 1925), might be considered extracellular products. The surface coat, which is firmly attached to the underlying outer membrane of the egg, is not an indispensable part of the egg or the epithelia derived from its surface (Holtfreter, 1943). Like the hyaloplasmic layer of sea urchin eggs (Figure 23-10), the surface coat remains on the surface during cleavage of the egg.

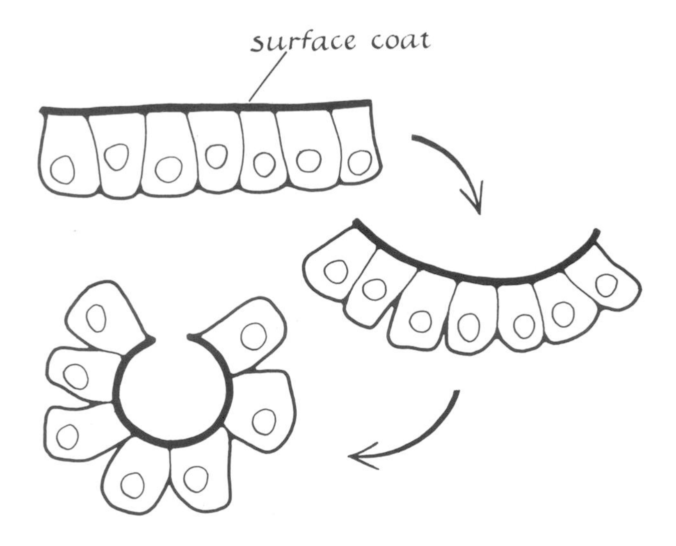

Figure 23-11. *Diagrams illustrating changes in position of explanted amphibian embryo cells when the surface coat to which they are attached contracts.*

The surface coat consists of a densely pigmented, contractile, rubber-like substance, as indicated by its contraction to close a wound made by puncturing an uncleaved egg. The surface coat on isolated groups of cells from a gastrula stage contracts; this causes the relationships of the cells to change (Figure 23-11).

The surface coat's behavior suggests that it functions in integrating the cell population of the embryo and probably in morphogenetic processes such as gastrulation. Recall (Chapter 13) that an unfertilized, aging, and noncleaving frog egg exhibits morphogenetic surface movements simulating gastrulation. Perhaps the surface coat is involved in these changes. However, as noted by Holtfreter, the coat apparently functions in controlling but not in causing gastrulation and neurulation movements.

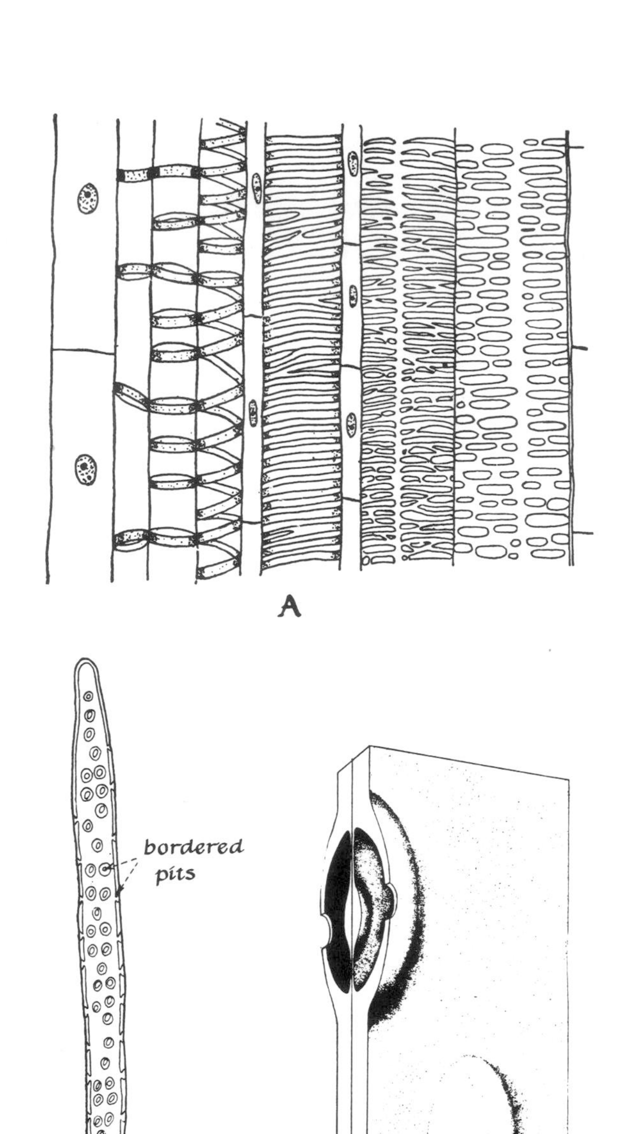

Figure 23-12. *A, left, first xylem to be formed in a stem; middle, next xylem to be formed; right, still later xylem to be formed; B, a single pitted tracheid from pine wood; C, diagram showing the structure of a bordered pit. A from* Botany: Principles and Problems *by Sinnott. Copyright 1963 by McGraw-Hill Book Company. Used with permission of McGraw-Hill Book Company. B and C from Holman and Robbins, 1940, courtesy John Wiley and Sons, Inc.*

Plants

As noted earlier, cellular differentiation in plants involves the formation of the cell wall. The primary framework of the wall in most higher plants is cellulose. In fungi it is chitin and in some algae it is β, 1-4 linked xylan (Clowes and Juniper, 1968). The cell walls of many plants also contain other constituents,

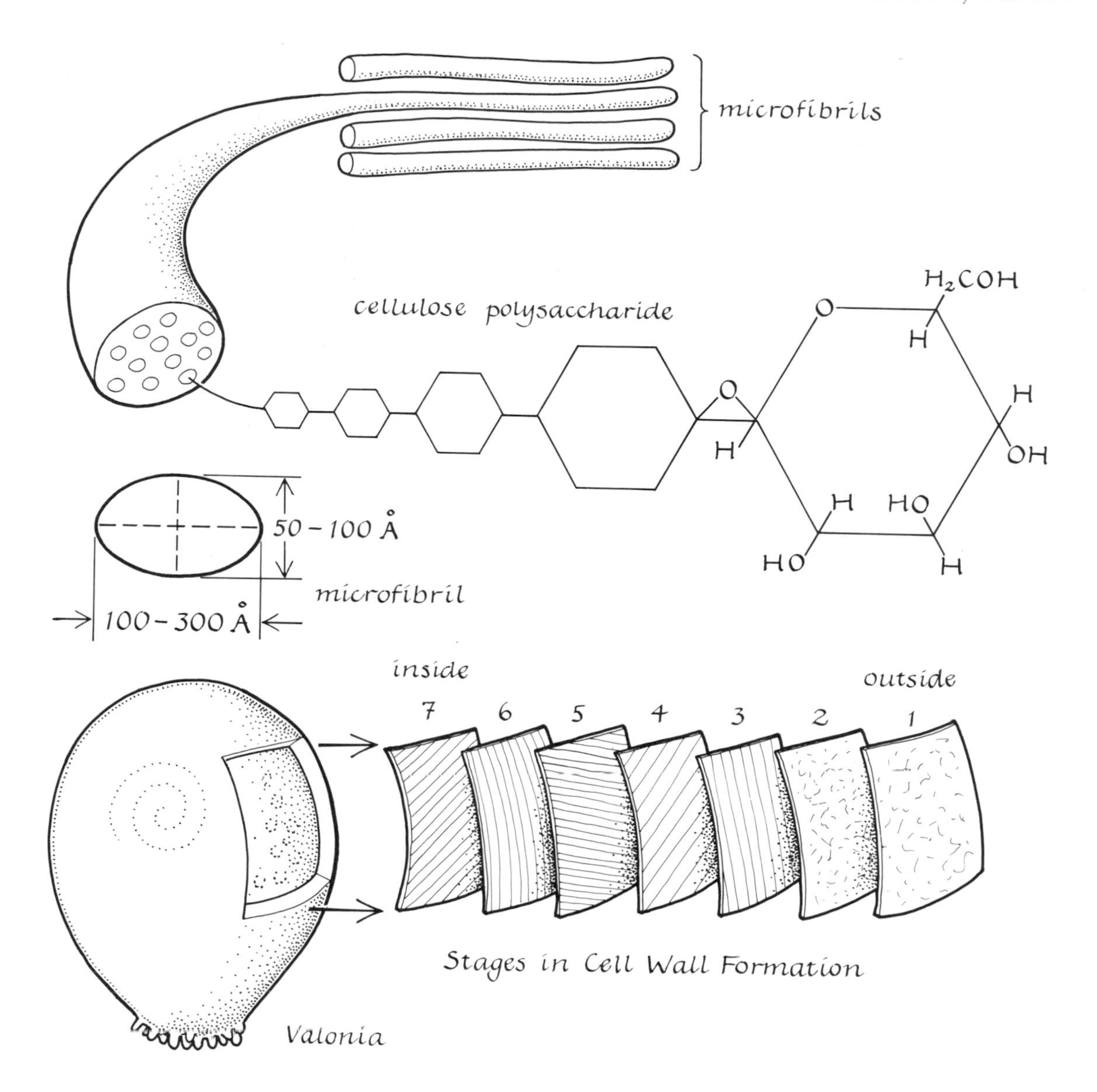

Figure 23-13. *Top, the chemical structure of cellulose, and bottom, stages in cell wall formation in* Valonia.

such as lignin, hemicelluloses, and pectin-like substances. In Chapter 11 we noted the unique composition of the cell wall of the pollen grain. As the plant cell grows its shape changes, and this shape is related to the deposition of cellulose microfibrils in special orientations in the primary wall formed during early growth of the cell. This deposition is followed by the laying down of cellulose fibrils and other substances constituting the secondary wall on the inner face of the primary wall. The patterned distribution of cellulose microfibrils in the wall appears to determine the direction of cell growth and thus of cell shape. The wall then yields to internally generated pressure in different directions. As the wall yields, the cellulose microfibrils slide past one another; this results in a reorientation of the fibrils, which tend to lie parallel to the direction of growth. Understanding plant form thus requires understanding what controls the orientation in the deposition of the cell wall materials.

Patterns of fibrillar distribution are often complex, consisting of the formation of bars, loops, helices, and networks. For example, in the differentiation of xylem cells in the growing stem, the arrangement of fibrous elements in the walls of the first xylem cells formed differs from the arrangement of the fibrous elements in the walls of xylem cells formed later (Figure 23-12A).

Note in Figure 23-12B and C that the walls of *tracheid* cells that are formed later in development have thinner, specialized areas called *pits*, which are arranged in *pit fields*. Pit fields in the wall of one cell are frequently aligned with those in an adjacent cell; this indicates that the wall-building processes in neighboring cells are correlated and that cellular interactions are probably involved.

Although noncellulose substances, such as waxes in leaves and fruits and sporopollenin in pollen grains, are arranged in definite patterns in the cell wall, cellulose apparently functions as a template in guiding their deposition.

The guiding forces in cell wall formation appear to involve microtubules, the endoplasmic reticulum, and cytoplasmic streaming within the cell (reviews of Preston, 1961; Heslop-Harrison, 1968; and Newcomb, 1969). Formation of patterns in cell walls is probably guided in part by template and in part by self-assembly mechanisms. Deposition of wall constituents can occur at a distance from the surface of the protoplast, which might suggest self-assembly. However, if self-assembly does occur, its spatial pattern is probably guided by some kind of template, possibly consisting of a definite arrangement of microtubules adjacent to the growing cell wall. What controls the arrangement of the microtubules thus becomes an important question. Undoubtedly the genome controls these patterns, but exactly how this is done remains to be discovered.

The structure of cellulose and stages in formation of the cell wall of the unicellular alga *Valonia macrophysa* are illustrated in Figure 23-13 (Preston, 1961). Note in the figure the alternate direction of alignment of cellulose fibrils in the successive layers of the cell wall and the diversity of alignment in different regions of the wall of the alga.

Conclusion

Extracellular products function significantly in the development of form in many organisms. In many organisms the form of the individual is the form developed by its products, and the cells function primarily as the source of material. In addition, extracellular products often influence the behavior of the cells which have produced them.

References

Bell, E. 1959. A new approach to some problems in experimental embryology through the use of ultrasound. In H. Quastler and J. J. Morowitz, eds., Proc. First Nat. Biophysics Conf., pp. 674–682. Yale University Press, New Haven, Conn.

Bjorksten, J. 1968. The crosslinkage theory of aging. J. Amer. Geriatrics Soc. **16:**408–427.

Bonner, J. T. 1952. Morphogenesis. Princeton University Press, Princeton, N.J.

Clowes, F. A. L., and B. E. Juniper. 1968. Plant Cells. Blackwell, London.

Donnay, G., and D. L. Pawson. 1969. X-ray diffraction studies of echinoderm plates. Science **166:**1147–1150.

Edds, M. V., and P. R. Sweeny. 1961. Chemical and morphological differentiation of the basement lamella. In D. Rudnick, ed., Synthesis of Molecular and Cellular Structure, pp. 111–138. Ronald Press, New York.

Goetinck, P. F. 1970. Analysis of skin and limb development in the chick embryo. Bull. Inst. Cell Biol. **11:**13–14.

Gray, J. 1925. The mechanism of cell division: I. The forces which control the form and cleavage of the eggs of *Echinus esculentus*. Proc. Camb. Phil. Soc. Biol. Sci. **1:**164–188.

Grobstein, C., and J. Cohen. 1965. Collagenase: effect on the morphogenesis of embryonic salivary epithelium *in vitro*. Science **150:**626–628.

Hauschka, S. D. 1968. Clonal aspects of muscle development and stability of the differentiated state. In H. Ursprung, ed., The Stability of the Differentiated State, pp. 37–57. Springer-Verlag, New York.

Hendy, N. I. 1945. Extra-frustula diatoms. J. Roy. Micr. Soc. Ser. III **65:**34–39.

Heslop-Harrison, J. 1968. The emergence of pattern in the cell walls of higher plants. In M. Locke, ed., The Emergence of Order in Developing Systems, pp. 118–150. Academic Press, New York.

Holtfreter, J. 1943. Properties and functions of the surface coat in amphibian embryos. J. Exp. Zool. **93:**251–323.

Howard, F. L. 1931. Life history of *Physarum polycephalum.* Am. J. Bot. **18:**116–133.

Jones, W. C. 1954. The orientation of the optic axis of spicules of *Leucosolenia complicata.* Quart. J. Micr. Sci. **95:**33–48; 191–203.

Jones, W. C. 1955. Crystalline properties of spicules of *Leucosolenia complicata.* Quart. J. Micr. Sci. **96:**129–149; 411–421.

King, L. J. 1969. Biocrystallography—an interdisciplinary challenge. Biol. Science **19:**505–518.

Konigsberg, I. R., and S. D. Hauschka. 1965. Cell and tissue interactions in the reproduction of cell type. In M. Locke, ed., Reproduction: Molecular, Subcellular and Cellular, pp. 243–290. Academic Press, New York.

Korner, W. E. 1952. Untersuchungen über die Gehäusebildung bei Appendicularien (*Oikopleura dioica* Fol). Zeit. Morph. Okol. Tiere **41:**1–53.

Lenhoff, H. M. 1968. Some prospects for investigating Hydra cells *in vitro.* In M. M. Sigel, ed., Differentiation and Defense Mechanisms in Lower Organisms, pp. 33–48. Waverly Press, Baltimore.

Liang-Ho, Su. 1965. The ontogeny of silicate crystals. Scientica Sinica **14:**113–119.

Needham, A. E. 1964. The Growth Process in Animals. Sir Isaac Pitman, London.

Newcomb, E. H. 1969. Plant microtubules. Ann. Rev. Plant Physiol. **29:**253–288.

Nissen, H. U. 1969. Crystal orientation and plate structure in echinoid skeletal units. Science **166:**1150–1152.

Picken, L. 1960. The Organization of Cells. Clarendon Press, London.

Preston, R. D. 1961. Cellulose-protein complexes in plant cell walls. In M. V. Edds, ed., Macromolecular Complexes, pp. 229–253. Ronald Press, New York.

Richards, G. 1951. The Integument of Arthropods. University of Minnesota Press, Minneapolis.

Rudall, K. M. 1955. The distribution of collagen and chitin. Symp. Soc. Exp. Biol. **9:**49–71.

Smith, G. M. 1933. Freshwater Algae of the United States. McGraw-Hill Book Company, New York.

Thompson, D. W. 1942. On Growth and Form. Cambridge University Press, London.

Verzar, F. 1963. The aging of collagen. Sci. Amer. **208:**104–114.

Wessells, N. K., and J. H. Cohen. 1968. Effects of collagenase on developing epithelia *in vitro:* Lung, ureteric bud, and pancreas. Devel. Biol. **18:**294–309.

Wilson, O. T. 1929. The colonial development of *Navicula rhombica.* Am. J. Bot. **16:**825–831.

Wise, S. W., Jr. 1970. Microarchitecture and deposition of gastropod nacre. Science **167:**1486–1488.

24 *Developmental Basis of Behavior*

In a thought-provoking paper entitled "That Was the Molecular Biology That Was," Stent (1968) concludes that "... it is also possible that the study of the nervous system is bringing us to the limits of human understanding, in that the brain may not be capable, in the last analysis, of providing an explanation for itself." Stent further suggests that "... students of the nervous system, rather than geneticists, will form the avant-garde of biological research." If these ideas are accepted, development of the nervous system constitutes a crucial aspect of developmental biology.

Neuroembryology is very complex, and we shall describe only some of its general features. Our primary concern is in the development of the neuronal circuitry underlying relatively simple behavior patterns, mostly of the reflex type. Sperry (1951, 1965, 1968), Jacobson (1966), Szekeley (1966), Hughes (1968), and Hamburger (1968) treat this problem in more detail.

Hamburger emphasizes that the development of the nervous system and the genesis of behavior are inseparable, since the latter emerges from activities of the former. Sperry (1965) concludes, "Behavioral networks are organized in advance of their use, by growth and differentiation primarily, with relatively little dependence on function for their orderly patterning. The center of the problem has thus shifted from the province of psychology to that of developmental biology." We shall not be mainly concerned with the role of function, which remains largely unresolved. Instead we will focus on how the central nervous system becomes connected with end-organs in the developing body and how regenerating nerves reestablish their original end-organ or central synaptic connections.

Development of the Neuron

Modern neuroembryology began with the brilliant work of Ross Harrison, who demonstrated in 1907 that a nerveless limb bud could be produced in frog tadpoles by excising the right or left half of the spinal cord at the level of the limb bud. When the nerveless limb bud was grafted to a host tadpole in place of or just behind the host limb bud, it acquired a normal set of nerves.

Other investigators had claimed that the peripheral nerves developed from preformed structures (protoplasmic bridges or cell chains) in the limbs, but the fact that a nerveless limb could be produced by removing part of the spinal cord at the limb bud level suggested to Harrison the incorrectness of this claim.

In 1910 Harrison performed the crucial experiment, in which the contribution of preformed structures was completely eliminated. Into a drop of frog lymph on a coverglass, Harrison placed a fragment of adult frog spinal cord. The coverglass was then inverted over a depression slide and sealed to it. Thus the technique of tissue culture was born. By repeatedly observing the culture Harrison found that processes grew out into the lymph from cell bodies within the spinal cord fragment. The outgrowth is diagramed in Figure 24-1.

Harrison's studies confirmed the *neuron theory,* which states that the nervous system consists of individual cellular units, neurons, whose processes collectively constitute the peripheral nerves. Thus, the nerves revealed by anatomical dissection are cables consisting of many processes, each of which is an outgrowth from a nerve cell body in the central nervous system or in a ganglion. Harrison's pioneering experiments have been confirmed many times (see Hughes, 1968). A differentiated motor neuron covered by sheath cells is illustrated in Figure 24-2.

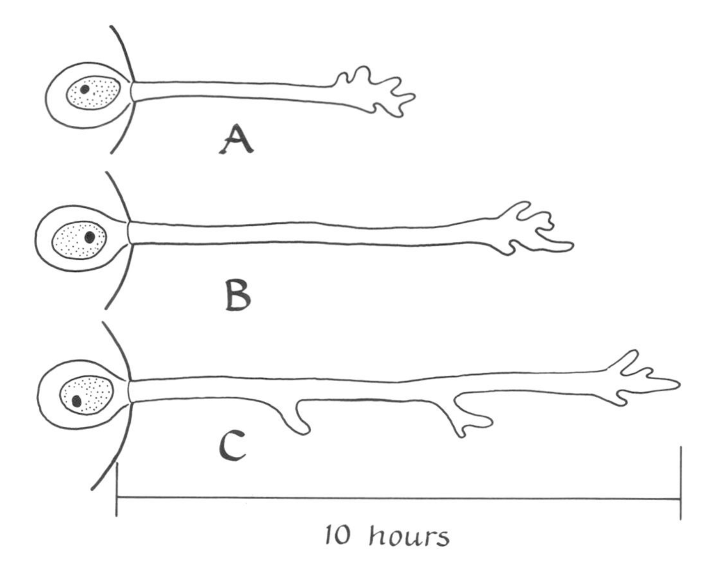

Figure 24-1. *Diagrams illustrating the outgrowth of the nerve cell process from the cell body located in an explanted piece of the spinal cord of the frog.*

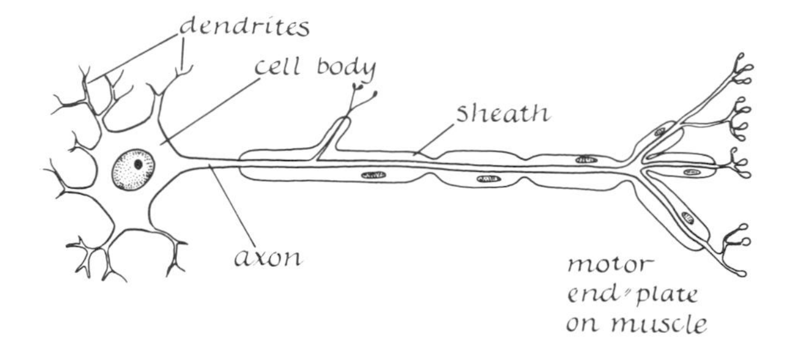

Figure 24-2. *A differentiated motor neuron with the axonal fiber covered by sheath cells.*

Regeneration of Nerves

In many animals, including man, the peripheral motor and sensory nerve processes can regenerate after the nerve is cut. The general features of this are shown in Figure 24-3.

After cutting, the distal portion of the nerve process (for instance, an axon) degenerates. At the cut surface of the proximal part, a relatively thin fiber grows and elongates until it contacts an end organ such as a muscle. Gradually, the thin process increases in diameter. If the thin nerve fiber is constricted, it continues its elongation but the part distal to the constriction remains thin. On the proximal side of the constriction the diameter of the fiber increases normally as a result of synthesis of protoplasm in the cell body (Weiss, 1959).

Sometimes nerve regeneration results in reestablishment of the original nerve-end organ connection In other cases, no end organ or the wrong end organ is innervated. The latter can occur after accidental severance of nerves. This raises the problem of the relationship between peripheral nerves and end organs.

Regeneration is similar to the outgrowth of nerves during embryonic development. In both instances the distal tip of the fiber crawls out along a substrate. Like an ameba, the nerve fiber cannot move unless it contacts a substratum. In some cases, the degenerating, distal part of the fiber provides this.

Nerve-End Organ Connections

After Harrison and other workers demonstrated that the peripheral nervous system is an outgrowth from the brain and spinal cord or associated cranial and spinal ganglia, the problem arose as to how the large number of peripheral nerves are guided to their end organs during development. This problem

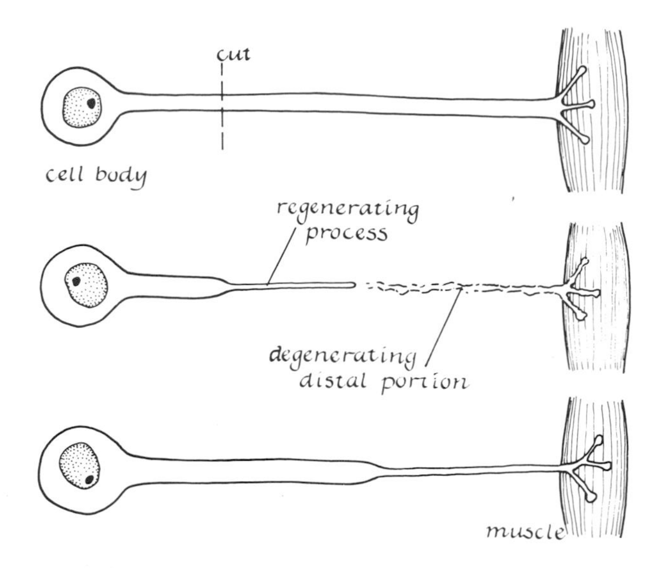

Figure 24-3. *Stages in regeneration of a cut nerve cell process.*

was most extensively investigated by Weiss (1955), who found much evidence for the role of preformed and oriented structural or ultrastructural elements in guiding the processes that made contact with them. Although a substratum is necessary for outgrowth, the orienting influence of structures *in vivo* remains obscure. Weiss and other neuroembryologists have presented evidence that neither motor nor sensory end organs exert any specific attraction for the outgrowing nerves. For example, it has been repeatedly shown that either initially developing or regenerating motor nerve processes can functionally innervate almost any muscles in their path. However, sensory nerves cannot functionally innervate muscles nor can motor nerves functionally establish connections with sense endings in the skin.

The great length of some peripheral nerves in the adult animal (for example, from the spinal cord to the foot in man) is not a problem faced by the embryo. The distance between the central system and sensory or motor end organs is far shorter in the embryo at the time outgrowth occurs (about two and one-half to three days of incubation in the chick). According to Weiss, first a pioneering fiber makes contact with an end organ. This is followed by the outgrowth from the central system of additional fibers which use the pioneer as a guide. Finally, as the embryo increases in size and the limb buds elongate, the end organ to which the nerve fibers are attached moves outward, "towing" the nerve along.

Anatomical dissection of adults within a species reveals a fairly constant one-to-one relationship between anatomically named nerves and end organs. The problem thus is whether the outgrowing nerves of the young embryo differ from one another qualitatively before contacts with different end organs are made. That is, are the neurons differentiated in correspondence to the end organs they will contact? Most evidence for at least a tentative answer to this question comes from studies involving the transplantation of limb buds to abnormal locations on the bodies of salamander (Detwiler, 1936) and chick embryos (Hamburger, 1939). For example, when the forelimb bud of a newt *(A. punctatum)* embryo at a tail-bud stage is grafted behind the limb bud of a host embryo it becomes functionally innervated by spinal nerves that normally never innervate a limb bud (Figure 24-4).

The newt forelimb is normally innervated by spinal nerves 3, 4, and 5. The grafted limb becomes innervated by nerves 6 and 7 and a branch from nerve 5 (Detwiler, 1936). A forelimb bud grafted to the head of a salamander larva can be functionally innervated by nerves growing out of the fifth cranial ganglion, which normally supplies the snout and upper and lower jaws (Detwiler, 1936). However, the pattern of nerve branching in the functionally innervated grafted limb is not necessarily like that in the normally innervated limb. Piatt (1952) found that the anatomical pattern of nerves in the forelimb of *Amblystoma punctatum* is influenced by the source of the nerves. Thus it cannot be concluded that all the peripheral nerves are exactly equivalent before they contact end organs. This is even more true in respect to sensory and motor nerves, neither of which can functionally innervate the opposite kind of end organ.

Most studies indicate that the peripheral nerves (or cell processes) growing out from the spinal cord during early development are relatively unspecialized and only become completely specified as a particular sensory or motor nerve (or cell process) after contacting an end organ. We may now examine some evidence for neuronal specification by end organs (see reviews of Sperry, 1951, 1965; Weiss, 1955).

Influence of End Organs

To explain the homologous function of grafted limbs, Weiss (1936) suggested that an end organ might impose specific chemical changes in the nerves that connect to it. This phenomenon, called *modulation,* is based upon normal muscle coordination in the limb in the presence of abnormal nerve-muscle connections. Alteration of the peripheral connections apparently has a feedback effect causing synaptic connections in the central system to change so that a typical pattern of neuronal linkage is restored.

There is also some evidence that end organs in the skin specify sensory nerve fibers (Miner, 1951). In frog tadpoles a strip of trunk skin from the mid-dorsal to near the midventral line was cut free, rotated 180°, and replaced (Figure 24-5A).

A transverse section through the operated level is shown in Figure 24-5B. Note that anatomically dorsal cutaneous nerves reinnervate ventral skin located on the back on the operated side of the tadpole, and that anatomically ventral nerves innervate dorsal skin located in the belly region on the operated side. After metamorphosis, stimulation of the dorsal or ventral regions of the rotated strip of skin resulted in erroneous responses as indicated in Figure 24-5C. These results show that the anatomically dorsal cutaneous nerve branch has become specified subsequent to the operation as a functional ventral cutaneous nerve. The anatomically ventral nerve has become specified as a functional dorsal nerve. Fibers of the anatomically dorsal nerve branches form central reflex connections appropriate for ventral nerve branches and vice versa. The quality of the cutaneous nerves thus appears to be imposed upon them by their connection with end organs in the skin, and the specificity of the fibers seems to be progressively and irreversibly fixed as the tadpole develops into an adult. Training cannot correct the misdirected responses of the frog to tactile stimuli.

Jacobson and Baker (1968) have repeated and extended the above experiment. In addition to rotating a strip of skin 180° from dorsal to ventral (inverting the dorsal-ventral axis) in frog tadpoles *(R. pipiens)*, these workers also rotated 180° a strip of skin along the backs of other tadpoles (inverting the anterior-posterior axis). In this latter experiment, stimulation after metamorphosis of various areas of the rotated back skin resulted in normal limb movements by the frog. That is, response was not

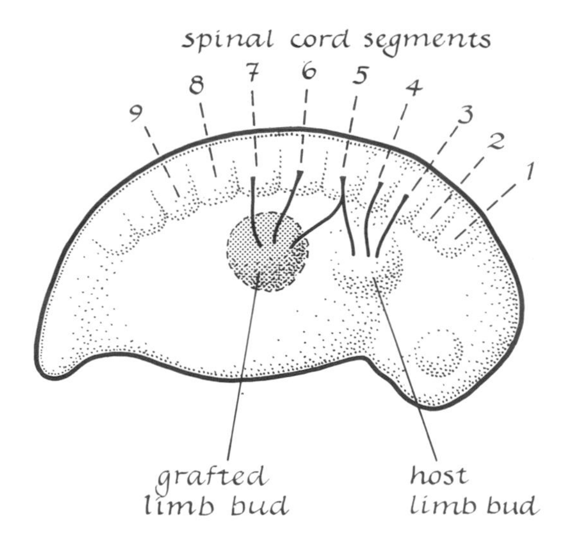

Figure 24-4. *Diagram illustrating the innervation of a limb bud grafted posterior to the host limb bud of a salamander embryo.*

misdirected away from the point of stimulation. A reason for the latter result appears to be that receptive fields in rotated dorsal skin overlap, but dorsal and ventral receptive fields in rotated dorsal-ventral grafts do not. The receptive fields were mapped with a recording electrode placed on the cutaneous nerve fibers during stimulation of skin areas. The shapes of the receptive fields were found to be determined mainly by the kinds of nerves entering them, not by the skin into which the nerves grow. These investigators concluded that each region of the skin has a local property that determines the specific central connections of its nerves. This is the conclusion reached by Miner, who found that the sensory nerves and their connections acquire their specificity by interacting with locally different end organs in the skin.

Once the peripheral cutaneous nerves have become specified they retain their *local sign*—that is, their skin area specificity. Sperry (1951) showed this by crossing the cutaneous nerves of the left hind foot into the right hind foot of fourteen- to twenty-six-day-old rats. This was done before the animals had much experience in localizing cutaneous stimuli. Following regeneration and recovery of function, stimulation of the right foot caused the rats to withdraw and lick the left foot, at the same time pressing the right foot more strongly against the offending stimulus (Figure 25-6).

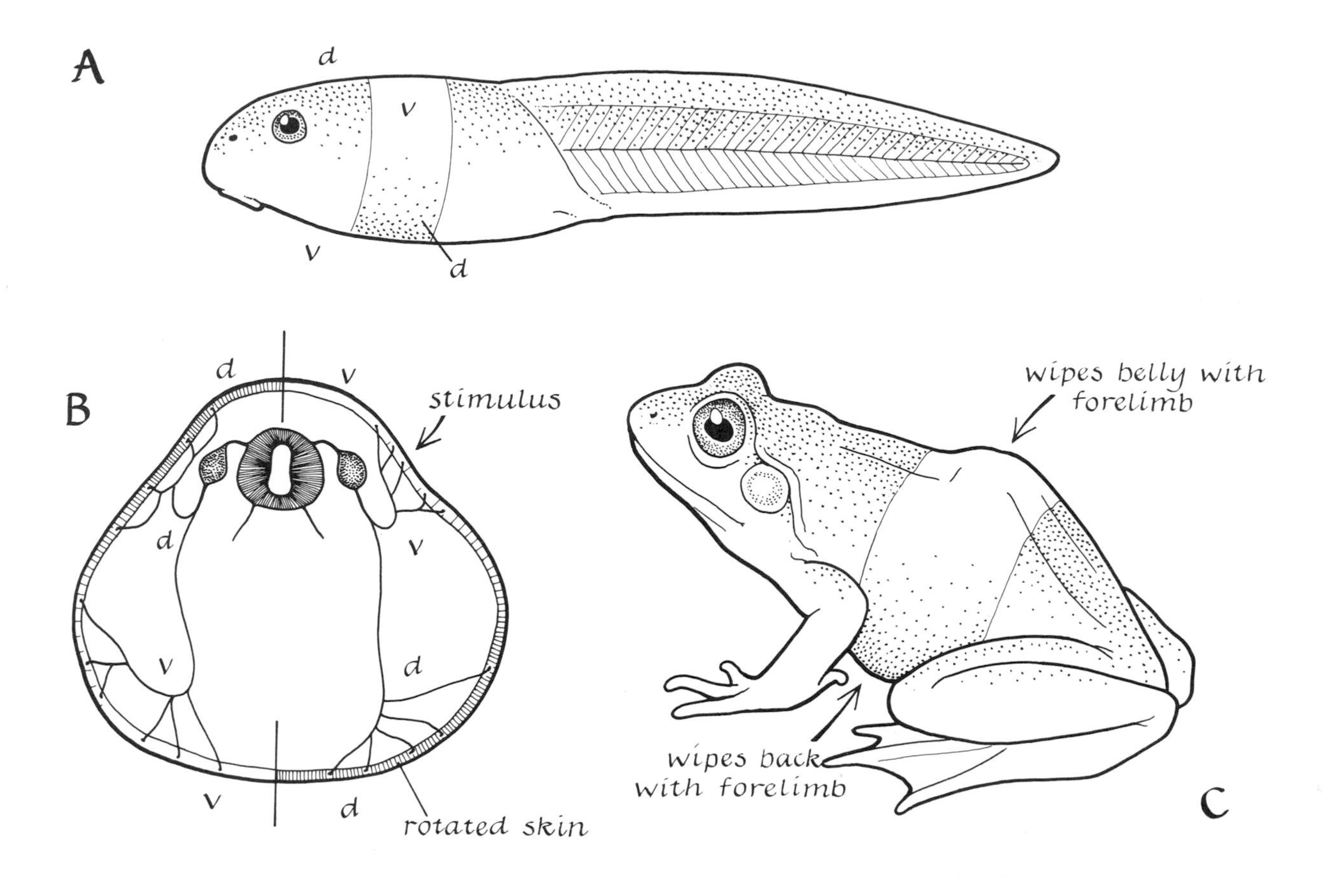

Figure 24-5. *A, a frog tadpole in which a strip of skin has been cut out, rotated 180° and replaced; B, diagram of a cross section through the trunk region in which a strip of skin on the left side had been rotated; C, response to stimulation of adult frog developed from the tadpole with rotated skin strip.*

Again, the maladaptive reactions were not correctable by training, which suggests that cutaneous sensation and its localization are built into the nervous system during embryonic development and are not a product of experience or training. This is supported by further experiments in which the right and left *ophthalmic* nerves to the snout of the frog tadpole were crossed. The metamorphosed frogs wipe the right side of the snout when the left side is stimulated and vice versa. Here it is apparent that the ophthalmic nerves are already specified as right or left in the tadpole. It is thus not the right or left side of the snout where the regenerating ophthalmic nerve terminates but the right or left fifth cranial ganglion into which the impulse enters that determines the motor response.

Misregeneration of nerves severed accidentally in man results in abnormal motor responses or the erroneous referral of sensation to the region originally innervated by a cutaneous nerve. Thus, once a nerve has become specified in relation to a particular end organ, stimulating the nerve via the wrong end organ or in the absence of any end organ is sensed as coming from the original end organ. Amputees, for instance, feel that the lost limb is still there when the skin over the stump is stimulated. Erroneous sensation or motor responses following misregeneration of nerves in man can rarely be corrected by training.

In general, differentiation of the entire nonneural organism is apparently stamped upon the nervous system and represented in a condensed form within the central nervous system. During development the

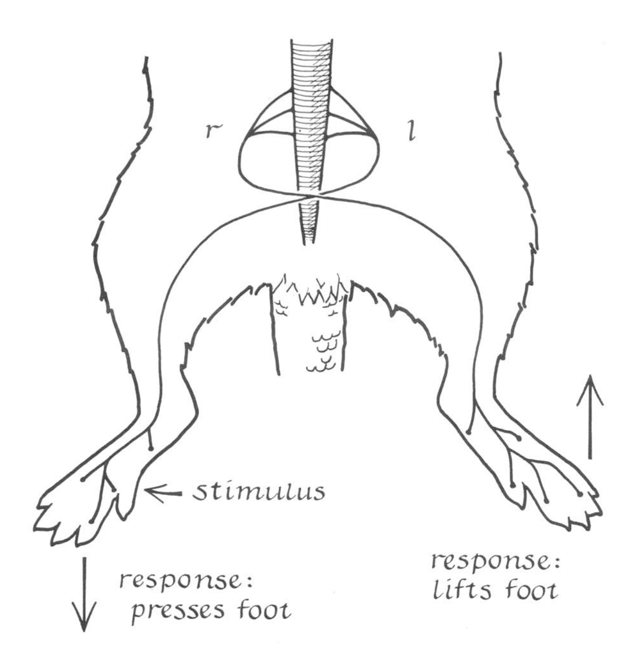

Figure 24-6. *Improper response of a rat in which the cutaneous nerves to right and left legs had been crossed.*

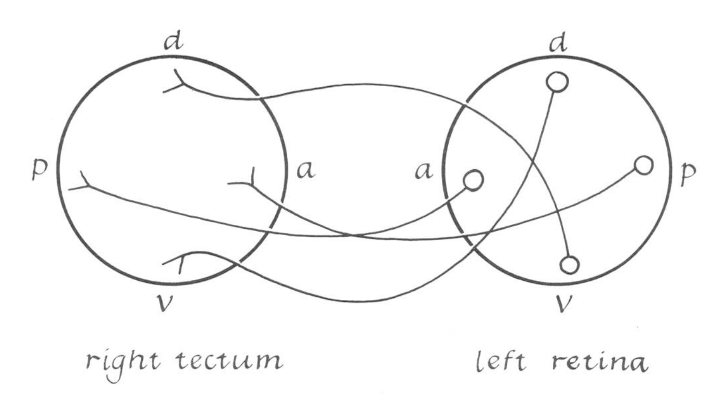

Figure 24-7. *Simplified diagram showing the general relation between dorsal, ventral, anterior, and posterior regions of the left retina and corresponding regions in the right optic tectum.*

central system seems to acquire a point-for-point copy of the entire outer surface of the body, internal surfaces, muscles, and other organs. However, differentiation within the central system is also occurring and results in what appears to be a parallel, independent, and corresponding specification of central regions prior to their connection by nerve processes with end organs. An example of this is the development of corresponding neuronal patterns in the retina of the eye (embryologically, a part of the brain) and the visual center in the midbrain.

Retinal-Visual Center Relations

The optic nerve of amphibians, after cutting, is capable of functional regeneration with good visual recovery (Matthey, 1925; Sperry, 1951, 1965). Functional regeneration occurs in salamanders, frogs, and toads in both larval and adult stages; it also occurs in young teleost fishes. The cell bodies of optic nerve fibers are located in the ganglion layer of the retina. During embryonic development the nerve processes grow through the optic stalk (which connects the

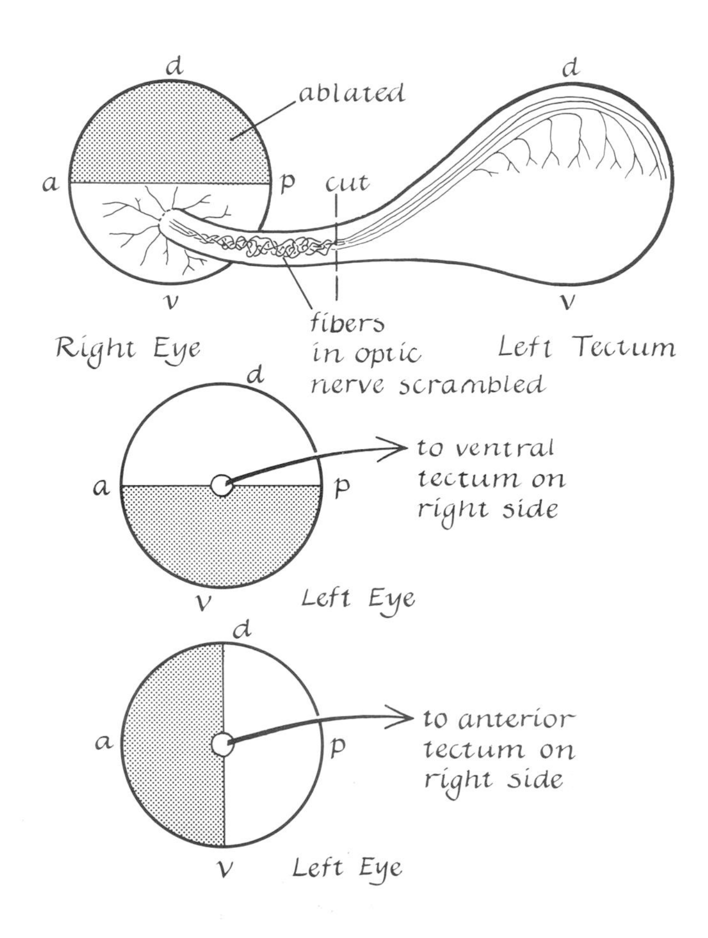

Figure 24-8. *Diagrams of the Sperry's retinal ablation experiments on the adult goldfish.*

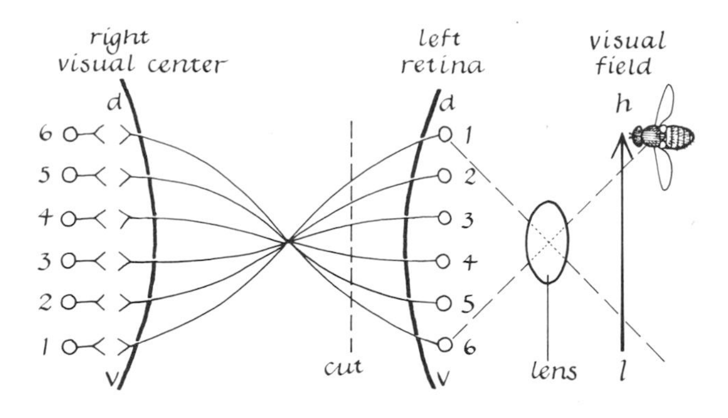

Figure 24-9. *Diagram of the relationships between the visual center, the optic nerve processes, the retina, and the visual field in* Triturus *and* Rana.

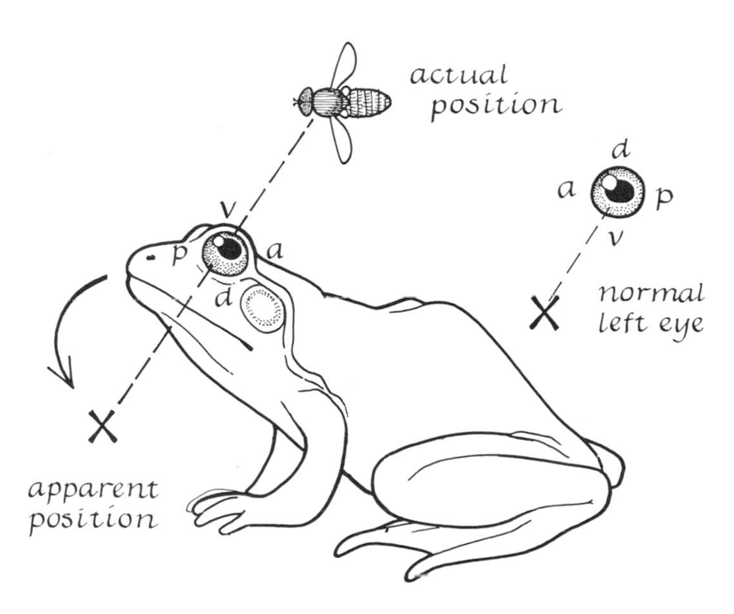

Figure 24-10. *Erroneous response of a frog in which the left eye had been rotated 180° after cutting the optic nerve.*

optic vesicle to the forebrain) into the optic center *(tectum)* of the midbrain, where synaptic connections are established with other neurons. In submammalian vertebrates the optic fibers cross completely at the optic chiasma from the right or left eye to the contralateral side of the brain. Electrophysiological mapping of the optic tectum has revealed the detailed relationship of retinal neurons and tectal neurons (Jacobson, 1966, 1968; Gaze, 1967). Figure 24-7 is a greatly simplified diagram of the general relations between dorsal, ventral, anterior (nasal), and posterior (temporal) regions of the left retina and the corresponding regions of the right optic tectum. These relations are established during embryonic development.

Neuronal connections can be reestablished following ablation of various segments of the adult goldfish retina by regeneration of the optic fibers back into the corresponding regions of the tectum (Sperry, 1965). Thus after ablating the dorsal half of the right retina and cutting and scrambling the optic nerve fibers, the nerve fibers from the ventral half of the retina connect only with neurons in the dorsal portion of the tectum. This is shown in Figure 24-8A. In Figure 24-8B and C, optic fibers from the undestroyed dorsal or posterior half of the retina regenerate and reestablish their normal, original connections in the tectum.

Using mostly the adult amphibians, *Triturus* and *Rana*, Sperry (1951) repeated, confirmed, and extended the experiments of Matthey. These experiments and their results are briefly summarized as follows. The normal but simplified relationship of the visual center, the optic nerve processes, the retina, and the visual field are illustrated in Figure 24-9. As indicated, an object high in the visual field is focused by the lens on the ventral portion of the retina. Though the image is thus inverted, the object is seen by the animal, as indicated by its response, as noninverted. The one-to-one relation of the optic fibers and the tectal neurons is denoted by the numbers 1 through 6. After cutting the optic nerve from the left eye, the eye was rotated 180°. Following return of vision after regeneration, behavior tests revealed that the visual field of the animal was rotated 180°. The animal responded as if it saw everything upside down and reversed front-to-back. This is shown in Figure 24-10.

As indicated in the figure, the image of the fly is focused on the posterior-dorsal quadrant of the retina of the 180°-rotated eye. In an unrotated left eye, an image focused on the posterior-dorsal quadrant would be located in position x. The operated frog thus sees the fly with its left eye in position x and jumps toward this spot to catch it. The same behavior occurs when the left eye is rotated without cutting the nerve. Training could not correct the erroneous response.

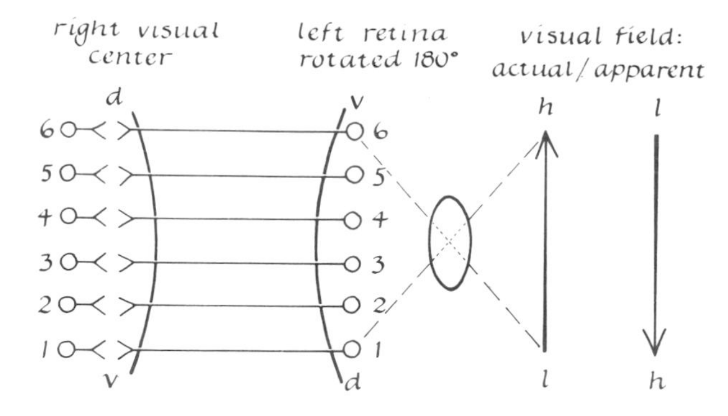

Figure 24-11. *Diagram illustrating reestablishment of the original synaptic connections between specific optic nerve fibers and specific neurons in the optic center.*

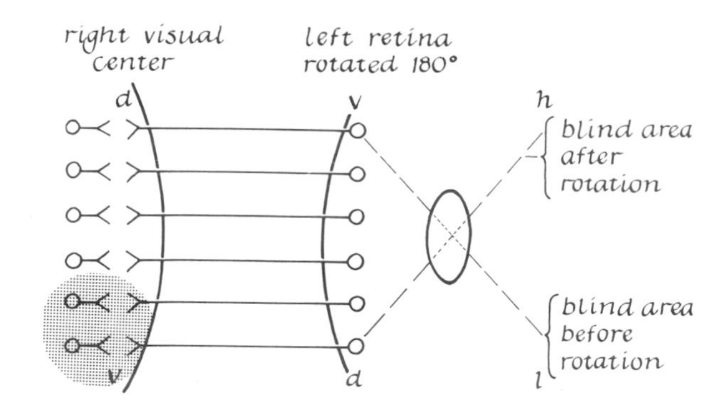

Figure 24-12. *Selective reconnection of retinal neurons and optic center neurons following the production of localized lesions in the optic center. Cutting the optic nerve and rotating the eye 180° results in 180° rotation of the position of the blind spot in the visual field.*

The results of the foregoing experiment indicate that optic fibers arising from different retinal loci differ from one another in quality (denoted by numbers 1 through 6) and that the original, specific synaptic connections in the optic tectum (numbers 1 through 6) are reestablished. This is diagramed in Figure 24-11. Compare this with Figure 24-9.

The following experiment further demonstrates the selective reconnection of retinal neurons with tectal neurons. Sperry found that a localized lesion in the optic center resulted in a distinct blind area in the animal's visual field. After cutting the optic nerve and rotating the eye 180° in such animals the position of the blind area was found to be rotated 180°. This is shown in Figure 24-12.

Eliminating the optic chiasma in adult frogs and forcing the nerves to regenerate into the ipsilateral visual centers causes the animals to respond as though everything viewed through one eye were being seen through the other (Figure 24-13).

As shown in the figure, the image of the fly impinges upon the left retina; nonetheless, the fly is seen by the frog as being at position x.

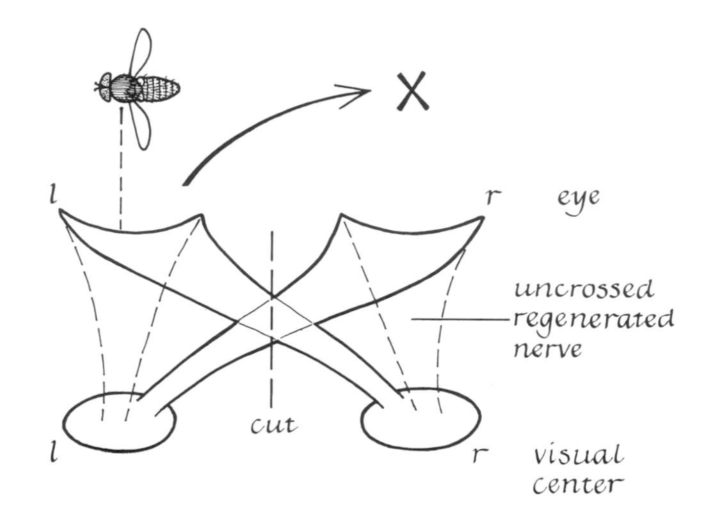

Figure 24-13. *Diagram illustrating the method of eliminating the optic chiasma in a frog and the erroneous response of the animal after regeneration of each optic nerve into the ipsilateral visual center.*

Progressive Specification of Retinal Neurons

During embryonic development of amphibians there is a progressive and parallel specification of retinal and optic tectum neurons in accord with their positions in the retina and tectum, respectively. Connections made by the ingrowing optic nerve have been determined by recording electrical responses from stimulated retinal neurons at their termination in the optic tectum (Jacobson, 1966, 1968). At a late tail-bud stage of *Xenopus*, before connections between the retina and tectum have been

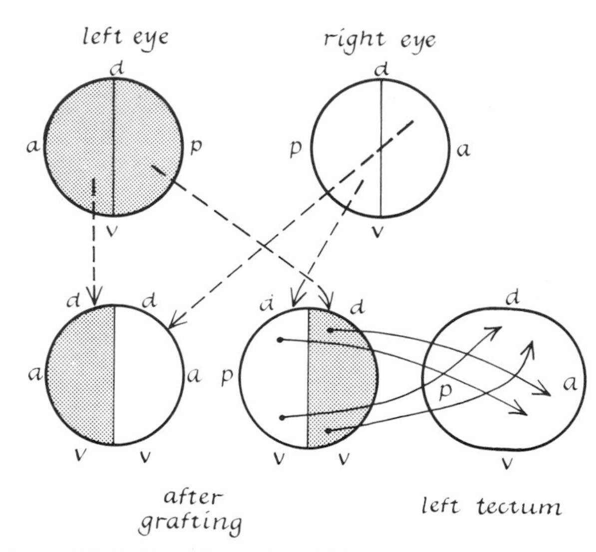

Figure 24-14. *Experiment in which compound eyes were produced (anterior-anterior or posterior-posterior halves of the retina were fused).*

formed, compound eyes can be produced by grafting the nasal (anterior) half of the retina of one *Xenopus* embryo in place of the temporal (posterior) half in another embryo and vice versa. This experiment and its results are shown in Figure 24-14.

As indicated in the figure, fibers from each half retina grew into the whole tectum—not only into the appropriate half (anterior in the illustrated result), as occurs when completely specified fibers of an adult regenerate. This indicates that each retinal neuron is not absolutely specified but is determined by its relation to its neighbors. There is thus no mosaic of specific neurons, but a gradient system of neuronal specificities. In half the retina of an early stage embryo, a complete but steeper gradient develops and connects with the whole tectum to establish the normal point-for-point relation of retina and tectum. This is, however, not the case after the neurons

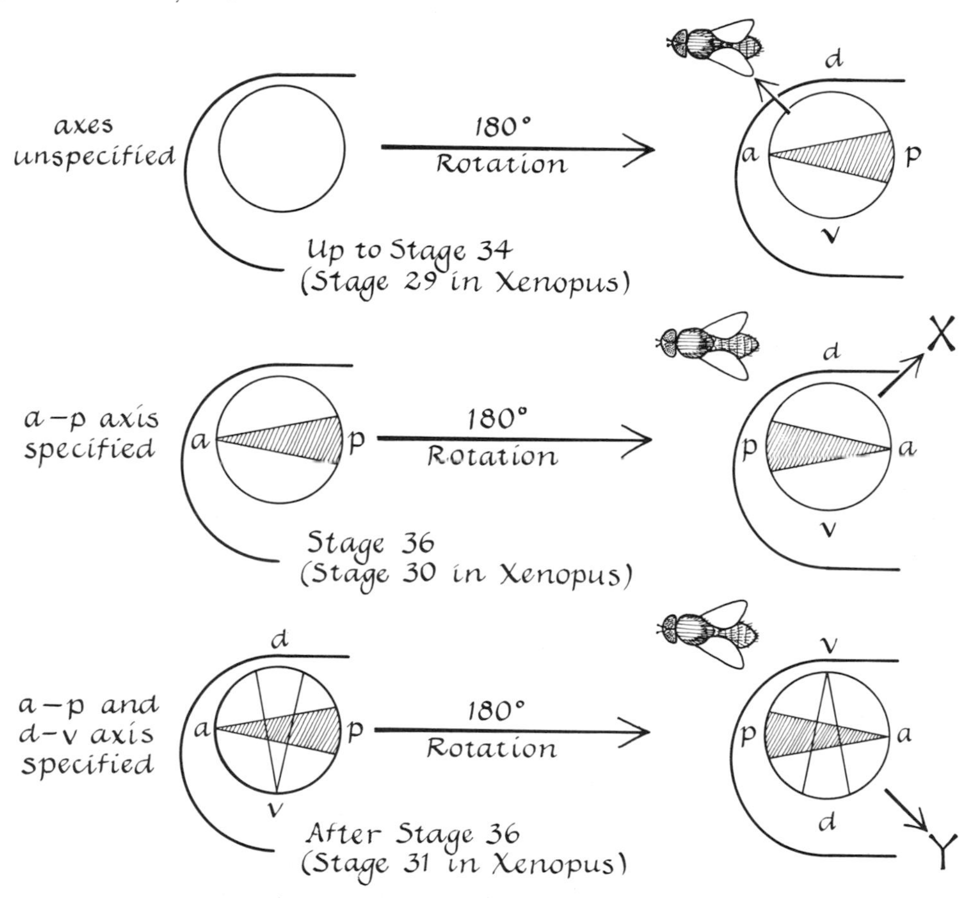

Figure 24-15. *Progressive specification of retinal neurons in* Ambystoma *and* Xenopus.

have been specified in an adult (see Figure 24-8).

Progressive specification of retinal neurons can be demonstrated by 180° rotation of the retina at different stages of development in *Amblystoma* (Stone, 1960), *Xenopus* (Jacobson, 1968), and other amphibians. After functional regeneration of the optic nerve, the animals' responses to visual stimuli were tested (Stone) or the retino-tectal relation was mapped electrophysiologically (Jacobson). Generalized results are shown in Figure 24-15.

Specification of the retinal neurons is independent of that in the optic tectum, since the two do not connect until after stage 38 in *Amblystoma punctatum* and stage 32 in *Xenopus* (both late tail-bud stages). In *Xenopus* this is about 20 hours before the outgrowth of axonal processes and 80 to 100 hours before the axons reach their terminal loci in the tectum. Note in the figure that the operations were performed prior to retinal-tectal connections. At an early stage rotation produces no defects in response to visual stimuli; later, however, the antero-posterior axis of the retina is specified, and so the animal sees everything reversed (Figure 24-15B). At a still later stage (36 in *Amblystoma* and 31 in *Xenopus*) the dorso-ventral axis of the retina is specified and the visual field is thus rotated 180°. Neurons in the tectum are also being specified during this period, since otherwise the visual field would not be inverted or rotated: the centers in the central nervous system would adjust to retinal inversion as they do in the embryos prior to stage 34 in *Amblystoma* or stage 29 in *Xenopus*.

Neuronal Basis of Behavior

The experiments outlined above show that the structure of the nervous system largely determines certain behavior patterns in fish and amphibians. The nervous system develops mainly in consequence of the reading out of entries in the encyclopedia of genetic information. As Jacobson (1966) has emphasized, at present we cannot ascribe the development of a single neuronal circuit to the action of use and experience. Segaar (1965) has reported a striking example of the control of behavior by the central nervous system. In some species of stickleback fish, the male builds a cylindrical nest and leads a female to it; she lays her eggs in it and leaves. The male releases sperm over the eggs and remains in the nest, fanning the eggs with fin movements. Males are aggressive, attacking other males and nongravid females (see Ingle, 1968). When a portion of the forebrain in an adult male fish was destroyed, the normal, aggressive sexual and parental behavior disappeared. Instead, the male built a flat, inadequate nest and dug pits in the sand at the bottom of the aquarium. Eventually the destroyed part regenerated and normal male behavior resumed.

Sexual behavior patterns are also influenced by sex hormones. Genetically female rats injected with testosterone (male sex hormone) shortly after birth exhibit male sexual behavior at maturity. The hormone apparently controls the development of neuronal circuits underlying sexual behavior; these then become stabilized, since the hormone need not be continuously present. The basic behavior pattern in rats is female in both genetic sexes; only under influence of the male sex hormone does the male type develop (Harris and Levine, 1965).

Especially in humans, but also in other animals, environmental influences, including training and learning, can modify behavior. A man's brain develops until he is about nine years old, and during this time exhibits a special kind of functional plasticity. Injuries to the nervous system are more easily compensated during this period than afterwards, and language skills are more easily acquired (Sperry, 1968). The mechanisms underlying this plasticity of the still-developing brain are unknown but probably involve neuron differentiation and growth of neuron processes.

There are numerous examples of the inheritance of behavior patterns in animals. Timidity, aggression, savagery, and defensiveness are inherited and thus have a clear genetically determined neuronal basis (see review of Krushinski, 1962). The nest-building habits (materials used and structure) of birds are, of course, genetically determined. How much the environment influences behavior also seems to depend upon the animal's genotype. This is a common observation in humans and is also true of dogs and many other animals. For example, German shepherds are more subject to environmental influences than are Airedales.

The extent to which functional activity of the nervous system during development might play a role in establishing neuronal circuits is not known (see

Hughes, 1968). Spontaneous functional activity begins in many embryos at relatively early stages—for instance, 3½ days of incubation in the chick (Hamburger, 1968) and two months in the human fetus. The movements of early chick embryos, extensively observed by Hamburger, consist of bending of the head at an early stage, followed by limb, eyeball, and eyelid movements. The activity begins with a few twitches at 3½ days and builds up until the embryo is in motion 80 percent of the observation time at 13 days. In every type of movement the behavior appears only after the underlying neuronal circuits have been completed. However, Hamburger's studies provide no evidence that adaptive neuronal connections are the result of selection by trial and error from a randomly interconnected and unspecified network.

Conclusion

The present state of development of neuroembryology permits only tentative conclusions about the mechanisms of neuronal specification and the formation of neuronal circuitry underlying even simple behavior patterns. Relatively simple behavior is probably based upon neuronal connections that are genetically determined and are not significantly influenced by experience or learning.

In general, the specificity of neurons, like that of other cells, develops gradually in the embryo. In peripheral nerve fibers connecting to muscles or sensory receptors in the skin the specificity appears to be at least partially imprinted upon originally unspecified nerve fibers by the end organs with which they connect. The overall result of this is a recording in the central system of a condensed copy of the entire nonneural portion of the animal's body. Thus, the central system appears to acquire its awareness of the outside world via its connection with receptor end organs and acquires its capacity for response by connection with motor end organs.

After experimental modification of central-peripheral relations (rotation of the retina, crossing peripheral nerves in amphibians and rats), the animal's behavior is misdirected and is not correctable by training. This indicates that specification of neurons, like that of other cells in the embryo, is essentially irreversible. The nervous system in adult subprimate animals is characterized by its lack of plasticity. The extent to which the human nervous system is plastic remains a challenge to students of neurology.

Function and training probably do not add or subtract any neurons or determine new synaptic connections. However, function before and after birth is probably important in maintaining neuronal pathways preestablished during early development and in continuing neuronal specification. Function (behavior) might even play a role in selecting pathways (Hughes, 1968). The human central nervous system, particularly in young children, has more plasticity than that of other animals. Behavior and learning appear to play important roles in the emergence of order within the cerebral nervous system, which continues to develop postnatally. As a working hypothesis we might assume that the final order attained in the brain, as reflected in behavior of an educated adult human, results from interaction between the inherited nervous system and the external environment.

References

Detwiler, S. R. 1936. Neuroembryology. Macmillan, New York.

Gaze, R. M. 1967. Growth and differentiation. Ann. Rev. Physiol. **29**:59–86.

Hamburger, V. 1939. The development and innervation of transplanted limb primordia of chick embryos. J. Exp. Zool. **80**:347–389.

Hamburger, V. 1968. Origins of integrated behavior. In M. Locke, ed., Emergence of Order in Developing Systems, pp. 251–271. Academic Press, New York.

Harris, G. W., and S. Levine. 1965. Sexual differentiation of the brain and its experimental control. J. Physiol. Lond. **181**:379–400.

Harrison, R. G. 1907. Experiments in transplanting limbs and their bearing upon the problem of the development of nerves. J. Exp. Zool. **4**:239–281.

Harrison, R. G. 1910. The outgrowth of the nerve fiber as a mode of protoplasmic movement. J. Exp. Zool. **9**:787–846.

Hughes, A. F. W. 1968. Aspects of Neural Ontogeny. Logos Press, London.

Ingle, D., ed. 1968. The Central Nervous System and Fish Behavior. University of Chicago Press, Chicago.

Jacobson, M. 1966. Starting points for research in the ontogeny of behavior. In M. Locke, ed., Major Problems in Developmental Biology, pp. 339–383. Academic Press, New York.

Jacobson, M. 1968. Development of neuronal specificity in retinal ganglion cells of *Xenopus*. Devel. Biol. **17**:202–218.

Jacobson, M., and R. E. Baker. 1968. Neuronal specification of cutaneous nerves through connections with skin grafts in the frog. Science **160**:543–545.

Krushinskii, L. V. 1962. Animal Behavior. International Behavioral Sciences Series, J. Wortis, ed.

Matthey, R. 1925. Récupération de la vue apres résection des nerfs optiques, chez le *Triton*. Compt. Rend. Soc. Biol. **93**:904–906.

Miner, N. M. 1951. Cutaneous localization following 180 degree rotation of skin grafts. Anat. Rec. **109**:326–327.

Piatt, J. 1952. Transplantation of aneurogenic forelimbs in place of the hindlimb in *Amblystoma*. J. Exp. Zool. **120**:247–285.

Segaar, J. 1965. Behavioral aspects of degeneration and regeneration in fish brain: a comparison with higher vertebrates. Prog. Brain Res. **14**:143–231.

Sperry, R. W. 1951. Regulative factors in the orderly growth of neural circuits. Growth Suppl. **15**:63–87.

Sperry, R. W. 1965. Embryogenesis of behavioral nerve nets. In R. L. DeHaan and H. Ursprung, eds., Organogenesis, pp. 161–186. Holt, Rinehart and Winston, New York.

Sperry, R. W. 1968. Plasticity of neural maturation. In M. Locke, ed., The Emergence of Order in Developing Systems, pp. 306–327. Academic Press, New York.

Stent, G. S. 1968. That was the molecular biology that was. Science **160**:390–395.

Stone, L. S. 1960. Polarization of the retina and development of vision. J. Exp. Zool. **145**:85–93.

Szekely, G. 1966. Embryonic determination of neural connections. Adv. Morph. **5**:181–219.

Weiss, P. 1936. Selectivity controlling the central peripheral relations in the nervous system. Biol. Rev. Camb. **11**:494–531.

Weiss, P. 1955. Neurogenesis. In B. H. Willier, P. Weiss, and V. Hamburger, eds., Analysis of Development, pp. 346–401. W. B. Saunders, Philadelphia.

Weiss, P. 1959. The nature of biological organization. In C. H. Waddington, ed., Biological Organization, pp. 1–21; 105–108. Pergamon Press, New York.

25 *The Emergence of Order in Development: Molecular and Ultrastructural Aspects*

A unifying aspect of development is the emergence of increasingly complex levels of order from order existing at lower levels of complexity. The developmental biologist, as well as the geneticist, is concerned with a twofold problem: how order is transmitted at various levels of complexity, as in reproduction, and how increasingly complex order emerges epigenetically, as in the transformation of a reproductive unit into an adult. Molecular order that is transmitted from generation to generation is primarily packaged in the genome of the nucleus, or its equivalent in prokaryotes, in the form of deoxyribonucleic acid (Chapter 3). As noted in Chapter 18, this order can also be transmitted in part in the form of products of gene action, as in maternal inheritance. In some instances—for example, in the ciliate cortex and in plant plastids—the order is transmitted as a nonnuclear template. We have also seen that molecular order in the form of supramolecular aggregates—structural organelles, cells, and even tissues—is also transmitted from parent to offspring (Chapter 9). Thus, reproduction may involve the transmission of order at any or all levels of complexity.

The epigenesis of levels of order presents basic and extremely complex problems of how higher levels of complexity arise from lower levels. Specifically, how does complexity at a molecular level transform into supramolecular complexity? We noted in Chapter 22 that each level of complexity functions in the context of the levels below it, but this does not explain how the entire system of levels in multicellular organisms arises from transmitted information. The differential expression of inherited information at one stage or another is clearly regulated by the expression of information at prior stages. Thus, normal development is largely an automatic consequence of the kind of information with which it begins and illustrates the *principle of inevitability.* The principle of inevitability, however, does not imply that development is completely automatic. As shown in Chapter 22, environmental agents influence the expression of inherited capacities. Nevertheless, programs of development, such as the sequence of stages along a path of differentiation, appear to be controlled by mechanisms internal to the system, not causally dependent upon the external environment. However, the composition of the environment can select which programs will be expressed. This chapter discusses how the principle of inevitability operates, from the molecular to the organism level.

We shall see how increasing complexity arises primarily as a result of multiplication of "ultimate units" (molecules) through reproduction, and assembly of the units into larger aggregates. Reproduction provides the necessary numbers of molecules; assem-

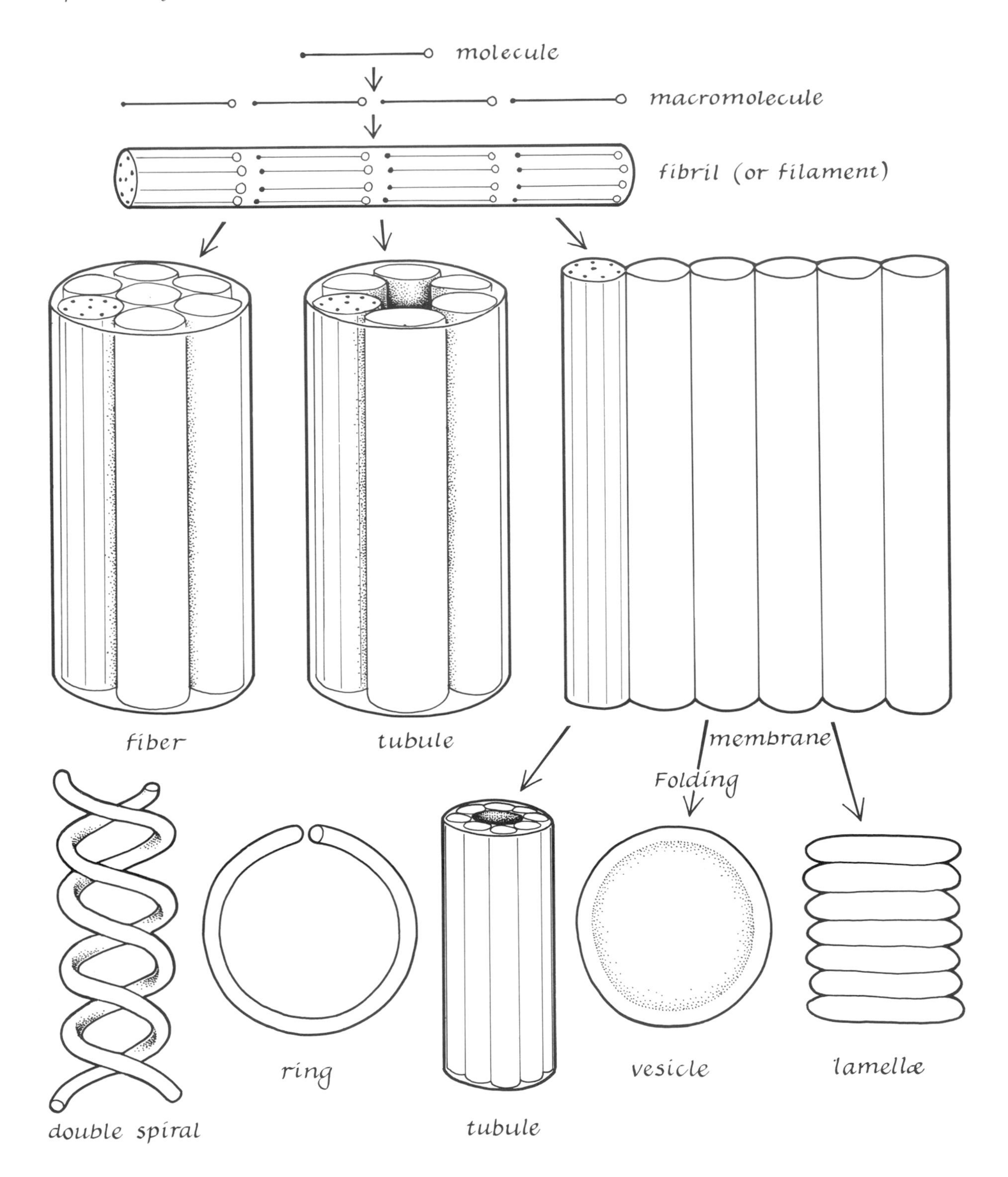

Figure 25-1. *Diagram illustrating the formation of various organelles by assembly of molecules, macromolecules, and fibrils.*

bly provides the means for arranging them into higher orders of size and complexity of pattern. Assembly may be accomplished either under guidance of a preexisting assembly of units (a template such as DNA) or more epigenetically by properties and interactions between the individual units which may be identical or heterogeneous.

Structural Continuity in Development

The electron microscope and new techniques in macromolecular (for instance, protein) chemistry have opened numerous vistas to the biologist. The molecular and ultrastructural aspects of development may now be studied with hope of eventually understanding how processes at these levels are basic to those at the cell and higher levels. The discovery that many patterns of organization at the molecular and cellular organelle levels are universal among organisms from the simplest microorganism to man is encouraging. In general, development from a reproductive unit recapitulates at the molecular and organelle level the evolution of molecular complexes from universally occurring molecules.

Studies that employ x-ray diffraction and electron microscopy suggest that development involves not only the reproduction of units varying in size and complexity but also the association of relatively simple units to form larger and more complex structures. This is what is meant by structural continuity (Reinert and Ursprung, 1971). The sequence in which units of increasing size might be built up from smaller units during development is approximately as follows:

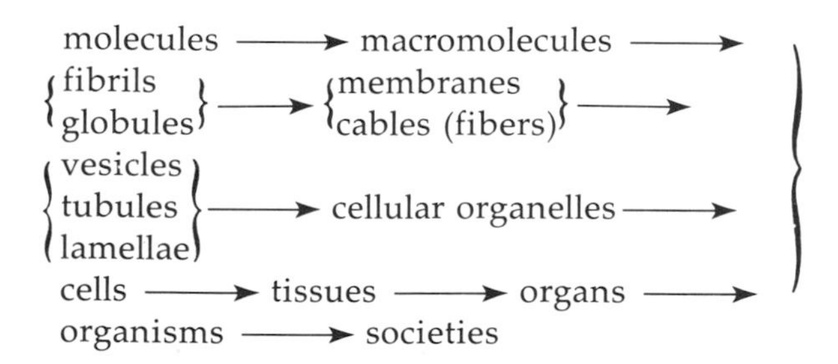

The formation of vesicles, tubules, lamellae, and other kinds of organelles from smaller units is diagrammatically and partly hypothetically illustrated in Figure 25-1.

Cellular organelles built up through association of fibrils or filaments include microtubules, spindle fibers, centrioles, cilia, and flagella (Porter, 1963, 1966). Those built up from membranes include cisternae of the endoplasmic reticulum, mitochondria, grana of chloroplasts, and the golgi vesicles (Mercer, 1962; Waddington, 1962; Thompson, 1964).

The attractive hypothesis that the basic molecular structure of membranes forming the endoplasmic reticulum, mitochondria, golgi, plasma, and nuclear membranes is very similar (Robertson, 1964, 1967) has not been universally accepted (Green and MacLennan, 1969). Nevertheless, all cellular membranes probably share a number of similar properties. According to Green and MacLennan, all biological membranes studied contain a mosaic of lipoprotein repeating units. Membranes thus appear to be assemblies of subunits which are joined to one another. Electron micrographic and x-ray diffraction studies indicate this, as does the fact that membranes can be reassembled *in vitro* from these subunits.

Some of the cellular organelles we have mentioned, such as mitochondria and chloroplasts, are transmitted intact to daughter cells during cell division; they replicate when the parent organelle divides. In other instances the parent organelle serves as a template upon which the new organelle is assembled; this appears to be true of centrioles, the basal bodies of cilia and flagella, and the mitotic spindle. Other ultrastructural components of the cell can apparently assemble themselves to some extent.

Ubiquitous Assemblies

Many organisms contain assemblies of molecules, macromolecules, fibrils, tubules, and membranes that are similar in design. Assembly of nucleotide molecules to form DNA or RNA is probably universal, as is the primary structure of proteins as a linear sequence of amino acids. Unit-type membranes, endoplasmic reticulum, microtubules, spindle fibers, centrioles, cilia, flagella, mitochondria, dictyosomes, nucleoli, and chromosomes are a few of the assemblies which, though not of universal occurrence, exist in many organisms.

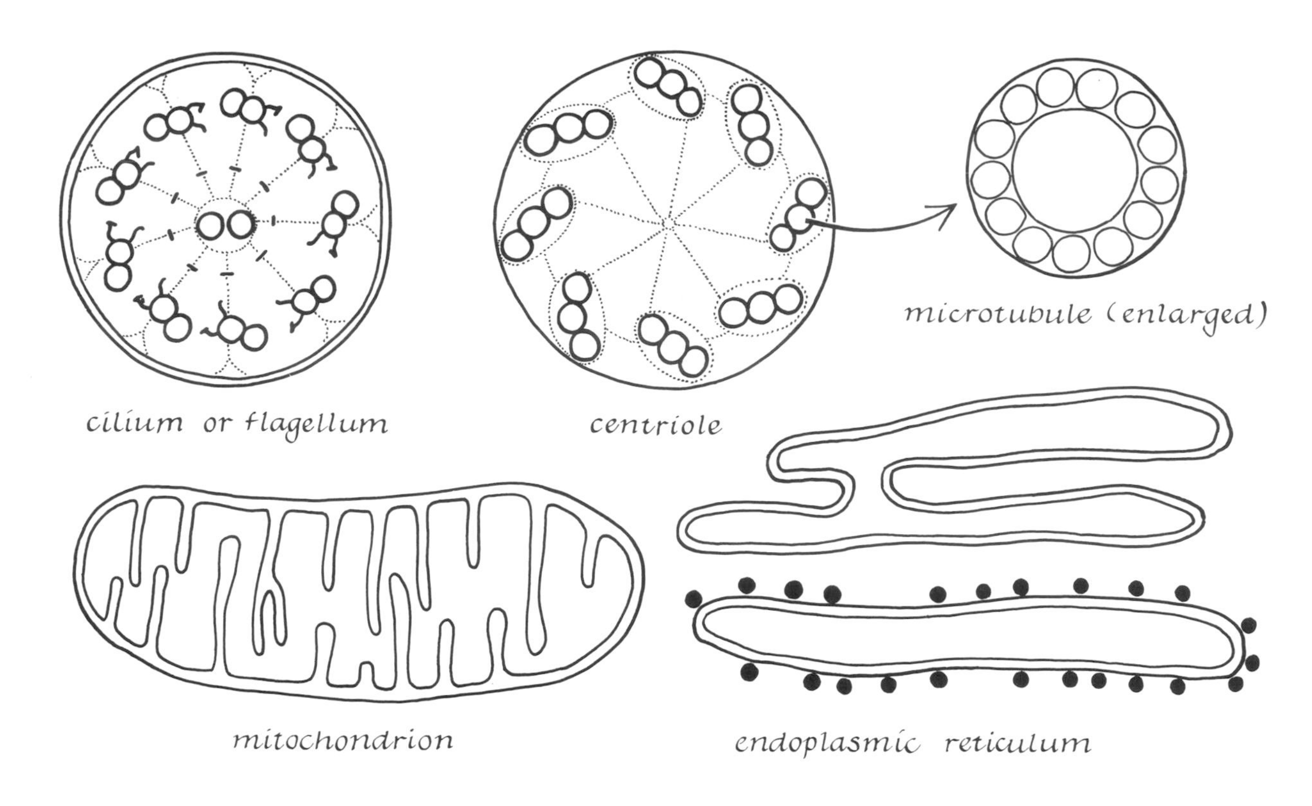

Figure 25-2. *Diagrams of the ultrastructure of widely occurring organelles as deduced from electron micrographs.*

Ultrastructure

The ultrastructure of several widely occurring organelles as deduced from electron micrographs is diagramed in Figure 25-2 (see Fawcett and Porter, 1954; Manton et al., 1956; Porter, 1966; Randall et al., 1967; Gibbons, 1967; Harven, 1965).

As shown in the figure, a cilium or flagellum, as seen in transverse section, consists of a pair of central fibers surrounded by nine double outer fibers. The fibers are tubular. One of each pair of outer fibers bears what are called arms. In all cilia and flagella that have been studied, the arms point clockwise when viewed from the base looking toward the tip of the structure. The central and outer fibers are connected by spokes (Figure 25-2) spirally arranged along the length of the cilium or flagellum.

Most cilia and flagella, from those in algae and protozoa to those in man, exhibit the 9-2 pattern of arrangement of fibers embedded in a matrix and enclosed by a membrane. However, cilia with a 9-0 pattern are found in virtually every tissue at some time during development in animal embryos. For example, such cilia are present in (usually inside of) heart cells of embryo and adult lizards, mice, and rabbits (Rash et al., 1969). These cilia, which are nonfunctional, have been found only in nondividing cells. They seem to appear when cellular differentiation is initiated. Rash and co-workers suggest that the mitotic centriole, which has the 9-0 pattern, functions like the ciliary basal body. In this case, instead of forming the spindle fibers, the centriole forms a cilium. These observations attest to the similarity of the centriole and basal body.

The 9-2 or 9-0 associative pattern of microtubules is present not only in cilia, flagella, basal bodies, and

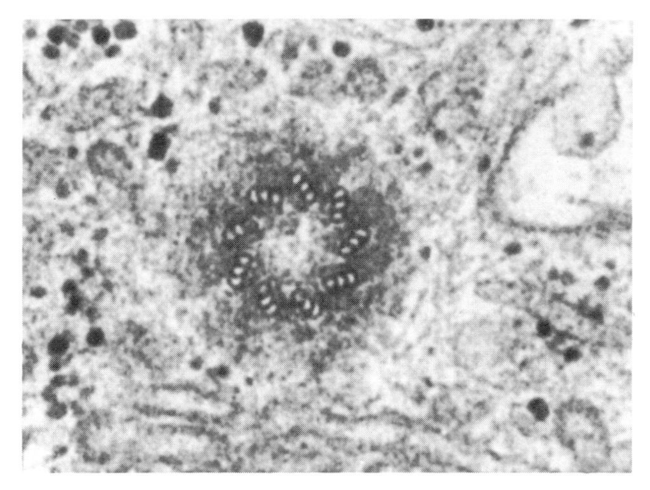

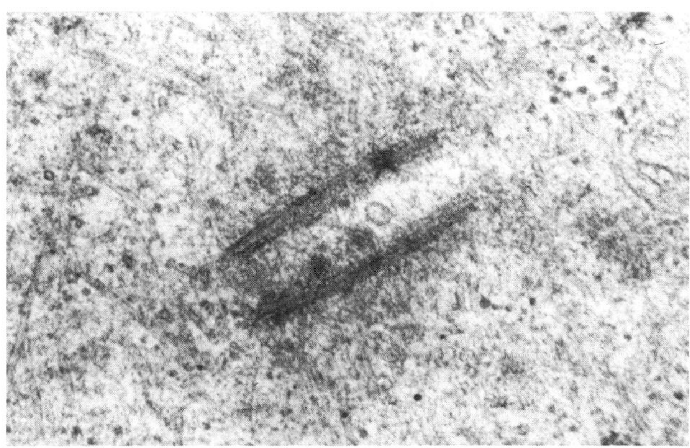

Figure 25-3. *Electron micrographs of cross (top) and longitudinal (bottom) views of centrioles from Chinese hamster fibroblast cells. The arrangement of microtubules composing the centriole is similar to that found in flagella. Top, × 132,000; bottom, × 110,000. Photographs courtesy Dr. E. Stubblefield from Jensen and Park, 1967. Courtesy of Wadsworth Publishing Company, Inc.*

centrioles, but also in the eye spot of the golden alga *Chromulina psammobia* and the retinal rods of animals (Fauré-Fremiet, 1958), including the third *(parietal)* eye of lizards (Eakin, 1970). This symmetry of nine, common to eucaryotes, shows the relationship of these organisms at the ultrastructural level.

A centriole (Figures 25-2 and 25-3) consists of nine outer groups of three tubules each. The triplets of tubules are connected to one another and to the axis of the centriole by a complex system of granular material. As in the cilium or flagellum, the arrangement of the triplets gives the centriole an enantiomorphic symmetry. However, the centriole, unlike the cilium or flagellum, lacks a pair of central tubules. The basal body from which a cilium or flagellum is formed also lacks the central tubules but otherwise is structurally very similar in arrangement of triplets of tubules to a centriole (Figure 25-8).

Microtubules (Figure 25-5) are apparently ubiquitous organelles in eucaryotic cells (Porter, 1966). They are the tubular components in cilia, flagella, centrioles, and mitotic spindle fibers (Inoué and Sato, 1967); they are present in elongating embryonic cells (Byers and Porter, 1964; Arnold, 1966) and in many other structures of protozoa, unicellular algae, and multicellular organisms (see review of Tilney, 1968; and Neucomb, 1969). As shown in Figure 25-2, the wall of a microtubule consists of 13 filaments (some cells have 12 or less) arranged to form a cylinder about 210 to 250 Å in diameter. There is evidence that the filaments are rows of globular protein units, each about 40 Å in diameter, assembled in a helical array (Porter, 1966; Tilney, 1968).

As shown in Figure 25-2, mitochondria are sausage-shaped structures composed of an outer and an inner membrane and the space enclosed by these. Infoldings of the inner membrane constitute the cristae in mitochondria of many cells. The basic structure of mitochondria is remarkably constant. However, minor differences in the pattern of the internal membrane of mitochondria appear to be characteristic of particular cell types (Fawcett, 1959).

The endoplasmic reticulum consists of flattened membranous vesicles *(cisternae)* and appears to constitute a continuous membrane system throughout the cell (Fawcett, 1959; Jensen and Park, 1967). Ribosomes frequently adhere to the outer surface of the cisternae (Figure 25-4). During cell differentiation, the endoplasmic reticulum often becomes larger and more complex.

Role in Development

The possible role of chloroplasts and mitochondria in cellular differentiation was discussed in Chapters 13 and 18, and the role of DNA and RNA molecules

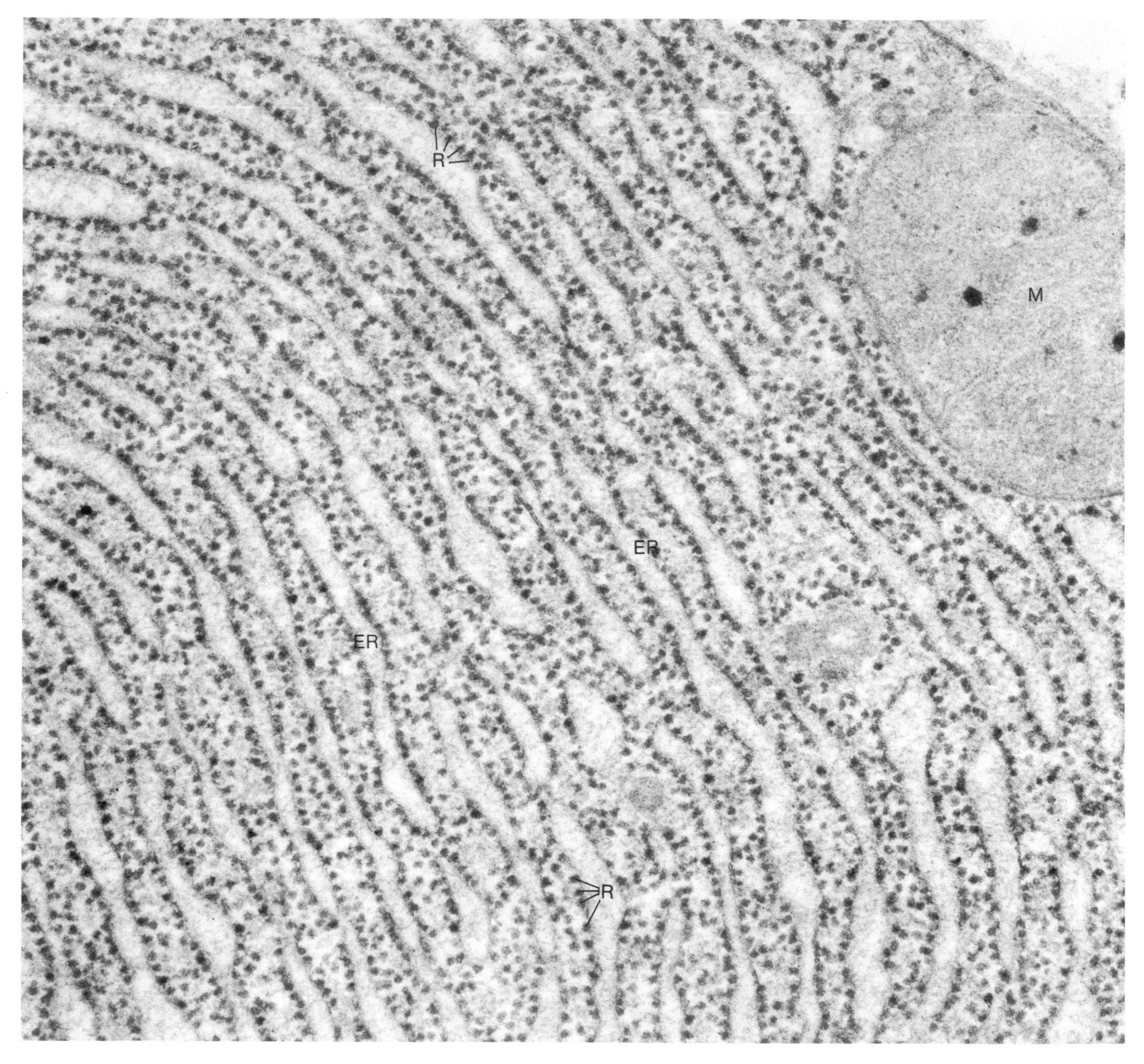

Figure 25-4. *Electron micrograph of endoplasmic reticulum and free and associated ribosomes in an exodrine cell of the rat pancreas. The quantity of endoplasmic reticulum,* ER, *and ribosomes,* R, *is typical of cells engaged in rapid protein synthesis. A poorly fixed mitochondrion,* M, *is also present.* × *66,000. Photograph by Dr. Palade from Jensen and Park, 1967. Courtesy of Wadsworth Publishing Company, Inc.*

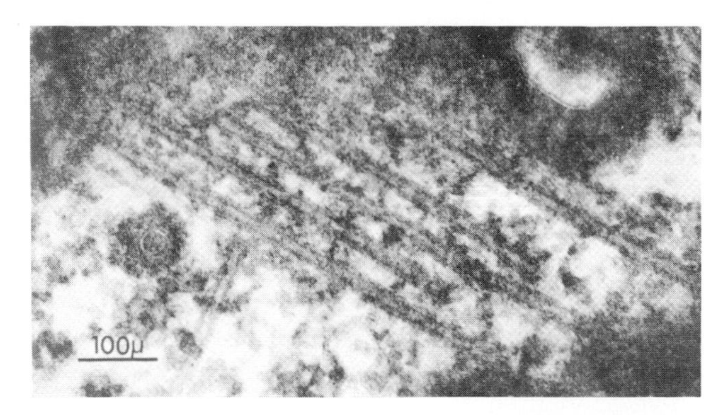

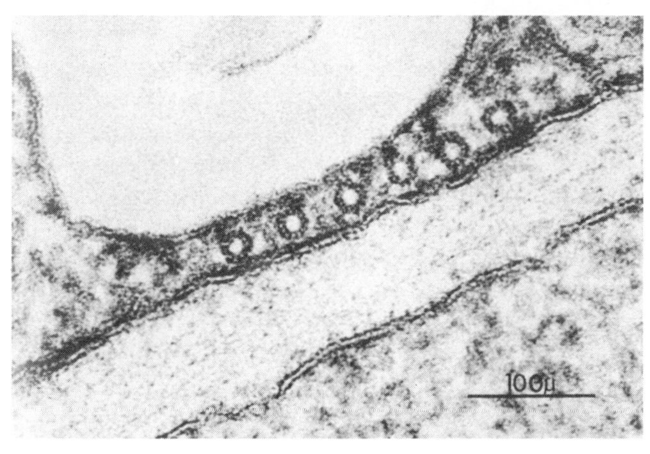

Figure 25-5. *Upper, electron micrograph of microtubules from a meristematic cell of a root tip of* Juniperus *(juniper). Lower, six microtubules from a meristematic cell of a root tip of* Juniperus. *Note that the wall of each microtubule is composed of a number (about 13) of globular subunits. Both photographs by Dr. Ledbetter from Clowes and Juniper:* Plant Cells. *Copyright 1968 by Blackwell Scientific Publications Ltd., Oxford.*

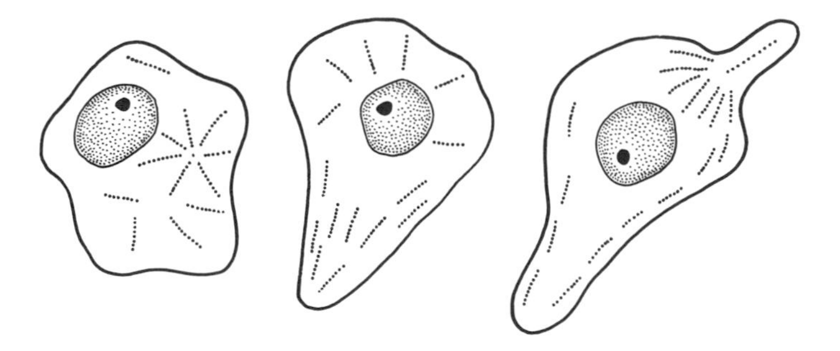

Figure 25-6. *Diagrams illustrating the relationship of the distribution of microtubules to changes in shape of a primary mesenchyme cell of a sea urchin gastrula.*

as the genetic material of the cell in Chapters 3 and 17. The centriole and/or the mitotic spindle participate in cell division in eucaryotes; the details of this participation are not completely understood. It is of particular interest that a program of changes in the distribution, disappearance, and reappearance of microtubules in the cytoplasm accompanies changes in cell shape in several unrelated plants and animals during development (Byers and Porter, 1964; Porter, 1966; Tilney, 1968; Newcomb, 1969). When the microtubules are randomly oriented or radiate from a central point, no specialized cell form is imparted. This is true of the newly formed, nearly spherical primary mesenchyme cells of the sea urchin gastrula (Figure 16-13). Later these cells become asymmetrical, correlated with orientation of the microtubules (Gibbins et al., 1968; Tilney, 1968). This is illustrated in Figure 25-6.

In plants, changes in shape of the generative cell during pollen grain germination (Chapter 11) appear to be controlled by the orientation of microtubules. The distribution of microtubules also seems to underlie asymmetrical cell divisions in the formation of root hair cells and stomatal guard cells in plant leaves (Newcomb, 1969).

Microtubular orientation occurs prior to the changes in cell shape in the developing lens fiber cells of both chick (Byers and Porter, 1964) and squid (Arnold, 1966). This is also the case in the developing bristles of insects, myoblasts of the chick embryo, and flagella (see review of Tilney, 1968).

These and other studies suggest that the patterned distribution of microtubules in the cytoplasm largely controls the form of a cell. This distribution is extremely orderly in cellular structures such as the *axopodia* of the protozoan *Actinosphaerium nucleofilum* (Figure 25-7). Note in the figure that the microtubules are precisely spaced and form two interlocking coils.

Basal bodies play an important role in forming flagella. Studies on the ameba *Naegleria gruberi* (Dingle and Fulton, 1966) and the alga *Chlamydomonas reinhardii* (Randall et al., 1967) demonstrate that the basal body disappears and reappears but is always present before the flagellum is formed. *Naegleria* becomes flagellated when transferred from a solid to a liquid culture medium (Chapter 22). Stages in development of the flagellum are illustrated in Figure 25-8.

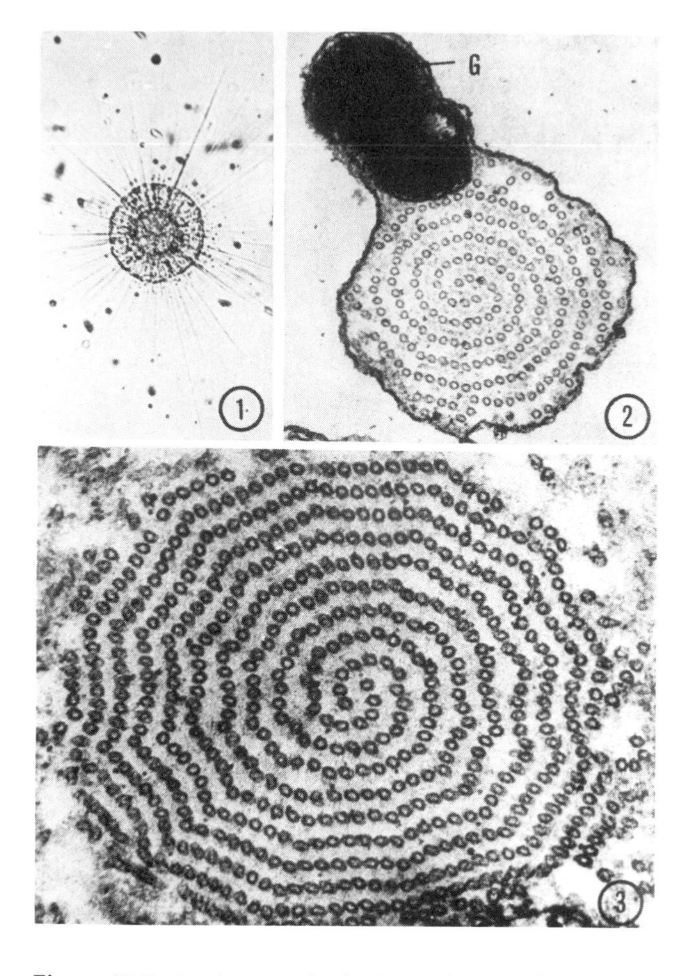

Figure 25-7. *A, photograph of a living* Actinosphaerium nucleofilum. *Axopodia radiate from the cell body. In each axopodium there is an axoneme (× 100); B, electron micrograph of a transverse section through an axopodium. The axoneme within the axopodium is constructed of microtubules arranged in two interlocking coils (× 60,000); C, cross section of an axoneme near the base of the axopodium (× 74,000). From Tilney, in* The Emergence of Order in Developing Systems, *p. 66. Copyright 1968 by Academic Press.*

In the amebae no basal bodies or flagellum structures could be detected. The first stage in formation of the flagellum is the appearance of the basal body with its long axis perpendicular to the cell membrane. Outer tubular fibers of the basal body elongate and extend into a progressively evaginating cell membrane. The two central filaments form and elongate to give the typical 9-2 organization (see Figure 25-2) within ten minutes after the basal body appears. A similar sequence of stages in flagellum development occurs in the water mold *Allomyces* (Renaud and Swift, 1964).

In *Chlamydomonas*, the flagella are resorbed prior to cell division, but the basal bodies persist. New basal bodies are also formed. In contrast, in the immotile zygotes of the sexual cycle (Figure 10-5), there was no trace of basal bodies (Randall et al., 1967). The elongation of the developing flagellum appears to involve sequential addition of material to the distal ends of the filaments. Autoradiography reveals that the labeled protein is added to the cut ends of protozoan flagella as they regrow (Rosenbaum and Child, 1967).

The elements of the flagellar structural pattern are probably assembled on the basal body, which functions like a template. On the other hand, the basal body itself can be assembled in the absence of a detectable template (Dingle and Fulton, 1966; Randall et al., 1967).

As seen in Chapter 3, DNA and RNA are constructed on preexisting templates. We have already mentioned that mitochondria and chloroplasts replicate by dividing and it is generally accepted that a new centriole is formed by the old, which serves as a template. Indeed, the structural continuity of cellular organelles is impressive. However, as noted in the introduction to this chapter, some organelles can be assembled from their subunits in the absence of preformed templates. Some evidence for self-assembly of cellular constituents follows.

Development Via Self-Assembly

Folding of Proteins

As we already know, the nucleotide sequences in DNA code for amino acid sequences in the polypeptide chain of proteins. In forming the functional protein, the polypeptide chain folds upon itself to produce a three-dimensional structure. This folding appears to result from a rapid and spontaneous interaction of amino acid side chains with each other and with the environment, without the input of additional genetic information. Most information about

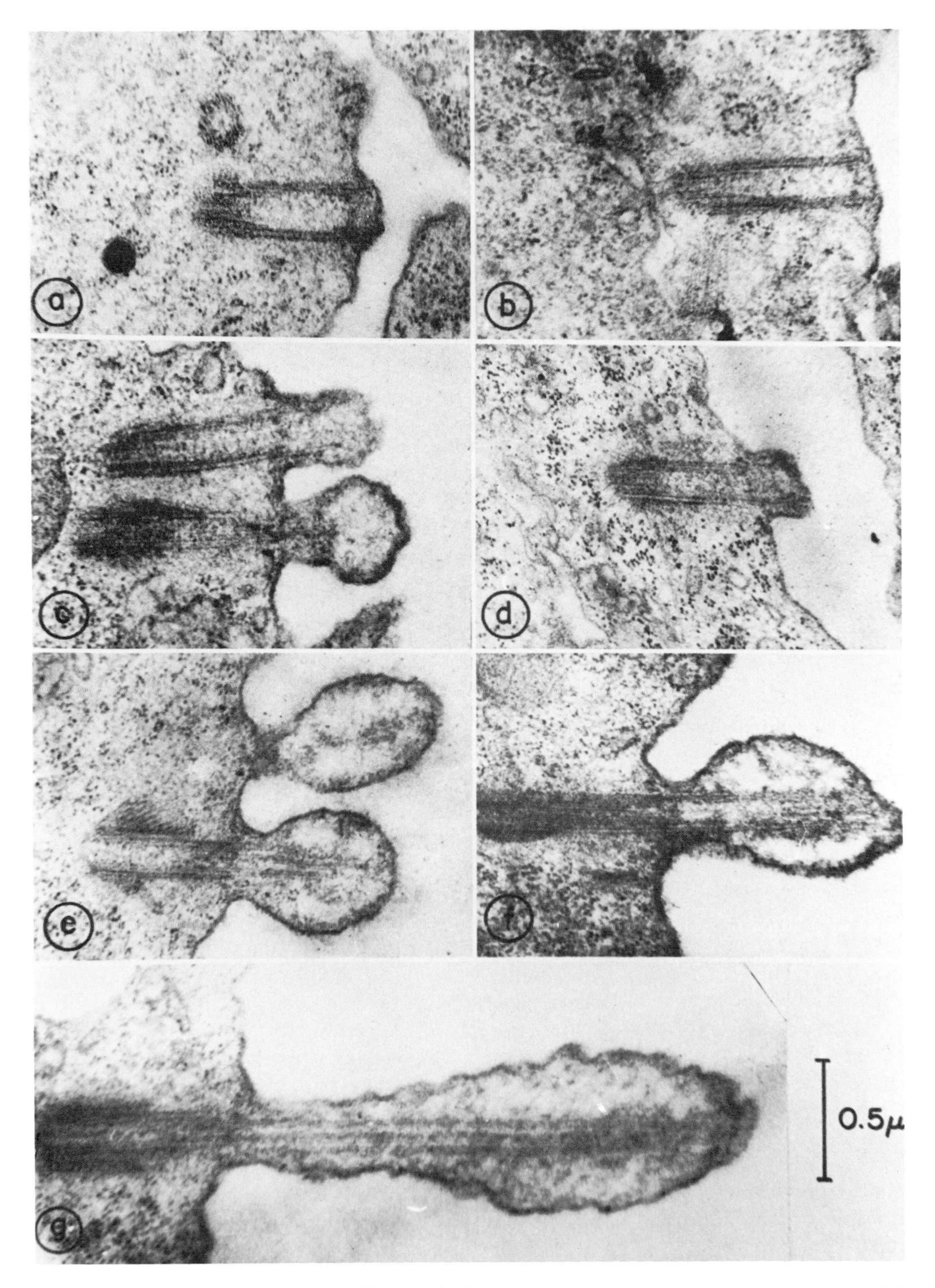

Figure 25-8. *Electron micrographs of longitudinally sectioned basal bodies and flagella showing the sequence of flagellum development in* Naegleria *(× 36,500). From Dingle and Fulton: Development of the flagella apparatus of* Naegleria, *in* J. Cell Biol., *Volume 31, copyright 1966 by the Rockefeller University Press.*

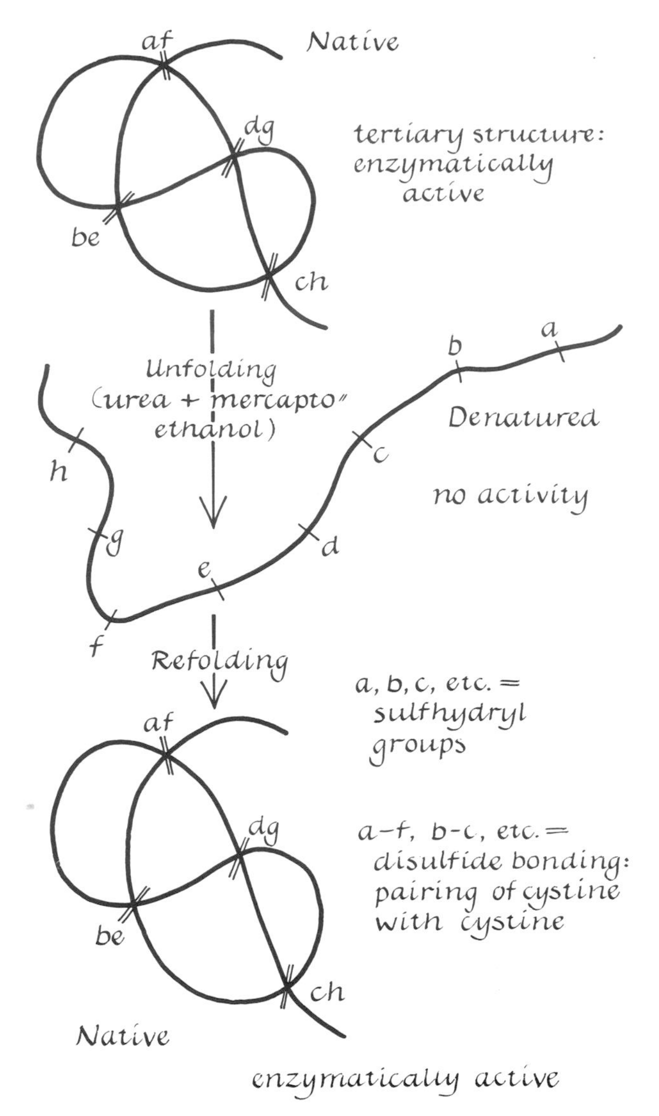

Figure 25-9. *Diagram illustrating the unfolding and refolding of a protein molecule in response to changes in the composition of the surrounding environment.*

the folding process has come from studying proteins that have disulfide bonds as cross links between amino acid pairs (Anfinsen, 1967, 1968). These proteins can be unfolded and refolded *in vitro* by changing the composition of the surrounding environment. This is shown in Figure 25-9. As indicated in Figure 25-9, the folding involves the formation of disulfide bonds during oxidation of the denatured form. With some proteins, such as enzymes, up to 100 percent functional activity is restored when the denatured form is oxidized. Note that only the native, three-dimensional form is functional. However, as Anfinsen has reported, large portions of many proteins are unessential for function. When parathyroid hormone (75 amino acid residues) is degraded to 45 residues, it becomes only 50 percent less active. The activity of papain (a protein-digesting enzyme from the papaya tree often used to tenderize meat) is maintained even when two-thirds of the molecule is removed experimentally. Synthetic adrenocorticotropic hormone (ACTH), which has 26 residues, is as active as native ACTH (39 residues). On the other hand, agents that break peptide bonds break up the sequence of amino acids in the polypeptide chain and destroy all functional activity. Thus, explicit information for forming the specific three-dimensional, active protein structure is coded in the amino acid sequence.

Collagen Synthesis *in Vitro*

A clear example of the self-assembly of macromolecules to form fibrils is the synthesis of collagen fibers *in vitro* (Gross et al., 1963). Collagen is a fibrous protein characteristic of connective tissue, cartilage, and bone (Chapter 23). The collagen molecule is about 2800 Å long and 14 Å in diameter. It is composed of three chains of amino acids coiled in a helical fashion and held together by hydrogen bonds (Figure 25-10). It behaves as if polarized. After synthesis in and release from fibroblasts, the molecules line up end to end and then side by side to form a collagen fibril. This is shown in Figure 25-10.

Note in the figure that the molecules overlap by about one-fourth of their length and are longitudinally associated "head" to "tail." This staggered arrangement is believed to give the collagen fibril its banded, 700 Å periodicity.

When native collagen fibers are dispersed to the molecular state, they will reassemble *in vitro*. The composition of the *in vitro* environment determines the pattern of the assembly. This is shown in Figure 25-11.

Acid treatment disperses collagen, but upon removal of the acid and warming, the molecules associate in the normal pattern. When the NaCl concentration is 0.35 M, fibrils with a 220 Å period form. Such fibers have been observed only in embryos and tissue cultures. At higher NaCl concentrations, nonstriated fibrils form. In the presence of glycoproteins, the dispersed molecules associate in a different

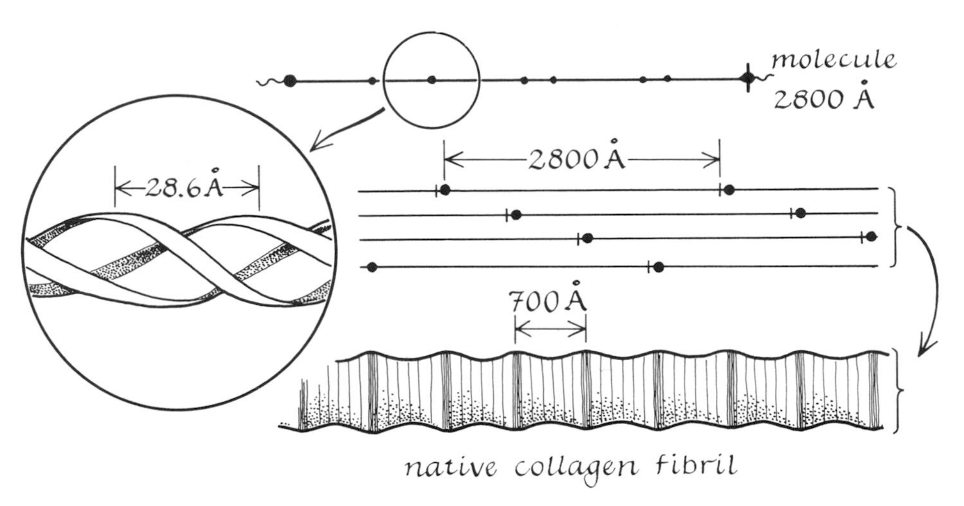

Figure 25-10. *Diagram of the assembly of collagen macromolecules to form a collagen fibril with 700 Å periodicity in banding.*

pattern to produce fibrils with a 3000 Å periodicity. Separate segments, very short fibrils, form in the presence of ATP.

Myosin

Myosin molecules associate *in vitro* to form long filamentous particles (Jakus and Hall, 1947; review of Harrington and Joseph, 1968). The filaments formed in 0.1 to 0.2 M KCl solutions are about 150 Å in diameter and from 2,500 to 20,000 Å in length; they are covered with surface projections in all but a central region. They correspond closely in structure to the thick filaments seen in longitudinal, thin sections of intact muscle. Slight changes in ionic strength and pH of the medium containing the myosin molecules influence the size of the regenerated filaments.

Microtubules

Microtubules in the axopodia of *Actinosphaerium* have been dispersed by treatment with antimitotic agents such as low temperature and colchicine (Tilney, 1968). In cells fixed during treatment no microtubules can be seen, but a few minutes after the living protozoan is returned to optimal conditions microtubules reappear and are associated in the characteristic array within the axopodia (Figure 25-7).

There is evidence that the subunits of the microtubule are conserved after disassembly and reused for reassembly. This is the case with the outer microtubules of flagella which, when dissociated to the subunit level, reaggregate to form microtubules (Stephens, 1968). Further evidence for reassembly of conserved microtubular subunits is the fact that the spindle fibers of the mitotic apparatus will reform in the presence of puromycin (an inhibitor of protein synthesis) following colchicine-induced disassembly (Inoué and Sato, 1967). It is not clear whether the microtubules contain all the information necessary to form their associations (for instance, the axoneme of an axopodium), or whether an orienting (Inoué and Sato, 1967) or initiating center (Porter, 1966) outside the assembly imposes the pattern.

Bacterial Flagella

Bacterial flagella are tubules about 140Å in diameter made up of 8 or 10 (depending on the species) longitudinal strands of globular subunits (Lowy and Hanson, 1965). The macromolecules constituting a flagellum have been dispersed by subjecting flagella preparations to an acid pH. The subunits are protein molecules with a molecular weight of from 20,000 to 45,000. When the pH of a solution of the flagellin molecules is raised to about 5.4, flagella are reconstituted (Abram and Koffler, 1964). Small pieces of native flagella act as "seeds" when added to a flagellin solution and initiate reassembly. The nature

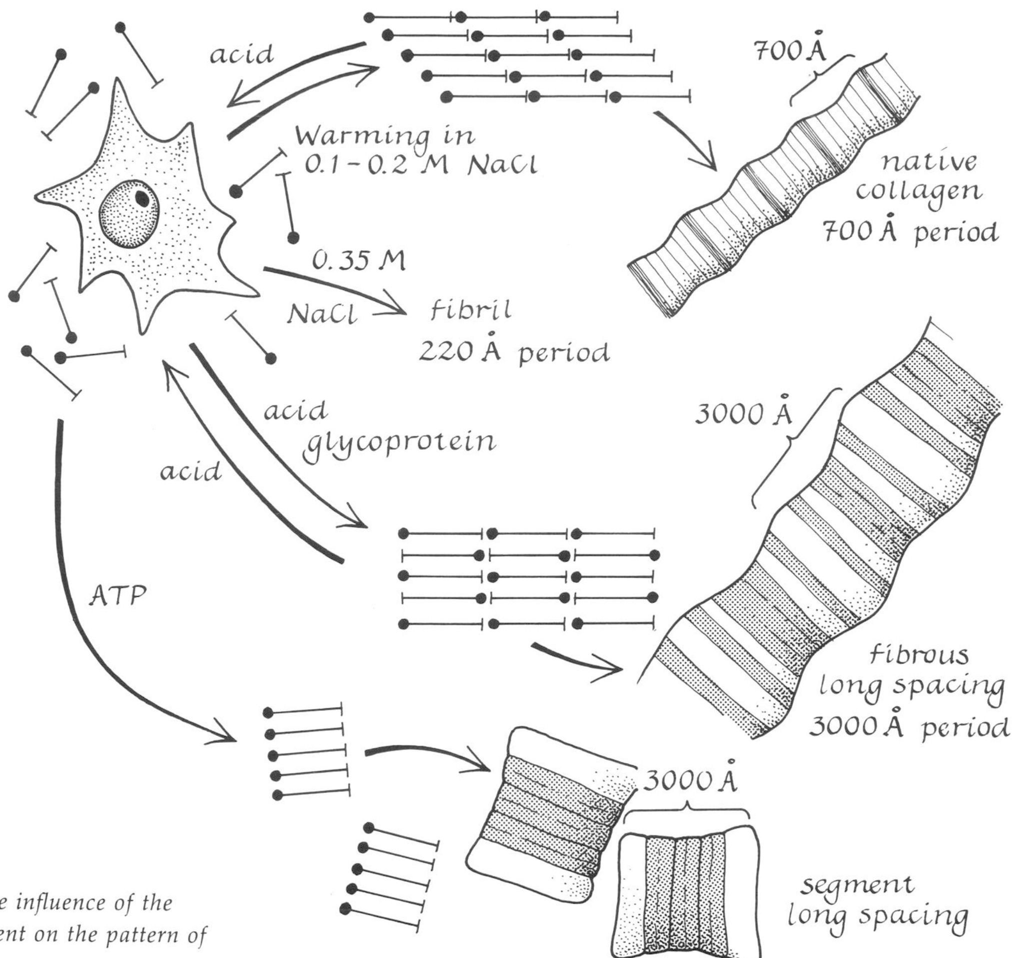

Figure 25-11. *Diagram illustrating the influence of the composition of the in vitro environment on the pattern of assembly of collagen molecules.*

of the seed flagellum controls the type of polymerization. If the seeds from a strain of bacterium possessing a curly flagellum are added to a solution of flagellin isolated from a normal *Salmonella*, the polymer is curly (Oosawa et al., 1966). Assembly of the subunits of bacterial flagella illustrates that although self-assembly can occur without a template, the presence of the latter can control the details of the structure formed. This appears to be the case with the flagella or cilia of eucaryotes as well.

Mitochondrial Membranes

Mitochondrial membranes can be dissociated into their repeating subunits, and these units can reassemble by hydrophobic recombination to form membranes (Green and MacLennan, 1969).

Other Organelles and Viruses

The subunits of a number of other cellular structures appear to assemble in the absence of detected templates. This has been reported for polyribosomes (Lerman et al., 1966), endoplasmic reticulum (Whaley et al., 1964), and mitochondria (Stoeckenius, 1966). Subjecting chick embryo tissues to low temperature (5 to 15° C) for six hours causes a crystal-like association of the ribosomes into sheets one ribosome thick (Byers, 1967). Partial reassembly of the flagella of eucaryotes has been reported, but in this case the incompletely disassembled axoneme of the flagellum appears to serve as a template (Gibbons, 1967).

Kellenberger (1964) and Pirie (1965) have reported the self-assembly of the protein coat of bacteriophages and the protein coat of tobacco mosaic virus (TMV), respectively. TMV can assemble in the absence of nucleic acid, but nucleic acid determines the helical structure and length of the virus. The phage's shape is determined partially by the structure of the subunits and partially by the DNA of the phage. In general, the development of phage from the DNA injected into the bacterium apparently involves template guidance as well as self-assembly of identical subunits into a symmetrical pattern (Anderson, 1967).

Experimental Reassembly of Components of a Cell

A living *Amoeba proteus* has been reassembled, using the membrane of one cell, cytoplasm from another, and nucleus from a third (Jeon et al., 1970). The basic procedure consists of (1) removing most of the cytoplasm and nucleus from one cell, (2) injecting cytoplasm from a second ameba to refill the cell, and (3) inserting a nucleus from a third cell. When components are from the same strain, most of the reassembled amebas divide and form clones normally. This technique promises to be valuable in studying the viability and compatibility of cell organelles from diverse sources.

Cells

Self-assembly of subunits is not restricted to the molecular and ultrastructural levels. As we have seen in Chapters 6 through 8, cells, as the subunits of tissues, organs, or organisms, can assemble by a natural aggregation (slime molds, some algae, annual fishes) or by aggregation after experimental dispersal (invertebrates and vertebrates). These aggregations are self-assemblies in that the properties of the cells determine the pattern of assembly.

Conclusion

Underlying the emergence of the order that characterizes developing systems is a sequential assembly of units beginning with molecules and ending with cells, tissues, and organs. The assembly of the subunits is controlled by template information in some cases and by the properties of the subunits in others. In the latter instances, the subunits, which are products of gene action, assemble into larger aggregates of specific design without the input of additional genetic information. This is what is usually termed self-assembly; it is described by Caspar (1966) as an "economical use of genetic information." Self-assembly, according to Caspar, "is possible for any structure whose design and stability are uniquely determined by the specific bonding properties of its constituent parts." Self-assembly typically involves apparently identical subunits and can occur *in vitro* in a number of structures (for instance, in collagen fibers, myosin, microtubules, bacterial flagella, and the protein coat of the tobacco mosaic virus). In other types of assembly, preexisting templates provide the information. However, in both template-guided and self-assembly, the assembly structure is primarily the inevitable consequence of information in the template or in the subunits themselves. Nevertheless, the environment of the assembling units influences the details of the assembly. We must not forget that an organism develops as the result of interaction between the genome and its external environment (Chapter 22).

The orderly nature of developmental processes and sequences is so obvious that we tend to overlook its importance in understanding development. In development, little is left to "chance," which is usually another word for ignorance.

References

Abram, D., and H. Koffler. 1964. *In vitro* formation of flagella-like filaments and other structures from flagellin. J. Mol. Biol. **9**:168–185.

Anderson, T. F. 1967. The molecular organization of virus particles. In J. M. Allen, ed., Molecular Organization and Biological Function, pp. 37–64. Harper & Row, New York.

Anfinsen, C. B. 1967. Molecular structure and the function of proteins. In J. M. Allen, ed., Molecular Organization and Biological Function, pp. 1–19. Harper & Row, New York.

Anfinsen, C. B. 1968. Spontaneous formation of the three-dimensional structure of proteins. In M. Locke, ed., The Emergence of Order in Developing Systems, pp. 1–20. Academic Press, New York.

Arnold, J. M. 1966. On the occurrence of microtubules in the developing lens of the squid, *Loligo peali.* J. Ultrastruct. Res. **14:**534–539.

Byers, B. 1966. Ribosome crystallization induced in chick embryo tissue by hypothermia. J. Cell Biol. **30:**C1–C2.

Byers, B., and K. R. Porter. 1964. Oriented microtubules in elongating cells of the developing lens rudiment after induction. Proc. Nat. Acad. Sci. U.S. **52:**1091–1099.

Caspar, D. L. D. 1966. Design principles in organized biological structures. In G. Wolstenholme, ed., Principles of Biomolecular Organization, pp. 7–39. J. and H. Churchill, London.

Dingle, A. D., and C. Fulton. 1966. Development of the flagellar apparatus of *Naegleria.* J. Cell Biol. **31:**43–54.

Eakin, R. M. 1970. A third eye. Am. Scientist **58:**73–79.

Fauré-Fremiet, E. 1958. The origin of the metazoa and the stigma of the phytoflagellates. Quart. J. Micr. Sci. **99:**123–129.

Fawcett, D. W. 1959. Changes in the fine structure of the cytoplasmic organelles during differentiation. In D. Rudnick, ed., Developmental Cytology, pp. 161–189. Ronald Press, New York.

Fawcett, D. W., and K. R. Porter. 1954. A study of the fine structure of ciliated epithelia. J. Morphol. **94:**221–281.

Gibbins, J. R., L. G. Tilney, and K. R. Porter. 1968. Microtubules in the formation and development of the primary mesenchyme in *Arbacia punctulata:* I. The distribution of microtubules. J. Cell Biol. **44:**201–226.

Gibbons, I. R. 1967. The organization of cilia and flagella. In J. M. Allen, ed., Molecular Organization and Biological Function, pp. 211–237. Harper & Row, New York.

Green, D. E., and D. H. MacLennon. 1969. Structure and function of the mitochondrial cristael membrane. Biol. Science **19:**213–222.

Gross, J., C. M. Lampiere, and M. L. Tasszer. 1963. Organization and disorganization of extracellular substances: the collagen system. In M. Locke, ed., Cytodifferentiation and Macromolecular Synthesis, pp. 175–202. Academic Press, New York.

Harrington, W. F., and R. Josephs. 1968. Self-association reactions among fibrous proteins: the myosin-polymer system. In M. Locke, ed., The Emergence of Order in Developing Systems, pp. 21–62. Academic Press, New York.

Harven, de E. 1965. The centriole and the mitotic spindle. In A. S. Dalton and F. Hagvenau, eds., The Nucleus, pp. 197–227. Academic Press, New York.

Inoué, S., and H. Sato. 1967. Cell motility by labile association of molecules. The nature of mitotic spindle fibers and their role in chromosome movement. J. Gen. Physiol. **50:**259–288.

Jakus, M. A., and C. E. Hall. 1947. Studies of actin and myosin. J. Biol. Chem. **167:**705–714.

Jensen, W. A., and R. B. Park. 1967. Cell Ultrastructure. Wadsworth Publishing Company, Inc., Belmont, Calif.

Jeon, K. W., I. J. Lorch, and J. F. Danielli. 1970. Reassembly of living cells from dissociated components. Science **167:**1626–1627.

Kellenberger, E. 1964. Morphogenesis of phage and its genetic determinants. In M. Sela, ed., New Perspectives in Biology, pp. 234–245. Elsevier, New York.

Lerman, M. C., A. S. Spirin, L. P. Gavrilova, and V. E. Golov. 1966. Studies on the structure of ribosomes: II. Stepwise dissociation of protein from ribosomes by cesium chloride and the reassembly of ribosome-like particles. J. Mol. Biol. **15:**268–281.

Lowy, J., and J. Hanson. 1965. Electron microscope studies of bacterial flagella. J. Mol. Biol. **11:**293–313.

Manton, I., B. Clarke, and A. P. Greenwood. 1956. Observations with the electron microscope on biciliate and quadriciliate zoospores in green algae. J. Exp. Bot. **6**:126–128.

Mercer, E. H. 1962. The evolution of intracellular phospholipid membrane systems. In R. J. C. Harris, ed., The Interpretation of Ultrastructure, pp. 369–384. Academic Press, New York.

Newcomb, E. H. 1969. Plant microtubules. Ann. Rev. Plant Physiol. **29**:253–288.

Oosawa, F., M. Kasai, S. Hatano, and S. Asakura. 1966. Polymerization of actin and flagellin. In G. E. Wolstenholme and M. O'Connor, eds., Principles of Biomolecular Organization, pp. 273–307. J. and H. Churchill, London.

Pirie, N. W. 1965. Biological organization of viruses. In G. E. Wolstenholme and M. O'Connor, eds., Principles of Biomolecular Organization, pp. 136–157. J. and H. Churchill, London.

Porter, K. R. 1963. Diversity at the subcellular level and its significance. In J. M. Allen, ed., The Nature of Biological Diversity, pp. 121–163. McGraw-Hill Book Company, New York.

Porter, K. R. 1966. Cytoplasmic microtubules and their functions. In G. E. Wolstenholme and M. O'Connor, eds., Principles of Biomolecular Organization, pp. 308–345. J. and H. Churchill, London.

Randall, J., T. Cavalier-Smith, A. McVittie, J. R. Warr, and J. M. Hopkins. 1967. Developmental and control processes in the basal bodies and flagella of *Chlamydomonas reinhardii*. In M. Locke, ed., Control Mechanisms in Developmental Processes, pp. 43–83. Academic Press, New York.

Rash, J. E., J. W. Shay, and J. J. Biesele. 1969. Cilia in cardiac differentiation. J. Ultrastruct. Res. **29**:470–484.

Reinert, J., and H. Ursprung, eds. 1971. Origin and Continuity of Cell Organelles, Springer-Verlag, New York.

Renaud, F. L., and H. Swift. 1964. The development of basal bodies and flagella in *Allomyces arbusculus*. J. Cell Biol. **23**:339–354.

Robertson, J. D. 1964. Unit membranes: a review with recent new studies of experimental alterations and a new subunit structure in synaptic membranes. In M. Locke, ed., Cellular Membranes in Development, pp. 1–81. Academic Press, New York.

Robertson, J. D. 1967. The organization of cellular membranes. In J. M. Allen, ed., Molecular Organization and Biological Function, pp. 65–106. Harper & Row, New York.

Rosenbaum, J. L., and F. M. Child. 1967. Flagella regeneration in protozoan flagellates. J. Cell Biol. **34**:345–364.

Stephens, R. E. 1968. On the structural protein of flagella outer fibers. J. Mol. Biol. **32**:277–283.

Stoeckenius, W. 1968. Structural organization of the mitochondrion. In G. E. Wolstenholme and M. O'Connor, eds., Principles of Biomolecular Organization, pp. 418–445. J. and H. Churchill, London.

Thompson, T. E. 1964. The properties of biomolecular phospholipid membranes. In M. Locke, ed., Cellular Membranes in Development, pp. 83–96. Academic Press, New York.

Tilney, L. G. 1968. The assembly of microtubules and their role in the development of cell form. In M. Locke, ed., The Emergence of Order in Developing Systems, pp. 63–102. Academic Press, New York.

Waddington, C. H. 1962. New Patterns in Genetics and Development. Columbia University Press, New York.

Whaley, W. G., T. E. Kephardt, and H. H. Mollenhauer. 1964. The dynamics of cytoplasmic membranes during development. In M. Locke, ed., Cellular Membranes in Development, pp. 135–174. Academic Press, New York.

26 *Pattern Aspects of Development*

The emergence of patterns is the most striking feature of development. Looking at development from the viewpoint of pattern formation unifies many of its features. The preceding chapter demonstrated how each level of complexity in a developing organism arises from a lower level of order. This process leads directly to the creation of pattern, which is order in space. Pattern is not simply an assembly of units; its significant feature is the orderly interrelation of its components. Earlier chapters provided many examples of how nonrandom distributions of units arise during continuous developmental activities in adults, during regeneration, and during asexual and sexual reproduction. We have seen how these interrelationships are established under control of preformed guidelines either inherent in the components, imposed upon them from outside, or controlled by progressively formed guidelines. However, the guidelines for pattern formation differ from those directing the synthesis of the materials out of which the pattern is made. For example, the guidelines for forming a cell, although participating in the formation of cellular associations and multicellular patterns, are clearly not the only guidelines controlling supracellular associations and patterns. Nevertheless, pattern formation is undoubtedly under genetic control, even though environmental influences can modify the details of genetic expression. As Goldschmidt (1955) emphasized, understanding pattern formation, which is the same problem in genetics as in experimental embryology, is understanding how a pattern in three-dimensional space results from four-dimensional processes (time being considered as the fourth dimension).

In this chapter we shall describe some patterns and suggest some mechanisms that appear to control their emergence. In many instances the evidence for mechanisms is incomplete, and in others it is lacking. If we only stimulate the reader's curiosity, we will have fulfilled our purpose.

Continuity of Patterns

As described in Chapter 25, patterns are continuous from the molecular to the cell-population level. The patterned associations of molecules that form macromolecules are the subunits of cellular organelles; organelles are the subunits of the pattern we call a cell; cells, the subunits of the patterns we see in tissues and organs, and so forth. Thus every pattern observed during development is the guideline or *prepattern* for the next pattern to be formed (see Ursprung, 1963, 1966). Although the complex pattern of

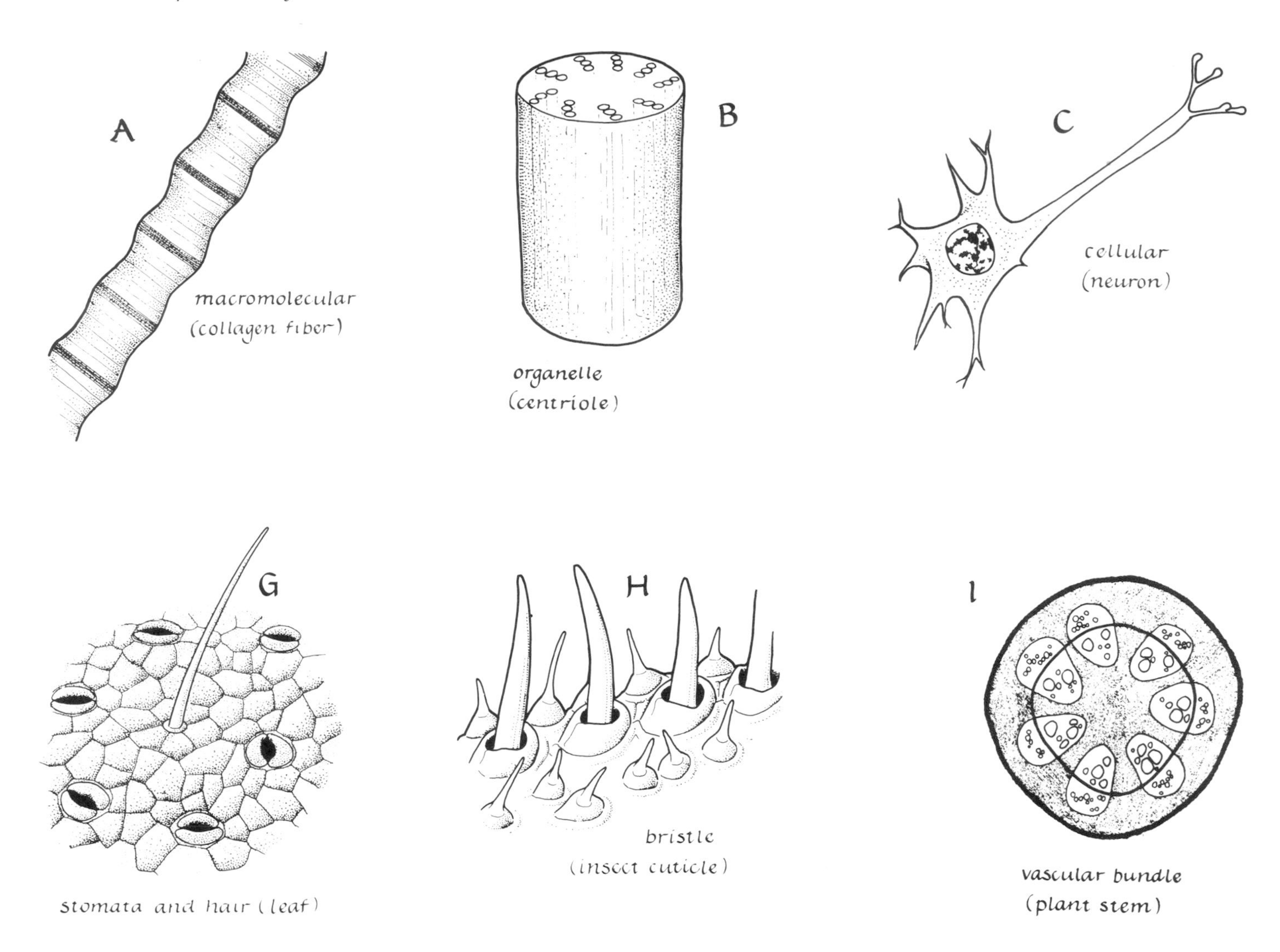

Figure 26-1. *Diagrams of the diverse patterns found in living organisms.*

an adult organism can be conceptually, and sometimes experimentally, analyzed into its subunits, the organism itself is a unit from the time of its inception to the time of its completed development. The building of design must be guided by some overall mechanism operating continuously during development. The continuity of organisms in reproduction and development suggests that they themselves are the mechanisms controlling the emergence of patterns which genetically characterize them. The fact that a pattern observed at one stage cannot be detected at an earlier stage does not mean that the genetic information for forming this pattern is not present at earlier stages. The orderly programmatic nature of development requires this continuity of emerging patterns.

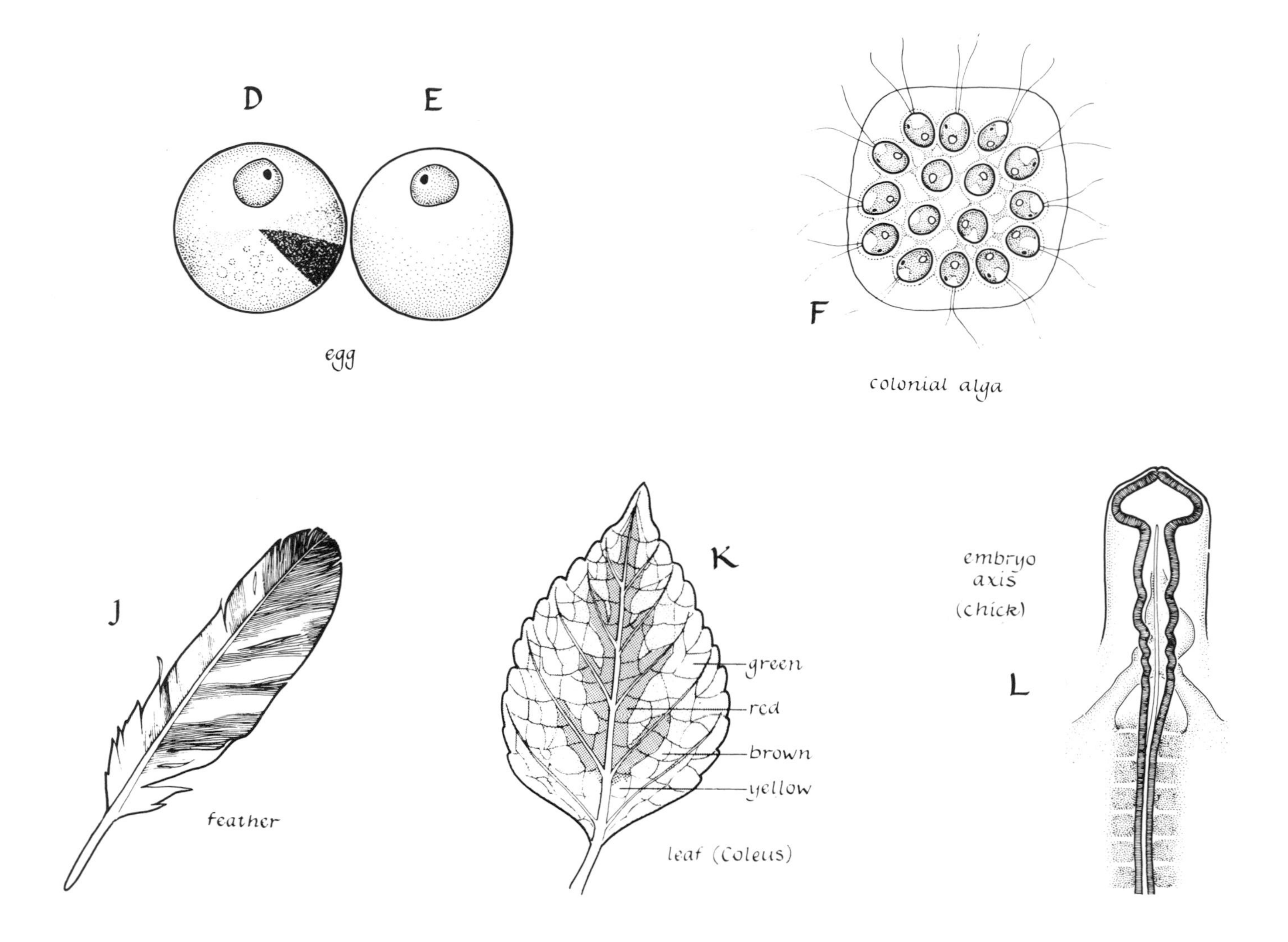

Some Selected Patterns

A few of the great variety of biological patterns are shown in Figure 26-1; these should illustrate the general problem of the genesis of pattern. As noted in Chapter 25, the end-to-end and side-by-side association of collagen molecules *in vivo* and *in vitro* results in formation of a collagen fibril exhibiting an orderly periodic banding pattern (Figure 26-1A). The association of microtubules in nine groups of three each constitutes the basic pattern of a centriole (Figure 26-1B). Cytoarchitectural patterns in eggs are either localized and mosaic or gradient in type (Chapter 13 and Figure 26-1D and E). Patterns of cellular arrangement in colonial algae are geometrically and numerically orderly (Chapter 7 and Figure

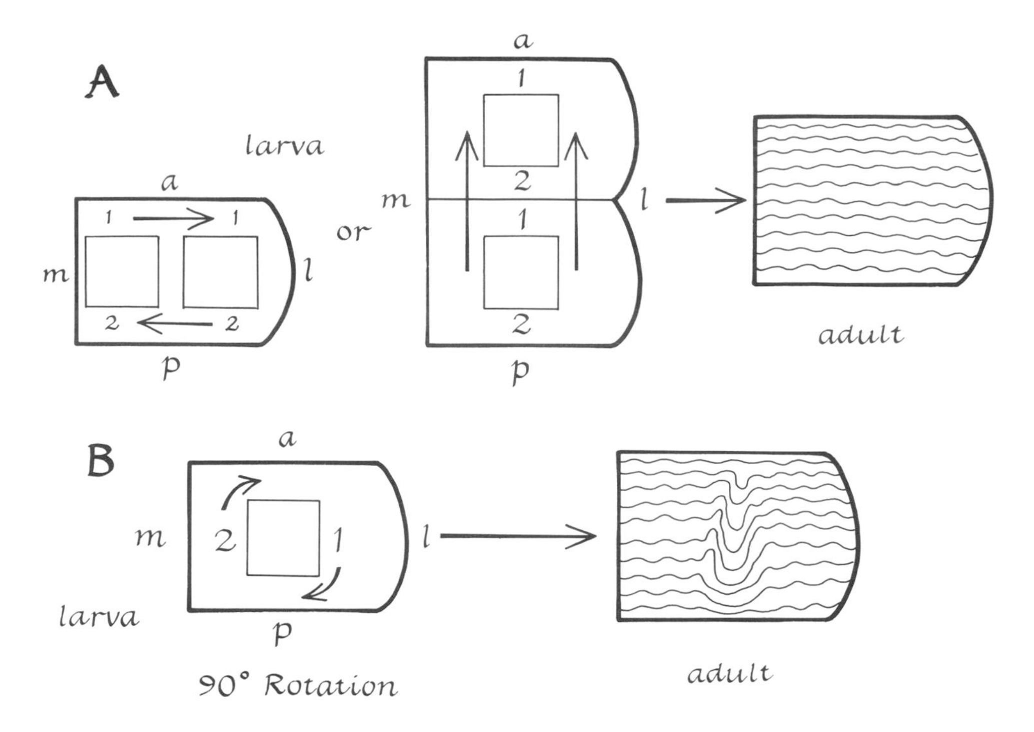

Figure 26-2. *Polarity in the prepattern of the integument in the bug* Rhodnius.

26-1F). A feature of many patterns in unicellular and multicellular organisms is the orderly spacing of the units. This is found in the spacing of stomata and hairs on plant leaves (Figure 26-1G) and insect bristles (Figure 26-1H), of vascular bundles in plant stems (Figure 26-1I), and of bands of black pigment in feathers (Figure 26-1J). Leaf patterns involving the distribution of chromoplasts are often very regular (Figure 26-1K). The axial structure of the vertebrate embryo (for instance, that of the chick) is complex and orderly (Chapter 12 and Figure 26-1L).

The Role of Preformed Guidelines

DNA Guidelines

Probably no pattern can arise in the absence of some preexisting guideline. This guideline can be either intrinsic to the units making up the pattern as in self-assembly, extrinsic in the form of a template (actually a prepattern itself), or progressively built up through activities of the units, such as release of cell products which constitute an influential internal microenvironment (Chapter 22). With some rare exceptions (Chapter 18), inherited DNA or RNA molecules seem to constitute the ultimate preformed guidelines for pattern formation. As described in Chapter 3, the linear association of nucleotides in the DNA helix (or in the RNA of some viruses) is the master template guiding the synthesis of enzymes and other proteins, which in turn control other developmental processes.

Egg Patterns

Several earlier chapters describe how gene action at one stage can establish conditions that control subsequent gene action. This is clearly the case when gene action during oogenesis establishes the pattern of egg organization, itself a prepattern for subsequent development. This prepattern, established through maternal inheritance, can sooner or later be modified by the male genome introduced at fertilization. Thus, the cytoarchitecture of an egg, whether visibly simple or complex (Figure 26-1D and E), is a patterned guideline specifying differential gene action in accord with the location of nuclei in the prepattern (Chapter 13).

Gradient and Field Patterns

In many organisms a quantitative gradient prepattern characterizes the early stages of development. The prepattern might consist of graded differences in concentrations of metabolites within an egg, spore, or unicellular organism. In multicellular systems, the prepattern might consist of graded differences in concentrations of cell products in the internal microenvironment. Graded differences in several metabolites could be manifest in all three spatial dimensions within the cell or cell population (Chapter 22). Since prepatterns (other than a template pattern) not only control but actually develop into patterns, the prepattern itself must progressively change and, along with this, the system of graded differences. A prepattern possessing a graded design is not rigid. An isolated part of the prepattern, provided it is representative of the whole, can reform or function like the whole prepattern. The result is termed regulation (Chapter 15). Such prepatterns are usually called *developmental fields* (Weiss, 1939). However, as Waddington (1966) has pointed out, there has been some confusion as to what a field is. It has been described as an area or group of cells giving rise to a specific structure such as a limb or eye, but capable of regulation at early stages. It probably matters little whether we define a field as a group of cells or as a set of microenvironmental conditions impinging upon the cells and controlling their behavior (Chapter 22). The essential point is that some prepatterns exhibit the properties of a field.

Field-type prepatterns exist at late as well as at early stages, and are even found in some adult organisms. For example, neurons in the retina and optic tectum are specified in a gradient system (see Chapter 24). Gradients in regenerative capacity exist in adult *Tubularia*, flatworms, *Acetabularia*, and liverworts (Chapter 5). In higher plants, gradients in auxins and other hormones constitute a type of prepattern controlling growth and differentiation (Chapter 19). Gradients are apparently universal in developing organisms (Child, 1941).

Some prepatterns exhibit a definite polarity of organization. An example is the prepattern underlying development of small folds or ripples in the integument of the abdomen in the blood-sucking bug *Rhodnius*. Exchange of grafts of integument within or between abdominal segments at the fifth instar larval stage demonstrates the stability of this pattern (see review of Locke, 1967). When grafts are exchanged as shown in Figure 26-2A, the normal adult pattern of ripples develops.

However, if a graft is rotated 90° relative to the anterior-posterior axis of the abdomen, a distorted adult pattern develops (Figure 26-2B). Thus, in order to obtain the normal adult pattern of ripples, the polarity of the graft must be maintained in alignment with that of the insect's body. These and similar experiments of Locke and others demonstrate that the larval epidermal cells are arranged in a gradient in which the cells become polarized in the anterior-posterior but not in the medio-lateral axis.

Complex preformed guidelines operate in asexual reproduction by fission or fragmentation: half or a fragment of the parent constitutes the prepattern on which the new half forms. This is clearly the case in *Microsterias* (Chapter 9) and in other types of asexual reproduction, such as zooid formation and budding.

Pattern in Primary Induction

In primary induction of the neural plate in amphibian embryos, the underlying chorda-mesoderm functions like a prepattern. The pattern in the chordamesoderm is apparently a gradient in concentration of diffusible inducing factors. One of these, the activating principle, is highest in concentration in the midline at the tip of the notochord and decreases laterally and posteriorly. A second factor, the transforming principle, which is responsible for the craniocaudal differentiation in the neural plate, is highest in concentration in the caudal region of the underlying mesoderm (Nieukoop, 1967; see also Chapter 21).

Genes and Patterns

There are many examples of gene mutations modifying patterns (Stern, 1955; Lewis, 1964). A number of these, including changes in the patterned spacing of bristles or hairs on the cuticle of *Drosophila,* have been studied. The problem here is how the mutant gene brings about these changes in bristle pattern. In the mutant called *achaete,* two bristles on each side of the posterior-dorsal part of the thorax are missing. In mosaic flies consisting of wild-type and *achaete* cells, the posterior-dorsal bristle develops when its normal site is occupied by even a small amount of wild-type tissue almost completely surrounded by *achaete* tissue. Conversely, when a small amount of *achaete* tissue occupies this site, no bristle develops (Stern, 1954, 1955). This is shown in Figure 26-3.

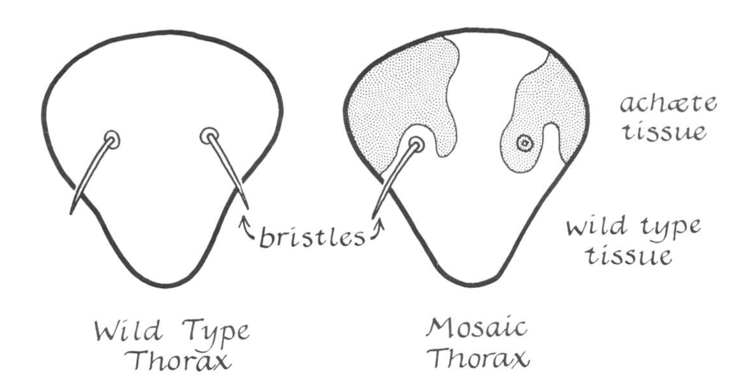

Figure 26-3. *Modification of pattern in bristle development by gene mutation in* Drosophila.

Stern interpreted these results to mean that the mutant gene does not change the prepattern that controls bristle distribution but changes the competence of the cells to respond to the prepattern. According to this hypothesis, a wild-type cell can respond to the prepattern by forming a bristle, whereas the mutant cell has lost this capacity.

Stern and his collaborator, Tokunaga, have suggested a similar explanation for the development of a secondary sex comb on the foreleg of a male *Drosophila* carrying the mutation "engrailed" (1961). In the wild-type fly only one sex comb develops. X-radiation of larvae, which increases the frequency of somatic crossing-over, produces flies having cells of two different genotypes. Phenotypically, these cells are wild type/wild type and yellow engrailed. The character "yellow" is a marker; wherever it appears in cells of a mosaic fly, it denotes that these cells also carry the mutation "engrailed." A secondary sex comb developed whenever yellow cells were present in the secondary sex comb area of the leg but not when wild type (black) cells occupied this area of the mosaic (Figure 26-4).

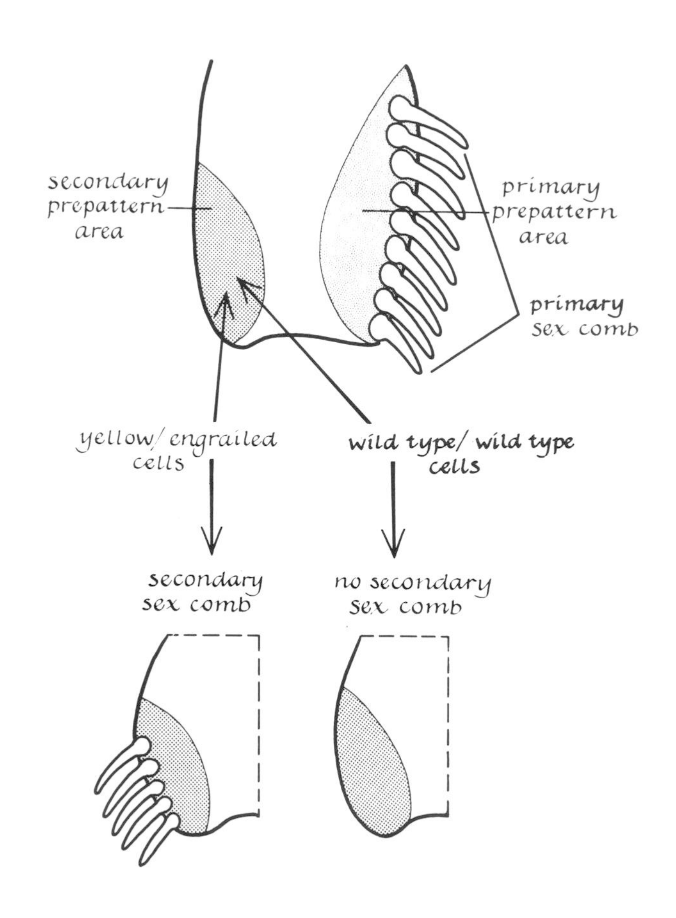

Figure 26-4. *Competence of cells carrying the mutation* engrailed *to respond to the secondary prepattern area on the foreleg of the male* Drosophila. *Note that wild-type cells do not respond to this prepattern.*

As in *achaete,* the prepattern underlying sex comb formation is the same in both wild-type and mutant flies. Cells carrying the mutant gene "engrailed" but not the wild-type cells are competent to respond when they occupy the secondary area of the prepattern in the mosaic fly.

Stern (1954) defines prepattern as "...a descriptive term for any kind of spatial differences in development." Although it is not known what the prepattern is a pattern of, it could consist of graded or localized distributions of many substances (for instance, oxygen, carbon dioxide, cell products). A prepattern of graded or localized differences pervading a population of cells of identical genotype could elicit different responses of the cells in accord with their positions. Recall the suggestion in Chapter 22 that the intercellular microenvironment helps control cellular differentiation. In genetic mosaics one can see how cells might respond differently to even a homogeneous kind of prepattern, but genetic mosaicism is probably not involved in cellular differentiation in normal development. Although the genome sets the limits of possible responses, the prepattern sets the specific responses of the genetic mosaic cells.

Epigenesis of Patterns

As noted above, all patterns arise during development under control of either preformed or progressively formed guidelines. Many normally developing systems contain progressively formed patterns resulting from the multiplication and differentiation, or from the aggregation of cells. Note, for example, the patterns illustrated in Figure 26-1C, F, G, J, K, and L (see also Chapters 4, 6 through 12, and 22). Under experimental conditions, a new pattern in a cell population can also be formed.

Induced Patterns in Plants

Recall how appropriate concentrations of an auxin and sucrose, when applied to the center of a mass of undifferentiated callus tissue *in vitro*, will induce concentric rings of phloem, cambium, and xylem cells (Chapter 22). Also recall that the experimentally altered shape of the shoot just below the apex determines the cross-sectional shape of the vascular bundle stele in ferns. In this case the cross-sectional shape of the shoot is the prepattern that controls the spatial differentiation of the stele. In Chapter 22, we noted that a cell population derived from one phloem tissue cell placed in liquid culture eventually becomes a whole plant.

Sometimes the pattern of cellular arrangement at early stages in division of the isolated phloem cell simulates that of embryo plants developing from eggs (Steward et al., 1961). Gravity induces the formation of a shoot, an indication of polarity, in these *in vitro* aggregates. This is demonstrated by the fact that aggregates grown in liquid media and rotated continuously around a horizontal axis (a technique that neutralizes gravitational stimuli) fail to develop shoots, whereas those placed on a stationary agar medium do so. The gravitational stimulus probably induces a polarized distribution of hormones to which the cells respond differently (see Chapter 19). Polarity of the pattern is thus imposed upon the system by external influences, but its radial symmetry is an automatic consequence of graded inside-to-outside differentials in concentration of cell products (see Figures 22-3, 22-4, and 22-5).

Induced Pattern in Chick Embryos

One can observe the epigenesis of a complex pattern in a population of chick embryo cells that normally does not form such a pattern. This is the development of the bilaterally symmetrical axial pattern characteristic of a thirty-six-hour incubated embryo from an anterior half or from a right or left half, fourth, or eighth of the unincubated blastoderm. The fragment must be explanted *in vitro* on an agar medium containing an extract of the egg contents (Spratt and Haas, 1960). Recall (Chapters 12 and 15) that the embryo body axis normally develops from the posterior marginal zone. Some of the results are illustrated in Figures 26-5 and 26-6.

In all but the youngest unincubated blastoderms, axial patterns form only in pieces that include the marginal zone. An axial pattern begins to form when a sheet of associated mesodermal and endodermal cells moves from the marginal zone toward the original center of the blastoderm. The movement of the lower germ layers across the underside of the relatively immotile upper layer induces a neural axis in the upper layer. The neural axis always is located in that radius of the blastoderm that lies above the axis of the movement pattern (Figure 12-31); this can be any radius irrespective of the posterior radius along which the embryo body normally forms in a whole blastoderm. These and later studies (Spratt, 1966) showed a one-to-one correspondence between the presence, position, and number of movement patterns and the presence, position, and number of primitive streaks and embryo axial systems.

The prepattern in the explanted pieces appears to be the direction (radius) in which the lower layers move. This does not correspond to any preformed pattern of bilateral symmetry, since movement can occur in any radius. However, in the intact blastoderm, the ring-shaped marginal zone is a bilaterally symmetrical gradient in cell population density (highest in the posterior part and decreasing around the ring to the anterior part). The ring could be considered the prepattern controlling the position of the embryo axis in normal development. Correlated with the gradient in cell population density is a gradient in embryo-initiating capacity, highest in isolated posterior pieces and decreasing gradually around the ring to its anterior part (Figures 26-6 and 26-7). Nevertheless, any portion of the ring can become an embryo-forming center. The marginal zone thus exhibits some properties of an embryonic *field*—a collection of cells, any portion of which, within a minimum size limit, can form the structure normally developing from the whole collection.

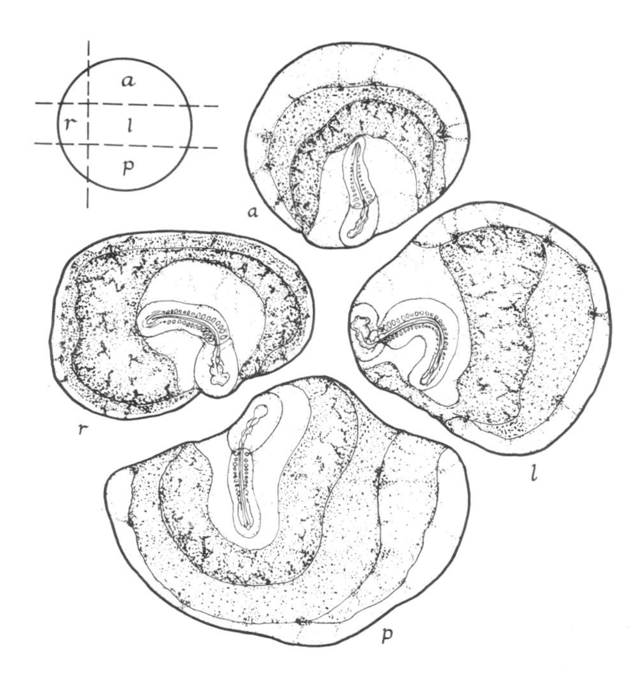

Figure 26-5. *Epigenesis of a complex pattern (the body axis of a chick embryo) in anterior,* a, *right,* r, *and left,* l, *pieces of an unincubated blastoderm explanted in vitro. The upper left diagram illustrates how the blastoderm was cut into pieces. From Spratt and Haas,* J. Exp. Zool., *Volume 145, p. 113, 1960. Courtesy Wistar Press.*

Figure 26-6. *Development of eight types of quarters isolated from unincubated chick blastoderms and explanted in vitro. The number and frequency in percent of embryonic axial systems are denoted in the diagrams* a *and* b. *The arrows in* c *denote the pattern of lower surface cell movements in anterior and posterior quarters. From Spratt and Haas,* J. Exp. Zool., *Volume 145, p. 117, 1960. Courtesy Wistar Press.*

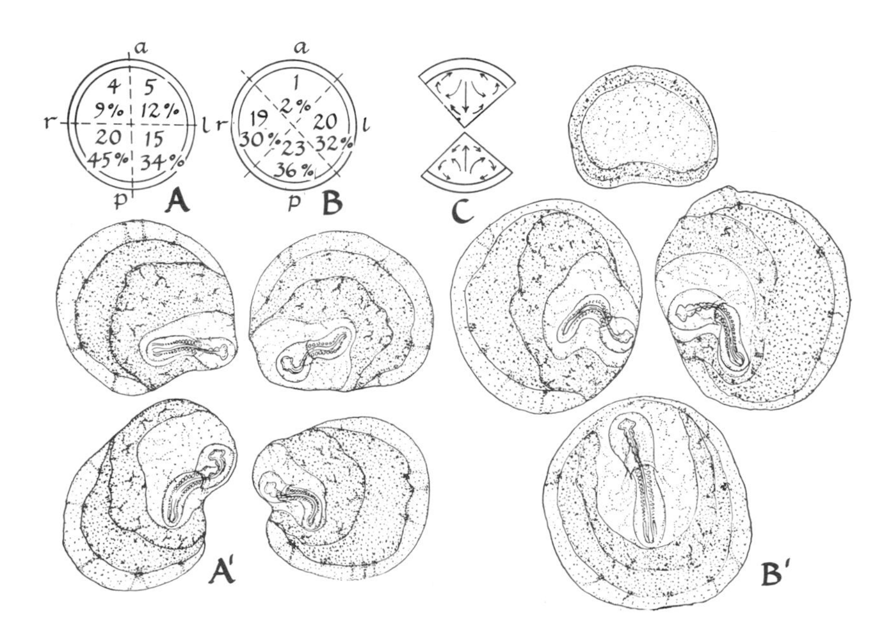

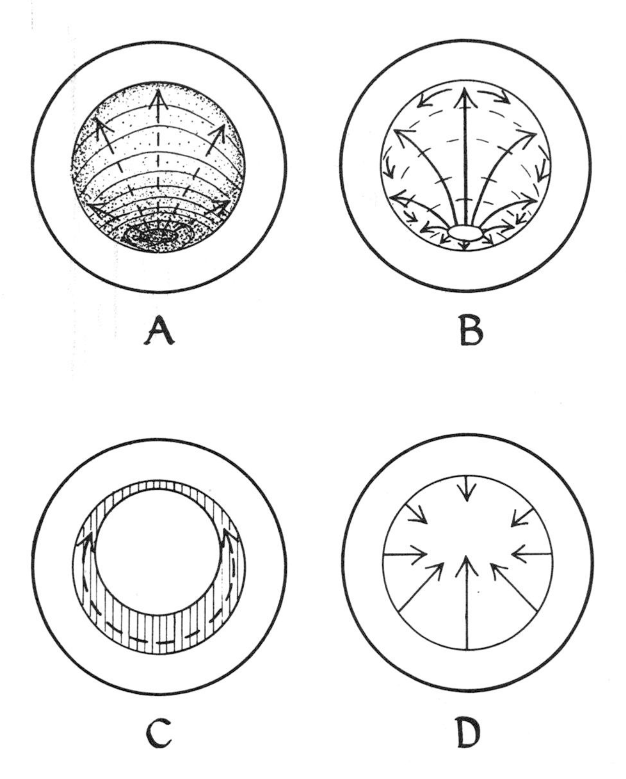

Figure 26-7. *Diagrams illustrating patterns in the unincubated chick blastoderm. A, gradient in cell population density; B, lower surface movement pattern; C, gradient in marginal zone growth potential; D, gradient in embryo body forming capacity. From Spratt and Haas,* J. Exp. Zool., *Volume 145, p. 135, 1960. Courtesy Wistar Press.*

Figure 26-8. *A, a one day old male White Leghorn chick showing area of black down feathers covering entire wing and part of breast produced by grafting to the wing bud of a host embryo at 78 hours a piece of head skin ectoderm from a Barred Plymouth Rock embryo of the same age; B, the same chick 18 days after hatching; C, the same chick 31 days after hatching. Note barred pattern in contour feathers of the wing and breast. From Willier and Rawles, 1940,* Physiological Zoology, *Volume 13, University of Chicago Press.*

Induced Patterns in Feathers

Patterns involving the distribution of melanin pigment granules in feathers arise epigenetically as a result of *melanocytes,* which deposit the granules on the developing barbs of the emerging feather just above the collar (Chapter 4). When melanoblasts derived from the neural crest of one breed of chicken embryo are grafted to the base of the wing bud of another breed at seventy-two hours of incubation, the wing feathers of the hatched chick have a pigmentation pattern like that in wing feathers of the donor breed. If the donor breed has barred (black and white striped) feathers, the host will develop barred feathers even when it is genetically a solid black or white breed (Willier and Rawles, 1940, 1948). This is illustrated in Figure 26-8.

Willier and Rawles demonstrated that the genotype of the melanocyte controls the melanin distribution in a feather. However, the region of the host (wing, back, breast) into which the melanoblasts are grafted or migrate modifies the details of the pattern. Thus, in the kind of experiment cited, the width of the black bar is much smaller in breast than in wing feathers of the white host; this is also true of feathers of the barred donor breed. The final pattern is the

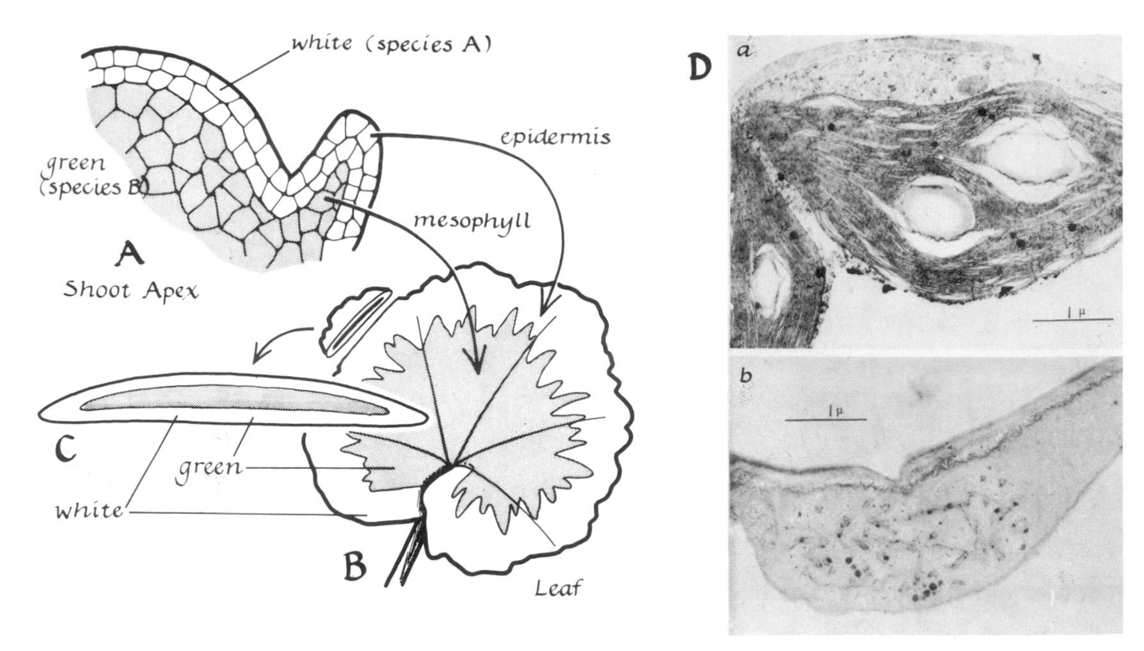

Figure 26-9. *A, diagram of a portion of a shoot apex in a geranium* (Pelargonium zonale) *chimaera; B, diagram of a leaf of the chimaera; C, diagram of a cross section of the leaf; D, electron micrographs of,* a, *normal, and,* b, *abnormal, chloroplasts in a chimaera of* Pelargonium zonale. *D from Clowes and Juniper:* Plant Cells. *Copyright 1968 by Blackwell Scientific Publications, Ltd., Oxford.*

result of an interaction between the genotype of the melanocytes and the kinds of feathers they enter. The prepattern in these cases is primarily a rhythmical production and/or release of melanin by the melanocytes, the underlying mechanism of which is not understood (see Ursprung, 1966). Pigment patterns also occur in amphibians. In some animals, following transplantation of neural crest cells from one species to another, the pattern of melanophore distribution can be either donor or host in type depending upon the species (see reviews of Twitty, 1949; Wilde, 1961).

Leaf Patterns

Variegation in plant leaves poses a great challenge to the student of pattern development. In some plants, variegation involves the segregation of white, green, or other colored plastids during mitosis in the leaf primordium (see Chapter 18). There seems to be no orderly mechanism for segregation, and the result is a very irregular flecking or spotting pattern, the density of which depends partly on when segregation occurs during leaf development.

Leaf patterns of various types also develop in plants containing cells of two different genotypes. The two cell types can arise by somatic mutation or as a result of a mixture of cells from different species. These mixtures are produced by grafting the shoot or bud of one species onto the stem of another. A shoot that is a chimaera can develop from the meristematic tissues at the graft union. The *periclinal* chimaera is particularly significant (Sinnott, 1960; Clowes and Juniper, 1968). The outer cell layers of the plant are derived from one graft partner and its entire inner part from another. When the epidermis cells that proliferate at the margins of leaves on the shoot apex are genetically white and the inner cells (the *mesophyll*) are genetically green, the leaf margins are white and the central regions are green. This pattern, found in some geranium chimaeras, is stable because cell layers in the apical meristem are discrete (Figure 26-9A and Chapter 4).

In other geranium chimaeras, which have a genetically green epidermis and genetically white inner layers, the opposite kind of pattern forms. The green epidermis cells that proliferate at the leaf margins to produce green mesophyll do not produce enough chlorophyll in the central region of the leaf to mask the underlying white tissue.

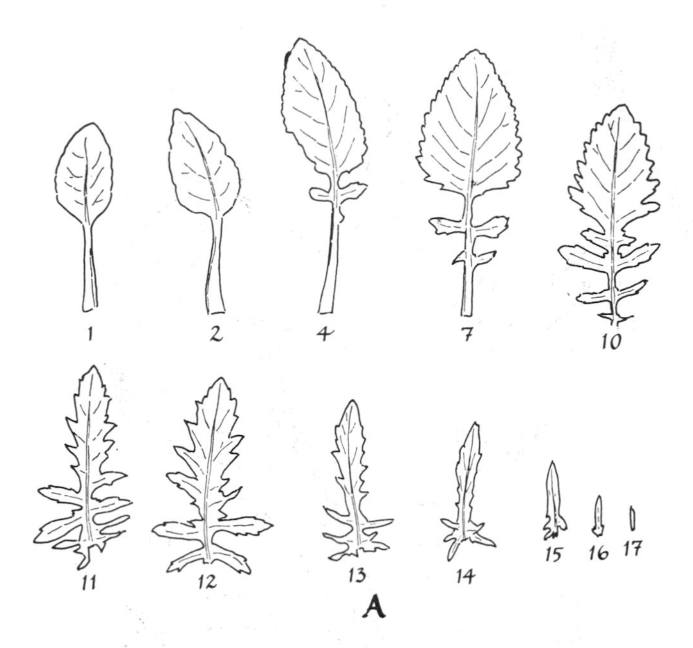

Figure 26-10. *Differences in the shape of leaves from apex, 1, to base, 14, on a flowering shoot of the groundsel* (Senecio aquaticus). *The small leaves, 15-17, are bracts in the floral region. From Wardlaw:* Morphogenesis in Plants, *1968. Courtesy Methuen and Company, Ltd.*

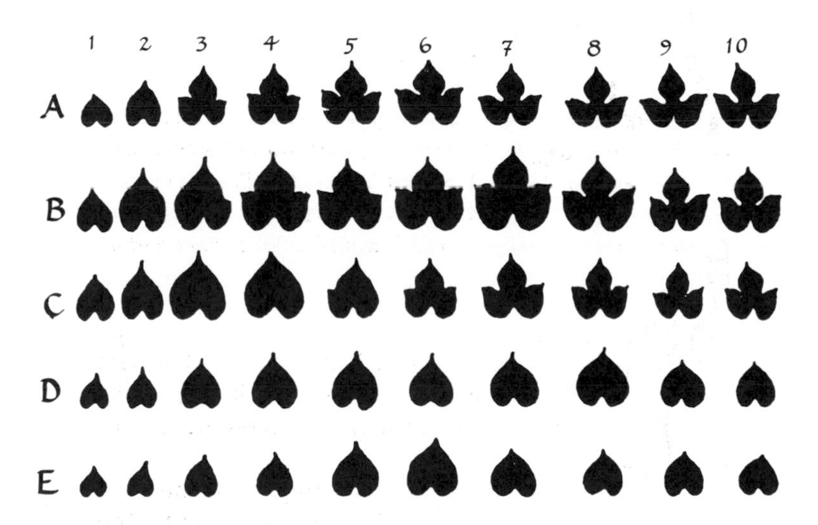

Figure 26-11. *Examples of leaves, 1-10, of the morning-glory* (Ipomoea coerulea) *grown in A, full daylight; B-E, increasing amounts of shade. From Wardlaw:* Morphogenesis in Plants, *1968. Courtesy Methuen and Co., Ltd.*

These chimaeric leaf patterns appear to form largely as a result of the manner in which the leaf primordium develops into the leaf. Recall that the leaf primordium is folded off the sides of the apical meristem and that its surfaces are continuous with the surface of the promeristem (Figure 26-9 and Chapter 4). The leaf primordium grows (elongates and expands) mainly at the apex (Sinnott, 1960). The tunica (epidermis) cells at the apex and the adjacent lateral edges of the young leaf are meristematic regions that add cells to the leaf margin. As noted above, if the epidermis cells are genetically white, the margins of the leaf will be white. If the underlying corpus (mesophyll) cells are genetically green, the central region of the leaf (except for the upper and lower epidermis) will be green (Figure 26-9B and C).

Also of interest are the regular patterns of plastid distribution in leaves of plants that are not chimaeras, such as those of Coleus (Figure 26-1K). Such patterns are regular in basic design from one leaf to another and probably are responses of the leaf-forming cells to local and different microenvironments. Genetically identical proplastids might respond to different environments by developing into different kinds of chromoplasts or by dividing at different rates. Mutant and wild-type plastids might be differentially sensitive to environmental differences, perhaps in a way similar to the manner in which engrailed mutant cells respond to the sex comb prepattern. In either case, we have to consider a prepattern in the leaf. Hypothetically, this could be a gradient type of microenvironmental pattern in the distribution of hormones and/or nutrients. The fact that many leaf patterns are geometrically associated with the pattern of vascular bundles or veins seems to support this hypothesis. In very young leaves, the basic pattern appears in miniature, and expands and becomes more complex as the leaf grows.

An equally challenging problem is understanding the control of leaf shape. In many plant species, the leaf shape is characteristic enough to identify the species. However, the shapes of leaves on a single plant can vary with the basal to apical position of the leaves, that is, with their developmental sequence (Figure 26-10).

Changes in leaf shape are also induced by changing environmental conditions such as temperature, light (Figure 26-11), and nutrient supply. The shape of a leaf is probably determined by the nutritional and hormonal status of the apex (which affects the shape of the leaf primordium), the number and orientation of cell divisions, and amount and distribution of cell enlargement (see review of Wardlaw, 1968).

How leaf shape is controlled by the genome is unknown, but it clearly involves differential growth of the leaf primordium. Differential growth is important in shaping organs or the whole body in many organisms, but we cannot explain the underlying pattern-controlling mechanism.

Mechanism of Spacing in Patterns

Many patterns are characterized by the orderly spacing of the subunits—for example, stomata in leaves, leaf primordia in shoot apexes, bristles in insects, bars in feathers, and feather follicles in the skin of birds. Generally, within a rather constant area surrounding a subunit, no other subunit will develop. A frequently invoked explanation for spacing is the production of an inhibitory substance that diffuses out of the developing subunit and prevents initiation of another within an area in which the concentration of the inhibitor is effective (Figure 26-12A). This has been postulated to explain spacing of leaf primordia (Chapter 4), stomata (see Leick, 1955), and hairs on leaves (Zimmerman et al., 1953). However, the presumed inhibitory substances have not yet been isolated or identified.

Auxins applied to the surface of a young leaf almost completely inhibit stomatal formation but not leaf growth (Bünning and Sagromsky, 1948). Bünning believes that a hormone, possibly an auxin, produced by the young stomatal cells diffuses and inhibits differentiation of other stomata in the surrounding region. If this is the mechanism, one must postulate that already differentiating stomatal cells are insensitive to the inhibitor.

An alternative explanation for spacing is that the subunit uses precursor substrates relatively rapidly, causing these to become depleted in the surrounding area. Because of the low concentration of substrates, no other subunit can develop within this area (Figure 26-12B). This has been postulated to explain spacing of bristles, ocelli, and other chitinous structures of the insect integument (reviews of Sondhi, 1963, and Wigglesworth, 1966), and bars in feathers (Ursprung, 1966). However, the developing subunits have not yet been shown to deplete necessary precursors.

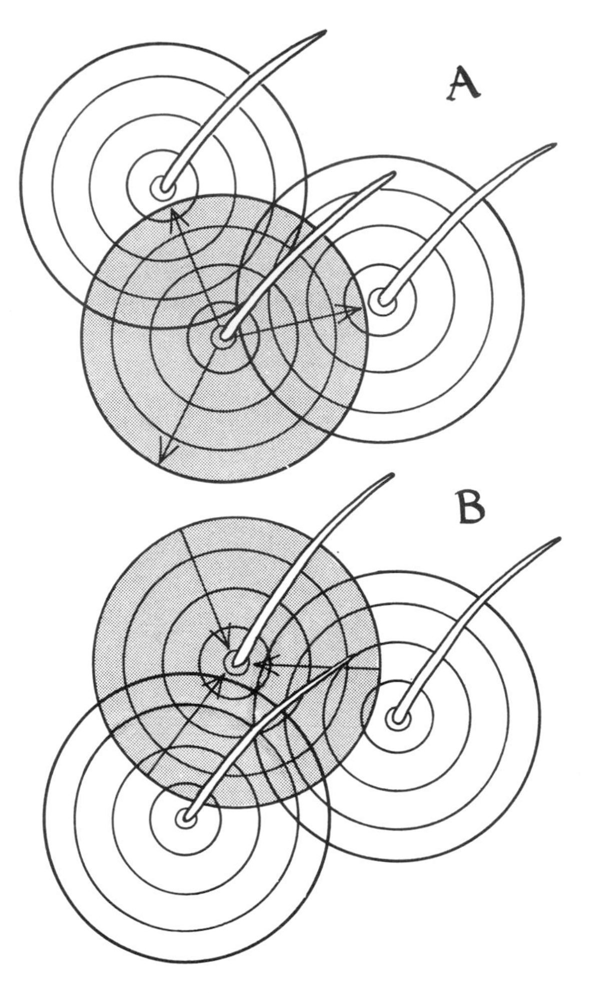

Figure 26-12. *Diagrams illustrating alternative explanations of the spacing of hairs on leaves. A, outward diffusion of an inhibitor from the base of one hair prevents initiation of new hairs within the outer circle; B, relatively rapid utilization of substrates by the hair prevents initiation of new hairs within the outer circle.*

Irrespective of the mechanism by which the subunits are spaced, we are still confronted with the problem of the nature of the prepattern underlying the spacing. Perhaps the first subunit arises at the center of a graded developmental field, and is followed by formation of others at a definite distance from the first and from one another in a progressively outward

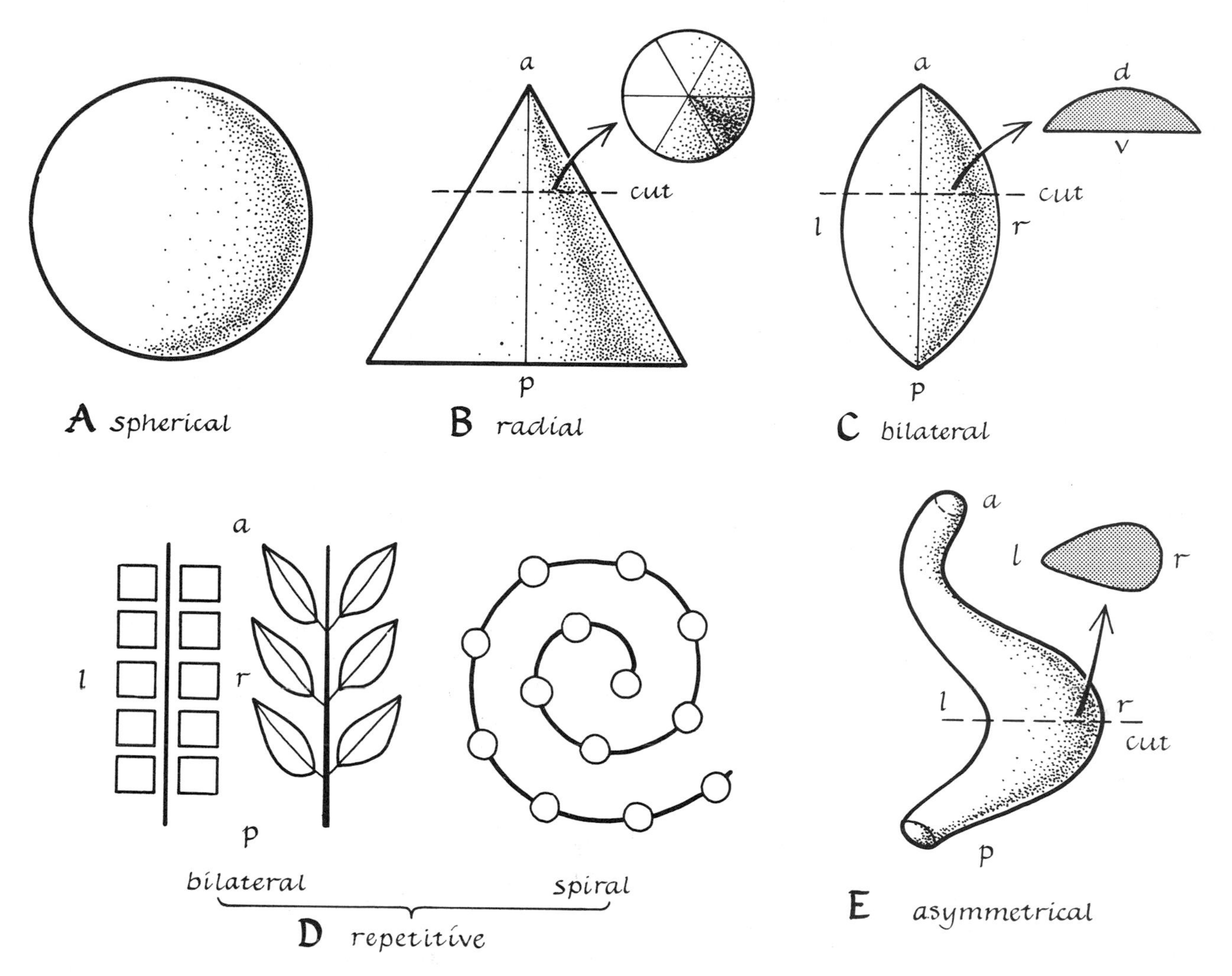

Figure 26-13. *Diagrams of basic types of symmetry.*

direction. Without such a progressive formation of subunits, it is difficult to see how orderly spacings could arise. If all subunits develop at the same time from a cell sheet, then spacing could probably be explained by postulating an underlying and inducing prepattern, but the origin of the latter would still remain a problem.

Symmetry of Patterns

A clear manifestation of organic pattern is the orderly spatial arrangement or symmetry of the components. The patterns of most organisms and their parts can be classified into one or more of the basic types of symmetry shown in abstraction in Figure 26-13.

Probably no organisms are spherically symmetrical, since this would entail that they had no polar axis. Even *Volvox* has a polar axis, and this is radially rather than spherically symmetrical (Chapter 7). Radial symmetry is common in plants and its modification, biradial symmetry, is typical of animals such as ctenophores and starfishes. Bilateral symmetry is common in both plants and animals and is usually accompanied by dorsoventral differences.

Multiple or repetitive symmetry occurs in two forms: (1) bilateral and repeating arrangement of the components, such as leaves or somites, on opposite sides of a polar axis, or (2) spiral arrangement of parts, such as leaves on the stem or blastomeres of a snail egg, around an animal-vegetal polar axis. As a number of examples in earlier chapters (for instance, in Chapter 13) have indicated, these symmetries develop from corresponding symmetries or prepatterns in the structure of the egg or other reproductive unit.

Asymmetric patterns (Figure 26-13E) also appear to be based on a corresponding asymmetry in the reproductive unit. Spemann and Falkenberg (1919) suggested that the bilateral asymmetry in the viscera of amphibian embryos and adults has its basis in a corresponding asymmetry in the microstructure of the egg cytoplasm. The viscera and heart of other vertebrates are also asymmetrical. The early development of the chick heart shows this very clearly: the ventricular loop bends toward the right in an embryo viewed from the dorsal side (Figures 12-35, 12-36, and 26-13E). Normally, the heart is formed by fusion of a right with a left group of mesoderm cells (heart areas) lying on each side of the anterior portion of the primitive streak. The heart mesoderm cells in the two areas aggregate to form tubes which fuse in the midline as the head of the embryo folds up from the remainder of the blastoderm. If the two tubes are prevented from fusing, the right tube bends toward the right and the one on the left bends leftward. After fusion in normal development, both tubes bend toward the right and constitute the ventricular loop. This asymmetry of the heart appears to be derived from an asymmetry of the whole blastoderm, as indicated by experiments in which the right heart area of an explanted stage 4 blastoderm (Figure 12-33A) was excised and replaced with a left heart area. The result was development of a heart with the normal curvature of the ventricular loop (Salazar and Spratt, 1971). The right and left halves of the blastoderm are thus different, and this difference constitutes a prepattern controlling heart development.

Asymmetry has been detected in chick blastoderms as early as the unincubated stage (Figure 12-29). When the anterior half of an unincubated blastoderm explanted *in vitro*, ectoderm side against the culture medium, is folded over the posterior half, the embryo body axis develops almost exclusively from the left side (Eyal-Giladi and Spratt, 1964; Eyal-Giladi, 1969, 1970). This is diagramed in Figure 26-14.

Asymmetry is also present in the convergence movements of the upper layer cells during streak formation in chick and duck blastoderms (Lepori, 1969), and morphologically in the displacement of the primitive pit toward the left in stage 5 blastoderms (Chapter 12). The geometry of the prepattern underlying these asymmetries and that of the heart is indicated by the following experiments. When unincubated chick (Spratt and Haas, 1960; Spratt, unpublished studies) and duck (Lepori, 1969) blastoderms were cut into quadrants as shown in Figure 26-15B, the asymmetry of the heart was normal in respect to the antero-posterior axis of the embryo which developed in each piece.

We can speak of anterior-posterior and right-left axes of whole blastoderms, but the experiment illustrated in Figure 26-15B also shows that these axial qualities and their expression in the behavior of the cells are not radially positioned but, as Lepori says, "unroll according to counterclockwise curved lines" (Figure 26-15C). The orderly and predictable asymmetrical development of the heart in isolated pieces is thus based upon a prepattern in the blastoderm, the nature of which is only suggested by the diagram of Figure 26-15C. The molecular basis of this kind of prepattern, which must exist in the cytoarchitecture of the egg, is unknown.

Reformation of Pattern After Disruption

In Chapter 8 we saw that experimentally dissociated single cells of embryonic organs can reaggregate, sort out, and reconstitute the original tissue architecture of the organ. Even dispersed cells of a whole amphibian embryo reaggregate, sort out, and approximate the normal arrangement of germ layers. In the reaggregation experiments, the properties or affinities of the cell types control the sorting out and reconstitution. No guidelines external to the aggregate seem to be involved, so the original pattern is formed strictly autonomously.

Ursprung (1963, 1966) has described another example of pattern formation following disruption of the pattern-forming cell population. Wild-type cells of the wing imaginal disc of *Drosophila* were dissociated to the single-cell state and mixed with similarly dissociated cells of wing discs of a genetically marked mutant (*yellow–multiple hair*). The mixed cells were allowed to reaggregate *in vitro* for a short

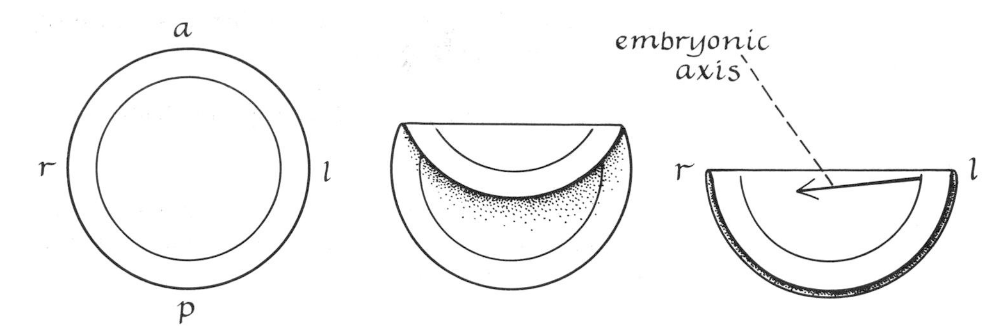

Figure 26-14. *Diagrams of an unfolded and folded (right) unincubated chick blastoderm explanted in vitro, illustrating early asymmetry in respect to initiation of the embryonic axial system.*

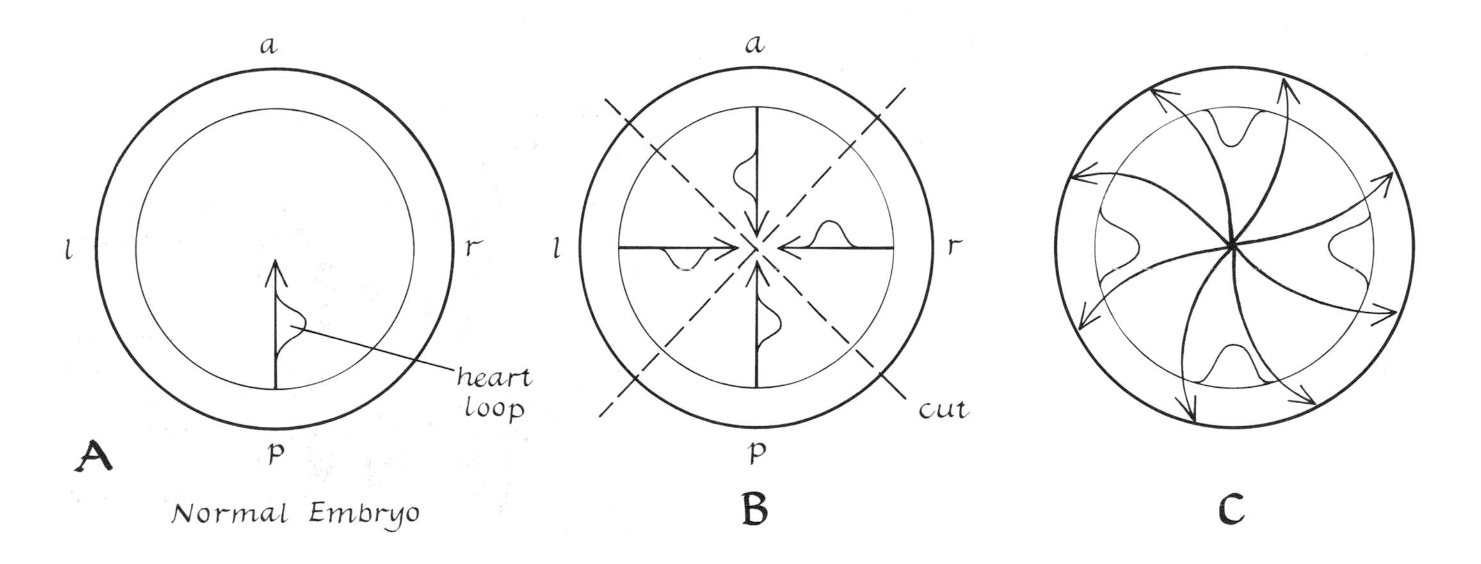

Figure 26-15. *A, bending of the heart loop toward the right in an unoperated chick blastoderm; B, normal asymmetry of the heart in isolated explanted quarters of an unincubated blastoderm; C, diagram of the asymmetrical prepattern hypothetically underlying asymmetry in heart development. The arrows in A and B denote the anterior-posterior axis of the embryo body.*

period and were then injected into a host larva. Recovery of the implant after metamorphosis revealed the formation of a mosaic wing lamella on which a normal bristle pattern had developed (Figure 26-1H). This consisted of rows of regularly spaced bristles, some of which were mutant (yellow), while others were wild-type in color (black). The pattern was thus shown to be composed of cells originating from different imaginal discs. Apparently, the bristle-forming stem cells of both discs are sorted out and find each other by cellular affinity, presumably by cell surface self-recognition. This implies that the bristle-forming stem cells in the imaginal discs were determined when the experiment was done, but how they become lined up and spaced properly during the sorting out is not known. Sorting out of like cell types has not been observed.

Conclusion

Mechanisms controlling pattern formation during development are largely unknown, yet the emergence of patterns constitutes an obvious and overall aspect of development. A striking feature of development is the continuity of patterns, in which the pattern at one stage or at one level of complexity is the prepattern guiding or setting the limits for the formation of the next pattern. No pattern arises in the absence of preexisting guidelines; these may be preformed, template-type prepatterns or prepatterns that are progressively built up from activities of the units making up the pattern. The prepattern can be field-like, consisting of graded differences in concentration of metabolites, hormones, or other cell products. On the other hand, it can consist of sharply localized differences in cytoplasmic composition, as in mosaic eggs.

In final analysis, all patterns are formed under control of the organism's genome. There is evidence that the genome of a cell in genetic mosaics can determine the cell's response to a prepattern. More probably, a prepattern, formed under control of many genes, controls later selective activity of identical genomes and causes a visible and more complex pattern to form. How prepattern-forming genes operate is not known.

References

Bünning, E., and H. Sagronsky. 1948. Die Bildung des Spaltöffnungsmusters in der Blattepidermis. Zeit. Naturforsch. **3b**:203–216.

Child, C. M. 1941. Patterns and Problems of Development. University of Chicago Press, Chicago.

Clowes, F. A. L., and B. E. Juniper. 1968. Plant Cells. Blackwell, Oxford.

Eyal-Giladi, H. 1969, 1970. Differentiation potencies of the young chick blastoderm as revealed by different manipulations: I and II. J. Embryol. Exp. Morph. **21**:177–192; **23**:739–749.

Eyal-Giladi, H., and N. T. Spratt, Jr. 1964. Embryo formation in folded chick blastoderms. Am. Zool. **4**:428 (abstr.).

Goldschmidt, R. B. 1955. Theoretical Genetics. University of California Press, Berkeley.

Leick, E. 1955. Periodische Neuanlage von Blatt-Stomata. Flora (Jena) **142**:45–64.

Lepori, N. G. 1969. Sur la genèse des structures asymmétriques chez l'embryon des oiseaux. Monitore Zool. Ital. **3**:33–53.

Lewis, E. B. 1964. Genetic control and regulation of developmental pathways. In M. Locke, ed., The Role of Chromosomes in Development, pp. 231–252. Academic Press, New York.

Locke, M. 1967. The development of patterns in the integument of insects. Adv. Morph. **6**:33–88.

Nieukoop, P. D. 1967. The organization centre: II. Field phenomena, their origin and significance. Acta. Biotheoretica **17**:151–177.

Rawles, M. E. 1948. Origin of melanophores and their role in development of color patterns in vertebrates. Physiol. Rev. Camb. **28**:383–408.

Salazar, J. S., and N. T. Spratt, Jr. 1971. Influence of the right and left halves of the chick blastoderm on the corresponding cardiogenic areas in respect to formation of the bulboventricular loop. Devel. Biol. (in press).

Sinnott, E. W. 1960. Plant Morphogenesis. McGraw-Hill Book Company, New York.

Sondhi, K. C. 1963. The biological foundations of animal patterns. Quart. Rev. Biol. **38**:289–327.

Spemann, H., and H. Falkenberg. 1919. Über asymmetrische Entwicklung und situs inversus viscerum bei Zwillingen und Doppelbildungen. Arch. Entw. Mech. **45**:371–421.

Spratt, N. T., Jr. 1966. Some problems and principles of development. Am. Zool. **6**:9–19.

Spratt, N. T., Jr., and H. Haas. 1960. Integrative mechanisms in development of the early chick blastoderm: I. Regulative potentiality of separated parts. J. Exp. Zool. **145**:97–137.

Stern, C. 1954. Two or three bristles. In G. A. Baitsell, ed., Science in Progress, pp. 41–84. Yale University Press, New Haven, Conn.

Stern, C. 1955. Gene action. In B. H. Willier, P. Weiss, and V. Hamburger, eds., Analysis of Development, pp. 151–169. W. B. Saunders, Philadelphia.

Steward, F. C., E. M. Shantz, J. K. Pollard, M. O. Mapes, and J. Mitra. 1961. Growth induction in explanted cells and tissues: metabolic and morphogenetic manifestations. In D. Rudnick, ed., Synthesis of Molecular and Cellular Structure, pp. 193–246. Ronald Press, New York.

Tokunaga, C. 1961. The differentiation of a secondary sex comb under influence of the gene *engrailed* in *Drosophila melanogaster*. Genetics **46:**157–176.

Twitty, V. C. 1949. Developmental analysis of amphibian pigmentation. Growth Suppl. **13:**133–162.

Ursprung, H. 1963. Development and genetics of patterns. Am. Zool. **3:**71–86.

Ursprung, H. 1966. The formation of patterns in development. In M. Locke, ed., Major Problems in Developmental Biology, pp. 177–216. Academic Press, New York.

Waddington, C. H. 1966. Fields and Gradients. In M. Locke, ed., Major Problems in Developmental Biology, pp. 105–124. Academic Press, New York.

Wardlaw, C. W. 1968. Morphogenesis in Plants. Methuen, London.

Weiss, P. 1939. Principles of Development. Henry Holt, New York.

Wigglesworth, V. B. 1966. Hormonal regulation of differentiation in insects. In Cell Differentiation and Morphogenesis, pp. 180–209. North Holland Publishing Company, Amsterdam.

Wilde, C. E. 1961. The differentiation of vertebrate pigment cells. Adv. Morph. **1:**267–298.

Willier, B. H., and M. E. Rawles. 1940. The control of feather color pattern by melanophores grafted from one embryo to another of a different breed of fowl. Physiol. Zool. **13:**177–199.

Willier, B. H., and M. E. Rawles. 1944. Genotypic control of feather color pattern as demonstrated by the effects of a sex-linked gene upon the melanophores. Genetics **29:**309–330.

Zimmerman, W., D. Woernle, and L. Warth. 1953. Genetische Untersuchungen am Pulsatilla. V. Die Entwicklung von Haaran und Spaetöffnungen bei Pulsatilla. Zeit. Bot. **41:**227–246.

Part Five:

Principles of Development

27 *Summary: Basic Guidelines and Principles of Development*

This book has surveyed the field of developmental biology, which we have defined as encompassing all biological activities involving progressive changes in structure and function. The purpose of the text has been twofold: to present in broad perspective the material of developmental biology—that is, the diversity of developmental patterns—and to suggest the nature of mechanisms that control the program of changes. Since the book is introductory, no attempt has been made to summarize and catalog all of our information about patterns of development or to analyze in depth what appear to be the underlying and controlling mechanisms. The subject of developmental biology is so vast that the amount of information and critical discussion allowed in this book had to be severely limited. It is hoped that what has been presented has been sufficient to stimulate the interest and curiosity of students. Nothing included should be considered as dogma, and many of the conclusions are clearly tentative.

The book may have helped those interested and competent in restricted and special aspects of development to relate their special interests to the much broader problems of development. It can comfort the specialist to see how his circumscribed knowledge fits into a larger edifice of knowledge.

This final chapter attempts to summarize the meaning of developmental biology by formulating what appear to be its general and fundamental principles, applicable to all kinds of developmental processes and patterns. These principles are presented as experimental and conceptual targets open to attack in the interest of truth.

General Guidelines for Development

As noted in Chapter 1, development proceeds under control of guidelines (informational and directive influences) specified by the genome. In final analysis, development is the synthesis of something new. The information specifies and limits the synthetic capacities of the developing organism and also specifies the spatial and temporal patterns of synthesis. Thus, the guidelines specify what can be synthesized, when it can be synthesized, and where in the developing organism the synthesis will occur. The genome contains the ultimate information for development in the form of a DNA, and rarely (for instance, in some viruses) an RNA code. The use of this inherited information—its replication, transcription, and translation—is development.

Preformed Guidelines

Some of the guidelines, such as the information packaged in the nucleus or nuclei of a reproductive unit, are directly inherited, preformed, and continuously present through generations of organisms. The cytoplasm of the cell or cells constituting the reproductive unit contain other important guidelines. The latter guidelines constitute the basic framework or overall pattern of development and apparently attain the highest degree of complexity in the cytoarchitecture of eggs. In general, these cytoplasmic guidelines are built up under control of information present in the nucleus while the reproductive unit forms. However, as we have seen, in some organisms the cytoplasmic guidelines have become relatively autonomous and independent of direct nuclear influences. The importance of cytoplasmic guidelines resides in their control over use of information in the nucleus.

Progressively Formed Guidelines

Epigenetic guidelines are those which are progressively built up as the reproductive unit develops. These guidelines are an automatic consequence of the action of the preformed guidelines and are therefore as exact and orderly in their development as is the progressive use of the preformed guidelines. In multicellular organisms, the progressively formed guidelines are an internal and heterogeneous microenvironment, to which cells of the developing organism respond differently in accord with their positions. The result is cellular differentiation.

Both preformed and progressively formed guidelines probably operate in all patterns of development, but in some, such as the more mosaic patterns, preformed guidelines outside the nucleus have a greater influence than in others. In the more regulative patterns—for instance, in the development of an isolated tissue cell of a plant and to a large extent in the development of many single-celled reproductive units—progressively formed guidelines play the dominant role.

Although mechanisms for controlling development of an organism's parts (organs, tissues, cells, organelles) lie outside as well as inside the parts, the guidelines for normal development are exclusively within the organism as a whole. A developing system is causally closed; it is not directed by the outside environment. The environment can modify, block, even initiate or select, but it cannot direct the sequence of developmental transformations.

Principles of Development

1. Conservation of Genetic Potentiality

All cells of a developing organism contain the same genetic information. Except in special cases, such as that of the nonnucleated mammalian erythrocytes, genes are not lost when cells specialize. Cells differ not because of qualitative differences in information content but because of differential utilization of the information. Quantitative differences in information content do exist (for instance, in salivary gland cells in insects) but these are probably the result rather than the cause of cellular differentiation. Conservation of genetic potentiality must not be confused with conservation of developmental potentiality; the latter has been demonstrated in several types of somatic plant cells but remains to be demonstrated in animals.

2. Fate and Capacity

The capacity for manifold expression of genetic potentiality is characteristic of early development. Although cells or other parts of a developing organism have specific fates in normal development, these fates can be experimentally changed during early stages.

3. Regulation and Regeneration (Repair)

The capacity for manifold expression of genetic potentiality enables early stages of many organisms to repair naturally incurred or experimentally produced defects. It sometimes enables part of the organism to reconstitute the whole organism. Some adult organisms retain the capacity to repair defects incurred by injury (regeneration). Regulative or regenerative capacities are not considered mechanisms of normal development.

4. Progressive Restriction of Capacities: Cellular Differentiation

In normal development, the manifold capacities available for expressing genetic potentiality at early stages are gradually reduced until each cell can express only one capacity. The multiple capacities possessed by most or all cells at early stages are parceled out to cells that express a single capacity. This specialization, which we call differentiation, results in discrete and distinct cell types.

5. Egg Organization

Probably all eggs exhibit a cytoarchitectural pattern—a pattern of graded or localized differences in biochemical properties of the cytoplasm. This cytodifferentiation constitutes a heterogeneous intracellular environment to which the nuclei of the cleaving egg respond differentially. The egg's organization is thus a basic guideline for differential gene action.

6. Environmental Influences (Intercellular and External)

Although an organism's genetic constitution determines its developmental capacities, no organism ever expresses all its capacities simultaneously. The capacities are expressed in an orderly sequence under control of influences impinging upon the genome. The spectrum of environments extending from that immediately surrounding the genome to the outside world, and including the cytoplasmic environment of the nuclei in cleaving eggs, constitutes these influences. The composition of these environments selects which inherited capacities will be expressed at any one time and in any particular part of an organism. This is what is meant by the principle of environmental influence.

7. Gradients and Fields

Early gene action during oogenesis results in a patterned distribution of gene products within the cytoplasm of the uncleaved egg and/or in the cell population at a later stage. This pattern is typically a quantitatively graded distribution, although more localized distributions do exist. An internal microenvironment develops as a result. This environment is important because it controls the differential response of genes in accord with the position of the nuclei. Differential gene action occurs in a geometrical framework consisting of a gradient in time plus gradients in the three spatial dimensions. The resulting expression of genetic capacity is spatially and temporally orderly.

8. Cellular Interactions

Cells of the increasingly large population arising during development of multicellular organisms influence one another profoundly. These influences are primarily mediated by products released from the cells. Sometimes the cells or groups of cells interact, thus influencing one another's development. In other instances, one cell or cell group acts upon another, causing a change in the development of the latter while remaining itself essentially uninfluenced. Such influences are called inductions. The influence of one cell group upon another can restrict the capacities of the latter, and is thus a mechanism for cellular differentiation.

9. Stability of the Differentiated State

With rare exceptions, cells that have attained a final state of specialization remain specialized. This is clearly true in the adult organism, where cells of one type do not transform into another except in rare cases. However, under experimental conditions in which the microenvironment surrounding the cells is changed, the stability of some kinds of differentiated cells is a matter of degree. In theory, the conservation of genetic potential could provide the basis for dedifferentiation followed by redifferentiation along a new path.

10. Integration

In multicellular organisms, a large cell population builds up during development, a population so unified that it constitutes a single organism. Cellular associations in colonial organisms illustrate the varying degrees of unity that have evolved. The division of labor among the cells necessitates their staying together. Interdependence within the cell population is a result of cellular specialization and is basic to the development of multicellularity.

11. Developmental Centers

In many organisms, developmental processes are initiated in and spread out from a localized region within the cell or cell population. Such a developmental center is characteristically a dominant or organizer center that influences surrounding cells and is relatively uninfluenced by them.

12. Differential Growth

Differential growth underlies differential increase in size of parts and organs in developing organisms. Developing and adult plants and some animals contain localized growth centers. Especially in plants, differential growth, including the direction of cell division, helps shape parts, organs, and whole bodies. Differential growth is probably less important in animal than in plant development. In animals, it is partially supplanted by cell movement and cell death.

13. Programmatic Gene Action

Action of genes at one stage results in establishment of conditions that control gene action at the next stage, and so on, resulting in a gene relay system. In reproductive patterns, this system begins to operate when the reproductive unit forms. In patterns of regeneration, the system starts operating as soon as the part is lost. In patterns of continuous development, the system operates continuously. Thus development is the sequential use or expression of inherited capacities. Not only is information inherited, but also a built-in program for orderly expression of this information. The molecular basis of this inherited program remains the most important and challenging problem in developmental biology.

14. Inevitability

Development results inevitably from the spectrum of subunits involved, molecular to cellular. The emergence of order results from the association of the subunits. The design of the association often depends upon properties of the subunits as well as the context of the association. Inevitability is basic to orderliness, which is a prime feature of development. Orderliness makes possible our eventual understanding of the developmental phenomena. There is little, if any, opportunity for the operation of chance, which is usually invoked to mask ignorance of the causal factors involved.

15. Maintenance of Polarity in Regeneration

The original polarity of the organism is maintained during regeneration of a missing part. This is true of both plants and animals. This is not considered a process of normal development, because it applies only in the organism's response to injury.

Interrelation of the Principles of Development

Although all cells of a multicellular organism have the same genetic capacities or potentialities (1), typically in normal development each cell expresses only one of these capacities—its fate (2). If the normal fate is suppressed experimentally, some other capacity is expressed, as, for example, in regulation or repair (3). In normal development a cell's capacities are progressively restricted (4). This restriction can result from the cytoplasmically different environments surrounding the nuclei of the cleaving eggs (5) and later from intercellular environmental influences impinging upon the cell (6). In early development the environmental influences are mostly quantitatively graded (7). Later the influences may be relatively specific, as in the case of cellular interactions or inductions (8). However capacities are restricted, the result is cellular differentiation or specialization (4). Once a cell has become specialized it tends to remain stable in its differentiated state (9). Cellular differentiation within a cell population involves a division of labor among the cells and leads to an integration or unification of the cell population (10). In many organisms a dominant group of cells arises and exerts control over surrounding cells (11). Such developmental centers often consist of more rapidly growing or precociously specialized cells (12).

All of the above activities are the consequence of programmatic gene action—that is, result from the operation of a gene relay system (13), which is an inevitable consequence of the genetic composition of the organism (14). These interrelations are shown below.

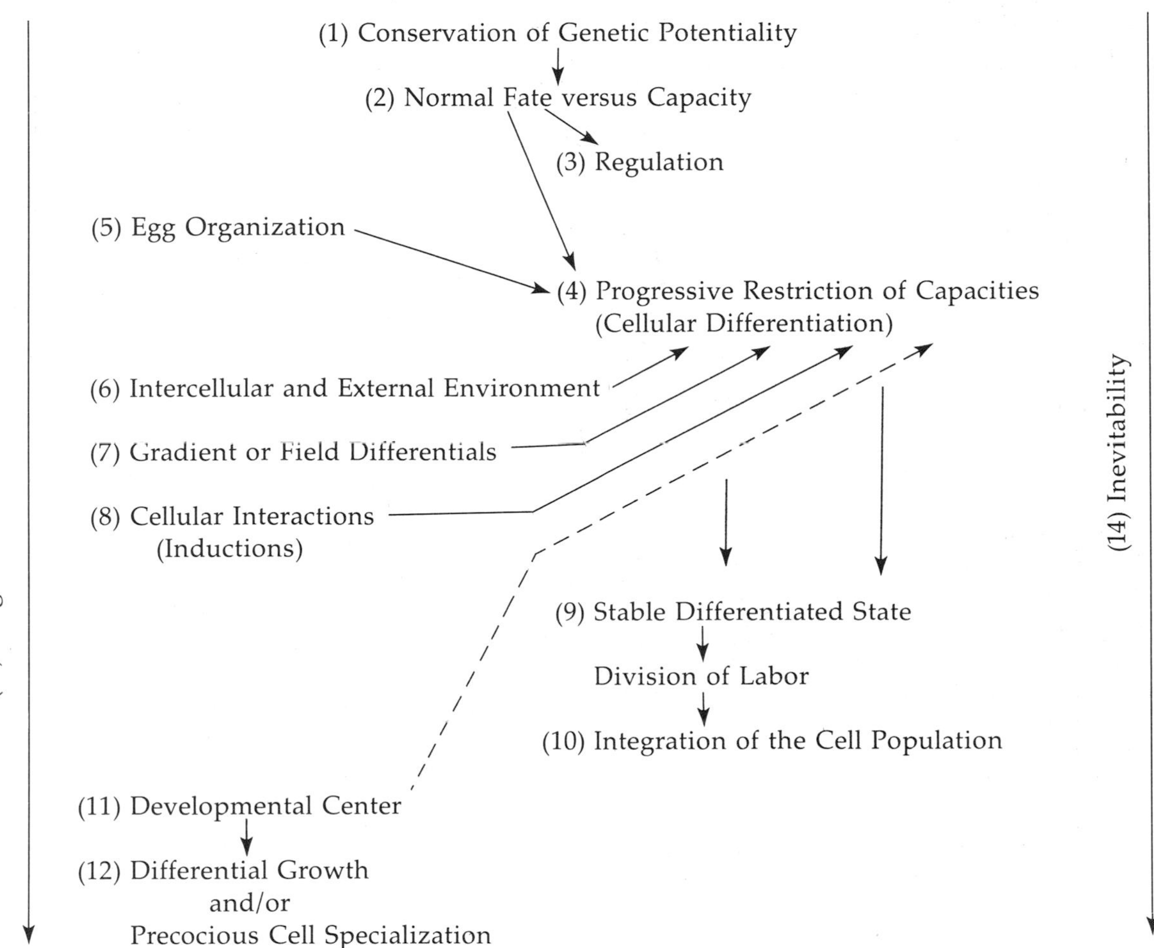

Subject Index

D

E

Species Index